T ... CE

JUN   2012

THE
FOURTH SOURCE

Effects of Natural Nuclear Reactors

Robert J. Tuttle

Universal-Publishers
Boca Raton

The Fourth Source: Effects of Natural Nuclear Reactors

Universal-Publishers
Boca Raton, Florida
USA • 2012

ISBN-10: 1-61233-077-0
ISBN-13: 978-1-61233-077-8

www.universal-publishers.com

Front cover: False color image of Venus by radar mapping by *Magellan* (NASA/JPL/USGS); Spiral Galaxy M74 with a supernova photo courtesy of Isaac Newton Group of Telescopes, La Palma/Simon Dye (Cardiff University)); Pu'u O'o, Hawai'i (photo by the author); and natural color image of Earth from space by *Apollo 17* mission (NASA/JSC/Apollo 17 crew).

Back cover: Image of M51, the Whirlpool nebula, and its companion, by the Canada-France-Hawaii Telescope (© 2002); photo of author (by the author), December 2007.

Library of Congress Cataloging-in-Publication Data

Tuttle, Robert J., 1935
    The fourth source : effects of natural nuclear reactors / Robert J. Tuttle
      p. cm.
    Includes bibliographical references and index.
    ISBN 978-1-61233-077-8 (pbk. : alk. paper) – ISBN 1-61233-077-0 (pbk. : alk. paper)
    1. Radiation sources. 2. Radiation, Background. 3. Extraterrestrial radiation. I. Title.
    QC476.S6T88 2012
    551.3'9—dc23
                                                                                            2011045154

PROLOGUE

Science progresses by making mistakes. Sometimes abruptly, these mistakes are corrected, and science proceeds on. Understanding the history of science is essential if we are to understand what we understand. The corrected errors may start with a flat Earth, as the center of the Universe, with the stars being distant campfires in the sky. That fit the earliest observations of our ancient predecessors and was consistent with the earliest theories. Observation and theory continued to evolve, each correcting the other. Now, do we know everything, except the smallest details, the insignificant fringes, or is it only what is beyond the fringes that we don't know? Like every generation prior, are we muddling along, burdened unknowingly by an abundance of unseen errors? What are those errors, what will the next big corrections be? This book makes an attempt to answer some of those questions. I hope the answers will be as exciting for you as they have been for me, and raise as many more new questions.

Some of this must remain as an exploration. There are problems I have not been able to resolve but that I think are important enough for further consideration.

v

**for
Zoey and Ian,**

**because of
June**

---

*The real voyage of discovery consists
not in seeking new landscapes
but in having new eyes.*

Marcel Proust

*This is the greatest adventure of mankind:
To find something that was never known before,
or understood.*

Irving Stone

*And the end of all our exploring
Will be to arrive where we started
And know the place for the first time.*

T. S. Eliot

Table of Contents

List of Figures

Preface

In ancient times, when people looked at the skies and wondered about their world, the high priests told them stories to explain what they saw. We call those stories myths and legends. In modern times, when people look at the skies and wonder about their world, scientists tell them stories to explain what they see. We call those stories theories and models.

As myths and legends are questioned, and as both the questioners and the answerers gain knowledge, the stories become more clearly based on facts, and approach more closely to the reality that they try to explain. But sometimes, because an effect is overlooked or misinterpreted, the stories go off-course and the people are led astray. That seems to be so for our current stories of the Universe, from the Earth and its fellow planets, to our Solar System neighborhood, to the most distant reaches we can imagine.

Our Universe has been described in myths and legends invoking many mysterious forces, but we now explain it by theories and models using just three major sources of energy: gravity, nuclear fusion, and electromagnetism. Because of our limitation to these three energy sources, our explanations are often incomplete and confused, sometimes overwhelmingly complex, and we must invent phenomena never seen.

Our theories have become increasingly creative and ingenious, calling on unobservable objects and processes, such as black holes, neutron stars, and inflationary cosmology, to explain our Universe. We struggle with inconsistencies and confusion in the stories of our world. While our concepts of reality are shaped by our perceptions, our perceptions are directed by our concepts. In some cases, the currently accepted models have become so firmly entrenched as to constitute "paralytic paradigms". These concepts are so strongly held and so severely control our thoughts, that contrary or alternate theories cannot be considered, and contradictory observations are not recognized.

Three sources of energy have long been known: gravity, that controls the motions of all bodies, heavenly, Earthly, and human; nuclear fusion, that provides the power to light the stars; and electromagnetism, that carries energy throughout the Universe and holds the atoms together. Our current explanations of the Universe are based on these three sources.

Things could be simpler... Something was missed along the way that can clarify much of the current confusion. That something is nuclear fission. We have been trying to explain effects that are caused by nuclear fission without including the process of nuclear fission in the explanations. Including nuclear fission as the fourth source of energy brings unity and clarity to our stories of the world, from the planets of the Solar System to the galaxies of the Universe. This book provides some new solutions to old, solved problems.

The omission of nuclear fission from our understanding of the Universe is an interesting philosophical question. You probably know very little about nuclear fission, other than that it is a very energetic process, used in weapons and power plants, and you probably know little about nuclear weapons and nuclear power plants. We have a concept that nuclear reactors must be very complicated, because those that we make are complicated, but Nature ran reactors in the bottom of a river delta for 600,000 years, 2 billion years before we discovered nuclear fission. So, how hard can it be, with the right stuff, and enough time?

Excluded from academia by anti-war sentiment, hidden from curiosity by extreme governmental secrecy, nuclear fission has languished in an intellectual limbo. It is well known, well understood, by its practitioners, but it is a will-o'-the-wisp to outsiders.

My research, leading to a new understanding of the Universe by including the effects of nuclear fission, began in 1989 as a simple personal search for a better understanding of the impact of natural radioactivity on our world and our lives. It became a self-guiding tour, with a life of its own, in 1993, as I began to discover possible effects due to nuclear fission, and how poorly many of these observations were explained by present theories. Each discovery led naturally to more. At times, I was startled by the weakness of arguments that decided fundamental questions of reality, and how those decisions have not been re-examined as our knowledge and our observational techniques have improved. On the other hand, it is impressive that the fundamental observations were made so well, reported so honestly, and kept so accessible, that a stranger to the specialists' worlds can still glimpse reality clearly. The results of this research have been quite contrary to many of our current models, but provide the key to understanding our knowledge more completely.

The reader of this book will be faced with a pair of basic questions:

> 1. If what I propose is true,
>> is it likely to be important?

and

> 2. If what I propose is important,
>> is it likely to be true?

Each reader's answer to these questions will prove whatever worth this book may have.

As I came to an understanding of the presence and probable source of excess energy in the early eons of Earth, my search for its source took me into space, into our Solar System and far beyond. And then, surprisingly, I was brought back to Earth, with a clearer understanding of the puzzle of evolution, and how and when we became us. What was expected to be, and intended to be, a short article in a technical journal grew to be something that was too large, too broad, too interdisciplinary to fit into any journal. Instead it has become a book, for the curious members of *Homo sapiens sapiens* to discover their world.

This is not so much a science book as it is a book about gaining knowledge and understanding. That knowledge and understanding is the right of every citizen of this planet, and should not be restricted to an elite of scientists, by virtue of being written in an arcane language. Here, there are few equations (except in a special Appendix), no integrals, no calculus. There are graphs and plots. These are like maps to help you see where we are, and where we might be headed. This book is filled with observations and ideas, with the hope that you will gain knowledge and understanding. The math and the science are there, but behind the scenes, available if you wish to discover them, but not demanding your attention. The equations are embedded in the physics. If you need the equations, you probably already know them. If you don't already know them, you probably don't need them. You can find them in the references, if you want. In the references, I have tried to provide readily available publications that are acceptable to the specialist and others that are more accessible to the novice. We are all novices in most fields. While the artists of the Renaissance ground their own pigments and formulated their own paints, fabricated their own brushes and canvasses, you do not need to do that to appreciate a Rembrandt or a Reubens or a Titian. The Universe is a beautiful masterpiece and I want you to enjoy this book about it.

Since this is not a science book, I have tried wherever possible to write in ordinary English, translating and interpreting as much of the official technical language and jargon as I can. The experts will not approve, and will use this as evidence that I don't even know what I'm writing about. I'm sorry. I think it is more important for more people to understand what they are reading about, than to present the information in an officially satisfactory but obfuscated way. By avoiding technical terminology, interpreting and explaining in ordinary language, I hope to gain in accuracy of understanding what may be lost in precision and correctness. Perhaps this book can be read without a dictionary in your lap.

This book is an exploration, rather than a declaration, and in some cases the reader must be left to choose the best answers.

To offer a solution from a foreign field often appears to be an appeal to magic, or more bluntly, to be crackpot science. If the foundation for an explanation is not known, recognized, or understood, that explanation may appear to be nonsense. Nuclear fission and radiation physics are foreign fields to most researchers, just as their fields are foreign to me. In those foreign fields I can only offer the insight of an outsider, glimpses of ideas that are based on knowledge that is unknown to the resident experts. Every expert in every field that this book addresses knows more about his field than I do. My only advantage is that I haven't known so much that is wrong.

Until you have read this book, you may not be familiar with many of these problems, which leads to my quip that "those who understand the solution don't know the problems, and those who know the problems don't understand the solution." The purpose of this book is to get those two populations, each understanding part of the situation, together.

To a large extent, this book is a story of catastrophe. In the all-too-often either/or world of science, Catastrophism lost out to Uniformitarianism. Most of the Earth has certainly been formed and shaped by the slow, steady progress of processes that can be seen in action somewhere on the Earth at the present time. Still, some catastrophes can be recognized and accepted, rare as they have been. In our current view, large volcanic eruptions are local catastrophes. Recent eruptions are insignificant compared to major eruptions in the distant past, which were truly global catastrophes. The mass extinctions, in which nearly all life on Earth was terminated, or at least altered irreversibly, must be viewed as catastrophes, whatever the cause. Impacts by asteroids or comets are now accepted as one possible catastrophic cause.

Natural physical processes are proposed here to explain, by catastrophe, what has not been explainable by uniformity.

The basic framework for studying the structure and form of the Earth was established on the assumption that the Earth is cooling and contracting (shrinking). The contraction led to the formation of the mountains of the continents as "wrinkles". In spite of evidence to the contrary, for example, the expansion of the ocean floor at mid-ocean ridges, contraction has continued to be an underlying concept. Mountain-building has been transferred to the contraction caused by collision and subduction of crustal plates. The apparent success of this Plate Tectonic Theory has blocked consideration of alternatives.

Some of the ideas in this book are soft and speculative. I must overwork the word "suggest" to convey the sense of possibilities worth exploring, rather than dogmatic facts to be accepted. I take that as my privilege as a visitor, an intruder, an explorer in fields where I lack the specialists' knowledge and authority. Lacking the authority to establish ideas as facts, I shirk the responsibility to present these ideas in a fully rigorous manner. I hope that there is enough rigor where it is appropriate that the promise of the soft ideas will be recognized.

One of the symptoms of crackpot science is the solution of problems in widely unrelated fields, often by the use of strange and obscure techniques. At first glance, that criticism might be very easily applied to this book. But it is important to recognize that this book does not try to explain all things as the result of nuclear fission. Many (in fact, most, by far) of our observations are well explained by fully developed theories and need no help, from fission or anything else. And there are other problems that seem to be beyond the help of nuclear fission, such as the initial origin of life. (We will make occasional detours to examine these.) This book attempts to explain those observations that appear to me to be poorly explained at present, and are clearly related to the effects of nuclear fission. Some of the detours will insist on intruding, but I hope they help rather than hinder our understanding. Moreover, everything in the book is only physics. I have simply applied the methods of physics to problems that became unveiled once the possibility of natural nuclear fission reactors became clear. These methods and knowledge are straightforward and well-founded. I have invented little that is in this book. Mainly, I have interpreted and integrated other researchers' ideas by the use of the methods of physics. What I present here is no more complicated, inherently, than gravity. Things fall. Reactors happen naturally, natural reactors sometimes explode.

In the 1950s, Immanuel Velikovsky wrote some outrageously speculative books, based on what he saw as poorly explained events that were described in myths, legends, and religious writings [Ginenthal 1995]. He invented a mechanism to explain these observations, but the mechanism was judged, quite objectively, to be a crackpot idea. (Unfortunately, his books were rejected much more emotionally and his perceptions were thrown away.)

In the first part of the 20th Century, Alfred Wegener collected observations that showed many of the continental coastlines had at one time been adjacent to each other, but had drifted apart [Wegener 1928]. His mechanism for this "Continental Drift" was also quite unreasonable, physically, and his entire concept was repeatedly rejected. Drifting blocks of continents was scientifically unacceptable, until oceanographers discovered spreading ridges in the middle of the oceans, and invented the theory of Plate Tectonics, where continental and oceanic scum is carried across the globe by giant convection cells in the mantle. Just as the mechanisms invented by Velikovsky, Allan and Delair, Carey, and Van Flandern, were physically implausible for their observations, so Wegener's drifting continents was also crackpot science. Unfortunately, his collection of observations was also rejected as not requiring consideration or explanation. When the scientific establishment came across a surprisingly convincing observation, ocean spreading ridges, Plate Tectonics was developed to accommodate the drift of continents.

Over 400 years ago, Galileo saw mountains on the Moon, spots on the Sun, handles on Saturn, and stars around Jupiter, at a time when these were known to be impossible nonsense. Alfred Wegener proposed that continents drifted through the oceanic crust, to explain the undeniable matching of the continental shelflines and geology. The meaning of his observations was denied for 50 years. Edwin Hubble showed that the spiral nebulae were outside of the Milky Way Galaxy, and therefore were vast Island Universes of stars, just like the Milky Way Galaxy, because Galileo had said the Milky Way was made of innumerable stars. Immanuel Velikovsky said that Venus was extremely hot and that Jupiter had a magnetic field, because Venus had been ejected from Jupiter just thousands of years ago.

Pioneers and crackpots, missionaries and explorers. How can we separate the wheat from the chaff if we can't tell them apart?

Time proved Galileo to be mostly correct. We have had to reinterpret some of his words, but that was done so gently it is hardly noticed. His use of the telescope for direct observation of the heavens is the foundation of modern astronomy. Wegener had to propose a physically unrealistic process for Continental Drift, that the continental plates moved through the surface of the oceanic crust. His fundamental observations finally won acceptance, and it was decided that oceanic plates dived beneath the continental crust. Velikovsky built his ideas on religious writings and invented mechanisms that were thoroughly impossible to explain the observations. He may have seen some truths, but the fallacies of his explanations drew such vehement attacks, it was not acceptable to explore what he had found. Just as the mechanisms that were invented to explain the observations of Wegener and Velikovsky and others, such as Carey, Allan and Delair, and Van Flandern, were physically implausible, their work became crackpot science. Scientists tend to reject an observation if a reasonable explanation for its cause cannot be provided. "If we can't imagine how it happens, it must not happen." That is bad science, but it happens over and over. That puts great pressure on an observer to provide an explanation of the process. Often, the observations have great merit, but the forced explanation is so artificial that the entire story fails to convince. In this book I present a mechanism, the formation and action of natural nuclear fission reactors, that has been omitted from consideration in our exploration of the Universe, and I show that this mechanism can explain many existing observations better than our present stories. I hope that I have also shown that this mechanism is physically reasonable and plausible.

Part of this voyage of discovery was a stumbling attempt to find out what had gone wrong in astronomy and cosmology, for we have certainly gone astray. Some of these paths have been clearer than others: remarkably, as I was struggling to prepare a decent sketch of the Milky Way Galaxy and the Solar System, for I am a poor artist, I found that one had already been done, by Cornelius Easton, in 1900. Other signposts and way-markers required more careful reading.

Many of our most important astronomical explanations invoke monsters that have never been seen, and worse yet, can never be seen. In some of these cases, I can only act as a tour guide, pointing out where nuclear fission has had effect, and saying "Isn't that interesting, probably the result of ..." and leave the detailed exploration of the sidetrack as an exercise for the reader. Hopefully, someone who has greater resources to investigate these possibilities will become interested in exploring these suggestions. More detailed developments of the application of the fission process in explanations of observations are welcomed, sought, encouraged, and requested.

The result of this research has been to produce a revised view of the Universe, and our place in it, that rather resembles a fisherman's net in terms of intellectual structure. It is a mixture of evidence for the effects of natural nuclear fission reactors and suggestions as to how these natural reactors explain various observations. Facts are connected and cross-connected in a somewhat untidy manner at present, and that makes an orderly presentation difficult. The distinction between evidence and a suggestive explanation may at times be somewhat fuzzy. Turns must be taken and backtracks made in order to cover the territory fully. This voyage has been a search for patterns in the observations, searching for and trying to understand what causes underlie similarities and differences. The reader's patience in dealing with such an exploration is sincerely requested. But science itself is a net, with weak meshes and strong meshes holding the structure together. Sometimes these meshes get distorted, unknowingly disconnected from reality, but representing a reasonable consistency of their own.

It should be clear from the start that this book is barely a beginning and that much more work still needs to be done. But it opens doors to a view of a new Universe, which you are welcome to visit and study; after all, you live here, too. Many details have been left as exercises for the reader. I hope that you will welcome the challenge.

In science, we have to name something before we can understand it, but naming by itself does not guarantee understanding. "Quasars" and "gamma-ray bursters" are clear examples. These names describe the observations, but do not explain what the underlying objects are. This book will describe several objects from a new viewpoint, from an understanding of nuclear fission. I will try to give them names that provide a new understanding.

The basic concept used in this research is that if the results of a process provide an explanation for observations, then the observations provide evidence for the existence of the process.

Science appears to progress by a combination of the two opposing theories of science. Francis Bacon proposed that science observes and formulates a theory to explain the observations, while Karl Popper considers that science formulates a theory and observes to test the predictions. For example, we are assured of the existence of gravity, even though no specimen of gravity has ever been isolated, inspected, weighed, measured, or its color determined. It is observed that things fall to the ground, and gravity explains this, and how things fall to the ground, and more. The process of gravity is not perfectly understood, and current theories do not perfectly explain all the observations. Yet the principle of the process of gravity is deeply ingrained in our theories of the Universe. Its explanations are essential to our understanding of a multitude of observations. So too, for electromagnetism and nuclear fusion. This book presents evidence that nuclear fission must also be included with these other three sources in order to prepare a correct model of our Universe.

There are several ways in which science uses knowledge to understand our world. We may know how something can happen, such as why icebergs float in the ocean: ice is less dense than water, and gravity forces the lighter to float on the heavier. We may know how something cannot happen, so that airline pilots never need keep watch for icebergs at 37,000 feet, even over the North Atlantic in the winter. Sometimes, however, we simply don't know how something could happen, and presumptuously claim that therefore it can't happen. If we don't know how something cannot happen, we may at least be more open to the possibility that it could happen.

Development of this book progressed by two considerations:

1. If nuclear-fission chain-reactions could cause a certain effect, have the consequences of that effect been observed, and particularly, are they poorly explained at present?

2. If an observation is poorly explained at present, could it have been caused by the effects of nuclear-fission chain-reactions?

Some of what is presented here is clearly speculation. That speculation is intended to show directions in which good research should be able to make good observations, and either confirm or refute the proposed possibilities. Our existing paradigms, an expanding Universe, Plate Tectonics, stellar galaxies, gradual evolution, are so deeply embedded in our science that it may only be from some sweet accidents that the ideas in this book are recognized, proven, and accepted.

This book reexamines several concepts in modern science that most of us, certainly this author, thought had been clearly settled long ago, if I was aware of them at all. I found that the settlement was often not that clear, and was determined more by the personal strength of ego of the participants than by factual reasoning. There are many things in our science that are well founded, well considered, and well settled. Those successes have led on to the remarkable success that science has shown in explaining most of our world. However, to insist that long-settled concepts should not be reexamined, as some critics will, is to pretend that science is thoroughly correct, at this moment at least. History proves this wrong.

A mistake often made in scientific controversies, is to structure them as though they were debates in logic, with pre-established choices for conclusions, argued much as a flamboyant trial lawyer might, with victory won by the loudest voices. Specious arguments, apparently plausible but truly misleading, are built by changing words without changing meaning.

For many of those involved in a scientific controversy, being right is important to their careers, assuring grants, tenure, and publication. For us, understanding our world is more important. Unfortunately, the either/or structure that most controversies must create, for the sake of identifying a "winner", means that most of us, and science, lose. When any scientist wins or loses, science is the poorer, possibilities have been eliminated.

An unfortunate aspect of many scientific controversies is that one side will specifically define the other side's position, and then proceed, quite successfully, to demolish that self-defined position. This battle with a strawman is always gloriously won by the constructor of the strawman, who then claims that he has defeated the other idea's champion. No prejudice is intended here. The tactic works both ways, for new and old, and is distressingly prevalent.

There are many arguments in science, between scientists, few dialogues, and almost no joint explorations of differing ideas. Unfortunately, for us and for science, many scientific arguments are produced by strong opinions based on the firm support of personal attitudes.

In the arguments (hardly a debate) between the Creationists and the Evolutionists, much effort has been spent on proving that the other side is wrong. Little effort is spent in attempting to extract an understanding of reality from what is known. Both sides distort the interpretations and explanations of the other, to discredit the foundation of the enemies' understanding. This detracts from the integrity of the two groups and distracts from our efforts to understand our world.

This work offers reconciliation between religion and science. Miraculous, catastrophic events have occurred, as told by many religions, that were caused by understandable processes, as studied by science. Re-united, these two ways of thought may help us to more completely understand our world, spiritual as well as physical. Perhaps religion was created by people at a time of crisis, to serve as a means for preserving their knowledge through a foreseeable disastrous time. Now, we see science as the finder of knowledge, becoming religious. Science has its defenders of the faith, who seem to feel that (nearly) all that can be known is already known, and that only they will find any new knowledge themselves. In other times, in other cultures, these intellectual protectors would have sought out the heretics and burned them at the stake.

This is seen most often in conflicts between true believers and debunkers, both of whom seem to take their status as proof of their correctness. Attempting to find common agreement among the true believers and the debunkers is like trying to be comfortable with your feet in the fire and your seat in the snow. In fact, the positions and arguments of both sides become so extreme it is difficult to determine which side are the debunkers and which are the true believers.

A shortcoming, sometimes a fatal flaw, of science is the need to make assumptions in the absence of knowledge. Often we simply do not know what we do not know. Usually the assumption is simply a continuation of what we think we know, either by observation or by theory, or by imagination. The folly is then that these assumptions become firmly embedded in our theory, our theory shapes our interpretations, our interpretations determine our observations, and what we assume becomes what we know. When an assumption, no matter how apparently secure, is used as a foundation for knowledge, the uncertainty and the potential for error must be recognized as an integral part of the assumption and everything derived from it.

Science often must accept what has been decided in the past, to form a foundation for progress into further discoveries and understanding. We do not need to prove $2 + 2 = 4$ over and over, when we balance our checkbooks. But because the foundation determines the shape of the growing structure, a false foundation leads us astray. Our foundations are always incomplete because they must constantly be rebuilt of new material. This book presents some new material for a sounder foundation.

This book describes a world that is quite different in some respects from the world described over the last few centuries. It uses knowledge from the most recent 80 years, knowledge of the nuclear age that was unimaginable to the founders of our sciences. Like many voyages, this exploration will have a few side trips. These are excursions to consider topics that do not seem to relate directly to natural nuclear fission reactors, but seem to be essential parts of the story. There are two major side trips which we must take, to explore some complicated effects. Even though these effects do not seem to relate to nuclear fission, their consequences are so important they demand investigation. These are the expanding Universe (it isn't), and the expanding Earth (it is). The standard view of these problems became locked into our science before enough was known for us to be aware of the possible alternatives.

This book addresses the possibility that there are significant fundamental errors in our understanding of the Universe. These errors developed because of perceptions that were expressed long before our more complete knowledge of the details of the physical world. Theories of the Universe were cast in place before nuclear fission was known, before antimatter had been recognized, and were shaped by the attitudes of our ancestors.

Specifically, it will be shown that our concept of the distance scale of the Universe is wrong because of a misunderstanding of the fundamental technique for measuring distances, Plate Tectonics is wrong because a fundamental fact in the formation of the Solar System was overlooked, and our understanding that the Sun is a star led to the misleading belief that all stars are suns.

Specific cases provide evidence for these errors, when scientific judgment led our ideas astray. An incorrect estimate of a perturbing mass in the Solar System, based on the wrong mass for Pluto, led to the rejection of the perturbation analysis. Assumptions that some observations of Supernova 1987A in the Large Magellanic Cloud were meaningless and could be ignored prevented a fuller understanding of that event. Careless and prejudicial review of the measurements by van Maanen, showing physically real rotational motions in nearby spiral galaxies, justified rejection of those measurements. Cleaning out the errors will lead to a fuller understanding of the place where we live.

In this new view of the Universe, there are no miracles, and no matters of faith must be invoked. This new Universe consists of the straightforward effects of well known processes, and no strange monsters are needed to explain our observations.

In a cultural world of experts and specialists, I claim to be one only in the fields of nuclear fission, radioactivity, and radiation. This has come from my education, at Caltech, and my work experience, at Atomics International and Rocketdyne, for a while a combined division of Rockwell International Corporation, and later of The Boeing Company. I am a stranger to the other fields into which this research has intruded, and therefore a layman, as you readers are also in many of these fields. This work has benefited immeasurably from the cooperation and help that I received from many experts, even though my research led me away from accepted models. For this, I am deeply grateful.

Perhaps, in these days, the only danger I may face, unlike Galileo and Giordano Bruno, is the annoyed rejection by colleagues who are unable to face a new Universe. It is not proper in science to tell everybody else that they're wrong. Most of this book does that. I worry only that this book will be seen as attacking problems that are not recognized with explanations that are not understood. At the risk of seeming melodramatic, I am comfortable with the words of the philosopher Immanuel Kant, in his *"Universal Natural History and Theory of the Heavens,"* (1755): "I have ventured, on the basis of a slight conjecture, to undertake a dangerous expedition; and already I discern the promontories of new lands. Those who will have the boldness to continue the investigation will occupy them, and may have the satisfaction of designating them by their own names." [Crowe 1994]

I also wish to express my appreciation specifically to G. Subbaraman, F. C. Dahl, R. T. Hammock (and Lydia, Zoey, and Ian), and T. E. Dix for assistance in completing this work, and to several technical libraries, especially those of the California State University, Northridge, and of Rocketdyne, for providing essential stores of knowledge, the raw material for this study. The SAO/NASA Astrophysical Data System Abstract Service was especially valuable in retrieving scientific publications through a long range of time. I must express my appreciation to the past and present editors (F. W. Walker, J. R. Parrington, and F. Feiner, and now E. M. Baum, H. D. Knox. and T. R. Miller) of the General Electric *'Chart of the Nuclides,'* for providing an invaluable road map to the nuclear world, and to Katharina Lodders and Bruce Fegley Jr. for doing the same with the Solar System. I wish to thank researchers around the world for their generous cooperation. I particularly appreciate the efforts of Gene Watson to keep my mind open to other people's new ideas. Valuable comments were provided along the way by Ray McGinnis, Don Erway, and Duane Doty.

Original graphs were prepared using PSI-Plot from Poly Soft International, and ProbPlot from Rad Pro Calculator. The book was prepared in Microsoft WORD and most of the analytical interpretations were done with Microsoft EXCEL. The Internet and email service were essential aids.

This research has made use of the NASA/IPAC Extragalactic Database (NED) which is operated by the Jet Propulsion Laboratory, California Institute of Technology, under contract with the National Aeronautics and Space Administration.

Reviews of the partially completed manuscript by Harry Pearlman, Earl Curtis, Carolyn Mallory, Gail Kennedy, and Arden Wray are greatly appreciated. This work benefited from the intellectual climate established by the people of Atomics International and Rocketdyne. Some of the ideas that I had thought were my own, and new, I found to have been preceded by the thoughts and work of Paul Kuroda, Robert Driscoll, and Marv Herndon.

This book has some grim parts, and some hard parts, but mostly, it should be fun. Do we live at a special moment in time, a special place in space? The Copernican Principle says not, but for us personally and collectively, it must be special. Personally, this is the only time we have here. Collectively, humans did not exist more than a few million years ago, and Earth, now, is the only place we know of where we can live. What we see around us must be special, and rare, and will be different from what was before and will be hence. This is a special time and place, and we must learn from it, while we can. We are all more or less aware of our surroundings, our environment, our world. Awareness leads to observations, observations produce data. Data can yield information, and information gives us knowledge. With knowledge we can achieve understanding, and understanding is the foundation of wisdom. Our world seems to need some more wisdom, and we seem to be the only creatures capable of providing it.

The Universe is far different from what we have thought, and far more fatally dangerous than we have imagined. As our knowledge and understanding grow, we are forced to become the caretakers of Life.

I hope that what may at first be seen to be foolish misinterpretations of our established observations will be accepted as more nearly correct reinterpretations that will lead us to a clearer understanding of our world.

Because this has been a very personal, almost solitary, exploration for me, I will write the book in the first person singular, almost as an expedition logbook. When we have finished this book's voyage, we will still not know all that is right, but we will know less that is wrong. Sometimes, our knowledge grows to force us to leave behind our old world, and enter a new world. You are now entering the future, a new world newly known. Please embark on a voyage of discovery, with new eyes open to a new view of our Universe. I hope that you will enjoy the voyage.

Moorpark, California
January 22, 2012

I. FOUNDATIONS

You have to learn a lot
to know a little.
Charles de Secondat

There is no such thing
as useless knowledge.
Virginia Trimble, 1996

Introduction

It is expected that nearly everyone who chooses to read this book will be a layman in some or many of the fields that we need to explore. Because of this, I have tried to write for the non-specialist throughout. I hope that specialists will forgive my treatment of their fields in this manner, for the benefit of improved understanding of what I have written in fields in which they have no special knowledge. I have probably failed most in this attempt in my own fields, falling prey to the specialist's assumption that what is common knowledge is commonly known.

There are some major obstacles to understanding that must be overcome before we begin our voyage. These obstacles are in the form of technical terminology and tools of the trade, that intrude and obscure the meaning of what has been discovered and understood. Much of this work deals with things that are nuclear, and that word is only slightly removed from unclear. It is hoped that this introduction to terms and tools will help to make this voyage of discovery easier. While the emphasis will be on the usage exercised in this book, some alternative forms will also be discussed, since independent review of the research literature by the reader will provide powerful confirmation of this discovery. There are times when we will struggle for a word, searching for the right word when there may be none.

A helpful book for this purpose is *"Planet Earth,"* by Cesare Emiliani [Emiliani 1992]. It is about as long as this present book, and is a science book, and so forms a fine foundation for understanding what is to come. Perhaps most importantly for the spirit of this voyage (which is grim at times), Emiliani plays with our vast store of knowledge. Knowing about our world should be fun. He describes science as resting on three legs: facts, figures, and theories. To that trio must be added thought, for it is thought that links the other three and helps us to tell one from the other. (Truly, that is a recurring theme in Emiliani's book.)

The reference list includes both technical literature and writings intended for the general public, often with some duplication. This was done to promote accessibility to all, and particularly across the compartmental walls of specialized science. Excellent articles and books exist for the generalist, and the authors of those pieces have worked hard to make the matters understandable to those readers without the specialist's knowledge of the field. (An unfortunate accident of this present work being done by someone somewhat outside the research establishment is that many of the references are to secondary sources, and so only indirectly credit the individuals responsible for the research. For those unintended slights, I apologize in advance.)

Sometimes, even often, this independent review will require the reader to develop an independent conclusion, different from that provided by the researcher, an alternative interpretation of the meaning of the observations. This same approach should be applied to this book, as well.

Consistency, Errors, Corrections, and Contradiction

Science is a continuing effort to achieve consistency of our models and theories with the reality of the Universe, and often with very small and limited parts of it. We attempt to achieve this by identifying errors, correcting them, and avoiding contradiction.

An important measure of our success is self-consistency: are the models and theories consistent with each other? It is difficult to imagine contradictory theories that are both consistent with a single reality. (General Relativity and Quantum Mechanics may be excellent examples of this problem; they cannot agree.) Astronomy has a rich history of developing theories that have been self-consistent, and even consistent with observations, as a result of careful theoretical work. But these were not valid representations of reality. However, each new theory brought us closer, in some ways, to the truth.

In some cases, it will be found that the observations are generously consistent with more than one model, and the value of consistency in recognizing reality vanishes. Consistency is an important test, but can be a dangerous trap that predetermines the outcome of analytical thought. The exercise of consistency and its ready appearance can hide errors.

Observations that don't match our expectations often carry the grim label "discrepancy." Initially, such a disagreeable measurement provides the observer with considerable personal trauma: observations are supposed to be better than before, not different! Indeed, a discrepancy may be the result of poor technique, instrument malfunction, or any of many other artifacts that may be identified and corrected, often to the observer's embarrassment and dismay.

However, many unresolved discrepancies survive, some to become part of our history, others to be dismissed, hidden, and forgotten. These small problems slip from our consciousness, surviving only in the ever-older, out-of-date literature, ignored.

Centuries ago, discrepancies in the observed positions of the five known Solar System planets (other than Earth) led to the invention of orbital epicycles. In time, more, and more complicated, epicycles were required for the elimination of the observational discrepancies, until finally it became clear that a drastically different model of the Solar System could be applied. The Sun is at the center! Unresolved discrepancies are Nature's way of saying, "You just don't understand me." Now, in our current astronomy, the Sun has lost its place at the center of the Universe, but discrepancies persist. Where is the missing mass? What is the Dark Matter? Where is the heat?

Discrepancies in the orbital positions of Uranus led to the discovery of Neptune [Burgess 1991, Moore 1995], and then Pluto [Asimov 1991, Reeves 1997, Stern and Mitton 1998]. Small, uncertain discrepancies in the orbital positions of Uranus and Neptune suggest other discoveries [Littman 1988, Watters 1995], but, as the discrepancies appear to fade with time, our interest fades, and the opportunity may be lost [Standish Jr. 1993, Quinlan 1993]. Perhaps we just don't understand. Least Squares, Myles Standish's method of analysis, smears the residuals over the orbit, and makes the perturbations average out. Least Squares gives the best result for the chosen model, but does not prove that the chosen model is correct.

Discrepancies in the precession of Mercury's orbit led to a confirmation of Einstein's General Theory of Relativity. Few similar opportunities have been found, but the observations of two pairs of binary stars, DI Herculis and AS Camelopardalis, contradict this proof [Naeye 1995]. Is our theory wrong? Perhaps not, even the conventional prediction fails for DI Herculis. We may be applying the theory improperly.

In the theory of the expanding Universe, the Hubble Constant has so far been found to be far from constant [Trimble 1996]. Different measurements, as good or as bad as they may be, give different results. The results conflict with other results, and the Universe seems to be younger than its oldest stars. With great effort, this discrepancy seems to have been conquered. Is our theory wrong? Perhaps, but at present, most efforts are directed toward reducing or eliminating the discrepancies. This is a one-sided approach that risks hiding the truth.

The best observed star, our Sun, continues to present discrepancies. Why do we not observe enough neutrinos, where did all the angular momentum go, how did it lose so much of its lithium, why doesn't it have enough beryllium? Is our theory wrong? Perhaps, perhaps not. We might just have our assumptions wrong.

SN 1987A, the best-observed supernova yet, produced a glorious collection of discrepancies [Goldsmith 1990, Marschall 1988, Murdin 1990, Woosley and Weaver 1989, Keishner 1988]. A neutron star/pulsar was expected [Lindley 1988], found and lost [Kristian *et al.* 1989], and rejected [Anderson 1990]. Some of these have been authoritatively dismissed as having no significance or reality, because we don't understand them [Bahcall 1989; Mann 1997; Cumming and Meikle 1993]. That misses the point. A discrepancy cries out for attention, *because* we don't understand it.

A discrepancy that survives easy resolution should be treasured, protected from arbitrary dismissal. These are the clues that lead to a new understanding of our world. Nature is saying: "Pay attention. You just don't understand me."

Numbers

Since our discussions must range from the submicroscopic to the cosmological, and consider the energy of a single electron and that of stars, we need a way to manage the numbers that represent values that cover an extreme range. For that purpose, so-called scientific notation will be used whenever normal numbers do not easily cover the span. This uses a number followed by a multiplying factor of 10 raised to a numerical power. The power explains how many places to move the decimal point, and the sign, positive (usually omitted) or negative, shows in which direction. In most cases in this book this format will be used where it will be clear that a positive (or unsigned power) will be associated with a very large number, while a negative power will represent a very small number.

As an example, the average distance from the Earth to the Sun can be represented by these values:

1 Astronomical Unit or 149,597,900,000.0 meters or 1.495979×10^{11} meters.

Since scientific notation also allows us to present numbers without using more digits than required for the purpose, or reflected by the accuracy with which the value is known or exists, this distance might be represented approximately by

1.5×10^{11} meters.

Towards the other end of the scale, we could consider the wavelength of green light as

0.00000556 meters

or

5.56×10^{-7} meters

However, to make a rule, break a rule, so wavelengths of visible light will usually be seen as nanometers, 556 nanometers for green light, or with the historical unit, the angstrom, 5,560 angstroms in this case.

Units will be used that are natural to the field being explored, with some additional clarification for particularly obscure situations. In most cases, these will be the metric units ("SI" for Systeme Internationale) consisting of the kilogram for mass, meter for length, the familiar second for time, with metric fractional units for short times and metric multiples of the year for long times. Usually, long time-spans will be so long and vague that the refinement of what kind of year is meant, calendar or mean or leap or Solar or sidereal, will be irrelevant and will be ignored.

In some cases, ordinary English (American English, here) has special names for large numbers. These are,

1,000	thousand	10^3
1,000,000	million	10^6
1,000,000,000	billion	10^9
1,000,000,000,000	trillion	10^{12}
1,000,000,000,000,000	quadrillion	10^{15}

For simplicity and familiarity, I'll use these named numbers whenever suitable.

Mass and Energy

A bewildering variety of units have been developed to describe quantities of mass and energy. Because different units are natural to the particular uses and are so widely used in the technical literature, we will use several different units here.

The reference unit for mass is the kilogram, equal to about 2.2 pounds. The gram is one-thousandth of the kilogram, and is roughly one-thirtieth of an ounce. The atomic mass unit is approximately the mass of a proton or a neutron, and is 1.66×10^{-27} kilograms. (A very small mass.)

The reference unit for energy is the joule, representing the energy released at a rate (power) of 1 watt for 1 second, and therefore the joule is 1 watt-second of energy. This is similar in concept to the unit used for billing electrical service, the kilowatt-hour. A kilowatt-hour is the energy released at a rate of 1,000 watts for 1 hour, or 3,600 seconds. Thus, a kilowatt-hour is 3,600,000 watt-seconds (or joules) or 3.6×10^6 joules.

One of the great accomplishments of Albert Einstein was the establishment of the equivalence of mass and energy. His truly famous equation states:

$$E = mc^2$$

The energy content of matter is equal to the mass times the square of the speed of light. This lets us calculate how much energy would be released if a certain amount of mass were destroyed, actually, converted into energy, since the energy can also be converted into matter.

1 kilogram = 3.5×10^{42} joules

1 atomic mass unit = 931 million electronvolts

The energy content of light varies with the frequency or wavelength of the light, so that light with higher frequency (shorter wavelength) has more energy. Light comes individually packaged as photons, and

1 photon of blue light

(wavelength of 450 nanometers, frequency of 6.66×10^{14} cycles per second, or hertz) = 2.76 electronvolts

1 photon of red light

(wavelength of 700 nanometers, frequency of 4.28×10^{14} cycles per second, or hertz) = 1.77 electronvolts

Notice that the unit of energy for a photon of light (electronvolt) is a million times smaller than that used for an atomic particle (million electronvolts). The unit for frequency, the hertz, is just the same sort as is used for radio broadcasting as the kilohertz or the megahertz. Red light has a longer wavelength (and lower frequency) than blue light, and this extends throughout the range of light, or electromagnetic radiation. Higher energy X-rays and gamma rays have shorter wavelengths (and higher frequencies) than do the colors of visible light, and visible light has shorter wavelengths (and higher frequencies) than does infrared light, microwaves, and radio waves. The long-wavelength end of the spectrum is always termed "redder" even when it is far beyond red, and the short-wavelength end is "bluer", even though it may be far past blue. Otherwise, all electromagnetic radiation is the same, regardless of its energy or name. Its interactions with matter will be greatly affected by its energy.

In the nuclear world, a unit derived in the early days of research, the electronvolt, is the supreme unit. It is equal to 1.6×10^{-19} joules. We will tend to use the metric multiples of the kilo-electron-volt (1,000 electronvolts, keV) and the mega-electron-volt (1,000,000 electronvolts, MeV).

Energy can be related to temperature in many cases, usually in the absolute or kelvin scale, as the cold of space (3 kelvin, or simply 3 K) and the heat of the Sun (5,600 kelvin). Standard body temperature is 310 kelvin, water normally freezes at 273 kelvin and boils at 373 kelvin.

Cerenkov radiation is a special form of light produced by high-energy electrons, traveling faster than the speed of light in a transparent medium. This radiation is completely independent of temperature. Plasma Cerenkov radiation is a low energy (radio-wave) radiation produced by electrons in a plasma [Cohen 1961], such as the cloud of protons and electrons that would exist around a nuclear fission detonation. The spectrum is roughly opposite that of the optical Cerenkov effect, increasing towards longer wavelength. The spectrum has a low frequency cutoff and fades away in intensity at short wavelengths. For a typical plasma frequency of 10^8 cycles per second, the wavelength is on the order of 3 meters.

Nuclear Physics

Nuclear physics is old "new physics", invented to explain an accumulating variety of observations that did not fit into the classical ideas of matter. Isaac Newton had described (1672) how light was actually a combination of many colors, the spectrum of the rainbow. Nearly two hundred years later, Kirchoff and Bunsen found that different chemical elements emitted (and absorbed) light at very specific and distinctive wavelengths, or colors. The patterns of these "spectral lines" could not be explained with the old model of an atom as a single indivisible object, devised by the Greek philosopher Democritus (460-370 BCE). The situation became further complicated by the discovery of the electron by J. J. Thomson in 1897. The atom contained individual discrete points of negative charge, so neutral atoms must also contain positive charges.

Alpha particles (from the Curies' isolation of natural radioactive elements, discovered by Henri Becquerel, because of Wilhelm Roentgen's discovery of X-rays) were found by Rutherford, and by Geiger and Marsden, to be scattered from their original paths, in passing through air or mica or gold foils, in a way that theory could not explain.

In a remarkable interplay of experimental observations and theoretical deductions, Rutherford defined the nuclear atom as a tiny, massive point of positive charge, encircled by negative electrons constituting the major volume of the atom. Nuclear Physics was born in 1911.

The smallest particle that represents a consistent chemical element is an atom. The center of an atom is the nucleus, smaller by a factor of 100,000. It contains the mass of the atom and determines the nuclear behavior. The deepest details of the nucleus are not important here. We will deal only with a few simple particles that make up the nucleus and are involved in the nuclear reactions that we need to consider. Just as the elements have different chemical properties and actions, different forms of atoms, that we will call isotopes or nuclides, have different nuclear properties and actions.

The important terms for our consideration are

proton, shown as p, or as ^1H, to indicate a normal hydrogen atom, when the atomic character is dominated by the nuclear character. It has a single positive electrical charge, and a mass of 1.7×10^{-27} kilograms. The number of protons in a nucleus is called the atomic number, and determines the chemical behavior of an atom.

electron, shown as e^-, or as β^- when its production in nuclear decay is more significant than its nature as a charged particle. It has a single negative electrical charge, exactly equal to the charge of the proton, but opposite, and a mass 1,836 times smaller than the proton. The number of electrons around a nucleus determines how the atom will react in a chemical sense.

neutron, shown as n, combines with protons to form the various types of nuclides with different mass and different nuclear properties. It has no electrical charge. The mass of the neutron is slightly greater than that of the proton.

nucleus, the combination of protons and neutrons that determines how the atom will react in a nuclear sense. Depending on the particular nuclide, it will be shown as the chemical symbol, which identifies how many protons are contained in the nucleus, and the mass number, which is the number of protons and neutrons combined in the nucleus. The simplest, ^1H, to more complicated, uranium-235 (^{235}U), are identified in this way. In any radioactive decay or other nuclear reactions, the combined number of nucleons (the protons and neutrons) or "baryons" stays unchanged. Nuclides of the same chemical element, but with different mass numbers are known as isotopes.

isotopes, which are identified in the shorthand way as ^1H, ^2H, and ^3H for three isotopes of hydrogen, for example. These may also be spelled out as, for the hydrogen isotope with 2 nuclear particles (1 proton and 1 neutron), hydrogen-2, or given a special name, deuterium in this case. History happens more than history is planned, so in some of these cases, proper names have been given, such as protium (rarely used except when specific identification of this kind of hydrogen is needed), deuterium, and tritium, for the three hydrogen isotopes, with their own special symbols, p, d, t when the particle nature is important ("protons", deuterons", and "tritons"), and H, D, T when the nuclear nature is important. (Note the glorious freedom from consistency that nuclear physicists maintain. The intent is usually clear from the context.)

ions are atoms that have lost (or gained) one or more electrons around the nucleus. This results in a net electrical charge (atoms are normally electrically neutral because of the balance between equal numbers of protons and electrons), either positive or negative. These are usually shown by the chemical symbol and a charge sign indicating how many electrons the atom has gained (-) or lost (+) (!). Thus, ionized hydrogen will be shown as H^+, while oxygen that has chemically gained two electrons will be shown as O^{2-}. Other specialized notations are sometimes used and will be discussed when needed.

photons are the carriers of energy that comprise the electromagnetic spectrum, which includes in our ordinary experience, radio waves, microwaves, infrared light, visible light, ultraviolet light, and X-rays and gamma-rays. Starting with radio waves at the low energy end of this list, the higher the frequency, the shorter the wavelength, and the greater the energy. For radio waves, the frequency is considered the characteristic parameter; for microwaves through ultraviolet light, the wavelength is used for identification; and for X-rays and gamma rays, the energy is the typifying measure. Photons have no mass or electrical charge.

neutrinos are massless and chargeless particles that were postulated in order to balance certain radioactive decay observations. Their effects have been observed independently, and so they have become quite well established in our stories and are extremely important in understanding the Universe. In an effort to explain the "Solar Neutrino Problem" and some other anomalous observations, current theories propose that neutrinos have very small masses, and change form, or "oscillate" as they travel.

antiparticles: while it is thought that each particle has its own antiparticle (in some cases, itself), which has several exactly opposite traits, here we need be concerned with only four:

positrons are the positive version of the electron, and are produced in certain forms of radioactive decay and high-energy interactions;

antineutrinos are the antiparticle for the neutrino and are also specifically produced in certain forms of radioactive decay.

The proton and neutron also have antiparticles.

Occasionally, we must speak of

baryons, "heavy particles", the protons and neutrons and other matter made of these components, and

leptons, "light particles", the electrons and all other less massive particles.

Since the concept of chemical element and isotope is so important to our discoveries of the Universe, some additional time should be spent on these concepts. The primary building blocks of the nucleus are the positively charged protons and the uncharged neutrons. The number of protons in a nucleus determines what chemical element it is, and its chemical characteristics, and is called the atomic number, Z. The combination of protons and neutrons determine its nuclear characteristics. The combined number of protons and neutrons gives the nucleus its mass number (A) and we use this to distinguish between special forms of any particular chemical element, the isotopes.

Radioactivity

Radioactivity, the spontaneous release of energy from the nucleus in the form of various nuclear particles, was discovered in 1896, and since then has become a major component in our exploration and explanation of our Universe. Several types of radioactivity are known, and each has particular characteristics. Radioactive decay may take place over an extremely wide time scale. (The word decay was taken from the fading away of the rate of energy release observed by the early researchers, not because something is rotting.) The time scale is represented by the "half-life": the time for one half of the radioactive atoms to change to the next form. This may range from less than a microsecond (a millionth of a second) to 1.4×10^{17} years, or longer. The ultimate death of the Universe may rest on the possible radioactive decay of the proton, with a half-life longer than 10^{33} years. (That is a very long time.) The form of decay is characterized by the particle emitted, if one is, using the historical names of alpha (α), beta (β), and gamma (γ) for the observed radiation. Later discoveries added electron capture, positron emission, and spontaneous fission to the known forms of decay. Slight differences in nuclear energy between the original nucleus and the final nucleus and emitted radiation, cause the decay to occur. Nature seeks the lowest energy state available.

The characters of these different decays are markedly and significantly different:

alpha decay releases a particle composed of two protons and two neutrons from the nucleus of the radioactive atom. This combination of nuclear particles is exactly the nucleus of a normal helium atom and as the alpha particle loses energy from its ejection, it collects two atomic electrons and becomes, indeed, an atom of helium. (Later, it will be important that we recognize that this helium is the isotope helium-4.) The alpha particle is emitted at a specific energy, or several different energies for a particular decay, and this is distinctive for each radioactive nuclide. Alpha decay is often followed by emission of gamma rays (γ), that are distinctive of the decay. Emission of the alpha particle leaves the new nucleus with a reduction in the number of protons (the atomic number) of two, and a decrease in the mass number of four. The alpha particle can be shown symbolically by the Greek letter alpha, α.

beta decay increases the atomic number of a nucleus without changing the mass number, by creating an electron in the nucleus, in changing a neutron into a proton, and ejecting the electron as a historically named beta particle. The beta particles have a broad range of energy, with a clearly defined maximum energy that is specific to the decay. Initially, the new atom will have a deficit of one electron, because of the increase in the number of protons, and will have a net positive charge of one unit. (An interesting conservation rule of this game is that there can be no net change in the number of a class of particles that includes the electron. This rule applies to any nuclear reaction, including beta decay, and to the Universe as well. This problem of the creation of an electron, and the broad energy distribution of the beta particles, is solved by the existence of the antineutrino, which, by being an antiparticle, can balance the electron. That is, at the moment of decay, both a particle (the electron) and an antiparticle (the antineutrino) are created. This leaves the net number of these light particles (leptons) unchanged.) For every beta particle emitted, an antineutrino is emitted. The beta particle can be shown symbolically by the

Greek letter beta, β. When it is important that this be recognized as the normal, or negative, beta particle, a negative sign can be added: β^-. Gamma-radiation generally follows beta-particle decay.

electron capture is the opposite of beta decay: an atomic electron is captured by one of the nuclear protons, which changes into a neutron. Since a lepton (the electron) has been destroyed, another lepton (a neutrino) must be emitted, or an antiparticle lepton (an antineutrino) must be absorbed, in order to keep the number of leptons balanced. Symbolically, electron capture can be indicated by EC. Electron capture is also called inverse beta decay.

positron decay is another opposite to beta decay and is often referred to as positive beta decay. (To keep things fair, beta decay is sometimes referred to as negatron decay.) Positrons have a broad range of energy, as in beta decay, with a clearly defined maximum, and produce characteristic gamma rays at 0.511 MeV when they combine with ordinary electrons and annihilate each other. As the positron is an antiparticle lepton, a neutrino is emitted at the moment of decay, to balance the number of leptons, or an antineutrino may be absorbed to induce positron decay. The positron can be shown symbolically by the Greek letter beta with a positive sign, β^+, to distinguish it from the normal beta particle.

Nuclear Reactions

The science of nuclear physics developed by using energetic nuclear particles to probe the nucleus. Initially, research was limited to use of the alpha particles produced by the decay of the natural radioactive elements discovered by Henri Becquerel in 1896 and studied intensively by Marie and Pierre Curie in Paris. As electrostatic particle accelerators were developed in 1932, other nuclear particles could be used, with much greater intensity and control over the energy. Since the particle energy was selected by adjusting the voltage of the accelerator, and depended on the number of electronic charges, the unit of the electronvolt (eV) was a natural unit for nuclear energies.

The nuclear particles used for this research in the early days (about 1900-1940) were generally limited to natural alpha particles and the identical artificially accelerated helium-4 nuclei (fully ionized atoms), and ions of hydrogen (^1H or protons, p); heavy hydrogen, deuterium, (^2H or deuterons, d); and super-heavy hydrogen, tritium, (^3H or tritons, t). Since both the particles aimed at the nuclei being studied and those nuclei themselves have positive electrical charges, the repulsion of like charges makes reactions very improbable and tremendous numbers of particles and target nuclei must be used. The concept of a unit to represent the probability of an interaction between an incoming particle and a target nucleus was developed as the cross-sectional area of a sphere representing the effective size of a nucleus which the particle would impact. For charged particle reactions, the probabilities could be represented with a unit on the order of 10^{-28} square centimeter.

A special type of nuclear reaction studied with the particle accelerators became known as nuclear fusion, because it involved the combination of a particle and target nucleus that were comparable in size or even identical. Thus, two deuterons could be combined to form a helium nucleus:

$$^2H + {}^2H \rightarrow {}^4He + \gamma.$$

(Actually, two gamma-rays are released, each with about 0.87 MeV.)

As it became known that the Sun was composed of hydrogen and helium almost exclusively, fusion reactions were seen to be the long-sought source of the Sun's energy, and fusion was seen as the energy of the stars. Because fusion reactions involve the interaction of similarly charged particles, considerable energy must be brought into the reaction, to overcome the like-charge repulsion. This input energy is supplied in the laboratory by the accelerator, and in nuclear fusion weapons and the Sun and stars by high temperature. Nuclear fusion resulting from interactions at high temperatures is called thermonuclear fusion.

In 1932, Chadwick discovered the neutron as an uncharged nuclear particle with a mass about equal to that of the proton. This particle was first produced by exposing beryllium to alpha particles from naturally radioactive elements, but more controllable reactions using the accelerators were soon developed. Artificial radioactivity was discovered in 1934, by Irène and Frédéric Joliot-Curie, by bombarding targets of beryllium or boron with alpha particles.

This provides an opportunity to describe the manner in which nuclear reactions can be written, in both a "chemist's style" and a "physicist's style":

For alpha particles on beryllium, producing a neutron and leaving a carbon atom,

$$^9Be + {}^4He \rightarrow {}^{12}C + n \qquad \text{or} \qquad {}^9Be(\alpha,n){}^{12}C.$$

The physicist's style will be particularly useful because the parenthetical statement, meaning "alpha particle in, neutron out" implies a general (α,n) reaction that can be applied to a variety of target nuclei, other than beryllium-9, as in this case. For historical reasons it might be proper to restrict the name "alpha" to only those helium nuclei emitted by alpha decay from radioactive atoms. However, it is useful to use this name for any fully ionized helium-4 nucleus that is involved in nuclear reactions. So, we may use "alpha fusion" when we mean "helium burning". (Of course, it isn't really "burning", but nuclear fusion.) For more complicated reactions, these equations can be strung together:

$$^{58}Fe(n,\gamma){}^{59}Fe(n,\gamma){}^{60}Fe(EC){}^{60}Co(\beta^-){}^{60}Ni$$

which shows natural iron-58 absorbing a neutron to make iron-59 which is allowed to absorb another neutron before it can decay, to produce iron-60 which decays by electron capture, with a moderately long half-life, producing radioactive cobalt-60, which decays rather quickly, leaving nickel-60 as the end result.

Further, this form provides a means for expressing many other general reactions, as;

(p,n) proton in, neutron out

(d,n) deuteron in, neutron out

(α,n) alpha in, neutron out

(n,γ) neutron in, gamma-ray out (neutron capture)

(n,p) neutron in, proton out

(n,α) neutron in, alpha particle out

(n,2n) neutron in, two neutrons out.

These will be developed further as the need arises.

Charged particles, the proton, deuteron, and alpha, are repelled by the positive charge of a nucleus. To force a charged particle to interact with a nucleus, it must be given a large amount of energy. Even then, the chance of reaction is very small.

The neutron is uncharged, so there is no repulsive force from a target nucleus, and the neutron can quite easily penetrate the nucleus and interact. This led to interaction probabilities or cross-sections that were 10,000 times greater than those for charged-particle reactions, 10^{-24} square centimeter in size rather than 10^{-28} square centimeter, which resulted in the exclamation that neutron cross-sections were, relatively speaking, "as big as a barn!" Thus, the unit of interaction probability, or cross-section, for nuclear reactions induced by neutrons became the barn, equal to 10^{-24} square centimeter.

Hans Bethe found in 1936 that the energy production in the Sun (and by association, in stars) could be accomplished by nuclear fusion, either by direct proton-proton fusion or by the catalytic C-N-O cycle. Eventually, it was realized that proton fusion could not produce the wide range of elements that we see in our world. That problem was solved comprehensively (though not necessarily entirely completely) by Geoffrey and Margaret Burbidge, Willy Fowler, and Sir Fred Hoyle in 1956-57 [Burbidge *et al.* 1957]. That work described several processes that were able to build heavier elements, beyond oxygen, silicon, and iron. The process that forms uranium and the heavier elements (essentially all the elements beyond iron), is a neutron-capture process in supernovae. This was named the r-process, for "rapid" neutron capture. Fowler shared a Nobel Prize for this research, in 1983.

Starting with the observed uranium-235 content in natural uranium at present, 0.72%, and the half-lives of uranium-235 (0.7 billion years) and uranium-238 (4.5 billion years), we can calculate the isotopic composition in the past. (This is subject to the assumption that only constant radioactive decay has changed the abundances. This assumption seems right in most cases. The only major deviations I am aware of are in the ore bodies of the fossil nuclear reactors at Oklo (less uranium-235) and the meager uranium found in some meteorites (more uranium-235). Bursts of neutrons can significantly alter this view, and remarkably, drive both richer and poorer uranium towards a value, 0.74%, that is very close to what we consider normal today, 0.72%.

At 4.55 billion years ago, the assumed age of the Earth, the uranium-235 content was 24%. That is, in current Nuclear Engineering terminology, it was 24% enriched uranium, an enrichment appropriate for the construction of what are called "fast breeder reactors".

The calculated uranium-235 content in the uranium produced in a supernova is 55-59%. The uranium is produced by the r-process, in which neutrons are released by the collapse of the stellar core, and the neutrons are absorbed by nuclei out to the neutron drip-line, before beta decay can occur. (As neutrons are added to a nuclide, the binding energy for each next neutron becomes less, until the nucleus can no longer hold on to any more. The next neutrons just "drip" off. There was some promising research begun on absorption beyond the drip-line, the prompt process, with nuclear weapons detonations supplying the neutrons, but that ended with the test ban [Becker 1993].) The uranium we have now would have had an enrichment of 60% at 6.4 billion years ago [Ozima 1987, Harwit 1988, Rolfs and Rodney 1988], so we think that the supernova that made the star-stuff of which we are made happened 6.4 billion years ago. Considering the estimated age of the Earth as 4.55 billion years, based on the apparent age of the materials in the meteorites, this age for the supernova seems "reasonable". There are other "clocks" based on other radioactive pairs, and these clocks give variously conflicting ages, suggesting that something is happening that we don't understand. Some measurements are consistent with an age of 6 billion years. These have been rejected, or ignored, or corrected, as they do not fit with our accepted age. If the asteroids, and the meteorites derived from them, are the result of a nuclear fission detonation, the meteorite age represents the time of that detonation and is only a lower limit to the true age of the Solar System. The event that led to the setting of the meteorite clock had no relation to the initial formation of the planets of the Solar System, except that it occurred somewhat after the formation of the now-departed asteroidal planet. (I'll call that planet by the name "Asteria". Asteria was a titan and they aren't here anymore, except for some moons of Saturn.)

Nuclear Fission

Just over a hundred years ago X-rays were discovered, and natural radioactivity, a process unthought of that radically changed our ideas of our world. Einstein, relativity, quantum mechanics, an expanding Universe, and then the Great Depression and World War II. Sandwiched in there, nuclear fission was discovered. Not only previously unimagined, but unimaginable: the atom was the smallest indivisible part of matter. But in 1939 scientists started splitting atoms. (It actually had been done earlier, but the data were not understood and were misinterpreted.) The nucleus became a weapon, ending one war and preventing the next. Secret and feared, nuclear fission became the mystic knowledge of a very few researchers, mostly those who were willing to tolerate overwhelming governmental bureaucracy, secrecy, security, and control. Very few university professors fit this pattern and nuclear fission never made it into the academic mainstream where most scientific research is conducted. The most energetic process that we know of is not included in any of our theories of the Universe. Energy so powerful it can light a city, and so intense it can destroy a city. Before I started on this voyage of discovery, I had no idea that such a source of energy should be included in our view of the world. Now, in retrospect, I am amazed that it has been neglected for so long.

Some atoms can be made to fission, split, into two parts, releasing millions of times as much energy as a chemical reaction, such as burning wood or exploding TNT. That is what makes nuclear reactors so useful for power production (compared to burning wood), and nuclear weapons so effective in destruction (compared to exploding TNT). Those atoms occur in nature, and we work hard to make them suitable for use in reactors or bombs. Nature has other ways of working, and this story is about the effects of natural nuclear fission reactors.

Nuclear fission, in this context, is the process in which a heavy-element nucleus, such as uranium-235, uranium-238, thorium-232, plutonium-239, and many others, splits (fissions) into two roughly equal parts, either spontaneously as a mode of radioactive decay, or as the result of an induced increase in nuclear excitation energy, most commonly in modern experience, by the absorption of a neutron. Since neutrons are released from the fissioning nucleus in this process and become free to be absorbed by other fissionable nuclei, the possibility of a self-sustaining (self-perpetuating) chain-reaction exists. This chain-reaction is the process that is used, in a controllable manner, in our nuclear fission reactors, and in a self-destructive manner, in a nuclear weapon. When a reactor has established a self-sustaining chain-reaction, it is said to be "critical" or to have achieved "criticality". In the approach to a detonation, it is "supercritical".

Nuclear-fission chain-reactions of this sort, mediated by the neutrons produced by the fissions induced by neutrons, do not require input energy as do the nuclear fusion reactions that power the stars and make H-bombs (thermonuclear weapons) work. Thus, while nuclear-fission chain-reactions can create tremendous temperatures, no elevated temperature is needed to promote a chain-reaction. The high temperature produced by gravitational collapse of a stellar mass is irrelevant in nuclear-fission chain-reactions, and nuclear reactors can exist in quite small objects, as evidenced by nuclear power reactors and nuclear weapons. The chain reaction process, mediated by the extra free neutrons released in fission, is shown in Figure 1.

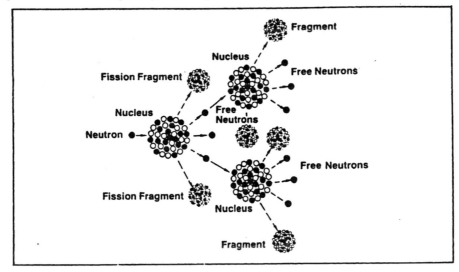

Figure 1. Nuclear-fission chain-reaction.
Conceptual diagram of the process of a nuclear-fission chain-reaction. Free neutrons released in each fission contribute to propagating the chain of fissions by inducing fission in additional nuclei; the fission fragments convert nuclear energy into kinetic energy, heat [Assistant Secretary for Nuclear Energy 1983].

Fission reactions release considerably more energy, per reaction, than do fusion reactions, by a factor of about 10 to 40 depending on the reactions being compared. Because the neutron-induced chain-reaction fissions discussed here are the result of nuclear excitation by an un-charged particle (the neutron), rather than the interaction of charged nuclei as is the case for stellar fusion reactions, the fission reactions proceed more rapidly than do the fusion reactions and do not require high ignition temperatures. The consequence of these differences is that large amounts of energy may be released in short periods of time from relatively little material, compared to the stellar masses and high temperatures that are required for nuclear fusion reactions. A nuclear reactor can operate at any temperature, depending upon removal of the heat that is generated.

Fission reactions, and the related absorption of neutrons by non-fissioning nuclei, leave distinct indicators in the form of the resulting isotopic compositions that differ from those established by the synthesis of the elements in stars.

Neutron-induced nuclear fission was discovered, after a series of misses, in 1938 [Hahn and Strassman 1939, Meitner and Frisch 1939, Hahn 1958]. Soon afterward, the development of nuclear fission reactors resulted in the first artificial self-sustaining nuclear-fission chain-reaction, at the University of Chicago, on December 2, 1942. The first man-made nuclear reactor had been made "critical".

The discovery of nuclear fission was announced in 1939 [Hahn and Strassman 1939, Hahn 1958, Sime 1998], resulting from irradiation of uranium by neutrons produced by exposing beryllium to the alpha particles from radium, the (α,n) reaction just described. The discovery was made by recognizing the nature of the radioactive products from the fission reactions. These radionuclides became known as fission products, a term that now also includes the stable nuclides ultimately resulting from radioactive decay. Nuclear fission had actually been achieved earlier by Enrico Fermi in Rome, but the experimental results were misinterpreted for several years as indications of neutron capture producing elements heavier than uranium. Similar misperceptions had also delayed the discovery in Germany. The assumptions through which the researchers viewed their results blinded them to the possibility of nuclear fission. They thought they had found what they were looking for, based on these assumptions, because their search was limited to what they had expected to find. Before its discovery, fission was unimaginable. Lise Meitner, Otto Hahn, and Fritz Strassmann had stubbornly picked at their results until the truth became clear. The experimental evidence eventually overcame their theoretical expectations. In an interesting dichotomy of sciences, the chemists of the research were convinced that barium, lanthanum, and cerium (fission fragments) had been produced. At first, the physicists could not accept that the "indivisible" atom had been split. Physicists agreed with the chemists a month later.

It has been discovered that various isotopes of thorium, neptunium, plutonium, and others at the end of the periodic table are also more or less easily fissionable. Some of these undergo fission spontaneously as a form of radioactive decay. Nuclear fission produces a variety of nuclear particles that will be crucial to this story. In addition to the fission products described above, which originate as intensely energetic fission fragments, neutrons and gamma rays are released; the fission products emit beta particles, gamma rays, and antineutrinos in their decays. The relative amounts of the fission product nuclides are different for each different fissioning nuclide, and can be used to identify these nuclides long after other fission effects have vanished.

Nuclear fission can occur in isolation as a form of radioactive decay called spontaneous fission, and can be induced in individual atoms by neutrons, protons, and other nuclear particles, and even by gamma rays. When this happens in uranium and many similar nuclei, some neutrons are released. These neutrons can be captured by other nuclei, causing those nuclei to fission, starting a "chain reaction". This is the process that is used in nuclear reactors and the atom bomb. The chain-reaction process is crucial in understanding our world. The observation that neutrons were released in nuclear fission reactions that had been induced by neutrons led to the concept of a continuing chain-reaction, in which succeeding fissions were induced by the neutrons released by fission.

Absorption of a neutron by a nucleus of the proper sort, leads to a highly energized, excited nucleus which then splits into two roughly equal fragments. Neutrons are released in this process and can propagate the reaction in a chain of fissions until either the fuel (the fissionable nuclei) or the neutrons run out. The fission fragments are highly radioactive, and decay quickly to less radioactive (longer half-life) or stable fission product nuclides. Gamma rays, X-rays, beta particles, and antineutrinos are also released as a result of the fission process.

Uranium and plutonium are easily fissioned by low-energy neutrons, as a result of the nuclear excitation energy produced by the absorption of a neutron. For some isotopes (uranium-233, uranium-235, plutonium-239, plutonium-241), the neutron need bring no other energy than its binding energy in the newly produced nucleus. Other isotopes (uranium-238, plutonium-240, and thorium-232) require additional energy that can be supplied by the kinetic energy of the incoming neutron.

For low-energy fission, the fissioning nucleus most often breaks into two rather unequal halves, one with a mass number of about 86 to 103 atomic mass units, and the other with about 128 to 146 atomic mass units. Several neutrons are also emitted, usually 2 to 4, and these neutrons are what keeps the chain reaction going in a nuclear fission reactor [Gönnewein 1991, H.-H. Knitter *et al.* 1991].

For higher energy neutrons, the production of more nearly equal-mass fission fragments increases. In the fission of uranium-235 by thermal neutrons (essentially zero kinetic energy), the peak-to-valley ratio is approximately 500:1. (The valley is where the two fission fragments have equal masses.) For neutrons with 14 MeV of kinetic energy, the ratio is only 5:1.

Further increases in excitation lead to increasingly symmetric fission, and a greater yield of fission neutrons. This will produce a preponderance of fission fragments with mass number near 112, with the release of approximately 12 fission neutrons. This great increase in the yield of neutrons will contribute to far more fissions during a detonation. As the effective "temperature" of the fission neutron spectrum increases with increasing fission energy, the yield of fission neutrons (per neutron absorbed) also increases. This further increases the fission rate, and so, further increases the temperature of the detonation. These are positive feedback mechanisms that will lead to a detonation in a confined body of uranium.

These fission neutrons are emitted with energies ranging from a small fraction of an MeV to above 5 MeV, with very nearly a Maxwellian distribution at an average energy of 2 MeV, typical of a temperature of 1.6 x 10^{10} K. As a side issue of some importance, free neutrons in the vacuum of space will decay by beta emission, producing protons and electrons. The protons will have an energy distribution that is close to that of the decaying neutrons, that is, a temperature of about 1.6 x 10^{10} K, 16 billion degrees kelvin. Neutrons that have been slowed down, or "moderated", by scattering off light-weight nuclei, will contribute little energy to the protons, and those protons will have the energy distribution resulting from the recoil in the beta decay. The proton energy distribution is shown in Figure 2, while the corresponding electron (beta particle) energy distribution is shown in Figure 3.

In a nuclear weapon, deliberately designed and constructed to detonate, the fission chain reaction continues and grows in intensity until the energy released in the nuclear detonation exceeds the inertial confinement that was produced by the initiating chemical explosion. Yields of purely fission weapons may be tailored by the design of a "tamper" material around the fissioning core, to confine the detonation for greater or shorter times.

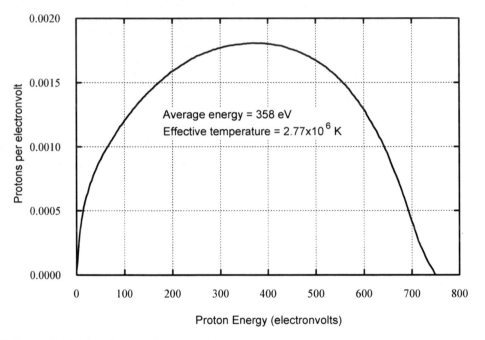

Figure 2. Energy distribution of protons from neutron decay.
The energy distribution of protons resulting from the beta decay of low-energy free neutrons in space. The protons have an effective thermal temperature of 2.77 million degrees kelvin. Note that the energy scale is in electronvolts.

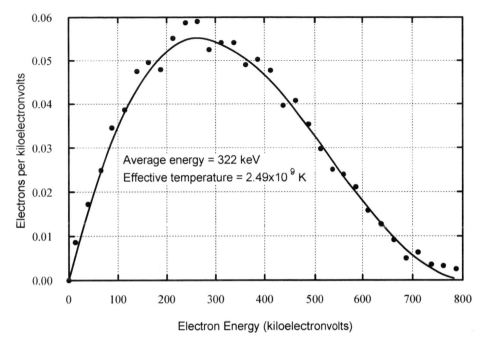

Figure 3. Energy distribution of electrons from neutron decay.
The energy distribution of electrons (beta particles) resulting from the beta decay of low-energy free neutrons in space. (Data points are from [Ito 2007]). The electrons have an effective thermal temperature of 2.49 billion degrees kelvin. Note that this energy scale is 1,000 times as great as for the previous figure.

Since the fission fragments have initial kinetic energies of 70 to 90 MeV, an intense detonation, where fragments interact with high-energy neutrons before the fragments have lost much energy, will experience very high excitation energies. As high-energy fragments collide with high-energy neutrons, further fissions will occur. If a detonating mass is so firmly confined that the detonation progresses nearly to completion, the final fission fragments will be moving in a gas that is largely neutrons, at a temperature on the order of 20 MeV, or about 1.5×10^{11} K (150 billion degrees kelvin). The fission fragments have kinetic energies in the range of 60 to 100 MeV (and will also be in an excited state, internally), so that interaction energies may be in the range of 40 to 120 MeV. This will lead to further fissioning of the fission fragments, producing lighter elements. Since this goes beyond normal fission, we might call it "superfission". This further fissioning will produce fragments leading to the iron group of nuclei (around iron-56). Fissioning of these nuclei, which might be called "ultrafission", will lead to nuclei in the sodium-magnesium-aluminum-silicon group (around magnesium-26). Even lower mass nuclides may be produced by a continuation of this process, perhaps named "ultimate fission".

A special feature of the fission reaction, designated ternary fission for the production of three nuclear fragments rather than the more usual two, produces the low-mass trace elements that are not easily produced by the Big Bang nucleosynthesis or by stellar processes. In particular, ternary fission produces some tritium (hydrogen 3, ^{3}H), which decays to helium-3 (^{3}He).

In developing these explanations, several unique characteristics of nuclear-fission chain-reactions should be recognized. These are:

1. Release of large amounts of energy, over a long or short time-period, independently of the ambient temperature. This energy is expressed as heat, neutrons, X- and gamma-radiation, beta particles, antineutrinos, and residual radioactivity.

2. Large time-integrated neutron fluxes (neutron fluence) with energy distributions ranging from essentially zero ("thermal") to 20 MeV, and possibly higher. In free space, with little material to absorb the neutrons, radioactive decay of the neutrons produces clouds of protons and electrons. These will recombine, producing the distinctive hydrogen spectral lines, as seen by the astronomers.

3. Creation of fission-product radionuclides and stable nuclides, production of induced radioactivity, and modification of the isotopic composition of existing elements.

The lack of inclusion of the distinctive nature of these characteristics is revealed in the difficulties of many poorly explained observations. Consideration of the nuclear-fission chain-reaction process in explaining a multitude of observations leads to explanations that are unified, understandable, and consistent. The possibilities discussed in this book are offered as those effects that appear most clearly to result from the nuclear-fission process.

Nuclear fission is the most energetic process in our Universe, except for the ultimate destruction of matter by mutual annihilation with antimatter, and can occur at any temperature. Nuclear fusion is less energetic, and requires extremely high temperatures and pressures, as we think exist in the hearts of stars.

Nuclear Fission Reactors

The continuing fission chain-reaction, and its control, form the basis for the nuclear reactors used in research, electrical generation, naval propulsion, and radioisotope production. In its uncontrolled form, the chain reaction provides the detonation energy of nuclear weapons. In a nuclear reactor, the chain reactions proceed at a generally steady rate, releasing neutrons, producing radiation and radioactive isotopes of various sorts, and releasing relatively large amounts of heat. It is that heat energy that is used in commercial nuclear power plants to generate steam and electricity. That heat is removed from the reactor itself by pumping large amounts of a heat-transfer medium (usually water) through the core of the reactor. In a nuclear weapon, the nuclear material is forced into a "supercritical" configuration (usually by compression by a high-energy explosion) so that the chain reactions increase rapidly and greatly until enough energy has been produced in the form of heat that the compression is overcome and the detonating nuclear material expands in a tremendous explosion.

In the average nuclear fission reaction, more than enough neutrons are released to maintain the chain reaction, generally about 2.5 or more neutrons per fission. This number varies from fission to fission, is somewhat dependent on the energy of the incoming neutron, and is different for the different types of fissioning nuclei. Over half of these fission neutrons must be lost, by non-fission absorption or by escaping ("leaking") from the reactor core, or the reactor power will increase rapidly. On the other hand, if too many neutrons are lost, the reactor will shut down and its power will quickly decrease. These effects provide feedback by which the power level of the reactor is controlled. When the conditions of a reactor are such that the rate of fissions continues at a steady pace, the reactor is said to be "critical" or that a state of "criticality" has been achieved. In the nuclear reactor industry, this is a very desirable condition, it is what makes reactors run, and for the initial startup of a new reactor, it is almost as exciting as "It's a girl!" or "It's a boy!"

On this basis, a nuclear-fission chain-reactor, a "uranium pile", was constructed in Chicago, and commenced operation, "went critical", on December 2, 1942. CP-1, Chicago Pile Number 1, was built of graphite blocks and natural uranium. The researchers ran out of uranium metal, and had to bring the reactor to critical by use of uranium oxide lumps. The total amount of uranium required was 11,659 kilograms (12.8 tons).

The uranium found in nature, in uranium ore on the Earth, consists primarily of two long-lived radioactive isotopes. These are uranium-238 and uranium-235. (A very small amount of uranium-234 is also present, in radioactive equilibrium with its parent, uranium-238. This isotope has important uses in radioactive dating.) In the production of heavy elements in a supernova, more uranium-235 is produced than uranium-238. However, the uranium-235 decays more rapidly and at the present time is only a very small fraction, 0.72%, of natural uranium. Of the two important uranium isotopes, uranium-238 has 3 more neutrons than uranium-235 has. When uranium-235 absorbes a neutron, more nuclear binding energy is released than for uranium-238, and this often leads to splitting the atom, fission, and a great release of energy, and some additional neutrons.

According to the currently accepted Big Bang theory for the creation of the Universe, there was no uranium at the beginning of time. As the initial fireball of creation cooled and the Universe filled with stars, there was only hydrogen, some helium, and trace amounts of the elements lithium, beryllium, and boron. the stars gradually produced heavier elements by nuclear fusion, up to iron, and abruptly produced much heavier elements by neutron absorption in supernovas. As successive generations of stars produced successively more heavy elements, an equilibrium may have been reached over the age of the Universe, with uranium decaying (or being consumed) at about the same rate as it is produced. We think that the ratio of iron to uranium in the Universe might be about 10 million to one [Allen 1976]. In the crust of the Earth, uranium is more plentiful, actually a thousand times as abundant as gold, and the ratio of iron to uranium is about 10,000 to 1 [Weast 1980]. The crust of the Earth is the only place where we can sample and analyze comprehensively. For the Sun, we see only the surface, and the meteorites that we can test may have catastrophically lost most of their uranium.

Fortunately for our ability to investigate the chain-reaction process in nuclear reactors, the tiny fraction of the easily fissioned uranium-235 is enough to permit the operation of a chain-reaction with moderators that do not absorb neutrons, such as graphite (pure carbon), heavy water (D_2O, water consisting of hydrogen-2 and oxygen), and beryllium. The moderator slows the high-energy neutrons produced in fission to "thermal" energies, where the cross-section for fission in uranium-235 is far greater than most other absorption cross-sections.

Ordinary water (H_2O) can also be used as a moderator, and this is the basis for the "Light-Water Reactors" (LWR) that comprise the majority of power reactors and naval reactors currently in use. However, the ordinary hydrogen (hydrogen-1) in water absorbs neutrons too readily to permit the use of uranium with its normal uranium-235 content (0.72%). In the competition for neutrons, the hydrogen atoms have the advantage over uranium-235 simply because of the much greater numbers of hydrogen atoms required to provide moderation.

In nuclear power reactors, this obstacle is overcome by using uranium in which the uranium-235 content has been enriched to 3 to 4%. In the reactors of the nuclear navies, the uranium has been enriched to 93 to 97%.

In the early days after the discovery of nuclear fission in uranium, it was realized that the chain-reaction could grow in intensity so rapidly as to create a powerful explosion, if nearly pure uranium-235 could be obtained [Peirls 1989].

This idea launched the atomic-bomb projects in Britain and the United States, initially from fear that Nazi Germany would produce such weapons. The Manhattan Project in the United States developed two methods for enriching uranium. Initially, high-throughput mass–spectrometers, called calutrons, were used, separating the two isotopes on the difference of their different charge-to-mass ratios. Enrichment by gaseous diffusion was developed next. The compound uranium hexafluoride (UF_6) is a gas at moderate temperature (above 56 C). Molecules of this gas that contain a uranium-235 atom move more rapidly, due to thermal energy, than the slightly heavier molecules containing uranium-238. This promotes the diffusion of uranium-235 through thin, porous ceramic barriers, so that gas on the far side of the barrier contains slightly more uranium-235 than does the input gas. Many stages of this enrichment process are required to increase the uranium-235 content significantly. Huge gaseous diffusion plants were built in Tennessee and Ohio to produce the highly enriched uranium needed for the first atomic bomb, nicknamed Little Boy, used at Hiroshima. After World War II, these enrichment plants continued in service to produce highly enriched uranium for weapons and the nuclear navy reactors, and low enrichment uranium for the civilian nuclear power program.

During the Manhattan Project, it was realized that use of the isotope plutonium-239 might permit construction of a more powerful bomb. This isotope had been produced in 1940, by way of its progenitor neptunium-239, from the neutron irradiation of natural uranium. It was the first truly artificial element discovered by laboratory research [Seaborg 1958]. During and since World War II, plutonium-239 is produced in large nuclear reactors that convert uranium-238 into the more easily fissioned isotope. Modern fission weapons utilize a combination of uranium-235 and plutonium-239.

Enrichment methods using lasers and high-speed centrifuges have been developed more recently, and offer advantages of efficiency and scale. Power reactors convert some of the uranium-238 into plutonium-239 and a modest fraction of the nuclear power is generated by fission of plutonium.

Nuclear fission reactors may be grouped into two distinct types: "thermal" reactors in which the neutrons promoting fission are "thermalized" or slowed down, moderated, to nearly thermal equilibrium with the reactor materials, and "fast" reactors in which the neutrons retain much of the kinetic energy that they were born with in fission. Thermal reactors use moderating materials of low atomic mass, such as light water (H_2O), heavy water (D_2O), or graphite to thermalize the fast fission neutrons by multiple scattering collisions. Thermal reactors generally require less fuel to achieve a critical mass, but are more susceptible to being halted in the chain-reaction by so-called fission-product poisoning, the absorption of neutrons by high-cross-section nuclides produced by the fission process. Fast reactors require greater amounts of fuel to achieve criticality, but are immune to the buildup of the thermal fission-product poisons. Nuclear weapon fission triggers (and the A-bombs dropped on Japan at the end of World War II), function briefly as fast reactors, until they are disrupted by the excess energy released. There are over 400 nuclear power reactors in use around the world for the generation of electricity [IAEA 1999], and several hundred more in service for propulsion of the nuclear navies. There are horrendous numbers of nuclear weapons.

There is evidence that natural nuclear fission reactors exist, in the past and present, on the Earth, in the Earth, and throughout the Universe, and provide a fourth source of energy, that has been previously overlooked in the development of our theories and models of the Universe. Because nuclear-fission chain-reactions produce effects and phenomena that are distinctly different from the consequences of the three standard sources of energy that are considered in our explanations, these existing explanations have failed to accurately describe our world.

Nuclear fission is the most intense source of energy we know (other than mutual annihilation of matter or the supposed capture of mass by hypothetical black holes), the quickest, and the only source whose action is automatically adjusted by the material. If the Earth is living, as Gaia, natural nuclear fission reactors are its life force.

Natural Nuclear Fission Reactors

Nature does nothing uselessly.
Aristotle

*If something is not expressly forbidden,
it is imperative.*
Murray Gell-Mann, from J. R. R. Tolkein

It is widely recognized that the major source of energy in the present Universe, and in the development of the Solar System, is due to thermonuclear fusion in the interior of stars, including our Sun. However, the complementary process of nuclear fission has not been included as a significant contributor to the conditions of our world, and has been overlooked in attempts to explain a wide variety of observations [Herndon 1992, Herndon 1993, Herndon 1994, Tuttle 1994, Herndon 1998, Herndon 2008].

Fission reactions exceed the energy, power, and speed of fusion reactions, and have no requirement for excessive temperature or density. Energy is generated, new elements and isotopes are produced, and various forms of radiation are released by fission. In a philosophical sense, it seems strange that the nuclear-fission chain-reaction process would have served no function in the Universe, other than for mankind's advantage as a highly technological source of power and of destruction.

In a philosophically reasonable Universe, such an important process would be intimately involved in routine operations, just as gravitation, nuclear fusion, and electromagnetism are. And, just as our knowledge of these processes has come to us slowly, it may be time now, though late, to recognize the effects of nuclear fission in nature. As it was not possible to properly understand stars until nuclear fusion was known, it would not be possible to properly understand the Universe until nuclear fission became known. Perhaps the secrecy associated with nuclear fission and weapons of war has made this understanding come slowly. As the 19th Century changed to the 20th Century, it was not possible to understand what was known then, until we came to know and understand fission. So far we have failed to integrate our knowledge of fission, and we still do not understand the Universe. Examples of our lack of understanding are abundant, in the Earth, the Sun, the planets, and the Universe.

The element uranium consists of several radioactive isotopes, and these decay at different rates. The two major isotopes, for our story, are uranium-235 and uranium-238 (U-235, U-238; ^{235}U, ^{238}U). We think that uranium-235 was produced more abundantly than uranium-238 in the supernova explosion that is thought to have made the material that ended up in the Solar nebula that led to the formation of the Solar System. Uranium-238 does not decay as rapidly as uranium-235 does, so at present, uranium consists of 99.28% uranium-238 and 0.72% uranium-235. Uranium-235 is easily fissioned; fission of uranium-238 requires high-energy neutrons. Uranium in the Earth is the result of rapid neutron capture in the later stages of a supernova. The uranium produced is rich in uranium-235, approximately 41% [Ozima 1987, Harwit 1988, Rolfs and Rodney 1988], but because of the more rapid decay of uranium-235, the present composition and abundance of uranium on Earth has been reduced from the initial levels to the present value of 0.72%. The reactors in use in the United States and in many other countries must use "enriched" uranium, in which the fraction of uranium-235 has been increased from the natural 0.72% to about 3 to 4%. At the time of formation of the Earth, the fraction of uranium-235 was about 24%; at 2 billion years ago, it was 3.63%, and that made the natural nuclear fission reactors at Oklo, Gabon, Africa, possible. It is predicted that the ratio of uranium-235/uranium-238 produced in a supernova is 1.42 [Ozima 1987] to 1.6 [Harwit 1988], or 1.93 or 1.24 to 1.42 [Rolfs and Rodney 1988]. If the uranium currently in the Earth were the result of a single supernova, the stellar explosion is estimated to have occurred 6.4 billion years ago, for an initial ratio of 1.47. If less time had elapsed from supernova to formation of the Earth, or if greater relative amounts of uranium-235 had been formed in the supernova, the nuclear fuel would have been correspondingly richer. A further confounding possibility is that the relative abundance of uranium-235 observed in surface material at the present time could have been modified from what it should have been, as a result of consumption, or production, of this isotope in the fission process.

Mass is converted to energy, and various forms of radiation are released. Uranium-235 is easily fissioned; uranium-238 takes high-energy neutrons to do it. Some complications in estimating uranium isotopic concentrations long ago will be explored later.

Since the uranium was richer in the past, about 24% uranium-235 at the supposed time of formation of the Earth, it had been suspected that natural chain-reactions might have occurred in favorable situations in the geologic past, in rich uranium ore bodies. Effects of these natural reactors were sought after, and apparently found [Orr 1949] in the form of isotopic depletion of the uranium-235 to approximately 0.32%, at a time when experimental apparatus for these measurements was extremely crude. When Wetherill and Ingham commented on the interference with measurements of spontaneous fission that would be caused by neutron-induced fission [Wetherill and Ingrham 1953] in 1953, the possibility of self-sustaining chain-reactions two billion years ago was recognized.

This possibility was predicted anew and in detail by Kuroda [Kuroda 1956] who also foresaw the possible confounding effect this would have on dating geologic deposits by uranium/lead chronology. Kuroda recognized the great probability of natural nuclear fission reactors occurring in rich deposits of uranium mixed with water, such as pitchblende (hydrated uraninite, $UO_3 \cdot UO_2 \cdot nH_2O$), in early times, when the shorter-lived isotope uranium-235 was much more abundant. A homogeneous mixture of pure uranium and water must have a uranium-235 content exceeding 0.94% in order to achieve criticality [Thomas 1978]. Present natural uranium has a uranium-235 content of 0.72%, so natural nuclear fission reactors of this type are no longer possible. A brief history of some of the development of the idea of nuclear reactors in the Earth's crust has been provided by Kuroda [Kuroda 1975]. Discovery of fossil nuclear reactors in ore deposits at Oklo, in Gabon, confirmed Kuroda's prediction [Bouzigues *et al.* 1975]. The early speculations and suspicions of natural nuclear fission reactors were not widely recognized, and even after the discovery of the fossil reactors at Oklo, the predictions of the existence of these reactors did not become generally known.

In 1972, French uranium analysts noted that some of the uranium won from ores from Gabon, the former French Equatorial Africa, was slightly deficient in uranium-235. Indications of a possible occurrence of this effect were reported as anomalous isotopic fractions in uranium from the French mills that were processing uranium ore [Bouzigues *et al.* 1975]. Confirmatory evidence of natural nuclear fission reactors in uranium ore deposits in Oklo, Gabon has been presented in detail [IAEA 1975, IAEA 1978, Brookins 1978, Morrison 1998, Petrov *et al.* 2006, Meshik, Hohenberg, and Pravdivtseva 2004]. Ancient reactors have also been found at nearby Okelobondo [Naudet 1977] and Bangombé [Nagy 1994]. As research into the Oklo reactors was completed, 17 individual reactors had been discovered.

In Gabon, in western equatorial Africa, about 1.7 to 2 billion years ago, at a time when the uranium-235 content of uranium was much greater than now, fission chain-reactions consumed a significant fraction of the available nuclear fuel, and produced an ore body that is locally depleted in uranium-235. These reactors may have operated at a power level of 100 kilowatts. In a river delta on a boggy lake-bottom, in deposits of rich uranium ore, a collection of natural nuclear fission reactors operated for hundreds of thousands of years [Cowan 1976]. This was made possible by a special combination of events: uranium minerals were deposited in a sufficiently concentrated form at a time when the uranium-235 content was high enough to permit a chain-reaction to develop with normal water moderation. That was a special period of time on the Earth. Earlier, when the atmosphere was lacking in oxygen, uranium could not be easily transported in solution, so massive deposits of uranium could not be formed. Later, by a billion years or so, the uranium-235 content would be too low for reactors to develop, no matter how massive the uranium deposit.

Research showed that the uranium deposit had operated as several independent nuclear reactors for a time estimated to range from 100,000 to 2 million years. Approximately 6 metric tons (6,000 kilograms) of uranium was fissioned. Eventually, consumption and decay of the uranium-235, and accumulation of fission product poisons, the "ashes" of nuclear fission, shut the reactors down. Searches for similar reactors elsewhere have not found any, other than the locations in and near Oklo, but there are no deposits of uranium ore known to be older than 2.3 billion years. (That may not be too significant, since there are not very many deposits of anything older than 2 billion years.)

At the time of the reactors in Gabon, the uranium-235 content, or enrichment, is assumed [Bryant *et al.* 1976] to have been over 3%, based on extrapolation of the current content for surface uranium to that time. (An exact calculation gives the uranium-235 content as 3.672 weight percent of the uranium.) At the time of the formation of the Earth, the enrichment is similarly assumed to have been about 24%. For this high an enrichment, it would seem almost inevitable that similar natural reactors would have been common on the surface of the Earth. These reactors would have been clustered into localized sites, isolated from each other, as those in Gabon seem to have been, and the effects would have been limited due to the strong control mechanisms provided by evaporation of the water and accumulation of fission-product poisons. However, perhaps concentrated deposits of uranium were not able to form so early.

Formation of the Oklo deposits resulted from prolonged and intense weathering, which has occurred only twice, during the middle Proterozoic (about 2 billion to 1 billion years ago), forming the Oklo deposits, and during the Permian (290 to 245 million years ago), which was too late for water-moderated reactors to form near the Earth's surface because of the decline in the uranium-235 content [Apt 1976]. Scattering of the high-energy fission neutrons by hydrogen nuclei in the water slows them to so-called "thermal" energy, essentially in thermal equilibrium with the surrounding material, just as the atoms in air are in thermal equilibrium with each other. At low energies, neutrons show a high probability for being absorbed by uranium-235 nuclei (a high "cross-section") and causing fission. A chain-reaction can develop only when the uranium-235 content is great enough to overcome competition for the neutrons by other absorbers, including hydrogen. At the time of the Oklo reactors criticality was easily achieved. In the chain-reaction process, these reactors produced plutonium by neutron absorption in the major uranium isotope, uranium-238. While some of this plutonium also underwent neutron-induced fission, much of it naturally decayed to nuclides indicative of the parent isotopes: plutonium-239 to uranium-235, plutonium-240 to thorium-232, and plutonium-241 eventually to bismuth-209. At low powers, plutonium-239 decays to uranium-235 more often than it fissions, so the reactor is dominated by uranium-235 fissions.

The fossil natural reactors in Gabon, at Oklo, Okelobondo, and Bangombé, are intriguing, for many reasons, and at least for their uniqueness. Why only there, so far as we know? There have been systematic searches for other uranium deposits depleted in uranium-235 or enriched in fission-product elements, but no others have yet been found. Gauthier-Lafaye gives some reasons for this uniqueness [Gauthier-Lafaye 1997]. The original ore deposit must have formed before about 1.8 billion years ago, for the uranium-235 fraction to be high enough for criticality, and the deposit must have been preserved geologically, for us to find it. Other conditions, such as concentration, shape, moderation, and lack of strong neutron absorbers, must also be suitable. It is likely that in earlier times, when the uranium was even richer in uranium-235 than for the Oklo reactors, criticality disrupted ore deposits in the process of forming, and none survived. The isotopic composition of this uranium, the composition that we consider to be "normal", may have been determined by the conversion ratio of these primitive reactors, the rate at which uranium-238 is converted to plutonium-239, as the uranium-

235 is consumed. Given sufficient time, the plutonium-239 decays to uranium-235, concealing the action of these ancient reactors.

The possible existence of natural reactors of this sort had been predicted by a Japanese émigré, Paul Kuroda, in 1956, and some others earlier. Natural nuclear fission reactors involved in planetary development, as I will discuss, have been described by Marvin Herndon [Herndon 1992, Herndon 1993, Herndon 1994, Herndon 1998, Herndon 2008] and Robert Driscoll [Driscoll 1988], but the standard theory can only explain why such reactors are not possible [Stevenson 1995].

So, it may just be luck, with long odds against natural reactors, that the uranium we find is all the same isotopic composition. This may also be a sampling artifact. Only concentrated uranium, in ore bodies, may be subjected to accurate isotopic mass analysis. Dilute uranium, from reactors that have consumed most of their fuel or were disrupted, would be overlooked.

A search for other natural reactors was reported by a team from LANL (then LASL) in the second Oklo conference in 1977 [IAEA 1977], and the project continued (at LASL) till the end of 1980. Nothing was found. However, George Cowan, (part of that team) had reported on slight anomalies in uranium ore from New Mexico and Texas [Cowan and Adler 1976]. Pitchblende ore from the Belgian Congo, deposited about 800 million years ago when the uranium-235 content was approximately 1.6%, shows a slightly greater amount of that isotope than is natural at present. This has been interpreted as an indication that natural reactors may have operated then and there, functioning as breeder reactors and producing slightly more fuel than they consumed [Cowan 1976, Maurette 1976]. Plutonium-239 that is not fissioned during the operation of the reactor decays naturally to uranium-235, making an observable increase over the normal abundance of 0.72%.

While there has been some speculation that massive uranium ore deposits could not have survived formation when the uranium-235 content was much higher [Gauthier-Lafaye 1997, Draganić *et al.* 1993, Kuroda 1975] it may be that none could have formed. One of the (rather speculative) reasons considered for why there are no uranium deposits from before 2.3 billion years ago, is that the uranium was so rich then that most of the reactors self-destructed, dispersing the uranium so that there was no future interest in exploration as a potential ore deposit [Draganić *et al.* 1993, Kuroda 1975] and so, no isotopic measurements would be made. Actually, that is rather appealing, in terms of some geologic structures called crypto-volcanoes. Diatremes may be the most significant of these "hidden" volcanoes. They look like volcanoes that blew their tops, but are volcanoes only skin-deep, and have no observable magma chambers or channels. The oldest known rich uranium deposit is at Elliot Lake, Canada [Nagy 1991], a 2.3-billion-year-old deposit of uraninite. Only suggestions have been found in other ore deposits, but it may be significant that no uranium ore bodies older than 2.3 billion years are known.

Otherwise, there is one other instance of promising interest, that found with (now) incredibly crude instrumentation. In 1949, J. B. Orr reported [Orr 1949] some measurements using thermal neutron-induced fission in uranium extracted from the mineral thucolite from the Besner Mine in Canada and from uranium in uraninite from the Ruggles Mine in New Hampshire. His measurements suggested a uranium-235 content of about 0.32% in the thucolite, which is about the lowest values found in Oklo ore. This seems to have been overlooked, as I don't know of any follow-up.

A possible observation of interest here is the finding of differences in the yield of the fission product xenon-136 from spontaneous fission of uranium-238 [Ragettli *et al.* 1994]. The xenon-136 trapped in crystalline phases containing dilute (zircon and monazite) and concentrated (pitchblende, uraninite, and coffinite) amounts of natural uranium was measured and found to give results for $\lambda_{sf}{}^{136}Y_{sf}$, the product of the decay constant λ and the yield Y, that were 17% lower in the concentrated minerals than in the dilute.

Uranium-238 decays both by alpha emission (release of a helium-4 nucleus) and by spontaneous fission, producing a variety of fission products. One of the fission products is xenon-136, which can be extracted from the uranium mineral and identified by mass spectrometry. Neutrons are released in the fission and these will be absorbed by uranium-235, causing fission (releasing more neutrons) or by uranium-238, producing plutonium-239, which decays to uranium-235. Any externally produced pulse of neutrons will exaggerate this effect.

In a very dilute uranium-bearing mineral, essentially all the neutrons will be absorbed by nuclei other than uranium, and so this effect will be absent. In concentrated uranium minerals, the effect will be amplified by the action of (short) fission chain reactions. The overall effect will be to reduce the production of xenon-136 (and other fission products) by the spontaneous fission of uranium-238 in rich uranium minerals, but make no change in the dilute minerals. That is the effect that was observed by Ragettli *et al.*

Draganić [Draganić *et al.* 1993] estimates that the potential for natural (near-surface) reactors was about 100 million reactor sites like the Oklo deposit, each with an average of 5 reactor zones per site. However, the great erosion that occurred about 2.5 to 2.0 billion years ago and the presence of chemically oxidizing and reducing environments that promoted the concentration of rich uranium ore deposits, seems to have been special circumstances. Perhaps only a very few natural nuclear reactors occurred near the surface and we were fortunate to find those in Gabon. For the rest of Nature's reactors, we must look deeper and farther.

Review of a wide variety of observations suggests that nuclear-fission chain-reactions in planets and stars occur almost universally. The action of these natural nuclear fission reactors has major consequences for the development of planets, stars and other astronomical objects, and our understanding of the Universe. These observations suggest the existence of other natural nuclear reactors, fast breeder reactors, in the core of the Earth, and in all planets and stars, all bodies that formed from the starstuff thrown out by exploding supernovae.

Herndon has shown, in a manner similar to that of Kuroda, that an on-going nuclear-fission reactor in the core of the Earth, rather than in ore deposits at the surface, is possible and likely [Herndon 1993], that it can provide the energy necessary to drive the Earth's geomagnetic dynamo, that similar reactors in the giant planets Jupiter, Saturn, and Neptune are the probable sources of the excess energy in these planets [Herndon 1992], and provide ignition for newly formed stars [Herndon 1994]. The possibility that nuclear reactors could develop within a planet, and result in a detonation that could disrupt the planet, as will be discussed in detail in this book, was described several years earlier by Driscoll [Driscoll 1988]. The possibility that the energy released by natural nuclear fission reactors in the earliest age of the Earth contributed to its formation, along with many other sources of heat, has been suggested, but no further study seems to have been done [Rogers 1993]. The possibility of these reactors continuing operation indefinitely as "breeder" reactors, producing new fuel as old was consumed, has been calculated by Hollenbach and Herndon [Hollenbach and Herndon 2001]. Like Kuroda's predictions, these ideas have had little notice.

The Faint Young Sun Paradox [Sagan and Mullen 1972] proposes that our Sun was not bright enough in the beginning to keep water from freezing all over the Earth. In view of the evidence for liquid water at early times, this is a serious problem. It is made even more obstinate by the White Earth Scenario, which considers that the Earth would have been such a white, bright snowball that it would have reflected so much of the Sun's warmth that the Earth would not have thawed yet.

I found that natural nuclear fission reactors in the core of the Earth, from the beginning of its formation, could have provided the energy needed to keep the water from freezing.

Natural nuclear fission reactors within the Earth (somewhat like those found in 1972 at Oklo, Gabon, West Africa) could have produced the supplemental heat that was needed to keep the Earth from freezing over during the first 3 billion years, when the Sun was not bright enough to keep water liquid on the face of the Earth. This provided an answer to the Faint Young Sun Paradox and the White Earth Scenario. I then found that there was evidence for the action of these reactors, and a very natural process for their original formation and continued operation.

It is likely that a reactor system exists in the Earth, with central reactors in the inner core, which modify global climate by heat production, and peripheral reactor cells, associated with the sources of mantle plumes at the core-mantle boundary, that modify local climate. In the Earth, the reactors are concentrated primarily at the core-mantle boundary and in and on the inner core. Reactors cycle between these locations, passing repeatedly through regions that are chemically oxidizing or chemically reducing. Natural nuclear fission reactors in the center of the Earth and at the core-mantle boundary could have provided the energy necessary to keep the Earth from freezing over before the Sun was warm enough.

Uranium became concentrated in the supernova ejecta cloud even before the formation of the Solar nebula, and concentrated further as the planets formed, contributing greatly to the evolution of the Solar System. Chemical, electrostatic, and mechanical processes in the pre-Solar cloud can concentrate uranium to the extent needed to form natural nuclear fission reactors, in planets and the Sun, and similarly in other stars. The consequences of these reactors have led to features of our world that cannot be reasonably explained by conventional means, and awkward theories have been developed and prevail in our view of the world. The failure to imagine the possibility of nuclear-fission chain-reactions within the Earth and on a planetary and stellar scale has resulted in the continued omission of this process in developing an understanding of our world. Consideration of the effects of nuclear fission, as the fourth source of energy on a large scale, provides a clearer view of the Universe.

Planetary reactors of this sort may act in a relatively stable mode for an extended period of time, although with pronounced fluctuations in power, producing climate and weather changes. Eventually reactors may reach a stage in which this stability is lost, and a nuclear fission detonation occurs. It appears that nuclear detonations produced the asteroids in our Solar System, as fragments of a detonated planet between Mars and Jupiter, ejected the crust of Mercury, dehydrated (and possibly sterilized) Venus and Mars, created the icy rings of Saturn, tipped Uranus on its side, destroyed many of the large natural moons, produced the fragments of the Kuiper Belt, and formed what we call the Milky Way Galaxy. Structural and isotopic characteristics of these planets are explained by nuclear fission processes. Many features of meteorites are explained as the result of the effects of the planetary nuclear detonation that produced the asteroids. (Interestingly, this process of nuclear fission planetary detonation to form the asteroids was considered to be the only way in which a planet could be blasted apart, but was rejected because a way to initiate the detonation could not be imagined [Napier and Dodd 1973]. There are some natural ways, but these were not recognized at the time.) There is evidence that these reactors continue in operation in the Earth, in the giant planets Jupiter, Saturn, Neptune, but not Uranus, and in the volcanistic moon of Jupiter, Io. (Perhaps even Europa and Ganymede still have nuclear life. Only Callisto seems to be completely dead.)

A reactor that is too small cannot act as a reactor. It will simply be a mass of radioactive material. It may heat itself by self-absorption of its own radiation, but it will scarcely be more active than that. Also, a reactor that is too large cannot stay together. It will overheat and break apart, if it can. This constrains the existence of natural nuclear fission reactors to a small range of sizes close to the well-defined critical mass for the composition and shape. Therefore, wherever the conditions for natural nuclear fission reactors occur, there will be many reactors, gathered in groups and scattered about.

The size of natural reactors, in the Earth's core, can be estimated from calculations of the critical mass of uranium-iron systems. The potential number, like Draganić's estimate for surface reactors (500 million), is very large. This number can be more precisely estimated from the power and energy deficit from 4.5 billion years ago to 1.6 billion years ago, keeping the average power of each reactor below that which leads to plutonium detonation. Uranium enriched to 24% uranium-235 is a nice nominal material to make a fast breeder reactor, and the Earth must have formed with a multitude of fast breeder reactors in its core. As uranium oxide, these reactors would have needed about 600 kilograms of uranium each. As elemental uranium dissolved in iron, the critical mass ranges from about 60 kilograms to about 5,000 kilograms of uranium. These reactors would be "bigger than a breadbox, smaller than a car". There may have been enough uranium at the time of formation of the Earth to make 100 trillion (10^{14}) natural nuclear fission reactors. Such efficient utilization of a rare resource is unlikely, but a much smaller number would have sufficed to overcome the faint young Sun.

There are many observations of effects in our Solar System and in our Universe that can be better explained by considering the consequences of large-scale, self-sustaining nuclear-fission chain-reactions, natural nuclear fission reactors, than is provided by our current understanding, which has been limited to the use of more conventional processes in interpreting observations. Inclusion of this process in explaining a variety of observations is discussed here: further exploration of the details requires application of the research tools and knowledge of specialists in the various fields. Review of these observations suggests that nuclear-fission chain-reactions in planets and stars occur almost universally. The action of these natural nuclear fission reactors has major consequences for the development of planets, stars and other astronomical objects, and our understanding of the Universe. These observations suggest the existence of other natural nuclear fission reactors, fast breeder reactors, in the core of the Earth, and in all planets and stars, all bodies that formed from the starstuff thrown out by exploding supernovae.

Nuclear-fission chain-reactions may start the fusion reactions of stars [Herndon 1994], before gravitational heating has increased the internal temperature to the necessary degree for fusion ignition to occur, and may cause many of the difficult-to-interpret observations of stellar variability and composition. An ongoing contribution by nuclear fission to energy production in stars, such as our Sun, might explain the apparent deficit in the Solar neutrino flux, by bypassing the p-p fusion, the first step in the fusion reaction. This allows a solution without applying mass to the neutrinos, which is not required by the very successful Standard Model of Particle Physics. Other effects must be studied to understand this problem. Many features of the recent supernova in the Large Magellanic Cloud, SN 1987A, including the two antineutrino bursts and the "mystery spot" and the light halos, are understandable in terms of detonations of asteroidal fragments, rather than the collapse of a supergiant star.

The geomagnetic field can result from the electron flux produced in the inner and the outer core of the Earth, by the intense gamma radiation of these reactors.

Natural nuclear fission reactors can induce the various isotopic anomalies seen in helium, neon, and many other elements, that are poorly explained otherwise. The tiny mystery of unnatural radiogenic halos in minerals is explained by the actions of natural nuclear fission reactors.

There are no natural nuclear fission reactors on the surface of the Earth at the present time, and we are protected from the nuclear radiation of the deep reactors by the great thickness of the rocky mantle, which absorbs all high energy radiation. Only antineutrinos, from the decay of the fission products, can escape to the surface of the Earth. If the cross section (probability) for the antineutrino-induced electron capture reaction presented in Appendix A is correct, even those antineutrinos will be absorbed and re-emitted, producing a glow, rather than a direct stream of radiation from the deep reactors.

The actions of such reactors explain a variety of otherwise awkward problems, such as the Faint Young Sun Paradox and the White Earth Scenario, the presence of helium-3 on Earth, and the formation of the asteroids.

Feedback

Dynamic systems usually involve feedback mechanisms that determine the behavior of the system. So-called negative feedback is usually desirable: it stabilizes a system, reducing the effects of small changes and maintaining steady performance. It is built into many engineered systems, such as aircraft and most nuclear reactors, as an inherent property of the design. For many types of equipment, a human operator is part of the controls. For example, a driver adjusts the speed of a car in response to changes in conditions, hills, Highway Patrol cruisers, and other observed situations.

Or the overall effect of the feedback can be to emphasize, amplify, and exaggerate any change. This is positive feedback, and is experienced when a microphone and a loudspeaker are too close together, and the squeal builds until its intensity is limited by some part of the system, or until some part of the system breaks. A wildfire often creates its own winds, fanning the flames into a firestorm, until all fuel is consumed.

Both these types of feedback are important in nature, and systems can change from negative to positive feedback, or the reverse, or even alternately, as the systems evolve.

In technological systems, such as commercial nuclear power reactors, negative feedback inherent in the physics and the design of the system provides remarkable stability. Nuclear electric generating plants are run continuously for a year or more, with only minor adjustments of the controls made occasionally to maintain a constant flow of heat and electricity. Accidents, such as at Three Mile Island and Chernobyl, can occur only when the human operators take drastic and inappropriate actions, and drive the system from its realm of stability into one of positive feedback. In well-designed systems, such as the Three Mile Island reactor, the realm of stability is large and its boundary is resistant to inadvertent crossing. In poorly designed systems, or those put into inappropriate use, as Chernobyl was, the realm of stability is small, the boundary weak, and the consequences of crossing the boundary are severe.

In nature, for the natural reactors we will explore in this book, it appears that crossing the boundary into the positive, self-destructive realm is done naturally. When this happens, a nuclear detonation results.

Science, a thoroughly technological system, is thought to be, and is intended to be, a self-correcting system. However, at times it too crosses the boundary into positive feedback, with theory upon theory (or epicycle on epicycle, as in older astronomies) compounding simple errors into gross distortions of our view of the world.

The negative feedback from those researchers who say our current scientific interpretations are wrong is dismissed and rejected as error itself. Those researchers, such as Halton Arp and Geoffrey Burbidge in astronomy, Robert Gentry in nuclear geology, Sir Fred Hoyle and Tom Van Flandern in cosmology, and many others, are excluded and cast out, becoming dissidents and heretics. We will reconsider their ideas along this voyage.

Detonation (with superfission and ultrafission)

In our nuclear weapons, the fission detonation is the result of a complicated sequence of events involving the violence of high explosives and the precision of electromechanical devices more delicate than a wristwatch. The conditions that permit the detonation can exist, in a nuclear weapon, for only a brief fraction of a second. Nature proceeds more slowly, and in a more natural manner.

As additional nuclear fuel accumulates in a planetary reactor, and as the thorium-232 and uranium-238 are transformed to more easily fissionable nuclides (uranium-233 and plutonium-239), the reactor can reach a point, in terms of temperature, size, and composition, where the effects controlling the reactor power change from negative feedback, providing stabilization of the fission rate, to positive feedback, leading to a rapid power runaway, resulting ultimately in a nuclear detonation. Such detonations of individual reactors would induce detonations of neighboring reactors in a chain-reaction of chain-reactions, and the planet (or satellite or planetoid) would disintegrate.

Conceptually, the most important parameter relating to the behavior of a nuclear reactor is the criticality constant (usually indicated by k). Sometimes called the "multiplication factor", it represents the ratio of the neutron population in a reactor at succeeding time intervals. If the criticality constant is exactly 1.0, the neutron population is constant, steady, and so the fission rate neither increases nor decreases. If the criticality constant is less than 1.0, the chain reactions will fade away and come to a halt. The only production of neutrons and fissions will be by so-called "source multiplication", in which neutrons arising from spontaneous fission, or nuclear reactions involving the natural alpha radioactivity of the material, are multiplied by sequential fissions, until the chain of fissions eventually comes to an end. A reactor can be large or small, running at high power or low. If its criticality constant is equal to 1.0, it is an operating reactor.

The stability and the time response of a reactor with a criticality constant close to 1.0 is determined by the release of delayed neutrons following the beta decay of a few of the radioactive fission products. These delayed neutrons amount to a fraction of one percent of the prompt fission neutrons, but provide the necessary time lag permitting easy direct control (in our industrial reactors), and facilitate inherent feedback control of nuclear reactors. If the criticality constant exceeds 1.0 by more than this small fraction, typically 0.4% to 0.7%, the time response approaches the time scale taken by the prompt neutrons from birth in fission to absorption or escape from the system. The reactor is said to be "prompt-critical" since the chain reactions continue without need for the delayed neutrons, but are propagated by the "prompt" neutrons alone. In reactor systems of the sort discussed here, as in the fission trigger of nuclear weapons, this time is on the order of 0.01 microsecond (10^{-8} second), and the fission rate can increase faster than any mechanical effects that might provide control. If the criticality constant becomes greater than 1.0 by a few percent, a detonation will occur. This is normally not possible, except in explosive devices, because feedback during the process of increasing the criticality constant limits the rate of increase and the magnitude of the increase. However, changes in the nuclear properties of a natural nuclear fission reactor can occur that will induce large and rapid increases in the criticality constant, and will lead to a detonation.

Several effects come into play in shifting the chain-reaction from a stable condition to a runaway detonation, and the detonation progresses through three phases.

As plutonium-239 becomes a significant fuel material, from conversion of the uranium-238 and depletion of uranium-235, the Doppler reactivity effect becomes positive [Hetrick 1971], thus providing positive feedback, accelerating the fission chain reaction.

As the rate of the reaction and the neutron flux increase, the intermediate nuclides uranium-239 and thorium-233, produced by neutron absorption in uranium-238 and in thorium-232 and leading (after two beta decays) to plutonium-239 and to uranium-233, become immediately available for neutron-induced fission.

As the local temperature increases, the fraction of neutrons extending above the fission threshold energy for uranium-238 (and thorium-232) increases, and more of these nuclei become immediately available as fuel. As the effective temperature of the neutrons increases, the yield of fission neutrons, per neutron absorbed, also increases. This further increases the fission rate, and the rate of energy release, and so the temperature increases more. This is positive feedback, and drives the detonation to nearly complete consumption of any fissionable atoms.

If a detonation progresses to completion before disruption, the final fission fragments will be moving in a gas that is largely neutrons, at a temperature on the order of 20 MeV, or about 1.5×10^{11} K. The fission fragments have kinetic energies of about 60 to 100 MeV, so that interaction energies may be on the order of 100 MeV. This will lead to mutually excited coulomb fission [Oberacker *et al.* 1979]. The fission fragments will also have internal excitation energy, and this will promote further fission, producing lighter elements, such as iron. I will call this fission of fission fragments "superfission", and it contributes to the elemental and isotopic abundances that we can observe. Fission of the iron-group elements will produce still lighter elements such as sodium, magnesium, and aluminum. This further fission of fragments warrants the name "ultrafission". Fissioning of iron and lighter nuclides requires energy, rather than producing it, but there is plentiful energy available in a fully contained fireball.

In a manner quite analogous to the super-prompt-critical behavior of a pulsed reactor, or a nuclear weapon, the chain-reaction will increase rapidly. Because of the extreme pressure and inertia of the surrounding planetary material, the fissioning region will remain together (unlike a nuclear weapon) until essentially all of the available nuclear fuel is consumed. In a very brief moment of time, extremely high temperatures will be produced. Fission neutrons that are emitted with a few million electronvolts of energy will be "thermalized" to higher energies. These higher energy neutrons will be effective in producing many of the unusual nuclides found, and can induce fission of all sorts of elements. The fissioning energy can be so great as to produce sequential fissions, "superfission" resulting in atoms of the iron group, and "ultrafission" resulting in atoms of sodium-magnesium-aluminum as fission-products. The fate of the planet then depends on the magnitude of the energy release relative to the energy required to disintegrate the planet. If the energy released in the nuclear detonation is great enough to overcome the planet's material and gravitational cohesiveness, the planet will explode like a cannonball. The outer crust will be blasted away by the shock wave and the rest of the planet will be fragmented by the explosive pressure.

An asteroidal impact could trigger a detonation by the compression of operating ("critical") reactors, and increasing the criticality constant. Compression increases the density, and the criticality constant increases with approximately the square of the density [Paxton and Pruvost 1987]. An increase in density of 1% would increase the criticality constant by 2%, well above the prompt critical threshold.

Estimates of the energy required to disintegrate a planet, and the energy available in the form of nuclear fuel, are shown in Figure 4. The energy required for disintegration is estimated as equivalent to that released during accumulation but not differentiation [Tozer 1978]. The energy available is based on a presumed average concentration in the planetary core of 1 part per million uranium and 3 parts per million thorium. (These must be taken as just good guesses.) Specific values for each of the planets, and for possible postulated planets: "Asteria" for the asteroids, and "Planet X" for a planet beyond Neptune, are shown explicitly in this figure.

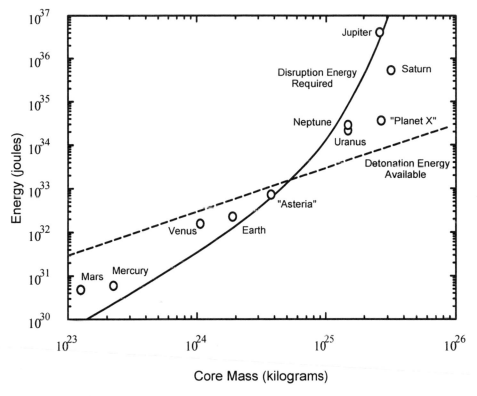

Figure 4. Energy required for disruption of planets.
Energy available from complete fission of uranium and thorium considered to be present in the cores of the planets, compared with the energy required to disrupt a planet and disperse the fragments to a mathematical infinity. The dots show the energy required to disrupt each planet. The smooth curve gives a representative value for the required disruption energy, the dashed line gives the energy available from complete fission of uranium, at 1 part per million of the core mass, and thorium, at 3 parts per million of the core mass.

For the cases in which the energy required for disintegration of the planet is less than the energy available, with the current choice of fuel content, a disintegrating detonation is possible. For those cases in which the energy required is greater than that available, the planet is likely to survive the detonation as a whole body, but with severe modification of the character of the original planet.

Under certain conditions reactors within a planet could develop an autocatalytic power increase that would lead to a detonation. This detonation process was expressed in a totally destructive disruption of a planet that had originally formed between Mars and Jupiter, resulting in the planetary fragments we now identify as asteroids. There is evidence in the meteorites that demonstrate effects of this nuclear detonation. Also, I found that the planet Mercury is thought to have lost half its mass by shocking off its outer material as the result of a presumed impact, but I think this was caused by a less-than-totally-disruptive detonation. I found that Venus had been volcanically resurfaced at about 500 million years ago (let us say "at the start of the Cambrian"). The carbon dioxide atmosphere was created by conversion of oxygen and nitrogen to carbon by the Giant Dipole Resonance reaction with gamma rays, and by disruption of carbonate rocks. An internal nuclear detonation could have overheated Venus to produce the massive global volcanism; the intense neutron radiation resulting from such a detonation would have dissociated most of the water, with a nuclear scattering bias leaving only some deuterium. I found that the glaciers and oceans of Mars were melted, vaporized, and destroyed at 180 million years ago (again, let us say "at the time of the Mt. Kirkpatrick/Palisades flood basalts on the Earth").

The detonation of nuclear weapons has been described in government handbooks [LASL 1950, DOD 1962, DOD and ERDA 1977; each edition reflects a different emphasis. For the present purpose, the earliest edition is best].

At the moment of detonation of a nuclear weapon, the central temperature exceeds 1,000,000 K. Then, a fireball of ionized material is formed, which radiates at a blackbody temperature of 300,000 K at 100 microseconds after initiation of the nuclear detonation. The ionization is sufficiently great that the fireball is opaque to the gamma radiation emitted within it, and this radiation is further converted to thermal energy and radiated as a continuous spectrum.

As the fireball expands and cools, it becomes transparent to the gamma radiation, and gamma emission from the intensely radioactive short-lived fission fragments shines forth.

The fission process in a nuclear weapon detonation ceases when the pressure produced by the fission energy release exceeds the inertial pressure created by the high-explosive trigger, and the fissioning material starts to expand. This terminates the chain-reaction by permitting the fission neutrons to escape from the fuel volume. At the instant of detonation, this volume is extremely small, on the order of a few centimeters in diameter. Only a few kilograms of nuclear fuel are needed to release a tremendous amount of energy in the brief instant of a detonation.

Considering the greater confinement of the nuclear detonation by the inertial mass of a planet, and the much greater amount of nuclear fuel available to be consumed, these effects must be scaled up greatly to match the power of a planetary disruption.

The gamma-ray emission detected from a disrupted planet should show a rapid increase as the planetary fragments expand and reveal the central fireball, followed by a decrease related to the cooling of the fireball and the decay of the short-lived gamma-emitting fission products. The energy spectrum will be dominated by the gamma rays from these fission products, with energies around 1 MeV. If the iron core of the planet is rich in oxygen, the 6.13-MeV gamma ray following the decay of nitrogen-16 should be seen. (This is a common, short-lived (7 seconds) radionuclide in commercial water-moderated nuclear power plants and in nuclear navy propulsion units.) As the short-lived activity dies away, the spectrum should be dominated by the 0.0586-MeV gamma rays from iron-60 and the 1.17 and 1.33-MeV gamma rays from cobalt-60, and high-energy gamma rays from iron-59. If planetary disruptions by nuclear fission detonation are responsible for the gamma-ray bursts, these details should be observable.

This may explain why there is so much iron in space. Iron is supposed to be made by helium-burning and supernovas, but with enough input energy, fission ("superfission") produces 4 iron atoms per fission:

$$^{238}U + n + 90 \text{ MeV} => 4 \, ^{56}V + 15n, \text{ and the } ^{56}V \text{ beta decays to } ^{56}Fe.$$

This may be a two-step process, with nearly symmetrical fission fragments being fissioned. Nearly symmetrical fission produces palladium and silver, for Gerry Wasserburg's puzzle.

This would happen in a glob of uranium, tightly surrounded ("tamped") by other material, in a planet or planetoid or satellite, so that the fissioning uranium got to an equilibrium temperature of about 170/2 MeV (85 MeV of kinetic energy per fragment, or a temperature of 6.6 x 10^{11} K, 660 billion degrees kelvin).

The planetary detonation is a chain-reaction of chain reactions. A planet (or planetoid or asteroidal fragment) contains many critical and subcritical masses. Reactors that are larger than critical masses cannot exist for long, and will be disrupted into smaller regions. As one critical mass goes supercritical and detonates, it triggers neighboring fuel-rich regions, and the detonation spreads throughout the body, until sufficient energy has been released to disrupt the object, or until all the available fuel has been consumed. This detonation process can proceed only deep within a planetary (asteroidal) size object. Reactors near the surface will be able to control power in a stable manner by simple thermal expansion. The reactors that we construct for power generation cannot detonate.

Astronomy

As a science that ascended from mythology, astronomy carries a broad and diverse culture, and focuses on a large variety of objects for study. The roots of astronomy in the ways of philosophy and the traditions of astrology show in our continued use of the astrological symbols for the Sun, Moon, and planets; our preservation of the Roman deities in the naming of newly discovered planets such as Uranus, Neptune, Pluto, and Ceres; and use of the ancient constellations as a basis for finding our way in the sky. Johannes Kepler, one of the founders of modern astronomy, cast a horoscope for Hans Hanibal Hütter von Hüttershoffen, born in 1586 [Misch 1999]. Astronomy has its head in the skies and its feet in the past. In more recent times, the philosophers have turned into theoreticians, and the astrologers of the past have become observers. However, some would say that the ancient dedication to dogma, and opposition to heresy, have continued [Ratcliffe 2007, Arp 1987].

Astronomy struggles with investigations of a three-dimensional world in a two-dimensional construct, in which only the positions on the spherical shell of the sky can be measured directly, for most astronomical objects. As with nuclear terminology, a listing of these objects helps to keep the players straight.

stars are hot spheres of gas, shaped and confined by gravity, and heated by thermonuclear fusion reactions. The Sun is a star. Stars exist in a large range of sizes, temperatures, and types, and are distributed over an extended range of distances in space. Special categories have been created for those stars that do not completely meet these criteria, such as brown dwarfs, white dwarfs, and neutron stars.

planets are generally bodies of solid or liquid and gas, shaped and confined by gravity, but without an internal source of heat from thermonuclear fusion. In the traditional sense, there are only and exactly nine (or eight) planets, Earth and its fellows in the Solar System (except that Pluto has been demoted, leaving us with only eight planets). However, it should be conceivable that other planetary systems, similar to our own or quite different, might exist elsewhere, with planets orbiting other stars, their suns. More planets might form abundantly in isolation, as brown dwarfs and smaller, from the smallest wisps of interstellar gas that are able to cool, condense, and adhere. To keep these other objects clear, we call them extrasolar planets.

nebulae (the somewhat awkward Latin plural for nebula) are extended collections of gas and dust that glow with induced or reflected light or may be dark. In a breakthrough decision by astronomers in the 1920s, spiral nebulae were determined to be at great distances from the Earth, and this has set the scale of the Universe ever since.

galaxies are considered to be collections of stars and other astronomical objects that belong with them, including planets, nebulae, interstellar gas and dust, and all sorts of other objects. Our Solar System has been assumed to exist in a spiral galaxy, the Milky Way Galaxy, that is not uniquely different from a large number of other spiral galaxies.

The possibility of an error in the decision that spiral nebulae were spiral galaxies at great distances will arise naturally in our exploration. The conclusion will be that these objects are instead relatively near, as was thought a hundred years ago, and not at all "Island Universes" composed of stars like our Sun. This will affect our view of the Milky Way Galaxy as well.

Astronomers report distances in a variety of unrelated, historically derived units. In some cases, these reflect the method of measurement. The major units of linear distance are

1 kilometer (km)	=	1,000 meters (m)
1 Astronomical Unit (AU)	=	1.5×10^8 km
1 light year (LY)	=	9.5×10^{12} km = 63,240 AU
1 parsec (pc)	=	3.1×10^{13} km = 3.2616 LY
1 kiloparsec (kpc)	=	1,000 pc
1 megaparsec (Mpc)	=	1,000,000 pc.

The only direct measurements of distance in astronomy come from radar and laser ranging within the Solar System. All other measurements, even trigonometric parallax, which pretends to be an absolute measurement, are relative. (In this regard, the term "parallax" has come to signify some types of distance measurements that are totally removed from the original trigonometric method. Thus, there are spectroscopic parallaxes and statistical parallaxes and dynamical parallaxes, and hypothetical parallaxes. Beware.)

Brightness, luminosity, and rate of energy output are expressed in several forms, a bit bewildering, but based on the history of the science.

For objects that can be seen or imaged in visible light, the magnitude scale is used. This scale is based on a ratio of 100 in apparent brightness being divided into 5 equal factor steps. In this way, each step in magnitude represents a factor of 2.512... in brightness, this number being the fifth root of 100. The scale runs backwards to the layman's normal concepts, with the brightest objects having negative magnitudes, and progressively dimmer objects having magnitudes that are progressively larger, positive numbers. Historically, this developed from early astronomers establishing a scale based on visual impressions, starting with the brightest stars "of the first magnitude", just dimmer stars "of the second magnitude", and so on.

Magnitudes may be determined for specified energy or wavelength ranges, or for a specific response function. The absolute magnitude is the apparent brightness (magnitude) of the object at a standard distance of 10 parsecs (32.6 light years). The apparent visual magnitude of the Sun, the brightest object in the daytime sky, is −26.74. Its absolute visual magnitude (at an imagined distance of 10 parsecs) is 4.83. The apparent visual magnitude of the planet Jupiter averages −2.4, and is bright in the night sky, easily visible to the naked eye, but its absolute visual magnitude (at 10 parsecs) is 25.68. At 10 parsecs, Jupiter could be just barely detected by our largest telescopes.

An absolute bolometric magnitude has been defined so that the zero of that scale represents 3.0×10^{28} watts.

In radio astronomy, the unit of intensity in common use is abbreviated as mJy, standing for the millijansky, honoring Karl Jansky, the founder of radio astronomy. In conventional units, 1 mJy is equivalent to 10^{-29} watts per square meter per hertz, that is, the rate at which energy passes through a unit area, per unit frequency of bandwidth.

Luminosity may also be expressed as watts per square meter, when the energy distribution can be integrated into one measure.

Geology

In this book it will seem that geology is where astronomy gets down to Earth. The methods and theories of geology can, are, and should be applied to our studies of other planets as well. Earth as a planet, perhaps one of a multitude, is an important concept here. As a planet, Earth includes

an **atmosphere**, the thin interface between us and space that provides the air we breathe, the source of weather and modifier of climate

an **ocean**, great puddles of water that present great puzzles

a **crust**, the outermost solid material, importantly different in continents than under the ocean a **mantle**, the major bulk of the Earth, extending directly from below the crust to the core

the **core**, consisting of an outer core of molten iron and an inner, central core of solid iron.

These shells and layers are shown in simplified form in Figure 5. As in all simplifications, it should be realized that reality is much more complicated. A clear blue sky is really filled with condensations of many different forms, and the wind in it blows in different ways; the crust of the Earth has mountains and valleys of different rocks; the oceans have their own forms and styles and flavors. And so, our faint glimpses of the mantle and the core hint at significant complications. These complications, these departures from smooth uniformity, make the world work.

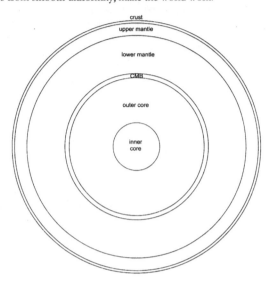

Figure 5. Schematic arrangement of the modern Earth.
The regions of the Earth, as if it were a perfect sphere. An approximate rendering of the internal structure of the Earth. The innermost region is the solid core, thought to be mostly crystallized iron. The next region outward is the fluid outer core, considered to be molten iron. A thin layer, D", separates the core from the lower mantle at the core-mantle boundary. The lower mantle extends above this to the upper mantle. These differ in density, possibly in mineral phases, and seismic velocities. The last layer is the lithosphere, the oceanic and continental crust, separated from the upper mantle by a slushy asthenosphere. On top of all this are the oceans and the atmosphere.

Based on the radiometric ages of meteorites, it is thought that the Earth (and the rest of the Solar System) formed about 4.5 billion years ago and, in a manner still subject to discussion, differentiated fairly quickly into these basic structures. The Earth continues to be geologically active, with volcanoes and earthquakes providing glimpses into the unreachable lower tiers of our planet. Our current model of the Earth explains the relative motion of the continents and ocean floors on the basis of the theory of Plate Tectonics. This hypothesis uses colliding and sinking plates of the crust to explain the observation of Continental Drift, as described by Alfred Wegener as early as 1915 [Wegener 1928]. Formed at a time when geological thought pictured formation of mountain ranges on the surface of the Earth as a consequence of the cooling and contraction of the interior producing a wrinkling of the crust like a drying apple, this theory was based on the observation that the continental shapes fit together well at the edges of the continental shelves. This lead to the interpretation that the continents had previously been joined, resulting in the explanation that the continents, as tectonic plates floating on the mantle, had drifted apart. The driving force for this motion has been variously seen as thermal convection currents in the mantle, the push of midocean spreading ridges, and the pull of sinking subducted plates.

Geologists have divided the life history of the Earth into a hierarchy of subdivisions, on the basis that it is usually possible to distinguish older from younger even when it is not possible to specify the date. Usually, geologic deposits, layers of different rocks, form from the bottom up, younger rocks deposited on the older, and this can be used to designate relative ages. Sometimes, rocks are turned up vertically, sometimes even overturned, so this rule cannot be applied blindly. This provides a well-established relative dating scale. Determining the actual age of the Earth, geologic deposits, fossils, and archaeological sites has been more elusive. Considerable effort has been made to produce absolute geological dates, with some real success, but lurking errors abound.

While we currently recognize that the intervals often end in a catastrophe, such as the lava flows of flood basalts, or an asteroidal impact, the boundaries are conventionally set at the beginning of the next stage (shown as Age in this table in millions of years in the past, Myr), rather than at the ending of the previous. This may have some significance as the strata immediately above a clear marker of catastrophe may be definitely different, such as a cap carbonate, a laminated sandstone, or what is labeled a glaciomarine deposit but may be an impact breccia, but underlies by a considerable distance the first appearance of new fossils. In stratigraphy, distances (differences in elevation) are interpreted to represent time. Thus, one interval may end well before the next begins.

The boundaries between these geological intervals are set by international agreement, or are tentatively assigned. This table is taken from The Geological Society of America Geological Time Scale of 2009.

Interval	Age (Myr)
Present	0.00
Holocene	0.01
Pleistocene	2.6
Pliocene	5.3
Miocene	23.0
Oligocene	33.9
Eocene	55.8
Paleocene	65.5
Cretaceous	145.5
Jurassic	201.6
Triassic	251.0
Permian	299.0
Carboniferous	359
Devonian	410
Silurian	444
Cambrian	542
Proterozoic	2500
Archaen	3850
Hadean	4550

We live in the Present, in the Holocene, which began about 10,000 "radiocarbon" years ago (11,500 calendar years), at the end of the Younger Dryas, a time of strangely perturbed isotopic anomalies. In radiocarbon dating, the present is specified to be 1950.

Isotope Variations and Dating

The relative abundances of isotopes that we find in nature often vary in interesting, instructive manners. These variations have been interpreted in terms of ages, biological activity, separation of related chemical elements, and differences in temperature. All these processes can have the understood effect, but nuclear effects have been overlooked.

One of the strongest nuclear effects involves absorption of neutrons. The probability of neutron absorption, expressed as its cross section (usually given in the somewhat whimsical unit of the "barn", equal to 10^{-24} square centimeters per atom) varies greatly from nuclide to nuclide, over a range of more than 10^{10}, as shown in Figure 6. Some nuclides, such as helium-4, have a zero neutron absorption probability.

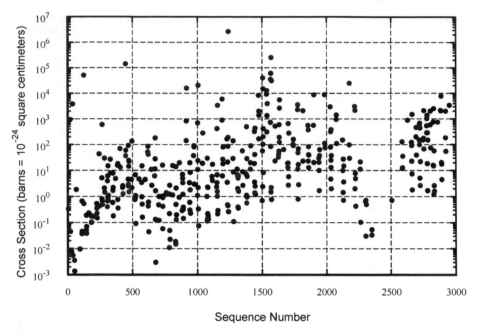

Figure 6. Neutron absorption cross sections.
Cross sections, leading to a different nuclide, shown in sequence from a nuclear data set at the National Nuclear Data Center, Brookhaven, New York. Cross sections greater than 10 (10^1) are often significant, particularly when a related nuclide has a very small cross section.

Isotopic dating methods usually rely on ratios between pairs of isotopes that can be related to each other. These methods assume that the individual behavior of the isotopes is not changed by unknown (nuclear) processes. Unfortunately, any significant flux of neutrons can seriously perturb the ratios, because one of the pair of isotopes has an absorption that is large compared to that of the other. Dating pairs listed by Harland *et al.* [Harland *et al.* 1990] are shown below, with the disturbing neutron cross sections. (The cross section is related to the probability of a neutron absorption reaction, and is given in the conventional unit, the barn, 10^{-24} square centimeter.)

Uranium-lead

^{235}U	681.4 barns
^{238}U	2.68 barns

(This pair is also subject to the further disturbance that a pulse of neutrons both consumes and produces ^{235}U (by way of transmutation of ^{238}U to ^{239}Pu, with decay of the ^{239}Pu to ^{235}U) and consumes ^{238}U.)

Rubidium

^{85}Rb	0.988 barns
^{87}Rb	0.34 barns
^{87}Sr	17 barns

(The rubidium isotopes have reasonably matched cross sections, but are produced with somewhat different yields, 1.319% and 2.56%, in nuclear fission. The nuclide ^{87}Sr will be depleted by neutron absorption.)

Strontium

^{86}Sr	1.04 barns
^{87}Sr	17 barns
^{88}Sr	0.0058 barns

(Neutron absorption will deplete the ^{87}Sr and increase the ^{88}Sr.)

Potassium

^{39}K	2.1 barns
^{40}K	30 barns
^{41}K	1.46 barns

(Neutron absorption will deplete the ^{40}K and increase the ^{41}K, the decay product ^{40}Ar will subsequently be less.)

Samarium

^{144}Sm	1.64 barns
^{147}Sm	57 barns
^{148}Sm	2.4 barns
^{149}Sm	40,140 barns
^{150}Sm	100 barns
^{152}Sm	206 barns
^{154}Sm	8.3 barns

(^{147}Sm has a yield of 2.25% in nuclear fission.)

Neodymium

^{142}Nd	50 barns
^{143}Nd	325 barns
^{144}Nd	3.6 barns
^{145}Nd	50.0 barns
^{146}Nd	1.49 barns
^{148}Nd	2.58 barns
^{150}Nd	1.04 barns

(^{143}Nd will be depleted by neutron absorption. Neodymium isotopes are produced in nuclear fission.)

Chlorine

^{35}Cl	43.6 barns
^{37}Cl	0.43 barns

(^{35}Cl will be depleted by neutron absorption. These isotopes have been used to study minerals from the Moon [Sharp *et al.* 2010].)

In all of these, neutron absorption will affect the age calculated from the analytical results.

Aside from dating, isotope ratios have been used to study variations in the material of various parts of the Solar System. The tungsten ratio ($^{182}W/^{184}W$) has been used to investigate the geochemical development of Earth's mantle, by comparison with meteorites, considered to be primitive material from the formation of the planets [see for example, Kleine 2011, Wilbold, Elliott, and Moorbath 2011]. This ratio is subject to disturbance from several neutron absorption reactions as implied by the differences in the thermal neutron absorption cross sections:

^{182}W	19.9 barns	
^{183}W	10.4 barns	
^{184}W	1.7 barns	
^{181}Ta	20.5 barns	(If tantalum is present, this will increase ^{182}W or produce ^{183}W with a subsequent reduction in ^{182}W)
^{182}Ta	8200 barns	(This reaction will result in less ^{182}W)

In a nuclear fission detonation, these reactions will result in a decrease in the ratio, as is shown in the meteorites compared to most of the Earth's mantle.

For light-weight elements, the mass differences between isotopes can have a large effect on fractionation, but again, nuclear reactions have been overlooked.

Hydrogen consists of two stable isotopes, and one radioactive form which occurs rarely in nature. The ratio of hydrogen-2 (deuterium) to hydrogen-1 can vary as the result of several well studied effects. Two nuclear effects have been ignored. Hydrogen-1 can absorb a neutron, converting it to hydrogen-2, increasing the deuterium to hydrogen ratio, D/H. In a water molecule, HOH, one (or both) of the hydrogen atoms can be a deuterium atom, DOH or DOD. The scattering cross section for neutrons on hydrogen-1 is about 6 times as great as for scattering on hydrogen-2. In an energetic neutron flux, neutron scattering from hydrogen-1 will break the HO bonds with 6 times the probability of breaking the DO bond with hydrogen-2. This will result in an enrichment of the bound deuterium in the water. If the water is subsequently disassociated, by ultraviolet light for example, the atmosphere will be enriched in deuterium. These processes have caused the apparent differences in the comets, Venus, and Mars, compared with the "normal" value for the Earth.

Carbon also consists of two stable isotopes, and the radioactive form, carbon-14, that is important in dating. The ratio of carbon-13 to carbon-12 is used for a variety of indications of climate and biological activity. However, several nuclear reactions can also affect this ratio. The following reactions will change the relative amounts of the two isotopes:

$^{12}C(\gamma,\alpha)^8Be$	Giant Dipole Resonance reaction destroys carbon-12
$^{13}C(\gamma,n)^{12}C$	Giant Dipole Resonance reaction destroys carbon-13, and creates carbon-12
$^{16}O(\gamma,\alpha)^{12}C$	Giant Dipole Resonance reaction destroys oxygen-16, and creates carbon-12
$^{17}O(\gamma,\alpha)^{13}C$	Giant Dipole Resonance reaction destroys oxygen-17, and creates carbon-13
$^{17}O(\gamma,n)^{16}O$	Giant Dipole Resonance reaction destroys oxygen-17, and creates oxygen-16
$^{18}O(\gamma,\alpha)^{14}C$	Giant Dipole Resonance reaction destroys oxygen-18, and creates carbon-14

A nearby gamma ray burst, though rare, could provide the high energy gamma rays that disrupt the various kinds of nuclei.

The variations in the ratios of two isotopes of iron can be used to study reduction/oxidation reactions [Homoky *et al.* 2009]. The ratio of iron-56 to iron-54 varies according to the amount of chemical recycling the iron has experienced. However, a nuclear reaction can also affect this ratio. Both isotopes have very similar neutron absorption cross sections, but iron-56 can be produced by neutron absorption in manganese-55:

$$^{55}Mn(n,\gamma)^{56}Fe \qquad 13.3 \text{ barns}$$

Some additional dating techniques are also described, for we need all the help we can get. Magnetic reversals and anomalies, and the ratio of strontium-87 to strontium-86, are more useful as means of correlations, but can be related to known absolute dates obtained by other techniques.

While relative dating, earlier or later, before or after, is reasonably well established and can often be correlated around the world, absolute dating in terms of a number of years, is much more difficult and rather uncertain in most cases. The methods involve isotopic ratios, magnetic polarity reversals and anomalies, and climatic cycles based on the dynamics of Earth's orbit and rotation. These require the assumption that the processes have not changed over the time period to which they are applied.

A series of estimates of the age of the Earth has steadily pushed the time of formation of our planet back in time. These have ranged from a few thousand years to a few trillion years, ample enough to include whatever the true value might be [Dalrymple 1991].

The development of each of these estimates has been based on the best knowledge of the culture, but most are now recognized to have been ridiculously wrong. The estimates, though properly worked out in a scheme that was acceptable at the time, were misled by lack of knowledge and understanding.

Perhaps the most famous estimate is that of Archbishop James Ussher, in 1650 in Ireland. By a detailed review of the chronology in the Book of Genesis, and an assortment of astronomical and calendrical cycles, Ussher arrived at a remarkably precise time for the beginning of all Creation. He declared that Creation began in the evening of October 22, 4004 BCE (actually, he gave 4000 BCE, but a revision to the date of the birth of Christ added the 4 years). Such a short time, only 6,000 years long, conflicted with the growing realization that mountains formed slowly and wore down even more slowly. Unless the world was created in nearly its present form, it appeared that hundreds of millions of years were needed to shape the surface of the Earth as we see it now. Thus began the either/or conflict between science and religion. If the world were created in final form, there would be no need for the immense amount of time needed by science. But if the world were created in final form, why were all the ancient forms created also? Science provided simpler, more complete answers to the development of the world than did religion, at least to scientists.

Scientific estimates were initially based on the assumption that the Earth (and the Sun) had been created as molten spheres and had then cooled to their present states. The estimates of William Thomson, later Lord Kelvin, carried the most impact, because of the great respect and honor accorded to this British scientist of the 19th Century. It is a bit distressing, but unfortunately typical, that the discussions between Kelvin, with his relatively short time estimate of roughly 30 million years, and the field geologists, who imagined hundreds of millions of years, were characterized as a "half-century-long feud" [Dalrymple 1991]. Those precise mathematical calculations (reminiscent of the precise Biblical accounting) were criticized on the grounds of unfounded assumptions. Indeed, the true flaw, regardless of the quality of the mathematics, lay in the problem that "we don't know what we don't know."

Kelvin's estimates were made before the discovery of radioactivity by Henri Becquerel in 1896. When radium was discovered, with its great heat output, the geologists were quite pleased. Radioactive decay provided the long-term internal heat they needed to justify an age of the Earth beyond 100 million years. However, that heat is truly insignificant, and somehow that fact has been overlooked after the solution was accepted. The release of heat by radioactivity in the Earth was seized upon as an explanation for the conflict between Kelvin's cooling calculation and the time required by the geologists for the Earth to develop. However enthusiastically that solution was seized, in retrospect it appears to fail, and the solution was accepted prematurely. The energy released by radioactivity in the Earth is now estimated to be only a ten-thousandth of the heat received from the Sun, and so seems to be as close to negligible as possible. We'll come back to this problem later.

While radioactivity fails to explain the heat of the Earth, radioactive decay has provided the best quantitative estimates of age in our Universe. Unfortunately, there may be serious errors lurking unnoticed.

The elements of uranium and thorium are radioactive and eventually produce stable isotopes of lead at the end of an extended decay chain. The amounts of lead, uranium, and thorium can be compared to estimate the age of the crust of the Earth. More accurate estimates can be made by comparing the several different isotopes that are involved.

That method led to the development of many different techniques that make up radiometric dating, on which are based most of our estimates of the age of the Earth, and various events in the life of the Earth, the age of the Solar System, and the age of the Universe.

Radiometric dating rests on two basic assumptions. We assume, with considerable experimental support, that radioactive decay proceeds with a constant time-rate, unaffected by age or other conditions. In the case of electron capture by beryllium-7 [Dalrymple 1991] clear changes in the decay rate have been produced by chemical compounds, pressure, and ionization. Recent research into this otherwise unanticipated effect has suggested effects due to variations in the distance to the Sun [Jenkins *et al.* 2008] and Solar activity [Jenkins and Fischbach 2009, Fischbach *et al.* 2009, Sturrock *et al.* 2011, Sturrock, Fischbach, and Jenkins 2011].

These changes are small and we are sure that they do not affect the radioactive decays used in dating. We also assume that the "clock" represented by the decrease in the number of parent (radioactive) atoms and the increase in the number of daughter (stable) atoms is not reset or otherwise changed by unseen effects. This assumption may be less secure, for we do not know what we do not know, and there are several ways in which the clocks can be affected. Our confidence in this assumption rests on the general success of radiometric dating to produce reasonable ages, in general agreement with estimates produced by other means and by our failure to imagine what might produce unnoticeable errors.

As an example, measurements of the small variations in uranium-235 concentration in meteorites [Brennecka *et al.* 2010] has shown that ages measured by the Pb-Pb (lead/lead) method may be in error by 5 million years or more. Admittedly, that is a small fractional error for ages in the billions of years, but when age differences are considered, it becomes significant. Further, that may be only the beginning of errors.

Radiocarbon dating is simple in principle, but it is difficult to discuss in detail without appearing to poke fun at the method. After all, its practitioners use a half-life for carbon-14 that is known to be wrong, refer to 1950 as "the present", and produce results that offer choices of several ages that may be centuries apart. All that aside, it is a serious method that can produce the best results we have available for a very important time in our pre-history. And its true flaws may be much more severe.

The method was first developed in the 1940s, primarily by Willard Libby at the University of Chicago, when he discovered that radioactive carbon-14 was present in modern organic compounds, but not in ancient material. The short half-life of the carbon-14 led to its complete decay in anything older than a hundred thousand years. New biological material has a concentration of carbon-14 in equilibrium with the concentration in the biosphere, atmosphere, and hydrosphere, while older material has less and less carbon-14 due to its radioactive decay.

Carbon-14 is considered to be produced in the upper atmosphere as a result of the absorption by nitrogen-14 of neutrons produced by cosmic ray interactions with oxygen and nitrogen nuclei. This was initially expected to be reasonably constant over time, but research showed considerable variations in comparison with tree ring dating, during the past 12,000 years. While the nuclear reaction is certainly known, I have not found any actual determination of the carbon-14 production in the atmosphere. (That nuclear reaction, involving neutrons released by testing of nuclear weapons in the atmosphere, seriously affected the concentration of carbon-14 after 1952.) The carbon-14 is almost identical, biochemically, to ordinary carbon, and so is bound into biological material as the plant or animal grows.

The complete reaction, for neutron absorption, is:

$$^{14}N(n,p)^{14}C(\beta^-)^{14}N.$$

Stable nitrogen-14 is converted to radioactive carbon-14 by absorption of a neutron and emission of a proton. The carbon-14 decays by emission of a beta particle (and an antineutrino), returning to stable nitrogen-14. Thus it would seem to vanish without a trace, unless the carbon-14 is sequestered in a form of carbon that is relatively free of nitrogen, and then it may leave an isotopic anomaly in the isotopic ratio of nitrogen-14 to nitrogen-15. This might occur in diamonds.

High energy gamma rays and cosmic rays release neutrons in the atmosphere. The neutron absorption reaction can occur anywhere that nitrogen, such as is contained in air or in ammonia for example, is exposed to a flux of neutrons, from whatever source.

An unconsidered source of carbon-14 involves the Giant Dipole Resonance reaction with gamma rays on oxygen-18. In this case, the reaction is:

$$^{18}O(\gamma,\alpha)^{14}C(\beta^-)^{14}N.$$

Sufficiently high energy gamma rays are released by gamma ray bursts, and this radiation, at some unexpectedly close distance from the Earth, would produce carbon-14 in any materials containing oxygen. That would include inorganic materials as well as biological compounds. The carbon-14 and the nitrogen-14 are identical to what is produced in the standard reaction.

A similar reaction involving nitrogen-15 also produces carbon-14:

$$^{15}N(\gamma,p)^{14}C(\beta^-)^{14}N.$$

If the cross section model for antineutrinos that is presented in Appendix A is correct, there may be another important reaction for the production of carbon-14. This is the antineutrino-induced electron capture reaction on nitrogen-14. This reaction is:

$$^{14}N(\bar{v}EC)^{14}C(\beta^-)^{14}N$$

If something like a gamma ray burst occurred relatively nearby, admittedly very rare, high energy gamma rays and antineutrinos would impinge on the Earth and disturb the assumed normal production of carbon-14.

The point of this section has been to alert researchers to the fact that for many of the critical isotopic ratio pairs, the two nuclides often have very different neutron absorption (and perhaps other nuclear reactions) probabilities. If nuclear reactions could have occurred in the material being studied, variations in the ratios may exist that are completely unrelated to the standard theories of fractionation.

Biology

Here, the concepts of life, species, mutation, and evolution are important. What is life? Life separates living things from the rest of the world. In most cases, life is easy to recognize, but difficult to define. Where did life come from? We have some ideas, but we really don't know. With such a foundation, biology begins. Life on Earth (and that is all we know, so far) exists in an amazing variety, divisible into five (or six) kingdoms, and three domains. These are still subject to change, but should not disturb us much.

Life consists of the simplest creatures that can only be seen through microscopes, consisting of single cells, to the most complex, trees and roses, whales and octopus, elephants and sharks, grasshoppers and us.

Individuals live and die, but as they live, they form populations which may continue for hundreds of thousands of years, and represent species which may survive for millions of years. Individuals are born and die. Populations grow and decline. Species originate and become extinct.

There is geologic evidence that life existed on Earth, in very simple forms, as early as 3.7 billion years ago, and has continued since then. Life has changed the world by producing oxygen, depositing carbon dioxide as carbonate rocks, and liberating hydrogen, which escapes to space. An oxygen atmosphere permits development of an ozone layer, which shields living organisms on land and in the surface of the sea from the intense ultraviolet radiation of the Sun.

Over two billion years ago, photosynthetic organisms used sunlight to feed themselves, and discharged oxygen as a waste product. Eventually, through the working of the Sun, the Earth, and the plants, the early atmosphere of methane, ammonia, water, carbon dioxide, and hydrogen, became an atmosphere rich in oxygen and nitrogen. Abruptly, the surface geology changed, iron ore and uranium ore were deposited, and the world was made ready for animals. While photosynthesis has been considered to be the major source of oxygen, an overlooked source is the breakdown of water vapor high in the atmosphere, by ultraviolet light from the Sun and high energy radiation from violent nuclear events. Water is continuously supplied to the surface of the Earth by volcanic transport of magma from the mantle, and in different atmospheres than we have at present water vapor may have easily reached the top of the atmosphere and been broken apart.

The source of the nitrogen that makes up about 80% of our present atmosphere has not been clearly explained. It is a major constituent of the atmosphere of Saturn's moon Titan [Lodders and Fegley Jr. 1998], perhaps as a breakdown product of ammonia.

The simplest, perhaps the most general, distinction between plants and animals is that plants take in carbon dioxide and release oxygen, while animals take in oxygen and release carbon dioxide. (That oversteps some of the boundaries a bit, but is generally correct.) An interesting difference is that the compounds of chlorophyll and hemoglobin, the source of energy for plants and the lifeblood of animals, differ mainly in the placement of a magnesium atom at the center of chlorophyll, and an iron atom at the center of hemoglobin. Surrounding molecules change slightly to accommodate the different ion sizes.

Plants prepared the world for animals by filling the atmosphere with oxygen. Animals, in a simplest form, became a possible form of life around 1 to 2 billion years ago.

A species is defined as a type of organism whose members interbreed but do not breed with organisms outside the species. This is obviously difficult to determine for extinct species, for which we have only fossil evidence of their bodily structure and some hints of lifestyle. Identification of species is generally based on the collections of morphological traits and lifestyle that are observed for each group of organisms, and the recognition that present organisms do not breed if the differences are too great.

In biology, living creatures (and dead living creatures) are labeled by the use of two names, a genus name and a species name (for example, *Tyrannosaurus rex* and *Homo sapiens*). Creatures included in a genus share similar characteristics and are related by descent from a relatively recent common ancestor. The common ancestor is the source of the traits shared in common. Within a genus, creatures are divided into species. The firmest definition of a species requires that individuals interbreed and produce healthy and fertile offspring. All cats, from house cats to lions and tigers, belong to the same family, and have clear similarities, but are divided into two genera, for large and small. The house cat is labeled *Felis domestica* (or properly *Felis catus*), while the larger "cats" are *Panthera leo* (lions) or *Panthera tigris* (tigers). The differences can be clearly seen in size, color, and form.

The formal names are derived (or constructed) from Latin, according to formal rules, and not all are as easy to recognize, by amateurs, as the cats. Within a species, there may be many varieties, such as the various breeds of the domestic dog, *Canis familiaris*.

The initial describer of a newly found fossil organism has the opportunity to name the organism, in the now traditional method of Linnaeus, through the several levels of nomenclature, down to the only ones we need to be concerned with: genus and species. This naming can be done with considerable latitude, according to the judgment of the namer and the constraint of the scientific community.

Creatures of some similarity but clear differences, as the house cat and the lion, may be grouped into a Family, and Families that share some significant characteristics are combined into Orders. Similar Orders make up a Class, which is a division of a Subphylum or Phylum. We now have four or five or six Kingdoms of life, derived from three Domains reflecting the fundamental origin of living organisms. These three Domains, Bacteria, Archaea, and Eucarya, are thought to have developed from existing simple life about 3 billion years ago. Those origins are so obscure it is difficult to tell the story with great confidence. Did life originate just once, or two or three times, in two or three slightly different ways?

The characteristics that gather many distinct types of creatures into an Order or a Class may contribute to success and to failure, ultimately to extinction. The dinosaurs thrived for 140 million years, yet all perished by 65 million years ago. Birds are now considered to be descendents of the dinosaurs, themselves dinosaurs that actually survived the great extinction. They are more likely to be descendents of the ancestors of the dinosaurs, and must have differed from true dinosaurs in significant ways to avoid complete extinction. Here, it may be useful to consider the true dinosaurs to be just those dinosaurs that went extinct by the Cretaceous-Tertiary boundary (the K/T extinction).

Individuals die, as a certainty of life, but *Genus* and *species* may go on indefinitely. While recent times allow us to identify the end of a species, extinction, by noting the death of the last known survivor, such as the last passenger pigeon or Tasmanian wolf, the great history of life lies in the fossil record of rocks. In the fossil record, we can only find that we no longer find evidence of a particular species. Close examination of the details, and the lucky existence of geologic markers, may permit us to identify the time of extinction closely. Because no dinosaur fossils are found in rocks that are more recent than a particular iridium-rich layer of clay at the boundary between the Cretaceous and the Tertiary Periods, we say that dinosaurs went extinct at the end of the Cretaceous. (There is some continuing controversy over this, fueled by the field and laboratory investigations by Keith Rigby on land, and Gerta Keller at sea [Sloan 1986, Keller 2004, Keller 2010].)

It is estimated that there are 8 million species alive on Earth at present. We have labeled roughly 2 million. A species ends with extinction. New species are created, in "speciation", but this may happen in several different ways. Speciation has been seen (in the fossil record) to occur before, during, and after a major extinction event. A major extinction event is often followed by a major speciation event, extending "upward" through the hierarchy of names to the level of Order, and even Class. Darwin had problems with the fossil record, and yet was able to see that creatures change, and so could propose his theory of evolution.

Individuals maintain populations and species by reproducing. The processes that produce offspring that are similar to their parents, thus maintaining the species, are coded in genes carried by the reproductive cells, and passed on to all the growing cells of the new individual. The science of genetics, started by the research of Gregor Mendel on peas in Austria in the 1850s, has grown to a maturity that allows us to describe much of the code. The genes are arranged in chromosomes, and the sets of chromosomes form distinct identifiers of species. While the genes are sequences of four nucleotide bases (labeled A, T, C, G), these always pair in only two ways, adenine with thymine and cytosine with guanine, so that the genetic code is a binary code, just like computers. (Or computers are coded just like genes.)

In a bit of linguistic confusion, terms of considerable importance in this discussion have been used differently for different purposes. While in most uses in this book, "radiation" will refer to ionizing energy resulting from nuclear or atomic/electronic effects, biologists have used the term to mean a rapid spread of life types, as in the "Ediacaran radiation." Mostly, by "explosion" we will mean a detonation, while biologists have established the "Cambrian explosion" as an abrupt proliferation of animal forms. And even "fission", usually meaning a splitting of the atomic nucleus in this book, may refer to "karyotypic fission", a splitting of the chromosomes in an organism's set of chromosomes, or karyotype.

Summary of Part I

This part introduces the tools and terms that are needed to explore our Earth, the Solar System, Life, and the Universe. Emphasis is placed on the features that are special to nuclear fission as it may occur in nature.

Consistency is both a goal and a trap, establishing incorrect theories. In contrast, discrepancies must be fully explored. It is what we don't understand that will lead us to knowledge and new understanding.

Different forms of expressing the numbers that we use to specify the small, medium, and large features of our world are discussed. I will use ordinary numbers, exponential notation (powers of ten), and named numbers.

The equivalence of mass and energy is presented, and various ways of expressing energy are described. Cerenkov radiation is introduced, as a light spectrum extending into the ultraviolet that is unrelated to temperature.

Some of the history and special features of the nuclear world are discussed. Antimatter, the opposite of our normal matter, is introduced. The process of radioactive decay, that opened our eyes to the nuclear world in 1896, is described. Nuclear reactions, those that naturally occur and those that we can use to experimentally explore nature, are presented and the terminology is explained. Nuclear fission, the splitting of the atom with a great (relatively) release of energy, in a controlled form and an explosive form, is shown to be an important part of our world, from the Sun and Earth, to the greatest reaches of the Universe. The release of neutrons from fission into space is shown to produce protons with a temperature distribution of 3 million degrees kelvin, and electrons with a temperature distribution of 2.5 billion degrees kelvin. The possibility of further fissioning of the initial fission fragments in subsequent reactions to produce the lower mass elements is proposed. Types of nuclear fission reactors for controlled nuclear fission chain reactions are discussed. Natural nuclear fission reactors, and the features of production of elements in supernovas, are described. The history of the discovery of the existence of natural reactors, in Africa, is presented.

The process of feedback, where the condition of a system provides a force to either control or disrupt the system, is shown to be important in the physical world and in the system of science.

The possibility of planets suffering naturally induced nuclear fission detonations is presented. Both disruptive detonations, such as produced the asteroids, and contained detonations, such as overheated the planet Venus, are possible.

Some of the development of modern astronomy is presented. Significant characteristics of astronomical studies are described. The present form and history of the Earth, as developed by geology, are described. The conflict between Plate Tectonics and Earth Expansion in explaining Continental Drift is discussed. Geological ages are listed, and the many pitfalls of isotope dating are described. Some of the effects of the Giant Dipole Resonance reaction, by gamma rays from astronomical detonations, are described. Historical attempts to date the Earth are described. Possible flaws in radiocarbon dating are suggested.

The study of life, evolution, mass extinctions and speciation events, is described.

In all of these matters, numerical, nuclear, astronomical, geological, biological, it is important that the reader be ready to search out alternative sources of explanation and description, beyond this book alone. This extra work will be needed to come to a full understanding of the stories that describe our world. As Ralph Waldo Emerson advised, " 'tis a good reader that makes the good book." With these terms and tools in hand or in mind, it is time to begin our voyage.

II. THE EARTH

There is only one Earth
in this world.
Zoey Hammock, 2010

Uranium and the Creation of Planets

The lamps are lit...

Begin at the beginning...
Lewis Carroll

Must one always begin at the beginning?
Much time could be saved if we could begin at the end.
A. P., as quoted by Marilyn vos Savant, 1999

So, let us begin at the end: the death of a star, exploding as a supernova.

Nearly all the heavy elements, beyond iron, are produced in the tremendous burst of neutrons set free in a supernova explosion. These neutrons are produced as the shock wave of the collapsing star forces electrons into the protons of the atoms of the core of the star. The neutrons are quickly absorbed by the existing nuclei, in what is called the r-process, more rapidly than radioactive beta decay can adjust the ratio of nuclear protons to neutrons, producing a mix of nuclei that gives us distinctive clues to the process itself. These newly produced heavy nuclei, and most of the rest of the star, are blown out (blasted) into the old, tenuous atmosphere of the star, which is still mostly hydrogen. As most of the new atoms are surrounded by 10^6-10^{12} as many hydrogen atoms (a million to a trillion), the new atoms will react to form hydrides whenever it is chemically possible. (Much oxygen is more tightly bound in carbon monoxide.) These reactions occur in the fresh material of the supernova cloud, while it is still moderately dense, and warm enough to promote the reactions. These are the materials needed to construct planets.

A neat trick happens as the cloud of new nuclei is ejected from the supernova. As the cloud cools, the ionized nuclei gather electrons and become elemental atoms, and then form chemical molecules. In the cloud, every new atom is surrounded by hydrogen atoms, and the first molecules formed, if chemically possible, are hydrides. Thus, carbon, nitrogen, and oxygen form methane (CH_4), ammonia (NH_3), and water (H_2O). Silicon forms silane (SiH_4). There is some adventurous research proposing that silane, accreted into a gas-giant primordial Earth, produced granites and quartzites [Collins and Hunt 1992]. These ideas have some appeal. Uranium reacts with the hydrogen to form uranium hydride (UH_3).

The uranium hydride crystallizes as the cloud cools further. As the uranium decays, by alpha and beta emission, the smallest grains develop negative electrostatic charges because of the excess removal of positive charges by escaping alpha particles (2 positive charges each). These negatively charged grains are attracted to larger, positively charged, grains of uranium hydride from which only the longer-range beta particles (1 negative charge each) can escape. The uranium hydride becomes, selectively, the fastest accumulating material in the condensing cloud. Since the uranium hydride is about the densest material in the cloud, these grains turn into the fastest growing, gravitationally and unselectively, globs. There are further interesting details which we will ignore right now, but these globs and the dust and gas in the cloud grow into planets with cores that are rich in uranium, at a time when the uranium was enriched to 24% (or more).

It is thought that these clouds of supernova ejecta eventually cool to very low temperatures, in the range of 10 K to 60 K. Nearly everything condenses and solidifies, except for hydrogen and helium and neon, which remain gaseous.

...the stage is set...

Nothing can be created from nothing.
Lucretius

In the cataclysmic collapse of a star, a supernova is born. The neutrons released in this collapse are so plentiful that successive neutron absorptions in the surrounding material occur so rapidly that radioactive decay plays a small role in determining the nuclides that are produced. This is the r-process [Burbidge *et al.* 1957], and nuclides beyond bismuth, including uranium and up to at least californium-254, are formed. The superheavy nuclides soon decay to lighter, longer-lived radionuclides, such as uranium-235 and uranium-238, and thorium-232, which accumulate beyond the abundance originally produced in the supernova explosion.

The shock wave produced by the collapse of the star propels the outer shell of the star into space, beyond the grasp of the remaining gravitational force of the stellar core. These clouds of newly produced elements disperse through interstellar space, expand, cool, contract, and the stage is set for the formation of new stars, and planets.

...let the play begin.

Judge not the play before the play is done.
Francis Quarles

Assuming an initial beginning for the Universe, as proposed by the Big Bang Theory, until a significant amount of the hydrogen and helium in the Universe had been processed through supernovae, there was little matter that could condense into solid or liquid planets. Hydrogen and helium, by themselves, make poor planet-building material. Add carbon, oxygen, magnesium, aluminum, silicon, and iron, and a pretty decent planet can be formed. With uranium added, a living planet is born. (An interesting aspect of our understanding that hydrogen and helium are the most abundant elements in the Universe, is that both of them are produced by uranium: protons from the beta decay of free neutrons from fission, and helium from the alpha decay of heavy elements.)

Uranium is specifically a product of the r-process of nucleosynthesis (neutron capture so intense and rapid that the distribution of elements and isotopes cannot be fully adjusted by beta decay during the process) in a supernova that contributed material to the formation of the Solar System [Rolfs and Rodney 1988, Spitzer 1963]. This uranium would have condensed out of the supernova cloud forming the pre-Solar nebula, and would have concentrated in the solids that accumulated to form the planets. The progress and decline of the uranium isotopes after their production in a supernova is shown in Figure 7. The relative abundances of uranium-235 and uranium-238 are shown, declining with time solely as a result of the easily calculated radioactive decay, and assuming that there are no effects that change this.

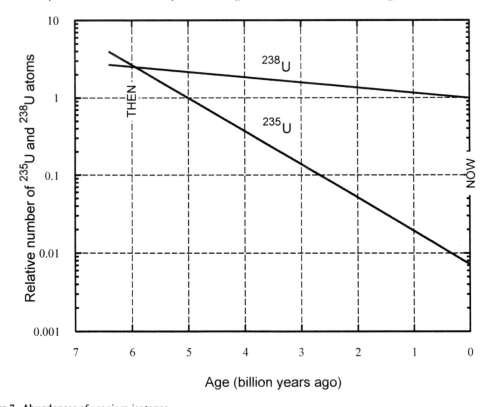

Figure 7. Abundances of uranium isotopes.
Uranium isotopes are produced in a supernova and then steadily decay. The uranium-235 decays more rapidly than the uranium-238. The uranium-235 enrichment at the time we think the Earth was formed (4.55 billion years ago) was about 24%.

Because of its shorter half-life, uranium-235 decays away faster than uranium-238, and fades from an overabundance in the supernova cloud to a trace material in the uranium of the present Earth. The "enrichment" of the uranium, representing the concentration of the easily fissioned isotope uranium-235, decreases from 60% in the supernova cloud to 24% at the formation of the Earth, to 3.63% at the time of the Oklo natural reactors, to 0.72% at the present time. It is this paucity of the nuclear fuel isotope that forced the development of gaseous diffusion plants and other devices for concentrating, or "enriching" uranium to the point at which nuclear fission chain reactions can be maintained, and nuclear reactors can be made to work. Fast breeder reactors can be constructed with uranium having more than about 5.6% uranium-235, although 25% is more in line with current technology. These reactors are "fast" because the neutrons released in fission are used to induce more fissions before they have lost much of their energy and slowed down to thermal equilibrium. They are "breeders" because more new fuel, in the form of plutonium-239, can be produced by neutron absorption by uranium-238, than is consumed by the fission of uranium-235. Thus, their operating lifetime can be extended far beyond that expected for our current electrical power generating reactors. At lower enrichments, down to about 1%, reactors including water as a moderator to slow down the fission neutrons to energies where the fission cross-section of uranium-235 is overwhelmingly greater than most other absorption cross-sections, can function. It is this type of reactor that was discovered to have formed in the rich uranium ore deposits of Gabon, Africa, and became known as the Oklo phenomena.

Herndon has pointed out that uranium compounds are denser than iron and have melting points (generally) that are higher than iron [Herndon 1993]. These compounds would be expected to precipitate from the molten iron produced during the differentiation of a planet, as the core formed, and because of their much greater density, especially at high pressure, would be concentrated at the center of the core by the action of gravity. Herndon cites earlier research that shows the density of elements at the pressures in the deep interior of the Earth are more determined by atomic number and atomic mass than by the chemical form and low-pressure crystal structure. Therefore, uranium, thorium, and related elements would be the densest elements, by far, in the iron core. The atomic number of uranium is 92; that of iron, 26. The atomic mass of uranium is 235 or 238; that of iron, 56 to 58. Accumulation of all the uranium in the core into a single giant reactor is prevented by the need for a reactor to be only slightly greater than its minimum critical mass. If that mass is significantly exceeded, the power will increase so rapidly as to disperse the material. Thus, there is no single reactor in the core of the Earth, but many.

A concentrating process could occur at the earliest stages of development of the pre-Solar nebula, even as the clouds of gas and dust from the supernova began to cool and slow. Diamonds and small grains of silicon carbide and silicon nitride have been found in material that is thought to be of interstellar quality and those compounds have been observed in interstellar clouds [Taylor 1992, Allamandola *et al.* 1993, Huss 1990], indicating that chemical combination and crystallization occur in diffuse gas in space.

The supernova cloud, containing freshly produced uranium, would consist primarily of hydrogen, and would provide a chemically reducing environment. The atom ratio of hydrogen to uranium would be approximately 10^{12} and the gas would be filled with electrons from the beta-decay of nuclides formed by the r-process, promoting chemical reactions. The uranium would easily react with the hydrogen, forming uranium hydride molecules (UH_3). The radical H_3^+, formed in the atmosphere of the star, has been found dispersed in the cloud of the supernova SN 1987A [Miller *et al.* 1992].

Structurally, the uranium hydride molecule is 3/4 of an equilateral tetrahedron, and these molecules would condense into small crystal grains by vapor-phase crystallization. Joining each new molecule leaves an open point of attachment for the next. Chains will grow and collect other chains to form a large dendritic clump [Beckwith 1994]. The uranium (and thorium) bearing grains would grow preferentially, aided by electrostatic effects [Glanz 1994] resulting from radioactive decay.

In the decay of radioactive uranium to stable lead, 10 more positive charges are emitted in alpha decays than negative charges that are emitted in beta decays. Because of this, very small grains of uranium- (and thorium-) bearing material (less than 1 micrometer in diameter, small compared to the alpha-particle range) will develop a negative static electric charge, with a potential up to a few million volts. Furthermore, the smallest grains (on the order of 10 micrometers) will receive some recoil energy from each alpha decay, and so will move erratically among the neighboring grains, inducing collisions that accelerate accretion and grain growth. The negative electrostatic charge makes this accretion more effective in adhering grains that have lost some electrons and so have a positive charge.

As the grains grow, a size is reached at which fewer of the alpha particles are emitted within range of the surface (10 to 20 micrometers) so that they can escape from the grain, and the static charge is instead produced by the longer range (up to a few hundred micrometers) beta particles and internally generated photoelectrons, resulting in a net positive charge on the grain.

As the grain then develops a positive charge at a potential up to a few million volts, it will attract more of its smaller, still negatively charged neighbors, those containing uranium (and thorium) and grow more rapidly than would inert grains of iron, silicon, carbon, and other more abundant elements. Calculations of this effect show that the attractive force of a 1-centimeter diameter grain of uranium hydride for very small grains, due to electrostatic charges, may be 20 times greater than the gravitational force. (A simple science demonstration can show how an electrified comb or plastic rod can lift a piece of paper against the Earth's gravitational attraction.) As these grains coalesce, the electrostatic charge will re-distribute and the growing piece will develop dendritically [Beckwith 1994], with branches and fingers reaching out to sweep in more material until it eventually collapses in on itself due to its own gravity.

Larger grains will experience self-heating from absorption of their own radiation, that further increases the effectiveness of accretion by impact, by melting ices or making the grain less brittle. As these develop into meter-sized objects, critical masses of uranium will accumulate and the object may melt due to the heat produced by the fission process, producing more segregation and further aiding accretion by making them "stickier" and more resistant to disruption by impacts. Eventually, sufficient heat would be generated to thermally decompose the uranium hydride, and nearly pure uranium would accumulate in the heart of the glob. These reactor globs would become electrostatically charged by the ejection of electrons from the radioactive fission products and by interaction of the gamma radiation, ejecting electrons by the photoelectric effect and the Compton effect. A reactor would be surrounded by a cloud of energetic electrons which would become attached to grains of silicate or other electrically insulating materials, and these grains would then be electrostatically attracted to the reactor glob, causing it to grow a mantle of non-reactive materials, or a "reflector" in reactor terminology. The gamma radiation from the reactor would also eject electrons from nearby grains and blobs of uranium, because of the high photoelectric cross-section of high-Z materials. These globs would become electrostatically charged, positively, and so would be repelled from the reactor glob. Thus, a critical mass of uranium would be self-limiting in size, and would not grow beyond the critical state by further accumulation of fuel materials. Stable operation of the chain reaction process would be assured by thermal expansion of the entire glob.

Pure hydrogen is an effective neutron moderator, and small globs of uranium hydride (UH_3) could form natural nuclear fission reactors in space. Calculations of the critical mass for a bare (unreflected) sphere of uranium hydride were done using the neutron transport-theory code ANISN/PC [Parsons 1988], with cross sections formed by use of COMBINE/PC [Grimesey *et al.* 1990]. The results are shown in Figure 8.

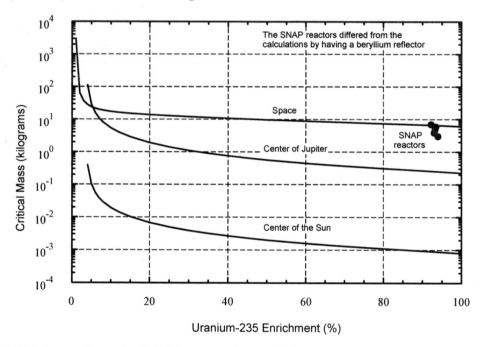

Figure 8. Critical masses for uranium hydride in space, Jupiter, and the Sun.
Critical masses of uranium hydride spheres, unreflected, in space. Globs of this sort may be critical down to uranium-235 concentrations as low as 0.94%, and subcritical blobs may be made critical by production of plutonium by neutron capture in uranium-238. Critical masses for several man-made space reactors are shown at 93% uranium-235. The increase in density of reactors at the center of Jupiter and the Sun reduces the critical mass.

Roughly 10 kilograms of uranium in this form will make a reactor, as shown by the critical masses of various SNAP (Systems for Nuclear Auxiliary Power) reactors, for powering artificial satellites. These reactors were fueled and moderated by zirconium-uranium hydride, using highly enriched uranium (93% uranium-235). The critical masses for these reactors are somewhat less than for the ideal spheres of the calculations because of their beryllium reflectors, that reduce the loss of neutrons to space.

Other globs may continue a very quiet existence, nearly unchanged with time except for the slow radioactive decay of the uranium. As these globs age, hydrogen will be lost and dispersed from the uranium hydride as a result of the radiation and heat released by the alpha decays, and helium will accumulate, also produced by the alpha decay. It seems likely that those globs that started to compact, as a result of impacts, electrostatic charge, or self-heating, would continue to do so to completion, while those that did not compact would, obviously, stay loose. Compact objects would lose hydrogen by thermal dissociation of the chemical bonds and diffusion of the molecular hydrogen. Helium produced by alpha decay of the uranium and its daughters would be trapped in the resulting massive material. In a loose object, the hydride form (and the hydrogen) might be retained while alpha particles, upon becoming helium atoms, would escape into space. Two types of uranium globs would exist: those rich in hydrogen and those rich in helium.

Those objects caught in the formation of a planetary system would form the cores of planets and, as the densest material, would remain there as the gravity field developed with further accumulation of material. Sizeable dense globs would then gravitate to the core of the newly forming planet as more material accreted and fused together. Planets sufficiently large will differentiate into a core, mantle, and crust. Oceans and an atmosphere may form, and life might eventually develop. The accumulations of uranium would become natural nuclear fission reactors within the core of the new planet. To take the Earth as a model of a fully differentiated planet formed in this way, these reactors may reside in two distinct regions of the core: in the solid inner core, where much of the uranium and thorium may have originally accumulated during differentiation of the newly formed planet, due to their great relative density; and in peripheral cells at the core-mantle boundary, where fresh uranium may be extracted into the core from minerals in the mantle. These separate concentrations of fission fuel may act as a coupled reactor system, with the power of each reactor being affected by that of its neighbors. Because of resonance interference in the neutron scattering cross-section of iron-56 at 27.9 keV, iron is extremely "transparent" to neutrons near this energy, and neutrons may travel several meters in iron before scattering out of this energy range [Weinberg and Wigner 1958]. This may promote the involvement of individual reactors in a region in a coupled, interacting system.

The uranium isotopic composition used for these calculations is based on extrapolating back in time to 4.55 billion years ago, when we think the Earth formed, with the assumption that the current ratio of uranium-235/uranium-238 is representative of the uranium present in the early Earth, after adjustment for normal radioactive decay.

The industrial safety practice of criticality control, developed in nuclear fuel and weapons fabrication plants, is based on the philosophy that if more than a few hundred grams of enriched uranium is involved in a process, a self-sustaining chain-reaction ("criticality") will occur, unless controls are established to prevent this. The quantity of uranium in the Earth, at an enrichment of 24% at 4.55 billion years ago, is estimated to have been about 6×10^{19} kilograms. That is a large amount and so this philosophy, that criticality will occur unless it is prevented, was applied to the research for this book. Since critical reactors require only 100 to 1,000 kilograms of uranium (and sometimes much less), all that is required is that enough uranium accumulate in a sufficiently concentrated form to create a critical mass. Critical accumulations of uranium in the Earth appear to be inevitable, and would result in self-sustaining chain-reactions, natural reactors like Oklo, on a planetary scale.

The empirical calculation of the critical mass for enriched uranium mixed with iron [Thomas 1978] has been justified in its industrial safety applications for fuel densities down to about 50% on the basis of experimental measurements. Because of its intentional conservative bias, this calculation should underestimate the amount of fuel required to achieve a critical condition. These calculations show that amounts of uranium (containing 24% uranium-235) in the range 60 to 5,000 kilograms of uranium in iron, could support naturally induced nuclear-fission chain-reactions in the Earth.

To better define the conditions leading to nuclear reactors in the Earth, calculations were also done using a finite-difference multi-group neutron transport computer code, ANISN/PC [Parsons 1988]. Multi-group cross-sections were formed using COMBINE/PC [Grimesey et al. 1990]. In these calculations, representative compositions have been considered and an adjustment was made for the higher density of material in the Earth's core. In these idealized calculations, the fission fuel was considered to be pure and in a compact spherical form. These assumptions may lead to a slight underestimate of the actual required critical mass. The results of these calculations are displayed in Figure 9, for uranium metal of various uranium-235 content, or "enrichment", at the surface for comparison with known critical masses, and adjusted for the increased density at the center of the Earth. Compression of the uranium by the pressure at the inner core/outer core boundary reduces the critical mass by about a factor of 4, producing reactors that are about the size of a large grapefruit to as small as a fist [Paxton and Pruvost 1987].

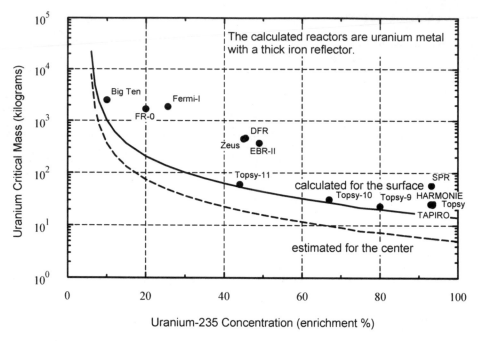

Figure 9. Critical masses of uranium on and in the Earth.
Critical mass of uranium in iron as a function of uranium-235 concentration, at the density on the surface and as expected for the center of the Earth. The critical masses of existing developmental nuclear reactors at different uranium enrichments are shown for comparison. Because of the inclusion of other materials in the industrial nuclear reactors, predominantly coolant, these critical masses are somewhat larger than the idealized natural reactors. Some were unreflected.

As the uranium-235 content decreases, as it does as a result of radioactive decay, increasingly large amounts of uranium are required to make a critical mass. These calculations show that a critical condition could exist for reasonable masses of uranium in iron containing more than 5% uranium-235, with critical masses ranging above 20,000 kilograms at the lowest possible enrichment. For example, at 93% uranium-235 (so-called "highly enriched uranium") the ideal critical mass (at high density) is 16 kilograms. At only 10% uranium-235, the critical mass in the same conditions is 1,000 kilograms uranium. For even lower uranium-235 content, a point is reached at which no amount of uranium can be critical, without a moderator. At uranium-235 contents below this value, a critical nuclear reactor is not possible unless light elements are also present as a moderator, such as the water in the reactors at Oklo, and hydrogen and helium in the Sun. Different compositions and fuels will show different values for this limiting enrichment.

Extrapolating back in time to the formation of the Earth, at 4.55 billion years ago, the uranium-235 content in the uranium would have been about 24%, essentially the same as the fuel in the Fermi 1 reactor, which had a critical mass of 2,000 kilograms of uranium, in an active volume of less than 0.4 cubic meters. That is a sphere less than a meter in diameter. (During this formation, if a reactor is larger than the critical mass for its specific composition and fissile content (uranium-235, plutonium-239, uranium-233), it will generate too much power, overheat, and quickly break up. Thus, there will be many small reactors, all somewhat similar in size.)

The critical masses of some existing research and development reactors are shown for comparison, and are considerably larger because of the dilution of the fissionable fuel by coolant and structural material, such as fuel rod cladding, in the industrial reactors. For example, actual reactors built at the surface of the Earth, TAPIRO and HARMONIE, at 93% enrichment, have critical masses of 25 kilograms, and BIG TEN, at 10%, has a critical mass of 2,500 kilograms.

These concentrations of uranium could be created at the center of the Earth by thermochemical action as a result of the various phases of these materials. Figure 10 shows the phase diagram for uranium in iron [Moffatt 1984], as it might exist in and around the inner core of the Earth, and other planets. At low concentrations of uranium, as the temperature decreases, iron solidifies and the liquid becomes richer in uranium. At high concentrations, the uranium will solidify and, because of its much greater density, is likely to gravitate towards the center of the Earth. Since the phase boundaries are subject to pressure effects that act as temperature effects, transport towards the center of the Earth is likely to concentrate the uranium further. The increased pressure at the center of the Earth, approximately 3.6 million bars (3.6 million times the pressure of air at the surface) [Poirier 1991], may elevate the temperatures of the phase boundaries by 5,000 K. Since the temperature at the Earth's center may be only 4,800 K [Stevenson 1981], phase changes may be induced as much by changes in pressure as by changes in temperature.

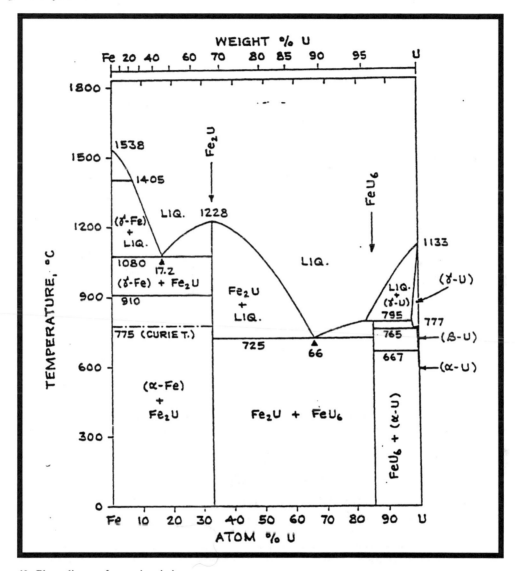

Figure 10. Phase diagram for uranium in iron.
Phase diagram of uranium in iron showing various phases subject to segregation by changes in temperature and, indirectly, pressure. High pressure within the Earth's core acts to elevate the temperature/phase boundaries [Moffatt 1984].

After differentiation of the Earth, processing of the mantle material at the core-mantle boundary is likely to extract uranium from the mantle into the core to produce local uranium-rich phases, containing 23% uranium oxide [Feber *et al.* 1984]. The phase diagram of uranium oxide in steel (which can be taken to represent the iron of the Earth's core) in Figure 11 shows how this could happen. For temperatures typical of the D″ layer and the core-mantle boundary in the Earth, 2,937 K to 3,800 K [Poirier 1991], small concentrations of uranium oxide (UO_2), on the order of a few percent, will exist in solution in molten iron (the right-hand side of the plot). If the local temperature decreases slightly, the uranium oxide will separate into an immiscible liquid phase and, if the temperature declines somewhat more, the liquid uranium oxide will precipitate as a solid. Thus, as temperatures fluctuate in the core-mantle boundary region, and as new mantle material is brought in contact with the molten iron core, uranium oxide would be extracted and concentrated, just as this happens near the surface of the Earth in the production of ore deposits.

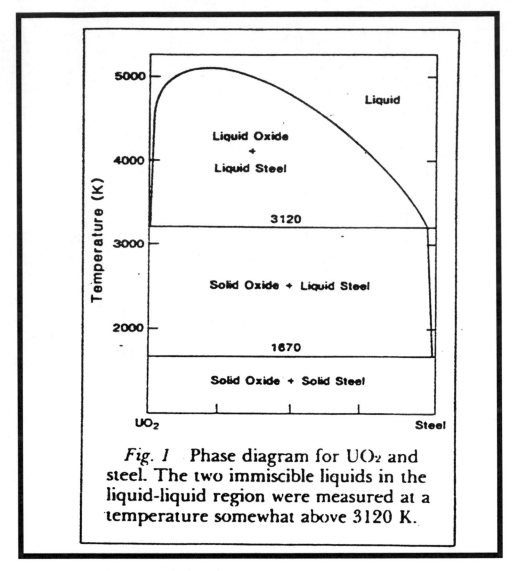

Fig. 1 Phase diagram for UO₂ and steel. The two immiscible liquids in the liquid-liquid region were measured at a temperature somewhat above 3120 K.

Figure 11. Phase diagram for uranium oxide in steel.
Phase diagram of uranium oxide in iron showing various phases subject to segregation by changes in temperature and, indirectly, pressure [Feber *et al.* 1984].

The critical mass for uranium oxide in iron at the core-mantle boundary and at the surface is shown in Figure 12. A concentration of pure uranium oxide (at an enrichment of 24%, the extrapolated uranium-235 content of uranium at the time of the formation of the Earth) would be sufficient to produce a reactor critical mass of about 600 kilograms uranium, in a volume of about 0.05 cubic meter, scarcely larger than a breadbox. Compression of the reactor material by the pressure at the core-mantle boundary will produce reactors with smaller critical masses, and smaller in size. Thus, a multitude of critical (and sub-critical) reactors could develop at the core-mantle boundary, and the power produced by these reactors would be quite significant in affecting conditions throughout the mantle and at the surface of the Earth. Local collections of reactors could provide the heat that drives the mantle-plume, or hotspot, volcanoes.

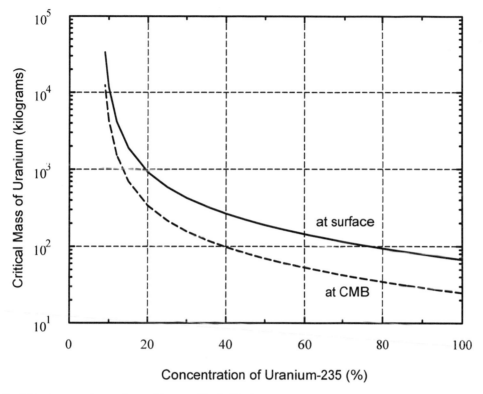

Figure 12. Critical masses for uranium oxide on and in the Earth.
Critical mass of uranium oxide in iron as a function of uranium-235 concentration, at the surface (dashed line) and adjusted to the density expected for the core-mantle boundary of the Earth (solid line).

In addition to iron, oxygen, and sulfur, the commonly considered major constituents of the Earth's core, a study of the solubility of hydrogen in hot iron at high pressure shows the likelihood of significant amounts of hydrogen in both the inner and the outer core [Fukai 1984]. That study showed that the ratio of hydrogen to iron developed during the accumulation and differentiation of the Earth could have been approximately 0.27 to 0.36. The presence of hydrogen is likely to have significant effects on the chemical concentration of uranium from the oxide. As a neutron moderator, it may also affect the action and development of the reactors, that may produce results quite different from a totally unmoderated system.

Because of its location, approximately 2,600 kilometers underfoot, and due to the intervening materials, it is somewhat difficult to explore the condition of the Earth's reactors by sampling the core, even more so the inner core. To borrow from Birch's high-pressure vocabulary, as quoted by Stevenson [Stevenson 1981], the inner core, by its solidity, might be supposed to be "pure iron", but indeed it might more likely be, as an accumulation of fission products and transmuted nuclides, "an uncertain mixture of all the elements".

Power stability will be controlled during this steady phase of the chain-reaction by the time-lag provided by delayed neutrons [Keepin 1965] and the negative feedback effects of increased neutron leakage from the reactor due to thermal expansion, and the prompt negative Doppler reactivity effect for uranium fuel [Hetrick 1971].

Scattered uranium produced in a supernova can accumulate into globs forming active reactors. These reactors ultimately provided heat for the planets, ignition for stars, and detonations for asteroidal fragments.

Solid accumulations of starstuff in space, forming much like our planets formed but without the benefit of a Sun, may need to be called planetoids, to make clear the distinction from our Solar System planets. Planetoids will contain natural nuclear fission reactors, deeply buried and confined. These reactors will occasionally make a transition from stable power to a self-destructive detonation. Peter Catalano discusses "rogue planets" as runaway planets that were formed next to a star but were ejected. [Catalano 1997]. I think planetoids may form by themselves, or in a company, but without a starsun at the end of the process.

These detonations may be observed in space as the gamma-ray bursts. The fragments of planetary detonations may, in many cases, be the objects that are interpreted as pulsars, QSOs and quasars, blazars, and other objects that are currently explained poorly. The old controversy over the nature of nebulae, that was resolved early in this century in favor of the "Island Universe" theory, may have been decided by mistake. The characteristics of the so-called galaxies are better explained as the consequences of relatively near and recent planetary detonations, than as vast collections of stars like our Sun. Many of the difficulties in understanding galaxies and quasars arise solely from the assigned great distances, and these difficulties disappear if these objects are recognized as nearby remnants of planets.

Expulsion of large quantities of iron, that had been accumulated in the cores of the disrupted planets, would have created a disperse fog of iron crystals that is the source of the Cosmic Microwave Background Radiation. Iron can be created as a fission product in detonations that reach extreme temperatures.

Home Planet Earth

The Earth is the planet on which we live, and it differs drastically from all the others in our Solar System, as they differ from each other. To understand the other planets, we should first study Earth. What is its structure? How did it form? What gives it the special character that supports life?

The structure of the Earth has been studied by many means. Only the very outermost skin of the Earth can be explored directly. We have direct knowledge of the atmosphere, the ocean, and the crust to less than 10 kilometers deep. Our knowledge of the lower depths, on down to the center of the Earth, comes from painstaking analysis of the seismic waves that pass through the Earth from earthquakes, measurements of the size and shape of the Earth and its gravitational field, and measurements of its magnetic field. These measurements are supplemented by samples of the deeper crust, brought up by volcanoes, without our quite knowing from where. Our interpretations are based on laboratory experiments that don't quite span the scale of conditions within the Earth. Everything we say about the Earth, beyond that, relies on what we think we know and our imaginations.

Seismology and a very few other measurements have given us as good a picture of the internal structure of the Earth as we have, possibly as good as we can ever get objectively. A cross section was presented in Figure 5.

Many theories of how the Earth formed have been developed. Most of these are based on the assumption that the Earth we see today closely represents the early Earth as it was formed. In those theories, the atmosphere may have changed somewhat, the continents have been rearranged, but this is just "remodeling". Exploration of the effects of natural nuclear fission reactors has led to a somewhat different concept of the Earth's formation, a much more dynamic process that agrees better with current discoveries. We will look at that in detail later, but for now will just consider the direct effects of the uranium that was produced in a nearby supernova explosion just before the birth of the Earth.

A solid core at the center, generally considered to be solid iron and nickel like some meteorites, is surrounded by a fluid outer core that does not transmit certain kinds of seismic waves. That fluid is contained by the mantle, which is thought to be much like the basaltic lava erupted by volcanoes. The boundary at the base of the mantle has been named the D″ layer because we don't know what else to call it, and appears to be extremely irregular, in terms of its surface (it has mountains and valleys the size of continents), its temperature (with hot and cold spots), and density (which fluctuates between the density of the outer core and the lower mantle) [Jeanloz and Romanowicz 1994]. The D″ layer is washed by a hot, chemically reactive liquid (molten iron) that is as fluid as water is at the surface of the Earth. It is hard to know what thermochemical reactions might be taking place at this interface. It is likely to be a very active region and may play a very important part in the workings of our world. We must rely entirely on seismology to study this region, and the various effects are difficult to sort out.

The mantle itself is divided into layers or shells, the lower mantle, the upper mantle, and the asthenosphere, all of which come to play in various discussions of the outer Earth. The mantle is covered with the crust, which may be oceanic material or continental material. Alfred Wegener imagined blocks of the continental crust floating through the oceanic crust, to permit the dispersion of the continents that has been labeled Continental Drift. The theory of Plate Tectonics explains this dispersion as a result of the motion of many crustal plates, with the edges of the oceanic plates being subducted, or sinking, beneath the edges of the continental plates. The theory of Earth Expansion explains the same dispersion as resulting from the breakup of an initially continuous continental crust as the Earth grew larger, and provides for the observations of subduction by the overflowing of the continental crust onto the oceanic crust. That might be called "superduction".

It has been thought that the Tibetan plateau had not reached its greatest height until about 8 million years ago, and that this uplift had strengthened the monsoon, causing more continental sediment to flow to the ocean. According to the Plate Tectonic theory, the Indian subcontinent caused the uplift of the Tibetan plateau by a collision that began 54-49 million years ago, and forced the plateau to rise above its maximum sustainable elevation, and the plateau began to collapse, spreading to the east and west. This collapse, or rather the initiation of the faults allowing it to spread in a manner aligned with the circumference of the Earth, has been dated at 14 million years ago [Coleman and Hodges 1995, Searle1995]. In Earth Expansion, the Himalayan block was forced upward by localized (deep) expansion, and crumbled, overflowing onto the subcontinent of India. This is superduction on dry land.

The crust is covered by ocean water in places, and all is enclosed in a gaseous atmosphere that allows us to breathe, see, talk and hear, and move with reasonable ease and grace. Beyond that is space, where we are beginning our first tentative explorations.

Keeping the home fires burning

Calculation of the Earth's thermal flux, and thus the extent, character, and origin of its internal heat, has played a major role in exploring the nature of our planet. In an effort to estimate the age of the Earth, William Thomson (later Lord Kelvin, for whom the principal temperature scale is named) calculated (in 1899) the rate of heat flow through the surface at the present time, assuming that the Earth had initially started at the temperature of the surface of the Sun. The temperature of the Solar surface is now thought to vary with depth in the photosphere from about 5,400 K to 6,200 K [Chapman 1992]. Using the then accepted structure of the Earth as a model, Kelvin estimated that the Earth was probably 25 million years old [Stacey 1969]. This short time, relatively speaking, compared to much longer estimates based on geological processes, stood as an obstacle to the development of modern geology until the heat released by the decay of natural radioactive elements, which were discovered in 1896, was recognized. This heat can be estimated to produce the currently observed heat flow, by suitable adjustments of the composition of the crust and mantle [Stacey 1969, Poirier 1991, Verhoogen 1980, Stevenson 2010]. However, it seems a bit unfair to use the observed heat flow to estimate the composition and structure of the Earth and then use these estimates to calculate the heat flow. And other problems remain. How could radioactivity, which we now calculate to contribute only a ten-thousandth of the heat supplied by the Sun, be considered anything more than negligibly trivial? Radioactivity does not solve Kelvin's problem.

The Faint Young Sun Paradox: Cool Sun, Warm Earth

The path of paradox
is the path of truth.
Oscar Wilde

Occasionally, science runs into a paradox, where our understanding conflicts with our knowledge. One of the most important is the Faint Young Sun Paradox [Sagan and Mullen 1972, Kasting, Toon, and Pollack 1988]. Resolving that paradox is essential in the understanding of the origin of life on Earth, the present condition of the planet, and our very existence. Many efforts have been made to explain this conflict [Kasting, Whitmire, and Reynolds 1993, Sagan and Chyba 1997] but they seem to have failed [Rampino and Caldeira 1994, McKay, Lorenz, and Lunine 1999].

The Sun evolves over time, and the Sun we see now, like the porridge in the tale of Goldilocks, is just right. However, in the beginning, it was too cold, and in the end, it will be too hot. How can something as bright and as hot as the Sun be too cold? Certainly, it has always been hot, since it became a star. High temperature is essential for the fusion reactions that power stars to take place. The problem is that the Earth is surrounded by bitterly cold space, at 3 K, and the Earth radiates away its heat almost as quickly as it receives it. At present, the Earth is covered by a blanket of air, the atmosphere, which contains the "greenhouse" gases, carbon dioxide, methane, and water vapor. Infrared absorption in the atmosphere prevents the loss of some of the heat that is absorbed from the Sun.

If the Earth were heated only by the Sun, and had no insulating atmosphere, the surface temperature can be calculated to be in the range of 248 to 255 K. (The freezing point of water is 273 K.) This range is based on assumptions that the Earth's albedo was in the range of 37% to 30%. (A lower albedo permits the Earth to be warmer.)

Briefly, the astronomers' model of the Sun shows that the Sun was much fainter (30%) at the beginning of the Solar System. The Sun has gradually expanded since then, and is currently providing adequate heat for a very livable Earth at the present time. According to our understanding of the Sun, it was only 72% as luminous as it is now when the Earth formed about 4.5 billion years ago. The deficit in heat is so great that climate-model calculations show that, with the present atmosphere, the surface water of the Earth would have become frozen at an early time after the Earth formed and would have remained so until approximately 1.6 billion years ago [Kasting *et al.* 1988], or perhaps even permanently [Gleick 1987, Manabe and Bryan Jr. 1985]. The temperature of the Earth when the Sun was new, again with no atmosphere but the same range of albedo, would have been in the range of 228 to 234 K. All these temperatures are well below the freezing point of water. With all reasonable modifications of climate models, this would leave the surface of the Earth below the freezing point of water until about 1.6 billion years ago. If the Earth had been that cold, it would have been covered with ice and snow, and the albedo would have been so high, it would not have thawed out yet. We know that did not happen from ancient sedimentary (hydrologic) deposits and temperature indicators, and that life developed long ago, presumably in warm, liquid water. That is the Faint Young Sun Paradox, for we know there was liquid water at least 3.8 billion years ago, yet our understanding of the Sun tells us that it could not have supplied enough heat.

If we assume that there was enough water on the surface of the Earth at an early time to cover the surface with snow and ice, the temperature would have been in the range of 121 to 181 K, because of the high albedo provided by snow. That is cold enough to freeze carbon dioxide, eliminating one of the greenhouse gases from consideration. If the Earth today were covered with snow, we can calculate the temperature without an atmosphere to be in the range of 132 to 197 K, which would keep water and carbon dioxide frozen solid. Even with the present atmosphere, and its assumed "greenhouse effect" of 33 to 40 K, the temperature would be less than 237 K. If the Earth had ever been frozen over, it would still be frozen over. And that is the White Earth Catastrophe [Caldeira and Kasting 1992]. If our understanding of the Sun and the Earth is correct, we should not be here to puzzle over these things. The Earth should be a bright, white, shiny snowball, drifting around a fruitless star. The Earth is somehow warmer than we can understand, and that is a paradox.

It should be clear from these discussions that calculations result in a range of answers, yet overall, the calculations all say the Earth should be frozen over. It has been proposed that this actually happened, as recently as 700 to 600 million years ago [Kaufman, Knoll, and Narbonne 1997, Hoffman and Schrag 2002, Walker 2003], and yet the Earth was able to recover, in spite of warnings about a White Earth Catastrophe. That is also a puzzle.

Since it is clear that these conditions did not occur (liquid water has been present since 3.8 billion years ago [Wicander and Monroe 1989] and the Earth is certainly not still frozen now), there must be a deficiency in the calculational model, and this deficiency has been generally considered to be the assumption of an atmosphere with the composition of the present atmosphere. Inclusion of somewhat greater amounts of carbon dioxide in the atmosphere than at present leads to predictions of warmer conditions, due to the greenhouse effect, or to cooler conditions, due to formation of CO_2 clouds [Caldeira and Kasting 1992], or hotter, due to formation of CO_2 clouds [Forget and Pierrehumbert 1997]. Alternate solutions involve a more rapidly spinning Earth in Precambrian time, with a 14-hour day, and higher CO_2 concentrations in the atmosphere [Earth 1994, Jenkins, Marshall, and Kuhn 1993] or very much higher CO_2 concentrations [Kasting and Ackerman 1986, Kasting 1993].

This situation can be considered in terms of heat input sources to the Earth (Figure 13). (The unit of heat production used here, the petawatt (PW) or 10^{15} watts, a quadrillion watts, is equivalent to roughly a million modern electrical power plants.) This figure shows the major sources of energy input to the Earth since the start of its formation, based on the conventional model for the formation of the Earth. This includes energy released by the gravitational infall of material in the form of dust and planetesimals (the spike in heat input at 4.5 billion years ago) as planet Earth was formed; the mineralogical separation of the early Earth into core, mantle, and crust [Tozer 1978]; energy deposited by the presumed impact formation of the Moon (the spike at 4.2 billion years ago) and by tidal friction [Stacey 1969] estimated here from formation of the Moon to 2.84 billion years ago, and then calculated for a representative ocean distribution [Burns 1985].

The actual deficiency may be considerably greater than that used for comparison here: evidence in the isotopic ratios of chert, a rock composed primarily of silicon and oxygen, shows that average temperatures were well above freezing in the distant past [Knauth and Epstein 1976], possibly ranging up to 52 to 70 C, at a time when the Sun's heat couldn't melt ice on the Earth, which would require temperatures just above 0 C. Moreover, this deficiency is not just a matter of climate, a surface thing that can be solved by surface changes. The problem is more than skin-deep. Convection deep in the mantle of the Earth, well below the thin surface crust, has been found to be ten times as great as at present, during the Late Archaen period, 2.7 billion years ago [Blichert-Toft and Alberède 1994].

In these calculations, the Moon is assumed to have been formed by a single cataclysmic impact by a Mars-sized object according to our current theory [Newsom and Taylor 1989], and to have stabilized in an orbit at 7 Earth radii from the Earth's center, at 4.25 billion years ago. Estimated values for Lunar tidal friction at 0.66 billion years ago [Sonnett *et al.* 1988, Katterfel'd 1962] are shown for comparison.

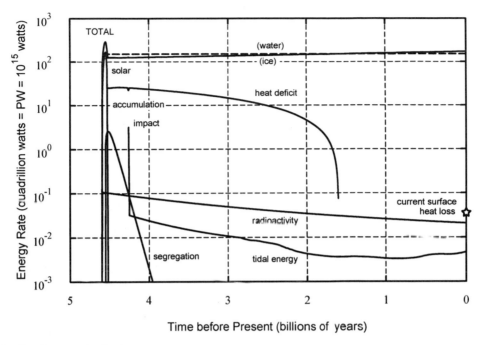

Figure 13. Heat inputs to the Earth.
Heat inputs to the Earth: Energy released by accumulation, structural differentiation, radioactivity, sunshine, impact formation of the Moon, and the dissipation of the tides. The total heat input is less than that required to melt ice from immediately after the accumulation phase until about 1.6 billion years ago. Rocks that were deposited in liquid water are known from 3.8 billion years ago. No recognized source of heat approaches the magnitude of this deficit.

The energy released by the radioactivity of long-lived radionuclides is calculated here for potassium-40 (225 ppm), and the uranium (0.0152 ppm) and thorium (0.0543 ppm) chains [Kargel and Lewis 1993]. The small heat release due to rubidium-87 (0.0001 quadrillion watts) has been neglected. Verhoogen [Poirier 1991, Verhoogen 1980] calculates a somewhat greater production of heat by radioactive decay than here, 0.0242 quadrillion watts, from slightly different abundances for the radioactive elements, but this is still not sufficient to answer the need for heat to keep water liquid in the past. At present, the heat released by continued cooling of the core, slow crystallization of the inner core, and settling of solidified particles of iron in the outer core, is estimated to be about 0.0036 quadrillion watts [Poirier 1991].

In this figure, a constant line at about 150×10^{15} watts (150 quadrillion watts, 150 PW) indicates the energy input required, for an atmosphere with the present composition, to maintain a surface temperature above 0 C, the nominal freezing point of water. The heat input estimated in this figure falls short of the heat required for liquid water from soon after formation to 1.6 billion years ago. However, it appears that liquid water was present at the surface of the Earth as soon as it was possible for water to condense, and its geologic effects have been found in rocks dating to 3.8 billion years ago. The deficit in the heat required for liquid water from 3.8 billion years ago to 1.6 billion years ago may be estimated based on the calculations shown in this figure. The total deficit amounts to approximately 10×10^{32} joules (J).

It is not possible to achieve adequate heat to maintain the warmth of the Earth from the sources that have been identified so far, even by delaying the development of the energy dissipation of the lunar tides. This source of heat is strongly dependent on the assumptions for the tidal model. The tidal model assumes the same characteristics of tidal friction as exist today [Stacey 1969] and at 0.66 billion years ago [Sonnett *et al.* 1988]. At present, this heat is calculated to be 0.00472 quadrillion watts. Absence of oceans and seas from the early Earth could have reduced the tidal friction to the point that the energy would have been released more gradually, and become available over the succeeding 2 billion years ago. A totally global ocean might have released less tidal heat than at present. However, there is not enough energy available for the Moon to have kept the Earth warm before the Sun was able.

Carefully averaged values of measurements of the heat flux from the solid Earth [Stacey 1969] show a global heat production rate of 0.042 quadrillion watts. Lunar tidal heating contributes 0.00472 quadrillion watts at the present time, and so can be neglected. Volcanic eruptions, on land and under the sea, release locally intense heat, but it is not clear if this heat is included in the determination of the global average. This global heat-flow is slightly more than that produced by the radioactive decay, estimated above, by 0.0125 quadrillion watts. It has been speculated that this difference may be made up by residual radiogenic heat [Kargel and Lewis 1993].

Heating of the early Earth by radioactivity, with standard estimates of composition, is too small to make up this deficit by a factor of 1,000; tidal heating by the Moon is similarly too small. Even if all the primordial heat from formation of the Earth had escaped to the surface during the first 3 billion years, that would have only contributed a few percent of the warmth needed to keep water liquid.

According to the climate modelers, the reduced amount of heat received from the Sun in the beginning would not have been sufficient to keep water from freezing on the face of the Earth. The Sun was not warm enough to keep water liquid until about 1.6 billion years ago. We know that the Earth had abundant liquid water throughout its history: the oldest rocks, about 3.8 billion years old, were initially formed as sedimentary rocks deposited in water. Further, the White Earth Scenario says that if the Earth had frozen over, it would have been such a bright, white snowball that the Sun's heat would be reflected. The Earth would not have thawed out yet. I found that there were no known sources of heat that could compensate for this deficit. Neither tidal friction nor radioactive decay, recognized as ongoing sources of heat independent of sunshine, provide nearly enough heat to affect these conclusions in the slightest.

After I learned of the Faint Young Sun Paradox and the White Earth Scenario, I was faced with a puzzle. How could I re-design the Earth so that it could keep warm until the Sun heated up? Climate modelers had failed. No reasonable amounts of greenhouse gases could thaw a frozen Earth. The modelers had to add a dummy source of heat to make their results agree with what we thought the Earth was like in long distant times. In order to match climate models with known climate conditions, the climatologists have had to add an artificial source of heat, often as warm humid air across a boundary. This suggests that there is an unaccounted-for source of heat, needed to produce the warmth of the Earth.

Since the discovery of natural radioactivity a hundred years ago, it was recognized that radioactive decay would produce some heat to warm the Earth. So, I tried to increase the amount of radioactivity in the Earth in order to get more heat. But no reasonable amount of radioactivity could make up the deficit. The Sun was too weak, the Earth was too cold, and we simply couldn't get here from there. Then, in one of my calculations, I put a large amount of uranium into the inside of the Earth, into its core, where most researchers think there is none. Uranium is naturally radioactive and produces much of the radioactive heat in the Earth, but even this added amount of radioactivity was not enough. However, I recognized that there was so much uranium in the Earth's core in this model that it would have acted like nuclear reactor fuel, and natural nuclear fission reactors would have been formed.

Nuclear Reactors and the History of the Earth

The possibility that the energy released by natural nuclear fission reactors in the earliest age of the Earth contributed to its formation, along with many other sources of heat, was suggested, but no further study seems to have been done [Rogers 1993].

More recently, and independently of the work that has led to this book, Herndon has shown, in a manner similar to that of Kuroda, that an on-going nuclear-fission reactor in the core of the Earth, rather than in ore deposits at the surface, is possible and likely [Herndon 1993], that it can provide the energy necessary to drive the geomagnetic dynamo, and that similar reactors in the giant planets Jupiter, Saturn, and Neptune are the probable sources of the excess energy in these planets [Herndon 1992]. The possibility that nuclear reactors could develop within a planet, and result in a detonation that could disrupt the planet, as will be discussed in detail in this book, was described several years ago by Driscoll [Driscoll 1988]. Unfortunately, these ideas did not receive adequate exposure, recognition, or acceptance.

In our standard theory of the creation of the Universe, the only elements available at the start are hydrogen, helium, and a small amount of the other lightest elements. These elements are not the stuff that planets are made of. Heavier elements must be made in other ways.

In the last seconds of its existence as a star, an exploding supernova produces most of the elements we know of, much beyond iron. The star spreads out into space, to become starstuff, of which planets and people are made. The cloud of new material cools, reacts, combines, and eventually becomes ready for the formation of a new star, possibly a new planetary system.

Well-known chemical reactions in the supernova cloud that supplied the raw material for the Solar System produced a variety of chemical compounds, mostly with hydrogen. One of these compounds, of particular importance to our story, is uranium hydride. With subsequent crystallization of uranium hydride, selective electrostatic attraction resulting from radioactive decay, and gravitational forces between the densest grains in the protostellar nebula, uranium-rich cores would form in the planets.

Natural uranium, on Earth at the present time, contains only 0.72% uranium-235, the easily fissioned isotope that fuels our commercial power reactors. However, since more uranium-235 than uranium-238 is formed in a supernova explosion, and uranium-235 decays more rapidly than uranium-238, the uranium-235 content at the time that the planets formed amounted to 24%. That is roughly the "enrichment" used for the design of modern fast breeder reactors, that produce more fuel isotopes than they consume. Such reactors, formed inevitably by condensation of the protostellar nebula into planets, could operate for billions of years, supplying the excess heat that we observe, but cannot account for. These are the natural nuclear fission reactors that have kept the Earth warm when the Sun could not.

These reactors would be small spheroids less than a meter in diameter, each containing a few hundred kilograms of uranium. Operating at an inherently stable power on the order of 1 to 1,000 watts each, these natural reactors could have overcome the deficit in heat needed by the Earth (the maximum deficit is calculated to have been 25.94 quadrillion watts (PW) at 4.38 billion years ago), if only 2 parts per million of the Earth's core mass were uranium concentrated into reactors. Herndon estimates that the core of the Earth contains about 4 parts per billion, with about 1.1 parts per billion in the lower mantle [Herndon 1993]. The Earth's crust has 1.7 to 4 parts per million [Lodders and Fegley Jr. 1998, Weast 1980]. The crust is the only part of the Earth where we have observational knowledge. To overcome the lack of heat from the Sun, over billions of years, the core needed to have about the same concentration of uranium as we know the crust has.

However, geophysicists still tend to think of the core of the Earth as generally smooth, quiet, and homogeneous. Conventional geophysics explains that there is no way in which uranium can accumulate sufficiently in the core because of its initial extremely low concentration and the effects of entropy and statistical mechanics in preserving this dilute condition [Stevenson 1994]. However, as will be presented in the balance of this book, evidence of the consequences of these reactors is so pervasive that the challenge becomes to discover how uranium could accumulate, rather than to explain why it can't. Just as changes of phase cause water vapor to concentrate in our sky, and rain to the ground to form pools of water, phase changes in the mantle and core concentrate uranium in ways that result in natural nuclear fission reactors. Uranium may rain down from the mantle, through the fluid outer core. Pools of reactor fuel would collect on the surface of the solid inner core and drive the heat-transfer system of the Earth.

At that, I started to recalculate the amount of heat produced in the Earth from radioactive decay, using a different assumption for the concentration of uranium in the core, and I found that I had "put" in the core a far greater amount of uranium than would be permitted in a reactor fuel fabrication plant. That is, the amount and type of uranium I assumed to be present 4.5 billion years ago would almost certainly have resulted in natural nuclear fission reactors. Moreover, I found that natural nuclear fission reactors deep within the Earth could provide the missing energy. According to our understanding of how elements are formed in supernovae and then condense to form stars and planetary systems, such reactors would be inevitable. All planet-sized bodies would form with a core of uranium-rich nuclear fuel. Considerable evidence from a variety of observations supports this suggestion. That understanding drove all that came after.

The discovery that natural nuclear fission reactors might exist within the Earth came to me unintentionally and was quite a surprise. I later found that this had been suggested previously by Robert Driscoll [Driscoll 1988] and by Marv Herndon [Herndon 1992].

The importance of nuclear-fission chain-reactions within the Earth first arose in considering the apparently discrepant heat balance of the Earth, during its early history. Trying to increase the heat from radioactivity in the Earth, I found that adding more uranium to the iron core, to increase the radioactive heating, resulted in relatively huge amounts of uranium, still without enough heat from radioactive decay. If this uranium were able to accumulate into globs of a few hundred kilograms of uranium each, at the beginning of the Earth, when the enrichment (uranium-235 content) was about 24%, fast breeder reactors would have formed. I didn't see any assuredly effective way these accumulations would have been prevented, so I went looking for indications that such reactors actually had operated in the Earth.

Since each nuclear fission reaction releases 190 MeV of energy (neglecting the antineutrino energy), a total of 3×10^{43} fissions would be required to make up the energy deficit and prevent the Earth from freezing. This would consume 1×10^{19} kilograms of uranium-235, or about 2 parts per million (ppm) of the Earth, by mass. (Of course, the fissioned mass remains inside the Earth, as fission-product atoms.) At 4.5 billion years ago, the uranium content of the Earth, based on the current crustal composition, adjusted for radioactive decay, is estimated to have been 10 ppm. The Earth's core has usually been considered to be relatively devoid of uranium because of the absence of uranium in nickel-iron meteorites and an expected incompatibility of uranium with iron. Recent experiments [Feber et al. 1984] have shown miscibility of uranium oxide in iron up to 23%, and its observed absence from the iron meteorites appears to be a direct consequence of the process of their formation by a nuclear-fission detonation and will be discussed later in this book.

Because of the conversion, by neutron absorption, of the less fissionable nuclides thorium-232 and uranium-238 into the easily fissioned nuclides uranium-233 and plutonium-239, respectively, the additional content of these other two nuclides gradually becomes available as reactor fuel. In addition, some of the fission neutrons have energies above the fission threshold of thorium-232 and uranium-238 and cause fission of these nuclides directly. All these nuclides can be considered to be suitable fuel, under various conditions, for a planetary reactor.

The fission chain-reaction proceeds so long as there is adequate nuclear fuel and the production of neutrons meets or exceeds the losses due to capture in non-fission reactions or by escape from the reactor region. For large systems, the loss by escape is small. Since generally more than two neutrons are produced in each fission, the fuel can be quite dilute in systems with low parasitic capture, such as a system consisting mainly of iron, oxygen, and uranium.

In addition to being a globally significant source of heat, operation of these reactors within the Earth may cause numerous other effects. The next sections consider effects resulting from the intense gamma radiation of these reactors, influencing the geomagnetic field, and the localized production of heat in causing hotspot volcanism and controlling local climate.

Nuclear reactors are generally pictured as being rather gigantic, from pictures of industrial power plants, and the well-publicized accidents at Three Mile Island and Chernobyl. But the reactors themselves are small, smaller than a gasoline tank truck. Nature's reactors could be smaller still, "bigger than a breadbox, smaller than a car", and they could keep running, and producing heat, for billions of years. (The "fossil" reactors that were found in the uranium mines of Oklo, in Africa were somewhat different from the ones in the Earth's core and so were able to operate for less than a million years.) So a few trillion (!) of these little reactors, scattered throughout the Earth, operating at about 1,000 watts each, could have provided the heat the Earth needed, that the Sun could not supply. That would have required that about 13 parts per billion uranium in the core of the Earth to have been arranged into critical reactors. Clusters of subcritical reactors can function as reactors by absorbing neutrons that escape from each other.

So, I had found a possible solution to the Faint Young Sun Paradox and the avoidance of the White Earth Scenario. Was there any evidence that I was right? In a search that was like going to the store without knowing what you want, but hoping you'll recognize it when you see it, I browsed the libraries. I found the evidence, just a few bits, then more and more as I learned what to look for and where to find it.

Helium, that makes the blimp fly, and heliarc welding, and birthday balloons, is plentiful in the Sun, rare in the Earth. All of the helium that the Earth might have had in the beginning was lost to space as the Earth formed, and all it has now was produced by the natural radioactive decay of uranium and thorium. But there is another kind of helium, called helium-3 for the number of particles in its nucleus, and there should be no helium-3 in our Earth. Yet gases from volcanoes that draw lava from very deep in the Earth, like the Hawaiian volcanoes, have helium-3. Nuclear fission produces helium-3, and here it was in the lava that came from where I thought the natural reactors might be. Actually, the helium-3 is produced as a radioactive form of hydrogen, called tritium. (The same tritium that makes a glow-in-the-dark watch dial glow.) Tritium changes somewhat quickly to helium-3, so what we find in the environment must have been made recently, and tritium itself had been found in the gases of a Hawaiian volcano [Östlund 1985]. Tritium in the lava could only come from natural nuclear fission reactors heating up the lava of the volcano.

As Earth's climate changes, in response to varying heat sources and losses, we think that glaciers come and glaciers go, and ice-sheets grow and shrink. A variety of effects may combine to affect the temporary climate on the Earth. Cyclic combinations of different aspects of the Earth's rotation and orbit around the Sun, known as the Milankovitch cycle, are thought to be major contributors to the glacial cycle [Broecker and Denton 1990]. The glacial cycle is also thought to be influenced by the composition of the atmosphere [Owen and Cess 1979, Newman and Rood 1977, Berner 1990], the location of continents [Crowley et al. 1987, Crowley and Baum 1991], ocean circulation [Chandler 1993], continental uplift [Molnar and England 1990], and the distribution of lakes and rivers [Yemane 1993, Valdes 1994]. However, the possibility of a varying output of heat from the Earth's reactors has been omitted from these considerations, and global glaciation may have been a mistaken idea, after all.

The fossil remains of a young and rapidly growing forest from the Late Permian (about 290 to 250 million years ago), in a polar region of ancient Antarctica, has been found and has been interpreted to show a notably warm climate [Taylor et al. 1992]. Growth was rapid during periods of sunlight, and there is no evidence of frost damage.

The remains of dinosaurs have been found in Australia, at Dinosaur Cove, suggesting adaptation to long dark periods [Vickers-Rich and Rich 1993]. At the time that these dinosaurs lived, Australia was close to Antarctica and Dinosaur Cove lay well within the Antarctic Circle, providing several months of darkness each winter. However, these winters may not have been bitterly cold. The Dinosaur Cove region shows evidence of a volcanic hotspot, and at the time of the polarsaurs, might have been associated with the current Mt. Erebus hotspot. This additional subterranean heat might have kept Dinosaur Cove temperate all year long, and thus the climate was controlled by supplemental heat rather than by latitude. At present, the Mt. Erebus hotspot may be providing the little extra heat that makes ice sublime in the three "inexplicably dry" valleys of Antarctica [Bond and Siegfried 1990].

Approximately 3 million years ago, the East Antarctic ice-sheet was largely melted, even though temperatures were only slightly warmer than at present [Barrett et al. 1992]. This seems to have been a localized condition that may be associated with the Mt. Erebus hotspot.

Fossil fragments of a dinosaur have been found in Early Jurassic rock on Mt. Kirkpatrick, only 700 kilometers from Mt. Erebus [Hammer and Hickerson 1994]. This may have been due to localized volcanic heat, or it may have resulted from a much denser atmosphere distributing heat more uniformly over the Earth.

Similar evidence for arctic dinosaurs is found in the North Slope area of Alaska [Gore 1993], a region of winter darkness and presumed cold. However, this area is also associated with a volcanic hotspot, the Bowie hotspot, one that has been traced, seismically, down to the D" layer at the core-mantle boundary [Nataf and VanDecar 1993, Vidale and Benz 1993].

The factors that have been considered by the climatologists have failed to produce predictions for the glaciation that was thought to occur in the Ordovician period (440 million years ago) when the atmospheric concentration of the greenhouse gas carbon dioxide was about 13 times the present value, or the ice-free mid-Cretaceous period when the carbon dioxide level was essentially the same as at present [Barron 1987, Kasting 1992, Barron *et al.* 1981].

A reasonable source of the heat needed to maintain liquid water and otherwise unexpectedly warm climates throughout Earth's life is available in the form of natural nuclear fission reactors deep inside. What evidence is there for this possibility and what additional effects might it explain?

Volcanoes, the Earth's Heat Transfer System

… volcanological terminology is a mine field
into which one treads perilously.
Peter Francis, *"Volcanoes"* 1993

While Lord Kelvin considered only heat conduction through the solid Earth as a means for transporting heat from the interior of the Earth to the crust [Stevenson 2010], one of the major means for the transport of heat is by convection in the form of magma flow, leading to volcanic action at the surface and igneous intrusion below. In calculating that the Earth could be only 25 (or 100) million years old, Lord Kelvin might have arrived at a much shorter life had he considered volcanoes.

Magma generally originates in the mantle, or at the core-mantle boundary. Melting can be caused by a local increase in temperature, a reduction in pressure, or addition of water.

Volcanoes of all sorts are very unevenly distributed. Most of the volcanoes that we give names to are clustered in strands that run raggedly around the Pacific Ocean as the "Ring of Fire". Roughly ten times this amount of volcanic activity occurs along the midocean spreading ridges, out of sight, and those eruptions are rarely named or noted. The Arctic, separated from the Pacific basin by a considerable distance, has only one volcano, while Antarctica has many, lying along the Pacific coast. Greenland has none, while neighboring Iceland owes its existence to numerous volcanic eruptions. Northern Europe has no volcanoes. In the New World, North and South America have no volcanoes east of the great mountain chain running along the western edge of the continents.

Material erupted from volcanoes provides us with the only true samples of the Earth below its outer crust. Unfortunately, the samples are certainly much modified in the process of delivery.

On Earth, volcanoes consist primarily of several distinct types: those resulting from the subduction of one crustal plate beneath another (or its burial under an overriding, expanding crust), and "hotspot" or mantle plume volcanoes. A mantle plume is a convection current from deep in the mantle, perhaps resembling a plume of smoke in still air, in some cases rising from the boundary between the molten core and the lower mantle, known as the D" layer [Francis 1993, Coffin and Eldholm 1993, Anderson *et al.* 1992, Walker *et al.* 1995].

The Hawaiian Islands and the trailing Emperor Seamounts, extending from the northeast corner of Asia to the middle of the Pacific Ocean, are considered to be the prime example of volcanic eruptions from a mantle plume, a rising column of hot magma from deep in the mantle. Seismic tomography has been used to trace the mantle plume as a narrow column of higher temperature rock deep into the lower mantle, as deep as 1,500 kilometers [Wolfe *et al.* 2009]. Studies with less spatial resolution showed the column extending downward to 1,900 kilometers depth and further to the core-mantle boundary at a depth of 2,891 kilometers [Montelli *et al.* 2004, Li *et al.* 2008].

An additional form of lava eruptions, the flood basalts, seems to differ in cause and form from normal volcanoes.

Some volcanoes spout or flow, others explode. These differences seem to depend on the viscous strength of the magma. The flowing lava has been able to release most of the volatile compounds, water and carbon dioxide, on the way to the surface. Some of this gas provides the propelling force for the high-shooting fountains that are sometimes seen. Explosive volcanoes have much thicker magma that contains the gases until the last moment, when the lava explodes, much like a steam explosion, when the pressure is abruptly released. Kilauea flows, Pinatubo exploded. Explosive volcanoes spread evidence of their existence around, and so some very old, very large eruptions can still be found.

The La Garita caldera in Colorado has been explored by Peter Lipman, and shows that 5,000 cubic kilometers of magma was erupted in a single blast. That is roughly 1,000 times as large as the Pinatubo eruption of 1991 [Lipman, Dungan, and Bachmann 1997, Lipman 1997]. An interesting aspect of this eruption is that it was immediately preceded by a flow of lava and a momentary quiescent period.

Evidence of a still larger, much older eruption has been found by Warren Huff [Huff 199?]. Chemically distinctive clay beds were formed from ash that was erupted roughly 450 million years ago, that extended from northeastern North America to northeastern Europe. Huff estimated the volume of ash to have been about 285 cubic miles (1,300 cubic kilometers).

Hotspot volcanoes and mantle plumes

A mantle plume is a long-term volcanic feature associated with a specific location in the Earth. The prime example is the volcanic activity on the island of Hawaii, where the current volcano is trailed by progressively older volcanic structures extending to the northwest, and back in time to 70-80 million years ago. This hotspot volcano can be traced back in a trail that takes an abrupt bend that had early been cited as evidence that the Pacific plate had changed direction. Now, later research shows this change in direction could not have happened. The plate tectonic theory explanation for the Hawaiian Islands (and Emperor Seamount) chain is that a relatively fixed mantle-plume hot-spot (whatever and however that is) is overridden by a tectonic plate that changed direction about 42 million years ago. There is no explanation of what produces the mantle plume, other than a self-circular statement that it is caused by an upwelling of hot material, possibly from the base of the mantle. Mantle plumes seem to be associated with flood basalt eruptions (like the Siberian Traps and the Deccan Traps).

The root of a mantle plume is formed by a cluster of natural nuclear fission reactors stranded at the core-mantle boundary, producing overheated rock, magma that travels to the surface to erupt as a hot spot, mantle plume volcano.

The source of these reactors is likely to be in a dimple on the surface of the inner core. Such a dimple would be formed by the shockwave of a major impact, spalling off a small amount of the solid surface. These dimples would accumulate uranium and plutonium as it rained down from the core-mantle boundary. As a sufficient thickness developed, the accumulation would become a critical reactor and begin to heat up, relative to its surroundings. This would reduce its density, causing it to float off the dimple and assume a more spherical shape, increasing its reactivity, and so its power and temperature will also increase. The reactors will then float to the core-mantle boundary. These will be the root heat source of the mantle plume, reactors that produce the high ratios of helium-3 to helium-4 [Basu 1993]. Reactor fuel material will continue to accumulate in the various dimples on the inner core, and will replenish the heat sources at the roots of the many mantle plumes, gradually migrating over the surface of the Earth, as the inner core changes orientation relative to the crust.

In the expanding Earth/natural nuclear fission reactor model, the shockwave from an impact causes a spall to be ejected from the surface of the inner core, leaving a small depression there, like a dimple on a golf ball, but kilometers wide. (There may be 30 or so dimples on the inner core, each the source, the root, of a mantle plume. These dimples had been foreseen by Driscoll [Driscoll 1988].) Shockwaves from the same impact disrupt the farside of the Earth, like the "weird terrain" on Mercury and the Moon, opposite giant impacts, and if the rock is hot enough, it erupts in a massive flood basalt eruption. (These differ greatly from what I usually think of as volcanoes. The surface of the Earth is ripped apart and magma flows from the tears. After the lava flows are removed by erosion, the roots remain as dike swarms [Ernst and Baragar 1992, Heaman, and Tarney 1989].) The rock may be heated due to a boost in the power of the natural reactors resulting from antineutrinos released in the nuclear detonation that produced the fragment that impacted the Earth. Things tie together.

If a metallic uranium reactor created a temperature increase enough to become buoyant, it could float from the inner core to the core-mantle boundary where, cooling, it would have chemically reacted to become uranium oxide, and cooled enough to become denser and sank, to become metallic again at the inner core, as it reached a higher temperature. Herndon says the temperature at which uranium oxide equilibrates with metallic uranium is 3,571 K [Herndon 1992] for oxidizing conditions that existed in the Solar nebula. That temperature is reached just above the bottom of the mantle [Stacey 1969].

The convection column impinges against the bottom of the mantle, the core-mantle boundary, the D" layer, and melts the rocks. The magma rises as a mantle plume. Just like a gas flame under a pot on the stove, the convection column spreads beneath the mantle. In spreading, it cools. In cooling, the uranium oxidizes and collects in innumerable blobs of uranium oxide in the fluid of the outer core. It cools and becomes denser, and sinks. On reaching the surface of the inner core, it loses its oxygen, and collects in small puddles of metallic uranium. Perhaps the inner core has a surface like a golfball, with dimples, albeit irregularly spaced. At the melting of the rocks of the bottom of the mantle, geochemical processing occurs and some uranium oxide is extracted from the rock into the outer core.

Uranium that rains down from the core-mantle boundary, as reactors get stalled and dissipate there, creating the mantle plume, eventually flows into one of the dimples on the core. (Uranium may also be extracted from the minerals of the mantle, adding fuel to the reactors in the outer core.) Dimples created in the surface of the inner core by the shock waves of impacts will accumulate uranium, as the densest material in the fluid outer core. An accumulation will increase until it reaches the critical mass for that shape, and it will start to generate nuclear fission heat. As the thickness of uranium in a dimple increases, a critical mass (in that geometry) is reached and a reactor is born. As the reactor generates heat, it becomes less dense and starts to rise, becomes spherical and therefore gains reactivity to compensate for its loss of density, and streams to the core-mantle boundary, where it produces a new mantle-plume eruption, along the hot-spot path. The spherical shape is more reactive so that the fission power will increase, the reactor will expand more until it reaches stability, and continue to rise through the outer core [Driscoll 1988]. Descending uranium may be very rich in uranium-235 because of the conversion (in the previous reactors) of uranium-238 to plutonium-239, which decays to uranium-235. Thus, the fluid outer core performs pyrochemical reprocessing, supplying new fuel to the reactors forming on the surface of the solid inner core.

The peripheral reactor cells may be self-perpetuating for a period of time, rapidly extracting uranium from the mantle as a result of the increased local heating. This heating would drive a local convection column and could establish a deep mantle plume, making a hotspot volcano. Such localized heating could produce warm climates in locales where modern predictions using computer models predict bitter cold, such as the warm Wyoming winters deduced by Wing and Greenwood [Kerr 1993] from fossils found in an area that was over the Yellowstone hot spot.

During the passage of the initial magma from its source to the surface, mantle material is entrained in the plume [Hart et al. 1992], diluting and obscuring the original elemental and isotopic character of the source material. However, these hotspots provide the best messages from the depths of the Earth, in some cases as far as 2,600 kilometers down.

Research has shown that mantle convection velocities were 10 times greater at 2.7 billion years ago than at present [Blichert-Toft and Albarède 1994]. This suggests much greater concentrations of heat at great depths than exist in the current Earth.

Further, the structure of the Slave Craton in Canada shows that during the time period of 2.6 to 1.9 billion years ago the thermal gradient in this area was 2 to 4 times greater than present values, implying a major subterranean supply of heat in this region [Grotzinger and Royden 1990]. This has been judged to be a local effect, since the Kaapvaal and Superior Cratons do not show this condition.

A hotspot volcano formed at the juncture of the present continents of South America and Africa, about 125 million years ago [Francis 1993, White and McKenzie 1989]. This hotspot, with an age set precisely from the age of the Paraná flood volcanism as 133 ± 1 million years [Renne et al. 1992] (or 134.7 ± 1 million years [Thiede and Vasconcelos 2010]) is apparently associated with expansion that caused the drift of Africa away from South America. Strangely, there were only minor species extinctions. However, this time marks the first appearance of placental mammals, the direct line of descent to us.

The South Atlantic Anomaly, an area with markedly reduced magnetic field strength [Sherrill 1991, Stassinopoulis 1970], is centered over the South Atlantic remnant of this hotspot. It is tempting to speculate that both the hotspot volcano and the South Atlantic Anomaly are the result of the operation of a powerful peripheral reactor cell that has left a long-lasting imprint on the outer core, that extends into space. Further, this anomaly lies above (by 2,500 kilometers) the coldest area yet found near the level of the D" layer [Anderson 1990], which may indicate that the nuclear fuel in the region has been depleted and that little fission power, and insignificant electron flux to bias the geodynamo, is currently being produced in this region.

Small mantle plumes have been identified that seem to be too small to maintain their heat content sufficiently to reach the surface [Langmuir 1990]. However, in a "China syndrome" in reverse, hot material from a reactor cell may carry its own heat source, in the form of fission-product radioactivity and neutron-induced radioactivity. In 1967, W. K. Ergen proposed that the material of an overheated nuclear power reactor could melt its way through the containment building floor and through the Earth to the opposite side. Ralph Lapp used "to China", to invent the term "China Syndrome", which formed the basis for a movie. This bit of speculation became the basis for much fiction, unintentionally, and intentionally misleading. The energy released by radioactive decay of these unstable nuclides would replenish the heat lost by conduction to, and by the entrainment of, the surrounding mantle material.

Flood basalts, large igneous provinces, oceanic plateaus

The Siberian Traps and the Deccan Traps are the remains of vast deposits of layers of lava, eruptions that came, not from volcanoes in the common sense, but from distributed fissures, cracks in the surface of the Earth. As the lava flows were eroded, they formed stair-step hillsides, which took their name from the Swedish word *trappa* for stair-steps. These eruptions, and similar ones scattered over the Earth, were far greater than any that we have experienced in our modern volcanoes, and they erupt in a different way. The Siberian Traps eruptions [Renne and Basu 1991] occurred at the same time as the greatest known mass extinction, that between the Permian and the Triassic. The Deccan Traps eruptions [Basu et al. 1993] occurred at the end of the Cretaceous, when the dinosaurs, and many others, died. The Central Atlantic Magmatic Province (CAMP) [Marzoli et al. 1999] and the Karoo eruptions, the largest continental flood basalt yet found, are also notable flood basalt eruptions. These eruptions are shown in time, in Figure 14.

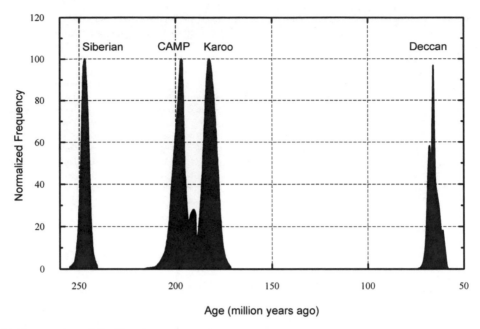

Figure 14. Four continental flood basalt eruptions.
Eruption intensities for the Siberian, CAMP, Karoo, and Deccan flood basalt eruptions.

The lavas of the Siberian Traps show two pulses of eruptions (Figure 15), which may be correlated with two apparent impacts about the end of the Permian, the worst mass extinction yet recognized [Renne and Basu 1991].

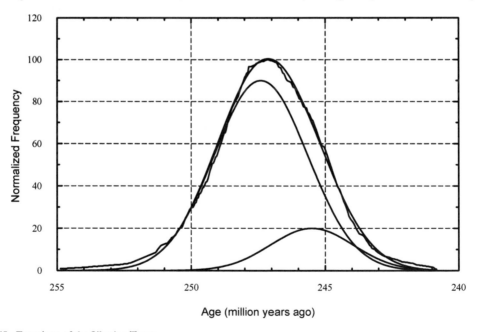

Figure 15. Eruptions of the Siberian Traps.
The geochronological ages of the Siberian Traps lava flows, decomposed into two Gaussian pulses. These roughly correspond to the two stages of mass extinction at the end of the Permian. The observed lava ages are shown by the jagged line; the two Gaussian distributions and their sum are shown by the smooth curves.

Eruptions of the Central Atlantic Magmatic Province (CAMP), related to the end-Triassic mass extinction and variously dated at 208 to 199.6 million years ago, are shown in Figure 16. These eruptions were associated with a large negative excursion in the carbon-13 isotopes [Whiteside *et al.* 2010]. (A negative excursion for carbon-13 is similarly a positive excursion for carbon-12, which may result from the conversion of oxygen-16 to carbon-12 by the Giant Dipole Resonance reaction from the gamma flash of a nearby astronomical nuclear fission detonation.)

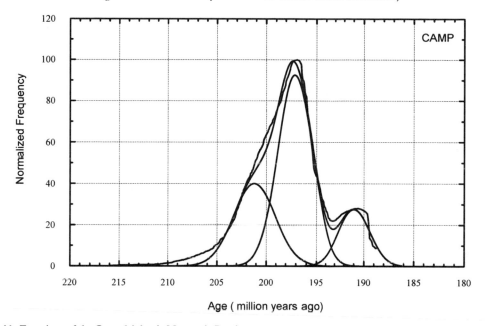

Figure 16. Eruptions of the Central Atlantic Magmatic Province.
A deconvolution of the ages of the lava flows of the Central Atlantic Magmatic Province (CAMP), showing three pulses, possibly from three separate impacts.

These eruptions produced the greatest area of flood basalts known, dwarfing even the Siberian Traps, which erupted 50 million years earlier. The mass extinction at the end of the Triassic Period [Smith 2011] clearly shows the action of the processes described in this book, initiated by the nuclear fission detonation of a large natural moon, killing by radiation, followed by impacts by the fragments of the disrupted moon, that caused expansion of the Earth and started and boosted the flood basalt eruptions. The extinction was essentially instantaneous and occurred just before the flood basalt flows (and therefore before the impacts), as shown by the "fern spike" overwhelming the pollen plants, just below the basalt flows. This showed a great resurging predominance of ferns, which propagate by radiation-resistant spores, as do fungi and algae. A similar fern spike has also been found at the end of the Cretaceous Period.

The extinction has been precisely dated at 201.4 million years ago, and a nearby crater at Rochechouart in France has a date of 199 to 203 million years ago, attracting interest in exploring an impact cause for the extinction [Smith 2011]. The ages of the lava flows in the CAMP flood basalts (Figure 16) show three strong initiations, presumably by impacts, one of which might be Rochechouart. The only currently identified impact crater that seems to be a good candidate is that of Manicouagan, in Quebec, Canada. The other might be found at the opposite side of the Earth from the CAMP eruptions, on a smaller, expanding Earth.

Extinction on land was greater than in the sea, showing the protective shielding of the water. The gamma radiation flash killed life on land, less severely in the oceans. The radiation-resistant ferns and similar land life repopulated the land, and impact-induced expansion changed the living conditions so that the previously successful land animals could no longer survive and went extinct. Those impacts also started and boosted the Camp basalt flows.

All this, even speculatively, is remarkable progress from our knowledge from the 1990s, when little was known of the end of the Triassic [Smith 2011, Raup 1991, Courtillot 1999]. A search for, and accurate dating of impact craters appropriately located on the opposite side of the Earth may help to fill in the story.

The eruptions of the Karoo Continental Flood Basalts began soon after (Figure 17), and may have been triggered by impacting fragments from the same detonation.

In an expansion event, resulting from the shockwaves causing a relaxation in a local supercompressed region, the near-surface pressure will be reduced by the decrease in gravity. This reduction in pressure will cause more melting, for a longer time.

There is evidence for related flood basalt intrusions, offset in time, in the Scourie dyke swarm in Scotland at 2.418 billion years ago and the Hearst-Matachewan swarm in Canada [Heaman and Tarney 1989] at 2.452 billion years ago.

An impact generates shock waves in the base material and these shock waves travel through the body of the planet, as compression waves and as shear waves. Sankar Chatterjee has depicted the paths of seismic shockwaves resulting from an impact on the Earth, and considered some complex possibilities. [Chatterjee 1995, Chatterjee and Rudra 1996].

Shockwaves passing directly through the center of the Earth arrive at the far surface, the antipode, first. This produces an uplifting effect. For the Caloris impact on Mercury the uplift has been estimated to be as great as 1 kilometer. When a compressive ("pushing") wave is reflected from a free surface, a rarefaction wave is returned, with a local intensity approximately twice that of the pressure wave. This induces a tension in the rock that tends to pull the crustal rock apart [Boslough *et al.* 1994].

When compression waves arrive at the far side that have travelled the long way around, reflecting from the crust and the core, reinforcement occurs as the waves meet, doubling their pressure, and then as the waves pass, the rarefaction is doubled, pulling the broken crust apart horizontally. This produces great rifts in the crust. This may occur over a fairly large area. If there is a localized region nearby that is close to melting, the extra energy and depressurization will push the region to melt and erupt through the fissures.

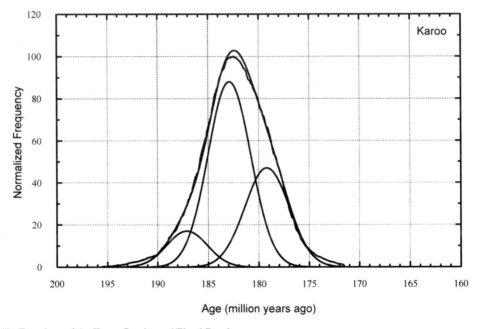

Figure 17. Eruptions of the Karoo Continental Flood Basalts.
A deconvolution of the ages of the lava flows of the Karoo flood basalts, showing three pulses, possibly from three separate impacts.

The Deccan Traps (Figure 18), near and at the end of the Cretaceous Period, show four distinct pulses [Courtillot 1988], which can be related to two known and two suspected/proposed impacts. The dimple on the Earth's solid inner core (for the subsequent mantle plume) is initially aligned with the flood basalt eruption (though one does not cause the other) and the point of impact. Ideally these three locations lie in a straight line, but the Earth doesn't always act in an ideal manner. With subsequent impact/expansion events, and due to differential rotation of the core, the mantle plume (anchored at its base to its dimple on the core) drifts away from the flood basalt site. Because of the continental drift caused by expansion, no place is ever opposite any other place ever after.

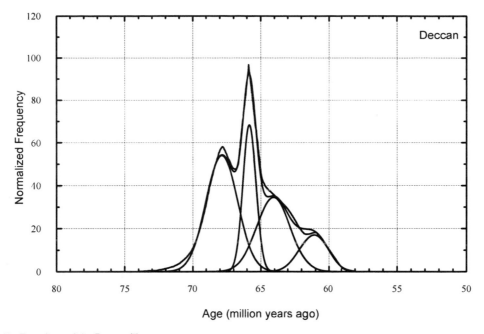

Figure 18. Eruptions of the Deccan Traps.
A deconvolution of the ages of the lava flows of the Deccan Traps, showing four pulses, possibly from four separate impacts, Manson, "Shiva", "Pacific", and Chicxulub.

Spreading-ridge volcanoes

In the 1960s, various observations of the seafloor led to the discovery of midocean ridges, and the realization that new oceanic crust was being formed and spreading apart at these ridges [Menard 1986, Erickson 1996, Birkeland and Larson 1989]. Initially, it was thought that this new crust was pushing its way against the distant continents, but further study showed that the new crust was being pulled away from the ridges.

Spreading ridges extend in an intricate network around the world, amounting to 40,000 to 65,000 kilometers, and providing 90% of the volcanic activity on Earth. The Earth seems to be the only planet in the Solar System where this behavior can be clearly identified.

For volcanoes that erupt under the sea, it is impossible to estimate what fraction of water accompanying the lava, and spouting out of hydrothermal vents, is new water, actually adding to the volume of the oceans. It is estimated that an amount of water equivalent to the entire volume of the oceans passes through these hot regions in 8 million years [Ross 1995]. Since it is assumed that the volume of the oceans has remained essentially constant for most of Earth's life, it is assumed that all of the water passing out of the midocean ridges is recycled water [Judson and Kauffman 1990], and that no new water is being added. At the estimated rate, water would fill the oceans to overflowing. This overlooks the loss of water (as hydrogen) to space, with an increase in the oxygen content of the atmosphere, and the effect of Earth Expansion in enlarging the ocean basins. The abrupt drops in sealevel, matching the impact cratering record (shown in Figure 26), are clearly apparent in the Exxon sealevel curves. These drops in sealevel are proposed here to result from impact-induced expansion events. Dry land volcanoes erupt considerable amounts of water, the most plentiful volatile in magma generally forming 1% to 7% of the mass [Frankel 1996], and it seems unlikely that the undersea versions would have less. Katterfel'd (quoting A. P. Vinogradov 1959) reports an estimate of the total amount of water in the mantle as about 100 times the current content of the world oceans [Katterfel'd 1969].

Island-arc and Backarc Volcanoes

The volcanoes that form the Ring of Fire surrounding the Pacific Ocean basin are all of a similar type, but different names are applied according to their location. The innermost island volcanoes are island arc volcanoes, while the next row out from the ocean basin are backarc basin volcanoes. These are concentrated at the western margin of the Pacific. The backarc basins form the Sea of Okhotsk, the Sea of Japan, and the East China Sea. In some areas, the continental crust has been torn apart to the extent that the underlying oceanic crust forms the bottom of the sea. This poses a strange situation. The oceanic crust is being pulled apart at the midocean spreading ridges, and on the distant side of the ocean, the continental crust is being stretched to breaking, being pulled toward the midocean spreading ridge. That would happen on an expanding Earth.

The eastern rim of the northern Pacific, the western coast of North America, is quite similar in form to the western rim. The major difference seems to be that the North American rim is higher in elevation. The Basin and Range region, between the Sierra Nevada (Cascade Range) and the Rocky Mountains, matches the Sea of Japan, except it has not been stretched and torn to the extent that the underlying oceanic crust is exposed. If that were to happen, the Cascade Range would become an island chain much like the Japanese Islands from Sakhalin to Kyushu. In addition to this similarity, the basin portion of this region closely resembles the hilly and lineated terrain seen at the antipodal locations to the giant impacts on the Moon and Mercury.

Explosive Eruptions

Mount St. Helens, El Chichón, Pinatubo, and Krakatau are notable recent explosive eruptions. Even larger explosions happened in the past. About 7,700 years ago, Mount Mazama exploded and left behind a caldera that filled with water to become Crater Lake, Oregon. About 28 million years ago, a volcano exploded and left the La Garita caldera and ash deposits covering nearly 3,000 square kilometers (1,000 square miles) of southwestern Colorado [Lipman, Dungan, and Bachmann 1997, Lipman 1997]. Further back in time, only the largest explosions left sufficient evidence to be found at present. In the Ordovician period, about 450 million years ago, an explosive eruption scattered ash from northeastern North America to northeastern Europe [Huff 199?].

Nearly simultaneous eruptions in a local region have been noted for particularly explosive eruptions [Simkin and Siebert 1994]. These were not associated with interconnecting magma chambers, but suggest that some external cause affected a wide region. Globally, no significant correlation between eruptions has been seen. Unzen, in Japan, erupted so violently that 43 persons were killed, in June of 1991. Just a week later, Pinatubo erupted equally violently, in the Philippines. (These eruptions were about 4.4 years after the explosion of the supernova SN 1987A in the Large Magellanic Cloud.) Two volcanoes connected to the same magma chamber, Vulcan and Tavurvur, in Papua New Guinea, erupted simultaneously in 1994 [Williams 1995, Williams 1996].

Krakatau, in 1883, erupted between 6.5 and 9 cubic kilometers of magma in the hottest lava ever studied, at 500 C, possibly up to 650 C [Mandeville *et al.* 1994]. The amount of heat discharged to the surface (at a heat capacity of 840 joules per kilogram-degree kelvin for basalt) was 1.134×10^{19} joules, or about 2,700 megatons TNT equivalent, to put it in the terms of nuclear weapons. Additional energy was released in the form of kinetic energy of the eruption, which had a range of up to 15 kilometers.

Explosive eruptions often occur abruptly, with little precursory activity, and release great amounts of energy. This suggests an abrupt supply of energy as might occur from a detonation of a cluster of reactors at the core-mantle boundary, while other volcanoes, such as Kilauea in Hawaii, erupt steadily with flows for many years, suggesting a steadier source of energy.

El Niño

El Niño is a local weather disturbance that has global consequences [Nash 2002, Zimmer 1999a, Webster and Palmer 1997, McPhaden 1999, Suploe 1999]. The warm El Niño brings floods and droughts to distant parts of the world. This is the warm phase of an irregularly varying weather pattern with El Niño (warm) and La Niña (cool), coupled to the Southern Oscillation. The full disturbance has been labeled the El Niño/Southern Oscillation, or ENSO.

El Niño has become a familiar part of life lately. The prevailing trade winds, established by the global circulation of the atmosphere, normally push warm surface water across the Pacific Ocean, from South America to Indonesia. That flow of warm water and warm air establishes the weather and climatic conditions we call "normal". (El Niño often occurred near the Christmas season and was named by the Peruvians, whose fishing industry it seriously affected, for the Christ Child. Its climatic opposite form is called La Niña.)

The standard model for the fluctuations in weather is that the trade winds die down and warmer water accumulates in the eastern Pacific. However, the process may be just the reverse. Increased volcanic activity along the spreading ridges of the eastern Pacific, the East Pacific Rise, supplies warm water to the surface, rather than the usual cold upwelling that normally supports the fishing industry. This may heat the surface air which rises as a curtain that blocks the trade wind flow, interrupting the westward wind, holding the increasingly warmer water in place. The warm rising water then accentuates the effect. This produces the climate disturbance we call El Niño.

When the volcanic heat production subsides, the trade winds return and push the remaining warm water across the central Pacific. Eventually, with less warming from the volcanoes along the sea bottom, the upwelling water that must replace the migrating surface water along the coast of Peru is increasingly cool. Many weather patterns then reverse from the El Niño condition, and we experience "La Niña", the little girl child.

In El Niño, a plume of warm water flows westward from the northwest corner of South America, above the midocean spreading ridge, the East Pacific Rise. The rising warmth produces a rising wall of warm air in the eastern Pacific which blocks the westerly flow of wind, stalling the trade winds. This plume of warm water is shown in Figure 19, in the upper frame. Helium-3 is associated with this water, coming from the eruptions of the East Pacific Rise, as shown in the lower frame of the figure.

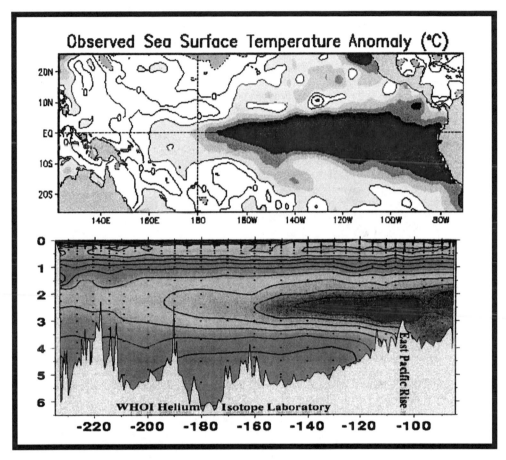

Figure 19. Sea surface temperature and helium-3 concentrations.
Plots of the sea surface temperature anomaly in the equatorial Pacific Ocean during the 1997-98 El Niño, (upper frame), and the helium-3 concentration in water above the East Pacific Rise, along Latitude 10N (lower frame). The sea surface temperature is from NOAA, the helium-3 is from WOCE at Woods Hole Oceanographic Institute.

In the view of this book, the heated water and the helium-3, as the product of tritium (hydrogen-3) released in ternary fission, come from natural nuclear fission reactors that power the eruptions of the East Pacific Rise.

If the warm water of El Niño was produced in varying amounts by submarine volcanic activity, this might affect atmospheric pressure patterns, which drive the winds, that usually give us "normal" weather. Other indicators of climate distributions and volcanic activity appear to show strong associations between these effects. I found that some other researchers had seen those associations ahead of me. To set the scene, Figure 19 shows the Sea Surface Temperature Anomaly in the equatorial Pacific Ocean and helium-3 from the East Pacific Rise. (The helium-3 plot is an elevation profile with depth in kilometers. It was measured along latitude 10°N, during Spring 1989. That was a time when El Niño was starting to recover from a relative low.)

There is a remarkable similarity in location and extent between the El Niño warm water pool in the eastern Pacific and the seemingly unrelated helium-3 from the volcanic sources at the East Pacific Rise, one of the mid-ocean spreading ridges. If it is recognized that El Niño results from a release of excess heat, and that the helium-3 is a signal of increased reactor power heating the magma under the spreading ridges, the relation becomes clear.

Dan Walker at the University of Hawaii, Manoa, had found that there is a strong correlation between El Niño and earthquakes near Easter Island, just off the East Pacific Rise [Walker 1988, Walker 1995]. My work agrees with that. Paul Hatchwell of the University of Salford has briefly commented [Hatchwell 1998] on the correlation of El Niño with world seismicity. My work for two additional seismic areas (Galapagos and Azores) appears to support that also. There is a similar, but different, pattern in the North Atlantic Oscillation Index used by James Hurrell at NOAA-Boulder and Mike McCartney at WHOI.

The disturbance oscillates between warm El Niño and cool La Niña with brief periods of "normal". Large amounts of heat are involved in the warm El Niño, and significant changes in temperature occur. In the 1952 El Niño, the sea surface off the coast of California heated as if it had received more than four times as much solar heat as it actually did [Nash 2002]. This is a remarkable amount of heat, with no identified source. A similar change occurred in the 1997-98 El Niño. In October 1997, during the strongest El Niño, the sea surface temperature in the eastern Pacific rose by more than 5 C [Webster and Palmer 1997] so we must ask "Where's the heat?"

These weather conditions are summarized in a numerical index developed by Klaus Wolter at NOAA-CIRES Climate Diagnostic Center, the Multivariate ENSO Index (MEI). Working at the Earth Systems Research laboratory of the National Oceanic & Atmospheric Administration, Klaus Wolter has developed an index combining measurements of pressure, temperatures, winds, and cloudiness to account for the changes that make an El Niño or a La Niña [Wolter and Timlin 1993, Wolter and Timlin 1998]. A recent plot is shown in Figure 20, from data downloaded from
http://www.cdc.noaa.gov/people/klaus.wolter/MEI/mei.html.

As the Index becomes progressively more positive, sometimes exceeding values of 2 or 3, the El Niño pattern becomes more extreme. As the Index declines, sometimes as low as -2, La Niña appears.

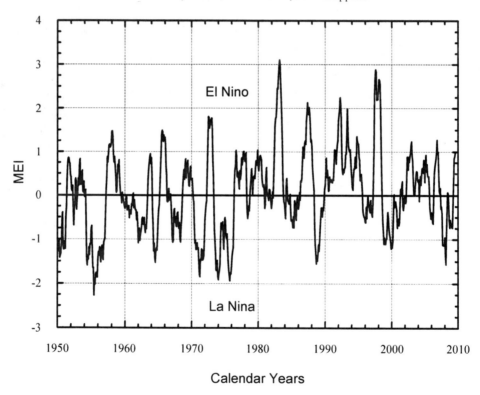

Figure 20. The Multivariate ENSO Index, MEI.
The Multivariate ENSO Index, from Klaus Wolter, showing its change in value with El Niño and La Niña conditions. Strong positive values signal a strong El Niño, strong negative values show a strong La Nina.

An explanation of the process was developed by Jacob Bjerkenes [Neelin and Latif 1998]. The pattern develops as a coupled cycle in which sea surface temperatures in the equatorial Pacific Ocean cause the trade winds to strengthen or slacken and the winds, in turn, drive changes in ocean circulation that produce changes in the sea surface temperature. This envisions a self-contained oscillating system, without any external driving force, other than the essentially constant solar heat. The system switches states when it is ready.

Strong earthquakes have been proposed as the events that start an El Niño [Walker 1988, Walker 1995, Walker 1999, Hatchwell 1998]. Explosive volcanic eruptions have been related to the start of an El Niño [Adams, Mann, and Ammann 2003, de Silva 2003], even though the aerosols discharged to the upper atmosphere cause surface cooling [Kerr 1992a].

Earthquake tremors are known to be associated with the activity of several well-studied volcanoes, and contribute to our knowledge of how those volcanoes work. Volcanic activity, strong eruptions, may be triggered by distant earthquakes [Linde and Sacks 1998]. Large volcanic eruptions were found to occur within one day following large (magnitude 8 or greater), but distant earthquakes. Strangely, no close earthquakes, less than 117 kilometers (70 miles) distant seemed to be related to the eruptions, suggesting that the effect may be more global than local. Slightly weaker earthquakes (magnitude = 7 to 7.9) were just as effective as the stronger earthquakes. (The statistical accuracy of this comparison is not very great, so this should be taken as a suggestion only.)

It may be that the two types of activity, volcanic eruptions and strong earthquakes, are independently responding to some other process within the Earth, and volcanoes simply take a little longer to react than does cracking rock.

There also seems to be a connection between earthquakes and volcanic eruptions that is much less immediate, but is global in nature. These events seem to reflect energy released within the Earth. The volcanoes seem to take somewhat longer to respond.

Earthquakes and volcanic explosions indicate energetic activity within the Earth.

Further, the oscillatory behavior is driven by episodic and intermittent boosts in the reactor power by distant supernova explosions.

The association between supernova brightness through the years and the Multivariate ENSO Index (MEI) is shown in Figure 21.

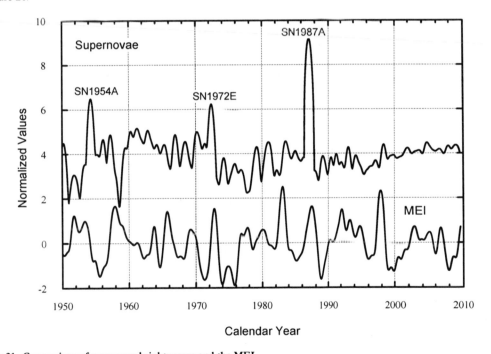

Figure 21. Comparison of supernova brightnesses and the MEI.
Comparison of the action of supernovae (derived from reported magnitudes) and El Niño (as shown by the MEI). The supernovae in 1972 and 1987 seem to have had immediate effects on weather as well as effects delayed by 10 years. Supernova 1954A shows a less clear relation.

The connections are weak and variable and more suggestive than convincing. Responses may be prompt, in about a year, and delayed, in about ten years, to judge from the plot. The response must be non-linear and self-limiting. Comparing a logarithmic forcing function (the supernova magnitudes) with what may be a mostly linear action (the MEI) is at best an awkward comparison.

Irregular forcing (variable in both time and intensity, and possibly location as well) of a quasi-periodic system (with its own poorly determined periodicity and phase) is more likely to produce erratic results than a clearly defined correlation. All that appears to be true here.

If the supernova/ENSO connection is real, it appears that the only messengers capable of carrying the signal are the neutrinos and antineutrinos released by supernova detonations. These travel at the speed of light and arrive at the Earth at the same time as the light flash of the supernova, or sooner. They may penetrate deep into the Earth, or pass completely through with no interaction.

The Maunder minimum is a period of time in which the Sun was free of visible sunspots. This extended from about 1645 to about 1715 [Eddy 1976]. This corresponded with the coldest part of the Little Ice Age, when the Thames River in London froze over to the extent of allowing sellers to set up stalls to display their goods. On the other side of the world, in the Indo-Pacific warm pool, proxies for sea surface temperature also show an extended decline and rise in temperature, spanning the Maunder Minimum, and the Little Ice Age [Oppo, Rosenthal, and Linsley 2009]. An absence of sunspots was associated with a drop in global temperature. It is suggested here that sunspots are caused by nuclear reactors deep in the Sun and result in the transport of additional heat to the surface.

Climate Change

Climate is what we expect,
weather is what we get.
Unknown

If supernova explosions drive the weather pattern of El Niño, they should also drive climate change. This was explored by comparing the global temperature from 1880 to the present (from the Goddard Institute for Space Sciences at http://data.giss.nasa.gov/gistemp) with a binned, smoothed, and offset indicator of supernova brightness. This is shown in Figure 22, where the Goddard temperature data have been smoothed with a pseudo-Gaussian with width ±3 years, and the supernova brightnesses have been binned into yearly bins, smoothed in the same way, normalized, and shifted in time to 18.57 years later. Comparing shape to shape, the subtle changes in temperature match the variations in supernova brightness quite well, with some variance in the alignment time.

Technically, this comparison suffers from the same shortcoming as the El Niño comparison: the supernova brightness is a logarithmic function, while the temperature variable is linear. This suggests that the response of the Earth to signals from the Universe is non-linear and self-limiting.

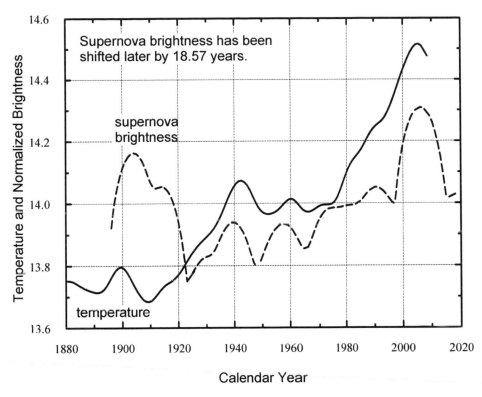

Figure 22. Surface temperature of the Earth and supernova brightness.
Comparison of supernova brightness (dashed line) and global surface temperature (solid line). The temperature data have been Gaussian smoothed, ± 3 years, and the smoothed and normalized supernova brightness has been shifted later by 18.57 years. These temperature data are from the Goddard Institute of Space Science, and the supernova data are from the Sternberg Astronomical Institute.

It should be noted that temperature reconstructions rely on different sorts of data, processed in different ways. These interpretations produce controversy and reveal uncertainties that are often not recognized.

To go further back in time, for the past 2,000 years, it is necessary to use proxy indicators for the temperature. This has been done by Anders Moberg at Stockholm University and several colleagues, using a variety of proxies, including tree ring thickness and oxygen isotope variations [Moberg 2005]. This time series, Figure 23, showed a generally constant temperature with fluctuations, until about the year 700. This time span (0 to 700) included three historically recorded supernovae [Clark and Stephenson 1977], with possible upward spikes in the temperature. Following the second bump, the temperature began to rise until it reached a maximum at about the year 1000, in the middle of the Medieval Warm Period. The temperature started a clear decline after the supernova of 1006, but reversed at the supernova of 1054.

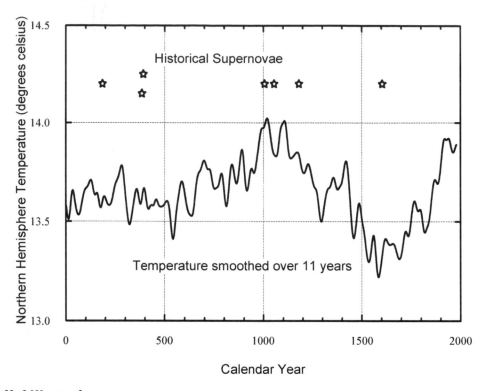

Figure 23. 2,000 years of temperature.
A temperature reconstruction over the past twenty centuries, from [Moberg *et al.* 2005] smoothed over 11 year periods. Supernovae historically noted by the Chinese are shown by stars.

Isotopic Evidence for the Earth's Reactors

Nuclear-fission chain-reactions result in the creation of distinct fission products and modify the isotopic composition of existing elements. Consideration of nuclear-fission chain-reactions in planets appears to provide better explanations of a number of isotopic and elemental anomalies in the Solar System. The steady phase of the chain-reaction may be compared to the s-process in stars [Burbidge *et al.* 1957], with heavier elements forming by neutron absorption, and similar isotopic effects may be expected. The detonation phase will produce distinctly different effects. Some explanations follow.

The clearest evidence of the action of natural nuclear fission reactors in the Earth, I think, comes in the helium-3 (from tritium produced in ternary fission) found in the magma from mantle-plume volcanoes and midocean spreading ridges. There is also evidence in isotopic variations of carbon, nitrogen, oxygen, strontium, and others that show the effects of neutron interactions. I don't know if anyone has ever looked at the uranium composition in these lavas. Uranium in lavas may be so dilute and locked in place that it has never had much effect on surface uranium deposits.

Helium, discovered initially in 1868 by studies of the spectrum of the Sun, is present in small amounts in the Earth's atmosphere, trapped in rocks and minerals, and in gases emitted from volcanoes. This helium, helium-4, was discovered in the laboratory in uranium minerals in 1895.

The diffusional lifetime of helium (helium-4) in the Earth's atmosphere is 20 million years. Because of the lesser mass of the other isotope of helium, helium-3, and the consequent higher thermal velocities, the lifetime of this isotope is expected to be about 1 million years. All the primordial helium, both helium-3 and helium-4, has long ago been lost to space, particularly during the formative period of a molten Earth with essentially no atmosphere. Because of the high diffusivity of helium (helium-4) in the atmosphere, it would seem unlikely that any un-degassed reservoir of this gas (including primordial helium-3) remained after differentiation of the Earth [Matsuda *et al.* 1993]. Because of its lighter atomic weight, helium-3 should have been lost even more rapidly and completely. However, for lack of any obvious source of this isotope, some controversy has developed over where the helium-3 comes from.

The currently observed amounts of helium (predominantly helium-4) are due to alpha decay of uranium and thorium and their decay products. In this mode of radioactive decay, an alpha particle, a helium-4 nucleus, is emitted from the parent nucleus and becomes an atom of helium upon attracting two electrons to balance the nuclear charge.

The other isotope, helium-3, is found in magma from deep-rooted volcanoes. The Earth should have no helium-3. It would have been lost during planetary formation, even faster than helium-4, and not be produced (in any appreciable amount) ever after. Helium-4 on Earth is from the alpha decay of uranium and thorium, while there is no recognized source of helium-3. But helium-3 is found, in distinctly different amounts, in lava and magmatic gases.

In the transmutation by radioactive decay to stable lead, one atom of uranium-238 ultimately produces eight atoms of helium-4. However, helium-3 cannot be produced in this manner. One possible method for producing helium-3 on Earth has been proposed as a two-step nuclear reaction [Mamyrin and Tolstikhin 1984]: an alpha particle emitted by a uranium or thorium atom, or a more energetic alpha from a daughter of these long-lived nuclides, reacts with a light-weight nucleus, such as beryllium, oxygen, fluorine, aluminum, or silicon, which then emits a neutron. The neutron is captured by the nucleus of an isotope of lithium (lithium-6) present in the surrounding rock, which then splits into a hydrogen-3 nucleus (commonly called "tritium") and a helium-4 nucleus. While the neutron reaction is surprisingly probable, due to the large capture cross-section of lithium-6 for low-energy neutrons, the neutron-production reaction is very improbable, with generally less than one neutron being produced per million alpha particles emitted, for most common target elements. Thus, the ^{3}He/^{4}He ratio due to this process will not be greater than about 1×10^{-6}. Measurements with helium from ancient granite has shown a ratio of 1 to 3×10^{-8}, and more detailed calculations of this process provide good agreement, although the method fails in isolated cases by factors of 40 and 2,000 [Mamyrin and Tolstikhin 1984].

In contrast to this expected ratio (1 to 3×10^{-8}), helium extracted from the magma of Iceland [Mamyrin and Tolstikhin 1984] and Hawaii [Kurz and Hart 1982, Honda *et al.* 1993], Yellowstone [Kennedy *et al.* 1985], and from the East Pacific Rise [Lupton and Craig 1981], the Deccan flood basalts in India [Basu *et al.* 1993], and the Galápagos [Graham *et al.* 1993], has shown ^{3}He/^{4}He ratios up to 3.3×10^{-5}. This is 24 times the ratio of helium isotopes found in the lower atmosphere [Mamyrin and Tolstikhin 1984], and a thousand times greater than the ratio found in granite.

Another proposed explanation of the high ^{3}He/^{4}He ratios is based on the settling of interplanetary dust particles that contain helium-3, without loss of the entrapped gas, onto the Earth's surface and subduction of these particles into the mantle [Anderson 1993]. These particles must pass through the zone where the magma for the subduction arc volcanoes is formed, without significant loss of helium, since these magmas do not show high ^{3}He/^{4}He ratios. The helium (helium-3 and helium-4) is then presumed to be released deep in the mantle to become part of a mantle plume magma.

This extraterrestrial source is contradicted by studies on the helium and neon ratios in basalt, deep-sea sediment, the Solar wind, and cosmic dust [Allègre *et al.* 1993]. These studies conclude that the neon can come from cosmic dust, but that the helium-3 must come from a deep un-degassed reservoir in the lower mantle. More recent studies on helium and neon in interplanetary dust particles have shown that neither helium nor neon can survive the subduction trip [Hiyagon 1994, Craig 1994]. It also appears that the core is not capable of retaining noble gases [Matsuda *et al.* 1993] and so, presumably, outgassed long ago. The core is not likely to be a reservoir for primordial helium-3 at the present time.

Another possible explanation has been linked to the observation that helium-3 appears to have been produced preferentially in high altitude Hawaiian lavas, but not in shielded lavas, indicating cosmic-ray production. However, the cosmogenic helium-3 was a small fraction of the total [Marti and Craig 1987].

Spontaneous fission, of uranium and thorium isotopes, and of plutonium-244 as well, has been considered as a possible source of the helium-3 (by the decay of hydrogen-3 produced in ternary fission), but the low fission rate, compared to alpha (helium-4) decay, is not adequate to produce the high isotopic ratios of helium-3 to helium-4 that are observed [Mamyrin and Tolstikhin 1984]. This possible source was also found to be unsatisfactory because of the high ratio of xenon-136 to helium-3 that would result from fission. High ratios are not observed in the volcanic gases. However, production of the helium-3 by fission, initially in the form of hydrogen-3 in the core, as discussed below, with the high mobility of the hydrogen-3 transporting the yet-to-become-helium-3 away from the source, would leave the xenon behind.

Thus, the likely source of the excess helium-3 is the production of hydrogen-3 (tritium) in ternary fission, in which a third fragment is released coincidentally with fission [Wagemans 1991, Hyde 1964, Halpern 1971]. While helium-3 itself is not produced in ternary fission, the hydrogen-3 decays by beta emission to helium-3 with a half-life of 12.33 years. The yield of hydrogen-3 in ternary fission is about 0.00012 atoms per fission and the high mobility of tritium in hot (or molten) iron assures widespread distribution. Ternary fission also produces ^{4}He, at about 0.002 atoms per fission, so ^{3}He/^{4}He ratios as high as 6×10^{-2} could be achieved, if all the helium came from ternary fission. However, much helium-4 is produced by the alpha decay of uranium and its daughters, and so this ratio would be limited to lower values.

How does that relate to nuclear fission? In about 1 fission per 8,000, a triton (nucleus of tritium, hydrogen-3) is emitted, along with the normal pair of fission fragments and neutrons. Tritium decays, with a half-life of 12.33 years, to helium-3. If the core of the Earth is heated by nuclear power, and the mantle-plume volcanoes are powered by reactors at their bases, tritium and helium-3 will be entrained in the magma, and surge on ahead in the gases.

Tritium itself, in the gas phase, has been detected by air sampling on Mauna Loa in Hawaii, as Kilauea erupted nearby [Östlund 1985]. Research related to the possibility of nuclear fusion reactions of deuterium (hydrogen-2) occurring at very slow rates in special geologic settings [Jones *et al.* 1989] reported that Östlund, at the University of Miami, has found evidence [Östlund 1985] of apparent atmospheric emissions of tritium (hydrogen-3) in gaseous (elemental) hydrogen, not as water vapor, from eruptions of the Mauna Ulu volcano on Hawaii, from the data of a tritium-monitoring station on Mauna Loa, about 40 kilometers away. An alternate explanation that might be suggested for the observed tritium in the volcanic gas is that ocean water carrying tritium from the testing of nuclear weapons has intruded into the magma, and it is this tritium that is detected following the eruptions. But this seems unlikely for the following reasons: deep water, below about 50 to 100 meters, has a negligible tritium concentration, due to the mixing with deep, old water in which the tritium has decayed. On the other hand, if surface water, with a tritium concentration of about 60 picocuries per liter, were the source, the observed concentration would require a water content of only 0.08 gram per cubic meter, much less than the saturation concentration of water in air (17.118 grams per cubic meter at 20 C), which lends some credibility to this idea. (Showing the great sensitivity of theses measurements, 60 picocuries per liter is about 2 disintegrations per second.) However, the tritium from the volcano was found in the elemental hydrogen component of the atmosphere. No significant increases were found in the water vapor component.

This observed tritium release has been interpreted as evidence of the occurrence in the volcano of the fusion reaction $^2H(d,p)^3H$, where two deuterium nuclei (^2H and d) fuse to form a normal hydrogen atom and a tritium atom (p and ^3H), with a significant release of energy, approximately 4 MeV. Since the half-life of tritium is only 12.33 years, the appearance of tritium in volcanic gases from ternary fission in the Earth's core would require relatively rapid movement of the magma. Magma velocities may be on the order of 1 meter per year [LaTourette *et al.* 1993] to as fast as one or two meters per second [Frankel 1996], and so in some cases tritium produced at the core-mantle boundary, roughly 2,900 kilometers down, would survive the trip, to be detected at the surface. However, the central conduit of a well-established plume might carry the tritium more rapidly, especially if it were in a gaseous phase as observed for sulfur dioxide [Wallace and Gerlach 1994]. Some melts rise directly to the surface without stopping on the way, as at Mount Etna [Frankel 1996]. Similarly, flow of highly volatile material through porous regions or in channels [Iwamori 1993] could bring the tritium to the surface quickly. Transport of magma through the crust has been reported to be as fast as 4 meters per second [Kelley and Wartha 2000]. If that rate of rise were maintained from the base of the mantle, magma would be transported in as little as 8 days, a short time compared to the half life of tritium, 12.33 years. A similarly rapid rise of magma was found for the Chaitén volcano in Chile [Castro and Dingwell 2009]. Magma traveled from a storage reservoir about 5 kilometers deep in about 4 hours. It may be that magma travels faster than has been ordinarily thought.

Using a nominal magma flow rate (10 cubic kilometers per year) [Schilling 1991] and temperature difference (1000 K) [Francis 1993], and assuming that all the heat is produced by fission at the D" layer and that the heat and all the tritium is transported to the surface, the observed concentration of tritium at Mauna Loa could be produced if the transit time were about 160 years.

In some volcanoes that have the appearance of being hotspot volcanoes, it has been found that the melting temperatures are below average and that melting is the result of excess water: they are "wet-spots" [Bonatti 1994]. These volcanoes show $^3He/^4He$ ratios that are lower than the midocean ridge basalts and the true hotspot volcanoes, which suggests that the melting occurred without the added heat of mantle-plume reactors.

An isotope of the noble gas neon, neon-20, is easily produced in the Sun, by the fusion of five helium-4 nuclei. Like helium-3 and helium-4 from the primordial nebula, this gas too should have been lost from the solid Earth [Matsuda *et al.* 1993, Hiyagon 1994]. However, in association with helium that is relatively rich in helium-3, high ratios of neon-20 are found. This isotope can be produced by neutron absorption in sodium-23 with emission of an alpha particle.

The isotopes of strontium show some variability related to the source of the material [McDermott and Hawkesworth 1990]. This is interpreted as resulting from the origin of the strontium, whether it comes from weathering of mountainous terrain or from the deep ocean. Much of the strontium-87 comes from the decay of rubidium-87, with a 48 billion-year half-life. Separation of strontium chemically from rubidium will result in a lower ratio of $^{87}Sr/^{86}Sr$ than for material in which the two elements remain in contact. Separation of rubidium from the mantle into the crust is thought to have occurred early in the history of the Earth, and this richer concentration of rubidium would have resulted in higher ratios of $^{87}Sr/^{86}Sr$ in crustal materials than in strontium released directly into the oceans by submarine volcanoes and hydrothermal vents. Thus, an increase in the erosion of the crust will result in enhanced transport of relatively high-ratio strontium by the continental rivers into the ocean, and an increase in the ratio of $^{87}Sr/^{86}Sr$ in the marine sediments produced at that time [Richter *et al.* 1992].

However, in general, the ratio is found to be lower than would be expected, with less strontium-87, based on the relative abundance of rubidium and strontium. Modeling of the effect of erosion on the isotopic ratio notably fails to show the strongest reductions that have been observed in this ratio.

The strontium isotope ratio has been measured and reported for a number of deep-sea sediment cores, as it exists in biologically deposited calcium carbonate [DePaolo and Ingram 1985, Capo and DePaolo 1990, Dia *et al.* 1992, Clemens *et al.* 1993, Edmond 1993, Froelich 1993]. Biochemically, strontium acts like calcium, and so is tightly bound in the carbonate deposits. These results, from a variety of oceanic sites, are shown in Figure 24. A smoothed curve has been drawn to show significant changes more clearly.

If rubidium were mixed uniformly and consistently with strontium, regardless of the locale, the radioactive decay of rubidium-87 to strontium-87 would cause the ratio to increase at a nearly constant rate, probably approximating the rate shown for the time period from about 40 million years ago to 15 million years ago. Attempts to match these variations by varying the erosional input of crustal strontium to the ocean have failed [Capo and DePaolo 1990, Richter *et al.* 1992]. Comparison of the strontium isotope ratios and ratios of germanium to silicon provides opposite indications as to the effect of chemical weathering [Kump 1993].

An alternate way to reduce this ratio involves neutron absorption by strontium-87, which converts it to strontium-88, the predominant isotope of strontium. The possibility of this process has so far been overlooked in studies of the strontium ratios. The cross-section for thermal neutron absorption by strontium-87 is 16 barns (1 barn = 10^{-24} square centimeter), which is unusually high. Thus, it is likely that variations in the $^{87}Sr/^{86}Sr$ ratio would result from absorption of neutrons produced by the reactors at the bases of the mantle plumes. Changes in the neutron flux at presumed volcanic sources of the oceanic strontium would produce changes in the isotopic ratio. Changes in the $^{87}Sr/^{86}Sr$ ratio could indicate the operating status of the Earth's reactors, and provide a measure of the internal heat generation during past geologic ages.

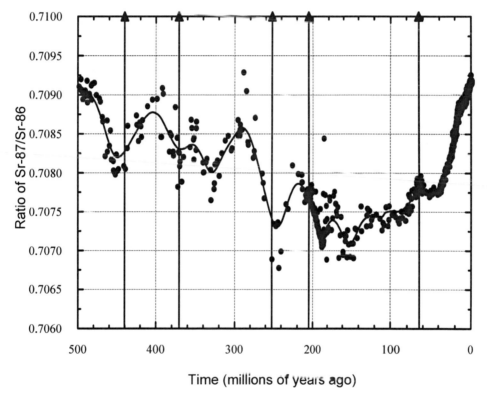

Figure 24. Strontium ratios and extinctions.
Variation of isotopic ratio of $^{87}Sr/^{86}Sr$, measured in sedimentary marine carbonates, with time. Individual measurements are shown as solid dots; a smoothed curve has been drawn to more clearly show the major trends. Principal mass extinctions are shown [Raup 1991].

This ratio has been related to what are thought to be occurrence of ice ages, and so warrants careful study. The association with the ice ages has usually been explained as an effect of increased weathering of surface rocks during warm periods, but measurements in the Bay of Bengal show that this may not be true [Burbank *et al.* 1993] (even if ice ages had occurred).

If this proposal of neutron absorption is correct, and the neutrons are produced by nuclear-fission chain-reactions in the Earth's core, the heating effects of this additional power should be apparent. Considering the evidence shown in Figure 24, increases in the ratio indicate reduced reactor power, and decreases in the ratio correspond to increases in the heat-production rate. While the general trend of the ratio is down, indicating a long-term contribution to the Earth's heat, the record is punctuated with periods when the ratio increases rapidly, perhaps indicating almost complete shutdown of the reactors.

These ups and downs correlate interestingly with major events in the geologic record. For example, the Cretaceous period (65 million years ago to 140 million years ago) was significantly warmer than the present, and the Earth was globally ice-free [Barron *et al.* 1981]. Climate model simulations fail to show this, and additional, unknown sources of heat must be invoked to explain the warm climate. During this time, the strontium isotope ratio was nearly constant, indicating destruction of the strontium-87 by neutron absorption at about the same rate that it was being formed by radioactive decay. Where's the heat? Heat production by the Earth's natural reactors could have warmed the climate, and fission neutrons would have affected the strontium ratios.

In a contrary manner, in the Ordovician period (450 million years ago), when the ratio was increasing at its steepest rate, suggesting shutdown or dormancy of the reactors and little production of internal heat, the Earth is thought to have been cold and glaciated, in spite of the carbon dioxide concentration in the atmosphere being about 13 times its present level [Crowley *et al.* 1987].

Recognizing that steep declines in the $^{87}Sr/^{86}Sr$ ratio may indicate a high production rate of heat by reactors in the Earth's core, and steep increases in the ratio may indicate a quiescent period in which little heat is produced, and regions of the Earth, or the entire Earth, may cool, it is interesting to compare periods of mass extinctions [Wicander and Monroe 1989, Shoemaker and Wolfe 1986, Muller 1986] with the pattern presented by this indicator of heat production.

The five principal mass extinctions within the last 500 million years [Raup 1991] are marked in Figure 24, with guidelines to show the relation to the strontium ratio curve. With the exception of the extinction event at the Cretaceous/Tertiary boundary, which is generally considered to have been caused by the effects of an asteroid impact [Alvarez *et al.* 1984, Sharpton *et al.* 1993], the principal mass extinctions coincide with steep increases in the strontium ratio, when abrupt chilling in the Earth's climate might be expected to have occurred. A "spike" in the strontium ratio at the K/T boundary is believed to have resulted from chemical erosion due to acid rain from the impact increasing the flux of continental strontium [Geodigest 1992, Meisel and Pettke 1994, Hess, Bender, and Schilling 1986, Vonhof and Smit 1997]. The ratio declines immediately after this perturbation.

As will be explored later, the mass extinctions are more likely to have been caused by major impacts, just as the dinosaur extinction was, rather than an independent change in climate.

Radiogenic Halos

There is a considerable history of research on radiogenic halos, small spherical regions of crystallographic damage associated with naturally occurring alpha radioactivity, in diamonds, micas, and other minerals. [See Gentry 1974a and Armitage 1992 for references.] Radiohalos are formed in these minerals as a result of the radiation damage, consisting primarily of lattice dislocations and defects, caused by the high energy alpha particles emitted by various members of the chains of naturally radioactive isotopes originating with uranium (uranium-235 and uranium-238) and thorium (thorium-232). Heavy charged particles are required for this damage to occur; beta radiation does not produce halos. Because of the distinctly different alpha energies of these isotopes, and the increased ionization produced at the end of the well-defined ranges of these alpha particles, these halos are equally distinctive and well defined.

Clear examples of the complete uranium-238 chain have been found, and these examples show an appropriate density of fission tracks from natural spontaneous fission decay of uranium-238, which competes with alpha decay. However, other examples seem to show only the halos caused by polonium-210, the last alpha-emitting member of the uranium-238 chain, or only the halos of polonium-210 and polonium-214, or of polonium-210 and polonium-214 and polonium-218, which might be considered to be the last sequence of alpha emitters in this chain. Ion microprobe mass spectrometry has found an excess abundance, compared to natural lead, of the isotope lead-206 [Gentry 1974b]. The microprobe analysis showed an absence of normal lead, and a slight presence of bismuth-209, the only stable isotope of bismuth. Other "fuzzy" halos show about half the size of the complete uranium halos [Gentry 1992]. "Dual-ring dwarf" halos have been found that cannot be explained as the result of alpha radioactivity [Gentry 1973].

The difficulty in interpretation that these halos pose is that the half-lives of the polonium isotopes are far too short to permit their existence, unsupported by their radioactive parents, in a natural geological material [Gentry 1992]. The half-lives range from 164 microseconds to 138 days. Therefore, it is not possible for these isotopes to exist without the longer-lived parents, nor is it possible for them to have been transported from another location of production.

"Not possible for these isotopes to exist ...", except by production *in situ* as a result of a brief, intense burst of neutrons. Neutron absorption by bismuth-209 produces bismuth-210, which beta decays to polonium-210, and by successive absorptions, the higher isotopes are produced. (Bismuth-210 also produces lead-206 as a result of alpha/beta decay.) Giant halos may result from long-range alphas emitted from metastable states in the bismuth isotopes, produced by rapid sequential neutron capture. Some small halos, otherwise not identified, may show the effect of nearly total fission of a uranium inclusion. The range of fission fragments is about half that of alpha particles, and shows a dual nature as a result of the mass yields being concentrated in a light peak and a heavy peak. Such intense bursts of neutrons could be produced by self-terminating power excursions of small, nearby reactor cells.

Interesting examples of halos in coalified wood from the Colorado Plateau provide evidence of the recurrence of bursts of neutron power over geological time periods [Gentry et al. 1976, Gentry 1992]. In uranium mines in this region, which contains major uranium ore concentrations, coalified (not petrified) logs are found embedded in the sedimentary deposits. These deposits have been found in the Jurassic, Triassic, and Eocene formations. The wood of these logs contains numerous halos, many of which are compressed as a result of compaction by the overburden of sediment, after formation of the halos. The deposits show indications of rapid, catastrophic burial, as in a flood. In some cases, the compressed (flattened) halos are superimposed with normal spherical halos. The ratios of lead to uranium (lead-206 to its "ancestor" uranium-238), which should increase with the age of a deposit, are exceptionally small. These effects are understandable as the result of an intense burst of neutrons, while the trees were still alive and standing, and the wood was still soft.

The intense burst of neutrons, induced by a distant event in the Solar System or beyond, would have caused a brief surge of fissions in the uranium of the underlying soil, releasing a significant amount of heat. If this is speculated to occur in a snow- or ice-covered environment, as is the case in the region at the present time, during winter, large amounts of water would be released, possibly in conjunction with disruptive steam explosions in the water-soaked soil. Trees would be violently dislodged and carried down in flood streams. Uranium deposits in the wood of the trees, extracted biochemically from the soil by the roots while the trees were growing, would fission (uranium-235) and be transmuted by neutron absorption, uranium-238 to plutonium-239, plutonium-240, and plutonium-241. The fission fragments would produce radiation damage in the still normal wood at the instant of the burst, and this would subsequently become the compressed halo as the tree was buried. Plutonium-240 emits an alpha particle at 5.17 MeV, plutonium-239 at 5.16 MeV, which may be indistinguishable from the 5.3-MeV alpha of polonium-210, and would slowly produce the spherical halo. The small amount of lead detected in the inclusions may result from loss of radon (radon-222) by diffusion in the coal, which is likely to be much less retentive of this noble gas than crystalline minerals are, or because the decays stop at thorium-232 and uranium-233 and uranium-235.

The intense radiation would produce the coalification (or carbonization) of the wood, by breaking the hydrogen-carbon bonds and releasing the hydrogen.

Coalified trees, as charcoal in Paleolithic settlements in France, have been found, with distinct deformations that are judged to have been produced while the wood was hot, prior to use [Théry et al. 1995]. Wood that had been converted to charcoal, possibly by this process, has been found at two Paleolithic settlements in France. The coalified wood appeared to have been "mined" by the Paleolithic people, for use in fires. The deformation of the wood cells without fracture was found by the researchers to be possible only with heat, or a long-term compression. This appears to have been the same process that produced the coalified wood of the Colorado Plateau, half a world away.

Other halos, in a diamond [Armitage 1993], are approximately one-half the radius expected for alpha radiation, and could have resulted from nearly complete fission of uranium inclusions in the diamond. The range of heavy and light fission fragments in diamond is about 7.0 to 8.4 millimeters. Since fully developed uranium alpha-particle halos should have an outer ring radius of about 29 millimeters, these smaller halos have been labeled "embryonic" [Mendelssohn et al. 1979]. Some of the halos in diamonds show "etched-out centres" [Mendelssohn et al. 1979], suggestive of the total consumption of a tiny inclusion of uranium.

Plutonium-244

Additional indication of the occurrence of a brief burst of neutrons on Earth in the geological past is the detection of plutonium-244 in the processing liquids of a cerium ore, bastnäsite, from Mountain Pass, California [Hoffman et al. 1971]. This nuclide has a half-life of 82.8 million years, and assumption that the plutonium-244 originated in a supernova significantly prior to the formation of the Earth requires unreasonably large concentration factors for plutonium from that source relative to cerium in the ore. Similarly, influx of plutonium-244 in cosmic rays requires an unacceptable concentration factor for rock that is considered to be Precambrian in age, older than 550 million years. This nuclide can be produced by rapid sequential neutron absorption in uranium-238. If this production occurred in the more recent past, as a result of an intense burst of neutrons, lesser quantities are required for the residue to still be detectable at present. Comparison of the amounts of plutonium-239, plutonium-240, and plutonium-244 detected in the extract, and reasonable production ratios of these isotopes by sequential neutron absorption in uranium-238, suggests the possibility that the plutonium was produced by a brief burst of neutrons within the past 20,000 years, as is suggested here for the Younger Dryas, at 12,900 years ago.

The Geomagnetic Field

So many dynamos!
Jon Agee

The Earth and several other planets in the Sun's family have significant magnetic fields. These magnetic fields are thought to be actively self-generated in the past and at the present time, by a dynamo action, powered by heat within the electrically conductive core. In this way, the planetary magnetic fields differ from permanent magnets and the possible magnetic fields of asteroids. However, a dynamo cannot initiate a magnetic field: it must have a magnetic bias in order to further generate the field. The field will then be maintained, unless conditions change. Like Jon Agee's palindrome quoted above, Earth's dynamo seems to reverse easily.

Studies of the spreading ocean floor have shown that the polarity of the Earth's magnetic field has switched back and forth between the "normal" polarity that we now experience, and a "reverse" polarity. Either polarity is generally aligned with the Earth's rotational axis. These reversals can be used to provide relative dating for geological strata and paleoarchaeological deposits that might otherwise be undatable. Reversals have occurred some 27 times in the last 4.5 million years alone.

Natural nuclear fission reactors offer a source of heat (speculated in the standard theory to be from crystallization of iron at the surface of the inner core), a biasing magnetic field (for which the standard theory has no clue), and a mechanism for reversing polarity (by changing the sense of the bias. Again, nothing in our current theory helps with that). There are computer simulations of the magnetic field but I think there is no set of equations that provide any understanding of the mechanism. All I can do at present is point to the occurrence of a self-induced magnetic field in a sodium-cooled nuclear reactor, and the failure to produce a dynamo in a non-radioactive liquid sodium simulation, and the success where a weak bias field was used.

A magnetic bias is required to initiate the dynamo action, and a change in this bias is required to initiate the instability that leads to a field reversal. Since the magnetic field seems to exist only in either of the two polarities, the changes that induce these reversals might also result from switching between two specific modes of behavior in the core. The magnetic bias that creates the geodynamo may be supplied by the electron flux from the nuclear reactors. Each reactor cell in the Earth's core will act as a local source of electrical current, due to the intense flux of gamma radiation, which will propel electrons away from the center of the reactor by the photoelectric and Compton effects. These current sources may contribute to and control the Earth's dynamo that establishes and maintains the geomagnetic field. The polarity of this bias is directed by the Coriolis force acting on the individual electrons emitted from the reactors, differently according to whether the reactor is at the core-mantle boundary or at the center of the Earth. Changes in the location and intensity of the reactors may induce the reversals in the geomagnetic polarity that have been observed in the remanent magnetism of rocks.

Once established, the magnetic field will increase to a particular value dependent on the characteristics of the system, and can only be reversed if a perturbation occurs that is sufficiently strong to overcome the main dipole field [Cox 1968].

Effects involving the location of the reactors, the intense electron flux produced by the gamma radiation from the reactors, the differing conductivity of adjacent materials, and the electron path-curvature induced by the Coriolis effect, could directly lead to establishing the initial excitation of the geodynamo. The Coriolis force results from the Earth's rotation, and causes the magnetic axis to be aligned with the rotational axis. Briefly, electrons ejected from the material composing and surrounding a reactor, by the action of high-energy gamma rays, would traverse a curved path due to the Coriolis effect, and the compensating return current would close a loop, creating a microscopic magnetic field. This field would induce greater curvature in the paths of succeeding and accompanying electrons, and eventually the macroscopic field would become sufficiently developed that it would form the basis for dynamo action in the circulating fluid outer core. Current loops would be preferentially curved in one direction for the electrons ejected from the central reactors and oppositely curved for those ejected from the peripheral reactors. Variations in the intensity of the outer reactors, relative to the central reactors, would lead to changes in the dipole and non-dipole fields, and to polarity reversals. These polarity reversals will always occur from one polarity direction to the opposite polarity, rather than to any intermediate direction, because of the two different effects of the reactors, based on their location in two regions, at the center or at the core-mantle boundary.

The Coriolis effect will cause electrons that are directed outward from the center of the Earth to move in a generally curving path with a clockwise sense, while those directed inward from the core-mantle boundary will have a general tendency to curve in a counter-clockwise sense. To balance the radiation-induced flow of electrons from within and near the reactor, electrons will return from the outer regions, in the highly conductive iron. These brief and individually minuscule current loops will create a magnetic field which could establish the geomagnetic dynamo. Current loops formed by the central reactors are predominantly in the clockwise sense, since most of the electrons are emitted outwards. Current loops from electrons emitted outwards from the reactor cells at the core-mantle boundary are suppressed relative to those produced by inward-directed electrons because of the low electrical conductivity of the mantle rock, compared to the liquid iron of the core [Wahr 1990]. In the model proposed here, the dynamo is biased by the strongest electron emitters among the several reactors, and then becomes self-energized. While the individual field contributions are small, in their multitude there is strength.

Thermal gradients created by the power output of the reactors drives the rotational motion of the liquid core, and electrical and magnetic gradients energize the dynamo. The polarity of the geomagnetic field will be determined by which reactor region is dominant, and will switch as the relative intensities of the local fields ebb and flow with reactor power. It is likely that one polarity will be associated with mantle plume (hot spot) volcanism, due to heat from core-mantle boundary reactors, while the opposite polarity will be associated with more global conditions, resulting from the general heating produced by the central reactors.

Herndon has calculated that energy produced by nuclear fission in the center of the Earth is adequate by itself to provide the power needed to drive the geomagnetic dynamo [Herndon 1993]. He further suggested that alternate on-again/off-again operation of a natural nuclear fission reactor at the center of the Earth would lead to disruption of the convection within the outer core, resulting in the observed reversals of the geomagnetic field. (However, recall that the physics of nuclear reactors precludes the existence of a single, super reactor. This is better approximated by a cluster of coupled (interacting) reactor cells.) Here it is suggested that the geomagnetic field polarity is more actively determined by the alternating dominance of the central reactors and those located at the core-mantle boundary.

In spite of extensive theoretical studies and proofs of fluid dynamos and the clear evidence provided by the Earth's magnetic field, an experimental fluid dynamo has not yet been deliberately created for controlled study. However, in a nuclear reactor in Beloyarsk in Sverdlovsk, in the former Soviet Union, an unwanted and unexpected dynamo was born [Mukerjee 1995].

The Soviet/Russian nuclear reactor BN-600 is a liquid-metal-cooled fast breeder reactor (LMFBR). This reactor uses molten sodium metal to remove heat generated by nuclear fission in a relatively small nuclear reactor core. Neutrons are absorbed by the fuel atoms to produce fission, before they have lost much of their original energy, so this is called a "fast" reactor. (High-energy neutrons go faster than low-energy neutrons.) It produces more fuel, as plutonium, than it consumes, so it is called a "breeder" reactor. At full power operation, it produces approximately 1,500 million watts (megawatts, MW).

During the initial full power start-up of this reactor, a self-excited magnetic field developed. [Kirko, Telichko, and Shrinkman 1983]. This possibility had been considered earlier, in a theoretical study [Pierson 1975]. That study concluded that electromagnetic self-excitation was unlikely in LMFBRs (Liquid Metal Fast Breeder Reactors). It further noted that a large liquid sodium pump had been operated with "no observed difficulties". Presumably, that indicated that no self-generated magnetic field was observed. However, neither the theoretical study nor the practical operation of the sodium pump involved the intense radiation field of the operating nuclear reactor. Thus, in the real reactor, the effects of high-speed electrons traversing the liquid-metal coolant may have induced the self-excited magnetic field.

The Soviet nuclear reactors used for submarine power use molten bismuth/lead alloy as the heat transfer medium, the "coolant" for these reactors. Several of the Soviet nuclear submarines have sunk or been disabled under conditions never clearly explained. (While that in itself is not surprising, for a secretive nation, only two similar failures have been experienced by the U. S. submarines, which use high-purity water, a non-conductor, as a coolant. The sinking of the *USS Thresher* was well explained as being unrelated to the reactor. The loss of the *USS Scorpion* was not so clearly explained, but was probably not due to problems with its reactor.)

The reactors of the high-powered radar satellites that were used by the Soviets to track submerged U. S. nuclear submarines also used a molten metal, lithium, to transfer heat from the nuclear fuel to the electrical generating system. This space-based surveillance of U. S. naval activities, so vital to the defensive strategy of the Soviet Union, was "one of the most error plagued elements of the Soviet space program" [Hart 1987]. Of the two dozen or so satellites of this sort that the Soviets operated, at least two, Cosmos 954 and Cosmos 1402, failed from malfunctions of the electronic command and control systems. Normally, at the end of the useful life of these satellites, the power system, including the reactor with its radioactive fuel, is moved to a high-altitude orbit with a sufficiently long orbital life that the fission-product activity can decay before the reactor re-enters the Earth's atmosphere. Cosmos 954 re-entered prematurely and scattered pieces over northern Canada on January 24, 1978. Cosmos 1402, launched on August 30, 1982, malfunctioned in orbit. The reactor could not be separated from the spacecraft and was incapable of being transferred to a safe parking orbit. The spacecraft reentered over the Indian Ocean on January 25, 1983 [Curtis 1990]. The reactor finally reentered and burned up on February 7, 1983, over the South Atlantic Ocean [Tag's Broadcasting Services 1998, Leifer *et al.* 1987].

The one reactor that the U. S. launched into space, S10A-FS4 ("SNAP10A-Flight System 4"), ended operation early, after 43 days of operation, rather than the intended 90 days, apparently shut down by command of its automatic control system. The explanation for this shutdown was that the Lockheed-built voltage regulator had malfunctioned. The malfunction, however, appears to have been forced on the regulator by the reactor. In anticipation of a somewhat greater loss in reactivity than originally planned, the reactor was started up (in orbit) running a bit hotter than designed. This eventually overwhelmed the capability of the regulator, and the reflectors were ejected, shutting down the reactor [Johnson 1995]. This satellite was then moved to a "parking" orbit, where all the short-lived fission products will decay before the reactor finally returns to Earth.

Perhaps, in all these cases of poorly explained reactor system failures, a nuclear-reactor dynamo created magnetic fields with stray electrical currents that intruded and interfered with the electronic systems and caused the fatal failures.

In the S10A-FS4 reactor, at least, rotary motion of the coolant was deliberately induced by use of helical wire wraps around the fuel rods. The wire wraps swirled the coolant along the length of the fuel to improve heat transfer. The Coriolis effect resulting from this swirling, and the electron flux generated by the gamma radiation of the fission reactions may have produced a self-excited dynamo that disrupted the electrical system of this satellite, resulting in the premature shutdown.

Permanent magnets of different shapes have slightly different shape fields. These are all dipolar in general form. A bar magnet produces a rather elongate field, while a ring magnet produces a rather squat field. A spherical magnet produces a field much like that of the Earth.

The Earth's magnetic field traps energetic charged particles from the Sun and space and sorts them according to charge, energy, and momentum. The Van Allen belts of electrons and protons have been known since 1958 by measurements with *Explorer 1* [Van Allen 1990], and a recent similar belt of heavy ions from space has been discovered by the *SAMPEX* satellite [Mewaldt *et al.* 1994]. Because of the greater energy and momentum of these heavier particles, they are able to penetrate the Earth's magnetosphere more deeply than the other particles, and show details of the geomagnetic field not revealed by the other belts. The greatest intensity of high energy oxygen ions shows as a streak across the South Atlantic, between the tips of South America and Africa [Cummings *et al.* 1993]. This streak passes directly over the Tristan de Cunha hotspot [Molnar and Stock 1987], and Gough Island [Francis 1993] [see also O'Connor and le Roex 1992], and the deep structures suggested by the tails of South America and Antarctica, that curl around at the South Sandwich Trench.

Planetary magnetic fields may help us to understand the planets themselves. Planetary fields are shown below (as magnetic dipole moments), with the observed amount of heat radiated, compared to the solar heat received. (Uranus radiates just slightly more heat than it receives; Neptune radiates nearly two times more than its solar heat.)

Planet	Magnetic Field	Excess Heat
Mercury	3.0×10^{-7}	none measured
Venus	$<3 \times 10^{-8}$	none measured
Earth	0.61×10^{-4}	nil
Mars	0	none
Asteroid (Gaspra)	0	none
Jupiter	4.3×10^{-4}	0.67 x solar
Saturn	0.21×10^{-4}	0.78 x solar
Uranus	0.23×10^{-4}	0.06 x solar
Neptune	0.133×10^{-4}	1.61 x solar

The magnetic fields of both Uranus and Neptune are significantly offset from the planet center and are misaligned with the rotation axis [Wagner 1991]. Perhaps in these two planets, the dynamo is overwhelmingly dominated by a peripheral reactor cell, and this produces a field distinctly different than would be produced by a central reactor. The observable magnetic fields of the planets are remarkably similar, ignoring those without magnetism. The lack of excess heat in Uranus is remarkably different from the other giant planets.

If natural reactors at the core-mantle boundary and on the surface of the inner core control the polarity of the geomagnetic field, in opposite directions, movement of the reactors from one region to the other will induce a polarity reversal. If most reactors become stalled in either region, staying thermally buoyant at the core-mantle boundary, or taking a long time to re-accumulate in the dimples on the inner core, the geomagnetic field will remain constant. This was seen during the "superchron/superplume" period of the Cretaceous, with no changes in polarity from about 124 million years ago to 84 million years ago.

That was a period in which the Earth was particularly warm, suggesting that the reactors became stalled at the core-mantle boundary and generated considerable excess heat that easily made its way to the surface.

The Snowball Earth

Since Louis Agassiz invented the Great Ice Age as an explanation for the world-wide deposits of sand, gravel, and bones that had once been blamed on Noah's Flood, global glaciation has been a useful explanation of similar deposits. Periods of glaciation have been described throughout geologic time, primarily based on the appearance of greatly disturbed sediments, laminated layers with erratically scattered "dropstones", fragments of rock that do not belong in the local neighborhood. These deposits are often followed, "capped", by a layer of carbonate, a cap limestone or a cap dolomite depending on composition.

The possibility of an early global ice age had been suggested by Brian Harland in 1964 as a result of his work on providing a world-wide geological map. He noticed that underlying the Ediacaran strata (about 570-590 million years ago) there were worldwide deposits of "tillite", the till and drift previously associated with the flood deposits supposed to have been left by Noah's Flood.

Global glaciation and the obvious recovery (or we wouldn't be here) from an ice-covered world has been well developed by Paul Hoffman [Hoffman and Schrag 2002] and Alan Kaufman [Kaufman, Knoll, and Narbonne 1997]. In this hypothesis, distinctive geologic deposits, special isotope ratios, and the occurrence of an anomalous "spike" of the rare element iridium, are linked together in a detective story that paints a picture of a world covered with ice, continental ice sheets, and oceans frozen over. Truly, a bright white Snowball Earth had been produced [Walker 2003]. This is the White Earth Catastrophe, from which no recovery should be possible, as most of the solar heat would be reflected by ice and snow. The global temperature should have been so low that all water vapor froze onto the surface. Water vapor, carbon dioxide, and methane are helper gases in the greenhouse effect. Recovery is proposed to have ultimately occurred as a result of millions of years accumulation of volcanic carbon dioxide.

The variations in the carbon isotopes -12 and -13, and the strontium isotopes -86 and -87 have been used as diagnostic markers. The carbon isotopes have a clear signal, benefitting from intensive analysis, but the strontium isotopes appear to be more erratic than systematic in this time span. The carbon isotopes may offer a firmer basis for interpretation. The ratio of carbon isotopes, carbon-13 to carbon-12, is changed by the preferential production of carbon-12 by the Giant Dipole Resonance reaction on oxygen-16 by the gamma rays from a nearby astronomical detonation. This additional carbon-12 is dissolved into the ocean water as carbon dioxide, reducing carbonate deposition. A major impact will induce expansion of the Earth, releasing carbon dioxide from the ocean by several mechanisms, and the cap carbonate will be abruptly precipitated. This carbonate will be richer in the light isotope, carbon-12, as is observed.

Measurements of the remanent magnetization of Neoproterozoic rocks associated with what have been interpreted as glaciogenic (diamictites, a poorly sorted breccia, and "ice-rafted" dropstones), have shown that these were concentrated near the equator. This was consistent with the distribution of continents at that time (640-580 million years ago), from where they have been moved by continental drift.

Working on the assumption that Snowball Earth glaciation had actually occurred, and that extraterrestrial dust falling down from space through the atmosphere would accumulate in the ice, a group looked for excess iridium in the glacial strata as an indicator of this dust. Thinking that a rapid melting of the global ice sheet could produce a concentrated layer of iridium that would show the length of time of the glaciation, the group did a detailed analysis, from below the diamictite that is taken as evidence of the glaciation, to above the overlying cap carbonate [Bodiselitsch et al. 2005, Kerr 2005].

The Snowball Earth researchers did find strong iridium anomalies for the Marinoan glaciation (about 635 million years ago) and the Sturtian glaciation (about 710 million years ago). Peak concentrations from 0.5 to 2 parts per billion were found, in a very concentrated layer near the base of the cap carbonate, above the diamictite. Interpreting this anomaly as a steady accumulation, Bodiselitsch et al. used estimates of the influx of extraterrestrial dust to estimate time spans of 3 to 12 to 41 million years for the Marinoan glaciation. For the Sturtian, only an estimate of thousands to millions of years was given.

There was no discussion of the possibility that the iridium, the most definitive diagnostic indication of impact, might have resulted from a major impact that had also deposited the diamictite, and caused the expansion of the Earth that precipitated the cap carbonate.

The Impact of Iridium

Here we must digress a moment for a view of the history of the iridium anomaly. In 1976, Walter Alverez, a geologist, asked his father Luis, a physicist, if there were a way to determine the accumulation time of a layer of apparently ordinary red clay, only a few centimeters thick at the K/T boundary, when the dinosaurs died, as he found it in a geologic section near Gubbio, Italy [Alvarez et al. 1980, Alvarez 1997, Alvarez and Asaro 1990, Alvarez 1987]. The approach decided upon was to measure the amount of extraterrestrial (meteoritic) dust that had steadily accumulated with the sediment in the layer, from the continuous influx of dust from space. This could be done by using neutron activation analysis to determine the amount of the rare element iridium in the dust that had accumulated. Actually, the analysis only identifies stable iridium-191, by neutron activation to radioactive iridium-192 followed by gamma-ray spectrometry to identify the specific radionuclide decay.

While a very small concentration had been expected, about 0.1 part per billion, the first analysis showed 3 parts per billion, and refinement of the method produced 9 parts per billion. Subsequently, this iridium anomaly has been found in the special clay all around the world. It has been generally accepted as a definitive marker for the K/T boundary. Further, it has become an essential diagnostic for the impact of an extraterrestrial object.

An iridium anomaly, just like the one at the K/T boundary, has also been found near the Acraman impact structure in Australia [Gostin, Keays, and Wallace 1989]. This impact is thought to have occurred 580 million years ago.

Kaufman, Knoll, and Narbonne specifically identify two global glaciations in their work, the Rapitan and the Ice Brook, occurring about 755 million years ago [Kaufman, Knoll, and Narbonne 1997]. These events are marked by deposits of tillite, 400 meters and 200 meters thick, respectively. The tillites appear to be overlain by cap carbonates. Intriguingly, the Rapitan deposit contains an iron-rich layer more than 100 meters thick. Hoffman and Schrag recognize the cap carbonates as "strange" and also comment on the large sedimentary iron deposits. Cap carbonates are distinctly unique unto themselves, and differ from "normal" carbonate deposits.

Since the early 1900s, the distinctive deposits of till (coarsely graded sediments) and diamictite (poorly sorted conglomerates or breccias) have been used to identify widespread glaciation. Features that have been seen in current mountain glaciers have been used to confirm the identification. A list of 13 criteria for identification of tillites and diamictites was presented by L. J. G. Schermerhorn in 1974, and has formed the basis for such identifications since. Schermerhorn recognized that these features could be produced by processes other than glaciation. About half of these criteria seem to have been applied in the practice of identifying ancient glaciations [Oberbeck, Marshall, and Aggarwal 1993b].

However, in 1993, Verne Oberbeck, John Marshall, and Hans Aggarwal published a detailed description of the clear similarity between geological deposits and features that are known to be caused by large impacts, and those that had been adopted as definitive for glaciation [Oberbeck, Marshall, and Aggarwal 1993a, but also see Young 1993, and Oberbeck, Marshall, and Aggarwal 1993b]. By studying the record of identified impact structures on the Earth, cratering mechanics, and the cratering rate, Oberbeck, Marshall, and Aggarwal showed that there should be a considerable thickness of impact-produced deposits of poorly sorted conglomerates, of the sort that are identified as tillites and diamictites that are commonly interpreted as being of glacial origin. Schermerhorn's list was explicitly addressed [Oberbeck, Marshall, and Aggarwal 1993b] and examples of known impact deposits were described that fit those criteria. Three definitive criteria for identifying an impact event, not discussed in the cited papers, are shock effects in mineral grains, glassy spherules (that turn to clay), and the presence of an anomalous concentration of the rare element iridium.

Oberbeck, Marshall, and Aggarwal declared that it might be possible that some of the deposits interpreted as indicating widespread glaciation had instead been caused by impact, and had been misinterpreted. That misinterpretation would cause a great distortion in our knowledge of the distant past and of the workings of the Earth. It is a speculation of this book that all deposits interpreted as indicating widespread glaciation have been misinterpreted, and the deposits were the result of major impacts rather than global glaciation.

In 1996, Michael Rampino reported on studies of deposits in Belize that were produced by the Chicxulub impact in the Yucatan, over 200 to 300 miles away [Rampino 1996, Ocampo 1996, Pope 1996, Fischer 1996, King Jr. 1996, see also Rampino 1994, Gostin, 1986, and Williams 1986]. After summarizing the results of his research on the impact deposits in Belize, Rampino stated:

> "Moreover, these unusual deposits may have a significance that goes far beyond the events at the Cretaceous/Tertiary (K/T) boundary. Similar deposits occur in the geologic record, and they have been interpreted as the products of ancient glaciation primarily because of the presence of large boulders and scratched and polished stones resembling those found in deposits of the most recent ice age. The discovery that impact debris can have analogous features casts doubt on the interpretation of some of these deposits as glacial in origin. In reality, they may be impact debris, or the products of debris flows generated by other violent processes. Further study of the Belize material and other known impact debris will enable geologists to develop better criteria for distinguishing impact ejecta from true glacial deposits in the geologic record."

In considering the dropstones to be indicators of ice-rafted debris, it should be recognized that at the Ries Crater, lightly shocked rocks that had been near the surface were thrown nearly 200 kilometers from the impact site [Hofmann and Hofmann 1992 quoted in Melosh 1995]. Such rocks would be seen as dropstones and, as Rampino remarks, such deposits are not distinguishable from glaciomarine deposits that are used as indicators of global glaciation.

This book proposes that all so-called glacial deposits indicating ice ages, including "the most recent ice age", have been misidentified and actually were produced by major impacts. In the case of the Great Ice Age of Louis Agassiz, effects of the impact itself were modified and compounded by the "violent processes" associated with tremendous amounts of ice (water, ammonia, and methane) contained in the impactors, and creating a great flood.

The hypothesis of creation of glacial deposits by a major impact involves starting with events before the impact. A brief narrative can give the basic ideas needed to explain all the effects.

Consider a large natural satellite of one of the outer planets, the same type as Io, Callisto, Ganymede, Europa, Titan, Triton, and our own Moon. If reactors in the core of the satellite approach the point of detonating, by any of several possible mechanisms, and then detonate, several effects may result. Immediately, the flash of gamma radiation will convert some of the oxygen in Earth's atmosphere into carbon-12, causing an increase in the partial pressure of carbon dioxide. This will inhibit the deposition of carbonates in the ocean, by acidifying the water. The detonation will increase the temperature of the Earth slightly, just as supernovae do at present, acting both on the Earth and the Sun. The warmer conditions will promote greater photosynthetic action, resulting in an increase in the carbon-13 content of biological deposits. (That is seen in the gradual increase in the carbon-13 ratio just before a glaciation deposit.) The impact itself will spread impact debris over great distances, to be interpreted as diamictites and tills, with great boulders to be seen as dropstones. Large regions will be destabilized by the seismic agitation caused by the impact. In some cases, great systems of dikes will be opened in the crust on the opposite side of the Earth, allowing a flood basalt to start. A dimple on the surface of the inner core will be formed by the shockwave, and will serve as the root for a future mantle plume volcano, initially associated with the flood basalt location. Expansion of a local region of the mantle (or core) will cause the farside to bulge, displacing ocean water, which will flow across the continents as a megatsunami, reverberating several times around the world. The expansion will reduce atmospheric pressure (by reducing surface gravity and by increasing the area over which the atmosphere is spread). That will lower the calcite compensation depth, promoting immediate precipitation of the cap carbonate, with the light carbon composition caused by absorption of the newly produced carbon-12 into the ocean water. This will be accelerated by the outgassing of the ocean, by heating by the impact debris, and by the agitation from the impact and "sloshing" of the ocean. Iridium from the impactor will be deposited along with the cap carbonate. The world will then settle down to a new form, slightly larger than before, with lower atmospheric pressure, slightly less surface gravity, and new life forms may find the new world very suitable, while old life forms die out.

This is a multistep process, and examples may be found in the geologic record, if looked for diligently, in which only some of the steps occur. That makes identification and interpretation all the more difficult.

Banded Iron Formations

Banded Iron Formations are enigmatic deposits rich in iron, comprising the current richest iron ore deposits commercially exploited. Among the oldest rocks, formed mostly from 2 to 2.5 billion years ago, Banded Iron Formations are massive deposits of commercial iron ore, consisting of laminated layers of iron-rich/silica-poor, and iron-poor/silica-rich intervals [Emiliani 1992]. These formations continue to be well-explained enigmas. Kurt Konhauser has been quoted as saying: "Despite the fact that people have been looking at their formation for a century or more, we still don't have a really firm handle on where they formed in the oceans, how they formed and what they're telling us about the composition of the oceans or atmosphere at that time." [Perkins 2009a]. They start abruptly in the geologic strata and end equally abruptly. They are chemically precipitated (like the cap carbonates), and have laminations, alternating iron-rich and iron poor, with cherts and jasper, of greatly varying thicknesses.

The simplest story is that the Earth's atmosphere (and ocean) prior to the deposition of the Banded Iron Formations had very little oxygen. Iron was dissolved in the oceans in its soluble (chemically reduced) form. As oxygen produced by photosynthesis accumulated it oxidized the iron to an insoluble form, the iron precipitated (alternately with jasper and chert) and produced the Banded Iron Formations. This general explanation as the rapid deposition of soluble iron in an abruptly oxygenated ocean, is related to the Great Oxygenation Event about 2.4 to 2.0 billion years ago. The iron in the ocean water is oxidized to an insoluble form and is quickly deposited. Significantly, these formations also occur in association with the deposits that are used to show a Snowball Earth glaciation.

However, with the apparent association with the so-called glaciogenic deposits that have been interpreted as showing global glaciation, we must ask about possible impacts. The amounts of iron in these various deposits is consistent with the iron carried to Earth by a small to medium sized asteroid, on the order of the impactor blamed for the death of the dinosaurs at the K/T boundary. Hot iron (melt and vapor in the case of an impacting iron asteroid) can react directly with water to produce the observed oxide deposits [Taylor 1921]. Sloshing of the global ocean following the impact and the likely impact-induced expansion could produce the laminations.

In an effort to examine the history of atmospheric oxygen, researchers studied the chromium isotopes in Precambrian Banded Iron Formations [Frei et al. 2009, Lyons and Reinhard 2009]. The isotope ratios (chromium-53 compared to chromium-52) from deposits older than about 1.8 billion years appear to cluster into groups with relatively little variation, compared to ratios from deposits around 0.5 billion years ago. The older ratios are distinctly lower in chromium-53 than the more recent. The older ratios form groups with ranges on the order of ± 0.1 per mil, while the recent values range from +0.3 to +5.0 per mil [Frei et al. 2009 Supplementary Information, +4.9 per mil in the text]. The process considered for understanding the atmospheric oxygen levels is a multistep process. A surge of photosynthesis produces oxygen which accumulates in the atmosphere, oxidizing manganese in soil, which acts as a catalyst in transforming insoluble trivalent chromium (Cr(III)) to hexavalent chromium (Cr(VI)), which is soluble and is carried to the seas. In ocean water, iron reduces the hexavalent chromium to the insoluble trivalent form, which then precipitates, to remain in the geological deposit for our eventual studies.

However, it might be simpler to consider that the chromium came to the Earth, and the oceans, by way of impacting asteroids ("It Came From Outer Space!", literally). Fossil meteorites with chromite ($FeCr_2O_4$) were found, relatively abundantly (by two orders of magnitude compared to the recent infall) in Ordovician (480 million years ago) limestone deposits in Sweden [Schmitz *et al.* 1997, Schmitz *et al.* 2003]. This ancient infall occurred over a period of about 1 to 2 million years.

Considering the possibility that the asteroids were produced as the result of the nuclear fission detonation of a planet ("Asteria"), and that fragments of those asteroids come to Earth and its oceans, this is a clear example of the modification of isotopic ratios by neutron absorption in the detonation. The cross sections for neutron absorption for the nuclides involved in these studies are:

chromium-50	15.4 barns
chromium-52	0.86 barns
chromium-53	18.6 barns
chromium-54	0.41 barns
vanadium-51	4.94 barns.

The critical ratio is between chromium-53 and chromium-52. The cross section (absorption probability) for chromium-53 is significantly greater than that for chromium-52. Neutron absorption will reduce the isotope ratio by preferentially depleting the -53 isotope, and may also increase the -52 isotope if any vanadium is present. The isotope ratios of chromium from asteroidal fragments produced by a nuclear fission detonation would have been modified in just the way that is observed in the Banded Iron Formations.

Phosphoria Basin

Phosphorus has been found in exceptionally high concentrations in some of the Neoproterozoic iron formations of about 635 to 750 million years ago, compared to earlier and later deposits [Planavsky *et al.* 2010, Filippelli. 2010]. This has been proposed to have resulted from increased weathering of the continental surface following the Snowball Earth glaciations.

Instead of post-glacial weathering, however, the great influx of phosphorus may have come in the form of a large, phosphorus-rich impactor. (The body of Jupiter's moon Io might serve as an example of a phosphate-rich object, along with much silica and sulfur.) While not situated at the right time, the Phosphoria Basin, in the northwestern United States and dated at about 250 million years ago, may represent such an impact [Stephens and Carroll 1999, Hein 2004, Carroll *et al.* 1998]. The end of the Permian, at about 251 million years ago, has been considered to have resulted from a major impact [Kaiho *et al.* 2001], although the Phosphoria Basin was not suggested as the impact site. (These studies were done in southern China.)

A palinspastic reconstruction of the basin (adjusting for crustal deformation since its creation) [Tisoncik 1984] shows a multi-ring structure with an outer diameter of about 290 kilometers and a central peak located at Black Pine Peak, elevation 2,861 meters, latitude 42 N, longitude 113 W. The basin contains more than six times the phosphorus currently in the global ocean. Impact by a similar object around 750 million years ago could have provided the phosphorus (and the iron, and the chromium) found in the Banded Iron Formations of that time.

Science says "global glaciers"; the observations say "impact".

Oceanic Anoxic Events

Oceanic Anoxic Events occur in geologic time when the oxygen content of the ocean is somewhat abruptly reduced. The conditions of Oceanic Anoxic Event 1a (OAE1a, the earliest such occurrence for which we have observations, about 120 million years ago) have been briefly described by Erba and colleagues [Erba *et al.* 2010]. These conditions provide a list of likely effects produced as a consequence of a nearby astronomical detonation, such as a large moon of an outer planet:

1. Carbon dioxide concentration in the ocean increased by a factor of roughly 20 to 30 times that at present, as a result of conversion of atmospheric (and surface ocean) oxygen to carbon by the Giant Dipole Resonance reaction.
2. The carbon isotope curve ($\delta^{13}C$) showed an abrupt negative excursion, from production of carbon-12 by the Giant Dipole Resonance reaction in oxygen-16.
3. The oxygen concentration in the ocean decreased, because of destruction of atmospheric oxygen by the Giant Dipole Resonance reaction and the reduction in atmospheric pressure due to impact-induced expansion of the Earth.
4. There was severe global warming, due to the boost in power in the Earth's reactors.
5. Eruption of a flood basalt (Ontong Java Plateau) was initiated, by a major impact on the opposite side of the Earth.
6. Initiation of a mantle plume hot spot making the seamounts tracking to the southeast from the Ontong Java Plateau, by the shockwave creation of a source dimple on the surface of the inner core.

7. Enhanced organic material burial occurred, as a result of the killing of surface organisms by the gamma-flash of the detonation and the increased sedimentation of impact-generated debris.
8. Surface water was acidified, both by the increase in carbon dioxide and by nitric acid formed by the gamma-flash (and ultraviolet radiation) from the detonation.
9. The sizes of various coccoliths were reduced without a change in species, as a result of the stunting effect of the temporarily increased radiation.

Thus, a cursory exploration of a single geological occurrence shows at least nine associations of effects due to a nearby nuclear fission detonation.

Continental Drift, Plate Tectonics, and Earth Expansion

If we are to believe in Continental Drift,
we must forget everything which has been learned,
and start all over again.
R. T. Chamberlin 1928

While this book intends to focus primarily on the effects of natural nuclear fission reactors, the story of the Earth is so complex and interconnected that a side-plot must be explored at this point. The geological arguments over Continental Drift must take us aside, and hopefully, not astray. We must face the conflict between the theory of Plate Tectonics and the theory of Earth Expansion. I will use the label "Continental Drift" to refer to a theory of *observations*, in this case, the apparent movement of continental blocks from an originally contiguous configuration to the dispersed situation we find today. Plate Tectonics and Earth Expansion are theories of *processes*, that produced the situation we find today.

As seen at the end of the previous section, Earth Expansion, as a well developed alternative to Plate Tectonics, plays an essential role in the explanation of what have been perceived as glacial events. Instead of global glaciation, the abrupt, impact-induced expansion of the Earth produced impact breccia dropstones, diamictites and cap carbonates, megatsunamis, carbon-isotope excursions, and a newer world for new forms of life, with extinction for the old forms. As will be seen in the next section, if the Solar System planets formed from the outside in, as I think they must, then the earliest Earth must have been a massive gas giant planet with a supercompressed kernel. All subsequent actions, the loss of the massive atmosphere, blasted away by the newly formed Sun; the loss of the molten silica ocean, as it boiled away into space; and the congealing of a global continental crust, as the surface cooled; resulted in a naturally expanding Earth, producing the Continental Drift that is observed.

The observations of Continental Drift developed from the similarity in the shape of the coastlines of opposing continents, facing each other across oceans, which can be seen on any globe and has long been noticed. The idea that the opposite coastlines of the Atlantic Ocean seemed to match each other, as if the continents had once upon a time been joined together, was first suggested by Abraham Ortelius in 1596, similarity between the coastlines of South America and Africa had been noted by Sir Francis Bacon in 1620, and was demonstrated in 1858 by Antonio Snider. The match of the continental coastlines, done casually, is quite convincing. The more carefully it is done, however, the more problems appear. If the continents had once been connected together, as better geological and paleontological mapping suggested ever more strongly, the continents must have drifted apart, to be so scattered as they are today. This seems to be an idea that cried out for attention, yet for years most of the scientific effort it attracted was spent in rejecting it. The possibility that continents had moved from their original (or at least earlier) positions was opposed by those geologists who thought that the Earth had "always" been in its present condition, the Uniformitarians. Movement of continents was also opposed by those geologists who thought that the Earth had been much hotter earlier, and that thermal contraction had shaped the continents and mountains.

The continents are surrounded, at elevations below present sea level, by continental shelves that extend out to sea with a slope generally similar to that of the dry land. At a greater or lesser distance from the present coastline, these shelves end in abrupt drops, that would be called sea cliffs if they were now above the sea. These cliffs drop down to the ocean floor. Submarine canyons cut through the edge of the shelf, as if it had previously been above sea level, and rivers had run through it. These canyons are tentatively explained as the result of turbidity currents, muddy flows with the cutting power of rivers. Monterey Canyon is cut into granite, and greatly resembles the Grand Canyon of the Colorado, which may have been cut by dry-land rivers in as little time as a few million years. While some of these canyons are associated with modern rivers, others begin at considerable distances from the shore, with the intervening distance filled and covered by soft sediments. The shelves are covered with a mix of sediment types, some resembling what would be laid down on dry land, and the turbidity currents are just cleaning out the soft sediment that filled in the canyons.

The match between continental plates is best at the edge of the continental shelves, at a depth of 2,500 to 3,000 meters below current sea level. That is where the granitic (or sial) continental crust ends. The match (on our current globe) is not as good as would be expected, with gaps and overlaps, and there have been other observations that impeded the acceptance of

Continental Drift. This refitting of the continents leaves a giant ocean covering the rest of the globe. There is no other evidence of the existence of such a huge ocean, it is simply needed to fill a void produced by our assumptions of a constant-size Earth. Remarkably, the mismatches in the coastal fits can be eliminated by making the fits on a much smaller globe, one with essentially no ocean basins, a globe uniformly covered with continental crust.

This obvious similarity of the coastlines, geologic structures, and distributions of fossil life led Alfred Wegener to propose, in 1915, that the currently existing continents had originally been a single supercontinent that had split apart and its parts had drifted away on the face of the globe [Wegener 1928].

Wegener began his scientific career as a planetary astronomer in Germany in 1905. He soon switched to meteorology, and eventually became a leading character in the new field of geophysics, the study of how the Earth works. Putting together observations from a variety of different scientific fields, Wegener argued that the continents had once been joined and then drifted apart. To explain these observations, Wegener proposed that the continental blocks had drifted through the oceanic crust like icebergs through the ocean. This is a physically unreasonable mechanism because of the obvious rigidity of rock, and that prevented acceptance of his ideas for 40-50 years.

Both the observations and the proposed mechanism were lumped together as "Continental Drift" and were generally rejected by established geologists during Wegener's lifetime. (He died during an expedition in Greenland in 1930.) Wegener saw an effect, the congruence of the continental shorelines, and had to propose an implausible mechanism, the drifting of continents, to explain the observations of Continental Drift. He was rejected, because the mechanism was not considered possible. While most of the observations were valid, and should have been convincing enough for further exploration, the mechanism of "granitebergs drifting through basaltic seas" suffered from at least two fatal flaws. The rocks of the surface crust are too hard, stiff, and brittle, to permit continental blocks to move over thousands of kilometers. Further, there was no known force to propel the granitebergs on their drifting travels. Wegener's Continental Drift ran aground on an incompetent theory. Continental Drift remained an idea waiting for a reason, for many years after his death.

The idea of Continental Drift was kept alive by a few lukewarm supporters during the next 30 years. Although long opposed, Wegener's observations eventually led to the theory of Plate Tectonics, which then became the mainstay of current geological thinking. In the 1960s, the rapid growth of undersea exploration and extensive mapping of the seafloor gave a reason for the idea. Perhaps the most amazing discovery of this exploration was the extensive, globe-girdling network of midocean ridges, first discovered in the Atlantic Ocean. These ridges were found to consist of frequent underwater volcanic eruptions of magma, and were flanked by generally parallel and symmetrical patterns of alternating magnetism in the cooled lava crust. This was easily recognized as evidence that new oceanic crust was being produced by the eruptions along the midocean ridges. The spread of new oceanic crust could force the continental crust apart. Continental Drift was real after all. With the identification of numerous crustal plates, defined by continental margins and midocean ridges, and the hypothesis of slab subduction, the theory of Plate Tectonics was born [Anderson Jr. 1971, Menard 1986, Le Grand 1988, Stewart 1990, Sullivan 1991].

If oceanic crust is spreading out from the midocean spreading ridges, where does it go? Somehow, the old crust must vanish to make room for the new. At least, that is a requirement on a world of constant size, for the world cannot otherwise hold both the old crust and the continually produced new crust. The key to making the present version of Plate Tectonics work is in letting old oceanic crust sink, as a slab, below the surrounding continents. This was originally proposed to explain the absence of any old oceanic crust, older than 200 million years in the Atlantic, where seafloor spreading is most clearly shown. It was thought that the new crust formed at spreading ridges pushed the old crust into the mantle beneath the continents. However, it was found that the newly produced oceanic crust was actually being pulled away from the spreading ridges, and so the theory developed that the slabs were sinking at the subduction zone and pulling the crust behind them. The material of the subducted slabs becomes assimilated in the mantle and is recycled in giant convection cells in the solid mantle. The viscosity that permits this flow to occur in solid rock is estimated from the dynamics required to let the slabs sink by the flow of the mantle rocks. That is, subduction is assumed to occur, and the viscosity required to permit that is estimated, and then used to show that the mantle has a viscosity that will allow subduction. In rebuttal, Herndon has argued against the development of thermal convection cells in the mantle on the basis of heat transport [Herndon 2008, Herndon 2010].

The discovery of the mid-Atlantic ridge, a volcanic spreading ridge running the length of the Atlantic Ocean and flanked on both sides by a symmetric array of stripes of rock with alternating magnetic polarities was one of the key points in the final acceptance of the dispersion of the continents, Continental Drift as Wegener had proposed it. These rocks had developed their magnetization as they had been extruded from below the Atlantic Ocean floor in response to the widening of the Atlantic basin. As a spreading ridge produced more sea floor, and as the continental plates moved apart, oceanic crustal plates were sinking into the mantle as the continental plates overrode them. The descent of the oceanic crust below the continents assists in the movement of the continents. This plate subduction is an essential part of the theory, as it explains why there is no seafloor older than 200 million years. The theory of Plate Tectonics says that large crustal plates are moving about, generally away from the mid-ocean spreading ridges, and the oceanic plates get driven down under the continental plates, but somehow drag them along so that mountains form. That would seem to be about as physically unreasonable as drifting granitebergs. How can convection currents in the solid mantle be transferred across the "slushy" asthenosphere to provide the driving force for movement of the plates?

The theory of Plate Tectonics has become the pre-eminent theory of the development of the Earth's crustal arrangement, and the foundation of modern geophysics. Perhaps because it resolved a problem of such long standing, the theory was enthusiastically applied, and has generally been so successful, that there has appeared to be no need for alternative theories. Plate Tectonics was a premature success, and pre-empted the consideration of its ramifications and alternative theories. How was it that the spreading ridges were not seen to be giant rips in Earth's crust where it was being torn by expansion?

In this theory, the crust of the Earth consists of several, perhaps 20, crustal plates, continental and oceanic, that gradually move across the face of the planet, driven by convection currents in the mantle and the subduction of cool oceanic crustal plates into the mantle. A layer called the asthenosphere, composed of a semi-slurry of not quite molten rocks, provides the separation between crust and mantle that facilitates the motion of the crustal plates. The uplift of the Himalayas by the collision of the Indian plate is often shown as a great success of Plate Tectonic theory. The force and energy required to lift great mountain ranges are extreme. Can that force be transmitted through the crust/asthenosphere/ mantle? I think these concerns also apply to the Sierra Nevada, the Rockies, the Andes, and the Alps. There is evidence for elevation of a part of the Alps from a depth of 300-400 kilometers [Dobrzhinetskaya, Green, and Su Wang 1996]. That seems hard to imagine for plate tectonics. After the end of the Permian Period, the same land animals existed in Asia and in India. This is strong evidence that India wasn't going to crash into Asia some 200 million years later to form the Himalayas. That shows that the Himalayas arose from within.

The subduction zones can be deduced from the occurrence of earthquakes at active continental margins, such as the Ring of Fire, and seismic interpretation for the waves from distant earthquakes that pass through these zones. The seismology cannot tell us if the slabs are sinking or being pushed, or if something else is happening, just that some structure exists that is a source of earthquakes, and is different from the surrounding crust. Strangely, the spreading ridge along the eastern part of the North Pacific Ocean passes under the western coast of North America. The volcanic energy of the spreading ridge is expressed in the volcanoes of the Cascade Range, on continental crust.

Antarctica is surrounded by active oceanic ridges, so new crust is continuously spreading towards it, but is not being subducted. So, the ocean floor must be spreading away from Antarctica, but it is also assumed to be spreading away from Africa, and so the ocean floor is trapped between a large continent and a very large continent, with nowhere to go. If all the continents are moving away from Antarctica, the Arctic Ocean basin should be compressed and being made smaller. However, the Arctic Ocean is under tension, and is being pulled apart, getting larger. Similarly, the continents are converging into the Pacific Ocean basin, making that basin smaller, but the Pacific Rim, surrounding the basin, is getting larger.

An alternative theory (to Plate Tectonics) that answers these objections proposes that the Earth was originally much smaller than at present and the expansion to its current size has torn the continents apart and produced the spreading tears and trenches of the ocean floor [Carey 1976, Carey 1988]. This theory has suffered from the same obstacle that thwarted early adoption of the Continental Drift theory: no reasonable mechanism could be thought of that would provide the driving force for the expansion of the Earth, just as none could be thought of to propel the continents through the oceanic crust. Carey has proposed that expansion is driven by creation of matter at the center of the Earth, and other "gravitational null points", such as between galaxies, but this is not a physically acceptable theory. That is an objection to Carey's theory, but not to his observations.

As in Wegener's case for Continental Drift, a mechanism was proposed that is considered physically impossible. While matter can be created from energy (and is in laboratory experiments), it is created in equal amounts with antimatter, with which it easily and completely annihilates, returning to energy in the form of gamma radiation. However, Carey's creation of matter at gravitational null points is an invented explanation which violates one of the fundamental laws of modern physics, the conservation of mass/energy. He may be right, "laws" have been found to be wrong in the past, but right now there seems to be no basis for it other than wanting it to be so. Creation of matter alone (without a balancing amount of antimatter) would be a violation of very well established laws of physics. Carey proposed the continuous creation of matter at the center of the Earth as a means to maintain the surface gravity nearly constant, since the gravitational force at the surface would be related to the inverse square of the radius of the Earth. Carey invented a creation of matter wherever gravitational attraction is zero, like the center of the Earth and other planets, but this is contrary to current physics, we don't see it for our Moon, and I think it is no more realistic than Wegener's drifting graniteberg continents. Carey felt it important to maintain a reasonably constant value of surface gravity, and so did not want to keep the mass of the Earth constant from its formation. If the Earth expands, but its mass remains constant, the surface gravity will decrease. Like the present warmth of the Sun, our surface gravity seems just right. Gravity more intense in the past should have caused problems, or at least differences, that would be noticeable. Gravity on the surface of a smaller Earth, with the same mass, would be greater. That poses a problem for the dinosaurs and other huge ancient beasts, who were already heavier than should have been comfortable.

The rejection of the theory of Earth Expansion is primarily based on two reasons:

1. There is no known mechanism to make the Earth expand. This is the most fatal objection, although it is based on an erroneous principal, that if we don't know how something can happen, it can't happen. Sam Carey proposed creation of matter at "gravitational null points", but that goes against a fundamental law of physics. Someone else proposed that cosmic dust accumulates, but that doesn't add up. Another proposal was based on a declining value for G, but that doesn't seem to be and probably can't work either.

2. The most practical objection is simply, we already have Plate Tectonics, why do we need another theory? If we have a theory that seems to work pretty well, there is little motivation to learn and consider another theory, particularly if it is not mainstream.

Carey and a few others have presented voluminous amounts of not-quite-convincing evidence for this theory. Rejection of Earth Expansion (or Expansion Tectonics) appears to follow the pattern of rejection of Continental Drift. Since no acceptable mechanism for driving the expansion could be proposed, the interpretation that the Earth was expanding could not be accepted. The underlying problem is that if we can't imagine how a process works, we cannot accept evidence that indicates its workings.

For Earth expansion, there must be a long term driving force to supply the energy to expand, from the beginning of the solid Earth, as we know it. So, I have supposed that there is some other source of energy that forces the Earth to expand. I looked at what inherent mechanisms could account for the expansion of the Earth. Expanding from a radius of 3,000 kilometers to 6,371 kilometers (as a rough estimate) produces a volume increase of a factor of 9.57 I considered the increase in each possible effect that I could think of. The only one that looks reasonable is the decrease in density, by expansion from a super-dense gas-giant planet to what we now know and love. Only release of primordial compressional energy holds a possibility of being right: others require far too extreme variation in the parameters to be likely mechanisms. (It should be noted that Herndon has also proposed expansion of this sort [Herndon 2005, Herndon 2008].)

If the Earth had initially formed as a gas-giant planet, as is observed for the distant extra-solar planets, its inner regions would have been super-compressed compared to what we imagine when examining our present Earth. If the Earth formed by the gentle accretion of cold gas and sticky dust, the dense phases of minerals would have literally been frozen in place by the extremely low internal temperature resulting from accumulation of cold material in the cold of space, before the Sun ignited.

If the giant-planet model of formation is right, then the Earth should have formed as a much larger planet, perhaps 500 times its present mass. Its rocky core would have been greatly compressed. As its atmosphere and other volatile material were blown away by an increasingly active and hotter Sun, the pressure would have decreased, and the Earth would have expanded. Energy is needed to push one phase of rock, iron, or other minerals into another phase, even a less dense phase that is more stable at the lower pressure, and there must be an accommodation of the greater volume required. (Note that diamonds are a high-pressure phase of carbon, not stable at our low surface pressure, but "Diamonds are forever.") So, some excess compression would be locked into the Earth even while it lost most of its "weight". Major impacts, asteroids or comets, would cause jumps in expansion and we seem to see this in some of the ancient impacts. This connection is strongly suggested by the comparison of the cratering record on Earth and the drops in sea level seen in the Exxon Research curves, shown later.

Thus, an alternative mechanism for expansion of the Earth that does not seem to have insurmountable obstacles is based on the initial formation of the Earth (and other planets in the Solar System) as a gas-giant planet, like Jupiter, and like the giant planets being found around distant stars, and its subsequent episodic expansion is a result of regional decompression, induced primarily by major impacts. This possibility will be discussed in more detail when we explore how to form a Solar System. At this point we will consider how Earth Expansion fits some other observations. We will reconsider Earth Expansion in more detail later, in some other instances.

The inner (solid) core of the Earth may now consist of a phase that was stable at the very high pressure (and relatively low temperature) of the protoplanet, while the lower mantle may consist of phases that were stable at the decreasing (but still great) pressures that resulted as the molten rock boiled away from the surface. The outer (fluid) core may be the relaxed phase of the original supercompressed core. Our present atmospheres and oceans were then produced by generally continuous outgassing of inner regions of the planet, and by accretion of rock and water impacting from space.

The structure of the Earth exists in a tenuous stability with a touch of positive feedback. If the density of material in a shell of the Earth were to be reduced, and the material expand, the rock and water outside of this shell would move farther from the center of the Earth, and the force of gravity on this matter would diminish. This would reduce the pressure on the rock lying beneath, and in a region where the mineralogic structure changes to a phase of lower density, the initial expansion is amplified and further expansion ensues. This process may stop only because the reduction of gravitational pressure is a weak effect of the expansion. (The expansions seem to be triggered by major impacts, causing expansion of still dense material, where rarefaction waves from the impact are strong enough to relieve the intrinsic compression, and permit local expansion. Relaxation of the superdense phases requires energy and expansion room for the lattices to expand. The impact supplies both.) To make the Earth expand, an expansive force is created within the core or mantle, from the release of precompression of the mineral lattices during formation of the Earth. A major impact from an asteroid, for example, causes an abrupt increase in the expansive force by triggering the release of compression in a small region of the Earth. This force creates tension in the surface as the Earth expands. Although the expansion is limited to a small local region, and causes localized uplift above it, such as the Himalaya Plateau, Earth's large mass and gravity force it to maintain, or resume, a spherical form. This forces great movements of crustal material, and major redistributions of ocean water. The tension stretches the basins but also results in tears in the mantle and the crust. A tear in the mantle, from within, provides an easy path outward for magma and produces the midocean spreading ridges. A tear in the crust, at the surface, produces the deep trenches, such as the Peru-Chile trench. This tear allows the continental block to rotate, downward where the magma is flowing away into the mid-ocean ridge, appearing to make subduction zones and produce backarc basins, and upward where the trench is tearing apart, making the Andes in this case. Stretching of the oceanic crust in response to expansion of the Earth produces the globe-encircling tears that make the spreading ridges. Stretching of the solidified oceanic crust causes the poorly understood transform faults that proliferate as the continental blocks move apart [Gerya 2010].

Whenever an impact jolts the Earth, some of the deep minerals are triggered into transition to a less dense phase, and the Earth expands. It does not expand uniformly, but only in the inner core and in the lower mantle, and not in a spherically uniform manner. This forces blocks of lithosphere "up", which leads to the appearance of crustal plate subduction at the margins of continents and ocean basins, as the continental edge falls over onto the oceanic crust. No subduction of an oceanic "plate" is required, it is just covered over by the continental crust. The expanded and uplifted continental crust overflows onto the oceanic crust, leaving what appears to be slab subduction, but what might be better called "superduction". I think that is a significantly different mode of expansion than is generally pictured. The Earth is not expanding simply like a balloon, so that only points on the surface move farther apart. It is expanding unevenly, as portions of its mantle and crust are forced outward, perhaps expanding themselves as well, as portions move into lower gravitational pressure. This causes the continental (lighter) crust to override the (denser) oceanic crust, forming what are currently called subduction zones, because we have supposed that the oceanic crust was sinking below the continental crust. As continental blocks are forced up by the internal expansion of the inner core and the lower mantle, the edges of the continents fall down onto the oceanic crust, like rubble piles built past the angle of repose. Paradoxically, this gives the appearance that the oceanic crust is diving beneath the continental crust, because it is buried by the continental material. The oceanic crust is indeed being buried beneath the continents, as subduction proposes, but by the continents overriding the ocean floor, not by the oceanic slabs sinking into the mantle. On an expanding Earth, it is the continental crust that overrides the oceanic crust, but I think the structure is the same, and the relative motions are the same. Only the absolute motion, and the theoretical interpretation, are different. The theory of Plate Tectonics has provided a means for projecting back in time the positions of the continents, assuming a constant-size Earth. This has been computerized by Scotese and colleagues [Scotese 1997, Eldredge, Walsh, and Scotese 1997], and a plot of the positions calculated for 745 million years ago is shown in Figure 25.

It may be noted that this grouping of the continents results in a great ocean spanning, without interruption, nearly one-half of the world. This ocean, assumed to have existed, is assumed to have disappeared by subduction. There is no evidence of the existence of this ocean, ever. On a smaller Earth, shown in the lower half of the figure, the excess ocean is much smaller. (There is still room for improvement. Consider how well the continents would intermesh if wrapped around at the equator.) That is the Earth as proposed by Earth Expansion.

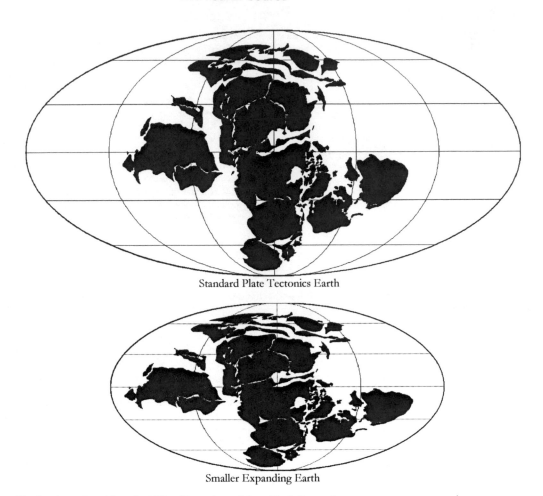

Standard Plate Tectonics Earth

Smaller Expanding Earth

Figure 25. Continental positions from Plate Tectonics and from Earth Expansion.
Positions of the continents calculated from Plate Tectonics [Scotese 1997], and placed on a smaller Earth according to Earth
Expansion.

By re-mapping the globe of the Earth, based on ages established for the observable ocean floors, James Maxlow
determined that the early Earth had a diameter equal to 0.27 of the present diameter [Maxlow 2005]. (That estimate
produces an average density that is not reasonable, so I will choose a somewhat larger Earth to start with, later.) His
analysis shows that the current expansion rate is 22 millimeters per year. He quotes several independent estimates of this
expansion rate:

Z. Garfunkel (1975)	20 mm/year
J. Steiner (1977)	20 mm/year
B. Parsons (1982)	23 mm/year
V. F. Blinov (1983)	19.1 mm/year.

It might be hoped that some sort of active measurements would be capable of determining the present expansion
rate of the Earth, if it really is expanding. One method that would seem to hold promise is the Very Long Baseline
Interferometry (VLBI) reported by Robaudo and Harrison [Robaudo and Harrison 1993]. However, the intent of that
study was to determine Plate Tectonic drift rates (horizontally, on a constant size Earth), and so the radial (vertical)
component of displacement was deliberately ignored:

"A further constraint on our solution was that the stations were not allowed to have any up-down
motion. A solution using baseline and transverse rates and allowing the stations to have three
independent velocities gave an RMS value of up-down motion over 18 mm/yr. This is extremely high
when it is realized that areas of maximum uplift due to deglaciation are moving at only 10 mm/yr or

less. We must expect that most VLBI stations will have up-down motions of only a few mm/yr. It therefore seems reasonable to restrict the vertical motion to be zero, because this is closer to the true situation than an average motion of 18 mm/yr." [pages 53-54]

It appears that the analysis was done using the method of Least Squares, with the basic assumption that there was no radial expansion. That method finds the values of specified parameters that best fit the model, whatever that model might be. If one of the three components of motion (radial/vertical, up-down, in this case) is omitted from the model, the variation in this component is simply accommodated in the values of the other two components, with an increase in the size of the residuals (differences between observed and calculated values). Considerable editing (selective rejection) of the data was done by the researchers and yet:

"However it was found that there were too many large residuals. For instance, out of the 512 available observed rates there were 40 weighted residuals greater than 2.5 when for a normal distribution there should only be about 6." [page 55]

Even after rejecting observations that differed from the Plate Tectonics model, there were nearly 7 times as many discrepant points as expected, relative to a constant size Earth. Adherence to Plate Tectonics led to overlooking strong signals that the Earth is expanding.

It should be noted that the unintentional result for radial expansion, 18 millimeters per year, is in good agreement with the expansion rate determined by Maxlow and those he quoted, 19.1 to 23 millimeters per year.

One objection to an expanding Earth is that the oceans should get shallower as the water is spread over greater surface area. That ignores the continual release of water from inside the Earth by volcanoes. And to turn the problem around, the standard story dismisses the water released by submarine volcanoes, about ten times as much volcanic activity as on land, supposing that it is entirely recirculated ocean water rather than new ("juvenile") water from the magma. Dry land volcanoes release water, but submarine volcanoes don't? The reasoning was that the oceans would overflow with that much water. On a constant size Earth the oceans would indeed overflow as more water is released, but not on an expanding Earth, constantly (or intermittently) accommodating additional water by increases in the size of the ocean basins, leading to abrupt drops in sealevel, followed by gradual rises. That is what the Exxon Research sealevel curves show.

A study of ocean surface salinity [Haug et al. 2001] has shown that low-salinity water flows from above the East Pacific Rise, one of the most active submarine volcanic areas, near the northwest corner of South America. This flow rivals that of the Orinoco and Amazon Rivers, into the Atlantic. There are no large rivers flowing from South America into the Pacific that could be the source of this nearly fresh water. This is new water from the magma released under the sea.

The Earth expands gradually, and episodically, as shown by the gradual encroachment of the ocean (transgression) and abrupt retreat (regression) in the geological record of rock strata [Haq, Hardenbol, and Vail 1987, Haq and Schutter 2008]. The abrupt drops in sea level show the same pacing as major impacts.

An impact creates a shockwave that unsticks a portion of superdense rock remaining from the original formation of the Earth as a giant planet. The Earth expands, unevenly, but retaining perforce a spherical form, oceans slosh as the gravitational equipotential surface changes shape and size, carbon dioxide is released from seawater by the reduction in atmospheric pressure, cap carbonates are precipitated (with special carbon isotopic compositions from the Giant Dipole Resonance reaction, $^{16}O(\gamma,\alpha)^{12}C$, from the gamma flash of the detonation that propels fragments of a planet or satellite helter-skelter into the Solar System), and whole orders of life collapse from the inability to survive in the new world. Surviving life forms, and new forms originated by radiation mutation, change by the "survival of the fittest" in response to the new conditions, and become the new kings of the hill.

Opponents of the theory of Earth Expansion say that it "denies the existence of subduction zones". This seems to be a disagreement over the mechanism and process of producing these zones. A sinking slab is very unlikely, as unlikely as Wegener's drifting granitebergs. The gravitational spreading of a continental block, forced up beyond its visible means of support, would produce what has been labeled as subduction. The overflow of expanding continental crust onto the oceanic crust can be demonstrated in any sandbox.

From many different reports that I have read, I think that the theory of Earth Expansion provides better explanations than does the Plate Tectonics theory:

1. Better fit of continents along the continental shelf, on a smaller globe,
2. Bilateral symmetry across ocean spreading ridges, from balanced tensional forces, due to expansion,
3. Definable Euler poles and angles for transform faults and ocean spreading,
4. Burial of oceanic crust at continental margins, what we otherwise call subduction,
5. No sludgy ocean sediments scraped off the oceanic crust as that slab slides beneath a continental slab,

6. The extinction of the dinosaurs by a reduction in oxygen partial pressure when the Earth expanded in response to four major impacts just before the K/T, thinning the atmosphere, and also decreasing the atmospheric buoyancy, so heavy dinosaurs collapsed,

7. The surge in mammal size just after the K/T, due to the reduced oxygen partial pressure,

8. The nature of Peter Vail's sealevel curves, showing a filling of ocean basins with a constant rate of increase of water volume (outgassing or comets) punctuated by abrupt drops when the Earth expanded,

9. The latitudinal drift of continents, from asymmetric expansion of a spherical Earth,

10. The abrupt change in direction of hotspot tracks, in coordination, when the center (or axis) of expansion changed.

Paleomagnetic data also provides evidence on the ancient radius of the Earth, but it is contradictory. One interpretation is that the Earth has contracted from 102% at 400 million years ago. However, van Hilten concluded (in 1963) that the paleomagnetic evidence "seems to indicate a noteworthy increase in the Earth's radius since the Carboniferous, the rate of which agrees roughly with the hypotheses of Carey and Heezen." I think, considering asymmetric expansion (while always keeping a spherical shape) and the possible overflowing of continental blocks, with most of the surface growth in the ocean basins, the choice of paleomagnetic latitude sites might seriously affect the outcome.

Joe Kirschvink at Caltech has recently proposed that a major effect in the change of life at the Cambrian was due to a slip of the Earth's crust [Kirschvink, Ripperdan, and Evans 1997]. A change in the moment of inertia of the Earth can cause a slip of the crust, around the fluid outer core, to produce a more stable rotation, with less energy. The change in the moment of inertia can be produced by an asymmetric expansion, so that crust and mantle slipped around the fluid outer core.

There is evidence for a great disturbance of the Earth in the recent past. ("Great" and "recent" are usefully vague.) The Moon's orbit, which should be aligned with the Earth's equator by tidal effects, is tilted by 5.15 degrees [Lodders and Fegley Jr. 1988]. This is one of the greatest misalignments among the more normal of the satellites in the Solar System. Also, the Earth's inner core, separated from the outer shell of rock by the low-viscosity fluid outer core, rotates more rapidly than the rocky shell does.

These growing pains can be accommodated by an expanding Earth, but not by an Earth of constant size and with slab subduction. With sufficient water, the early Earth should have been completely covered by an ocean. As the Earth expanded, the ocean would have been spread into shallow epicontinental seas, and then ocean basins would have formed along the continental tears, and dry land would have gradually emerged. (Perhaps that is why it took so long for life to occupy the land. There was no land to occupy.) As the Earth expanded, the ocean would have receded from the land as the ocean basins grew larger, perhaps only filling the basins outlined by the continental slope. Perhaps the oceans formed and defined this boundary as sea cliffs. There is a hierarchy of faint "beaches" at various depths on the ocean bottom, showing shorelines at various stages of expansion. This is also shown on the planet Mars, for only two expansion events.

This ocean recession could have been irregular and intermittent, and complicated by local subsidence and uplift. A further complication may have been episodic additions of significant amounts of water from volcanism and from cometary influx. Louis Frank at the University of Idaho thinks that all the oceanic water could have come to Earth by way of the small comets he discovered, and is still coming [Frank 1990]. Others think that the oceans were all delivered early in the history of the Earth, while still others recognize that large amounts of water are released by volcanoes. More might be recognized except that it is claimed that much of the volcanic water is recycled. Why? Because the oceans would otherwise overflow.

Global sealevels have been derived by Peter Vail and his colleagues at Exxon Research. These show abrupt drops in sealevel, marine regressions, followed by gradual rises, causing marine transgressions. Abrupt expansion of the Earth (by some small amount) would cause an equally abrupt marine regression, a drop in sealevel. If the Earth then stayed the same size for a while, continued discharge of water from the interior, by volcanic eruptions, would cause the sealevel to rise. A constant increase in volume of ocean water (a steady rate of volcanic discharge) into ocean basins with sloping sides would result in a sealevel increase that proceeds as the square root of time. That appears to be what the Exxon Research curves show. This nearly asymptotic increase in sealevel (and that asymptotic appearance would result from the flooding of continental basins) is interrupted by a marine regression, is resumed again, interrupted again, and so on. The abrupt drops in sealevel correspond to abrupt, but limited, expansions of the Earth. These expansions occur when major meteoroidal/asteroidal/cometary/bolide impacts shock the Earth so that some of the dense phases of the core and/or mantle are able to release some of their residual compressive strain and expand. The energy from an impact triggers partial phase changes. Peter Vail's sealevel curve shows that expansion was episodic, and I think the episodes or events might correspond in time with major impacts. The Earth expands episodically in response to major impacts and continuously in slow, silent earthquakes. That gradual expansion may be what is seen as crustal rebound following deglaciation. Vail's sealevel curves are not generally accepted, because there is no accepted mechanism to make sealevel rise and fall, in a global manner. Our standard assumption is that the ocean water has always been here, in constant volume. However, volcanoes are constantly releasing water, and so the oceans should eventually overflow. Unless the Earth expands.

In the Superdense Expanding Earth model, the Earth, originally formed as a massive gas-giant planet, was reduced to a supercompressed kernel by the loss of its thick insulating atmosphere and evaporation of the molten silica ocean, and has been expanding, intermittently, episodically, and geologically continuously.

The "onlap" curves showing the encroachment (transgression) of the sea onto the land, as sealevel rose, and withdrawal (regression) from the land, as sealevel fell, show the shape to be expected from a steady supply of water increasing the volume of the oceans in basins with sloping boundaries, interrupted by abrupt drops. In this book, the abrupt drops are interpreted as resulting from impact-induced expansion of the Earth, as shown in Figure 26. This figure shows the observed impact cratering record [taken from http://www.unb.ca/passc/ImpactDatabase/Age.html] and the sequence of sealevel drops. Adjusting the cratering record for the ratio of total area to land area, making the assumption that few craters from impacts in the ocean have been discovered, overlies the sealevel sequence very well, until about 200 million years ago.

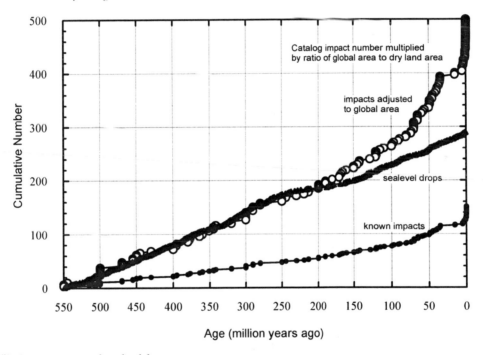

Figure 26. Impact craters and sea level drops.
A comparison of the accumulation of sealevel drops and expected impacts (adjusted to the whole global Earth from the known impact crater database). In any period of time, from the beginning of the Cambrian (550 million years ago) to about 200 million years ago, about the beginning of the Triassic, there are approximately equal numbers of expected impacts and observed sealevel drops. After that time, the impacts seem to be less effective in causing Earth expansion and sealevel drops.

Episodes of significant expansion occur in response to major meteorite impacts. As shown by the sealevel curves of Peter Vail and colleagues [Haq, Hardenbol, and Vail 1987, Haq and Schutter 2008], these expansions are individually on the order of a hundred meters. The ocean basin is refilled with volcanic water until the next jolt raises the land and lowers the water. While it has been generally assumed that all the water passing through oceanic volcanoes (submarine volcanoes) is recycled water, it does not seem reasonable that those volcanoes should be so different from land volcanoes, which emit considerable new water. There is some evidence that fresh water, not recycled water, is produced by undersea volcanoes. Above the East Pacific Rise, at the northwest corner of South America, there is a pool of low-salinity water at the surface [Haug et al. 2001] No large rivers drain into this pool, but it exceeds the size of the outflow pool of the Amazon River in the Atlantic. The association of low-salinity water with the volcanic activity of the East Pacific Rise suggests that this is juvenile water released from the mantle.

In addition to the episodic expansion, the nearly continuous expansion occurs in association with earthquakes as shown by the sealevel drops found in Sumatra [Sieh *et al.* 2008]. The Sumatra data show abrupt drops (generally and conventionally interpreted as local rises of the land) on the order of a hundred centimeters, about a hundred times less than the impact-induced drops.

It appears likely that the great Tohoku-Oki earthquake and tsunami of March 11, 2011, in Japan, was caused by continental superduction, of the far edge of the Okhotsk continental plate surging onto the Pacific oceanic plate. (Even though Plate Tectonics may be wrong, the plate concept and names are useful.) Along the shore, the ground surface was torn apart by tension cracks, and the surface subsided by about 3 feet (1 meter), rendering the tsunami-barrier sea-walls less effective than expected. A multitude of after shocks clustered along the coastline, out to the initiating quake and beyond, as the continental crust re-adjusted to its new configuration. This produced a massive scab of fractured rock, instead of a nearly linear delineation of a fault line.

This superduction by overflow of continental crust is shown clearly for the Tohoku-Oki earthquake [Heki 2011, Simons *et al.* 2011.] and also for the 2010 earthquake in Central Chile [Vigny *et al.* 2011]. Ground movements were determined by use of the Global Positioning System and so were referred to the location of the center of mass of the Earth. This avoids the ambiguity of under and down or out and over that is produced by the interpretation of geological structure determined by analysis of seismic signals.

The ground movements for the Tohoku-Oki earthquake are shown in Figure 27, and show a pronounced flow of the continental crust from the North American plate (containing the islands of Japan) onto and over the Pacific plate. This is also shown by sea-floor observations with surges as great as 24 meters [Sato *et al.* 2011].

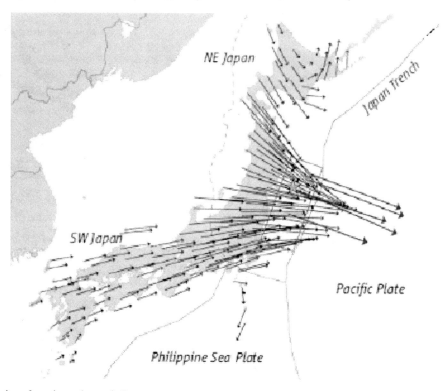

Figure 27. Slumping of continental crust in Japan.
Ground movements of northern Japan from the 2011 Tohoku-Oki earthquake [for details see Heki2011, and Simons *et al.* 2011]. The star indicates the epicenter of the main quake. This shows the superduction of continental crust onto the oceanic crust.

The overall trend of an expanding Earth is hidden by the continuous increase in ocean water due to the outgassing of the mantle by volcanoes.

Impacts are rare, regressions are rare, flood basalts are rare, and mass extinctions are rare. A time correlation, and global opposition between impact and eruption, ought to show a nice correspondence of these events.

Migration in latitude of the positions of landmasses as observed, can be accommodated by an off-center expansion. If a surface feature is embedded in the crust, and the subcrust under and around it expands circumferentially as the Earth expands radially, there will be a change in latitude. Assuming an impact in the northern hemisphere, and no radial expansion in the northward direction but a radial expansion southward, continental points will migrate to the north in proportion to their distance from the poles. Obviously, the land at each pole will not move, but land at lesser latitudes will move north in all cases. Once a dichotomy between land and sea was established, it would be promoted by subsequent impacts. Shockwaves, triggering relaxation, would be produced more intensely by impacts on land (predominately in the north) than in the ocean (predominately in the south), causing expansion to occur preferentially in the oceanic regions, from impacts on land.

But can we believe an Earth less than half the present size only 65-80 million years ago? If it had the same atmosphere, the air would have been spread over a surface area less than 1/4 the present and pulled down with a force 4 times as great. Atmospheric pressure 16 times the present would help explain the giant dragonflies with 3-foot wingspans and bodies too thick for diffusion of oxygen at the present pressure. It would have helped the pterosaurs to fly, rather than just climb trees and glide down. But how could dinosaurs stand up if they weighed four times as much because of the increase in gravity? The planet Venus, much the same size as the Earth, has an atmosphere roughly a hundred times as massive as ours. On a smaller Earth, that much air would have been nearly as dense as water, at the surface of our planet. The heaviest dinosaurs, the size of whales, were originally thought to live in swamps for buoyant support, but the evidence showed that they lived on dry land. A dense atmosphere would have supported them just as the ocean now supports giant whales. The greater partial pressure of oxygen would have given the dinosaurs the ability to live like warm-blooded mammals and birds now do, but would have suppressed the size of the early mammals. Too much oxygen makes mammals overheat fatally.

The rotation rate of the Earth has been considered to have been slowed by tidal friction with the Moon, but this energy loss is dependent on the configuration of the seas and may not be estimated well. Three estimates are used here, although there may be several more. About 4.25 billion years ago, a rotation rate of 5 hours per day was estimated [Touma, Wisdom, and Kuhn 1998]. If there were no tidal effect, and the distribution of material in the Earth remained relatively the same, that would correspond to a radius of 2,905 kilometers, compared to the present value of 6,371 kilometers. An estimate of 18.3 hours per day, for 900 million years ago [Sinnott. 1996] gives a radius of 5,548 kilometers. An estimate of 14 hours per day at about 0.6 billion years ago [Jenkins, Marshall, and Kuhn 1993] gives a radius of 4,867 kilometers.

Summary of Part II

Production and dispersal of heavy elements by a supernova detonation of a star is described. The subsequent chemically pure aggregation of uranium by crystallization and electrostatic attraction is discussed. Release of a cloud of debris from a supernova leads to the formation of new stars and planets. Uranium-235 is produced with sufficient abundance to form breeder reactors, converting uranium-238 into plutonium-239, and back to uranium-235. Because of its inherent density, uranium gravitates to the center of a forming planet. Uranium hydride could form natural nuclear fission reactors with less than 10 kilograms in space, or less than 1 kilogram as compressed at the center of Jupiter, or less than 0.01 kilogram (10 grams) as compressed even more at the center of the Sun.

Uranium accumulated in the Earth and formed natural nuclear fission reactors in the solid inner core and at the core-mantle boundary. Nuclear reactor calculations show that small amounts of uranium, tens to hundreds of kilograms, are sufficient to form reactors. Phase transitions of uranium and uranium oxide in iron can provide a reprocessing function to keep reactors operating as breeder reactors. Similarly small reactors of uranium oxide rise and fall in the fluid outer core.

Reactors deeply confined in a planet or moon can detonate. Detonation of a planet between Mars and Jupiter produced the asteroids. Many other effects can be traced to planetary detonations.

The structure of the Earth, as shells of different materials, is described. The Earth has a solid core at the center, assumed to be iron and nickel, surrounded by a fluid outer core considered to be of about the same composition, and separated from the lower mantle by the D" layer. The lower mantle transforms into the upper mantle which underlies the asthenosphere, which is considered to form a very weak layer between the mantle and the crust. The outer surface is the crust, oceanic basalt or continental granite, with the oceans and atmosphere above.

Continental Drift, Plate Tectonics (with subduction), and Earth Expansion (with superduction) are introduced. Lord Kelvin's calculation of the age of a cooling Earth fell far short of other estimates, and heat from radioactivity does not solve this problem.

The Sun was too small (faint) in the beginning to keep water from freezing on the surface of the Earth, and a Snowball Earth would not have thawed yet. Additional sources of heat were insufficient to prevent freezing.

A calculational attempt to put enough uranium in the Earth's core (where it is assumed to be absent) showed that there should have been sufficient uranium for natural nuclear fission reactors to develop. These reactors could have provided the supplemental heat to keep the Earth warm when the Sun could not.

Volcanoes are discussed as a means for transporting heat from inside the Earth to the surface. Various different types are described, with mantle plume volcanoes and flood basalt eruptions being most significant in this book. The function of nuclear reactors in powering the mantle plume volcanoes, and impact-induced expansion of the Earth in causing flood basalt eruptions are described.

The El Niño/La Nina climate variations and longer term climate changes are discussed and it is shown that these effects may be driven by antineutrinos from supernova detonations. Helium-3 (as would be produced by natural nuclear fission reactors) is shown to be associated with the warmer ocean water that causes El Niño.

Helium-3 in volcanoes is discussed, and the finding of hydrogen-3 (tritium) in volcanic gas is shown. Isotopic variations in strontium, possibly caused by absorption of fission neutrons, are related to mass extinctions.

Unexplained radiogenic halos in minerals are discussed.

The geomagnetic field is produced by dynamo action in the Earth, both biased by and powered by natural nuclear fission reactors. Changes in the locations and power of these reactors in the outer core cause geomagnetic reversals.

The Earth is now warm and relatively free of ice. It has a magnetic field. Strontium ratios are not increasing as steeply as they could. The climate system responds to the distant signals of supernovae. It is likely that the fission reactors that are proposed to have provided the life-giving warmth to the early Earth are still in operation.

A self-sustaining chain-reaction in uranium is not possible for uranium-235 concentrations in uranium of less than 5%, and our knowledge of the surface uranium composition indicates that the enrichment should be only 0.72% at present. However, after "billions upon billions" of years, the isotopic mixture deep inside the Earth is likely to be far different from that of the uranium and thorium found at the surface of the Earth. Evidence for the current state of the Earth's reactor must be sought elsewhere. Such evidence may lie in the detection of antineutrinos produced as a result of the fission process.

Theories of ancient global glaciation, a Snowball Earth, are discussed and are shown to be based on incorrect interpretation of iridium concentrations and sedimentary debris and dropstones. This and other considerations lead to the conclusion that all global glaciations, the Ice Ages, have been misinterpretations. The various associated effects resulted from radiation from a nearby astronomical detonation and from the impacts of the fragments produced by the detonation. Banded Iron Formations are shown to have been produced by the iron brought to the Earth in those fragments from space. Variations in chromium isotopes are shown to be due to neutron absorption in the detonation. The phosphorus associated with iron deposits is proposed to have come in impactors. The observations associated with oceanic anoxic events are shown to be related to the effects of a nearby astronomical detonation.

Continental Drift, Plate Tectonics, and Earth Expansion are further discussed, with a mechanism of decompression from formation as a giant planet proposed as the driving force that makes the Earth expand. The compression is partially and regionally released by the shockwaves of major impacts. A comparison of the position of continents in the distant past, as calculated for a constant-size Earth and for a smaller, expanding Earth shows a reduction in the size of the undiscoverable ocean that completely occupied a major portion of the surface. An inadvertent measurement of the current rate of expansion, by Very Long Baseline Interferometry, is shown to agree with several independent estimates based on Earth Expansion.

Ocean water volume is continually increased by the release of water from the mantle by volcanoes. The pattern of sealevel rise and drops shows a close relation to the pattern of major impacts, which are proposed to be the triggering events that cause expansion.

III. THE SOLAR SYSTEM
The Nature and Possible Formation of Our Solar System

The harmony of the world is made manifest
in form and number.
Sir D'Arcy Wentworth Thompson
"On Growth and Form", 1917

Long, long ago, in our thoughts, the Earth was the center of the Universe. The Sun rose in the East, and set at the end of each day, in the West. This seemed quite reasonable when the entire Earth was a flat Earth [Gribben 1991], as it seems to be when we simply look around about us. The Moon, and the tiny stars, and the roving planets also appeared, passed overhead, and vanished on a daily turn. The Moon clearly changed, sometimes full and bright, sometimes a dim sliver of a crescent, sometimes showing herself at night, at other times shyly sharing the sky with the Sun. The planets were points of light, like the stars, but took very complicated paths through the stellar patterns that we call the constellations. These constellations showed at their best at various seasons, each constellation coming into view and then no longer seen, through the course of the year. Certain constellations became identified as the Houses of the Zodiac, which, in early times and still into the present serve as the working material of astrologers.

Astrology was important in our early civilization, and the astrological observations of the planets and stars were carefully recorded. These observations were gradually used to refine our model of the Universe, and that model could then be used to make predictions about Solar and Lunar eclipses, and the various positions of the planets among the stars. As the observations became more precise, comparisons were made with the predictions, and the model was found to be flawed. In particular, changes in the positions of the planets (then limited to Mercury, Venus, Mars, Jupiter, and Saturn) were difficult to understand.

Just as the age of global exploration was opening, the astrological system was beginning to fail. The vernal equinox was off by ten days, full Moons were not predicted correctly, and Easter Day could not be foretold. The need to use stars for navigation on the ocean contributed to the transition from astrology to astronomy.

To correct these deviations, the orbits of the planets were made increasingly complicated, but more precise observations spoiled these adjustments. Finally, a change in viewpoint occurred. Perhaps the Earth was not truly the center of the Universe, perhaps the Sun was. Perhaps the Earth was round, perhaps the Earth rotated.

These ideas had arisen among the ancient Greeks, but had been opposed and discarded by the strong personality of Aristotle, a pupil of Plato, now 23 centuries ago [Pannekoek 1989]. (What a long time to carry an error!) It remained for Copernicus to put the old ideas together in 1512, providing a clear definition of the Solar System and setting the form of our present view of the Universe. Copernicus provided the details in the six books of *"De Revolutionibus"*. This work was published in 1543, and Copernicus died the next day.

Copernicus developed much of his Universe through reason, accepting what he thought was reasonable or natural, rejecting what was not. That was the way of the ancient philosophers, but Copernicus' insight was closer to the truth. Some of his details failed. His thinking was still directed by old ideas, but he introduced a new coherent model of the Universe that has survived to the present.

The success of the Copernican theory comes from its agreement with nature, rather than from a rigorous development. Yet, his theory has been tremendously successful. Most of what has been added since is in the form of objects and effects that Copernicus had no way of knowing. Copernicus did not know what he did not know. We still, and always, do not know what we do not know.

The history of astronomy is well told by Anton Pannekoek, in a manner that has the charm of a Dickens novel, a book to be read for the pleasure of reading, as well as for the understanding [Pannekoek 1989].

The Solar System has been known, in a sense, since the first thinking being saw the Sun. Our knowledge of the Solar System has grown since, and continues to grow, as we make new discoveries, and come to understand the old. Our concepts have changed, from a Sun going over and under a flat Earth that was the center of the Universe, to a system of small planets orbiting an ordinary star, somewhere in a not very special part of space. Early ideas die hard and fade slowly, and our present concept of the Solar System and how it was formed are products of the last few hundred years. The present concept has been built on thoughts of many researchers and philosophers that preceded our modern efforts, and their ideas have shaped ours.

As well as we know at present, the Solar System consists of a single Sun, orbited by 8 or 9 planets, with zones of fragmentary debris, gas, and dust. The uncertainty in the number of planets may seem odd, since we should certainly be able to count that many quite accurately. This uncertainty arises from the peculiar nature of Pluto. An intense search for an unknown planet resulted in the discovery of Pluto, and it was declared a planet before its characteristics were known. Depending on the criteria chosen to characterize a planet, Pluto may or may not be one, and it has recently been relabeled by the International Astronomical Union as a "dwarf planet", along with the larger asteroids. Pluto orbits the Sun somewhat like the other planets, a bit sloppily perhaps; it is shaped like a planet, but very small; it has its own moon, better than some others. It was discovered in a search for a planet, yet surely was not the one that was sought. It almost definitely did not form in the way the other planets formed. The planets have moons, and some of the fragments have moons. (But we have no indication that the moons have moons, and so on.) The whole collection is held together by gravity. The Solar System extends so far in space that a single figure cannot map it. The range of sizes is so great that a single figure cannot show them all.

Subdivision by size gives us some clues as to differences. The Solar System is truly dominated by the giant Sun, holding 99.8% of the mass. Its diameter is nearly a factor of 10 larger than the next largest, Jupiter. The planets Jupiter and Saturn are much larger than the next planets, Uranus and Neptune, and they all dwarf the inner four planets, Mercury, Venus, Earth, and Mars. The large moons are next in size, seven in total, remarkably similar in size and distinctly larger than the other moons. The large moons range in size from Jupiter's Ganymede, 5,260 kilometers in diameter, to Neptune's Triton, 2,706 kilometers in diameter. The Earth's Moon is roughly in the middle, 3,474 kilometers in diameter. Pluto fits comfortably within this range, at 2,304 kilometers, but is not a moon but a dwarf planet. All other moons are noticeably smaller, with the largest, Titania at Uranus, having a diameter of 1,580 kilometers. These small moons, and many other of the small big bodies in the Solar System, comprise a wide variety of objects that seem to share only small size in common. The two moons of Mars, the multitudinous asteroids, the many small moons of Jupiter, Saturn, Uranus, and Neptune, Pluto and Charon, and the objects in the Edgeworth-Kuiper Belt, are all of a similar size, shape, and character, as well as we can tell.

As with Pluto and the planets, we can clearly see a difference between the fragmentary satellites and the large moons. Even though the fragments orbit their planets like moons (with some differences), they are so small and sufficiently different in character as to make us suspect that there is a significant difference in their nature.

These fragments are scattered throughout the Solar System, clustered in the Asteroid and Edgeworth-Kuiper Belts, orbiting planets, and, as Pluto, pretending to be a planet.

Mercury, our Moon and the other large moons, and Mars, are similar in size, and have thin atmospheres or none. They all show the evidence of impacts on their surfaces, as even Earth does sparsely. Earth and Venus have been protected from many small impacts by the shield of their atmospheres, and small craters have been obliterated by weathering and volcanic activity. Mercury and the moons are completely covered by craters, from large to small. Mars generally has large craters, and very few small craters, the north and south hemispheres are different. Some of these impacts leave tangible evidence, in addition to the craters. Chunks of iron and rock have been found associated with craters and with bright meteors (shooting stars). These alien pieces may come from asteroids, comets, or other objects not yet identified. Some have come to the Earth from Mars and from the Moon.

Accurate distances are the key to understanding the Universe, and that is true for our Solar System as well. There are no easy ways to measure the distance from the Sun to the Earth, but thanks to Kepler's laws of orbital motion, only that distance (or the distance to any other planet) need be known to set the distance scale for the entire system. Attempts to measure the distance the Sun were not successful in any of the modifications. The method of horizontal parallax required the measurement of a small change in angle between the rising Sun and the setting Sun, and this was clearly impractical, given the great distortion caused by our atmosphere. Similar problems applied to the angular measurements for the planets. This method is an absolute method, as firmly true as the civil surveyors' triangulation. The observational difficulties were overcome by referring the angles to reference stars, but in principle, the absolute nature of the method was lost by this relative determination. Even this did not produce dependable measurements, and the true distance scale of our Solar System was not known until radar distances to Venus were measured. There will be more on this later.

For all this great variety in size, the planets show a remarkable regularity in their spacing from the Sun. This was noted as long ago as 1766, by Johann Titius of Wittenburg, Germany, when only five planets plus Earth were known. The idea was promoted by Johann Elert Bode, and so became known as the Titius-Bode Law.

This law, a simple mathematical formula for the average distance from the Sun for each planet, works remarkably well, even for the planets (and fragments) that were unknown at the time. The discovery of the planet Uranus by William Herschel in 1781 supported the law. Prediction by this law of a planet between Mars and Jupiter led to the discovery of the fragmentary asteroids, each and all too small to be a whole planet. Because of their place and behavior in the Solar System, much like Pluto, the asteroids are sometimes called minor planets.

The Titius-Bode Law is compared with the present distances of the planets in Figure 28. Distances were predicted by the Titius-Bode Law for the planets Uranus and Neptune and the asteroids before their discoveries, quite accurately for Uranus, less so for Neptune. Mercury seems to be a bit too far from the Sun. Neptune is a bit too close to the Sun. Pluto follows this trend of error, being at about the right distance as Neptune should be, but sometimes orbiting closer to the Sun than does Neptune. Like the asteroids, Pluto and Charon, put together, do not make a decent planet. Farther out, the fragments of the Edgeworth-Kuiper Belt may eventually be so well discovered as to identify the proper location of the next Titius-Bode planet. Because of its place in history, and in the present story, "Planet X" has been included as the tenth Planet.

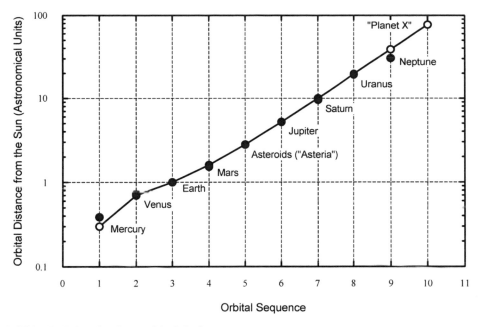

Figure 28. Titius-Bode Law for planets of the Solar System.
The Titius-Bode Law calculated for the planets of the Solar System. Current values are shown by solid dots, the calculated values by the open dots. The asteroids have been included at an average distance, and "Planet X" has been added to fill out the set. Only Mercury and Neptune show significant deviations from the Law, although Saturn, Uranus, and Neptune are systematically closer to the Sun than calculated.

We have come to know our Solar System well. It is composed of a variety of major and minor bodies, from the immense Sun to the tiniest grains of rock, iron, ice, and dust that we collect and call meteorites. Relative to our Earth, and the tiny distance from the Earth to the Moon, the Solar System encompasses a vast region of space, out to the edge of an imagined cloud of icy chunks left over from the very formation of the Sun and its planets.

A great change in the nature of the planets occurs at the asteroid belt. This divides the Inner Solar system from the Outer Solar System, the rocky planets from the gas-giant/ice-giant planets.

The distances are so great it is simply not practical to display the entire Solar System in one piece. Some outdoor representations have been made successfully, but even these are forced to allow the Oort cloud of comets to be somewhere vaguely off in the distance.

The planets of the Solar System, their satellites, the asteroids and kuiperoids, and the comets, are shown, simply by mass, in Figure 29. Here it is clear again, that the planets divide into two groups, the inner, rocky planets, and the outer, giant planets (and Pluto). Size divides these all. The Sun, of course, stands alone, dwarfing all the others.

The inner planets are clearly smaller than the outer planets, but even so, are much larger than anything else. (In this chart, each line is a factor of ten in mass. Earth is nearly a thousand times larger than Pluto.) The satellites divide into the large moons, 1 for Earth, 4 at Jupiter, 1 at Saturn, 1 at Neptune, and then 23 or 24 (counting Pluto's moon, Charon) little fragments, and many more as our explorations continue. The fact that little ol' Earth has a Moon as big as those of the giant planets should attract our attention. The asteroids and the kuiperoids match in size, as well as we can tell. The smallest asteroids look like re-located comets, and might be, since their orbits hardly belong with the rest of the asteroids. The Oort cloud, the storehouse of comets, is two decades in distance off scale. Since no more is known of the Oort cloud than people have thought to write about it, we may generally ignore it, and there seem to be reasons that it does not exist [Bailey, Clube, and Napier 1990].

Some of the groupings in the smallest objects may be an observational artifact. Small objects at considerable distances from the Earth and the Sun are hard to find. (The comets are detected as they approach the Sun and develop a visible cloud, the coma. After rounding the Sun, a comet returns to the distant part of its orbit, where it spends most of the time in its cycle.)

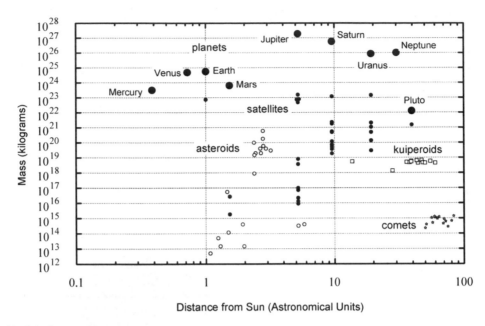

Figure 29. Solar System objects.
Masses of the known bodies in the Solar System. Pluto has been plotted as if it were a true planet, but its small size clearly shows it is an object of a different sort, fitting in more with the fragmentary satellites, asteroids, and kuiperoids. The rocky inner planets form a group, as do the giant planets, and the seven large natural moons cluster about 10^{23} kilograms. The asteroids, kuiperoids, and the fragmentary small satellites show a (probably) continuous size distribution down to extremely small (unobservable).

That is the portrait of our Solar System at the present time. In the development of theories of its formation, this portrait is accepted as accurate for all time, it is assumed that no significant changes have altered the character or number of the planets, satellites, and other objects through the last 4.5 billion years or so. Impacts have scarred the surfaces and gravitational forces have perturbed orbits, but generally the way it is, is the way it was. Our standard theories are designed and tailored to produce a Solar System that, in the beginning, was just like our Solar System is today. If we consider that things might not have always been as they are now, we open many other possibilities, and can approach closer to the truth.

Formation of the Solar System, the Earth as a Giant Planet

We can never really know how the Solar System formed, it is too long ago. It is unlikely that we will ever find another system forming, and that formation is too slow for humans to observe it. What we can hope to do is to develop a description of the conditions and mechanisms as accurately as possible and build an explanation.

From the best-accepted sets of evidence, we think that the Solar System formed about 4.5 to 4.6 billion years ago, starting with the Sun, and then the planets. Scientific explanations of that formation must be recognized as simply "best effort" attempts. Physically reasonable processes are described, based on the conditions that we observe today and the assumptions that we can infer for the distant past. For all the scientific effort exerted, we don't know how our Solar System was formed, and the only other planetary systems that we have studied don't fit our existing theories. We must be sure that we have considered carefully and knowledgeably, all the possible processes, mechanisms, effects, and phenomena, and their possible interactions. Some possibilities have been overlooked in our present theories, and will be described here.

Our standard theory for the formation of the Solar System is based on the rather direct approach of having simple things happen so that we have a Solar System like the present from the beginning of its formation. Briefly put, a cloud of gas and dust from a supernova collapses and forms a star. This star, our Sun, blows all the light elements out of the inner region of the remaining cloud, leaving mostly dust. The dust accumulates into planetesimals which then combine to form the planets of the inner Solar System, Mercury, Venus, Earth, and Mars. Just beyond Mars, the planetesimals were unable to combine into a single planet and have remained scattered, as asteroids. Farther out, where the gas of the original cloud still lingered, the gas-giant planets of Jupiter and Saturn, and the ice-giant planets of Uranus and Neptune formed. More planetesimals formed beyond these planets, two perhaps remaining as Pluto and Charon, the others composing the Edgeworth-Kuiper Belt, perhaps different in historical name only. Still farther out, remaining material coalesced into comets-to-be in the Oort cloud. For the next 4.5 billion years, very little happened to change or affect that original Solar System.

It is important for us to start our exploration a little bit earlier, and to consider a more dynamic, more extensive formation and development.

In our standard stories, there is one fundamental error, an omission, and it affects the whole formation story. We have assumed that the cloud of gas, dust, and planetesimals was cold and transparent to thermal (heat) radiation. It was certainly cold, by our standards, and so should have been transparent, because thermal opacity comes from free electrons, and there shouldn't be any in a cold dark cloud. But what was overlooked changes the conditions so much that the standard story won't work. The material in the cloud, fresh from a large supernova, was filled with radioactive elements. In the standard theory, ionization (and opacity) doesn't begin until the gas has reached about 30,000 K, and by then it is too late to retard the final collapse. But in reality, the radiation from radioactive materials keeps the cloud full of free electrons independently of the temperature. And that makes it opaque from the very beginning.

Radioactive elements decay and give off alpha particles (the bare nuclei of helium atoms), beta particles (free electrons in person), and gamma rays (high-energy photons). These radiations ionize the surrounding gas, producing one ion pair (an ionized atom and an electron) for every 32 electronvolts of energy, to judge from our experience with radiation in air. In the thin gas, the loose electrons range far, and serve ever ready to absorb a thermal photon, and re-emit it in any direction. This scattering of heat prevents the cloud from cooling as it would collapse. As the cloud would collapse to form the Sun, it heats up from compression, converting gravitational energy into thermal energy, and is self-supporting. The cloud, being radioactive, and ionized, and thermally opaque, cannot collapse to form the Sun. Something else must happen. The Solar System, the planets and the Sun, formed from the outside in.

Because of the ionizing power of the radiation from the radioactive decays, there is an abundance of free electrons, and free electrons scatter radiation of any energy. The gas is radioactive, from the production of unstable nuclides in the supernova. The radioactive decays release approximately 3 million electronvolts of energy each, on average. Ionization of an atom, the ejection of a bound electron from any atomic quantum level, requires on average approximately 32 electronvolts of energy (in air), so that each decay releases about 100,000 free electrons. The concentration of electrons accumulates until an equilibrium is reached where electrons and positive ions recombine at a rate that balances the production by radiation from radioactive decay. A rough calculation, using solar abundances and ignoring recombination, short-lived radionuclides, and the initial ionization of the cloud, shows that the radiation produces about 8 free electrons per 100 atoms in the first 2 billion years of the cloud's life.

Recombination occurs on an astronomical time scale, as evidenced by the HII (ionized hydrogen) regions seen in the Universe, which glow in lovely red hydrogen Balmer light, as the protons and electrons recombine.

The beginning of the story starts in space, with the explosion of a supernova approximately 6.4 billion years ago. All of the really heavy stuff in our world, the elements beyond iron, is produced by rapid neutron absorption, the r-process, in supernovae. These neutrons are produced as the shock wave of the collapsing star forces electrons into the protons of the atoms of the core of the star. The neutrons are quickly absorbed by the existing nuclei, in what is called the r-process, more rapidly than radioactive beta decay can adjust the ratio of nuclear protons to neutrons, producing a mix of nuclei that gives us distinctive clues to the process itself.

These newly produced heavy nuclei, and most of the rest of the star, are blown out (blasted) into the old, tenuous atmosphere of the star, which is still mostly hydrogen. As most of the new atoms are surrounded by 10^6-10^{12} as many hydrogen atoms, the new atoms will react to form hydrides whenever it is chemically possible. These reactions occur in the fresh material of the supernova cloud, while it is still moderately dense, and warm enough to promote the reactions. Much of what is produced is highly radioactive and decays by emitting beta particles and antineutrinos, or alpha particles, quite quickly, until it becomes a stable nuclide or a radionuclide with a long half-life, such as uranium-235 and uranium 238. These two nuclides have half-lives comparable to the age of the Solar System. Calculations of nuclear reactions in an exploding supernova show that the initial enrichment of the newly produced uranium is about 60%. As the uranium-235 decays more rapidly than the uranium-238, the enrichment decreases from 60% right after the supernova to about 24% at the formation of the Earth, considered to be 4.5 billion years ago.

What happens in the supernova ejecta is one of nature's neat little tricks. The uranium in the supernova ejecta is surrounded by hydrogen, as everything else is, hydrogen outnumbering everything else by a million times or more. As the supernova cloud cools, the uranium reacts with hydrogen to form uranium hydride (UH_3), which slowly crystallizes, forming small grains of (probably) high purity. As the uranium radioactively decays, the grains emit alpha particles (2 positive charges each) and beta particles (1 negative charge each). More positive charges escape from very small grains, leaving the grains with net negative electrostatic charges. The alpha particles have short ranges compared to the beta particles, so relatively more beta particles escape from larger grains, leaving them with net positive charges. Thus, large and small grains of uranium hydride will be selectively attracted and will grow faster than other materials in the cloud. (A home physics experiment showing how the electric field of an electrified comb (or similar object) can attract a piece of paper, against Earth's gravity, shows the strength of electrostatic attraction.) Uranium hydride is about the densest material in the supernova ejecta (11 grams per cubic centimeter), and so, with its electrostatic head start, forms gravitational centers and serves as the nucleation of planetesimals. The planetesimals grow to form planets, the uranium hydride loses its hydrogen as the temperature increases, and one of the densest elements, metallic uranium at almost 19 grams per cubic centimeter, concentrates in the core of the planet. The concentrated uranium makes fast breeder reactors and keeps the planet warm.

(An interesting aspect of our understanding that hydrogen and helium are the most abundant elements in the Universe, is that both of them are produced by uranium, protons and electrons for hydrogen from the beta decay of free neutrons from fission, and helium from the alpha particles of heavy element radioactive decay.)

It is thought that these clouds of supernova ejecta cool to very low temperatures, in the range of 10 K to 60 K, or 100 K [Kippenhauer and Weigert1990]. Nearly everything condenses and solidifies, except for hydrogen, helium, and neon, which remain gaseous.

There are several questions that a satisfactory theory of Solar System formation must answer. Existing theories have had varying degrees of success with these.

1. Why does the Sun have so little angular momentum, compared to the planets?
2. Why do the planets orbit in the same direction?
3. Why are the orbits in the same plane?
4. Why are the orbits so nearly circular?
5. Why are the planets spaced so regularly in distance from the Sun?
6. How do grains accrete?
7. How did the Sun ignite?

During the past few years, astronomers have indirectly detected the probable existence of planetary systems consisting of giant planets closely orbiting starsuns that seem to resemble our own Sun. This observation of distant planets, now numbering in the hundreds, with clear confirmation by observed eclipses, is somewhat predetermined by the method of detection: only a giant planet close to a starsun can be detected by this method. However, it demonstrates that our Solar System, with small, rocky "terrestrial" planets in the inner region, is not necessarily the standard of the Universe. This raises the possibility that the Earth might have been formed initially as a giant planet. It was then downsized by ablation by a freshly ignited Sun, and has been expanding, in fits perhaps, ever since.

Our current theories of the formation of the Solar System (essentially variants of a standard theme) all form objects from the inside (center) of a cloud of dust and gas, outward, after the Sun has formed and ignited. As the cloud shrinks, heat of compression is released. In the standard story, because the cloud is cold, it is transparent to the thermal radiation produced by this heat. The radiation escapes easily and the cloud continues to shrink, finally collapsing abruptly with enough presumed momentum to form a shockwave which ignites the Sun. Ignition of the Sun blows away the hydrogen and helium from the inner Solar System, causing the inner planets to be rocky dregs, the outer planets (Jupiter and Saturn most especially) to be rich in the gases. This is done so that the Sun may empty the inner Solar System of gas and so let the inner planets avoid becoming gas giants as they form.

A cold, transparent, nonradioactive cloud can collapse because the heat of compression and the ionization resulting from high temperature and the opacity that causes, are in the center of the cloud, and the outer region is still transparent to thermal radiation. The outer region can release heat and continue to collapse onto the central region, making a star. Out of a collapsing cloud of interstellar gas and dust, the Sun forms, and is left surrounded by a protoplanetary nebula, a cloud with about 5% of the mass of the Sun. The infant Sun heats the material of the inner Solar System, where the terrestrial or rocky planets will form, and drives out the gas (mostly hydrogen and helium). Then the planets begin to form, Mercury first, on out to Neptune. (Pluto seems to be different enough that it probably did not form in this process. It is classified as a planet or not mostly according to the classifier's personal attitude as to what is important about being a planet. This has now been made official by the International Astronomical Union, and Pluto is a dwarf planet.)

I think the standard theories of formation of the planets, in which the Sun is taken as forming first, are not reasonable. The central region of a collapsing cloud would be the hottest region in the cloud and the least likely to collapse of its own gravity, condense, and form the Sun. Instead, let us let the outside of the collapsing cloud cool, condense, form a planet, one zone at a time, and allow heat to escape from the now newly exposed surface of the cloud, unshielded from the cold of space. This process progresses inward, forming giant planets throughout and, last of all, out of chaotically moving gas atoms, the Sun.

Gravitational collapse is resisted by gas pressure. The gas pressure is determined by the temperature, which is affected by the heating rate, the surface-to-mass ratio, and the opacity. Radioactivity increases the heating rate over that generated by compression alone, and most importantly increases the opacity by producing free electrons by ionization, independently of the gas temperature. The cloud is opaque from the beginning, at any temperature. This effect decreases with time, as the radioactive nuclides decay. Two of the most significant, considering a time-scale of a billion years or so to form the Solar System, is the decay of uranium-235 eventually to lead-207, with a half-life of 0.7038 billion years, and the decay of uranium-238 eventually to lead-206, with a half-life of 4.468 billion years.

A problem with the standard theories of formation, in which the Sun is taken as forming first from a collapsing rotating cloud, is that the Sun has too little angular momentum. It is not rotating as rapidly as it should be, according to these theories. However, the central region of a collapsing cloud would be the hottest region in the cloud and the least likely to collapse of its own gravity, condense, and form the Sun. Instead, the outside of the collapsing cloud would cool, condense, and form a planet, one zone at a time. Heat would escape from the newly exposed surface of the cloud, unshielded from the cold of space. This process progresses inward, forming gas-giant planets throughout and, last of all, out of chaotically moving gas atoms, the Sun. The Sun has most of the mass, but the planets (primarily Jupiter) have most of the angular momentum.

To make a Solar System, start with a very large interstellar cloud (10 to 100 times the mass of the Sun), composed of gas (hydrogen, helium, some noble gases) and dust (silicates, rock stuff, hydrides, and "ices"), at a temperature of 100 K. This cloud may be in turbulent motion, mixing, fragmenting. Assume that some small parcel (about twice the mass of our Sun) establishes a unity, and breaks free of the original cloud. As the parcel separates from the original cloud, it comes under the effect of its own gravitation. (Its gravitational self-attraction was previously nullified by the external attraction of the balance of the original surrounding cloud.) The surface of this parcel cools further by evaporation of those molecules and atoms that are moving most rapidly. At this point, the random motions of the individual atoms, molecules, dust grains, and larger objects become converted into orbital motions around the center-of-mass of the parcel, each gravitationally attracted by a mass equivalent to the mass contained within a sphere with a radius equal to its distance from the center. That is, the outermost atoms and globs feel the attraction of the entire mass, while those near the center feel no gravitational attraction, and continue with their random motions.

Cooling of the bulk of the cloud is retarded by the production of heat by radioactive decay, and the increase in the opacity beyond what would result from thermal motion, by the ionization resulting from the nuclear radiation. This is where the standard story fails. The cloud is assumed to be cold and transparent, but the radiation ionizes the gases and releases free electrons, which make the cloud opaque to thermal radiation, regardless of temperature.

The free electrons scatter the thermal radiation, preventing the cloud from cooling except near the surface, where the stickiness of condensing methane and silane starts to form the kernels of planets. The planets form first, from the outside in, each with a retinue of large natural moons. The Sun forms as the last stage, when the cloud has become so small that cooling and collapse is possible. Ignition is not likely to result from any shock produced by the collapse, but rather by the energy and the neutrons released by nuclear fission deep in the core of the Sun. (Free neutrons in the Sun make it possible to bypass the proton-proton fusion, the first step in the fusion chain, and the slowest.)

With an opaque cloud, planets must form sequentially from the outside in, as the surface of the cloud cools and the material becomes sticky from the condensation of methane and silane. This is true accretion, and proceeds for each planet until the cloud has shrunk away from its gravitational sphere of influence, and then a new planet begins to form. As the forming kernel orbits the outside of the cloud, it accumulates material gently, extracted from the surface of the cloud. As material is captured from the cloud and as the cloud shrinks, the forming protoplanet loses its attraction, and the cloud is ready to form the next planet. (That accounts for the regular spacing expressed in the Titius-Bode Law.) Large natural moons form for each planet as tidal bulges induced in the periphery of the cloud by the protoplanet are captured by the newly formed planet. We now have seven large natural moons, one for Earth, four for Jupiter, one for Saturn, one for Neptune. In my view, each planet initially had several, without a unique impact event required to make one for the Earth, and no captures of planetesimals for the others.

This process explains our finding of giant planets in very close orbits around other stars like our Sun. I have to admit that this is weak evidence since those are the only kind of planets that we can find so far. This model of accretion is much like that proposed by Elder, that accretion of the planets was from the Solar nebula remaining in the form of a "dusty gas" throughout the accretion of the protoplanets, thereby avoiding the need to construct planetesimals [Elder 1986] In the present model, the dusty gas becomes a sticky dusty gas as it cools below 90 K, and planet formation proceeds from the outside in.

As the surface layer cools, this outer matter will begin to fall closer to the center of the parcel (what we could now call the pre-Solar cloud) and loses kinetic energy by inelastic collisions among the particles. These collisions heat the material, and this thermal energy is radiated away, at the surface, into space, or is absorbed by intervening matter in the interior of the cloud. Thus, the outer layer trades gravitational potential energy for kinetic energy, converts the kinetic energy into thermal energy and radiation energy, cools and continues to shrink. The interior of the cloud warms up. As the size of the cloud decreases, the force of gravity at its surface increases, and so faster atoms, and material at higher temperatures, can be contained. The higher temperatures lead to more rapid radiation of energy (as the fourth power of the absolute temperature) so that the exterior of the cloud can cool and shrink progressively more rapidly. The center of the cloud becomes progressively hotter, and there is essentially no ordered motion in the central region.

There are two distinctive differences between this model and the standard theories. The other theories have been tailored to produce the small rocky planets of the inner Solar System from the beginning. This is done by having radiation from the infant Sun drive the gas and volatiles away before the planets formed. The inner planets were then formed out of the residual solid materials. In those models, the giant planets that we have found close to their starsuns cannot be formed there. Their existence must be accommodated by other peculiar effects, such as paired encounters, or spiraling in, to make the present formation theories (as developed to produce our Solar System) agree with observed facts (of the distant solar systems). I think instead that, in my model, the center of the cloud is too chaotic and grows continuously hotter until the end, so that the cloud cannot collapse in the beginning and the Sun cannot form early. The existing theories simply use a Sun that was formed before the formation of the planets began. In my model, the Sun forms after the planets have been formed. In existing theories, the cloud continues to rotate about a cylindrical axis as it shrinks, and conservation of angular momentum predicts that the Sun should be rotating very rapidly, far faster than the currently observed 27 days. In my model, the central region of the cloud has very little angular momentum of its own. Most of the angular momentum is in the planets, as we observe at present.

As the cloud cools from the outside in, an outer shell or zone of gas and dust cools towards "ambient" temperature, perhaps 30 K. (It is important to note that in this model, since the Sun has not yet formed, very little radiant energy from inside the cloud warms the outside of the cloud.) Nearly all molecular materials condense and solidify, except hydrogen, helium, and neon. At this stage, the remaining gas passes through a very important transition. At about 40 K to 90 K, both methane (CH_4) and silane (SiH_4) condense, forming a tarry, gooey glue. The previously hard dry dust grains now are coated with sticky stuff. These grains agglomerate, forming nodules, probably of a very low density dendritic structure. Nodules combine, eventually forming planetesimals, at the same low temperature. Uranium is likely to have been concentrated at the centers of these nodules, and then accumulates in the cores of planetesimals. Planetesimals accumulate gravitationally, but with very low impact velocities (and energies) initially. Therefore, a protoplanetary core forms, of a broad mix of materials, at very low temperature, orbiting the outside of the cloud.

As the outer region of the cloud cools, dust grains adhere and build into planetesimals which collide and accumulate into the rocky core of a new planet. Most of the accumulation energy from these initial impacts is radiated away into the cold (3 K) space surrounding the cloud, and the forming planet stays very cold. The new planet attracts cooling gas from the surrounding cloud and develops a massive atmosphere, which then serves as an insulating blanket. Impact energy is then retained at the surface of the new planet. This heats the surface while the interior may still stay cool.

As the cloud progressively shrinks, the mass of material (gas and dust) available in each radial "feeding" zone increases. If the density of material in the pre-Solar nebular cloud varied with distance from the center-of-mass as $1/r^2$, which I think would be reasonable for gravitational collapse, then the amount of material available for planet formation at any distance would be proportional to $1/r$. (Density, $1/r^2$, multiplied by the circumference or orbital distance, $2\pi r$.) Considering the giant planets of our Solar System and the Sun, and especially the new discoveries of giant planets close to their starsuns, it appears that the planetary mass available is proportional to the inverse of the distance from the center of the cloud, that is, $1/r$.

This $1/r$ pattern appears to be supported by the giant planets that are being found orbiting close to stars that appear to be like our Sun. Again, this argument must be somewhat tempered by the fact that the current detection method can only find giant planets orbiting close to their starsun. However, there is significance in this simple affirmative answer to the question; do close, massive planets exist? In the accompanying graph, the Solar System is represented by the solid dots, the extra-Solar giant planets by the open dots. Masses of the extra-Solar planets are minimum values. The set of extra-Solar planets consists of individual systems, rather than a family of planets, like ours. (See Figure 30.)

If we use this $1/r$ rule with Neptune as a reference, the Earth would have been formed with a mass 515 times its present mass, roughly 2 times the mass of Jupiter. Most of this mass would have been hydrogen, helium, methane, ammonia, and water. Scaling from our model of Jupiter, the central pressure in this giant Earth would have been greater than about 22,100 gigapascals (a gigapascal is a billion pascals of pressure, about 10,000 times atmospheric pressure at the surface of the Earth), just slightly greater than Jupiter, but much greater than the present value of 364 gigapascals for the Earth. This much greater pressure would have created a great compression of the interior of the Earth.

A protoplanetary core forms, orbiting in the outer region of the shrinking cloud, of a broad mix of materials, at very low temperature. As the protoplanet grows, its surface temperature increases, due to the increasing impact energy of the accreting planetesimals. This impact energy is lost to space at first, but then is retained as a blanketing atmosphere (hydrogen, helium, trace noble gases, and volatiles) develops, held by the increasing gravitational attraction. The surface temperature increases further and newly accreted material begins to melt the surface. (As the protoplanet grows, the surface gets further away from the protocore. The protocore has been compacting during the time of growth; each impact produces shock waves which push the core minerals into denser phases, as those phases are stable at the increasing static pressure.)

As the surface temperature of the protoplanet exceeds the melting point of the iron minerals, the iron is reduced to metallic iron by the hydrogen atmosphere. This molten-slushy-solid mix of iron (and other dense elements) sinks through the molten silicates and accumulates in a shell around the solid, cold, dense protocore. Most elements and compounds that are as dense as iron will accompany it, whether they are "siderophile" or not. This material solidifies on contact with the cold protocore, until eventually, as sufficient hot materials migrate inward, the materials of the protocore begin to warm and melt. Being less dense than the iron, bits of the protocore begin to rise toward the surface, chilling and fragmenting the molten iron shell. These fragments, solid at first, sink to the center of the protoplanet. Eventually, as the last of the super-cold protocore floats toward the surface, a remnant molten iron shell begins to form around the core. However, it is not certain that the inner core is mostly iron, or even not cold. Francis Birch's research effort was directed at showing the core was not compressed rock, of the same sort as the mantle, and the shockwave experiments did that well. Those experiments do not support the idea that the core is made of iron.

As the protoplanet reaches its final size, at this stage of its development, it consists of a solid iron inner core, a molten iron outer core, a molten silicate mantle, and a very thick atmosphere of H_2, He, CH_4, NH_3, and H_2O.

The developing protoplanet will pull material from within its orbit and from outside its orbit, into a prograde orbital ring. The large natural satellites were formed by accretion of this material. The natural moons are natural miracles, the result of being at the right place at the right time. They are the moons labeled as "regular satellites" by J. A. Burns [Burns 1986], including our own Moon. In various ways they show similarities and we may suppose that those similarities (and some differences) and the great differences from the other moons may tell us about their origin. The forming planet causes a tidal bulge to develop on the surface of the cloud, and this bulge pinches off to form a large satellite. This continues until the surface of the cloud has been eroded to such a distance that the new planet can no longer attract material. The residual bulge is the starting point for the next planet. This sequence is what gave the planets their original even spacing, as shown in the Titius-Bode Law. The planetary orbits are formed in the same plane and orbiting in the same direction because of this tidal initiation of each next planet. Aerobraking, gas drag in the tenuous gas of the outside of the cloud forms the orbits into nearly circular paths.

As each planet sequentially accreted and extracted gas from its feeding zone at the outside of the cloud, the temperature of the gas declined because of the reduction in radioactive heat and the reduced opacity, and the increased surface-to-mass ratio. Further, the local density was increased by the gravitational attraction of the newly formed planets, forming concentrated masses. Eventually, when the radius had shrunk to less than the inner limit of Mercury's sphere of influence, the remainder of the cloud was available to make the Sun.

Further contraction of the cloud, and unshielding of the inner region as the planets were formed and scavenged the remaining gas and dust, allowed the ultimate condensation of the Sun. Unshielded, the inner region was able to radiate its heat, cool, and collapse. As the protoSun collapsed gravitationally, with relatively cool gas falling from considerable distance inward to the center, it over-shot the size at which the outer surface would radiate heat at the rate that it would soon be generated by nuclear fusion within the core. The shock wave of this first bounce drastically raised the central temperature and produced a great flash of heat that blew away the volatiles from the inner Solar System. Hydrogen and helium were blown a great distance, to be collected by Jupiter and Saturn, making them the gas giants. The "ices", methane, ammonia, and water, were collected by Uranus and Neptune, making them ice giants, a bit impoverished in gas. The rocky inner planets were stripped bare. (It should be noted that the giant exoplanets that we observe pose an objection to the formation process proposed here. Why weren't their atmospheres blasted off when their starsuns ignited? Perhaps their starsuns ignited more gently.)

The atmospheres of our present outer giant planets remained intact, as the heat at that distance was insufficient to strip the molecules away. Jupiter and Saturn gained some of the gas expelled from the inner planets. The asteroidal planet may have also, but there is so little left of this planet that we cannot say. As the Sun formed from the remainder of the cloud, with mostly random motion, it had little angular momentum. The major angular momentum was initially in the orbiting of the massive inner planets: protoMercury, protoVenus, protoEarth, and protoMars. Then most of this angular momentum was blasted out to the outer planets and beyond with the expelled gas and ice, leaving Jupiter as the carrier of most of the Solar System's angular momentum.

This abrupt ablation of the atmospheres from the inner protoplanets exposed surfaces of molten magma that were well above the boiling point (at zero pressure), due to the last stage of accretion. After the atmosphere was lost, rapid evaporation of the molten surface removed much of the remaining mass. Heat was transferred to the surface by rapid solid/liquid convection, so the base of the mantle froze early. Heat was lost from the surface by radiation and boiling. The downsized inner planets stabilized when the temperature dropped below the boiling point of the more common rock minerals, and then to the freezing point of quartz. Until the surface temperature fell below the melting point of quartz (about 1,900 K), materials that freeze at higher temperatures sank as solids towards the outer, molten iron core, where they floated, initially at still great pressure. This produced a solid mantle, predominantly of oxides of magnesium, aluminum, and calcium, surrounded by a quartz-rich crust. (Most of the hydrogen would have been lost in the intense ablation by the Sun, so that oxides became the stable forms. This cooling and sinking of oxides would have concentrated the last of the uranium and thorium as oxides at the base of the mantle.) The Earth's Moon would have been formed in the same manner as the other large natural moons of the planets, by retention of the tidal bulges formed in the surface of the cloud, towards the end of the formation of each planet. That would replace a unique event, a major impact on the Earth producing the Moon, with a commonplace process, the retention of neighboring accretions. The nearside of the Moon, tidally locked to face the Earth, would have been at a much higher temperature than the farside, and would have experienced the great lava flows that we see as the maria.

There should have been a lot of water around, formed from hydrogen, the most abundant element, combining with oxygen, the third most abundant element after helium.

By now, hundreds of planets around other stars have been found by detailed analysis of the star's spectral shifts. All of these planets have been huge, Jupiter-size or larger, and very close to their starsuns. To some extent, that is a result of the method of discovery: the method is most sensitive to large planets close to the star. However, it suggests that our theory of formation of the Solar System is not the only way to go, and is probably wrong.

I have compared the observed masses for a large number of extra-Solar planets with the masses of our planets, arranged according to distance from their starsuns, in Figure 30. I have then assumed that the inner planets of our Solar System have been drastically modified from their original form, but that the outer giant planets represent close approximations to their first forms. Then I fitted a straight line (on this plot) to the Sun and the four outer planets.

The straight line goes through the Sun (an artifact of the fit) and Neptune, which might be the planet least affected by changes after its formation. It seems to fit the extra-Solar planets we think we are finding around stars that we think are much like our Sun (the unfilled dots), but there is a severe selection effect in that most of the exoplanets are detected by the wobble of the starsun, and for distant planets to be detected, they must be larger. The exact fit (the solid line) depends a little on the choice of where to plot the Sun, but the slope looks generally correct. (The slope is actually -1.13, but for calculational purposes, -1 is good enough.) Thus, if planets formed before the Sun, or before it had blown away much of the nebular material, the inner planets might have started out as gas giants like the outer planets and the newly found extra-Solar planets, or even larger. If that were so, the Earth (the solid part) would have been much more densely compressed than at present. This would be true for the other inner planets also. We know something drastic happened to Mercury, losing half its solid mass, and maybe to Mars as well, small and with a puny core.

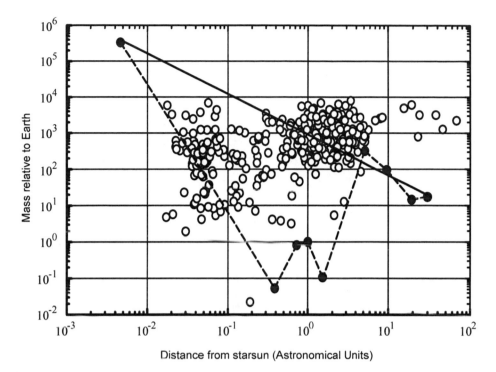

Figure 30. Planetary masses.
Masses of planets of the Solar System and the extra-solar planets that have been discovered around distant stars (exoplanets). The Solar System planets are shown by solid dots connected by the dashed lines, the exoplanets are shown by the open dots. A fit to the masses of the Sun and the giant planets gives a relation close to $1/r$, shown by the solid line. The exoplanet distribution shows the selection effect of the discovery method: at greater distances from the starsun, a planet must be more massive to be detected; close to the starsun, planets are more effective in causing a detectable "wobble", and more are detected. Using the $1/r$ scaling, relative to Neptune, predicts a mass for the protoEarth of 515 times the present mass of the Earth.

Using the proposed $1/r$ rule for the mass of newly formed planets, with Neptune as a reference, the Earth would have been formed with a mass 515 times its present mass, nearly twice the mass of Jupiter. The predictions of this rule for the original masses of the protoplanets, and the actual masses of the planets at present, are shown in Figure 31.

Taking Jupiter as a model, approximately 97% of the mass of the protoEarth would be atmosphere. The solid/liquid part of the new Earth was supercompressed. (Gaining mass beyond 4 Jupiter masses increases the density so much that the diameter of the planet actually shrinks.) At this point, I picture the new Earth to be composed of a core, which might actually have been cold, since it formed by accretion rather than by impact, exposed to the cold of space before it accumulated an insulating atmosphere. Surrounding the core might be a transition region, increasing in temperature outward, which is covered by a thick molten silica "ocean", melted by the retained heat of impacts. Silicon and oxygen are among the abundant elements. Boiling of this ocean is suppressed by the tremendous atmospheric pressure.

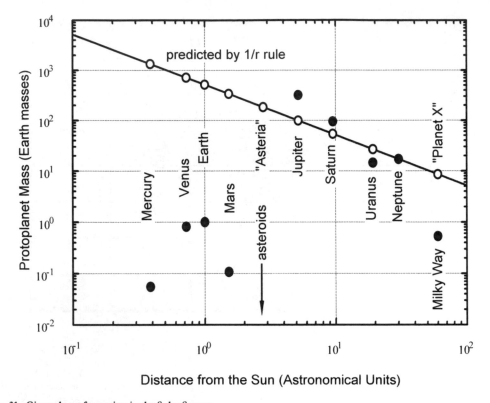

Figure 31. Giant planet formation in the Solar System.
Comparison of the predictions of the 1/r rule for the masses of the original protoplanets of the Solar System and the present masses. The predictions are shown by the open dots, present values by the filled dots. The value for the Milky Way is a corrected value from the analysis by J. L. Brady.

Most of this mass would have been hydrogen, helium, methane, ammonia, and water. The central pressure in this giant protoEarth would have been about 22,100 billion pascals (gigapascals, GPa), compared to the present value of 364 gigapascals. This much greater pressure would have created a great compression of the interior of the Earth. The central density would have been 58,146 kilograms per cubic meter (a specific gravity of 58.1), and the bulk density would have been 4,123 kilograms per cubic meter. A calculation of the variation of density is shown in Figure 32, compared with a model for Jupiter, and the PREM model (Preliminary Reference Earth Model) [Poirier 1991] for the present Earth.

Detection of an eclipse of its starsun by an exoplanet, WASP-18b, permits determination of the planet's mass [Heller *et al.* 2009, Hamilton 2009]. That planet has a mass of 10.3 times that of Jupiter (3274 times the mass of the Earth), but orbits its starsun (in an orbit that is nearly edge-on to our line-of-sight) at a distance of 0.02 Astronomical Units, about 0.0039 of the distance from the Sun to Jupiter in our Solar System. There it is thought to be spiraling in, with an estimated remaining lifetime that may be less than a million years. This is in some conflict with the estimated lifetime of its starsun, about 5 billion years.

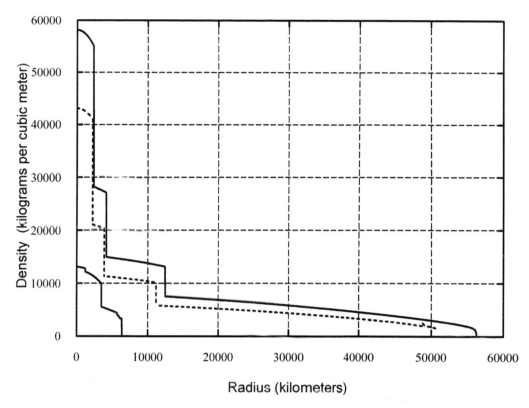

Figure 32. ProtoEarth, and present Earth and Jupiter.
Comparisons of the density profiles for the protoEarth, formed as a giant planet (upper line), a model of Jupiter (dashed line), and the PREM model for the present Earth (lower line).

Sometimes astronomers lose stars [Wagman 2003], comets, nebulae, asteroids, pulsars, and even planets. Not of our own Solar System, but the distant planets around other starsuns, that are so hard to find.

An example of this, showing the difficulty in confirming discoveries, and revealing how the analytical assumptions affect the outcome of the interpretations, is in the proposed discovery of a planet that might be able to have liquid water. This would make it possible for life as we know it to develop on a distant planet [Seven Days 2010a, Vogt, S. S *et al.* 2010]. This planet, 3.1 times the mass of the Earth, and another slightly larger, join four others previously reported by other researchers. It has a period of 36.6 days, a tenth that of Earth about our Sun, and has an estimated surface temperature of 228 K, just below the freezing point of water, because the starsun is dimmer than our Sun. However, immediately after this announcement, another group, analyzing a somewhat different set of data with different assumptions, declared that they could find no evidence of the potentially wet world [Seven Days 2010b, Kerr 2010d]. It will surely take time to resolve this dilemma, if possible, but it appears that the first group analyzed data with the assumption that the other (large) planets in the system have circular orbits, while the second team assumed that those orbits were eccentric. If the orbits have no eccentricity, other planets are found; if the orbits are elongated, no other planets exist. Is it is, or is it isn't? The uncertainty results from the use of different assumptions.

Another extra Solar planetary system that presents problems, that around the star HR 8799, has been criticized as "[That] solar system ... should not exist." [Close 2010]. It consists of four very massive planets (masses 7 to 10 times the mass of Jupiter) at exceptionally great distances, ranging from 14.5 Astronomical Units from their starsun to 68 Astronomical Units. (The derived distances are about those of Saturn, Uranus, Neptune, and the original Planet X in our Solar System.) While the large masses and great orbital diameters result from an error in the distance assumed for the system, 39.4 parsecs (128 light years), current theory has a difficulty in explaining the formation of the system, especially the innermost planet [Marois *et al.* 2010].

A system with a Jupiter-size planet that is more in keeping with the model proposed here, is the planet and starsun HD 70642 [Fischer 2009]. That planet is at half the distance of Jupiter from its starsun and has a mass that is twice the mass of Jupiter, as would be predicted by the present model.

A giant planet has been detected orbiting at about the Earth-Sun distance around a star that is considered to be a twin to our Sun, in size, temperature, and mass [Kürster *et al.* 2000]. The mass of this planet has been determined to be between 2.26 and 5.18 times the mass of Jupiter (718 to 1646 Earth masses), which is somewhat larger than the mass proposed here for the initial protoEarth (515 Earth masses). The planet is 0.925 Astronomical Units from its starsun, with an orbital period of 320.1 days, very similar to our planet.

However, this is not an entirely satisfactory part of the story. While it shows that a planet of the right size can form at the right distance, why wasn't the giant planet reduced to the size of the Earth when the star ignited as I have proposed for our Solar system? If it had, we couldn't have detected it. Perhaps some stars ignite gently, without the blast of energy necessary to blow away a massive atmosphere, at the distance of the giant planet. Perhaps planetary magnetic fields may develop early enough that the atmosphere is shielded from the starsun's stellar wind, and the planet retains its giant mass. (That is a problem with all the giant planets that are so close to their starsuns. Why weren't they downsized with the birth of their starsuns?) Any inner planets, corresponding to our Mercury and Venus, might have been reduced to sizes that are less than detectable. This remains an unresolved flaw in this version of Solar System formation.

When the Sun ignited, I assume (along with everyone else) that there was a great burst of radiant and particulate energy from the Sun, and that blasted all of the gases and volatiles out of the inner Solar System, stripping the inner planets of their massive atmospheres, perhaps transporting some of that material to the outer gas giants, otherwise simply out of the Solar System. (This is the same mechanism for making rocky planets as in our standard story, only the sequence is different.) With their atmospheres gone, almost all of the molten material forming the newly unshielded outer region of the inner planets boiled away into space, leaving a solid planet still held nearly as compressed by its stronger gravitational field (the same mass at a smaller radius makes a stronger field). The heat radiated from the surface of the Earth kept the nearside of the Moon hot, which is why the nearside is covered by melted plains while the farside is mountainous.

That left the Earth in a supercompressed state with a possibly cold interior (I don't know how important that is, or how likely), surrounded by a basaltic mantle and a silicic crust. While the interior is no longer constrained by the massive atmosphere and silica ocean, the material is locked into superdense phases by the mineral lattices. It can't expand because there is no room. The planet would have been in an unstable equilibrium between gravity and compression. Expansion would reduce gravity, and phase changes in the minerals of the core and mantle might have served as a ratchet to keep the expansion in effect. I think expansion is triggered by the rarefaction shockwaves from major impacts.

The bulk of the present Earth is composed of various phases of magnesium silicate ($MgSiO_3$), perovskite in the lower mantle, spinel in the upper mantle. At temperatures above 10,000 K and pressure greater than 1,000 gigapascals (10 million atmospheres), it is expected that this compound will be dissociated into magnesium oxide (magnesia) and silicon dioxide (silica) [Hirose 2010]. These conditions likely were present on the surface of the protoEarth, beneath the massive atmosphere. Magnesia is denser than silica, and melts and boils at higher temperatures. As the silica melted, the magnesia would settle below the surface, and a molten silica ocean would form.

Most gravitational encounters of a small object with a massive object (a planet) result in impact on the planet or ejection from the Solar System. Ejection is the result of the "gravitational slingshot" mechanism, which was used in the *Voyager* and *Pioneer* missions. These spacecraft are now leaving the Solar System, due to the gravitational encounters with Jupiter, Saturn, and Uranus. Impact is the result of the planets being larger than a mathematical point. The small object would prefer to zoom around the point, the center-of-mass, but runs into the surface of the planet along the way. That gravitational slingshot method has been used in space exploration by arranging planetary flybys to sling the space probe farther out (or sometimes in), gaining energy we could not provide by use of chemical fuels. Aerobraking, by barely diving into the outer layer of a planet's atmosphere, has been used to achieve orbits around Venus and Mars.

Energy must be dissipated in order for the small object to be captured into a stable orbit. This happened for the natural moons at the stage in planetary formation when the gas density in the planetary zone was just right for gas drag to dissipate the excess energy of the trajectory. At an earlier stage, the gas density was too great and the would-be moon became just another planetesimal impacting on the protoplanet. Later, not enough gas remained to slow the trajectory and the would-be moon would be ejected from the Solar System, or recycled by being returned to the cloud. In my model, the large satellites are also formed naturally around each new planet, so no freak impact was needed to make our Moon. It was made just like all the others. The extended thermal radiation of the Earth kept the nearside of the Moon hot so the maria formed on the nearside but not on the farside. As the Earth's Moon formed in much the same manner as Jupiter's and Saturn's satellites, it fits into that size range rather well. That replaces a unique event, a major impact on the Earth producing the Moon, with a commonplace process, the retention of a satellite formed by an orbital tidal bulge.

Thus, our Moon is the result of a natural process, just as the six other large moons were, and not a hard-to-explain freak impact by a planet too large to be there [Canup and Righter 2000]. The Moon's differences from the Earth resulted from its smaller size. Its developmental pace and end result were greatly affected by having a mass of only

0.00002 of its protoplanet. After the ignition of the Sun and the blow-off of the dense, pressurizing, and insulating atmosphere, the molten silica ocean of the protoEarth boiled into space. At an initial temperature approaching a million degrees K, it had ample thermal energy to evaporate most of its mass, as the escape velocity at the surface decreased as mass was lost, allowing the escape of the silica vapor to space. The surface was also losing heat by thermal radiation, and eventually, by latent heat of vaporization and thermal radiation, the surface cooled enough for the boiling and vaporization to stop. After time, the silica ocean froze solid, forming a new global continental crust.

There were probably more large moons, and we may play with the idea that Venus had one, although there is little evidence for that. Mars probably had one, and Deimos and Phobos are the fragmentary remnants. The outer planets have orbiting fragments as satellites. Orbital peculiarities may help assign groups.

As each protoplanet reached its final size, at this stage of its development, it consisted of a solid iron inner core, a molten iron outer core, a molten silicate mantle, and a very thick atmosphere of hydrogen, helium, methane, ammonia, and water.

As the final stage of cooling and contraction of the cloud, the protoSun can only contract at the rate allowed by cooling, by surface radiation and internal convection, and it is not clear that a temperature (15,000,000 K) sufficient to start proton-proton fusion at a significant rate will be produced. Even d-d fusion ("deuterium burning") requires a temperature in the range of 540,000 to 600,000 K to start.

After the protoSolar cloud had shrunk (isothermally) past the formation of the innermost planet Mercury, the surface gravity may be strong enough to cause the cloud to finally collapse to form the Sun. What caused ignition of the Sun is uncertain, although the standard story is that "the shockwave did it." This is somewhat similar to the ignition of fuel in a diesel engine, by the heat developed by the compression stroke. The compression must be irresistible for this to succeed, and this would not be possible in a cloud filled with the ionization from radioactivity. I have thought, and Herndon has proposed, that natural nuclear fission reactors formed and (he proposes) provided enough heat to raise the central temperature above the point where proton-proton fusion can occur, about 15,000,000 K. Since nuclear fission reactors can start with small but critical masses, and have no need for high temperatures to initiate or sustain the reactions, they provide the kindling fire needed to start stars [Herndon 1994]. I think that's only part of the story. Reactors release neutrons, which would be readily captured by protons (hydrogen) to form deuterium. The deuterium then fuses with a proton to make helium-3. That bypasses the initial step in the p-p fusion chain, which is the limiting reaction in the Sun's heat source, and is so slow it has never been observed. (Mixed with hydrogen and helium, a uranium critical mass with uranium-235 = 25-90% would be about 4 kilograms or less at zero pressure. If the central pressure compressed this by a factor of 10, the critical mass would be only 40 grams. According to the standard abundances, (uranium in the Sun is about 4×10^{17} kilograms), there should have been enough uranium to make 10^{17} to 10^{19} critical masses. Mixing of helium, which does not absorb neutrons at all but still acts as a good moderator, will make reactors possible with lower concentrations of uranium-235. Because of its great density, uranium would gravitate towards the center, and very dilute regions would become natural nuclear fission reactors. Oliver Manuel (University of Missouri at Rolla) has proposed that our estimation of the composition of the Sun has been grossly biased by the fact that we can only see the surface [Manuel and Hwaung 1983], and that gravity has differentiated the elements so that hydrogen and helium are most abundant at the surface, and iron is concentrated towards the center (as would be uranium).

When the Sun ignited, the inner planets were stripped bare, their massive atmospheres blown away. The atmospheres of our present outer giant planets remained intact, as the heat at that distance was insufficient to strip the molecules away. Jupiter and Saturn gained some of the expelled gas. As the Sun formed from the remainder of the cloud, with mostly random motion, it had relatively little angular momentum, as we observe. The major angular momentum was initially in the orbiting of the massive inner planets: protoMercury, protoVenus, and protoEarth. Then most of this angular momentum was blasted out to the outer limits and beyond with the expelled gas and ice, leaving Jupiter as the carrier of most of the Solar System's angular momentum.

The first major impact after the dissipation of the atmosphere (which had shielded the surface against impacts till its loss) caused local expansion towards the opposite side of the Earth, where the rarefaction effect of the shockwave was greatest. That expansion split the silicic crust (the enveloping global continent) and caused formation of a basaltic basin, which eventually became part of the ocean basins. Subsequent impacts caused more expansion, with uplift of continental crust, overriding, overflowing, subducting (or "superducting") the basaltic, oceanic crust at the margin of the continental blocks. Once oceans began to form from the accumulation of water from volcanic outgassing, subsequent impacts on hard rock (the continents) were more effective in converting impact energy into shockwaves than impacts on the oceans were, so that further expansion generally occurred opposite land, in the ocean basins. This resulted in mid-ocean ridges, and sea-floor spreading, and migration of the continents toward the North, as an accident of where the first impacts were. Expansion of the Earth gets rid of the giant ocean Panthalassa, of which no evidence can be found, no evidence exists, "because it was all subducted".

Uplift by expansion beneath the continental blocks brings deep (high pressure/high temperature) metamorphosed rocks to the surface, produces the broken strata that is seen near the surface, and raises sea floor deposits to the tops of mountains. The Tibetan Plateau, with the Himalayas crumbling at the leading edge, is an excellent example of the superduction of continental crust, onto continental crust in this case.

To complicate things there is a possibility that the Earth was actually formed 6 billion years ago, rather than the standard 4.5 billion years. This is just a guess at a possibility, based on the following:

We say the Earth was formed 4.5 billion years ago, and the Sun 4.6 billion years ago because,

1. The radiochemistry ages of the rocks from the sky, the meteorites, show an oldest age of about 4.55 billion years, and
2. we assume that these rocks come from the asteroids (a reasonable assumption, but I think that only 1 or 2 types of meteorites have been identified with asteroidal types), and
3. we assume that the asteroids are planetesimals remaining from the initial formation of the Solar System, because
4. we assume that a fully formed planet could not have detonated to form these fragments (this is wrong), and
5. we can't find any rocks on Earth older than about 4.4 billion years, so the Earth must be only a little older.

So the planets formed a little later (4.50 billion years ago) than the planetesimals (4.55 billion years ago) and

6. we assume that the Sun formed shortly before (4.6 billion years ago).

That all makes a pretty shaky foundation of assumptions.

So why 6 billion years instead?

1. Some diamonds on Earth were dated at 6 billion years ago [Zashu, Ozima, and Nitoh 1986]. This didn't agree with our standard age of the Earth (4.5 billion years), so it was decided that the age was in error due to an unusual concentration of argon-40 in the diamond at its formation [Ozima *et al.* 1989, also Dodson 1989].
2. The Sun has less lithium on its surface than a star only 4.6 billion years old should have [Balachandran and Bell 1998, Boothroyd, Sackmann, and Fowler 1991, News Notes 1991b]. Lithium is consumed by nuclear reactions and the predicted rate takes longer than 4.6 billion years. A study of stars with and without planets has suggested that the presence of orbiting planets may stir the convection zone of the star to the extent that lithium is mixed and destroyed [Israelian *et al.* 2009]. Since there are as many stars without planets that show low abundances of lithium as there are stars with planets, this is not entirely convincing. This point for the age of the Sun is weak because the presumed presolar abundance of lithium is derived from measurements on meteorites, as fragments of asteroids, considered to be remnant planetesimals. If the asteroids resulted from a nuclear fission detonation of a fully formed planet, the lithium abundance would be increased by the production of lithium by ternary fission. Lithium-6 is produced at a rate of 9.5 x 10^{-9} per binary fission, and lithium-7 is produced at a rate of 7.6 x 10^{-7} per binary fission [Wagemans 1991].
3. The Sun is more stable (less variable) than a star only 4.6 billion years old should be. Stability increases with age for stars [Lockwood *et al.* 1992, Foukal 1994, F. F. 1993].

Based on the fact that we don't really know when the Solar System formed, and there are 3 very weak contrary indications, I wouldn't be surprised if 6 billion years were right.

Herndon has also suggested how failure of the natural nuclear fission reactors to ignite stars that formed of material deficient in uranium-235, would lead to stars over a broad range of masses, that would have failed to ignite, and would be unobservable as stars [Herndon 1994].

Outside of our Solar System, there may be several classes of stellar formulations. Stars with adequate nuclear fission fuel will ignite and shine brightly. Others with inadequate fission fuel will gradually collapse into darkness, provided the residual radioactivity is sufficient to generate the opacity needed to slow the collapse. Some stars, with essentially no fission fuel or radioactivity could conceivably play out the classical gravitational collapse, with sufficient shock energy to initiate nuclear fusion.

This outside-in theory, with a radioactively opaque cloud, gives answers to the 7 requirements of a theory of Solar System formation, provides a basis for the Expanding Earth Theory, and explains some other observations.

The Sun

The Sun will not break the rules.
Heraclitus of Ephesius

What is the nearest star to us?
R. J. Tayler 1997

Of all astronomical objects, when it is seen to be an astronomical object, our Sun is the greatest. What is the nearest star? Our Sun. It is the gravitational anchor of the Solar System, and the primary source of energy for life. The Sun, our star, is shown in Figure 33.

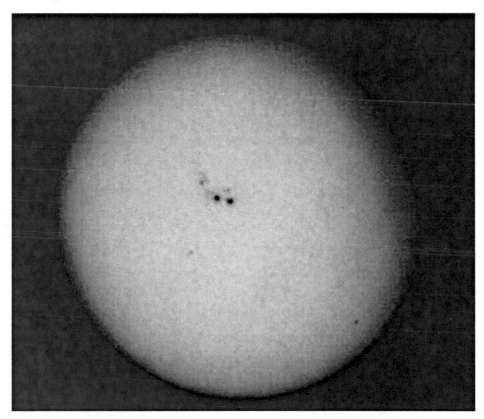

Figure 33. The nearest star to the Earth, our Sun.
A visible-light image of the Sun, showing a pair of sunspots (http://en.wikipedia/wiki/Sunspot, original in color).

The Sun has grown in our knowledge, as concepts changed and measurements improved. Its nature has changed, from a fire to a fiery rock to an inhabitable world, to a star with no known source of power, finally to our present understanding of a giant nuclear fusion reactor.

In 1929, the Sun and the stars were thought to be a thousand times older than our present ideas [Jeans 1929]. The Sun was thought to have been formed over 5,000 billion years ago. At that time (1929) the Earth was thought to be only 2 billion years old, somewhat less than our current idea of 4.5 billion years. The life of the Earth was a blink compared to the Sun, a thousand times shorter.

Attempts to understand the nature of the Sun's energy failed throughout history [Berry 1898]. Anaxagoras, in Greece in the fifth century BCE, thought the Sun was a ball of red-hot (white-hot) iron [Gribben 1991]. Combustion of coal equal to the Sun's mass could supply the Sun's heat for only a few thousand years [Jeans 1929]. The most energetic chemical reactions known could supply the Sun for no more than 3,000 years [North 1994]. The release of gravitational energy by infalling meteorites [North 1994] would require a supply of meteorites equal to the Sun's mass every 30 million years [Pannekoek 1961]. If the Sun itself were to be slowly contracting gravitationally [Brennan 1921, Berry 1898], it would have taken only 50 million years to reach its present size. William Thomson (later Lord Kelvin) carefully calculated that the Sun could supply its heat by shrinking the small amount of 50 meters per century, and would have a lifetime of 20 million years. Thomson's conviction in the completeness of physics in his day led to his confident defense of these calculations by qualifying them with "unless sources now unknown to us are prepared in the great storehouse of creation." And he was certainly right [Gribben 1991], unknown sources powered the Sun.

Throughout history, our science has always been incomplete, and yet its practitioners often fall into the trap of thinking that what is missing is trivial and will have little effect on their work as these last little pieces are discovered. Nuclear fusion was unknown, nuclear fission, radioactivity, even the nuclear atom, were yet to be discovered, totally changing physics. This situation was recognized in 1899 by Thomas Chamberlin, who stated that because of the unknown nature of the interior of atoms, it was possible to speculate that "No cautious chemist would … affirm or deny that the extraordinary conditions which may reside at the center of the sun may not set free a portion of this energy." [Gribben 1991] Later, in Jeans' time, even the newly discovered radioactivity failed, and annihilation of matter was suggested [Jeans 1929, Jeans 1931]. Perrier and Eddington found the right answer: fusion of hydrogen atoms to form helium [Jeans 1929, Pannekoek 1961]. However, this process could not provide the energy required for a Sun that was a thousand times older than our Sun.

Jeans had considered in 1904 that the annihilation of matter could power the stars. Einstein's famous equation $E = mc^2$, in 1905, allowed Jeans to calculate the energy released and the lifetime this afforded a star. These lifetimes were in agreement with what was thought to be the ages of stars at that time, so Jeans was assured that annihilation of matter was the source of the Sun's energy as well. (Indeed, annihilation of two protons provides 1,877 MeV of energy, while fusion of two protons provides only 0.425 MeV.) Because the stellar ages required such an extreme source of energy, Jeans was convinced that annihilation of matter was essential. As a mechanism, he proposed that an atomic electron would eventually merge with the nucleus and release a flash of radiation at the moment of mutual annihilation [Jeans 1931]. (This reaction would actually provide 938.8 MeV, because of the very small mass of the electron, but that was not known at the time.) Our knowledge of nuclear physics has progressed sufficiently for us to know this reaction is not possible. Our knowledge of the lifetime of the stars has progressed sufficiently for us to know this is not needed.

The lifetime of the Sun is estimated to be about 15 billion years from birth to death, and so the age problem has disappeared. Echoing Jeans' idea of hydrogen fusion, Hans Bethe and Carl Friedrich von Weisacker independently proposed in 1938 a carbon-nitrogen-oxygen (C-N-O) cycle, catalyzing the fusion of 4 hydrogen atoms into 1 helium atom [Karttunen *et al.* 1994]. Unfortunately, it was found that our Sun was not hot enough for this reaction to proceed at the required rate. Proton-proton fusion was also proposed by Bethe and Charles Critchfield [Gribben 1991]. Eventually, in the 1950s, the details of proton-proton fusion were worked out and a practical process for energy production in the Sun was understood. The nuclear fusion processes that power the Sun are reasonably well understood, and make up part of the "Standard Solar Model" [Bahcall and Ulrich 1988, Bahcall, 1989, Ulrich and Cox 1991, Bahcall and Pinsonneault 1992]. These fusion reactions produce elements heavier than hydrogen, and release energy in the form of kinetic energy of the reaction products, gamma rays (photons), and neutrinos.

The Standard Solar Models use the laws of gas dynamics and measured values for the nuclear reaction cross-sections (except for the fundamental proton-proton fusion, which has never been observed and must be theoretically calculated) to represent the Sun. These models are used to calculate the luminosity of the Sun, which we can observe to be 3.86×10^{26} watts. Adjustments are made to the calculations to obtain this value. A foundation stone of these models is that the Sun is in strict hydrostatic equilibrium, assuming that the energy production rate by fusion reactions in the Sun is at all times exactly equal to the radiation of energy from the surface, the luminosity. This assumption implies that this is a permanent state of the Sun, indeed, that a few million years ago, the production rate was just what is radiated today. This is considered to be a well-justified assumption because the Sun would adjust its size quickly if it departed significantly from hydrostatic equilibrium [Bahcall and Ulrich 1988]. However, if the Sun is instead in hydrodynamic equilibrium, with temperature, pressure, energy production and transport, and density distributions interactively responding to a disturbance, or repeated disturbances, our current model of the Sun might not be right.

If a disturbance briefly produces additional heat deep within the Sun, it takes a very long time for that energy to escape from the surface. Deep within the Sun, the photons released by the energy production are trapped by the electrons that have been stripped from the atoms. In the dense plasma of protons and electrons, a photon can scarcely travel a centimeter before it is scattered by an electron. This scattering transfers some energy to the electron and releases a new photon in a different direction from that of the initial photon. If a photon could fly directly from the center of the Sun to the surface, it would take only 2.3 seconds. Instead, the scattering and re-scattering of photons delays the escape of these photons and spreads the release of their energy over time. Some heat is transferred to the surface relatively quickly, some energy remains in the core of the Sun for astronomical times. This time spread is shown in Figure 34.

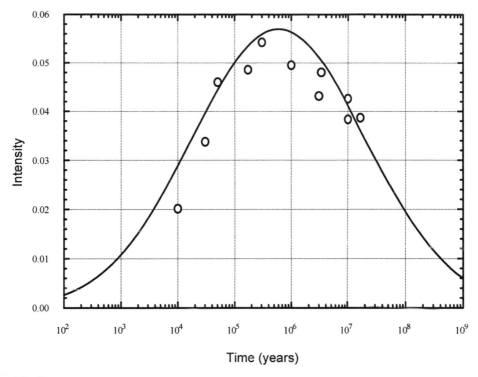

Figure 34. Escape of a pulse of heat from the center of the Sun.
The time required for energy produced in the center of the Sun to reach the surface varies over a long time scale and serves to average the power production. The indicated points are values gleaned from the literature, and the curve represents a log-Gaussian function as a consensus of the reported values.

Some energy begins to escape in less than a hundred years, most takes nearly a million years, and some takes more than the present age of the Sun. This figure shows that the heat leaks out over a time period spanning from 100 years to a billion years. There is no single answer, for the energy of the initial gamma ray is distributed into many lower-energy photons by many scatterings. Eventually, photons of infrared, visible, and ultraviolet light escape from the surface of the Sun, spread out in time over millions of years.

The complete release of energy is gradual and protracted, as shown in Figure 35. While the photons are trapped in the core of the Sun, the immediate effect of a brief power burst travels to the surface at the speed of sound, and this takes only 27 minutes [Kippenhahn and Weigert 1990]. The surface inflates slightly and radiates more energy. The Sun expands and becomes brighter.

The light and heat that the Earth receives from the Sun depends on the temperature of the Solar surface and the apparent size of the Sun. The energy flux, at the distance of the Earth, in watts per square meter, is termed the Solar Constant. This normally varies slightly, by roughly 0.1%, during the sunspot cycle. An increase in the number of the dark, cooler sunspots actually causes an increase in the rate of energy release. Each sunspot is surrounded by material that is hotter than average and this more than makes up for the cooler interior of the spots.

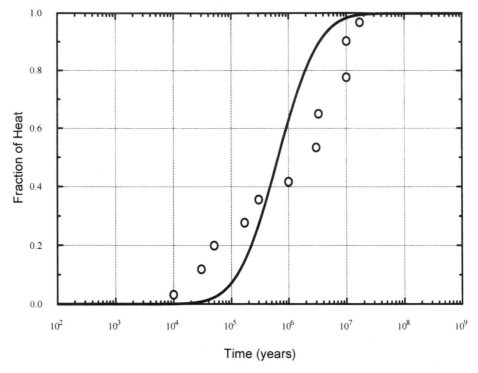

Figure 35. Cumulative heat escape from the center of the Sun.
The gradual release of energy from the center of the Sun. It takes 100,000 years for the first 10% to reach the surface and shine into space, and it takes more than 10,000,000 years for nearly all the energy to be released.

Uranium reactors in the Sun

Uranium is the densest material in the Sun, and will tend to accumulate towards the center under the influence of gravity. This may be more pronounced than usually expected because of the lack of convection in the inner part of the Sun. Both thermal diffusion and gravitational settling cause uranium to be transported toward the center of the Sun [Michaud and Vauclair 1991]. Since we see only the "top" layer of the Sun, and what it spits out in the flares and Solar wind, our knowledge of the innermost composition is very uncertain. Extremely small accumulations of uranium in a medium of hydrogen and helium will inevitably become critical reactors. Figure 36 shows how small these reactors might be, possibly needing less than a gram of 100% uranium-235.

Uranium reactors that formed in the earliest Sun, when the uranium-235 content was 24%, would act as thermal breeder reactors, converting uranium-238 into plutonium-239, which can act as a fuel material, or that would be separated out in a form of pyrochemical reprocessing that is sure to occur in the Sun. If the plutonium-239 formed by neutron absorption by uranium-238 were sufficiently dispersed that it was not able to form reactors by itself, it would decay by alpha emission to uranium-235, adding to the store of pure, 100% uranium-235, replenishing existing reactors or forming new ones. These reactors will not only contribute to the power required to keep the surface of the Sun shining brightly, the excess neutrons will jump-start the fusion chain, which is assumed at present to be the sole source of energy, and bypass the slowest step, proton-proton fusion. The fission process does not release the neutrinos that have been measured as an indication of the Sun's fusion process. Those neutrinos have been found significantly deficient, leading to the invention of neutrino oscillations, requiring that neutrinos of different flavors must have different masses, a violation of the Standard Model of Particle Physics, and the successful theory of Beta Decay. That invention, that invoking of a magical character for the neutrinos, is not necessary if fission provides the unaccounted-for power that the Sun reveals at its surface. Fission does release antineutrinos, but these are of relatively low energy and are detected only weakly by the Cerenkov neutrino detectors. This may contribute to the weak correlation with the sunspot cycle that is shown by these detectors.

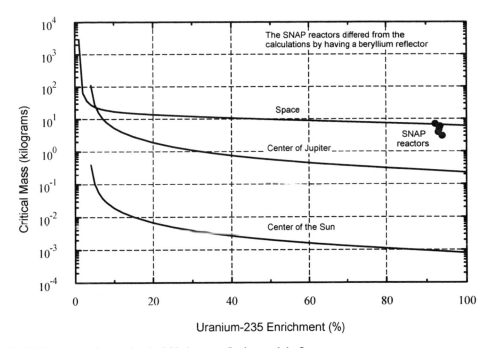

Figure 36. Critical masses for uranium hydride in space, Jupiter, and the Sun.
The critical mass of uranium reactors at the center of the Sun, extrapolated from known conditions, through uncertain calculations, to the extreme, unknown and unknowable conditions of the center of the Sun. Even with those great uncertainties, it is clear that extremely small amounts of uranium-235 would act as critical reactors.

Taking a very pessimistic estimate of 100 grams for the critical mass, corresponding to a uranium-235 concentration of less than 5%, there is so much uranium in the Sun that large numbers of reactors could form. The Sun is estimated to contain approximately 7×10^{17} kilograms of uranium. Doing simple (and silly) arithmetic shows that there could be as many as 7×10^{18} nuclear fission reactors in the Sun. However, in order to make a significant contribution to the Sun's energy production, as the sole source of energy, these reactors would have to produce power at an average rate of 55 million watts per reactor. While that is a small fraction of the power produced by a standard industrial nuclear power reactor, 3,000 million watts, it still seems too much to expect from something smaller than a pea. A very large, very dilute system would serve better. It appears to be possible for large dilute systems to detonate [Mather 1994]. In the 4.5 billion years that we think the Sun has been in operation, it has produced nearly 4×10^{43} joules (that is, watt-seconds). That energy release would require about 7×10^{29} kilograms of uranium to be fissioned, and our best estimate is only 7×10^{17} kilograms of uranium available, a shortage of a factor of a billion. To increase the amount of uranium, based on a speculation that we really do not know the composition of the interior of the Sun, would require that the Sun be 50% uranium. Clearly, production of all the Sun's energy by nuclear fission instead of nuclear fusion is not the solution to the Solar Neutrino Problem.

The Sunspot Mechanism

Ignoring the failure of natural nuclear fission reactors to totally solve the problems of energy production in the Sun, it is useful to consider a remarkable possibility in the formation of sunspots.

The way in which sunspots are formed has been an unexplained mystery since their discovery in ancient times, before scientists attempted to factually describe mysteries. We seem to be no closer today. However, some tricks of physics in conjunction with the multitude of fission reactors that are likely to be present offers an explanation. If two reactors in the Sun approach each other closely enough for one to absorb neutrons leaking from the other, and vice versa, the power of each reactor will increase. This increase will cause the reactors to heat and expand, preferentially in the regions nearest each other. That will force gas to be pushed out from the space between the reactors. The Bernoulli effect [Slater and Frank 1947] will then draw the reactors closer together, thus increasing the rate at which each reactor absorbs the other's neutrons. That will cause the power to increase, in a positive feedback. This will continue, accelerating until the reactors detonate, creating a superheated bubble of gas. The bubble will rise buoyantly, as it continues to heat by fission power, and will gain momentum which keeps it going towards the Sun's surface beyond the neutral buoyancy level. At that level, the bubble will begin to be cooler than the surrounding gas, but still rising by its momentum. It will break the surface, leaving a long void behind it, into which surface gas must fall. It will have carried up through the surface the magnetic field, making two sides of the sunspot different magnetic polarities. The excess energy released by fission, and the gas from lower, hotter regions that it has pushed ahead, will contribute to the increase in Solar Irradiance that is associated with the increasing sunspot cycle. These effects are all observed on the surface of the Sun. Since the number of sunspots in a year is not great, there are likely to be enough reactors that randomly come close enough together that this would produce the sunspots. The sunspot cycle would be caused by a regular variation in the concentration of uranium reactors through the Sun. Uranium dispersed at the surface would gradually settle towards the center, where it would be available to make new reactors and then randomly make more sunspots.

A further consequence of this mechanism of sunspot formation is that neutrons reaching the surface of the Sun will decay by beta emission, releasing protons and electrons. The protons will have average random kinetic energies of 358 electronvolts, which corresponds to an effective temperature of 2.8 million degrees kelvin. This is close to the temperature that is observed in the Solar corona. The corona is hot because of the decay of neutrons. Considerations are being developed for the measurement of the protons and electrons from beta-decay of neutrons from the Sun, associated with solar flares [Agueda *et al.* 2011].

Highly accurate measurements of the Solar Constant from the ground are not generally practical. Variations in cloud cover, transparency of the air, and changing thickness of the atmosphere along the line-of-sight to the Sun cause variations in the results. Complete measurements are not possible, because of the absorption of major bands of radiation by our atmosphere. This absorption prevents roughly half the Solar radiation from reaching the ground, and this absorption is quite variable.

These problems are eliminated by the use of instruments mounted on artificial satellites in the vacuum of space. Richard Willson has used measurements from four space missions covering 18 years to show that the Sun has actually brightened measurably, from the sunspot minimum in 1986 to the minimum in 1996 [Willson 1997]. A summary of his measurements is shown in Figure 37. These measurements show a sustained increase in the Solar Constant value, beyond the trend related to the sunspot cycle, starting in early 1987 and extending for most of the year.

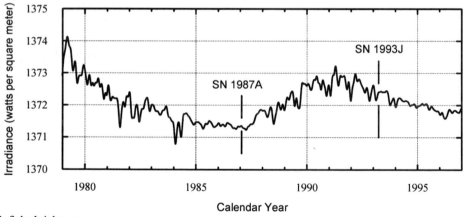

Figure 37. Solar brightness.
Solar irradiance, measured in space by ACRIM I and II, Nimbus-7, and ERBS. Daily values are combined and smoothed. Possible increases in the irradiance may occur after the supernovae, but that is not distinguishable from the variations.

The apparent size of the Sun has been successfully measured at the Mount Wilson Observatory, and these measurements show a small variation that follows the sunspot cycle closely [Ulrich and Bertello 1995]. These results, averaged over calendar years, are shown in Figure 38. A slight jump in the radius is shown in 1987, and possibly again in 1993.

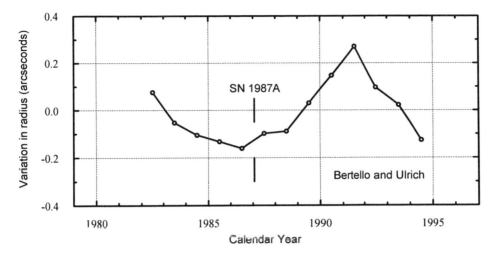

Figure 38. Solar radius.
Variations in the Solar radius measured by Bertello and Ulrich. A slightly noticeable increase in the annual average values is associated with SN 1987A.

If we see an unusual disturbance in the Sun, we should explore other disturbances in our neighborhood of the Universe. One of the most notable disturbances that was observed about that time was the astronomical detonation identified as a supernova, SN 1987A, in the nearby Large Magellanic Cloud. In these two figures the occurrence of SN 1987A is shown by the small time markers spanning the data curves. Comparison shows that SN 1987A occurred about the same time that the Sun increased in brightness and abruptly expanded, as small as these changes were.

As massive as the Sun is, we owe our lives to its tiniest particles, the protons and photons, electrons, and neutrinos. The neutrinos provide us with the deepest puzzle of all, for they let us look deep inside the Sun.

Neutrinos and the Solar Neutrino Problem

Neutrinos, they are very small,
They have no charge and have no mass,
and do not interact at all.
John Updike, *"Cosmic Gall"*,
Telephone Poles and Other Poems

You're telling me that neutrinos had no mass
10 years ago, but now they do?
Bill Minkler,
"Backscatter", Nuclear News July 1998

The neutrino is a particle with no charge and presumably no mass and apparently no magnetic moment, that is emitted in beta decay and shares the decay energy with the beta particle (which becomes an ordinary electron when it slows down). Lacking both mass and charge, the neutrino interaction with matter is extremely weak, just slightly more than "not at all", and most pass through the Sun and Earth with no indication of their passage. The neutrino was invented by a brilliant theoretician, Wolfgang Pauli, in 1930, to solve the problem of the continuous energy distribution of beta particles from radioactive decay [Pauli 1991]. This is one example of an invented solution that seems to have worked wonderfully. This particle (actually an antineutrino) was initially detected at nuclear fission reactors [Reines and Cowan 1953a, Reines and Cowan 1953b], and subsequently, detectors were built to search for neutrinos resulting from the radioactive decay of protons and fusion reactions in the Sun. Neutrinos produced in different nuclear reactions have different energy ranges, and the detectors likewise differ in energy response. Some will detect only antineutrinos, some only neutrinos, and some will detect both kinds.

The detection of neutrinos involves massive equipment and is a testimonial to the skill and perseverance of the researchers. Most neutrinos that are headed our way, regardless of the source, pass right through the Earth without any interaction at all. Antineutrinos may behave differently, as I discuss in Appendix A.

Antineutrinos produced by beta (negative electron) decay of fission products have an average energy of about 2 MeV. Neutrinos from the positron decay of boron-8 produced in fusion reactions in the Sun have energies up to 14.1 MeV, while those produced in the direct p-p fusion reaction have energies up to 0.420 MeV. Neutrinos from the electron-capture decay of beryllium-7 are emitted at energies of 0.383 and 0.861 MeV. Two types of detectors have been in operation long enough to provide significant information of some of the features of the neutrino field at the Earth: these are nuclear reaction detectors, such as the Homestake detector, known as the "chlorine experiment" [Kirsten and Wolfenstein 1991], and SAGE and GALLEX, and the water Cerenkov detector, Kamiokande-II [Bahcall 1989], which detects high energy neutrinos (and antineutrinos) by electron scattering.

The reactions that power our Sun are thought to be the following:

$$p + p \rightarrow d + \beta^+ + \nu \qquad\qquad *$$
$$p + e^- + p \rightarrow d + \nu \qquad\qquad *$$
$$d + p \rightarrow {}^3He + \gamma$$
$${}^3He + {}^3He \rightarrow {}^4He + p + p$$
$${}^3He + {}^4He \rightarrow {}^7Be + \gamma$$
$${}^7Be + e^- \rightarrow {}^7Li + \nu \qquad\qquad *$$
$${}^7Li + p \rightarrow {}^4He + {}^4He$$
$${}^7Be + p \rightarrow {}^8B + \gamma$$
$${}^8B \rightarrow {}^8Be^* + \beta^+ + \nu \qquad *$$
$${}^8Be^* \rightarrow {}^4He + {}^4He$$

Those reactions marked with a * produce neutrinos, at various different energies. The first reaction, p + p, is so slow it has never been observed in the laboratory. It is assumed to occur in the Sun at a calculated rate that makes the average proton take 9 billion years to contribute to the Sun's energy. After that, the reactions are relatively rapid. Deuterium fusion (d + p) occurs in seconds, ${}^3He$ fusion (${}^3He + {}^3He$) occurs in a million years. Neutrinos travel at the speed of light, scarcely affected by all the matter of the Sun, reaching the surface in 2.3 seconds and passing through the Earth 8.3 minutes later. Unlike the light from the Sun, which has diffused slowly outward for thousands to millions of years, neutrinos carry direct information about conditions in the core of the Sun, now.

The Homestake detector uses the conversion of a chlorine-37 nucleus to an argon-37 nucleus by the absorption of a neutrino, with subsequent collection and measurement of the argon-37. This is a different reaction than was used in the discovery measurements made at the high power nuclear reactors at Hanford and Savannah River. The radioactive fission-products, and most activation products made by neutron absorption, are neutron-rich and decay by emission of a negative beta particle, accompanied by an antineutrino. Because of the distinctness between neutrino and antineutrino, the two different particles cannot induce the same reactions, and so the Homestake detector is blind to fission reactor antineutrinos.

The Kamiokande I and II and Super-Kamiokande detectors use Cerenkov radiation produced by high energy electrons resulting from neutrino (and antineutrino) scattering. These detectors provide both energy and direction information, in addition to simply counting the events. It is sensitive to the high-energy antineutrinos produced in the β⁻ decay of nitrogen-16 which is produced by the (n,p) reaction with fast (fission) neutrons on oxygen-16. These antineutrinos have a maximum energy of 10.4 MeV but are very small in number. These detectors are sensitive to the neutrinos with energies above about 3 to 5 MeV, and are also sensitive to the highest energy antineutrinos from the decay of radioactive fission products.

Because of the weak interaction of the neutrino, these measurements require huge amounts of detector material and produce very sparse results. In terms of detection of Solar neutrinos, all detectors have shown a deficit: only about one-third as many neutrinos are detected as are expected. Evidence that even the few detected neutrinos actually come from the Sun is weak and ambiguous. A fit is made of the expected angular distribution to that observed, to eliminate the background of randomly directed neutrinos [Barszak 1998]. The background counts come mainly from antineutrinos resulting from decay of radon daughters and other radioactive nuclides and from cosmic-ray muon-induced spallation products [Barszak 1998]. Based on our knowledge of the energy produced by nuclear fusion reactions and the energy released by the Sun, it is possible to calculate the expected neutrino flux at the Earth. Stated in terms of the capture of neutrinos by material of the detector, a Solar Neutrino Unit (SNU) is 10^{-36} captures per second per target atom.

Neutrinos from the various fusion reactions have different energy distributions. The different detection mechanisms have different energy cross-sections or detection probabilities. The gallium detectors (SAGE and GALLEX) respond to all the types of neutrinos expected from the Sun, p-p, beryllium-7, boron-8 (but not to antineutrinos). The chlorine detector responds to the beryllium-7 and boron-8 neutrinos (but not to antineutrinos). The scattering detectors respond only to the boron-8 neutrinos (and high-energy antineutrinos), since an energy threshold cuts out the lower energy neutrinos.

The energy output of the Sun is known quite accurately. In our theories of the Sun, it is assumed that the energy production, due to nuclear fusion reactions in the core of the Sun, is exactly equal to the output at the surface, and has been for the last several million years. The neutrino flux indicates the present energy production rate, while the energy output represents the energy production rate 1 to 10 million years ago. At present, it is presumed that these are currently exactly equal. Since it takes thousands to millions of years for the energy produced by nuclear fusion in the core of the Sun to escape from the surface, those photons carry very little information about the core, and none for the present time.

Calculations of the heat production in the Sun are done using the Standard Solar Model. These calculations also predict the flux of various energy neutrinos produced in the fusion reactions that power the Sun. It was recognized that only neutrinos could tell us about the inner workings of the Sun. We asked the neutrinos to tell us the innermost secrets of the Sun. They told us, and we didn't believe them. We decided there was something wrong with the neutrinos.

All the detectors show that there are fewer neutrinos at the Earth than is predicted by the Standard Solar Model:

gallium	50.7%
chlorine	27.3%
scattering	39.1%

The gallium result shows that the fundamental reaction in the Sun, p+p fusion, is proceeding at only half what is needed to sustain the Sun's current energy output. The chlorine result says that the production of beryllium-7, which is very sensitive to the core temperature, is scarcely more than a quarter of what is expected. If the Standard Solar Model were right, the scattering result, which measures the boron-8 production and decay, should be the same as the chlorine result, or less. And yet this shows that proportionately more boron-8 is produced than is expected from the beryllium-7.

Clearly something is severely wrong. This is the Solar Neutrino Problem: where does the energy come from if it comes without neutrinos, or where did the neutrinos disappear to? This problem was discovered over thirty years ago, and proposed solutions still miss the mark and create controversy [Taubes 1994]. The accepted explanation at present is that some of the neutrinos become undetectable by mass-induced flavor oscillations into other types of neutrinos. This is not consistent with the Standard Model of Particle Physics, but detailed analyses have produced measurements of the parameters that agree with this idea.

Processes that produce energy in the Sun without producing neutrinos deserve some consideration. Since fission reactions release considerably more energy per reaction than do fusion reactions, and do not produce neutrinos, it might be hoped that fission reactions in the Sun could resolve the Solar Neutrino Problem. The deficit in neutrino detections recorded by these experiments, compared to the expected flux of neutrinos from the Sun, the "Solar Neutrino Problem," [Bahcall 1990] has been a puzzle for over thirty years. Not as many neutrino events are being detected, by about a factor of three, as would be expected from predictions of the Standard Solar Model, which assumes that essentially all the energy released in the Sun is due to fusion reactions. If a significant fraction of the Sun's energy production were the result of nuclear-fission chain-reactions, which release far more energy per reaction than does a fusion reaction and which do not produce neutrinos, this deficit would be a direct consequence. However, because of the small amount of uranium in the Sun, estimated to be about 2 parts per million by mass [Wagner 1991] it seems likely that only a small fraction, perhaps a few millionths, of the Sun's current heat may be produced by fission. What if there were more than 2 parts per million, and neutrons released by fission produced deuterons, promoting fusion? We don't know how much uranium is in the center of the Sun.

(Could the Sun contain sufficient uranium to assist in the generation of its heat? That Sun would gradually increase in size, as we think our Sun has, expanding from the Faint Young Sun that could not keep the Earth from freezing over, to our present Goldilocks Sun, just right for our form of life. This expansion would have resulted from the accumulation of helium produced by the alpha decay of uranium, and hydrogen from the beta decay of fission neutrons at the surface. It would have also been driven by the production of two fission fragment atoms for each uranium atom that fissioned. That hydrogen and helium, and hydrogen from thermal decomposition of uranium hydride within the Sun (and other light-weight elements) would collect at the surface, forming the material that we can spectroscopically analyze to determine the composition of the Sun. That analysis might be quite misleading. Some hydrogen fusion would occur, aided by neutron capture producing deuterium. Fewer neutrinos would be released, as we observe. Can we distinguish between this uranium Sun and the hydrogen Sun that we assume we have? Could a partnership of fusion power and fission power heat the Sun without producing the theoretically expected flux of neutrinos? That possibility seems worth exploring. The sunspots may be telling us that the reactors are there.)

The gallium and chlorine detectors can only detect neutrinos (actually only "electron" neutrinos). The scattering detectors are sensitive to all kinds of neutrinos, including the electron antineutrinos that are emitted in radioactive beta decay. Beta-decay radioactivity is produced abundantly in nuclear fission reactors, and that production formed the basis for the experimental verification of the neutrino. (Of course, Wolfgang Pauli really invented the antineutrino, but he called it the neutrino, before the physicists had clarified all the normal and antimatter types of particles.) Several experiments have been conducted to study antineutrinos produced at high-power nuclear fission reactors, Palo Verde in Arizona and KamLAND in Japan, and others.

Two methods are effective in detecting antineutrinos. The scattering method is used in the Kamiokande detectors, and the induced positron-emission method that was used in the original verification experiments. This method is the best for antineutrinos.

The Kamiokande results, which show there are too many boron-8 neutrinos compared to the beryllium-7 neutrinos shown by the chlorine results, might include nuclear fission antineutrinos.

The Solar Neutrino Problem is clear. We do not detect as many neutrinos, with the right energy distribution, as we think the Sun should be producing to keep its face shining brightly.

Since no acceptable modifications to the Standard Solar Model have corrected this problem, it has been decided, nearly universally, that the fault is in our description of the neutrino. There are three kinds of neutrinos, and three matching antineutrinos. The three types belong to reactions involving electrons, or mu-mesons, or tau-mesons. These had originally been proposed as completely massless, like photons, traveling at exactly the speed of light. If, however, any or all of these neutrino types (electron, mu, or tau) have mass, one form can change into another, and the other, and back, in a complicated oscillation. Since the detectors respond selectively to electron neutrinos, oscillations of this sort might reduce the number of electron neutrinos that reach the Earth from the Sun. No neutrinos are lost, the right number is produced, but some of them are temporarily invisible to our detectors as they pass the Earth. Other effects have been considered, including the decay of the neutrino, and a magnetic moment. None of these have been observed and must be very small. But after all, the distance from the Sun to the Earth is great, the Sun itself is huge, and the neutrino is a very tiny particle. Adjustments to the various theories can come close to solving the problem, with theoretical effects that are too small to measure. It just requires some new physics.

However, let us consider instead, changes to our concept of the Sun. Let us consider that it is not stable, moment by moment, but experiences brief hot flashes in its core, with the size and brightness of its surface waxing and waning in response. That is, the Sun is not in hydrostatic equilibrium but responds to the dynamics of its entire body. That is a condition that can be included in the Solar theory, but has been avoided by the assumption that the Sun is static. That change allows us to say that some time in the past, the Sun generated the heat we are enjoying today, but today the Sun may be generating heat at 50% of its current energy output. After all, that is what the gallium experiments show. If that reduced reaction rate, the p+p fusion, indicates a reduction in the present core temperature from what we calculate, the beryllium-7 production and decay rate would be reduced. That is what the chlorine experiment shows. The situation of the boron-8 signal is more complicated.

The boron-8 neutrino production rate depends on the number of beryllium-7 atoms and the temperature. At a lower temperature fewer beryllium-7 atoms will be converted into boron-8 atoms. Since most of the beryllium-7 atoms decay rather than fuse with a proton to make boron-8, this reduction in the fusion rate does not significantly affect the number of beryllium-7 atoms. However, beryllium-7 decays by electron capture and, in the core of the Sun, the distribution of electrons in the nucleus of beryllium-7 is markedly different from that of beryllium-7 in the laboratory. This leads to the beryllium-7 decay constant being almost 7 times less than is used in the Standard Solar Model. (This result may be an accident of normalization in the quantum-mechanical calculation, and should be investigated further.) This greater concentration of beryllium-7 would lead to a 7 times greater production rate of boron-8, except for the reduction in temperature. A further complication is that boron-8 decays both by positron emission (β^+) and electron capture (EC), which is favored energetically. The reference books are not clear on the fractionation between these modes, but the EC decay has been ignored in the Standard Solar Model.

Since the same ionization effect which retards electron capture in beryllium-7 should also be effective for boron-8, the positron decay may simply take priority and the electron capture would be negligible. The electron capture should have an associated neutrino at an energy of 14.939 MeV.

A still further complication is that the energy range of antineutrinos from nuclear fission approaches that of the boron-8 neutrinos. The scattering detectors are sensitive to both neutrinos and antineutrinos and so would not be able to discriminate between fusion neutrinos and fission antineutrinos coming from the Sun. Since approximately 6 antineutrinos are emitted per fission, fewer fissions would be required to mimic the boron-8 decays. It would take about 3×10^{34} fissions per second in the Sun to produce the same amount of fission antineutrinos as the boron-8 neutrinos. This would contribute about 0.002 of the Sun's observed energy output.

Neutrons emitted in fission in the Sun, generally 2 to 3 neutrons per fission, would be immediately absorbed by uranium, producing more fissions, or by protons, producing deuterium and moving past the first long step in the p+p fusion chain. That step is known to be possible only by theory, but could be replaced by the absorption of neutrons, without the associated production of neutrinos.

What could cause the Sun's hot flashes, and could they be intense enough and often enough to maintain its nearly steady output? (This is somewhat like an automobile engine, where brief, repetitive little explosions keep the engine turning.) Energy from a fusion reaction in the core of the Sun takes considerable time to reach the surface and escape into space, some to reach Earth as warming sunlight. This delay results because matter in the inside of the Sun is highly ionized, the initial photons have energies that are much greater than atomic binding energies, and the electrons scatter the gamma-rays emitted from the fusion reactions so that no photon has a direct path to the surface. The time delay has been variously estimated as 50,000 years [Phillips 1994], 1 million years [Hathaway 1995], and 10 million years [Gribben 1991]. Thus, the Sun acts as an averager of energy, delivering whatever energy might actually be produced in jolts, as a smooth flow of light and warmth.

Supernovae produce great bursts of neutrinos and antineutrinos, so great that our detectors responded to the 1987 supernova in the Large Magellanic Cloud, SN 1987A. In Kamiokande-II, it is estimated that 3 to 7 events were due to absorption of antineutrinos, a process that produces a neutron.

The Kamiokande-II detector contains 680 tons of water in the sensitive volume. The core of the Sun, where the subsequent proton-deuteron fusions are most likely to follow the creation of neutrons, contains 28% of its mass, or 5.6×10^{26} tons, equivalent to about 35×10^{26} tons of water. There should have been about 3×10^{25} antineutrino absorptions in the Sun, producing 3×10^{25} neutrons by conversion of protons to neutrons, making 3×10^{25} deuterons, starting 3×10^{25} fusion chains, and releasing 1×10^{13} joules of energy. This is trivial compared to the steady radiation of the Sun at 3.86×10^{26} joules per second. Clearly, supernova boosting of the Sun's energy production is not a significant factor.

Attempts to make the problem go away by making the neutrinos disappear before reaching Earth have been attempted, by endowing the electron neutrino (the type, "species", produced in the beta decay reactions accompanying nuclear fusion) with mass, but this seems to be unsuccessful [Morrison 1993].

The current approach to this problem is to look for new physics that gives mass to the neutrino and forces oscillations between the three different types of neutrinos, allowing the electron neutrinos from the Sun to fade away before they reach the detectors on Earth. While this at first seemed to be unlikely, it is now the accepted answer. This approach is based on Sherlock Holmes' dictum to eliminate the impossible, and accept the improbable. But this only works in situations where all the possibilities have truly been considered. Massy neutrinos are impossible in our standard physics, and so it has been concluded that the standard nuclear theory is wrong and we need new physics. We do not know what we do not know.

However, there is still hope... The following possibilities appear to offer adequate though novel, solutions to our neutrino problem. They may at least provide material for the experts to consider.

In the core of the Sun, where approximately half the mass is contained, electrons are stripped from most of the atoms and the gas is a plasma, consisting mostly of protons and electrons, at a temperature of about 15×10^6 K. In the core, and perhaps throughout the Sun, a sequence of reactions can occur that produce energy without producing neutrinos. Starting with an antineutrino, a proton is converted to a neutron by positron emission. The neutron quickly combines with a proton to make a deuterium nucleus, a deuteron. A gamma ray is emitted with 2.25 MeV of energy, and the nucleus recoils with a small amount of energy which is dissipated as thermal agitation, raising the local temperature slightly. In seconds, a proton fuses with the deuteron to make helium-3, again with the release of a gamma ray, but no neutrino. This reaction produces 4.5 MeV. The helium-3 may fuse with another helium-3 (12.86 MeV) producing helium-4 after releasing two protons (which directly heat the gas), but no neutrinos; or it may fuse with a helium-4 nucleus (1.59 MeV), releasing a gamma-ray, but no neutrinos; or it may fuse with a proton (10.16 MeV), with the release of a neutrino having a maximum energy of 18.77 MeV, which is easily detected by the Solar neutrino detectors, except for the fact that this reaction is thought to occur only in 20 chances per million [Bahcall 1990]. The reaction fusing helium-3 and helium-4 produces beryllium-7, which normally decays by electron capture with the emission of a neutrino, with 0.38 or 0.86 MeV energy, with a half life of 52 days. It is not clear how the electronic energy distribution of the plasma might affect this decay.

If a nearby astronomical detonation produces a burst of antineutrinos that induce electron capture in the protons of the core of the Sun, a surge of fusion power would be produced by the reactions just described. This energy would then diffuse out over an extended period of time. As the energy transported by photons is absorbed by atoms in the convective zone of the Sun, the additional thermal energy would increase the pressure and make the Sun expand. The neutrinos produced in this surge of power would be gone, at the speed of light, and would not be observed on Earth, hundreds of thousands to millions of years later. As this excess energy escaped from the surface, the Sun would cool to its normal state, and slowly contract.

It is estimated that the number of neutrinos and antineutrinos in the Universe are approximately equal (and approximately equal to the number of photons), and are much greater than the number of nucleons [Reeves 1994]. Since neutrinos are produced by nuclear fusion and antineutrinos result from nuclear fission, this equality of particles suggests an equality of reactions.

The neutrino, in all its guises, continues to puzzle and perplex us [Goldhaber and Goldhaber 2011].

The Inner Planets

The major planets of the Solar System clearly divide into two types, particularly if we ignore Pluto, as rocky and gas- or ice-giant, and the dividing line falls among the asteroids. Mercury, Venus, Earth, and Mars are relatively small, predominantly solid (rocky). They orbit the Sun in the Inner Solar System. Jupiter, Saturn, Uranus, and Neptune presumably contain small rocky kernels similar to the inner planets, but are predominantly thick, massive atmospheres, with internal pressures so great that the distinction between liquid and gas disappears as the center of the planet is approached. Actually, the core, or kernel, would be much larger than the rocky planets are because none of it boiled away from the outer planets. Keeping their atmospheres allowed them to keep the molten magma oceans. The planets in our current Solar System clearly divide into two sorts: the four small rocky planets close to the Sun (Figure 39), and the four giant planets farther out. The Asteroid Belt serves as a natural dividing line.

Figure 39. Images from space of the four inner planets.
Upper left, Mercury; upper right, Venus; lower left, Earth; lower right, Mars. These are portraits, not to the same scale. (All NASA images, originals in color.)

The inner planets vary remarkably in size, and this obscures some of the similarities (Figure 40). Two are tiny, Mercury and Mars, scarcely larger than Earth's Moon. Venus and Earth are nearly the same size, and are roughly ten times as massive as the small ones. Venus, Earth, and Mars have atmospheres ranging from massive to moderate to meager. As with many other things, the atmosphere of Earth is "just right", for us, just now. Similarities and differences between the inner planets give indications of the fission process in a destructive mode. Based on our current knowledge of these four, there are three dry, dead planets and one wet world, teeming with life. The outward appearances tell only part of the story, and should be considered in conjunction with the relative sizes and internal structure.

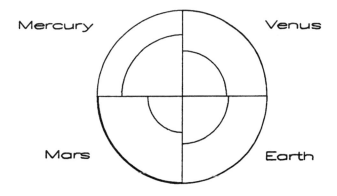

Comparison of internal structure

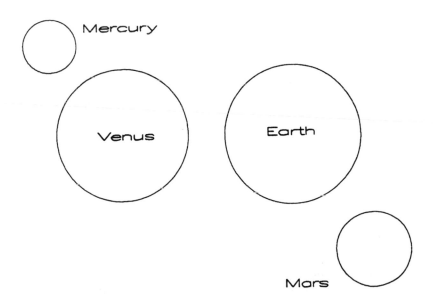

Comparative sizes of inner planets

Figure 40. Comparison of the four inner planets.
Earth and Venus are similar in size, as are Mercury and Mars. The cores differ greatly in relative size.

The ratios of the masses of the core (assumed to be iron) to the mantle and crust (assumed to be silicate rock) of the inner planets are:

Mercury	2.184
Venus	0.4014
Earth	0.4817
Mars	0.2134

It is clear that both Mercury and Mars are quite different, while Venus and the Earth are rather similar.

Having started at the center of the Solar System with the Sun, and bogging down only briefly in the neutrino morass, we can continue our explorations with the innermost planet, Mercury.

Mercury

Mercury orbits the Sun so closely that it completes its circuit four times while the Earth goes around once. So hot it has no atmosphere and tin, lead, and zinc would melt in the afternoon, but so bare that methane would freeze solid just before dawn [Strom 1987], it has ice at its poles [Nelson 1997].

One of our space explorers, *Mariner 10*, visited this planet and returned more information than we had ever known. (Now, 2010, *MESSENGER* is sending us newer news.) Mercury has a magnetic field, about 1% as strong as the Earth's. The planet is more spherical than the Earth or Mars, and is covered with a loose insulating surface regolith. It is denser than any other planet or moon except the Earth, and has a much lower ratio of silicate to iron. It appears to have lost a large fraction of its outer crust. It has been proposed that approximately half the mass of the protoplanet was shocked off by a giant impact [Benz, Slattery, and Cameron 1988].

On Mercury, one side of the planet is dominated by the Caloris basin, the largest known impact structure in the Solar System. The impact seems to have disrupted the crust on the opposite side into "hilly and lineated terrain" [Nelson 1997, Melosh and McKinnon 1988, Wagner 1991].

It is estimated that approximately 2.68×10^{22} kilograms of rock was deposited on the Earth and Moon during the Late Heavy Bombardment [Zahnle and Sleep 1996]. If the average of the Venus and Earth ratios of core to mantle and crust is taken to represent the "standard" for planets of this sort (0.4587), and that reasonable fractions of the ejected mass were captured by the Sun, Mercury itself, Venus, Earth, and Mars, in the following estimates:

Sun	0.50
Mercury	0.24
Venus	0.16
Earth	0.08
Mars	0.02

then the estimated mass deposited on the Earth and Moon from the sloughing-off of Mercury's crust is 2.53×10^{22} kilograms. This is 94% of the mass estimated to have been accreted during the Late Heavy Bombardment.

Mercury's surface is covered with sinuous cliffs and scarps, like the wrinkled surface of a prune, suggesting that the planet has shrunk since its surface was formed. One side of Mercury shows the enormous Caloris basin, an impact crater 1,300 kilometers in diameter, more than a quarter the size of the entire planet. Directly opposite lies "weird terrain", hilly and lineated, found only here, and in a similar occurrence opposite the Orientale and Imbrium Basins on the Moon, and possibly in the Great Basin of western North America on the Earth. The skin of the planet sloughed in response to the terrific impact, shattered and broke apart.

Mercury is covered with impact craters, much like the Moon. Neither were protected from small impacts by an atmosphere, like Venus, Earth, and Mars. Weather has never softened the surface on either body, and Mercury looks much like the Moon. There is a subtle difference. Mercury's surface looks looser, and there are no large seas of lava, the mare that we see on the nearside of the Moon.

The composition of the surface of the Moon has been confidently determined to be anorthosite (according to the text) or plagioclase (according to the figure caption) [Kerr 2009]. However, the crust of Mercury shows no anorthosite, just younger lava composed of entirely different minerals. The present surface of Mercury is likely to be a strange blend of the minerals that formed the earlier mantle and crust, as they returned from being ejected in a major detonation that shocked off the outer portion of the planet.

A cataclysmic event must have occurred in the Solar System soon after the formation of the Moon, after solidification of the lunar crust but while large amounts of magma were still near the surface of the Moon. Evidence for intense impacts on the Moon (and also on Mercury, with less ability to identify a time) is provided by the many craters of approximately the same age, and the large areas of lava-flooded plains, or mare, the lunar seas. This heavy bombardment could have been caused by a partially disruptive detonation in Mercury, in which the outer shell was blasted off from within.

Considering its size and compared to the other inner planets, the planet Mercury has an unusually high density. At present, Mercury is a small world, intermediate in size between the Moon and Mars. Yet its natural density exceeds that of all the other inner planets. The uncompressed (inherent) densities (in grams per cubic centimeter) of these bodies [Seeds 1992] are shown below:

Mercury	Venus	Earth	Moon	Mars
5.4	4.2	4.2	3.35	3.3

I found that astronomers think that the planet Mercury had lost approximately half of its original mass, judging from the size of its core. This is currently explained as the result of a tremendous impact early in the life of the Solar System. However, an internal nuclear detonation could have shocked off the outer crust of Mercury (without totally disrupting the planet) and left a remainder that looks much more like what we see than a reformed planet would. The debris from this detonation could have provided the massive objects that caused the Late Heavy Bombardment of the inner planets. (The impact ejection could have provided these objects also.) These objects are not easily accounted for otherwise.

It has been suggested [Benz, Slattery, and Cameron 1988] that the high density of Mercury is the result of a loss of most of the silicate crust from a differentiated (core/mantle/crust) planet, leaving only (or mostly) the mantle and the core. This is shown in the inner structure of Mercury as a relatively large core and thin mantle. The mechanism proposed for this loss is a major impact collision after development of the planet, that ejected the crustal material from Mercury [Benz, Slattery, and Cameron 1988, Cameron *et al.* 1988]. In this model, it is considered that Mercury was approximately twice as massive as at present. The ejected material was ultimately removed from Mercury's orbit by solar radiation pressure, or for debris in an extremely eccentric orbit, by Venus and the Earth, leaving only the denser interior material to reassemble as the current planet. To produce the absence of iron from the surface, the newly assembled planet would have to go through differentiation and segregation again.

An alternate explanation, based on the nuclear fission detonation process, is that the energy of the shockwave produced by such a detonation was just adequate to eject the silicate crust and much of the mantle of Mercury, leaving the core and inner mantle intact. After ejection from Mercury, the crustal material would be removed by solar radiation or attracted by Venus and the Earth. In this case, however, the disruptive energy would have come from inside the planet rather than from an external impact. From the thorough covering of Mercury's surface with impact craters, it would appear that this event happened somewhat before the period of Late Heavy Bombardment, that is, about 3.9 billion years ago. In fact, there is a suggestion that this ejection of Mercury's crust was the source of the objects that caused the many impacts associated with this period. Spectroscopic studies of the surface of Mercury and a laboratory specimen brought back by the Apollo 16 expedition from the Lunar Highlands show a remarkable similarity, as if the material were the same on both bodies [Taylor 1992]. Further, the Apollo 15 basalts, which have been identified with some uncertainty as volcanic rock that was subjected to a major impact, have been dated at 3.85 billion years ago [Ryder 1988]. The northern light plains on the Moon also resemble the highlands, and are judged to have been deposited by impact [Belton *et al.* 1994]. All this material might be debris from the shock-ejected crust of Mercury. In this explanation, the surface consists of loose rubble from the detonation, fallen back onto the surface.

Mercury's crust looks like the lunar highlands, but there are probably no volcanoes, the plains are not like mare on Moon, and there is no sign of erupted magma. Its thermal history is a problem. It can't shrink enough and it may be too hot inside. The absence of volcanism kept the heat in, and the regolith may be a deep layer of insulation. While basalt is erupted by most volcanoes, and covers the surface of the Moon, it is largely absent from the surface of Mercury. [Jeanloz 1995, Robinson and Lucey 1997].

This proposal is supported by the appearance of Mercury, as shown in the close-up photos taken by the *Mariner 10* spacecraft. The relatively smooth intercrater plains seem to be covered by a crumbled powder, rather than by solidified magma, as in the case of the Moon. Further, the giant impact that formed the Caloris basin and ring structure "shoved" the inner part of the planet through its loose crust, which slid in a loose, but not molten, manner from the antipode of the impact back toward the impact point [Thomas *et al.* 1988]. Mercury shows evidence of ancient saturation cratering of basins about 4.25 billion years ago. This phase ended with emplacement of the smooth plains about 3.8 billion years ago [Chapman 1988, Spudis and Guest 1988, Strom and Neukem 1988]. That is a time-scale that is consistent with our dating of materials and events on the Moon, corresponding to the Late Heavy Bombardment.

The surface of Mercury has been found to have less iron and titanium than the surface of the Moon, by comparison of radio mapping results [Astronomy 1994a1]. This is more consistent with ejection of the previous surface, as proposed here, than with disruption by impact and subsequent re-accumulation and differentiation, in a much smaller planet than before. In the giant-planet theory, Mercury was formed especially large, and so especially hot, from impact heating, and differentiation and segregation were even more effective than for the Earth. All the iron of the protoplanet sank to the core. No trace of iron in the surface materials has been found [Nelson 1997].

The magnetic field of Mercury is very weak [Zeilik and Smith 1987]. This suggests that internal heat generation has declined, and in the view of planetary fission, so has the reactor induced generation of electron currents in the core. The field may be facilitated by high internal temperatures retained by the insulating properties of a very loose, unconsolidated regolith.

Small deposits of water have been detected at Mercury's poles [Slade *et al.* 1992]. Water appears to have been an important constituent in the inner planets, even one as close to the Sun as Mercury, and its disappearance from all inner planets but Earth invites consideration.

Venus

... a riddle wrapped in a mystery in an enigma.
Winston Churchill, 1939

The planet Venus is often the nearest planet to the Earth, and is then the third brightest object in the sky, after the Sun and Moon. It plays the dual role of Morning Star and Evening Star and has been an important object and image in our astronomy and our astrology. It is nearly a twin to the Earth in size, but is viciously different in temperament. Its surface temperature is a uniform 735 K, above the melting points of tin, lead, and zinc. It is shrouded in a nearly opaque atmosphere of carbon dioxide, providing a surface pressure of 93 bar, a pressure found on the Earth only at a depth in the ocean of 800 meters. Its high clouds are highly reflective sulfuric acid. Sunlight penetrated sufficiently to the surface for the *Venera* landers to send back images. There is almost no water in its atmosphere. The thick atmosphere maintains the surface temperature uniform within 5 K from day to night, and from the equator to the poles. It rotates retrograde, "backward", a little slower (243 Earth-days) than it takes to go around the Sun.

The atmosphere of Venus is 96.5% carbon dioxide and 3.5% nitrogen. The surface pressure is 9.3 megapascals (93 bar) and the surface temperature is 735 K, above the critical points of both major constituents of the atmosphere and making the surface atmosphere a supercritical fluid.

Since Venus matches the Earth so closely in size and location, these drastic differences, discovered only recently, came as somewhat of a shock. Venus was once imagined as a lush, tropical sister of the Earth. Now we see it as a candidate for Hell.

The clouds of Venus are composed of droplets of sulfuric acid and water, much like the aerosol ejected high into our atmosphere by the eruption of Mount Pinatubo in 1991. Below the clouds, sulfur dioxide (SO_2) is abundant (180 ± 70 parts per million) but above the clouds, its concentration decreased from 0.090 ppm to 0.003 ppm during the *Pioneer Venus Orbiter* mission, from December 1978 to October 1992.

Carbon dioxide is known to be a greenhouse gas, acting as an insulating blanket, letting sunlight in, but trapping the radiant heat from the surface. This has become a standard model for a planetary climate gone extreme. It is supposed that the heavy blanket of carbon dioxide is able to trap this heat and build up a surface temperature 500 K degrees hotter than it would be without the greenhouse. However, the reflectivity of the clouds of Venus, the Bond (spherical) albedo, is so great that most of the solar energy is reflected. The Earth actually receives more solar energy at its surface than Venus does now. Sunlight must reach the surface for the greenhouse effect to work. Further, the magma that resurfaced the planet was molten at a temperature of 1,200 to 1,500 K, so it has been cooling ever since.

Venus was the tool used by Immanuel Velikovsky to reshape the Earth and Solar System in his ill-fated best-sellers of the 1950s [Velikovsky 1950, Velikovsky 1952, Velikovsky 1955]. Basing his research on a voluminous collection of myth, legend, and religious writings, he proposed that the planet Venus had been ejected from Jupiter a few thousand years ago.

This is generally recognized to be physically impossible (like Wegener's drifting granitebergs as an explanation of Continental Drift) and so his entire work was rejected and ridiculed, angrily, as crackpot science. A somewhat kinder and more rational response might be that he misinterpreted stories that had been repeated from long ago, and were much modified, as newsflashes from the front, and that the development of the Cult of Venus, even before Julius Caesar, misled his attention and his interpretation. Like many others, before and since, he also fell into the trap of being compelled to invent a mechanism for the observations. Unfortunately, his chosen mechanism, the ejection of one planet from another, faced too many physical obstacles. (In the case of this present book, by way of an apology, I "invented" a mechanism, natural nuclear fission reactors, although admittedly several others were there before me. Actually, I simply discovered the possibility of such a mechanism and I have just applied that mechanism to the observations, using what is known about nuclear reactors.)

The ancient myths, legends, and religions have many confusing references to Venus. Perhaps, for now, Velikovsky has done as well at assembling them as can be done, and proper integration must wait for another, more understanding and more scientific try.

I found that astronomers think that the planet Venus underwent a global resurfacing that wiped its face clean of any evidence of events prior to about 500 million years ago. This is not explained at present, only described. A nuclear detonation within Venus would have provided the excess heat that caused the overflowing of its volcanoes, would have produced the peculiar pancake domes, would have destroyed any existing carbonate rocks and converted oxygen into carbon for more carbon dioxide to form its currently hellish atmosphere. Neutron scattering would have preferentially broken hydrogen-1 chemical bonds, compared to hydrogen-2 bonds, and would have destroyed its water in such a way as to produce the high ratio of deuterium to normal hydrogen that we observe.

The current condition of the surface of Venus has been interpreted [Solomon 1993, Kerr 1993] as showing no areas (or almost none) that are younger than about 0.5 billion years (or 300 million years, [Strom *et al.* 1994]) and none that are older, that is, it appears that the planet was re-surfaced in an abrupt event, after which the tectonic and volcanic energy died, equally abruptly. This interpretation is controversial, and depends on difficult interpretations of surface features, but is supported by a study of topography and gravity data [Turcotte 1993]. Plate tectonic activity appears to be absent: the resurfacing was done by mantle plume volcanoes [Schubert 1991]. Observations suggest that there is no current volcanic activity on Venus [Taylor Jr. and Cloutier 1986]. These conclusions are subject to much argument, as the situation is so foreign to our experiences as to make common sense inapplicable [Kerr 1993b, Soloman 1993]. Resurfacing may have been in two events, like the beginning of the Cambrian (about 550 million years ago) and the Permian-Triassic extinctions (about 250 million years ago).

This gory glory of a volcanic inferno may have been echoed on a very small scale many years later. Indications of mantle plume volcanism, like those of Hawai'i, have been found [Smrekar *et al.* 2010]. The age of the eruptions has been broadly estimated to have been about 250,000 years ago. This is extremely recent compared to the resurfacing event at about 0.5 billion years ago, and quite different, as mantle plume volcanoes are different from flood basalt eruptions. (The global eruptions that resurfaced Venus resemble flood basalt eruptions, but probably resulted from the overheating of the planet by a nuclear detonation, rather than from a major impact as proposed in this book for normal flood basalt eruptions.)

At the end of the steady fission phase, the nuclear fuel in a planet is likely to be quite dilute, the reactor volume very large, the fuel isotopes consisting largely of plutonium and other unusual nuclides, and a runaway (positive feedback) stage is likely to develop. This will lead to a detonation. For Venus, as for Mercury, the energy was inadequate to disintegrate the planet, but was sufficient to melt the crust and mantle, destroying existing surface features, and creating an extremely dense new atmosphere.

The amount of lava erupted, as needed to cover the entire surface of Venus, has been estimated to be 100 million cubic kilometers [Kerr 1998a]. This leads to an estimate that it took 1.7×10^{26} joules to melt the magma. If that heat were provided by a nuclear fission detonation (actually, many distributed detonations throughout the planet), the amount of uranium fissioned would be about 2.5×10^{12} kilograms, or about 0.005 parts per billion of the planet's mass, a truly trivial fraction.

Venus is a remarkably smooth world: the NASA image constructors generally apply a vertical scale exaggeration of a factor of ten or more to the computer-generated perspective renderings of the *Magellan* radar mapping to make the Venusian "terrain" look normal to eyes accustomed to that of Earth. Lava filled the basins, volcanoes overflowed, mountains were softened. The lavas appear to have had very low viscosity, and flowed great distances. It has been suggested that some of these lavas were carbonatites, like the lava from Oldonyo-Longai, in Tanzania [Kargel 1997]. But, although the viscosity of carbonatites is low, so is the eruption temperature, and flows (on Earth) quickly congeal. Lava on Venus had to remain hot to continue a flow for great distances, and the surface had to be hot to permit this [Francis 1993].

Pancake Domes of Venus

More than a hundred circular, steep-sided flat-topped domes have been found on Venus by the *Magellan* radar mapping [Pavri *et al.* 1992, Stofan 1993, Burnham 1991, Head *et al.* 1991, Doody 1994]. These are distinctively different from any similar forms on Earth. A global nuclear detonation would cause near-surface deposits of uranium ore to explode, producing ejection craters, slump craters, and pancake domes, similar to what has been observed in the underground tests of nuclear weapons at the Nevada Test Site. A set of pancake domes on Venus, as imaged by the *Magellan* radar-mapping project, is shown in Figure 41.

Each dome is roughly 25 kilometers in diameter. It has been proposed that these flat circular structures are the result of highly viscous lava flows from the central vents [Cattermole 1994]. However, the overlaps of adjacent domes are not elevated, one on top of the other, but instead, show a shortening, non-circular distortion. An alternate explanation that has been suggested is that these domes resulted from processes that produced similar appearing domes under the sea on Earth [Bridges 1994]. However, the centers of the Venusian domes lie at about the level of the surrounding surface [Cooper Jr. 1993]. Therefore, gravitational transport of an outward flow of lava from a vent seems to be unlikely. A detailed comparison of the pancake domes with seafloor volcanoes on Earth showed that the differences were more significant than the similarities, and there could be no functional connection [Smith 1996].

This flatness is suggestive of the slump resulting at the site of an underground nuclear weapon test, when the bubble formed by the detonation deep in the rock collapses and a pit is formed. It is likely that these domes are the result of local shockwaves generated by detonation of subsurface uranium ore bodies. In uniform rock, the shockwaves would produce circular patterns on the surface, as has been seen with underground nuclear weapon tests. The overlapping edges of the pancake domes fit this explanation nicely. A shockwave arriving at the surface, somewhat late due to starting somewhat deeper than the others, in a region in which the rock has just been fractured by an earlier shockwave will be attenuated in the loose rubble, and will travel a shorter distance along the surface. The localized circular layer of rubble is formed as a result of the increase in the tensile stress when the spherical shock wave from the detonation is reflected from the surface at the rock/atmosphere interface. The surface rock is broken, jolted into the air, and falls back in a less consolidated form.

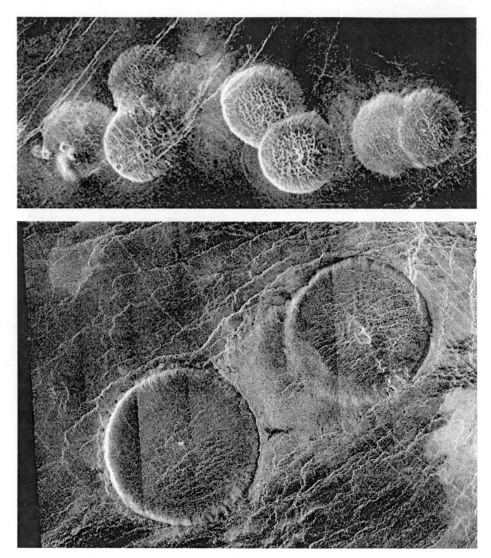

Figure 41. Pancake domes on Venus.
Pancake domes on Venus, imaged by radar mapping by the *Magellan* spacecraft. (JPL/NASA.)

On a much smaller scale, this effect is shown by the SULKY test in Nevada (Figure 42), where a nuclear weapon test device with an energy yield of only 0.092 kiloton of TNT equivalent, buried at a depth of 27 meters, produced a mound of rock rubble [DOD and ERDA 1977]. In this small detonation, only 5 grams of uranium was consumed in the process of nuclear fission, to make this crater.

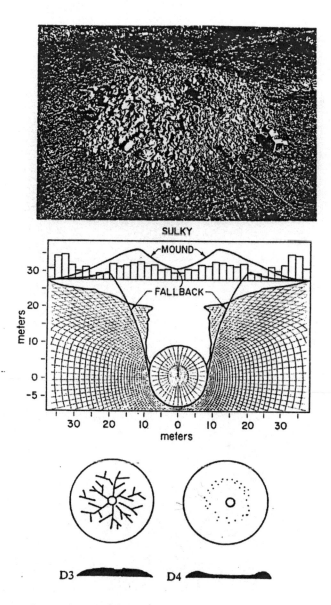

Figure 42. Rubble mound from underground detonation.
Rubble mound produced by underground nuclear detonation. SULKY test, Nevada Test Site, yield of 0.092 kiloton TNT at a depth of 27 meters. Note slight subsidence depression in center. The lower images show schematic diagrams of pancake domes on Venus [Cattermole 1994, after Guest *et al.* 1992].

More energetic nuclear explosions, even when buried deeper, produce ejection craters, with forms much like impact craters. This is shown in Figure 43 for the Sedan Crater in Nevada, and the Meteor Crater, in Arizona. The Sedan Crater was produced by a nuclear explosion with a yield of 100 kilotons, at a depth of 200 meters. The equivalent of only 6,000 grams of uranium/plutonium fissioned, in less than a microsecond (a millionth of a second) [Serber 1992, see also Knauth, Burt, and Wohlhetz 2005].

Scaling up the craters and mounds produced in underground nuclear weapons tests shows that fission of 20,000 to 2,000,000 kilograms of uranium in deep local concentrations would produce craters and domes similar to those shown by the

Magellan mapping of Venus. This is similar to the amount of uranium in the Oklo natural reactor ore body, 100,000 kilograms.

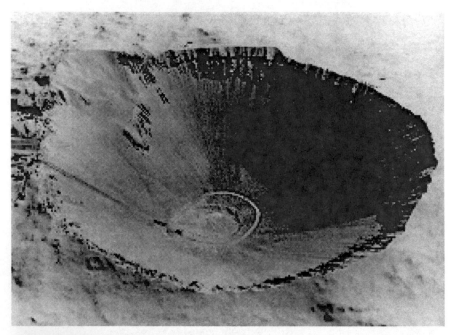

Figure 43. Sedan Crater (nuclear detonation) and Meteor Crater (meteor impact).
Sedan Crater, Nevada, a crater formed by an underground nuclear weapon test (U. S. Department of Energy), and Meteor Crater, Arizona, formed by an impacting meteor (D. Roddy, Lunar and Planetary Institute, original in color).

Volcanic features seen resting on top of lava flows could have been triggered at the same time, but magma for these vents reached the surface after the surface had cooled.

Evidence of intense neutron irradiation is found in the reduced amount of argon-40 detected in the present atmosphere of Venus [McGill *et al.* 1983]. This nuclide results largely from the radioactive decay of potassium-40, a rare isotope of the abundant rock-forming element potassium. The abundance of the element potassium on Venus has been found to be similar to that on Earth, but the argon-40 abundance is less than one-third that on Earth. Neutron absorption by potassium-40, with an absorption cross section of 30 barns (compared to 0.33 barns for hydrogen-1, for perspective), early in the life of the planet, would have reduced the amount of argon-40 that was produced. The isotope potassium-40 has such a small abundance (0.00017) that even complete loss would not be noticeable in an elemental analysis.

The ratio of argon-36 to argon-40 is greater than on Earth [Donohue and Pollack 1983]. In addition to the loss of potassium-40, and the consequent loss of argon-40, argon-36 will be produced by neutron absorption by chlorine-35. This isotope makes up 76% of elemental chlorine, and has an absorption cross section of 43.6 barns. Venus has more argon-36 than Earth, by a factor of 71; Venus has less argon-40 than Earth by a factor of 0.25. These differences show the effect of neutron absorption producing more argon-36 on Venus, from chlorine, and reducing the amount of argon-40 by early fission reactor destruction of potassium-40.

The magnetic field of Venus is weak or nonexistent [Zeilik and Smith 1987]. Here again, this suggests the absence of internal heating at present, and the complete shutdown of the planetary reactors and the dynamo.

Deuterium and Hydrogen

As we will see for Mars and the meteorites, the hydrogen isotope ratios, deuterium to hydrogen (hydrogen-2 to hydrogen-1), D/H, on Venus are different from the ratios commonly found on Earth. This is generally explained as an enrichment of deuterium, but conversely, it is actually a depletion of hydrogen-1. For technical purposes, the difference in the D/H ratios, labeled δD, (usually related to a standard such as Vienna Standard Mean Ocean Water (VSMOW) with D/H in this case equal to 155.76×10^{-6} and $\delta D = 0$), is reported in parts per thousand, or per mil (‰).

Indication of the earlier presence of large amounts of water has been found in the presence of deuterium (hydrogen-2, heavy hydrogen) in the atmosphere [Grinspoon 1993]. This is one of the most thoroughly measured properties of Venus. The ratio of deuterium to normal hydrogen ("protium") on Venus has been found to be 120 ± 40 times that on Earth [de Bergh 1991] or 150 ± 30 or 157 ± 30 or 138 times the Earth [Donahue *et al.* 1997]. As a result of the intense burst of fission neutrons and gamma radiation, water molecules would be dissociated into hydrogen and oxygen. The lighter normal hydrogen (hydrogen-1) would be easily lost to space, while the atomic oxygen would quickly react with the dust and debris ejected into the new atmosphere, making the newly formed crust unusually oxidized [Pieters *et al.* 1986] (just as Mars is). Since the neutron scattering cross-section of normal hydrogen is approximately six times greater than that of deuterium, molecules containing deuterium will be preferentially preserved, resulting in less loss of deuterium to space, compared to normal hydrogen. Some hydrogen-1 will also be converted to hydrogen-2 (deuterium) by neutron absorption.

Scattering of energetic neutrons breaks chemical bonds by the recoil of the struck nucleus. The scattering cross section for hydrogen-1 is about 6 times as large as for deuterium (see Figure 44), so the probability of scattering from hydrogen-1 is that much greater than for deuterium, and more H-C, H-N, and H-O bonds will be broken, than D-C, D-N, and D-O bonds. More hydrogen-1 atoms will be freed to be lost in space. Further, a neutron will impart more energy to a proton (hydrogen-1) than to a deuteron (hydrogen-2) because of the lesser mass of the proton. For scattering off a proton, a neutron can transfer all its kinetic energy to the proton, but can transfer only 89% of its energy to the deuteron. Energetic recoil protons will preferentially disrupt proton bonds (again because of the lesser mass of the proton) than deuteron bonds.

As the neutrons are slowed down by energy losses with each scattering, capture by hydrogen-1 may be significant. This directly produces deuterium, thereby increasing the D/H ratio.

These effects lead to an increase in the abundance of deuterium relative to hydrogen-1, due to the neutrons released in nuclear fission.

Bond strengths for hydrogen with carbon, nitrogen, and oxygen range from about 3.2 electronvolts to 4.4 electronvolts. From its average initial energy as a fission neutron, to below the bond dissociation energy, a neutron will break about 12 bonds. Most of these will be hydrogen 1 bonds and the recoil proton may be lost.

A planetary detonation produces an abundance of high energy neutrons, which preferentially dissociate H bonds compared to D bonds, resulting in a loss of hydrogen-1 and an apparent enrichment in deuterium.

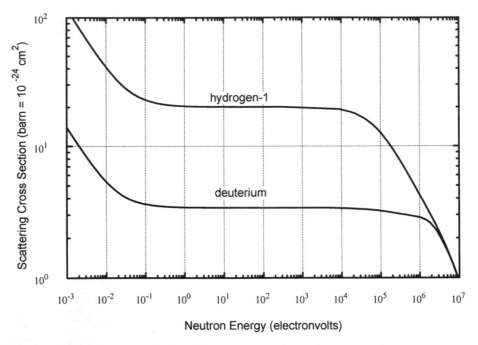

Figure 44. Neutron scattering cross sections for ordinary hydrogen and deuterium.
Comparison of the neutron scattering cross sections of hydrogen-1 and hydrogen-2 (deuterium), showing that scattering is 6 times as likely from hydrogen-1 as from hydrogen-2.

Our Earth and Moon

There's no place like home.
John Howard Payne, 1823

The home made of blue clouds,
I am grateful for that mode of goodness there.
Apache chant, translated by Harry Hoijer.

Of all the planets in our Solar System, we have studied Earth the longest, and the most, and should know it best. Our home planet serves as the standard of comparison for all the rest. We compare the elemental and isotopic compositions of the Earth's thin skin of air, ocean, and crust, with whatever we can find in the meteorites, the Moon, the Sun, and the other planets. We compare mass, distance, geologic activity of all the other planets with what we find on our home world, and our world is the best. It is the only world we know where we can live.

Unfortunately, the comparisons often fail because the differences are too great. We cannot say that Venus is a bit warmer than Earth because it is a little closer to the Sun. It is horrendously hotter and we must conclude that the greenhouse effect overwhelms. Nor can we say that Mars is a bit cooler than Earth because it is a little farther from the Sun. It is bitterly cold and we must conclude that the greenhouse effect failed. Plate-tectonic activity is not just a little weaker on Venus, because Venus is just a little smaller than Earth, it is absent on Venus. Similarly for Mars, Mercury, the Moon, all much smaller than the Earth, and no plate tectonics, no subducting slabs, no continental drift. Earth is still geologically active, while Venus and Mars (and Mercury and the Moon) seem spent. It was not always so, for most of the surface of Venus is covered by massive lava flows, and Mars has the largest volcano known in the Solar System, Olympus Mons. The ice ages of Earth are hard to explain when we see that the ice of Mars has vanished.

Earth is the Goldilocks planet, just right for us. The atmosphere of Venus is too heavy, Mars has too little. Venus is too hot, Mars is too cold. Both of our neighboring worlds seem dead, but may once have had life. Neither Venus nor Mars have oceans at present, although they might have had at an earlier time. Most of the Earth's surface is covered by oceans, and seems to have been for most of its life. Mars appears to have had glaciers that disappeared, and an ocean that covered half the planet. We can't find enough water so we don't know where the glaciers went, or the ocean. Venus has no moon; Mars has only two fragmentary moonlets. Earth has a single large moon, fifth in size among the seven large moons of the Solar System.

Like all the other planets of the Solar System, Earth is unique. Surprisingly, the Moon is not, but the Earth-Moon pair is. Home to mankind, home to the only life we now know. A planet with a mode of goodness that has let life thrive, and change, and wonder. Earth has been home to mankind, and all sorts of other weird creatures, and creatures that can hardly be called creatures, for most of its life. Now, we often puzzle over "why" questions, without quite realizing that we have barely begun to answer the "what" and the "how".

Earth is a rocky planet, as is clear beneath any cover of vegetation. In that way it is similar to the other three planets of the inner Solar System, Mercury, Venus, and Mars. Our deserts resemble the deserts of Mars; our great lava flows resemble the surface of Venus. Our Moon looks like Mercury, but in more details, its kinship to the moons of the outer Solar System shows through.

The best current explanation for the formation of the Earth/Moon system proposes a major impact on the Earth after its accumulation and differentiation [Newsom and Taylor 1989, Canup and Righter 2000.]. It has been assumed that suitable impactors, in the form of uncaptured planetesimals, were plentiful. However, in considering the formation of the Moon as the result of an impact with a "Mars-sized" object, with much of the Moon being formed by the mantle of the impactor [Taylor 1994], it is difficult to end with the right abundances of various elements [Drake 1990, Ringwood et al. 1990]. The problem is that a Mars-sized object is too small a planet to provide the right environment for its own differentiation to the degree required by the observed composition of the Moon. While later studies showed that it would take an impactor two to three times the size of Mars [Astronews 1998, Lissauer. 1997, Ida, Canup, and Stewart 1997], the simulations have been refined so that a glancing blow with a Mars-sized impactor does the job [Canup and Asphaug 2001].

Formation of the Moon in this manner is one of the miracles of modern astronomy. The character of the incoming object is somewhat obscured by naming it a "giant massive impactor". In theory it was a full-sized planet, as large as Mars. Where did it come from? Where did it go? It formed out of place, if the Titius-Bode rule is a guide, did its job, and vanished. These things can be accounted for, but it seems all a bit of a stretch. The simulations have been tuned to give the right answer, which is right if the process actually happened that way.

The Moon has more rock/less iron than Earth. In the natural formation of large moons proposed here, as part of the formation of giant inner planets, the silica ocean on the natural Moon might have stopped boiling sooner than Earth, and so kept more rock relative to its iron. The Lunar rocks are depleted in volatiles, enriched in refractories; Moon has less metallic iron, compared to Earth. Its orbit is more inclined, more eccentric, and it is larger, compared to the radius of its primary, than other satellites. The difference in inclination and eccentricity from what is expected may be from the Late Heavy Bombardment nearly 4 billion years ago, impactors from the shocked-off shell of Mercury, or this disturbance may have happened more recently, as recently as 11,300 years ago.

If the Moon formed naturally, as a satellite to a giant planet, as opposed to an unlikely freak impact, present differences from the Earth came about naturally. The Moon has less volatile material and less iron than the Earth, in proportion to its size. As the Moon was always smaller than the Earth, when it lost its atmosphere from the Sun's ignition, it lost more volatiles easily, since it had a weaker gravitational field. The difference in the iron content is not so much a deficiency in iron as a surfeit of rock. Since the Moon was smaller than the Earth, impact heating was less, and radiation and vaporization cooling were greater. The Moon froze over before it had lost as great a proportion of its rock as the Earth did. If the Moon had been tidally locked in rotation in orbit with the Earth from the beginning of its formation, as seems likely, the near side would have been hotter because of heat radiated from the evaporating molten silica ocean of the Earth. That would have resulted in more mare basalt flows on the near side than the far side. The center of figure/center of mass of the Moon is offset 2 kilometers to 19S, 194E. This could have been caused by a greater loss of material from the nearside, which would have been kept hotter by the shine from the boiling Earth.

The mare basalt of the Moon contains up to 14% iron, but the nearside highlands and large parts of the farside have virtually none. The surface crust of Mercury is also low in iron. Debris from its shock loss would be iron poor also, and would cover the Moon with iron-poor rubble. The Moon's mantle may have 10% iron, but the Earth's mantle has 20 to 30% [News Notes 1995].

During the Late Heavy Bombardment, clusters of multiple impacts stirred still molten lava in Imbrium Basin [Ryder 1988, Ringwood et al. 1990]. The nearside of the Moon developed great seas ("mare") of lava because the still boiling-hot Earth kept the nearside hot, while the farside cooled to solidity.

The surface of Mercury appears to exhibit the spectral characteristics of the mature mare solids on the Moon [Vilas and Melosh 1976, Matson 1977].

Mars

Mars, "the red planet of mystery".
Stuart J. Inglis,
"Planets, Stars, and Galaxies", 1961

While Venus hides its mysteries under a cloudy shroud, Mars lies beneath nearly cloudless skies, that are often filled with dust. Its mystery originated from being a bit too small, a bit too far away, to be seen clearly in our telescopes. At the limits of vision, imagination played tricks and offered hope of life on another planet. Better telescopes, space probes and landers, have eliminated many of the early stories about Mars, but mystery still abounds, and we hope for alien life again.

Mars has played an important part in our myths and theories. To ancient people, it embodied War, with its color of blood. It excited Percival Lowell, with the perceived possibility of civilized, highly technical, life. For Orson Welles, it provided an ominous base for an imaginary invasion of Earth. As our explorations provide more and better observations, Mars now provides us with planetary puzzles.

Interestingly, Mars is unexpectedly small [Wetherill 1992], and its iron core appears to be abnormally small compared to Mercury, Venus, and Earth [Zeilik and Smith 1987, Michaux 1967]: modeled core sizes for Mars range from 11.9% to 25.7% of the planet mass, compared to 32.4% for Earth [Longhi *et al.* 1992]. Furthermore, Mars is very deficient in volatiles, compared to Earth, generally having only 3% of the Earth's concentrations. In an ideal application of the model presented here for planetary formation, the early Mars would have had a mass equal to 0.43 that of the mass of the Earth. Instead, it has only 0.11 the mass of the Earth. Apparently something was lost in the transformation. If Mars had formed as a giant planet, but 0.66 times the giant protoEarth, it might have lost more material (like the Moon) because of lesser gravity and yet kept more rock compared to iron, because it cooled more quickly. It would have lost more volatiles, just like the Moon.

Mars seems to have formed a little differently than the other rocky planets. It is smaller than Venus or Earth, which are so similar to each other. Its core seems smaller, in proportion, than it should be, or it has more rock relative to its core. Its surface was clearly carved by flowing water, perhaps by glaciers as well, yet there is practically no water to be found. Enough liquid water was present to form clay minerals as recently as about 655-274 million years ago. With natural nuclear fission reactors providing heat from within, Mars could have been warm and wet like the Earth, in spite of a more severe Faint Young Sun problem, and would have been covered with water as we observe to have been likely, and perhaps covered with life as well.

If Mars had water, why didn't that water freeze solid when the Sun was too faint? Mars is farther from the Sun than the Earth is, and so the Sun seems 2.3 times fainter. Water frozen on and in the ground should be hard to lose, yet there is evidence for little water at the present time. The small amount of residual water is different from Earth's water, with a deuterium ratio 5.2 x greater than on Earth. Its surface deposits may be volcanic or sedimentary, or both, much like Earth, unlike Venus, where all the surface is volcanic. It has the largest volcano we know of, Olympus Mons, yet now it seems to be geologically dead, with no eruptions for 40 (or 30) million years. With plausible possibilities for past life, it now appears to be sterile. The atmosphere of Mars has 10 times as much carbon dioxide as Earth, but all the other air is gone. Yet, with 10 times as much greenhouse gas as the Earth, the surface temperature of Mars is no different from what it would be without any atmosphere at all. Mars has two fragmentary moons, one rough and rugged, the other soft and fuzzy.

Astronomers think that the glaciers and oceans on the planet Mars were destroyed about 180 million years ago. At present, this is only puzzled over. A nuclear detonation within Mars would have disrupted the frozen groundwater in such a way as to produce the observed chaotic terrain, melted the glaciers, disintegrated the water so as to lose the hydrogen, creating free atomic oxygen which oxidized the surface. The meager carbon dioxide atmosphere is the remainder of an atmosphere destroyed in the nuclear blast, with much oxygen converted to carbon by the Giant Dipole Resonance reaction.

Pancake Domes of Mars

Unusual hills that have been labeled as possible mud volcanoes, have been found by the HiRISE team at the University of Arizona [http://hirise.lpl.arizona.edu/PSP_008561_2205]. That research used images from the High Resolution Imaging Science Experiment on the Mars Reconnaissance Orbiter (MRO). One of these, next to a similar sized impact crater for contrast, is shown in Figure 45. Interestingly, the impact crater could possibly be a detonation crater, like the Sedan Crater (Figure 43), from the detonation of an ore body near enough to the surface that all the overburden was ejected. At least one Martian crater shows a pit in the bottom [Hartmann, year]. Some detonation craters at the Nevada Test Site show a pit at the bottom, where the floor of the crater has collapsed into the cavity produced by the explosion of a nuclear weapon.

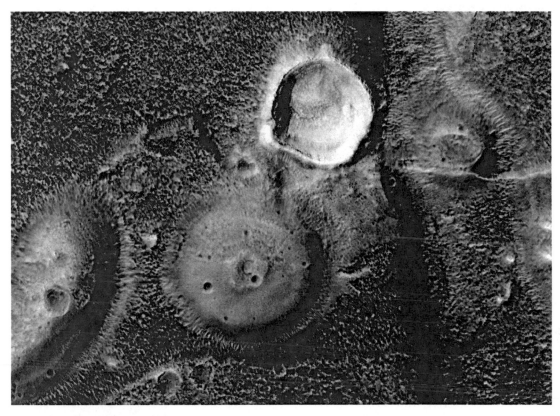

Figure 45. A pancake dome on Mars.
An image from HiRISE showing a pancake dome with a central pit or crater at the summit, adjacent to an impact crater. This dome is approximately 100 meters in diameter. (Department of Planetary Sciences, Lunar and Planetary Laboratory, The University of Arizona)

Typical mounds are described as circular, pancake-shaped domes with pits or craters at the summit. The domes are similar in color to the surrounding surface material, although subsurface material exposed by the impact crater is a different color. This discrepancy requires the speculation that the dome has been covered by a fine coating of dust to match its surroundings. The surface of the domes is finely fractured, considered to be consistent with fractured ice. The domes appear to be ringed by moats, encircling depressions that have been explained as showing surface collapse from the eruption of subsurface material, as a mud volcano [Tanaka 2005, Tanaka et al. 2003]. In some cases, the pancake domes occur as chains of pitted domes [see Figure 10 of Tanaka et al. 2003]. The pancake domes on Mars, imaged in visible light, bear a close resemblance in form to the pancake domes on Venus, imaged by the radar of the *Magellan* orbiter. If the form is the same, it might be suspected that the formation would be the same. Forced detonation of uranium ore bodies, accompanying the planetary detonation that destroyed the oceans, could have produced the pancake domes of Mars, in the same way as the pancake domes of Venus.

The circular, rounded domes with a finely fractured surface result from the shockwave from the local detonation rubblizing the surface. In uniform material this produces an inherently circular rubble pile with a clear boundary. The summit pit is caused by the venting of gas from the explosion. This occasionally happened at the Nevada Test Site (and probably at underground tests of the other nuclear powers). The surrounding moats were caused by the slumping of sub-sea sediments back into the explosion cavity.

Two planets, Venus and Mars, each with their oceans destroyed, show the rubble mounds of a nuclear test site, as pancake domes.

Oceans of Mars

One of the obvious problems with Mars is that it obviously has an ocean basin, the whole northern hemisphere, but it has no ocean.

The detailed topographic mapping of Mars shows two hemispheres, northern and southern, quite different. The northern hemisphere is smooth, much like the bottom of a shallow sea. The southern hemisphere is rugged and cratered. These two different landscapes are separated by a pair of apparent shorelines, as if the ocean had changed its depth. The two shorelines are not parallel, nor evenly level with the current surface of Mars. This has made acceptance of these terraces as ancient oceanic shorelines difficult. An explanation for this has been offered by Taylor Perron [Perron *et al.* 2007a, Kerr 2007, Perron *et al.* 2007b.]. By allowing the planet to roll in response to relocation of masses of ocean water, his analysis was able to reproduce the trace of the two shorelines while not disturbing other features of the planet. The *Science* report does not explain where and how the ocean water was relocated, and that needs a view of the Expanding Mars.

The ocean has been explained away, because quite obviously, there is no water there now. Where did it go, and how? The two shorelines don't quite match, and they aren't quite level. Plate tectonics has been invoked as a possible explanation [Sleep 1994], as has been a supergiant impact, at a glancing angle [Andrews-Hanna, Zuber, and Banerdt 2008, Marinova, Aharonson, and Asphaug 2008, Nimmo *et al.* 2008.]. That impact proposal says that the entire northern hemisphere is the crater from the impact.

The case for an ocean on Mars, at some time, has been strengthened by a study of river deltas. Those deltas are formed where rivers carry sediment into a lake, sea, or ocean, and the reduced flow velocity causes deposition of the sediment. From research on the deltas and river valleys of Mars, and topography from *Mars Observer Laser Altimeter* (*MOLA*), 52 river deltas have been identified [Di Achille and Hynek 2010a, Di Achille and Hynek 2010b, Kerr 2010a]. Of these, 17 appeared to be deposited at the edge of the northern lowlands, where an ocean might be presumed to have been. These deltas all lie at the same elevation around the planet, $\pm$ 177 meters (a range of more than 1,000 feet), and are judged by the researchers to represent river sediment deposits at the edge of an ocean, about 3.5 million years ago.

Large deltas would have been formed as a fully global ocean drained from the south into a newly formed basin in the north, as a result of impact-induced expansion.

The large range in elevation could result from unequal expansion of Mars induced by one of the early major impacts. Ross Irwin (quoted by Kerr in the online version [Kerr 2010b]) has raised the problem that some of the deltas and valley networks lie well below the proposed sea level. These could have been produced by the lesser flood resulting from flow into the lower basin formed by expansion induced by a later major impact.

By the model of planet formation and development proposed here, early Mars was a supercompressed planet, like the early Earth. Large impacts would have initiated great lava flows, like the flood basalts on Earth, and would have caused global expansions.

With the simple assumption that the local density of the planet-forming pre-solar cloud varied with distance from its center as $1/r^2$ (and the planetary masses were proportional to $1/r$), Mars should have a mass equal to 0.66 that of the Earth. Its mass is actually equal to 0.11 of the Earth, suggesting that this assumption is not correct for this region. (Other effects may have altered the mass accumulation and retention, and the smaller core of Mars needs explanation.)

In addition to the impact crater itself, major impacts should produce two effects at roughly the opposite side of the planet from the impact. Major lava flows should be started, and an expansion of some of the supercompressed inner material should result. The lava flows created the large igneous provinces seen on Mars, and the global expansions formed the northern lowlands, a basin that collected what might have been a fully globe-covering ocean. The eventual loss of the ocean water may have caused the roughly 3.3-kilometer offset between the center of mass and the center of figure [Smith 1999]. Crustal magnetism on Mars has been measured by the *Mars Global Surveyor*. A detailed analysis of the patterns shows a remarkable symmetry when reflected about a latitude of 1° N [Connerney *et al.* 2005]. This is consistent with the offset of the centers of figure and mass. This could suggest that the magnetizing field, a global dynamo magnetic field, was in existence when the northern ocean still had water.

On the other hand, the magnetic stripes that were found in the crust of the continental southern hemisphere, by *Mars Global Surveyor*, may have a different meaning. Some researchers have suggested that the stripes resemble the alternating pattern of magnetism straddling the midocean spreading ridges of Earth. They conclude that this indicates the existence of plate-tectonic movements on an early Mars. However, there are no apparent spreading ridges on Mars, and the pattern is in what might be labeled continental crust rather than "oceanic". All these differences suggest that what is seen on Mars is not what we see on Earth. The northern lowlands have no magnetism, and neither do the impacts, Hellas and Argyre, that might have produced the lowlands by impact-induced expansion. This suggests that the impacts and formation of the lowlands occurred after the dynamo of Mars had ceased operation.

I have put together what I have proposed as the effects of Earth Expansion induced by major impacts, and applied that to an early Mars. Mars has two oceanic shorelines, it has two major impact basins (Argyre and Hellas), and it has two major volcanic provinces (Tharsis Montes and Elysium Mons). The impact basins are nearly (now) opposite (antipodal) to the volcanic provinces, just as I have proposed that the volcanic trap eruptions (large igneous provinces) were caused on a less expanded Earth by major antipodal impacts, usually associated with major extinctions due to the change in living conditions. Hellas is opposite Tharsis Montes, and Argyre is opposite Olympus Mons. There are four impact basins identified on Mars, shown in Figure 46 with associated volcanic structures. There are four igneous provinces. These can be easily matched by inspection of the geologic map and selecting pairs that are roughly antipodal. These are:

Hellas (impact)	Tharsis (eruption)
	causing the first major expansion that formed the upper (outer) shoreline.
Isidis (impact)	Olympus Mons (eruption)
	no identifiable expansion.
Argyre (impact)	Elysium Mons (eruption)
	causing the second major expansion that formed the lower (inner) shoreline.
Utopia (impact)	lava flow that produced the southeast portion of the Tharsis plateau.
	no identifiable expansion.

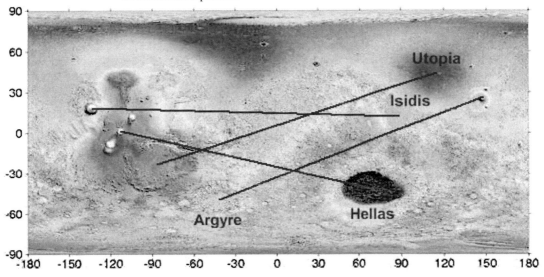

Figure 46. Mars Orbiter Laser Altimeter topographic map of Mars.
Connections between impacts and eruptions on Mars. The angular distances approximate 180 degrees, with impact and eruption being about antipodal.

Perron *et al.* do not discuss the obvious loss of the ocean water that formed the two shorelines. For this, I propose the destruction of the water by the dissociation of hydrogen and oxygen by energetic neutron radiation resulting from a nuclear fission detonation on a planetary scale. I think this detonation may have occurred about 180 million years ago. The hydrogen was lost to space and the oxygen reacted with the surface minerals, making the Red Planet red. (Produced as atomic oxygen by the breakup of water molecules, H_2O, much of the oxygen may have been left in a "superoxide" state, making it very effective in oxidizing minerals.) Some oxygen was converted to carbon, to form carbon dioxide, by the Giant Dipole Resonance reaction.

Convincing evidence for the early presence of large amounts of liquid (and frozen) water on the surface of Mars has been shown [Squyres and Kasting 1994, Kargel and Strom 1996, Zent 1996] and these raise the overwhelming theoretical obstacle to liquid water on that planet. That is the so-called Faint Young Sun Paradox, written about by Carl Sagan in 1972, a major problem for Earth, presumably insurmountable for Mars. If the Sun has been steadily expanding since it became stable 4.6 billion years ago, as our theories show, it was not bright enough for the first 3 billion years to keep water liquid on the surface of the Earth. Mars is 1.5 times as far from the Sun as is the Earth, and so receives only 43% as much solar heat. Thus, a Sun too small to melt the ice on Earth would have left Mars solidly frigid, or at least frosted, throughout its history.

Major glaciation occurred on Mars between 2 billion years ago and 0.3 billion years ago. A Martian "Little Ice Age" occurred between 300 million years ago and the recent past [Kargel and Strom 1992, Carr 1987, Baker *et al.* 1991]. Major water floods occurred on Mars, presumably after the Little Ice Age, and this was followed by volcanism so recent the deposits are still apparent from space, as yet unobscured by the blowing sand [Lucchitta 1987, De Hon and Pani 1993]. The end of the Martian Little Ice Age dates the nuclear detonation, which was the terminating event, resulting in the melting of the glaciers and the destruction of the water.

Catastrophic flood channels required the flow of enough water to form a global layer at least 0.5 kilometer deep. Volcanism and intrusive magma should have released enough for 3 to 6 kilometers more.

The evidence is strong that the Faint Young Sun Paradox is false, as presented by Kargel and Strom for Mars. The deficit in solar heat is significant, amounting to about 57% for Mars at the beginning of its life. The surface is not yet warm enough for liquid water to flow, but glaciers accumulated and marked the landscape, and water flooded through the canyons and over the plains. Therefore, some other effect must have compensated for the faint young Sun.

The mass of carbon dioxide on Mars is 23 quadrillion kilograms. Earth's carbon dioxide amounts to less than 2 quadrillion kilograms. The surface area of Earth is 3.5 times that of Mars, so our carbon dioxide is spread even thinner. The Martian greenhouse should be 46 times greater than Earth's. On the Earth, the assumed warming effect of the greenhouse is estimated to increase the surface temperature by about 40 K, from 247.3 K due to the Sun alone to a very livable 288 K (15 C, 59 F). On Mars, the temperature calculated for solar heating alone is 216.6 K, and the estimated average surface temperature is 214 K, nearly 60 degrees below freezing [Lodders and Fegley Jr. 1998]. On Mars, a much thicker carbon dioxide greenhouse than on Earth seems to have no effect at all.

While no generally accepted source of heat, nor modifications to the climate models, can overcome this deficit, a very natural source of heat has been overlooked: natural nuclear fission reactors. (The natural reactors in the uranium ore deposits of Oklo have been known for almost 40 years, but the aspects of natural reactors on a planetary scale have only been discussed in little-noticed papers by Driscoll, and Herndon, and myself.) Such reactors, operating from before the formation of the planets, could have provided the supplemental heat needed to keep our planet, and Mars, warm and wet, and provide the excess heat that is still evident in some of the other planets and satellites of the Solar System.

At present, there appears to be no surface water on Mars, but tantalizing hints keep coming back to us from our spacecraft explorers.

The "chaotic terrain" of Mars consists of broken, tumbled ground that often lies at the head of flood erosion features and is thought to have been the source of the floodwaters. The ground appears to have been abruptly disrupted by the eruption of groundwater [Taylor 1992, Carr 1981]. Here, it is proposed that ground ice in these locations was melted and partially vaporized and dissociated by the intense neutron flux, gamma radiation, and heat from the nuclear fission detonation. The dissociated water recombined violently in the ground, before it could escape, adding to the energy of the eruption. The atomic hydrogen and oxygen reacted rapidly and energetically, like Brown's gas [Jueneman 1998]. The erupted water then flooded across the plains and the ocean bottoms, emptied of their water by the radiation dissociation of its molecules.

Deuterium and Hydrogen

On Mars, as we saw for Venus and will see for the meteorites, the hydrogen isotope ratios, deuterium to hydrogen (hydrogen-2 to hydrogen-1), D/H, are different from the ratios commonly found on Earth. This is generally explained as an enrichment of deuterium, but conversely, it is actually a depletion of hydrogen-1. A planetary detonation produces an abundance of high energy neutrons, which preferentially dissociate H bonds compared to D bonds, resulting in a loss of hydrogen-1 and an apparent enrichment in deuterium.

In the atmosphere of Mars, the deuterium-to-hydrogen ratio is 5.2 or 6 times that on Earth [Donahue 1995, Owen *et al.* 1988], showing the effect of the greater dissociation of the water molecules and the loss of hydrogen to space [Owen 1992]. It is estimated that at least 99% of the water was destroyed and the hydrogen lost, to produce this ratio. This is about the ratio of H/D scattering, which would preferentially break H-O bonds compared to D-O bonds. The observed ratio of HD/H_2 is smaller by 11x than the HDO/HOH ratio. I think that is consistent with recombination of H^+ from neutron dissociation of water, into H_2.

Scattering of energetic neutrons breaks chemical bonds by the recoil of the struck nucleus. The scattering cross section for hydrogen-1 is about 6 times as large as for deuterium (see Figure 44), so the probability of scattering from hydrogen-1 is that much greater than for deuterium, and more H-C, H-N, and H-O bonds will be broken, than D-C, D-N, and D-O bonds. More hydrogen-1 atoms will be freed to be lost in space. Further, a neutron will impart more energy to a proton (hydrogen-1) than to a deuteron (hydrogen-2) because of the lesser mass of the proton. For scattering off a proton, a neutron can transfer all its kinetic energy to the proton, but can transfer only 89% of its energy to the deuteron. Energetic recoil protons will further preferentially disrupt proton bonds (again because of the lesser mass of the proton) than deuteron bonds. As the neutrons are slowed down by energy losses with each scattering, capture by hydrogen-1 may be significant. This directly produces deuterium, thereby increasing the D/H ratio. These effects lead to an increase in the abundance of deuterium relative to hydrogen-1, due to the neutrons released in nuclear fission.

Michael Carr has estimated an inventory of surface water equivalent to a depth of 1 kilometer before the D/H fractionation [Carr 1996]. That mass of water is about 1.4 x 10^{17} kilograms/kilometer of depth, or about 4.68 x 10^{42} water molecules for a depth of 1 kilometer.

The average number of neutron scatterings from an average fission energy of 2 MeV down to 5.18 eV (the energy of the H-O bond), is about 13. That takes 3.6 x 10^{41} fission neutrons, or about 1 x 10^{41} fissions, or 3 x 10^{30} joules, to disrupt 4.68 x 10^{42} molecules of water. That is about 1/2 the energy release required to disrupt Mars, according to my figure. It could take less, because of gamma-radiation and energetic recoil protons, so say 1-3 x 10^{30} joules. Since this is less than the energy required to disrupt the planet, it shows why Mars is still there, and is not just another asteroid belt. The hydrogen vanished into space, deuterium remained bound in water molecules, the oxygen was consumed by the minerals of the hot crust, and a thin residual atmosphere of carbon dioxide was left.

Measurements on meteorites that are thought to have come from Mars (SNC meteorites), showed variable enrichments of deuterium, compared to Earth's water [Watson *et al.* 1994, Jakosky and Jones 1994]. Further, amphibole phases in the meteorites contained only 10% as much water as expected for those minerals, suggesting that water was directly lost by the neutron scattering. This would occur by the neutron scattering breaking the H-O bonds, with diffusive loss of the free H.

On the surfaces of Mercury, the Moon, and Mars, impact craters (and some uncertain volcanoes) are clearly visible. Some craters can also be seen on Venus by use of radar, and some have been identified on Earth. Volcanic eruptions, and water flooding, work to erase the existing craters from previous impacts. The surface can be dated by assuming a steady, and known, rate of impact.

Large impacts are rare, so large craters are rare. Small impacts are frequent, and small craters are common. On Venus and the Earth, the atmospheres prevent the penetration and impact of objects below a certain size range. On worlds with little or no atmosphere, all incoming objects hit the surface hard.

Counts of craters as a function of size can be used to estimate the age of surfaces. A fresh lava flow may have no craters, one of intermediate age may have many small craters but few large ones, while an ancient surface may have so many small craters that they have reached an equilibrium state of overlap, and some medium and large craters are present.

From crater counts, it appears that volcanoes on Mars erupted from the earliest times, back to before 3.4 billion years ago, and progressively shut down. There seems to have been a boost in activity in the last half-billion years, with three volcanoes still, or newly, active at that time. The largest, Olympus Mons, was the last. Its final eruption may have been about 70 million years ago [Hartmann 1999].

The Tharsis Montes lava plains have been estimated to be about 1.4 billion years old, and direct radiometric measurement of the Martian (SNC) meteorites show ages of 4.5 billion years, 1.3 billion years, and about 170 million years [Hartmann 2000]. A crater count for Arsia Mons suggests an age in the range of 100 to 300 million years.

These ages are subject to some random errors due to the distribution of counts. They are also subject to systematic errors due to an assumption of a constant impact rate, after the end of the Late Heavy Bombardment at about 3.8 billion years ago.

High-resolution topography of the Martian surface by *Mars Global Surveyor* shows that the northern hemisphere is smooth and lower than the southern hemisphere, which is ruggedly cratered [Smith *et al.* 1999]. The northern lowlands are as smooth at the 100-meter scale as the abyssal plains of our oceans (where the gentle rain of deep-sea sediment falls to lie undisturbed) and are flatter than any other known surface in the Solar System [Kerr 1998b]. The northern lowlands appear to be edged by two "shorelines", suggestive of the existence of an ocean.

In a process of remarkable luck, sometimes an impact on Mars (or on the Moon, also) will blast fragments of the surface into space, where eventually, some are captured by the Earth and brought to ground. When found, these impact fragments can be identified as from Mars (or more properly, not from Earth, but we already knew that) by the similarity of various isotope ratios to what we have measured on Mars. Other planets may generally be ruled out as the source of these fragments. The gas-giant planets are unlikely sources because of their thick atmospheres, through which an impact fragment would have to pass, similarly for Venus. These atmospheres are too heavy to allow the escape of anything from the surface. Mercury is a possibility because of its low gravity and the absence of an atmosphere, but fragments are not so likely to escape from so close to the Sun, to reach the Earth. It's uphill all the way. (If Mercury lost a major fraction of its outer stony shell by a detonation early in the life of the Solar System, some of that material quite likely reached the Earth and Moon. Some very few fragments of that debris may still be in space.)

The Martian meteorites are labeled SNC ("snick") for the primary, initial recovery locations, Shergotty, Nahkla, and Chassigny, when it became evident that these meteorites shared similarities that suggested original sources on Mars. Now, most SNC meteorites are found in Antarctica, where the action of the continental icesheet serves to preserve and protect the meteorites, and present them for our collection.

The meteorites from Mars show evidence of the presence of mineral-bound water on Mars [Karlsson *et al.* 1992, McSween Jr. and R. P. Harvey 1993] as well as aqueous alteration [Gooding 1992, Gooding *et al.* 1991, Treiman *et al.* 1993]. Further, from examination of the surface features of Mars, it is evident that large amounts of water were present on this planet in the geologically recent past, also: canyons, flood plains, glacial remains, and other landscapes, give strong evidence of the action of quantities of water normally associated only with the Earth. Many of the glacial features give the impression that the glaciers simply and abruptly vanished.

While the chronology of the shergottites, the meteorites from Mars, is subject to some disagreement, a reasonable history can be constructed. According to samarium-neodymium (Sm-Nd) dating, the material from which these meteorites were derived crystallized as a cooling magma at 1.3 billion years ago, was subjected to a severe shock (30-45 gigapascals, 300,000 to 450,000 times the atmospheric pressure of Earth) at 0.18 billion years ago according to the crystalline rubidium-strontium (Rb-Sr) dating, and was exposed to cosmic rays in space for a period of 0.5 to 2.4 million years before reaching the surface of the Earth [Bogard *et al.* 1984].

These meteorites show evidence of a Martian magnetic field during the time of the cooling of the magma [McSween Jr. 1987] at 1.3 billion years ago. No significant magnetic field exists at the present time [Luhmann *et al.* 1992], and therefore the core dynamo has ceased to operate.

Considering the nuclear fission detonation process, these events may be interpreted as indicating that the original magma cooled slowly as basalt, deep below the Martian surface, allowing the mineral crystals to develop and grow, and to capture evidence of the Martian magnetic field, at 1.3 billion years ago. The nuclear detonation occurred at 0.18 billion years ago, severely shocking the basaltic material. In what may be a related event on Earth, the Dogger Epoch of the Jurassic Period began at 178 million years ago. Significant volcanism on Earth occurred during this time, Mount Kirkpatrick in Antarctica [Earth 1994], and the Palisades in North America. Mount Kirkpatrick lava has been dated to 177 ± 2 million years ago [Hammer and Hickerson 1994]. This eruption seems associated with lava flows that formed the Palisades along the Hudson River, in New York [Birkeland and Larson 1989], as part of the CAMP flood basalts [Marzoli *et al.* 1999]. Those flood basalt eruptions were a bit earlier than Mount Kirkpatrick, and were followed, at the time of the Dogger Epoch, by the Karoo-Ferrar flood basalt eruptions. These eruptions were probably initiated by the shockwaves from major impacts, on the opposite side of the Earth. These impacts could have induced expansion of the Earth that made it no longer suitable for the forms of life that had previously thrived, and produced conditions that favored new species. On Mars, the basaltic material was then brought to the surface in the violent volcanism resulting from the tremendous heat released in the core by the nuclear-fission chain-reaction. Between 2.4 million years ago and 0.5 million years ago, a meteoritic impact on Mars ejected some of this material, shocking it only weakly [Melosh 1993, Gratz *et al.* 1993] and propelling it into space. After a few million years in space, or less, these splinters fell to Earth.

The deuterium/hydrogen ratio of bound water in the SNC meteorites shows a variable increase up to roughly five times the terrestrial ratio [Watson *et al.* 1994]. A deficiency in water content, of about a factor of ten, was also found. These two observations show the effect of neutron scattering by water molecules, with the H-O bonds six times more likely to be broken, because of the greater (n,p) scattering cross-section, than the D-O bonds. Because of the high rate of diffusion of hydrogen in solids, recoil protons would be lost from the mineral grains by diffusion, resulting in a deficit of water and the increased oxidation of the Martian minerals.

The *Viking Lander* biochemical experiments on the surface of Mars in 1976 found high levels of oxygen in the Martian soil. This was expected to indicate the action of photosynthetic plants, but the gas chromatograph-mass spectrometer (GC-MS) found no organic compounds at the parts-per-billion sensitivity of the instrument. Organic compounds, from the accumulation of meteorites and comets, are found on the surface of the Moon at concentrations of a few parts per million, a thousand times greater than what was not found on Mars [Taylor 1992]. Organic compounds would be destroyed by the intense neutron flux accompanying the nuclear detonation, and further oxidized by the nascent oxygen released by the neutron-scattering destruction of water molecules.

An interesting situation exists with an Antarctic meteorite, the Elephant Moraine A79001 (EETA 79001) shergottite [McSween Jr. 1987]. The nitrogen isotopic composition in the carbonate-rich fraction of this meteorite does not match that in the glassy inclusions, which seem to have absorbed atmospheric gases while on the surface of Mars [Owen 1992]. This can be explained by intense neutron capture. These isotopic effects would exist if the glassy nitrogen were from the Martian atmosphere after the detonation, and the Martian nitrogen were enhanced in nitrogen-15 due to neutron absorption by nitrogen-14; and if the carbonate nitrogen were from neutron absorption by nitrogen-14, producing carbon-14, assimilated into the carbonate, and decaying back to nitrogen-14, producing nearly pure nitrogen-14.

Two anomalies related to the noble gas argon have been found on Mars [Owen 1992]. The total amount of argon is a factor of 16 less (per gram of rock) than the terrestrial value, and the ratio of $^{40}Ar/^{36}Ar$ is 10 times greater than on Earth. This reflects the neutron destruction of the argon isotopes, with proportionately more argon-36 being destroyed because of its relatively greater neutron-absorption cross-section, compared to argon-40. The cross section for argon-36 is 5.2 barns, for argon-40 it is only 0.66 barns. The ratio is roughly 8, rather close to the factor of 10 that is observed.

In the case of Mars, it appears that the detonation was sufficiently strong to disperse most of the atmosphere, dissociate most of the water molecules, and produce a temporary, extremely oxidizing atmosphere, while the crust was at a high temperature. With its nuclear fuel consumed in the detonation, the core of the planet would cool, the dynamo would run down, and the magnetic field would fade away.

Xenon and Iodine

Mars has an excess of xenon-129 [Gilmour, Whitby, and Turner 2001]. This would be produced by neutron irradiation (during the detonation) of the iodine chain as I discuss later.

The Asteroids

> *... fragments of a primeval planet,*
> *... blown to pieces.*
> Heinrich Wilhelm M. Olbers, 1807

If this were a children's book, we could play that the Asteroid Belt is Tearalong, the Dotted Lion, for it divides one part of our planetary system from the other.

Asteroids are "starlike". Most cannot be resolved into discs but appear as points of light, like the stars. They shine by reflected sunlight, and so have a spectrum in the visible range, like the Sun, like many stars. There are more than 30,000 identified asteroids with determined orbits [Bowell 2003]. Observationally they only differ from stars by having planetary motions in the sky that allow us to calculate their orbits. Most of these orbits lie between 2 Astronomical Units and 4 Astronomical Units from the Sun, and form the Asteroid Belt. Some asteroids pass closer to the Sun than the Earth does, and pose a potential threat of impact on the Earth. These are Near Earth Objects (NEOs) or Near Earth Asteroids (NEAs). Some, the "Trojans", orbit ahead and behind the planet Jupiter. Unlike the nearly circular orbits of the major planets, which all lie close to the plane of the ecliptic (Earth's orbital plane), the orbits of the asteroids are significantly eccentric and inclined. They are scattered all over their neighborhood. Figure 47 shows some indicators of the departure of these orbits from the plane of the ecliptic (semimajor axis x sin(inclination)) and from a circular orbit (semimajor axis x eccentricity).

The first asteroid, 1 Ceres (asteroids are numbered in the order of their discovery, and may be named also), was discovered accidently, just as a search for the planet predicted by the Titius-Bode Law was to begin. Giuseppe Piazzi discovered a slowly moving star on January 1, 1801, in Sicily. It might be a comet, by its nightly motion, or better, it might be the Bode's Law planet predicted for the gap between Mars and Jupiter, at a distance of 2.8 Astronomical Units from the Sun. The name Ceres was taken from the matron goddess of Sicily. Was this the planet predicted to orbit between Mars and Jupiter? Before more observations could be made, the object passed into the light of day and was lost. Carl Friedrich Gauss, one of the world's greatest mathematicians, invented an analytical method for determining the orbit from only three observations, so that the object would not be lost, as it moved into the daytime, sunlit portion of the sky. The calculated distance lay in the gap between Mars and Jupiter, at a distance of 2.8 Astronomical Units, where the Titius-Bode Law predicted a planet should be. Wilhelm Olbers found Ceres again on December 31, 1801, where Gauss' calculation predicted it would be. The present values for the orbit of Ceres, the largest of the known asteroids, place its average distance from the Sun at 2.767 Astronomical Units. This compares very closely with the Titius-Bode prediction of 2.8 Astronomical Units. (The Titius-Bode Law worked for the location of Uranus, discovered in 1781, after the law had been formulated in 1776, based only on the six planets known at the time. When applied to satellite systems, it also predicted Hyperion, one of the small moons of Saturn.)

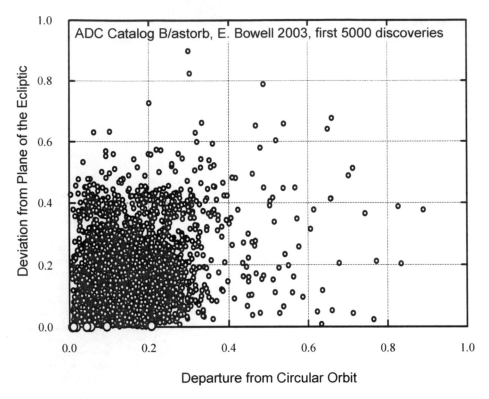

Figure 47. Asteroid orbits.
Departures from the plane of the ecliptic and from circular orbit, for the first 5,000 asteroids discovered. Data from the Astronomical Data Center at the University of Maryland. The eight major planets are shown as open circles at the lower edge of the plot. While Mercury has a significant eccentricity (0.2), all the planets lie close to the plane of the ecliptic, but the asteroids are scattered.

Even recently, asteroids have been lost. Apollo, Adonis, and Hermes were lost after discovery. Apollo and Adonis were relocated, but Hermes remained lost, until it was rediscovered in October 2003. Hermes is an Earth-crosser, a Near Earth Asteroid, and it had passed close to the Earth four times since it was discovered and lost, before it was rediscovered. Four close passes, without us being aware of its presence. Many close misses have occurred, with the interloping asteroid undiscovered until the last moment, or even later after the pass, from review of photographs taken a few nights earlier. If we had been hit, we probably wouldn't know anymore than the dinosaurs did.

Soon, other asteroids were discovered by visual searches (using telescopes), and astronomical photography eventually led to the discovery of hundreds and then thousands of these tiny objects. (Tiny is a relative term. Ceres, at 914 kilometers in diameter, is as large as France.) Now, over 249,568 asteroids have been sighted, and 30,769 have calculated orbits. The count changes rapidly, as automated telescopes and computer processing add new asteroids nearly daily (or nightly). Although Ceres is the largest known, it is much smaller than expected for a planet. It was clearly not the planet that had been sought. As more asteroids were discovered, Olbers (or perhaps Jan Oort) proposed that these were fragments of a single large planet that ruptured [Binzel, Barucci, and Fulchignoni 1991], blown to pieces by action of internal forces [Bailey, Clube, and Napier 1990].

There are only four large asteroids (1 Ceres, 4 Vesta, 2 Pallas, and 10 Hygiea) but there may be 1 million asteroids larger than 1 kilometer in diameter. All the fragments put together account for only 0.0004 of the mass of the Earth. The second-largest asteroid, 4 Vesta, roughly 529 kilometers in diameter and about 2.67×10^{20} kilograms, has been imaged by the *Dawn* spacecraft, the first to orbit an asteroid (Figure 48).

Figure 48. The asteroid 4 Vesta as imaged by the *Dawn* spacecraft.
This asteroid is the second largest in the Asteroid Belt (NASA).

Pallas was discovered in 1802, Juno and Vesta in 1807, and it was clear, from the star-like appearance, that these were not fully formed planets. William Herschel named them "Asteroids" from the Greek for starlike, because of their appearance.

Binzel says that the gravitational perturbations that cause groups, concentrations, and gaps to form in response to Jupiter's motion, prevented formation of a real planet from the planetesimals in the orbital zone between Jupiter and Mars. These perturbations produce shattering collisions making smaller pieces, rather than accretion, making larger rocks from small. Presumably, the rest of the material, perhaps 99.96% of the mass of the Earth or possibly much more, was lost to space, ejected from the Solar System, captured by the Sun, or impacted on the other planets. This could have happened so early in the life of the Solar System that there is no surviving evidence. (Some things may be unknowable, absolutely, and the best we can do is invent reasonable stories.) The Sun serves as a giant garbage disposal for comets at the present time, and it is likely that much of the debris was swept up in a short time by the Sun [Farinella *et al.* 1994].

However, as we receive samples of them in the form of meteorites, the asteroids seem to have been geologically differentiated, although even the largest are too small for significant gravitational, thermal, and chemical segregation. Most asteroids have irregular lightcurves, indicating an irregular shape or surface brightness, and most are so small as to be unable to gather themselves into spherical shapes [Kiel *et al.* 1997, Tonks and. Melosh 1992].

Spectroscopic comparisons with meteorites and Earthly minerals show that the asteroid Vesta has a crust that appears to be a surface lava flow [Gaffey, Bell, and Crruikshank 1989], while Ceres and Pallas have been extensively altered by water.

Obviously, until we sample an asteroid deeply, we can only see the surfaces, and so must draw our conclusions based on the surface dirt alone. A compositional gradient appears to exist relative to distance from the Sun. Asteroids most distant from the Sun have carbon and water on the surface. Those in the middle have less volatiles, less water. Those closest to the Sun have complex mineral assemblages. These may show the underlying material, while the others have been coated with whatever compounds were available and could condense on cold surfaces at temperatures varying with the distance from the Sun.

Two unusual asteroids, which may not be related to the Asteroid Belt, seem to be connected to Saturn. 944 Hidalgo orbits between Saturn and the Asteroid Belt ("inside" Saturn's orbit) while 2060 Chiron orbits between Saturn and Uranus ("outside" Saturn's orbit). Chiron is confusing, or at least confused. It has the orbit of an asteroid (almost) but shows almost the coma of a comet, and so is also listed as a periodic comet, 95P/Chiron.

There are four outstanding problems in the asteroids:

1.   Why are there fragments where a planet should exist?
2.   How did the fragments come to be geologically differentiated, like portions of a full-size planet?
3.   How did the fragments come to be magnetized like parts of a planet?
4.   Why do the fragments have the irregular shape of fragments, rather than the more rounded form expected for impact abrasion?

Several asteroids have been imaged by space probes. In addition, ground-based radar has produced some asteroid shapes. These show objects that are much like the two moons of Mars in appearance. This has supported the idea that the moons of Mars are like asteroids, but the reverse relation should be kept in mind. The asteroids may be fragments, like Deimos and Phobos.

The Near Earth Asteroids form three separate classes. The Amor asteroids approach Earth from outside its orbit; the Apollo asteroids cross Earth's orbit, and so approach from both outside and inside; and the Aten asteroids, that approach Earth from the inside [Kowal 1988].

Meteor Crater (Barringer Crater, Arizona) was caused by the impact of an iron object, a presumed asteroid (or fragment) about 60 meters in diameter, several trillion kilograms of mass, at a speed of 11 to 15 kilometers per second. (For comparison, a high-speed rifle bullet leaves the muzzle at 1 kilometer per second.) For an object of this size and density, very little energy is lost in its passage through our present atmosphere. That impact released the energy equivalent to a 15 megaton nuclear detonation, in a moment of time that is similar to a nuclear detonation.

Because of the possible consequences of such an impact, an automated search for objects with orbits within 1.3 Astronomical Units from the Sun (close to Earth's orbit at 1 Astronomical Unit) was started in 1995. The present best estimate of the number of such asteroids with diameters greater than 1 kilometer (a global threat to life on Earth) is 700 ± 230. The researchers estimate that 90% of the near-Earth asteroids may have been detected within the next 20 years [Rabinowitz 2000]. A defensive strategy has not yet been devised, much less implemented [Jewitt 2000]. Successful defense of life on Earth, perhaps all intelligent life in our corner of the Universe, will depend on a thorough understanding of the risks and hazards of cosmic impacts.

The Asteroid Belt forms the borderline between the rocky inner planets and the giant outer planets. In this section it is proposed that a fully differentiated planet had formed in that zone, as predicted by the Titius-Bode Law, and that an uncontrolled fission chain-reaction within that planet, and the consequent detonation, resulted in the formation of the asteroids, as fragments, as shrapnel.

The simple numerical rule of Titius and Bode [Seeds 1992] predicts that a planet should exist at 2.8 Astronomical Units (1 Astronomical Unit is the mean distance from the Sun to Earth, approximately 1.5×10^8 kilometers), between the orbits of Mars and Jupiter. There is no planet there, only fragments, and that raises the first problem of the asteroids. Why are there only fragments where a planet should exist?

The currently favored model for the formation of the asteroids involves gravitational disturbance of the forming planetesimals in orbit between Mars and Jupiter at the time the Solar System formed. It is thought that perturbations by Jupiter prevented accumulation of a complete planet, and ejected most of the mass in that region out of the Solar System [Wetherill 1992]. The planetesimals that had formed out of this material destroyed each other by mutual collisions, producing the tiny asteroids. Rather than being part of a cohesive, gravitationally bound planet, with a mass similar to that of the Earth or larger, these bodies are scattered through the Asteroid Belt. They have varying velocities relative to each other of up to 5 kilometers per second [Wetherill 1992] and collisions can cause further breakups. This theory raises some additional problems. If the fragments came from the breakup of planetesimals, how did the material come to be chemically, mineralogically, and geologically differentiated, like portions of a full-sized planet? This is seen spectroscopically with the asteroids themselves, and up close with the meteorites, which divide into stones, stony-irons, and iron-nickel, with several subdivisions along the way. How did the fragments come to be magnetized as if they had been parts of a planet? Why do the fragments have the irregular shape of fragments, rather than the rounded shape of impact abrasion?

The standard theory of formation of the asteroids is described well by Binzel [Binzel, Barucci, and Fulchignoni 1991], and the details of an alternative process, derived from computer simulations, were presented by George Wetherill [Wetherill 1992].

The eventual gravitational effects of Jupiter, on the individual asteroids, are clear at the present time. There are significantly depleted zones, the Kirkwood gaps, in the Asteroid Belt, that correspond to orbital period resonances with Jupiter [Binzel, Barucci, and Fulchignoni 1991].

Most meteorites come from the asteroids. In 1794, E. F. F. Chladni, the "Father of Meteoritics", suggested that the meteorites came from space, and had been produced by impacts or explosions of planetary bodies [Marvin 1994, McSween Jr. 1987]. This idea was seized by Wilhelm Olbers when he discovered the second asteroid, Pallas, in 1802 [Berry 1898]. For many years after the discovery of the asteroids, the idea that they had been produced by an explosion held favor [Inglis 1961]. James Jeans thought that a single, quite ordinary primeval planet had been formed, but was disrupted by passing through the gravitational field of Jupiter [Jeans 1931]. This is one of the methods of destruction considered by Napier and Dodd [Napier and Dodd 1973].

Instead of that single full-sized planet, many small bodies, the asteroids, have been found in that orbital zone, and others are somewhat scattered throughout the Solar System. The present total mass is a small fraction (0.0001 to 0.001) of the mass thought to have been available in that region for planetary accumulation [Wetherill 1992].

The theory that the asteroids had been produced by a disruption of a normal planet was finally dismissed by a detailed review of possible causes. A detailed review of possible disruptive processes had shown that gravitational forces, chemical reactions, and even nuclear fusion could not provide enough energy quickly enough to cause a planet to explode [Napier and Dodd 1973]. This review showed that nuclear fission, as in the Hiroshima A-bomb, could disintegrate a planet, but the reviewers could not imagine how that could come about and so rejected this possibility also. However, there is a very natural way in which a stable nuclear reactor, confined by the great mass of a planet, can progress to a condition in which a disruptive explosion is almost certain. Napier and Dodd considered chemical, gravitational, and nuclear sources of energy as possible sources for this explosion [Napier and Dodd 1973]. They concluded that there was no process capable of disrupting a planet. Chemical energy, such as the explosive power of TNT, was found to be far too weak to disrupt a planet. Gravitational forces from Jupiter are far too weak at the distance of the Asteroid Belt, where the parent planet might have formed. If a planet had strayed within the Roche Limit of Jupiter (inside Io's orbit), it would certainly have disturbed the currently existing system of satellites, and this is not observed. Three sources of nuclear energy were considered: radioactive decay, nuclear fusion, and a nuclear-fission chain-reaction. Radioactive decay is so slow that the energy released must be confined and stored until an explosion can occur, but the rock of a planet is not strong enough to do this for the time required. Fusion of ordinary hydrogen and deuterium was considered as the only possible fusion reaction fast enough and strong enough, but this reaction requires temperatures above 540,000 K, and such a high temperature cannot exist within a planet. They did discover that a nuclear-fission chain-reaction based on the naturally occurring uranium-235 was strong enough and fast enough to disintegrate a planet. However, they concluded that

> "... it is clearly improbable that, in nature, enormous numbers of sub-critical masses of uranium-235 could be assembled and brought together simultaneously within the planet."

Their thinking was limited by their knowledge that this rapid assembly method had been used for the uranium bomb, "Little Boy", that destroyed Hiroshima. In that bomb, a large mass of uranium-235 metal was explosively propelled into another large mass to make a super-critical mass. Triggered by an artificial burst of neutrons, the chain reactions quickly grew to the point of explosive disassembly. The disassembly terminated the fission reactions by dispersing the fuel. That bomb produced a yield of about 16,000 tons of TNT (5×10^{13} joules), and consumed about 700 grams of uranium.

I discovered a process for planetary detonation (to use a term that I intend as very general) after finding the probable existence of natural nuclear fission reactors in the Earth. (Natural nuclear fission reactors in uranium ore deposits had been predicted in 1957, and were discovered at Oklo in Gabon, in 1972.) My search for supporting evidence took me from the volcanoes of the Earth to the meteorites and then to the asteroids. I found that the original idea of the asteroids was that they were fragments of an exploded planet, but because no one could think of how a planet could explode, this possibility was rejected [Sagan 1980]. It was decided that the asteroids were the remnants of a planet that never formed, planetesimals that couldn't get together.

While an explosive nuclear-fission chain-reaction met the requirements for energy and quick action, Napier and Dodd could not imagine how it could occur in nature and so rejected the possibility, and decided that the asteroids could not possibly have been formed by an explosion. Indeed, such a rapid mechanical assembly cannot occur in the core of a planet, but that obstacle is overcome in nature by the dynamic changes that develop in the inherent behavior of extremely large critical systems. Rapid transformation of a subcritical mass to a supercritical mass to produce a nuclear detonation is a problem that nuclear weapons designers have solved, on a much smaller scale, by the use of high explosives to compress the fissionable material into a supercritical configuration. In the "Fat Man" bomb, which destroyed Nagasaki, high explosives were used to compress a plutonium core and confine it until sufficient fission energy had been released to overcome the inertial pressure of the bomb material and disperse the fuel.

Because the mechanism for such a planetary explosion could not be imagined [Wetherill 1992, Wood 1962, Sagan 1980, Napier and Dodd 1973], this idea for the formation of the asteroids was dropped from consideration. Regardless, there is abundant evidence from the mineralogical content and structure of the meteorites that they were once part of a large planetary body. Meteorites of a type previously presumed to represent primitive Solar System material have been shown to have veins of minerals that were formed from aqueous solutions, as might exist on a large, well-differentiated planetary body [Tomeoka 1990]. And the meteorites carry within themselves evidence of a nuclear detonation.

For planets of the size we are familiar with, Figure 4 showed that sufficient uranium fuel is available for non-disintegrating detonations that could release up to a range of 10^{30} to 10^{37} joules of energy, while disintegrating detonations might be in the range above 10^{32} joules. The disintegrating detonation is likely to display a greater observable energy release because of the dispersal of the solid fragments of the planet and the unshielding of the central fireball.

Following the rejection of the explosion hypothesis for the formation of the asteroids, it appeared that the gravitational perturbations of Jupiter might have been effective in preventing the formation of a single, complete planet: the "Phaeton" of Olbers and Daly [Wetherill 1992, Safronov 1969, McSween Jr. 1987]. However, if a single, complete planet, similar in size to the Earth, had formed in the asteroidal orbit, a sequence of events could have occurred that resulted in the cataclysmic disruption of that planet, sending fragments throughout the Solar System.

Reginald Daly, a notable geologist at Harvard University during the first half of the 20th Century, considered that the meteorites had "originated in the thorough fragmentation of a single planet, a former member of the solar system." He also considered it possible, even probable, that the asteroids had resulted from the same event, and were the source of the meteorites. He suggested that the "parental planet" had been disrupted by the explosion of gas driven out of solution as the minerals of the planet crystallized [Daly 1943]. That was two years before the first successful detonation of a nuclear explosion.

Because of the very small amount of material currently in the Asteroid Belt, while uncertain, Daly thought the parental planet ought to have been "somewhat smaller than Mars." He chose a planetary radius of 3,000 kilometers (compared to the 3,385 kilometers of Mars) with a 1,000-kilometer radius iron core. Its mass was 0.07 that of the Earth, compared to 0.108 for Mars. Daly thought 0.07 Earth was as much as it could be, because there is so little material there now.

Michael Ovenden determined a planet mass of 90 ± 5 of Earth, for a loss of the planet 16 million years ago [Ovenden 1972]. Ovenden analyzed the minimum-interaction-energy configuration of satellite systems and the Solar System, and found that our current arrangement of planets departs from this stable configuration [Ovenden 1972]. These departures could be explained by the prior existence of an additional planet, at the Asteroid Belt. He estimated that "Planet A" was 90 ± 5 times as massive as the Earth, and that it "suddenly disrupted" 16 million years ago, leaving only 0.1 of an Earth mass of material. He recognized the difficulty in "exploding" a planet.

According to Tom Van Flandern, the time of the disappearance of Ovenden's Planet A was estimated by use of a method of Meffroy from 1955. Ovenden later concluded that the method could not be applied for that purpose [Van Flandern 1978, Ovenden 1976]. Thus, Ovenden's 1972 research showed how the planetary orbits would adjust following the loss of a major planet, forming the Asteroid Belt, but could not correctly determine the time of its disappearance. I have proposed that Ovenden's Planet A was disrupted by a nuclear fission detonation at the time that has been incorrectly chosen for the formation of the Solar System, that is, 4.56 billion years ago, based on the many concordant radiometric dates determined from meteorites. The meteorites are the crumbs that we receive from the Asteroid Belt.

George Wetherill calculated the formation of a planet the size of the Earth at the distance of the Asteroid Belt, using the standard theory for the formation of planets in the Solar System, of accreting planetesimals in the presence of the early Sun [Wetherill 1989]. However, he concluded that this was an absurd result, considering the clear observation that there is no such planet in that zone at present, only fragments. He proposed several mechanisms to prevent the formation of a large planet. In the standard theory, it is necessary to assume that something different happened during planet formation in the Asteroid Belt than in the rest of the Solar System.

We might explore a way in which a nuclear detonation of such extreme energy as is needed to disrupt a large planet could occur naturally. As the asteroidal planet ("Asteria") accumulated material, and the core began to separate from the rocky material, steady and self-controllable nuclear fission chain-reactions would have contributed to the heating and mixing, and accelerated the segregation of the material within the planet by melting, creating high temperatures in the inner part of the planet. As the iron core solidified, it formed a lattice of crystals that trapped small grains of uranium oxide: the metallic uranium and dissolved uranium oxide continued to descend into the inner core. While many critical reactors will have accreted to form the planet core, in the form of uranium-rich globs, this segregation produced many more reactors. When an adequate concentration of uranium had accumulated, the nuclear fission chain-reactions began, intermittently at first, and then stronger and steadier. Deep within a planet, the reactors are confined by the great mass of material surrounding them. If the power in a reactor increases slowly, some expansion may allow the power to stabilize. If the power increases rapidly, the inertia of the planetary mass is too much to overcome, and the power excursion can continue to the point of a nuclear detonation. The detonation destroyed the planet.

As the planet was shattered by the shockwave produced by this detonation, portions of planetary material were exposed to a brief, intense neutron flux wave, traveling faster than sound in the solid material, faster than the shock wave, faster than the fragments of the expanding planet. Thermal neutrons with 0.0253 eV (at our laboratory standard of 300 K) travel at 2.2 kilometers per second, fission neutrons with 2 MeV travel at 19,560 kilometers per second. The time to traverse a planet of 7,000 kilometer radius is about 1 hour to less than 0.3 second. Most neutrons will be fast neutrons, but scattering will reduce their energy, slow their speed, and increase the diffusion time. Residual grains of uranium-bearing minerals were vaporized or melted, and the superheated iron around them flashed to vapor as the pressure was relieved, leaving bubble pocks and glassy grains in the fragments, as are found in the iron meteorites. The expanding cloud of iron vapor condensed on the cooler fragments, leading to the condition where all (nearly all) meteorites may be identified by their attraction of a magnet [Haag 1992]. Molten droplets of other materials were deposited on and throughout the fragments of the broken crust of the planet.

Fragments of this disruptive detonation moved to an extreme inner orbit and an extreme outer orbit, depending on whether the fragment original orbital energy was increased or decreased. Probably a major amount were neither increased nor decreased, and so occupied the original orbit, but with greater eccentricity than the orbit of the planet. Some of the energy of the detonation became kinetic energy of the fragments, giving them random velocities of about 3.3 kilometers per second. This agrees with the observation that mutual collision velocities in the Asteroid Belt are about 5 kilometers per second.

During the detonation phase, isotopic and elemental abundances were altered in a manner similar to the r-process in a supernova, or the prompt-capture process observed in weapons tests [Becker 1993]. The nuclear detonation of a weapon takes roughly half a microsecond [Glasstone and Dolan 1977]. The nuclear detonation of the reactors in a planet is likely to be complete in a few seconds. This may involve a chain-reaction of chain-reactions. Since the steady-state reactors are all separate, but may be interacting, individual reactors will detonate separately. Each detonation of a critical mass induces the detonations of its nearest neighbors, which in turn cause their neighbors to detonate until all the fuel is consumed, or the planetary body is disrupted, dispersing the remaining fuel.

Sodium (the stable isotope sodium-23) in the crust is irradiated by high-energy neutrons and is converted to radioactive sodium-22 which relatively quickly decays to neon-22; aluminum (aluminum-27) is converted to aluminum-26 which quickly (6.3 seconds) or slowly (0.74 million years) decays to magnesium-26; helium-3 is transmuted to hydrogen-3 or helium-4; the isotopic compositions of potassium and argon, and carbon, nitrogen, and oxygen, are altered; fission products are created; much of the uranium is consumed, with remnants of uranium-239 left to decay, after the neutron pulse fades away, to plutonium-239, which will ultimately decay to uranium-235. Potassium-40 is destroyed, resulting in a future deficiency in argon-40, and excess argon-36 is produced by neutron absorption by chlorine-35. In short, and in an extremely brief moment of time, many of the observed anomalies will be created and frozen in place.

The remanent magnetism (a term derived from magnetic remanence, and not to be confused with the similar word remnant) of a fully developed planetary core is shown in the magnetic field carried by the asteroid 951 Gaspra [Kivelson *et al.* 1993] as measured by the space probe *Galileo*. The magnetization of Gaspra, as indicated by interpretation of the *Galileo* measurements, is similar to that of the Earth, at present.

The possible existence of asteroidal satellites, small bodies in stable orbit around an asteroid, has been known since the observation of multiple occultations of stars by asteroids [Binzel and Van Flandern 1979] and apparent eclipsing binary lightcurves [Tedesco 1979]. These observations met with some disagreement at the time [Reitsema 1979, Van Flandern 1981], but seem to be confirmed in general by the image of the asteroid 243 Ida with an orbiting companion, Dactyl [Chapman 1994]. There seems to be some difficulty for an asteroid to capture a satellite [Kerr 1994a]. This might be more easily done in the presence of a magnetic field. Electrical currents would be generated within the new satellite by its motion through this field during the capture process. Excess energy could be dissipated by electrical heating. Some relative energy would be converted into rotation of each body by the magnetic torques. Ida has a magnetic field [Burnham 1994]. And Ida has a satellite, Dactyl, discovered by the *Galileo* space probe [Beatty 1995]. There are now known to be over a hundred pairs of asteroidal fragments, including Pluto and Charon, and others found in the more distant parts of our Solar System.

Magnetism might also attract small bits of material to form a covering, hiding the underlying surface and smoothing the contours, and reducing the measurable magnetic field. These objects would have a low density because of the low packing fraction of granular (particularly magnetic) material [Chapman *et al.* 1995]. Coatings of similar thickness would smooth a small body better than a large one. Alignment may be more secure by magnetism than by tidal effects. If we imagine two magnetized asteroids passing in the night, we find that the magnetic attraction is approximately the same as the gravitational attraction. However, the gravitational force is a conservative force, and does not dissipate energy, except by tidal friction. If only gravity affects the interaction of these two bodies, they will pass by each other, perturbing each other's motion. An initially parabolic or hyperbolic orbit will remain so, and the two objects will travel on their new paths, to completely separate from each other. A loss of energy is necessary to convert an open orbit (parabolic or hyperbolic) into a closed, elliptical orbit. The magnetic field can dissipate energy in two ways. Torque produced by the non-alignment of the magnetic axes will cause linear kinetic energy to be converted into rotational energy. The transverse force, that produces this torque, can amount to roughly half of the maximum attractive force [Craik 1995]. Electromagnetic induction will also consume energy by ohmic heating from currents induced in the bodies. This loss of kinetic energy between two magnetized asteroids will permit capture, producing asteroid-satellite pairs, like Ida and Dactyl. The newly closed orbit will evolve by loss of kinetic energy until a synchronous orbit is reached, with the two magnetic dipoles locked into alignment. Magnetic "tidal" friction is likely to be particularly effective for magnetic bodies which have become coated with a thick layer of magnetic dust. These small particles will wave back and forth and migrate over the surfaces as the relative magnetic fields change.

The Apollo and Amor objects are asteroids that pass close to the Earth and are too small for the force of gravity to be effective in modifying their original shape. The shapes of individual asteroids may be determined with reasonable accuracy during close approaches. Fujiwara has reported [Fujiwara *et al.* 1989] in a discussion of high-speed impact fragmentation, that one-quarter of these objects are far more elongated than any fragments produced experimentally by impact (collisional) fragmentation.

A similar condition is found for the Trojan group of asteroids, which are separated from the main Asteroid Belt by orbiting at the gravitationally stable Lagrangian points in Jupiter's orbit. The shapes of these objects, as estimated from the magnitude of brightness variations due to their rotation, are markedly more irregular, more elongated, than the asteroids in the main belt [Kowal 1988, Bell 1989]. It may be that the Trojans, sheltered from the bump and grind of the main belt, have more nearly retained their original fragment shapes. Such elongated shapes may be preferentially produced by the fragmentation of a spherical shell of a planet that was burst by internal pressure, in the manner of shrapnel resulting from the explosion of a cannonball. As we learn more about the possibility of mass extinctions by extraterrestrial impact, questions of how to defend life on Earth become more important. Methods to deflect a threatening asteroid hold the greatest promise. A near miss is better than any kind of hit. However, if the very character of the Earth is changed by a major impact, as seems indicated by our studies of the mass extinctions, shattering an asteroid to small pieces may be a necessary act of desperation. Many small fragments may then impact the Earth, some exploding in the atmosphere and devastating the ground below, as the Tunguska impact did. Actual ground impacts will destroy life for miles around. Limiting the size of the impactors to a small size may preserve the livability of the Earth for our kind of life, by preventing the expansion induced by a major impact. Such an expansion can change the character of the surface of the Earth beyond our ability to survive.

Rotating asteroids, reflecting unevenly the Sun's light, might give the appearance of variable stars, as shown in Figure 49 [Malcolm 2000, Pietrzyński *et al.* 2010]. (See also [Chang *et al.* 1981 and Sterken and Jaschek 1996].) Note that the plot for 729 Watsoonia shows two cycles, while the Cepheid plot shows only one.

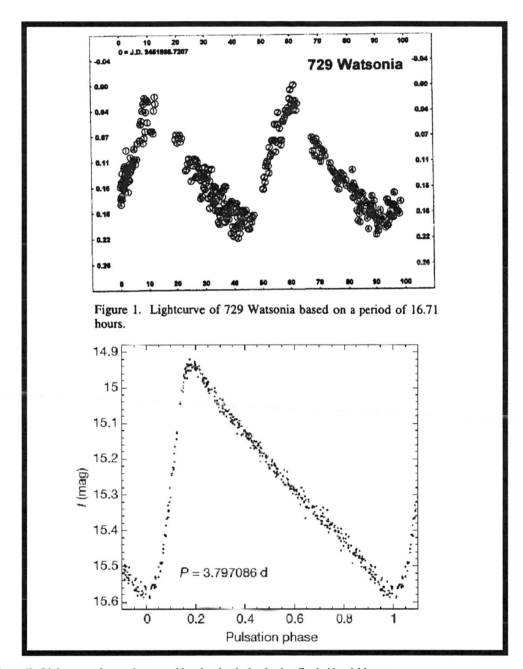

Figure 1. Lightcurve of 729 Watsonia based on a period of 16.71 hours.

Figure 49. Lightcurve of a rotating asteroid and a classical pulsating Cepheid variable star.
The lightcurve of a rotating asteroid, 729 Watsonia (upper plot) [Malcolm 2000], to show its similarity to the lightcurve of a variable star, a classical Cepheid in the Large Magellanic Cloud (lower plot) [Pietrzyński et al. 2010]. Two cycles are shown for the asteroid, only one for the Cepheid.

The asteroid, 729 Watsonia, is a cold, rotating rock reflecting sunlight at a distance of about 2.8 Astronomical Units from the Sun. The Cepheid variable, OGLE-LMC-CEP0227, is a red giant star in the Large Magellanic Cloud at an assumed distance of 157,000 light years.

Meteorites

Catch a falling star and
Put it in your pocket.
Vance and Pockriss
(with thanks to Perry Como)

... looking more like concrete than anything else.
John A. Woods

Meteorites are our grandest space samples, in most cases from the asteroids, some from Mars and the Moon. The best ages that we can establish for meteorites are taken to show the age of the formation of the Solar System, as meteorites are fragments of asteroids, and asteroids are left-over planetesimals that failed to form into a planet (according to our standard story). While the oldest securely determined age for a meteorite is 4.566 ± 0.008 billion years, by Pb-Pb (lead/lead) dating of a high temperature inclusion in the Allende meteorite [Carlson and Lugmair 2000], a starting time for the Solar System has been derived to be 4.571 billion years ago [Lugmair and Shukolyukov 2001].

This section will show that the dating of the meteorites, as fragments of asteroids, refers to a nuclear fission detonation that disrupted a fully formed planet, "Asteria", in the Titius-Bode orbit between Mars and Jupiter. This orbital space is now filled with tiny fragments showing evidence of segregation and differentiation, just like a large planet. The ages of the meteorites are not indicators for the age of the Solar System.

A precise terminology has been developed for meteorites. A "meteoroid" is an object in space which, streaking in a flaming path across our sky as a shooting star, appears as a "meteor", and when found on the ground is a "meteorite". Meteorites are samples from other worlds, brought to us by the dynamics of the Solar System, and subject to close examination in the laboratory. While most of the meteors we see, particularly in the periodic showers, are from dust left behind by comets, we think that most meteorites we find have come from the debris of the Asteroid Belt. Several trails of meteors have been analyzed to determine the orbits prior to their entry into our atmosphere and those orbits reach out into the Asteroid Belt. In addition, the surface spectrum of the asteroid Vesta matches the spectra of basaltic meteorites. None of our found meteorites are known to have come from comets, even though the trails of comets make our most spectacular meteor showers. Some meteorites have been identified as coming from Mars, and some have come from the Moon [Wood 1968, Wasson 1985, McSween 1987, Heide and Wlotzka 1995, Haag 2003].

Some meteorites are found immediately after their fall, announced by a fiery streak through the sky. Disappointingly, not all the big fireballs result in discoveries. Some are found many years after their fall, which may produce an impact crater in rare cases. Found meteorites are generally named after the locality where they were found.

While many meteorites resemble earthly rocks, or as John Wood says, concrete, some are quite different. Iron meteorites are composed of metallic iron, rarely found in its native state on Earth, and usually contain a few percent of nickel. The nickel-iron alloy is known to our metallurgists, but in the iron meteorites it has formed a crystalline structure that is estimated to take several million years of slow cooling to develop. It obviously has not been possible to produce such crystalline structure in our laboratories. The planes of the crystals intersect in the Widmanstätten pattern, which identifies the special crystalline lattice

Some meteorites, typified by the Pallasite meteorite, are a beautiful mixture of green glassy olivine, a common rock-forming mineral on Earth seen as the gemstone peridot, scattered through metallic iron. We have no clear understanding of how these were formed. The original meteorite of this type was discovered by Peter Pallas, a German naturalist, while travelling in Siberia. An iron meteorite had fallen in 1749 and was collected and preserved by the local inhabitants [Kouřimský 1993]. These belong to the "stony-iron" class, a mixture of metallic iron and various forms of stone.

A remarkably unusual class includes the stones, so-called "chondrites", which have chondrules, and achondrites, which do not. Chondrules are small (pea-size and smaller) spheres of once molten minerals that froze and congealed into a mass. Often the mass includes calcium-, aluminum-rich inclusions, labeled as CAIs, and a dusty matrix that holds everything together. Researchers remark on their uniqueness, like nothing known on the Earth. In fairness, nearly all the meteorites are unique, like nothing found on Earth, except meteorites. Some meteorites are basalt, like the lava erupted from Earthly volcanoes, and these give us some reassurance that we are dealing with truly geological objects, in spite of their outer-space origin.

Many of the meteorites that have been found on the Earth have, presumably, come from the Asteroid Belt. Many distinct components of meteoritic material have been identified and provide a variety of data on the composition and evolution of the Solar System and the parent bodies of the meteorites. Much of our knowledge of the composition of the Solar System, and our assumed knowledge of the Earth's core, come from studies of these meteorites. We base our determination of the age of the Solar System entirely on the meteorites.

Random collisions in the Asteroid Belt break off chunks of material and eject the pieces into new orbits. Gravitational perturbations by Jupiter and the other planets may cause some of these pieces to drift through the Solar System, eventually being captured by the Sun or planets, or being ejected from the Solar System into interstellar space. Some may simply take up other stable orbits and remain for an indefinite time. Very small pieces, the size of grains of sand or small pebbles, will burn up in our atmosphere on entering with a speed of about 15 to 40 kilometers per second. Larger pieces, up to about 10,000 kilograms, will lose most or all or their kinetic energy in the atmosphere and hit the surface with terminal velocity, a few hundred miles per hour, for stone and iron. These pieces can usually be recovered intact and provide the museums and collectors with fine specimens. Greater masses retain more of their initial kinetic energy and have the potential for creating impact craters, with possible severe damage from earthquakes, tsunami, heat, and ejected debris. At sufficiently great mass, a life-threatening event occurs, like the impact at Chicxulub that has been blamed for the extinction of the dinosaurs (and associated creatures) and other impacts that may have caused other major extinctions.

The existence of extraterrestrial impacts on Earth was established by studies of Meteor Crater, in Arizona. The crater is reasonably fresh in appearance, perhaps due to its location in a desert area, and is about 1,200 meters in diameter. It is estimated that the meteoroid (named Canyon Diablo) that formed the crater was approximately 100 million kilograms of iron, travelling at 15 kilometers per second, and released energy on impact equivalent to a 1 megaton nuclear explosion. While its age was estimated as about 20,000 to 30,000 years, based on weathering [Heide and Wlotzka 1995, Grieve 1990], and later radiometric dating has shown 49,900 years, the impact is mentioned in legends of the local Indians. Other, more recent impacts have been retained in local legends.

The Canyon Diablo meteoroid was completely shattered on impact, and only small pieces, of a distinctive iron-nickel alloy, have been found. The impact origin of the crater was initially rejected, but detailed research led to acceptance of this idea, and many more impact craters have been identified. We now know of at least 157 , and we're still counting [Grieve 2000].

The origin of at least 14 meteorites has been determined to be the surface of Mars, by the close match of several gases in the meteorites to the composition found on Mars by the *Viking* landers. While the match is not quite perfect, it has been convincing. Differences may arise from the great differences in time between the oldest Martian meteorites and the surface measurement on Mars.

Another group of meteorites has been identified as coming from the Moon. This identification is based on comparisons with the Apollo samples.

Meteorites from other planets require ejection of the material from the surface of the planet by an energetic impact, escape through the atmosphere, eventual capture by the Earth's gravity, and successful entry through the Earth's atmosphere, finally landing on land, not in the ocean. The meteorite must then survive for various times on the ground, and finally be recognized for the valuable prize that it is. Most of these planetary meteorites have been found on the icesheet of Antarctica, where they are preserved and easily seen.

Several meteorites have been analyzed for age by different isotopic methods [Lodders and Fegley Jr. 1998]. An average of these values gives 4.52 billion years and ages of that sort have been taken as the age of formation of the Solar System. This date is taken as the time of formation of the meteoroid, and in the standard theory of the Solar System, that would be the solidification of planetesimals. That solidification has been taken to be shortly after the formation of our Sun, according to the standard theory. Therefore, from these measured ages, the formation of the Sun is judged to have occurred 4.6 billion years ago. Various parts of meteorites have been used to establish the definitive composition of the materials that formed the Solar System, because of the assumption that these objects are remainders of primordial conditions, and have had little processing or alteration since the earliest times.

However, if many of the meteorites come from the asteroids, as seems reasonable, and if the asteroids are fragments of a planet that detonated some time after formation, as the previous section showed, all this knowledge becomes suspect. This age is definitely not the beginning of the Solar System, but the ending of the asteroidal planet. If this is true, there should be evidence in the meteorites themselves that shows the effects of a nuclear fission detonation. And that is what we find. Meteorites show the effects of natural nuclear fission reactors, and the consequences of the transition to a detonation.

If the asteroids are the fragments of a detonated planet, the meteorites date from that detonation and contain information about a fully formed planet, and the detonation that destroyed it. That information will be essentially unrelated to any conditions or age at the formation of the Solar System. Meteorites provide the best opportunity to examine what might once have been the core of a planet, disintegrated after formation by an explosive nuclear-fission chain-reaction.

The age of the Monahans meteorite has been determined to be 4.7 ± 0.2 billion years, within the uncertainty of the standard age for meteorites of 4.56 billion years, by the ratio of strontium-87 to strontium-86 and the radioactive decay of rubidium-87. The H group of chondrites define an age of 4.58 ± 0.04 billion years.

The Divnoe meteorite shows a K-Ar age of 4.67 billion years, but the Rb-Sr age is only 3.39 billion years, and it has a clear excess of xenon-129 [Shukolyukov *et al.* 1995]. Neutron absorption in strontium-87 will reduce the amount of this isotope, causing the age to be in error. These are signs of fission and neutron absorption.

Uncertain modification of the uranium isotopic content in nature has been noted as a confounding effect for dating methods. If the isotopic composition of uranium is altered in a small amount of matter at an early time, by destruction of uranium-235 by fission for example, or in the opposite sense by production through neutron absorption by uranium-238 and decay of plutonium-239, the U/Pb (uranium/lead) dating method may give an incorrect result. Similar modifications to the isotopic makeup of the various elements involved in the dating pairs used to estimate the age of the Solar System, the Earth, Moon, meteorites, and other objects, which can be caused by neutron absorption or created as fission products, will distort these age estimates. The current estimate for the time of formation of the Solar System, about 4.6 billion years ago, is based on element/isotopic analyses of meteorites. These results are generally uniformly in agreement, and so may all represent the same event. However, this event was related to the detonation of the asteroidal planet rather than to the formation of the Solar System.

The meteorites we find on the Earth, after a fiery passage through the atmosphere, consist of several types. Some of these types show that they are parts, tiny fragments, from a well-differentiated planetary body. For now, we may assume that nearly all the meteorites known have come from a single planet, and the detonation of that planet formed the asteroids. (Some meteorites have come from Mars [Wasson 1985], some from the Moon [McSween Jr. 1987], some may have come from Mercury [suggested, for the Martian meteorite ALH84001, Tuttle 1997, and rejected, Mittlefeldt 1997, but supported in general, Love and Keil 1995, Gladman and Coffey 2009]. Many may be the remains of comets. There are still other possibilities that we may want to consider. Most have come from the Asteroid Belt, and we can base our study on that.)

Many stony meteorites have droplets of once-molten iron splattered through their material, and all show a magnetism resulting from the condensation of iron vapor throughout the bodies. Many have once-molten droplets, chondrules, that have been assumed to represent a collection of the most primitive particles in the pre-Solar nebula. Some have small spheres of highly refractory elements, with low volatility, and they have lost the high volatility elements. These are the calcium, aluminum-rich inclusions, CAIs. Blowholes exist in iron meteorites that must have been produced in their original formation. They cannot have been produced by ablation during entry into our atmosphere.

The presence of water has been indicated by veins of hydrated magnesium sulfate ("epsom salt"), and by brine trapped in salt crystals [Cowen 1999, Clayton 1999, Zolensky *et al.* 1999]. Anders has speculated that liquid water could exist for long periods of time, deep within a large asteroid, if there were sufficient heat to raise the internal temperature to about 1,900 K (above "white heat" on the color scale of temperature [Hodgman *et al.* 1956]). So, where's the heat?

One of the significant problems the meteorites have brought to us, as representatives of the asteroids, is how such small bodies, all less than a few percent of our Moon in mass, could have differentiated into a solid iron core, a core-mantle boundary, a mantle, a crust with distinctive minerals, and liquid water, all in a colder part of the Solar System than we are in.

Water

Carbonate minerals have been found in several carbonaceous CI (I = one) chondrites, both as grains that might have formed as chondrules do, and as fragments suggestive of broken veins of mineral that were formed by water [Endress, Zinner, and Bischoff 1996]. Those meteorites, considered to be among the most primitive known, have dolomite, breunnerite, calcite, and siderite, carbonate minerals that typically form in water. A study of excess chromium-53 showed that the deposition had occurred less than 20 million years after the oldest refractory inclusions in the Allende meteorite, the oldest objects yet measured. The chromium-53 is assumed to have come from the extinct radioactivity of manganese-53, with a half-life of 3.7 million years. If the radioactivity originated in a supernova, perhaps 2.4 billion years before the formation of the Solar System, it is hard to understand how such a short half-life radionuclide could exist at the time of formation of the CI parent body. (This concern is also true for other short half-life radionuclides, such as aluminum-26, chlorine-36, calcium-41, iron-60, and iodine-129. While a supernova is far away, in space and time, these radionuclides could have been produced in a nuclear fission detonation of a fully formed planet, at about 4.55 billion years ago [Bogard *et al.* 1995].) While the neutron absorption in meteorites that was reported by Bogard *et al.* has been conventionally interpreted as being due to neutrons produced by cosmic rays in the meteroids while in space, the nuclear detonation of the single fully formed planet would have produced abundant neutrons throughout the planet. The major detonation is proposed here as taking place in the core. Uranium throughout the planet, including near-surface ore bodies as shown by the pancake domes of Venus and Mars, would have been fissioned, contributing to a widespread source of thermal neutrons.

Veins of poorly crystallized phyllosilicates (clay minerals) have been studied in the Yamato-82162 meteorite, found in Antarctica [Tomeoka 1990]. Sulphates and carbonates have also been found, indicating varied depositions of watery minerals.

There is some evidence for liquid water mingling with chondrules after their formation [Grossman 2000, Young 1999]. Residual decay heat from the fission products of a nuclear detonation could have provided the distributed internal heating required to keep water liquid, and promote thermal gradient flows.

Deuterium and Hydrogen

As on Venus and Mars, the hydrogen isotope ratios, deuterium to hydrogen (hydrogen-2 to hydrogen-1), D/H, are different from the ratios commonly found on Earth. This is generally explained as an enrichment of deuterium, but conversely, it is actually a depletion of hydrogen-1. Water and amino acids in the Murchison meteorite show an enrichment of deuterium, up to a factor of six, over terrestrial water [Pillinger 1987, Epstein *et al.* 1987]. Here, only the initial loss of the hydrogen isotopes released by the neutron irradiation would be expected: further action by atmospheric diffusion would not occur for bound water in the meteorite. (This meteorite, with amino acids, essential components of life, also shows "strange fibers" under the scanning electron microscope [Henbest 1992].)

For technical purposes, the difference in the D/H ratios are labeled δD (usually related to a standard such as Vienna Standard Mean Ocean Water (VSMOW) with D/H in this case equal to 155.76 x 10^{-6} and δD = 0), in parts per thousand, or per mil (‰). For the strong differences found in the meteorites, the direct ratio of the D/H ratio to the standard terrestrial ratio may be the most illuminating. These values are shown below for a variety of meteorite studies.

Comparison of D/H Ratios in Meteorites

Reference	observation	δD per mil	D/H (10^{-6})	compared to Earth	type
Mostefaoui 1998	Khohar	50	165	1.05	
Mostefaoui 1998	Khohar	200	188	1.20	
Mostefaoui 1998	Khohar	1500	390	2.49	
Remusat 2006	aromatic C-H	150	180	1.15	
Remusat 2006	aliphatic C-H	550	242	1.55	
Remusat 2006	benzylic C-H	1250	351	2.24	
Busemann 2006	GRO95577	19400	3173	20.24	hotspot
Busemann 2006	EET92042	16300	2691	17.17	hotspot
Busemann 2006	Al Rais	14300	2380	15.18	hotspot
Busemann 2006	Murchison	1740	427	2.73	hotspot
Busemann 2006	Bells	9700	1665	10.62	hotspot
Busemann 2006	Al Rais matrix	6200	1121	7.15	hotspot
Busemann 2006	Tagish matrix	8600	1494	9.53	hotspot
Busemann 2006	GRO95577	11780	1988	12.69	area
Busemann 2006	EET92042	13112	2195	14.01	area
Busemann 2006	Al Rais	6261	1130	7.21	area
Busemann 2006	Murchison	738	271	1.73	area
Busemann 2006	Bells	5702	1043	6.66	area
Busemann 2006	Al Rais matrix	867	292	1.86	area
Busemann 2006	Tagish matrix	3963	773	4.93	area
Alexander 2005	GRO95577	2973	619	3.95	bulk
Alexander 2005	EET92042	3004	624	3.98	bulk
Alexander 2005	Al Rais	2658	570	3.64	bulk
Alexander 2005	Murchison	712	267	1.71	bulk
Alexander 2005	Bells	3283	667	4.26	bulk
Messenger 1996	IDP	8000	1400	8.94	
Messenger 1996	IDP	20000	3266	20.84	
Messenger 1996	IDP	24800	4012	25.60	
Deloule 1995	Renazzo	4600	872	5.56	
Deloule 1995	Renazzo	3300	670	4.27	
Deloule 1995	Semarkona	1050	320	2.04	
Deloule 1995	Semarkona	20	160	1.02	matrix
Deloule 1995	Semarkona	345	210	1.34	matrix
Deloule 1995	Semarkona	470	230	1.47	matrix
Deloule 1995	Semarkona	2725	580	3.70	matrix
Deloule 1995	Semarkona	2210	500	3.19	matrix
Deloule 1995	Semarkona	2145	490	3.13	matrix
Messenger 1996	Butterfly IDP	2700	576	3.68	hotspot
Messenger 1996	Butterfly IDP	2900	608	3.88	
Messenger 1996	Butterfly IDP	12000	2022	12.90	
Messenger 1996	Butterfly IDP	50000	7930	50.60	

Messenger 1996	Butterfly IDP	1147	335	2.14	
Messenger 1996	Butterfly IDP	1906	453	2.89	
Messenger 1996	Butterfly IDP	2145	490	3.13	
Messenger 1996	Butterfly IDP	3	157	1.00	
Messenger 1996	Butterfly IDP	1155	336	2.15	
Messenger 1996	Butterfly IDP	11035	1872	11.95	
Messenger 1996	Butterfly IDP	1080	325	2.07	
Messenger 1996	Butterfly IDP	12007	2023	12.91	
Messenger 1996	Butterfly IDP	4122	798	5.09	
Messenger 1996	Butterfly IDP	5585	1025	6.54	
Messenger 1996	Butterfly IDP	1236	349	2.23	
Messenger 1996	Butterfly IDP	1376	371	2.36	
Messenger 1996	Butterfly IDP	5570	1023	6.53	
Messenger 1996	Butterfly IDP	929	301	1.92	
Messenger 1996	Butterfly IDP	2914	610	3.89	
Messenger 1996	Butterfly IDP	0	157	1.00	
Messenger 1996	Butterfly IDP	3580	713	4.55	
Messenger 1996	Butterfly IDP	80	169	1.08	
Messenger 1996	Butterfly IDP	678	262	1.67	
Nakamura-Messenger 2006	Tagish Lake	1800	437	2.79	glob
Nakamura-Messenger 2006	Tagish Lake	8100	1416	9.03	glob
Alexander, Fogel, Cody 2005	H/C = 0.17	5560	1021	6.52	
Alexander, Fogel, Cody 2005	H/C = 0.50	2321	518	3.30	

First of all, the great range in these values is remarkable, with D/H relative to Earth from 1.00 (just like Earth) to as high (in this data set) as 50 times the ratio on Earth. This happens even in the same sample (the interplanetary dust particle Butterfly [Messenger 1996]). Secondly, there are values equal to the ratio on the Earth, but none less than that. The ratio depends on the strength of the chemical bond [Remusat 2006], and the enrichment of deuterium is related to the loss of hydrogen [Alexander, Fogel, and Cody 2005].

Scattering of energetic neutrons breaks chemical bonds by the recoil of the struck nucleus. The scattering cross section for hydrogen-1 is about 6 times as large as for deuterium (see Figure 44), so the probability of scattering from hydrogen-1 is that much greater than for deuterium, and more H-C, H-N, and H-O bonds will be broken, than D-C, D-N, and D-O bonds. More hydrogen-1 atoms will be freed to be lost in space. Further, a neutron will impart more energy to a proton (hydrogen-1) than to a deuteron (hydrogen-2) because of the lesser mass of the proton. For scattering off a proton, a neutron can transfer all its kinetic energy to the proton, but can transfer only 89% of its energy to the deuteron. Energetic recoil protons will further preferentially disrupt proton bonds (again because of the lesser mass of the proton) than deuteron bonds. As the neutrons are slowed down by energy losses with each scattering, capture by hydrogen-1 may be significant. This directly produces deuterium, thereby increasing the D/H ratio. These effects lead to an increase in the abundance of deuterium relative to hydrogen-1, due to the neutrons released in nuclear fission.

Bond strengths for hydrogen with carbon, nitrogen, and oxygen range from about 3.2 electronvolts to 4.4 electronvolts. From its average initial energy as a fission neutron, to below the bond dissociation energy, a neutron will break about 12 bonds. Most of these will be hydrogen-1 bonds and the recoil proton may be lost.

The effect of the chemical bond strength is shown by the differences in the organic compounds [Remusat 2006]. The strongest bonds (aromatic) show the least increase in the D/H ratio, while the weakest bonds (benzylic) show the greatest increase, and the intermediate strength bonds (aliphatic) are intermediate. This bond breaking and loss of hydrogen-1 is also shown by Alexander, Fogel, and Cody [Alexander, Fogel, and Cody 2005], where the chondrules that lost the most hydrogen (H/C = 0.17) show the greatest increase in the ratio, while those that lost less (H/C = 0.50) showed a smaller increase in the ratio.

It is notable that there are no ratios less than 1.00, less than the standard value on the Earth. A distributional analysis (using ProbPlot [McGinnis 2009]) of the D/H ratios (Figure 50), relative to the Earthly value, shows that these data fit a log-Gaussian distribution very well. There are no gaps, and no standout groups. This suggests that all these values, as belonging to the same statistical population, were produced by the same process, acting with varying intensity on material that had, initially, a D/H ratio equal to that on the Earth. That initial value of the ratio shows as the few points at the foot of the distribution.

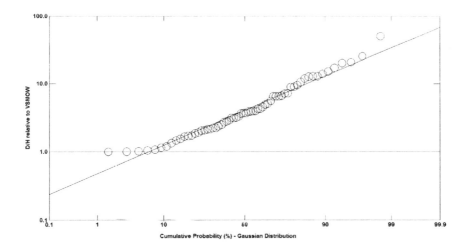

Figure 50. D/H ratios in meteorites.
A distributional plot of the D/H ratios in meteorites, normalized to Vienna Mean Standard Ocean Water. The data points fit a log-Gaussian distribution well, and show the initial ratio, just like the Earth's water, as a foot at the low end of the distribution.

A planetary detonation produces an abundance of high energy neutrons, which preferentially dissociate H bonds compared to D bonds, resulting in a loss of hydrogen-1 and an apparent enrichment in deuterium.

Helium-3

Helium-3 is a rare isotope of the light noble gas helium. It is produced in the fusion reactions that power the stars, but its life there is short. It can be produced by high-energy nuclear reactions that shatter and fragment individual nuclei into many parts. It is produced in nuclear fission as the decay product of tritium, hydrogen-3, which is produced in about 1 fission out of 8,000. Only when the products of many fissions are accumulated can there be a detectable amount of helium-3.

In explaining the helium-3 found in iron meteorites, it has been assumed that all the helium-3 was produced by high-energy cosmic rays shattering atomic nuclei in the meteorite while it was a small body drifting through space. Then, the amount of helium-3 found in a meteorite is used to estimate the length of time that the meteorite was exposed to cosmic rays. This is generally taken to be the time since its last disruptive impact with another asteroidal fragment.

The distribution of helium-3, and other noble gases, has been measured in several meteorites by Fireman, most notably in the Grant meteorite [Wood 1962] and in the Carbo meteorite [Fireman 1958, Mason 1962]. The observed distributions are generally a smoothly increasing function of distance from a presumed center of the meteoroid, relative to the surface before entry into the Earth's atmosphere. Heating and ablation of the surface material removes significant amounts, and changes the outer shape as a result of uneven erosion by atmospheric friction. Based on the assumption that the helium-3 was produced by nuclear spallation reactions in the iron of the meteoroid as it was in space, as some of the other noble gases probably were, this distribution has been explained as showing the attenuation of the cosmic-ray flux with passage into the depth of the meteoroid. In a spallation reaction, an extremely high-energy cosmic-ray particle, an atomic nucleus, often a proton, but possibly any kind, totally disrupts a nucleus in the target material and produces a variety of nuclear fragments, and other particles. Neutrons and neutral π mesons induce further nuclear reactions. These reactions have been observed to produce helium-3 as well as other lightweight nuclei. Values of the concentrations of helium-3 measured in Grant and in Carbo, as taken along a transect across the contour lines shown by Fireman, are shown in Figure 51.

These measurements provide strong evidence for the long-term action of nuclear reactors in the iron core of a planet. Reactors operating deep within the parent planet of the asteroids produced a uniform distribution of hydrogen-3 from ternary fission. Those hydrogen-3 atoms decayed to helium-3, to become locked in place. The abrupt burst of neutrons released in the detonation then re-converted the helium-3 to hydrogen-3. As mobile hydrogen (albeit superheavy), these atoms diffused away from the intense sources of heat, resulting from the fissioning of specks of uranium. These relocated hydrogen-3 atoms then decayed again to helium-3, leaving a fossil distribution which recorded the effects of the operation and detonation of natural nuclear fission reactors. The observed distribution does not match that expected for cosmic-ray production of helium-3, but does reveal the consequences of steady reactor operation followed by a nuclear-fission detonation.

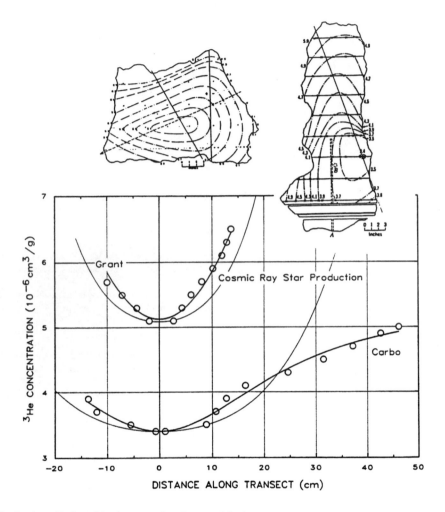

Figure 51. Distribution of helium-3 in the meteorites Grant and Carbo.
Sections of the Grant (left) and Carbo meteorites (right), showing contours of the concentration of helium-3. The graph compares the shape of the expected cosmic-ray interaction density, and the postulated temperature-induced diffusion of neutron-produced hydrogen-3, with the measured values. For this comparison, the measured values were picked off the figures along transects shown as the diagonal lines running from top left to bottom right. Because of this temperature-induced diffusion, the helium-3 distribution can be represented as a modification of an initially uniform distribution in small regions surrounding the fission sites, by a diffusional Gaussian distribution, with a diffusion constant varying with temperature based on distance from the origin. This distribution is represented in the figure by the smooth curves that pass through the Carbo and Grant data points. The contrast in the quality of fit between the two types of curves is noteworthy.

A smooth curve, which does not relate well to the observed distributions, is shown, based on the estimated "star" production rate as a function of depth [Goel 1962] with an original diameter of 102 centimeters for Carbo and 60 centimeters for Grant. (Because of the visual appearance of cosmic-ray spallation reactions in the photographic emulsions used to study them, these reaction events became known as "stars".) The contrast in the shape of the expected distribution of the cosmic-ray-produced helium-3 and the observed distribution is pronounced, and is most clearly shown by the results from Carbo.

An alternative manner of producing these distributions may be offered by the planetary fission process: during the relatively steady fissioning phase, hydrogen-3 produced by ternary fission diffuses rapidly through the hot, partly molten core, and is mixed to a uniform distribution within the core. As the hydrogen-3 decays to helium-3, which is much less mobile, and as the iron solidifies, this uniform distribution becomes frozen in place, as a uniform distribution of helium-3.

At the moment of the planetary disintegration, following the nuclear detonation, an intense wave of neutrons spreads outward from the detonation site, more rapidly than the detonation shockwave. These neutrons interact with the helium-3, converting some nuclei to hydrogen-3 and others to helium-4.

Simultaneously, nearly all the remaining nuclear fuel, uranium and thorium and any other fissionable nuclides available, is caused to fission by this intense flux of neutrons. Small specks of uranium, trapped in the crystalline structure of the iron, fission with the release of an intensely brief, intensely localized burst of heat. This heat diffuses away into the surrounding material, raising its temperature. Since the diffusion of hydrogen-3 atoms in iron is strongly temperature dependent, the newly produced hydrogen-3 atoms that are near the fission site diffuse away more rapidly than those that are more distant. The hydrogen-3 subsequently decays into helium-3, and this distribution is then frozen in place. (The departure of this distribution from spherical symmetry, clearly apparent from the contour-line plots of the original research reports, may still reflect the shape of the fragment as proposed [Wood 1962], but rather through the mechanism of cooling of the fragment, thus halting diffusion of the hydrogen-3, than by the attenuation of cosmic-rays.) The observations have been fit to a curve consisting of a Cauchy distribution for the temperature and a Gaussian distribution to represent the thermal diffusion, with a diffusion constant varying with temperature according to the distance from the origin.

If the amount of fissionable material present in a localized speck is greater than a few milligrams, the released fission energy will cause the surrounding iron to vaporize, with the size of the resulting bubble determined by the mass of the speck of fuel. Remains of spherical cavities have been found at the surface of some meteorites, with diameters up to 1 meter [Henderson and Perry 1958, Mason 1962]. In some cases, the cavity extends deep within the body of the meteorite. These are judged to have existed before passage through the Earth's atmosphere. A speck containing about 0.04 grams of fissionable material could release the amount of energy required to vaporize the corresponding mass of iron, not allowing for the increase in the cavity size resulting from expansion of the superheated iron vapor.

In the helium-3 studies of the Grant meteorite, the remnants of two surface cavities were found on the side of the meteorite that has the greatest concentration of helium-3. The material adjacent to these cavities showed a decrease in the helium-3 concentration, consistent with the process described, from conversion of helium-3 to hydrogen-3 with subsequent diffusion away from a central point, or several central points, followed by decay in place to helium-3. The cavities may be preferentially located at the surface of the meteorites as a result of their production contributing to the fragmentation of the parent body.

Many meteorites contain small (about 1 millimeter to 1 centimeter) spherical blobs that are thought to represent original unprocessed material from the Solar/planetary nebula [Wasson 1985]. These blobs are called chondrules (named from the Greek for granules) and the type of meteorite containing them is called a chondrite. The bulk "chondritic composition", elemental and isotopic, is generally considered to represent the starting composition for the Solar System. Isotope anomalies are found in some of these meteorites [Schramm 1985, Rolfs and Rodney 1988].

Formation of the chondrules, apparently universally, required a brief intense heating, flash heating, sufficient to heat the material to 1,900-2,100 K for only a matter of seconds and then quickly quench the molten droplets [Boss 1995]. Further cooling was slower [Ash 1994]. In the standard theory of planetary formation from planetesimals accreting in a Solar nebula, it has not been possible to identify the source of the intense but brief heating. Giant lighting flashes, shockwaves, and magnetic reconnection flares have been offered [Cameron 1995a]. A nuclear fission detonation would produce such brief intense heat.

Chondrules in Allende show olivine bars in glass, coated with metallic iron [Brearley 1997]. This iron plating could have been produced by condensation of the iron vapor from the detonation in the core of the planet.

The element neon is a noble gas found in small amounts in meteorites. Neon consists of three stable isotopes: masses 20, 21, and 22. The observed ratios of these three isotopes can be produced by neutron absorption in fluorine and sodium. Sequential neutron absorption in fluorine first converts fluorine-19 to neon-20, then produces smaller amounts of neon-21, and still smaller amounts of neon-22. While the cross-sections for these reactions are generally low, a peak occurs in the fluorine-19 capture cross-section at 25 keV, corresponding in energy to a pronounced minimum in the cross-section of iron. The neutron energy spectrum in a reactor consisting principally of uranium in iron has a pronounced peak at this energy.

Neon isotopic anomalies in which the neon-22 isotope is exceptionally high can also be produced by neutron absorption, in this case by a high-energy neutron reaction with sodium-23, which produces sodium-22 by an (n,2n) reaction (neutron in, 2 neutrons out). The sodium-22 decays by positron emission to neon-22, with a half-life of 2.6 years. The long (in this sense) half-life of 2.6 years would allow molten droplets of material containing sodium-22 to solidify and cool before much of the neon-22 is produced, thus locking it in place. This suggests the form of occurrence found in the chondrules of some meteorites. Since fluorine and sodium often occur in the same minerals, a variety of neon isotopic compositions can be produced. These different compositions can be further compounded by the different neutron energy dependence of the reactions: the $^{19}F(n,\gamma)^{20}F$ reaction, producing neon-20, has a resonance at a neutron energy of 25 keV, while the

^{23}Na(n,2n)^{22}Na reaction, producing neon-22, must have a neutron energy greater than 12.5 MeV. The neutrons are supplied abundantly by the nuclear-fission chain-reaction in the core of the planet.

Magnesium consists of three stable isotopes: masses 24, 25, and 26. In some meteorites an excess of magnesium-26 has been found, particularly in aluminum-rich regions, and this has been taken as indicating the presence of long-lived aluminum-26 (half life 740,000 years; a short-lived isomer has a half life of 6.3 seconds) in the gas cloud that formed the early Solar System [Wasserburg 1985, Hutcheon and Hutchison 1989, Boss 1995]. By positron decay (emission of a positive electron), this isotope produces magnesium-26. Production of this nuclide by the decay of long-lived aluminum-26 resulting from proton irradiation of pre-Solar material has been suggested [Lee 1978] and argued against on the basis of a failure of this process to produce an expected excess of lanthanum-138 [Shen *et al.* 1994]. The nuclide lanthanum-138 is expected to be produced by proton irradiation, but has not been found in excess in those materials where magnesium-26 is found in excess.

An alternate source is the production, in place, of both long-lived and short-lived isomers of aluminum-26 by a high-energy (greater than 15 MeV) neutron reaction with aluminum-27, by the (n,2n) reaction. In a planetary detonation, wherever there was aluminum (aluminum-27 only) and high energy neutrons, aluminum-26 would have been produced. Where the neutrons had lost energy, there would be no aluminum-26.

The distribution of a significant excess of magnesium-26 in a hibonite ($CaAl_{12}O_{19}$) grain in the Murchison meteorite has been carefully measured by use of an ion microprobe [Ireland and Compston 1987]. The excess of magnesium-26 (relative to its content in terrestrial magnesium) is closely correlated with the aluminum content, relative to the major magnesium isotope, magnesium-24. The excess of magnesium-26 in the rim of the sectioned grain (a surface layer surrounding the grain) is roughly one-half that in the center of the grain. This can happen by production of aluminum-26, by the reaction ^{27}Al(n,2n)^{26}Al. Because of the high neutron energy required for this reaction (15 MeV), the recoil of the aluminum-26 nucleus will be significant, and much of the product atoms near the surface of the grain (nearly one-half), will escape, resulting in a depletion of the final magnesium-26 concentration in this rim.

Excesses of magnesium-26 in many similar meteorites has been interpreted as showing that radioactive aluminum-26 existed in the early Solar System, and may have contributed enough heat of decay to melt the larger planetesimals [Cameron 1995b, MacPherson, Davis, and Zinner 1995].

Other isotopic anomalies in magnesium have been observed and can be explained as direct consequences of neutron absorption: an excess of magnesium-24 will be produced by neutron absorption in sodium (sodium-23), and excess magnesium-25 will result from neutron absorption in magnesium itself.

Potassium consists of three isotopes, two stable and one long-lived radioactive isotope. In terrestrial material, the radioactive isotope potassium-40 constitutes a very small trace of the element (0.000117), while the stable isotope potassium-39 constitutes over 93% of the element. In the Aroos iron meteorite and the Clark County meteorite, however, the isotopic composition is far different [White 1986]. In potassium from these two meteorites, the three isotopes have similar fractional abundances, ranging from 0.16 to 0.56. The observed abundances can be closely matched by calculations of neutron absorption, starting with the terrestrial abundances, with small adjustment of the neutron absorption cross-sections, which may reasonably result from differences in the neutron flux spectra.

Both negative and positive anomalies of chromium-54 have been found [Rotaru *et al.* 1992] in meteorites. The negative anomalies can result from neutron irradiation of vanadium, by the reaction ^{51}V(n,γ)^{52}V, the product of which beta-decays to chromium-52. Multiple neutron capture will produce small amounts of chromium-53 (from the decay of vanadium-53) and still smaller amounts of chromium-54 (from the decay of vanadium-54). Since chromium-52 is the reference isotope used in determining the anomaly, an increase in this isotope will appear to be a decrease (negative anomaly) in chromium-54. If the material is relatively poor in vanadium but contains chromium, neutron irradiation will reduce the chromium-52 content by the reaction ^{52}Cr(n,γ)^{53}Cr (which is a stable isotope) and will also increase the chromium-54 content, producing a positive anomaly.

Anomalies in the isotopic pair chromium-52 and chromium-53 have been observed and explained as the result of the decay of manganese-53 [Birck and Allègre 1988]. High-energy neutrons are effective at producing this radionuclide from iron-54, by the (n,np) reaction [Goldberg *et al.* 1966]. High-energy neutrons will be produced abundantly in a planetary detonation, in material that is intimately mixed with iron.

A search for evidence of extinct technetium isotopes in meteorites found an anomaly in the ratio of ^{99}Ru/^{101}Ru, which was considered to indicate the presence of technetium-99 at the time of condensation of the meteoritic material [Yin, Jagoutz, and Wänke 1992]. No finding of excess technetium-98, with a somewhat longer half-life, was reported. Neutron absorption in molybdenum will produce technetium-99, without producing any technetium-98. Tecnetium-99 is also a fission product with a high yield, 6.1% for uranium-235 fission.

Several isotopes of the element palladium are produced as fission products. Some of these isotopes are stable and some are radioactive decaying to the corresponding isotopes of silver with various different half-lives. The isotopic anomaly observed for silver in meteorites [Schramm 1985, Wasserburg 1985, Wasserburg and Chen 1981] can result from the decay of fission-produced palladium-107, with a half-life of 6.5 million years, to produce silver-107, and the decay of palladium-109, with a half-life of 13.5 hours, to produce silver-109. The variations that are possible form the basis for solving Gerry Wasserburg's favorite puzzle.

If the palladium and silver produced in this manner are not chemically separated, the ratio $^{107}Ag/^{109}Ag$ will be approximately 2.5, considerably greater than the terrestrial value of 1.07. If the two elements were chemically separated after the short-lived palladium-109 had decayed to silver-109, then the ratio in the palladium-rich phase could be as high as permitted by the efficiency of the separation, and would be proportional to the palladium content. The absence of evidence of other fission products in these meteorites is likely to be due to the loss of the more volatile fission products, relative to the highly refractory series of fission products that contribute to palladium and silver. In fact, the meteorites are described as showing "a strong depletion in volatile siderophile elements" [Wasserburg and Papanastassiou 1982]. (Siderophile elements tend to associate with iron in the molten phase. Lithophile elements associate with silicates. Chalcophiles associate with sulfides.)

Meteorites have been found containing xenon-129, which is considered to have resulted from the radioactive decay of iodine-129 (half life 15.7 million years) that is assumed to have condensed from a supernova cloud, in minerals rich in iodine [Anders 1963, Wasserburg and Papanastassiou 1982, Whitby *et al.* 2000]. In some cases, the xenon-129 is nearly pure, suggesting that all the radioactive iodine-128 (half life 24.99 minutes) had already decayed to xenon-128, which clearly would not condense with the iodine-129 in the chondrule, a 1-centimeter blob. Iodine-129 is a fission product with a yield of about 1%.

Xenon and Iodine

Xenon is a noble gas, with stable (and some unstable, radioactive) isotopes ranging from mass number 124 to mass number 136. It is produced abundantly in nuclear fission; about 20% of fissions produce a stable xenon atom. (Under certain circumstances this can be increased by the conversion of radioactive xenon-135 to stable xenon-136 by neutron absorption, because of the very large absorption cross section, 2.65 million barns.)

Neutron-rich fission fragments decay to more stable nuclides by beta-particle emission. When a stable nuclide is reached, the decays stop, and that nuclide becomes part of the isotopic pattern for the fission that produced the initial fragment. Since there is great variability in a process as chaotic as nuclear fission, this is a statistical pattern.

Some of the nine stable isotopes of xenon are blocked by stable nuclides in the beta-decay chain from a fission fragment and are not produced, almost at all. The isotopes 124, 126, 128, and 130 have essentially no yield in fission. For xenon, the last step in the decay chain is an iodine isotope. Iodine is volatile and, unless it is trapped in a solid, will remain in the vapor phase until it cools sufficiently to condense, at about 460 K, or react chemically. In a nuclear weapon, the detonation temperature is a few tens of millions of kelvins. In a confined planetary detonation, the temperature may reach a few tens of billions of kelvins (see the interpretation of the antineutrino spectrum from SN 1987A, in Appendix C). In dilute uranium, fissioned by the collateral neutrons from the detonation, there may be no sensible temperature increase.

The xenon isotopes in various meteorites (and in various parts) have been studied in precise detail [Wasserburg 1955, Reynolds 1960, Marti *et al.* 1989, Nichols Jr. *et al.* 1991, Kim and Marti 1992, Nichols Jr., Hohenberg, Marti 1992, Swindle and Burkland 1992, Gilmour *et al.* 1995a, Brazzle and Hohenberg 1995, Gilmour *et al.* 1995b, Ott 2000, Whitby *et al.* 2000, Gilmour, Whitby, Turner 2001].

Many studies show a variety of xenon isotopes, as is found in Earth's atmosphere, and as is produced by nuclear fission. Reynolds found an excess of xenon-129 (by roughly 50%) in the meteorite Richardton, and interpreted it as being from the presence of (extra) radioactive iodine-129 at the time of the formation of the meteorite material [Reynolds 1960]. Studying the Beardsley meteorite, Wasserburg found only an estimated upper limit at one-tenth the amount found by Reynolds [Wasserburg 1955]. Nichols found excess xenon-129 in the apatite grains of Acapulco, but none in the merrilite grains, showing how variable and chancy this research can be.

Iodine-129 has a half life that is long to us, 15.7 million years, but short compared to the age of the Solar System, and is produced abundantly (about 1%) in nuclear fission. The adjacent iodine isotope, -128, has a very short half life, 25 minutes. However, this isotope is not produced by fission, being blocked in the beta decay chain by stable tellurium-128. Both of the iodine isotopes decay to xenon isotopes. (All of the fission-product xenon isotopes are produced by beta decay from corresponding iodine isotopes, all but iodine-129 have short half lifes, ranging from minutes to days.) If the iodine is in the vapor phase, all the xenon that is produced by beta decay will be lost. Only after the iodine is formed into a solid body, a grain, by chemical reaction or condensation, will the xenon isotopes be trapped. For a fully formed xenon fission product distribution to form, the iodine isotopes must be trapped in less than a minute. (Stable xenon-136 will be most lost because of the 1.39 minute half life of its parent iodine-136.)

Reynolds found a significant excess of xenon-129, a slight excess of xenon-128 (which might have been produced from iodine-128 formed by neutron absorption in iodine-127), and a very minor deficit in xenon-136.

At the other extreme, Whitby found pure xenon-129, in halite crystals (common salt, sodium chloride) that were found in the Zag meteorite [Whitby *et al.* 2000]. This suggests that all the short half life iodine isotopes had decayed to xenon which was lost, before the iodine-129 was combined into the halite. The derived ratios of iodine-129 (at time of formation) to iodine-127 (at the present time) were very variable, with differences from the mean value amounting to as much as 30 standard deviations. Obviously, this was not a simple case. The estimated age of formation makes the salt among the oldest materials yet dated [Ott 2000].

In contrast to the production of xenon in a brief burst of a nuclear detonation, long operation of a nuclear reactor, as the natural nuclear fission reactors at Oklo, causes some of the iodine-129 to be transmuted to iodine-130, which then decays to xenon-130. Excess xenon-130 was found in the Oklo reactors [Meshik, Hohenberg, and Pradivista 2004].

Sources for xenon isotopes have been considered to be neutron fission of uranium-235, spontaneous fission of uranium-238, spontaneous fission of plutonium-244, and solar xenon [Nichols 1992, Marti *et al.* 1989, Kim and Marti 1992]. These sources do not seem to solve the problems, either for the individual meteorites or for the set. In an effort to measure formation ages, Brazzle and Hohenberg compared Pb-Pb (lead/lead) and I-Xe (iodine/xenon) ages [Brazzle and Hohenberg 1995]. For the Shallowater meteorite, the xenon ratios plotted on a straight line (an "isochron") as expected, but the data for Richardton and Bruderheim were completely out of line. Gilmour found the xenon ratios in ALH 84001 (a meteorite from Mars) were very variable and showed little relation to mixing between our air, the meteorite Chassigny (from Mars), and the Shergottite Parent Body (Mars) [Gilmour *et al.* 1995b]. The isotopes are not behaving the way we expect.

The ratio of iodine-129 to iodine-127 produced in a supernova is estimated to be $1.61/1.27 = 1.27$ [Cameron 1982]; a ratio of 1.4 is given by [Anders and Gravesse 1989]. The ratio of these isotopes at the time of trapping the xenon-129 is estimated to be about 1×10^{-4}; a mean value of 1.35×10^{-4} is given by [Whitby *et al.* 2000]. The radioactive iodine must decay by about 14 half lifes, losing the xenon, before solidifying. That is about 214 million years. Less time would result in retaining greater amounts of xenon-129, while more time would retain less xenon. In a planetary detonation, a relatively brief and intense burst of neutrons can produce a variety of iodine nuclides, in an absorption chain, each absorption making the next absorption possible, as in the prompt process, studied by use of nuclear weapons [Becker 1993]. This can be shown by a series of neutron absorptions, each occurring too rapidly for the beta decays to change the element. These absorptions start with the stable iodine-127, in place in minerals on and in the pre-asteroidal planet "Asteria". There is no involvement with a supernova cloud or the pre-Solar nebula, just a fully formed, solid planet. (Half lifes are shown above the chain, neutron absorption cross sections, in the nuclear unit "barn", equal to 10^{-24} square centimeter, shown below.)

$$\text{stable} \qquad \text{25 min} \qquad \text{15.7 Myr} \qquad \text{12.36 h} \qquad \text{8.02 d}$$
$$^{127}\text{I}(n,\gamma) \rightarrow {}^{128}\text{I}(n,\gamma) \rightarrow {}^{129}\text{I}(n,\gamma) \rightarrow {}^{130}\text{I}(n,\gamma) \rightarrow {}^{131}\text{I}$$
$$\text{6.15 b} \qquad \text{22 b} \qquad \text{30.31 b} \qquad \text{18 b}$$

This prompt absorption of neutrons will produce small enrichments of xenon-128 (as is shown by [Reynolds 1960]), xenon-129 (the isotope of main interest), and xenon-130, and -131 (as also shown by [Reynolds 1960]).

Several researchers have suspected that (natural) neutron irradiation caused some otherwise unexplainable effects [Bogard 1995, Nichols 1991, Kim and Marti 1992]. Their data permitted the derivation of a neutron fluence, the time integral of a neutron flux, that would produce the effects that were observed. Those interpretations are shown here:

Neutron Fluences in Meteorites

Observation	Reference	Neutron Fluence
		neutrons per cm^2
Excess ^{36}Ar, in Arapahoe, Bruderheim, Torino chondrites, and Shallowater aubrites	Bogard 1995	$0.2 - 0.6 \times 10^{16}$
Depletion of ^{149}Sm in LEW 86010 and Torino angrite	Bogard 1995	0.25×10^{16}
Chico	Bogard 1995	1×10^{16}
^{128}Xe from ^{127}I	Nichols 1991	30×10^{16} thermal $2 - 3 \times 10^{16}$ resonance
Forest Vale (metal) xenon from ^{235}U fission	Kim and Marti 1992	1×10^{16}
Adjacent to the Hiroshima atomic bomb detonation (16 KT yield)	Santoro *et al.* 2005	5×10^{16}

The similarity in the meteorite data suggests a widespread and common effect. The agreement of the meteorite data with the neutron fluence from a nuclear fission weapon [Santoro *et al.* 2005] supports the proposal made in this book that multitudinous natural nuclear fission reactors are distributed throughout the interiors of planets, and that at about 4.56 billion years ago, the reactors in the planet "Asteria" detonated, nearly simultaneously and concurrently, disrupting the planet and forming the asteroids. Thus, the well determined oldest ages signify the nuclear detonation, and do not give us information on the age of the Solar System, other than it is older.

Those elements that are lost from small grains and fragments that have been melted at one stage of the expanding detonation are regained by condensation in the next, so that the bulk composition is unaffected. All elements are retained in the bulk. Further evidence of the detonation comes from consideration of the recapture of volatilized elements from the surrounding saturated gas, suggesting a region 1,000 kilometers in extent [Desch 2006] or 150 to 6,000 kilometers in radius [Cuzzi and Alexander 2006]. That is the size of a planet. "Asteria" could have been about 8,000 kilometers in radius.

The H- and C-chondrites have approximately 20 times as much xenon, relative to the lighter noble gas krypton [Ozima 1986, Wacker and Anders 1984, McElroy and Prather 1981]. While this has been interpreted as indicating a deficit of xenon in the planetary atmospheres, it is more likely to be a surplus of xenon in the meteorites instead. The production of the observed concentration of xenon in the meteorites would require the nuclear fission of only 0.02 to 0.2 parts per billion (ppb or 10^{-9}) of uranium. Fission of uranium and plutonium can result in Xe/Kr ratios in the range of 6 to 16, not quite the number observed, but close.

The agreement between the xenon isotope distribution observed in several meteorites with that found for the spontaneous fission of plutonium-244 is remarkably good [Wasserburg 1985] and supports the explanation that plutonium-244 (with a half life of 80.8 million years) produced in a pre-Solar supernova provided this nuclide to the condensing Solar nebula. It may be more likely that this nuclide was formed in the nuclear detonation that disintegrated the asteroidal planet, in a similar manner as considered for the production of the plutonium-244 in the bastnäsite of Mountain Pass [Hoffman *et al.* 1971]. Formation of this nuclide is dependent on the burst of neutrons being intense and brief, as in a detonation. Many observations of evidence for radioactivity in the early Solar System have been explained as long-lived radionuclides produced in supernova explosions. The problem with this explanation is that several different supernovae are required, at different times, each producing the particular radionuclides without producing the others.

A model for production of many of the relatively short-lived nuclides (^{26}Al, ^{60}Fe, ^{129}I, ^{92}Nb, ^{53}Mn, ^{107}Pd, and ^{146}Sm) through the action of heavy cosmic rays colliding with hydrogen nuclei has been proposed [Clayton 1994]. This model gives results that are in reasonable agreement for some nuclides but are off by factors of 4 or more, to 100 for samarium-146 (^{146}Sm).

While details may differ depending on the choice of nominally continuous nucleosynthesis or a sudden "spike" of elements before solidification of the Solar nebula materials begins, the time before solidification as determined by several radioactive-stable nuclide pairs shows major conflicts:

Nuclide pair	time of supernova occurrence before solidification
$^{235}U/^{238}U$	1850 million years
^{244}Pu/spontaneous fission xenon	164 million years
$^{129}I/^{129}Xe$	100 million years
$^{107}Pd/^{107}Ag$	26 million years
$^{26}Al/^{26}Mg$	3 million years
$^{99}Tc/^{99}Ru$	2.9 million years
$^{41}Ca/^{41}K$	1.8 million years
$^{53}Mn/^{53}Cr$	1 million years
$^{22}Na/^{22}Ne$	10 years after!

It would appear that at least four to seven supernovae would be required to provide the raw material for the Solar System, and the condensation and solidification of material in the forming Solar System apparently 10 years before the supernova explosion does not seem physically realistic.

This problem has been recognized and an explanation has been suggested by Hubert Reeves [Allègre 1992]: every 120 million years, as the location of the Solar System is thought to pass through its spiral arm in the Milky Way, new supernova-produced material was added to the growing cloud. This material eventually coalesced into the Sun and solidified into its planets. However, the underlying problem is that the nucleosynthesis of each supernova did not interfere with that of the others. The supernova that produced plutonium did not produce uranium or palladium, and the production of manganese did not include aluminum. The observation of well-defined peaks in the abundances of r-process isobars also argues in a similar manner against multiple supernovae as the source of the elements [Mathews and Cowan 1990]. The evidence of the existence of sodium-22 with a half-life of 2.6 years, in the solidifying material from the supernova, is really insurmountable with the multiple supernova model.

The abundance of thorium in the atmosphere of the Sun has been measured [Lawler *et al.* 1990] and shown to be about 40% larger than in meteorites [Anders and Grevesse 1989]. Since it is unlikely that additional thorium has been produced in the Sun, beyond what was present at the time of the formation of the Solar System, this difference is likely to represent a deficit of thorium in the meteorites. Thorium will be consumed by nuclear fission induced by high-energy neutrons, as in a nuclear detonation, and this may explain the loss of thorium.

Uranium isotopic anomalies, related to variations in uranium concentration, have also been found in meteorites. Evidence of a nearly instantaneous fission reaction is shown in the isotopic composition and concentration of uranium found in certain meteorites. In Figure 52, the $^{235}U/^{238}U$ ratio is compared with the uranium concentration of small grains from a variety of chondritic meteorites [Arden 1977].

Two facts are immediately apparent: the uranium concentration in meteorites is very small compared to the rich African ores at Oklo, Gabon (and is less than the average concentration in the Earth's crust, 4 ppm), and the uranium-235 content of the uranium is significantly, and in all cases, greater than the uranium-235 content of natural uranium [Cowan and Adler 1976]. Low concentrations of uranium, where nearly all the uranium may have been consumed by fission reactions, have high values of the $^{235}U/^{238}U$ ratio, or enrichment, while the higher concentrations show more normal enrichment values. This can result from conversion of uranium-238 to plutonium-239 in a brief episode of neutron production, followed by eventual decay of the plutonium-239 to uranium-235. In contrast, the correlation of enrichment and concentration found at Oklo [Naudet and Renson 1975], which is also shown in the figure, shows a reduction in enrichment at high concentrations, where it is known that much of the uranium-235 has been consumed by fission.

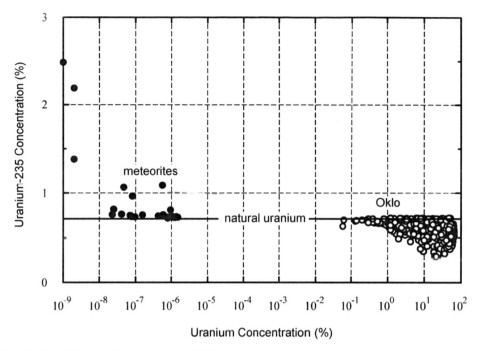

Figure 52. Uranium-235 content in meteorites and Oklo ore.
Variation of $^{235}U/^{238}U$ ratio with uranium content for meteorites and for Oklo ore bodies. The meteorite uranium shows a marked reduction in concentration and an unusual increase in uranium-235 content, as would result from an intense burst of neutrons in a nuclear detonation, by conversion of uranium-238 into plutonium-239, which subsequently decayed to uranium-235. The Oklo ores show the depletion of uranium-235 resulting from long-term operation of natural nuclear fission reactors.

At low neutron energies, so-called thermal energies, as are produced in water-moderated reactors such as the natural reactors at Oklo, the neutron absorption cross-sections (or absorption probabilities) are much greater for the destruction of uranium-235 by fission or by conversion to uranium-236, than for the conversion of uranium-238 to plutonium-239, which subsequently produces uranium-235, by alpha decay.

At higher neutron energies, such as at the 28-keV dip in the iron scattering cross-section, the reverse is true. For these higher energy neutrons, the cross-sections for uranium-235 and uranium-238 are more nearly equal, with conversion of uranium-238 to plutonium-239 occurring more frequently than the destruction of uranium-235. Thus, since the alpha decay of plutonium-239 produces uranium-235, irradiation of uranium atoms by high-energy neutrons leads to an apparent enrichment of uranium-235 in the remaining uranium, as observed in these meteorites. In a continuing chain-reaction, such as in the Oklo ores, the new plutonium-239 is "burned", fissioned, almost as soon as it becomes available from the decay of uranium-239 to neptunium-239 to plutonium-239, all of which have short half-lifes. There is not sufficient time for the plutonium-239 to alpha decay. In a brief burst, a detonation, all of the uranium-238 that was converted to uranium-239 will decay to plutonium-239, which will decay to uranium-235, enriching the remaining uranium. That is what is seen in the meteorites [Arden 1977] (Figure 52). The Oklo ores show a depletion in the uranium-235 content [Naudet and Renson 1975] while the meteorites show an enhancement.

Arden's measurements have not been well replicated [Tatsumoto and Shimamura 1980, Chen and Wasserburg 1980a, Chen and Wasserburg 1980b, Shimamura and Lugmair 1981, Lugmair and Galer 1992, Stirling, Halliday, and Porcelli 2005, Brennecka *et al.* 2010], and that might be because of the especially active three-step acid extraction Arden used. Progressive treatment with nitric and hydrochloric acids showed increases in the uranium-235 concentration. This suggests that the highest uranium-235 concentrations were bound in the most refractory grains. The most significant indication is that as the elemental uranium concentration decreases (as shown in Figure 52), the excess of uranium-235 increases. This would occur in a nuclear fission detonation, where the more uranium was consumed, the more uranium-238 was converted to plutonium-239, to ultimately decay to uranium-235, producing the observed excess. The detonation progresses and is over in a fraction of a second, avoiding the loss of plutonium-239 by fission, although some of the intermediate uranium-239 will be lost by fission. This contributes more neutrons to the reaction. The fragmentary and accreted material resulting from the detonation would be very mixed and heterogeneous on a grain scale.

While it was not possible to make a direct comparison with the other measurements because of the lack of elemental uranium concentrations, a distributional analysis (cumulative probability plot [McGinnis 2009]) suggests that the five highest uranium-235 concentrations shown by Brennecka *et al.* [Brennecka *et al.* 2010] are significantly greater than the Gaussian distribution shown by the lower 10 values. As with Arden's results, only increases in the uranium-235 concentration were found. Brennecka *et al.* found that this could introduce errors in the Pb-Pb (lead/lead) dating of 5 million years or more.

Lugmair and Shukolyukov commented that "…'normal' CAIs with ^{26}Mg excesses also usually possess isotopic anomalies in at least the iron peak elements with reasonably uniform excesses in the neutron-rich isotopes" [Lugmair and Shukolyukov 2001]. This would be expected in a confined planetary nuclear detonation, both from the effect of "superfission", the fissioning of the primary fission fragments, and by addition of neutrons by neutron absorption.

The isotopic abundance anomalies of neodymium in a FUN inclusion (Fractionation plus Unknown Nuclear effects) from the Allende meteorite and in silicon carbide that is considered to have originated independently of the Solar System [Ott 1993] have been compared, and shown to be greatly different. The isotopic abundances of the FUN inclusion are similar to the abundances produced by fission, possibly modified by rapid neutron capture. Here, the "unknown nuclear effects" are suggested to be the absorption of neutrons produced in an intense burst of nuclear fission power, in the detonation of the parent body of the meteorites.

Anomalous amounts of both barium and zirconium have been found, as isolated elemental anomalies, in meteorites [Mason 1962]. White calcium- and aluminum-rich chondrules have been found that contain notable amounts of the platinum group metals and hafnium and molybdenum [Blum *et al.* 1987, Wark 1987]. All these elements are produced in significant amounts by nuclear fission.

Shock-produced diamonds [Wood 1962] and shock-induced metallurgical changes have been observed in meteorites [Maringer and Manning 1962]. There appears to be considerable difficulty in forming diamonds under the conditions of pressure, temperature, time, composition, and external shock produced by impact on the meteorite parent body. No consideration appears to have been given to the possibility of internal shock as a diamond-forming agent. Such an internal shock would accompany the disintegration caused by a nuclear detonation.

The Widmanstätten pattern that is commonly observed in iron meteorites, that so distinctly sets them apart from terrestrial iron, is the result of slow crystallization of separate iron-nickel phases. This pattern has not been observed in any meteorite with greater than 14% nickel [Mason 1962]. There seems to be no metallurgical reason for this. This effect may be due to the nickel being present at the time of crystallization in the form of radioactive iron-60 that will eventually decay, through cobalt-60, to nickel after complete solidification has occurred, changing the ratios of iron to nickel. The metallurgical phases in meteorites are thought to not be in thermodynamic equilibrium [Massalski 1962, Moffatt 1984]. This condition of non-equilibrium would result from the gradual conversion of iron (radioactive iron-60 with a halflife of 2.6 million years) to nickel (stable nickel-60) after crystallization of the iron-nickel alloy. The iron-60 would be produced in a detonation by neutron absorption on iron-58, or by "superfission", in which the initial fission fragments are themselves fissioned into fragments with masses in the 55 to 65 atomic mass unit range. This could have resulted from production of a significant amount of nickel by the decay of iron-60.

All meteorites show magnetic effects, responding to a magnet [Krinov 1960, Haag 2003]. That is certainly not true for most Earth rocks. Something different must have happened. All meteorites contain metallic iron, often appearing to have been deposited from a spray of vapor that condensed on the small grains. Meteorites often show magnetic polarity [Krinov 1960] but remanent (or relict) magnetism is difficult to study [Cisowski and Hood 1991] and is rarely discussed. Since the outer part of a meteorite is likely to be heated above the Curie point on entry into Earth's atmosphere, the remanent magnetization will be lost from the surface, and the inner field will be shielded by the de-magnetized high-permeability nickel-iron alloy surrounding it. Further, it would not be expected that small objects, such as the planetesimals that are currently assumed to be the parent bodies of the asteroids, would have developed sufficient magnetic fields to impart remanent magnetism in the fragments of impact that we now see as the asteroids. This may explain why the magnetic field of the asteroid Gaspra [Kivelson *et al.* 1993] was such a surprise [Kerr 1993, Astronomy 1993, Sky & Telescope 1993d] when discovered by the *Galileo* spacecraft fly-by.

Existing remanent magnetization may be lost from a ferro-magnetic material by heating the object to a temperature above the Curie point. In pure iron this is 1,043 K; in iron-nickel it is 673 to 885 K, depending upon composition; unless the alloy phases are not in thermodynamic equilibrium, in which case [Hansen and Anderko 1958] it may be as low as 373 K.

In a planetary detonation, shockwaves exceeding the compressive and tensile strengths of the rocks shatter and pulverize the planet. The expanding fireball of iron vapor and fission products propels molten iron into the base of the mantle, breaking it apart, and mixing the molten iron with the peridotite. The volatile elements are vaporized, to condense as CAIs, and farther out, as the chondrules, "drops of fiery rain", and then coating all particles, large and small. These soft sticky blobs then run into the dust cloud created by shockwaves reflecting from the surface of the planet, and collect a dusty coat. Fractures in fragments would be filled with phyllosilicates as the fragments are propelled upward through the bottoms of lakes where clays had formed. Most of the detonation debris would be flung out of the Solar System, with some remainders to be scattered by the planets.

The Outer Planets

The outer planets, Jupiter, Saturn, Uranus, and Neptune, generally called "giants", are distinctly different from the inner, rocky planets, yet are somewhat similar among themselves. Structurally, these may be considered to consist of a rocky kernel composed of iron, rock, and ices, surrounded by an immense ocean and atmosphere of light gases such as hydrogen, helium, methane, ammonia, and nitrogen [Hubbard 1990]. These planets are shown, as "portraits", in Figure 53.

With the exception of Uranus, these planets radiate considerably more energy than they receive from the Sun [Zeilik and Smith 1987], somewhat more than would be expected from radioactivity if the entire planet were composed of the material of the carbonaceous chondrites [Hubbard 1990] (a rather poor comparison considering that the bulk of the planets is hydrogen and helium, and "ice", this comparison greatly overestimates the radioactive content). In particular, the weather activity observed on Neptune [Kinoshita 1989, Limaye 1991] is far more violent than expected for the small input of energy received by this planet from the Sun, less than two percent overall of the Earth's supply. This extra internal energy, at this late time after the formation of the planets, is difficult to explain on the basis of heating by contraction alone [Zeilik and Smith 1987]. Condensation of helium, falling as raindrops through metallic hydrogen, has been suggested for Jupiter. But the most severe objection to this idea is that Neptune radiates excess heat, while its twin, Uranus, does not. If the excess heat were a consequence of the formation of the planets, Uranus and Neptune should be the same. They are not, something is different. Herndon has also proposed that similar natural nuclear fission reactors in Jupiter, Saturn, and Neptune are the probable sources of the excess energy in these planets [Herndon 1992].

This excess energy could be produced by on-going nuclear-fission chain-reactions. In the case of Uranus, the steady phase may have been terminated by a detonation, consuming all the nuclear fuel, and halting any further production of heat.

Less than a hundred years ago, it had been thought that the outer planets, gas giants and ice giants, were miniature Suns, providing heat to their satellites. A more accurate model of the outer planets was devised by Sir Harold Jeffreys in 1923, with further refinements by Rupert Wildt, W. R. Ramsey, and (in 1980) W. B. Hubbard and J. J. MacFarlane [Hunt and Moore 1994]. Since a planet shines by reflected sunlight, its spectrum is much like that of the Sun, so this idea of miniature suns was not very farfetched, considering the knowledge of the time. Just as Jupiter's moons had been stars to Galileo, the giant planets had been stars to modern astronomy, until a new viewpoint showed that they could be cold, not hot, and a giant but greatly distorted version of our own home world, a planet.

The giant planets are aptly named. Each planet is huge compared to the Earth, in mass, diameter, gravity, and magnetism. Most of the mass is from gas, hydrogen and helium, and "ice", water, methane, and ammonia. Inside each is a rocky kernel, ranging from 4 to 10 times the mass of the entire Earth, each is different, and those differences tell their stories.

The outer planets, except for Uranus, radiate excess energy, compared to what is received from the Sun. (This observation is partly aided by the lower natural temperature at these great distances, and by our ability to observe them unhindered by glancing sunlight. If the Earth could be observed properly, with sufficient accuracy, we might find excess energy here as well.) The excess heat, more heat than is received from the Sun, is shown for each of these planets:

Planet:	Jupiter	Saturn	Uranus	Neptune
Excess heat (watts)	33.4×10^{16}	8.64×10^{16}	0.03×10^{16}	0.33×10^{16}

This remarkable observation stands out: three of the four similar planets emit more heat than they receive, but Uranus does not. This alone makes it unlikely that gravitational contraction produces the excess heat that is observed. Its twin sister Neptune radiates ten times as much heat. Uranus is also peculiar with a spin axis that lies almost in the plane of its orbit about the Sun; it rotates in reverse, like Venus; its magnetic field is at an angle of 60 degrees to its spin axis and is offset by roughly 8,000 kilometers (more than the radius of the Earth!) from the center of the planet.

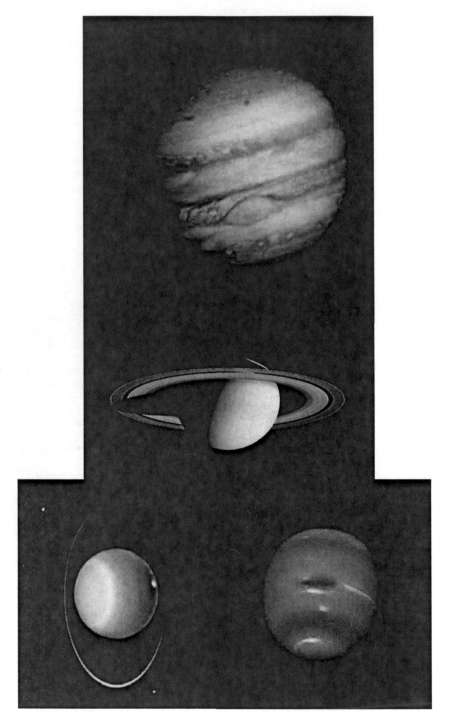

Figure 53. The four outer planets.
Jupiter, Saturn with glorius rings, Uranus with a scant ring, and Neptune, the Gas Giants of our Solar System (originals in color).

Uranus seems a bit too small, Jupiter and Saturn a bit too large, if we take Neptune as the standard and apply the 1/r rule for protoplanetary masses, as proposed in the section on forming the Earth as a giant planet. This is shown in Figure 54. By this interpretation, Jupiter gained approximately 220 Earth masses of material, while Saturn gained about 40. Uranus seems to have lost over 12 Earth masses, compared to the reference of Neptune. There was ample excess gas in the inner Solar System, that was ejected when the Sun ignited, to supply the amounts gained by Jupiter and Saturn. Uranus may have lost mass, both gaseous and solid ("ices") by the detonation of a large moon at about 80 million years ago, to end the Cretaceous Period on the Earth. A partial, internal detonation in Uranus, induced by the detonation of a large moon, could have ejected a large amount of solid and gaseous material, resulting in its tilt and offset magnetic field.

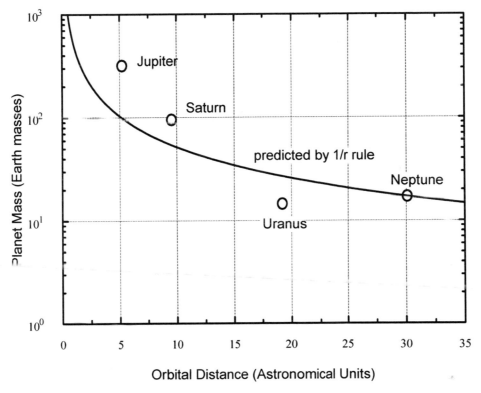

Figure 54. Giant planets in the Solar System.
The masses of the outer planets, in units of Earth masses, with the 1/r rule for protoplanet masses, developed for the formation of the Earth as a giant planet. Neptune is taken as the reference planet, as it is likely to have been the least disturbed. Jupiter and Saturn are significantly more massive than predicted; Uranus is similarly smaller, as if it had lost mass.

While three of the planets have large spots in their atmospheres, Uranus has none. Three of the four have barely detectable rings. Saturn has a glorious set of rings, clearly and obviously different from the others. Jupiter has four large moons, Saturn one, Neptune has one going around the wrong way. Uranus has only fragments, which the others have as well.

Uranus is just like the other outer planets, except that it has no excess heat, no large moons, may be a little lighter than it ought to be, a little wider than it ought to be, has a magnetic field that is more offset than the others, and a misaligned spin axis. Otherwise, it is no more unique than each of the others.

The giant planets are so large that nuclear detonations would be unlikely to disrupt them. However, the large moons could explode, leaving fragments (and rings of debris) orbiting their parent planets, and leaving marks on the surviving moons.

One of the lingering effects of the detonation of a satellite would be the heating of the planet and an inflation of its body. This appears to have happened for Uranus (a possible detonation of a satellite about 80 million years ago, leading to the end of the Cretaceous Period and the death of the dinosaurs), and Saturn (a likely detonation of a large moon about 12,900 years ago, leading to the end of the Pleistocene and the extinction of the megafauna, mammoths, mastodons, and others). Comparing the sizes of the giant planets, with Neptune and Jupiter as reference standards in this case, the sizes of Uranus and Saturn seem too large for their masses, as shown in Figure 55.

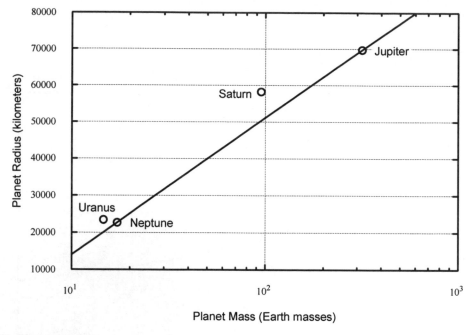

Planet Mass (Earth masses)

Figure 55. Giant planet sizes.
The apparent inflation of Uranus and Saturn, based on a comparison with Neptune and Jupiter.

The radius of Uranus appears to be 15.7% larger than it "should" be, the radius of Saturn appears to be 17.0% larger. To inflate Uranus by this amount would require 2.19×10^{33} joules, or about 18% of the energy needed to disrupt the then somewhat smaller planet. For Saturn, the energy for inflation would be 3.48×10^{34} joules, about 16% of the energy needed to disrupt that smaller planet. This energy would be deposited by thermal radiation from the fireball of the detonating satellite, by nuclear reactions, and by impacts of fragments. To lose 12 Earth masses from a larger Uranus would require about 2.25×10^{34} joules, about 54% of the energy to disrupt the entire planet. That would be enough to tip the planet on its side, and leave the atmosphere inflated.

In these hydrogen-rich planets, uranium hydride is likely to be the stable form of uranium in the core [Herndon 1992]. Herndon calculated the thermodynamic stability of uranium metal (U), oxide (UO_2), and hydride (UH_3), and found that the hydride was the stable form in conditions that are thought to represent those in the kernels of the gas-giant planets.

The critical mass for small spheres of uranium hydride in an assumed iron core of Jupiter has been calculated using the same methods as were used for reactors in the core of the Earth. The results of these calculations are shown in Figure 56. Natural reactors will develop where adequate concentrations of just a few kilograms of uranium accumulate. Similar conditions should exist in all the gas giant planets, and perhaps in their satellites as well.

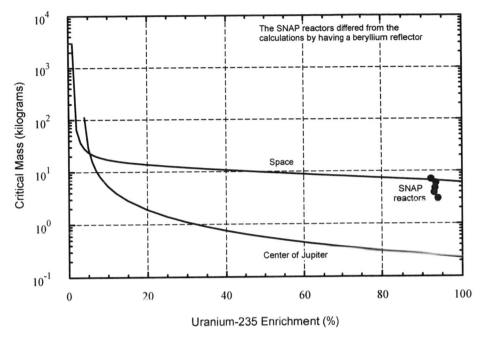

Figure 56. Critical masses for uranium hydride in space and in the center of Jupiter.
The critical mass in Jupiter is less than in space because of the considerable increase in density due to compression. For a uranium enrichment of 25%, the critical mass is only about 1 kilogram of uranium. The SNAP reactors (Satellite Nuclear Auxiliary Power) were developed by Atomics International for the Atomic Energy Commission.

Jupiter and His Moons

The composition of Jupiter is thought to match that of the Sun. What we can see of it, and its low bulk density (1,326 kilograms per cubic meter) shows that it is predominantly hydrogen and helium, like the Sun, although its measured helium content is slightly lower. Deep below the tops of the clouds the pressure and the temperature increase so that the gases become supercritical and no liquid surface exists, until the surface of molten silica is reached. As the pressure increases further, it is thought that the hydrogen is compressed to the extent that it becomes a metallic conductor. This has been extremely difficult to verify experimentally. An experiment with liquid hydrogen seems to have produces a metallic conductor phase [Nellis 2000].

The *Galileo* probe that descended into Jupiter's atmosphere on December 7, 1995 showed surprisingly strong winds that increased as the probe fell deeper into the atmosphere. This indicates that the weather on Jupiter is driven more by its own internal heat than by the energy it receives from the Sun [Kerr 1996a, Burnham 1996, Beatty 1996].

A suggested source of this heat is that helium becomes immiscible in hydrogen and condenses into droplets deep in the atmosphere where the hydrogen is converted to the metallic form, and releases gravitational energy as it falls toward the rocky kernel [Stevenson and Salpeter 1976]. This would also explain the reduction in observable helium content, compared to the Sun. However, the critical temperature for helium is only 5.25 K, which is exceeded throughout Jupiter. Above their critical temperatures, supercritical fluids are completely miscible, and have no surface tension, so raindrops would not form. Helium will remain as a gas, or a supercritical fluid. Unless helium is immiscible in metallic hydrogen, as is speculated, there would be no condensed phase, and no helium raindrops would fall. Jupiter's energy must come from some other process.

Moons

Jupiter holds four large natural moons, Io, Europa, Ganymede, and Callisto, discovered by Galileo in 1610, and 59 (more or less) scattered fragments, as shown in Figure 57.

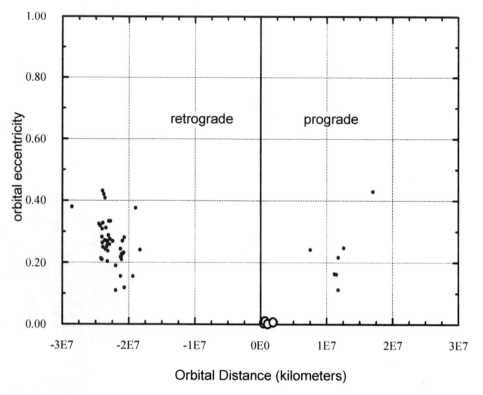

Figure 57. Moons of Jupiter.
The many moons of Jupiter, displayed by their orbital eccentricity, showing departures from a nearly circular orbit, and their orbital distances from Jupiter. A convention is used here of plotting the retrograde ("backwards") moons at negative distances from the planet. The four large natural moons are shown by large open dots clustered close to the planet, with very small eccentricities. The 59 (at present) other moons, all small fragments, are shown as small filled dots. Four small moons, likely to be fragments, are close to the planet and are hidden by the symbols used here for the large moons. The obscure notation for the distance scale indicates that the orbits range from -3x10[7] kilometers retrograde to 3x10[7] kilometers prograde.

Io

One of the satellites of Jupiter, Io, with incredible volcanoes, shows an anomalously high heat flow from its interior. This has been observed to be greater than 0.05 quadrillion watts (PW), approximately the same as for Earth, in a much smaller body. (*"The Planetary Scientist's Companion"* gives heat flux, and that is equal to 0.104 PW [Lodders and Fegley Jr. 1998].) This has been almost accounted for by calculation of the tidal friction induced by an eccentric orbit, maintained by its sister moon, Europa [Nash *et al.* 1986]. Io radiates more heat than it receives from the Sun and from tidal heating from Jupiter. The temperature of Io should be about 130 K, but volcanoes erupt at temperatures of 1,000 to 1,500 K [Francis 1993]. Hot spots have been observed at temperatures up to 654 K, centered on the eruptive volcanoes. Estimates of the lava temperatures range up to 1,613 to 1,723 K [Keszthelyi 2007]. Tidal heating has been proposed as the source of the eruptive energy [Frankel 1996], but the surface heat flow exceeded that by a factor of a hundred [Lodders and Fegley Jr. 1998]. Schubert, Spohn, and Reynolds estimated that tidal heating could keep Io molten for only about 300 million years [Schubert, Spohn, and Reynolds 1986]. The internal tidal energy dissipation has been calculated by analysis of orbital observations of Io (and the other three Galilean moons) about Jupiter, spanning 116 years [Lainey *et al.* 2009]. This was found to be $93.3 \pm 18.7 \times 10^{12}$ watts. This could produce a surface heat flux of 2.24 ± 0.45 watts per square meter. (This estimate may be compared with values for the Earth, which are $44 \pm 1 \times 10^{12}$ watts and 0.065 ± 0.0016 watts per square meter, respectively. The surface heat flux of the Earth is much smaller because of the much larger surface area. However, it is interesting that the total heat released from the two bodies is so similar.) The calculations have been refined, and constrained by orbital position measurements, and now produce results in agreement with the somewhat variable and incomplete surface heat flux measurements [Lainey *et al.* 2009]. Some caution is advised, as various determinations of a significant parameter in the derivation of tidal dissipation, the secular mean-motion acceleration, vary from −0.074 ± 0.087 to 4.54 ± 0.95. The value of Laney *et al.* (0.14 ± 0.01) is noticeably low and has a very small uncertainty. The major difficulties in these calculations involve estimates of the dissipation function and viscosity, and the history of satellite resonances, none of which we can know well. It was recognized that the range of viscosities required to produce the necessary tidal dissipation are too high to allow the transport of heat to the surface by convection. A different mechanism is needed. Something isn't quite right, yet.

An interesting sidelight on Io is its infrared spectrum, measured by Fink, Larson, and Gautier III [Fink, Larson, and Gautier III 1976], and shown in *"Jupiter"* [Morrison and Burns 1976]. The spectrum of Io is compared with a similar spectrum of a component of the multiple star Capella, as a reference standard. The similarity is nearly exact, as shown in Figure 58. The infrared spectrum of a planetary satellite in our Solar System, with a surface temperature in the range of 130 K to 1,723 K, matches almost exactly the spectrum of a Type G8III star with a temperature of 4,940 K, a bit cooler than our Sun at 5,780 K, at a distance measured by trigonometric parallax to be 42.2 light years. More properly, the spectrum of sunlight reflected from Io matches the spectrum of a very bright star. In the words of the astronomers, "A detailed comparison of the spectra of Io [lower] and α Aur [Capella Aa, upper] shows no features above the noise level that could be attributed to Io." [Fink, Larson, and Gautier III 1976, page 441] This would be a remarkable coincidence, that the spectrum of our Sun, reflected from Io, would so closely match, in detail, the spectrum of a different star. The point of all this is that an astronomical object cannot be determined to be a star on the basis of it having a stellar spectrum. If an object has a stellar spectrum, it may be a planet/satellite/asteroid shining by reflected sunlight. The problems with trigonometric parallax and our concept of the stars are discussed later in this book.

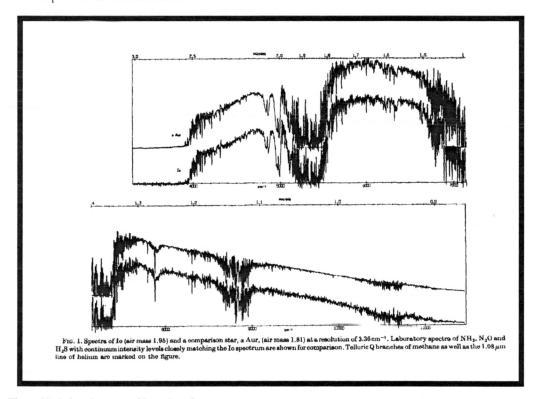

FIG. 1. Spectra of Io (air mass 1.95) and a comparison star, α Aur, (air mass 1.81) at a resolution of 3.36 cm⁻¹. Laboratory spectra of NH₃, N₂O and H₂S with continuum intensity levels closely matching the Io spectrum are shown for comparison. Telluric Q branches of methane as well as the 1.08 μm line of helium are marked on the figure.

Figure 58. Infrared spectra of Io and a reference star.
Comparison of the spectrum of reflected sunlight from Io (lower spectrum) and the spectrum of the G8III bright star, alpha Aurigae, Capella (upper spectrum). From Fink, Larson, and Gautier 1976. Some explanatory notes have been deleted.

Analysis of the observations of ultraviolet emissions from a cloud of hydrogen around Io showed a "huge escape rate of hydrogen" [Brown and Yung 1976]. Photolysis of ammonia (NH₃) has been proposed for the source of hydrogen, but the remnant nitrogen has not been detected. Neutrons in the vacuum of space decay into protons and electrons, and this could be the source of the observed hydrogen atoms. The magnitude of this source (10^{11} atoms per square centimeter per second for a total source of 4.2×10^{28} atoms per second) implies a fission rate (and fission neutron leakage) much greater (by 10,000 times) than what the excess internal heat suggests, with natural nuclear fission reactors as the source of heat. However, the source calculation involves a lifetime estimate that is 1,000 times shorter than the photoionization lifetime, so the discrepancy may be only a factor of ten. It is still difficult to understand how such a large flux of neutrons/protons could leak from deep within the satellite. The antineutrino-induced electron capture reaction (Appendix A) also produces neutrons and protons, but this reaction seems to be too weak to produce such a large source.

We don't know much about the other moons, except that Europa may have a liquid ocean below its surface. That suggests a temperature of close to 273 K just below a crust at 140 K.

Europa
The *Galileo* spacecraft returned remarkable images of Europa, showing an apparently resurfacing crust of ice. Clark Chapman remarked "Europa hasn't died. I'm willing to bet there are things happening today, but we haven't seen them yet." [Holden 1997]. Its surface shows evidence of rising heat from below, not certainly caused by tidal heating [Pappalardo *et al.* 1998]. It may have a subsurface ocean [Carr *et al.* 1998]. Without extra heat, it is estimated that this ocean should have frozen solid in 100 million years. Nighttime temperature anomalies require local heat flows of 1.1 watt per square meter, but tidal heating only gives an average of about 0.05 or less. Where's the heat?

Ganymede
Ganymede is the largest of the large natural moons in the Solar System. Emission of hydrogen from Ganymede has been found, as for Io. Plasma measurements found a loss of about 1.6×10^{22} protons per second [Frank *et al.* 1997]. Measurements of hydrogen Lyman alpha radiation produced an estimate of 6.1×10^{26} atoms per second [Barth *et al.* 1997]. These observations show an escape rate of hydrogen that is significantly less than that of Io, 4.2×10^{28} atoms per second. Dissociation of water (H_2O) ice has been proposed, but the anticipated outflow of oxygen ions was not found.

Ganymede was thought to have a subsurface ocean, presumed to be heated by radioactivity in a rocky core [Cowen 2000 citing D. J. Stevenson]. However, Ganymede has a distinct magnetic field [Kerr 1996b, Kivelson *et al.* 1996, Schubert *et al.* 1996], and tidal heating doesn't seem to be sufficient [Stevenson 1996]. The field may be produced by a molten core (where's the heat?), and there may be no subsurface ocean [Astronews 1997].

Callisto
Callisto has no magnetic field of its own, but may be surrounded by a field induced in a conducting ocean deep below its crust. The ocean may be liquid by virtue of dissolved ammonia acting as an antifreeze [Showman and Malhotra 1999]. Other mechanisms have been proposed to maintain the liquid ocean, such as a more rigid ice crust which would inhibit heat flow, and the presence of a heat source from radioactivity in a rocky core [Bennett 2001, Ruiz 2001]. Dissolved salts, or ammonia, contribute the electrical conductivity for the induction of the magnetic field.

Callisto is the most heavily cratered of Jupiter's moons, and appears to be cold, with no internal heat. There is a paucity of small craters. Callisto appears to have been partially segregated, with a normalized moment of inertia of 0.359 ± 0.005 [Showman and Malhotra 1999]. (A perfectly homogeneous sphere has a moment of inertia of 0.400.)

Titius-Bode Law for Satellites
The Titius-Bode Law, developed and as applied to the then five known planets, confirmed by the discovery of Uranus and leading to the discovery of the asteroid Ceres, guided the predictions of the location of an unknown planet, Neptune. It has been a successful rule, without an established theoretical foundation. Several variations have been proposed [Lynch 2003] and the concept of orbital regularity has also been applied to satellite systems. Since I have proposed that satellites may detonate and thereby disappear or leave fragments, the application of this concept to satellite systems, depending as it does on a sequential order, requires consideration of what moons that aren't there might be missing moons, detonated and gone from sight. Using a linear relation for the orbital distances of Jupiter's moons, and placing the existing moons in a sequence that provides a good fit, provides a very good fit.

The result of this analysis is shown in Figure 59. The existing moons have been placed in the following sequence:

0	fragmentary innermost moons
1	Io
2	Europa
3	missing moon
4	Ganymede
5	missing moon
6	missing moon
7	Callisto

A straight-line fit agrees quite well with the adjusted existing orbital distances and allows four moons to have been lost by detonation, one leaving fragments in its orbit. In this plot, the four existing large moons, with a selected sequence, are shown as solid dots; the proposed detonated moons are shown as open dots, taking the otherwise empty orbits. The innermost fragmentary moons are shown as bars across the axis at the 0 orbit. The unadjusted order is shown by small stars, and clearly does not fit the linear rule.

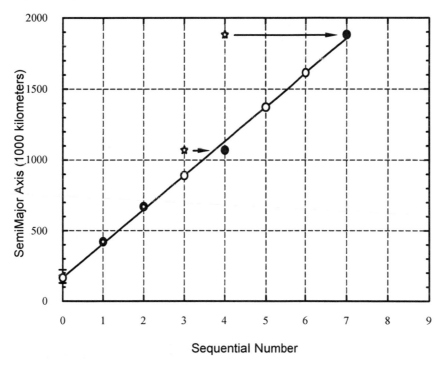

Figure 59. Large moons of Jupiter.
An application of the concept of regularity to the satellite system of Jupiter. The existing large moons have been placed in a selected order and are shown as filled dots. Missing moons, to fill in the gaps, are shown as open dots. The fragmentary innermost moons are shown as bars across the axis at the 0 orbit. Small stars show the unadjusted order, simply counting in sequence. After adjusting the sequential order for possibly missing moons, the orbital distances fall into a strictly linear progression. This shows, assuming that the regularity has some reality, that four moons are missing, and one left fragments behind.

This same pattern will be seen with the moons of Uranus.

Saturn and His Moons and Rings

Saturn is the next of the gas giant planets, perhaps the ideal image of other worlds in cartoons, with its remarkable rings. It is the second largest of the planets, less than one-third the mass of Jupiter, but nearly the same diameter. Compared to the other gas giant planets, with mean densities in the range of 1.318 (Uranus) to 1.326 (Jupiter) to 1.638 (Neptune), Saturn (at 0.6873) seems oddly inflated to have such a low density. This is particularly strange since Saturn appears to have more heavy elements than Jupiter, and its atmosphere is depleted in helium, which is thought to have sunk deep into Saturn. These effects should make a dense interior, holding the atmosphere closer. Saturn is cooler (95 K) than Jupiter (165 K), which should also increase the density and shrink its size. It radiates a similar amount of internal heat, approximately one-third as much as Jupiter. If mass counts for anything, Saturn's internal power relative to its mass (15.2×10^{-11} watts per kilogram) is nearly equal to Jupiter's (17.6×10^{-11} watts per kilogram).

While all the gas giant planets have been found to have rings of dust, only Saturn has glorious rings of ice. There is a suspicion that the ice is tinged slightly red by radiation damage [Eliot and Kerr 1984].

Saturn has a magnetic field similar to that of Earth, only one-tenth that of Jupiter [Hubbard and Stevenson 1988], but somewhat more closely aligned to its rotational axis. Saturn spins about its axis in 10.5 hours, almost as fast as Jupiter, and has the greatest oblateness, 0.097. The smaller magnetic field suggests that Saturn has less energy to drive the dynamo, in spite of its large amount of radiated power. Perhaps the problem is simply in the distribution of energy, and more of Saturn's energy has found its way out of the deep interior, to be radiated by the atmosphere.

Rings

While it appears that all the outer planets have rings, Saturn's rings are spectacularly more beautiful and extensive than the dusty little rings around the others. The rings of Saturn are considered to be short-lived compared to the age of the Solar System: if this is true, they are not permanent features [Mosquiera 1995, Dones 1991], and could not have been formed at the same time as the existing moons. Can it be that we are privileged to see these rings in the brief (astronomical) period of time in which they have their full glory, before they evaporate to tiny remnants? That could be so: if enough rare things happen, we should be treated to an occasional lucky glimpse.

The possibility that the rings might be young, perhaps just one-tenth as old as the Solar system (that is, about 450 million years) has been argued against. The major objection seems to be that "the generation of the entire ring through disruption of a Mimas-size (or larger) parent is unlikely on this time scale" [Cuzzi *et al.* 2010].

However, there are indications, as has been discussed here, that there have been several nuclear fission detonations in the Solar System, throughout its lifetime, and that some of those may have involved individual planetary satellites. In the case of Saturn's rings, a disruptive nuclear fission detonation of a large natural icy moon about 12,900 years ago may have produced the rings and altered the history of life on Earth.

Moreover, the difference in degree that we see in Saturn's rings, compared to the other planetary rings, may indicate a difference in kind, a difference in the manner of formation. While other rings may consist of magnetically captured material, Saturn's rings may be the debris of a detonated moon, a large moon like the icy moons of Jupiter, Europa and Ganymede.

The rings of Saturn are glorious, and have been puzzling ever since Galileo saw them and couldn't understand them. Now that we have had several spacecraft explorations, we know of several distinct rings. These have been labeled, in the order of their discovery, with letters of the alphabet, from the innermost to the outermost (so far): D, C, B, A, F, G, E. New, fainter rings continue to be discovered. It is a complicated system. The E ring is clearly established by material spewed out of the small moon Enceladus, one of the grandest surprises yet of the *Cassini* mission. (Another example of "We don't know what we don't know.") Only the C, B, and A rings have significant mass, and the B ring holds the most. The rings are shaped and shepherded by moons, large and small, some so small as to be called moonlets.

The C, B, and A rings appear to be a single structure that has been strongly affected by the sculpting effect of Saturn's moons [Esposito *et al.* 1984]. These may be three independent structures, as shown by compositional differences, or may have been sorted by tidal effects. They will be considered to be a single modified structure here.

Combined, these three rings contain about 31×10^{18} kilograms of material, a little less than the small moon Mimas, at 38×10^{18} kilograms. [Lodders and Fegley Jr. 1998 may have a unit error, showing gram values but labeling them as kilograms.]

The rings appear to be composed of a multitude of particles ranging in size from microscopic dust to meter-sized boulders. We can only examine the surface layer and this appears to be a form of water ice. There may be some rock dust, and the ice has a pinkish tinge, due to dust or from radiation damage of the crystalline structure of the ice or polymerization of organic chemicals [Cuzzi and Estrada 1998]. (Words can confuse. Radiation means different things to different astronomers, ultraviolet, infrared, or the 21-centimeter radiation sought by radio astronomers. Nuclear radiation, gamma-rays, beta particles, fast neutrons, can cause more damage on an atomic scale.)

The age of the rings has been a controversial problem. The simplest theory has been that the rings are material left over from the formation of the other moons (much like the theory that the asteroids are leftovers from formation of the planets), prevented from coalescing by the tidal pull of Saturn. The rings are inside the Roche Limit of Saturn, where a weak moon would be torn up, and a cloud of dust would be held apart. The rings should be 4.5 billion years old. However, meteoroid impacts should destroy the ring particles in 10,000 to 1,000,000 years [Durisen 1999]. The impact debris is likely to remain in the ring system, but the boulders should be reduced to grains. A further argument against a primordial origin is that Saturn's gravity and collisions among the particles would have prevented them from growing much larger than 10 meters across [Frank Spahn quoted by Cowan 2006b].

The rings are recent. Jeffreys [Elliot and Kerr 1984] described how a ring system of small particles would spread into a layer that was a single particle thick, in about a million years. Yet, at present, the rings reflect radar as if they were still at least several particles thick, and they brighten abruptly at opposition as a thick layer would. They may not yet be a million years old.

Approximate upper limits to the age of the main rings have been calculated ranging from 4.4 million years to 600 million years. The recent formation of the rings by a collision of a comet with a small satellite (6×10^{20} kilograms) has been studied [Ip 1988]. However, it was judged that this was so unlikely that the problem really remained a mystery. A common opinion holds that the rings may be about 100 million years old.

Here, it is proposed that a large natural moon of Saturn, with mass of 90×10^{21} kilograms (the average of the seven existing large natural moons) was disrupted by a nuclear fission detonation about 12,900 years ago. This detonation scattered debris throughout the Solar System, leaving only 0.03×10^{21} kilograms of the crustal ice of the moon in orbit around Saturn as the major rings. The icy crust of that moon might have held 40×10^{21} kilograms of ice, judging from what we know of Jupiter's moon, Ganymede. Thus, less than 0.1% of the debris would have remained around Saturn to form the rings. Nearly 40×10^{21} kilograms would have been captured by Saturn; much of the fragmentary debris would have impacted the other moons causing their asymmetrically decorated surfaces, or lost into the Solar System. For comparison, the oceans of Earth contain at present 1.4×10^{21} kilograms of water. The gamma radiation from the detonation caused the extinction of the megafauna on Earth at the end of the Pleistocene, and the impact of the ice debris, as a swarm of comets, caused the great flood ("Noah's Flood", but also known under many other labels), whose destructiveness has been interpreted as evidence of the Great Ice Age.

Only a few percent of the mass contained in one of Saturn's larger moons would have had to be captured in orbit about Saturn after a nuclear detonation of a moon to form the rings. While the internal material of such a moon might have been heated sufficiently to vaporize and completely disperse, the icy crust might have only been shattered by the shockwave, to be captured within the Roche Limit, unable to coalesce as a single body. Perhaps a moon was totally fragmented, and only small pieces remain. The small fragmentary satellites group like at Jupiter, as was shown in Figure 57. Titan rotates in 15.911 days, but orbits in 15.945 days. That is, its orbit is slightly larger than it should be. This would happen if an inner moon disappeared, and Titan drifted out.

Titan

Direct observations provide us with subtle hints of history. The *Huygens* probe carried to Saturn by the *Cassini* spacecraft, and released to descend through the atmosphere of Titan, carried a gas chromatograph-mass spectrometer. That instrument measured the elemental and isotopic abundances of gases in Titan's atmosphere [Niemann *et al.* 2005].

That atmosphere has 1.5 times the pressure as at the surface of the Earth, and is composed largely of nitrogen, much like on Earth, but with a significant amount of methane. Methane has a lifetime of only 20 million years, and so must be continually resupplied to the atmosphere [Lebreton *et al.* 2005].

This book proposes that there are bursts of neutrons, from nuclear fission detonations, that can leave changes in the material that has been exposed to the neutrons. That appears to be the case with regard to the isotopes of nitrogen on Titan.

Nitrogen has two stable isotopes, nitrogen-14 and nitrogen-15. The ratio of nitrogen-14 to nitrogen-15 in the air of Earth is 273; nitrogen-14 is much more abundant than nitrogen-15. In the atmosphere of Titan, the ratio is only 183. Something has removed one-third of the nitrogen-14 from the atmosphere.

Consider the following series of nuclear-chemical reactions resulting from exposure to neutrons in the gases of Titan:

$$^{14}N(n,p)^{14}C + 4H \rightarrow {}^{14}CH_4 (\beta) {}^{14}NH_3 + H$$

Translated into readable English, that means that when nitrogen-14 absorbs a neutron, it releases a proton, transforming into a carbon-14 atom, which is radioactive. Produced by the nuclear reaction, the carbon-14 atom is an energetic ion and reacts with four nearby hydrogen atoms, forming methane (as an example, other hydrocarbons are possible). The methane condensed into the pools of methane on the surface of Titan. The carbon-14 atom later decayed by beta emission, with a half life of about 5,330 years, returning to a nitrogen-14 atom. The nitrogen-14 atom is a low-energy ion, in a liquid puddle, and reacted with nearby hydrogen atoms to become ammonia. (Or, at low energy, it may simply retain three of the hydrogen atoms from the methane.) The ammonia, rich in nitrogen-14, remained in the surface material, and caused the observed depletion of nitrogen-14 in the atmosphere.

A notable difference in the ratio of deuterium to ordinary hydrogen (D/H) was also detected. The D/H ratio observed is 1.46 times that on the Earth, there is relatively more deuterium on Titan than on the Earth. This enhanced ratio may result from the preferential breakup of hydrocarbons by neutron scattering on hydrogen-1, which is 6 times greater than on hydrogen-2 (deuterium), with subsequent loss of the hydrogen-1, and by some neutron capture by hydrogen-1, forming deuterium.

Enceladus

One of the grandest surprises of the *Cassini* mission to Saturn was the discovery of plumes of water jetting from the southern region of Saturn's small moon Enceladus [Porco *et al.* 2006]. In spite of its small size and meager radioactive heat, and a lack of significant tidal energy, this moon shows excess heat, localized in the South Polar Terrain. It was estimated early that the South Polar Terrain emits 3 to 7 billion watts [Spencer *et al.* 2006], compared to the identified energy sources of about 0.5 billion watts [Porco *et al.* 2006]. The estimate has been refined by more detailed analysis of the data, so that Enceladus generates 15.8 billion watts of heat [Howett, *et al.* 2011, Kerr 2011], roughly equivalent to the heat produced by 5 commercial nuclear power reactors. This greatly increases the discrepancy between what can be accounted for and what is observed. Plume ejection of water requires heating the water ice to above 273 K, while the North Pole (in darkness) is at a temperature of 32.9 K. Fountains of ice spew from the South Polar Terrain and replenish the E ring.

Considering the possibility that Enceladus is a fragment of the large natural moon that detonated to form the major rings, and that this fragment managed to retain some of the nuclear reactor region of the detonated moon (this region must have somehow escaped destruction), ongoing operation of natural nuclear fission reactors could supply this excess heat. Alternatively, Enceladus may have been a fully formed large natural moon, just a lot smaller than the rest. This may also apply to Miranda of Uranus, which shows an unexplainable history of geophysical modification, requiring significant, unidentified amounts of heat.

The reactors could consist of individual units of uranium in water solution, but this would appear to raise a serious problem. The question arises as to how these reactors could function with the currently assumed low uranium-235 content, 0.72%. Uranium (uranium-235 and uranium-238) in water cannot form a critical reactor in ordinary conditions if the uranium-235 concentration is less than about 1% of the uranium (0.94%, [Thomas 1978]) and our best estimate of the present concentration, on Earth, is 0.72%. However, as shown in Figure 52, the uranium-235 concentration in the meteorites ranges up to 2.5%.

Some effects may result from the extremely low temperatures. At the very low temperature observed for the ice plumes, 77 K [Cowen 2006a], the moderated (thermal equilibrium) neutron energy is sufficiently low that the effective uranium-235 fission cross section is significantly greater than in ordinary conditions. Some other possibilities are: a methane moderator would be more effective than water, allowing a lower enrichment; for reactors operating over millions of years, some of the hydrogen-1 would have been converted to deuterium (hydrogen-2) which has negligible absorption; operation as breeder reactors could have produced sufficient plutonium-239/uranium-235 fuel that the reactor material would be richer than we expect; the low temperature reduces the thermal Doppler broadening of the uranium-238 absorption resonances, thus reducing the loss of neutrons.

This suggests that the geyser jets are powered by boiling liquid nitrogen (at 77 K) heated by natural nuclear fission reactors composed of uranium frozen into rock-hard water ice. Ice crystals near the surface are entrained in the expelled nitrogen gas and fall back to the surface as fresh snow, making Enceladus the most reflective object in the Solar System.

Considering this, and that there is little more reason to expect the uranium in a moon of Saturn to be the same as in the meteorites, I will look at a range of uranium-235 concentrations from 1.4% to 10%.

At 10% uranium-235, the minimum critical mass as a solution in water is 12 kilograms of uranium (1.2 kilograms of uranium-235) [Paxton and Pruvost 1987]. At the lower enrichment of 1.4%, the uranium mass must be at least 715 kilograms. These minimum critical masses correspond to volumes of 0.02 cubic meters (20 liters) to 0.22 cubic meters (220 liters). Taking a nominal value of 2.5% uranium-235, the minimum critical mass is 120 kilograms of uranium (3 kilograms of uranium-235) and the corresponding volume is 0.08 cubic meters (80 liters). To provide the excess power of the South Polar Terrain, nominally 6 billion watts, at the average power per reactor as was found for the Oklo reactors, 6 thousand watts, would require about a million reactors. This would require a mass of uranium exceeding 1.2 x 10^8 kilograms, sufficiently concentrated to form critical reactors. Compared to the total mass of Enceladus (1.08 x 10^{20} kilograms), this is 1 part per trillion (1 x 10^{-12}). Compared to a rocky core (which might be the remnant of the detonated moon's reactor region), with a mass of 0.61 x 10^{20} kilograms [Kargel 2006], this is 2 parts per trillion (2 x 10^{-12}). These are very small bulk concentrations, small even compared to the range seen in the meteorites (Figure 52), 10^{-11} to 10^{-8} (10 parts per trillion to 10 parts per billion). Localized concentration would result from melting of the ice and subsequent re-freezing, excluding the uranium from the new ice into the remaining solution.

To judge from the meteorite data, there may be 10 to 10,000 times as much uranium in Enceladus as is needed to constitute the natural nuclear fission reactors that could provide the excess power that is observed.

The system of Saturn provides abundant evidence for the existence and action of natural nuclear fission reactors.

Uranus

Uranus is the fourth largest planet of our Solar System, smaller in mass than its nearly twin Neptune, but larger in size. Uranus is a world out of kilter. It is tipped on its side and beyond, its rotational axis is at 98 degrees from its orbital axis. Its magnetic field is tilted in another direction and is offset from the center of the planet by a distance about half the width of the Earth. Compared to Neptune, it seems to have lost some mass, but is puffed up, inflated from the size it should be. It holds a small moon, Miranda, that appears to have actively resurfaced itself, with no identified source of heat. None of the other moons, large or fragmentary, have shown such recent activity, but the four larger moons all show significant resurfacing. Uranus is a puzzle.

The tilt may be from an impact, although "This does not sound very plausible, but it is difficult to think of anything better." [Hunt and Moore 1994]. Uranus has four moderate size moons, but no large moons. It is thought that the moons had to form after a tilting impact since they are equatorially aligned. Internal heating of Uranus, from a nuclear detonation that ejected some solid mass, or perhaps just enough overheating to blow off some atmosphere, could have been gentle enough that the moons could have stayed in place and slowly rotated around to the equatorial plane.

Uranus was discovered somewhat unintentionally by William Herschel on March 13, 1781, as he was conducting a survey of fainter stars. Its previous sightings by astronomers had been recorded as a star. Without repeated observations, the orbital motion that identifies it as a Solar System object would not be detected. In historical searches for records of Uranus before its discovery, several recorded stars were seen to be no longer there. Before it was recognized as a planet, Uranus had been recorded at least 22 times between 1690 and 1771, before discovery in 1781. Some of these observations were on sequential nights when its change in position should have forced a discovery [Hoyt 1980]. Some of the difficulty in identifying prior detections lay in the deviations of Uranus from its theoretical orbit, due to perturbations by the then-unknown planet Neptune [Alexander 1965]. Herschel was aided by the quality of his telescope and was able to recognize that the image was less point-like than a star should be. That led to his initial suspicion that the object was a new comet, and so he repeated his observations and detected its orbital motion.

The discovery of Uranus strengthened the acceptance of the Titius-Bode Law as representing the spacing of the planets in the Solar System, and led to the identification of the first asteroid, 1 Ceres [North 1995]. While several variations of the mathematical rule have been studied, and some of these can give spacings without room for asteroids, the underlying concept is that there is a regularity in the spacing, whatever its unknown cause might be.

Brunini proposed a recent impact to disturb the orbit, producing the perturbations that have suggested a more distant planet [Brunini 1992b]. This has been refuted by Tyson [Tyson *et al.* 1993], on the grounds that Brunini underestimated the energy required for such an impact to produce the observed residual perturbations.

Further suggestive evidence comes from the severe tilt of the axis of rotation of the planet Uranus. It has been considered that planetary rotation axes should align quite closely with the orbital axes, and that all orbits should lie in the same plane. Departures from these alignments have been taken to show the action of extremely massive impacts [Safronov 1969]. If it is accepted that the tilt of the Earth's rotation axis by 23 degrees is the result of the impact that formed the Moon (although this book suggests otherwise), a much more severe impact must be proposed for the re-orientation of the axis of Uranus to 98 degrees [Zeilik and Smith 1987]. Not counting Pluto, for the reasons discussed above, the orbital planes differ from the plane of the Earth's orbit by no more than 7 degrees.

The four larger moons also show cryovolcanic resurfacing. For Umbriel and Oberon, this is considered to have resulted from "geologically very late-occurring melts" [Croft and Soderblom 1991].

This might also be the case for Miranda. Surface features are more easily explained by Pappalardo and Greeley of Arizona State University as the result of deep convection upwellings [Astronomy 1994a2]. Small internal reactors could have driven these plumes.

The abundance of methane (CH_4) in the atmosphere of Uranus is about 30 times the solar abundance, while ammonia (NH_3) is depleted, at 0.01 times the solar abundance [Allison *et al.* 1991]. This suggests the action of nuclear reactions that convert nitrogen into carbon, to deplete the ammonia and enhance the methane. Even temporary conversion into radioactive carbon-14 could result in transport of what had been ammonia deeper into the atmosphere as methane. Re-conversion to ammonia by the beta-decay of the carbon-14 could leave the ammonia trapped and unobservable.

Voyager 2 showed the medium-size moon Miranda to be unexpectedly complex. While this moon shows canyons and resurfaced areas, and looks as if it had been catastrophically fragmented and then re-assembled, it is too small for the conventional heat sources (tidal and radioactivity) to have been effective. It has areas that appear to be ordinary volcanic flows [Smith *et al.* 1986]. The recurring question repeats: Where's the heat?

Miranda is remarkably similar in size to Saturn's moon Enceladus, with a diameter of 472 kilometers, compared to 498 kilometers, so the calculations presented previously for uranium and natural nuclear fission reactors in Enceladus apply as well to Miranda. Nuclear fission could supply the heat.

Miranda's surface shows two types of craters: those that have been mantled with a softening coating of dust, a kilometer thick, and those that appear to be fresh with sharp rims [Croft and Soderblom 1991, Greenberg *et al.* 1991]. Canyons and similar structures show the same mantling also. It has been proposed that the mantling is due to ejecta from the impact that formed Arden Corona, although it is not clearly certain that this was from an impact [McKinnon, Chapman, and Housen 1991]. The strongest evidence is that there are possible antipodal fractures near Elsinore Corona. In this book, it is suggested that the mantling material came from the detonation of a large moon about 80 million years ago, that precipitated the end of the Cretaceous and the extinction of the dinosaurs.

Cryovolcanic flooding, amounting to about 200,000 cubic kilometers of "magma", filled the supposed Arden Basin, forming the Arden Corona. Elsinore Corona is similarly interpreted as consisting of a cryo-flood of about 100,000 cubic kilometers.

Miranda shows the effects of apparent volumetric expansion amounting to global expansion of about 6%. The only way that such a large expansion could occur appears to be the melting of water ice, with the liquid water then refreezing and causing the expansion. But, "it is nearly impossible to melt H_2O in the first place, so refreezing of H_2O is problematical." [Croft and Soderblom 1991, page 622]. Again, Where's the heat?

Titius-Bode Law for Satellites, again

If a moon of Uranus had detonated, and is now missing, where did it belong, where is it isn't? As with Jupiter's moons, this was explored by use of the Titius-Bode Law concept of regularity in satellite orbits. A plot of the five somewhat large moons (about the same relation in mass to Uranus as the four large moons of Jupiter have to their planet) in terms of their distance from the planet in sequential order is shown in Figure 60. The Titius-Bode concept supposes a regular arrangement of orbital radii with sequential order. This analysis further supposes that some of the moons in the original sequence have been lost by detonation.

In this case, the existing moons have been placed in the following sequence:

0	fragmentary innermost moons (as for Jupiter)
1	Miranda
2	Ariel
3	Umbriel
4	missing
5	missing
6	Titania
7	missing
8	Oberon

Again, as with Jupiter, a straight-line fit agrees very well with the observed orbital radii of the existing moons, in the selected order. The existing moons are shown as filled dots, the missing moons as open dots, and the fragmentary moons as bars across the axis at the 0 orbit. The original sequence of the large moons is indicated by the small stars, simply put in the order in which they now stand. Four moons have been lost by detonations, and the innermost has left fragments in its orbit.

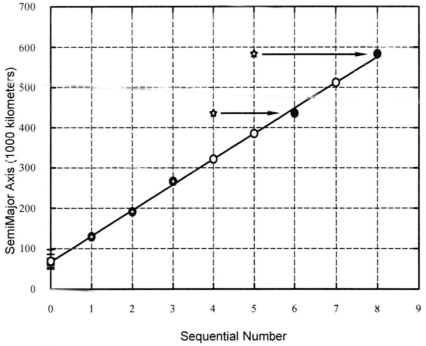

Figure 60. Larger moons of Uranus.

An application of the concept of regularity to the satellite system of Uranus. The existing large moons have been placed in a selected order and are shown as filled dots. Missing moons, to fill in the gaps, are shown as open dots. The fragmentary innermost moons are shown as bars across the axis at the 0 orbit. Small stars show the unadjusted order, simply counting in sequence. After adjusting the sequential order for possibly missing moons, the orbital distances fall into a strictly linear progression, as was seen for Jupiter. This shows, assuming that the regularity has some reality, that four moons are missing, and one left fragments behind.

It is thought that the cratering on the large moons was due to a cloud of objects somewhat chaotically orbiting Uranus, colliding with each other and fragmenting. This model has difficulties matching the age distribution, if the impactors were left over from the formation of the planet and the moons. Production of a cloud of chaotically orbiting objects at a late time, such as 80 million years ago, could satisfy these objections.

Alone among the five major satellites, Miranda has a significantly inclined orbit (4.22 degrees from the plane of Uranus' equator), surprisingly large for the innermost large satellite, with no identified perturbations. This might have resulted from a long history of geysering plumes, like Enceladus of Saturn, forcing the moon out of alignment. Perhaps the inclination was produced by the disruption resulting from the detonation of one of Uranus' other moons.

Greenberg *et al.* raise the recurring question: "At a distance from the Sun where the ambient subsolar temperature is only 85 K (for geometrical albedo 0.22), how could a body as small as Miranda have been heated enough to drive its apparently active geophysical evolution, to mobilize material in its interior, and to permit volcanic flow at its surface?" [Greenberg *et al.* 1991, page 705]. In other words: Where's the heat?

Accretion during formation of Miranda is estimated [Greenberg *et al.* 1991] to have raised the temperature of the outside to at most 84 K. Radioactive heating could have raised the temperature by 10 K to 20 K. If the ice were in the form of a clathrate with low thermal conductivity, the temperature could have increased by 140 K to 150 K. Initial tidal heating could have raised the temperature by 4 K. After becoming synchronous, the tidal heating could have raised the temperature by 10 K to 20 K. Taking all those sources of heat at face value leads to maximum temperatures in the ranges of 108 -128 K to 238-258 K, the latter range about to the melting point of water ice (273 K) but that requires extreme conditions.

Again, where's the heat? As shown for Enceladus of Saturn, natural nuclear fission reactors could have supplied the heat.

Neptune

After the discovery of Uranus, and after finding the asteroid Ceres at the right distance from the Sun, the Titius-Bode Law was used to help guide the search for another planet. This search was done mathematically from the deviations of the planet Uranus from its theoretical orbit. The Titius-Bode Law proved to be somewhat misleading in this case. The mathematical derivations produced two sets of orbits, one near where the Titius-Bode Law predicted, at 38.8 Astronomical Units, the other at about 30 Astronomical Units, where Neptune actually is. Both John Couch Adams and Urbain Le Verrier chose the farther of these two for their calculations, because of the Titius-Bode distance. The planet was discovered by Johann Galle and Heinrich D'Arrest at the Berlin Observatory, working from Le Verrier's predicted position, on September 23, 1846. Later historical studies showed that the planet had been observed before, most famously by Galileo in 1612 and 1614 [Hunt and Moore 1994]. Galileo failed to understand from its apparent movement that this might be a new planet, not just another star. That was true for several other observations before discovery. In addition to its sightings by Galileo, Neptune was recorded twice in May 1795, but the observed shift in its position was attributed to observational error. James Challis, who had been assigned by the Astronomer Royal George Airy to deliberately search for the planet predicted by John Adams, recorded it twice in August 1846, without realizing that he had seen a new planet. Three other sightings in 1845 and 1846 should have revealed the wandering nature of the star that became a planet [Hoyt 1980]. Telling a planet, of any sort, from a common star takes the follow up work of analyzing its position.

Of all the regular planets in the Solar System, Neptune is farthest from its proper place as predicted by the Titius-Bode Law. Neptune's actual orbit has a mean distance of 30.066 Astronomical Units, while the Titius-Bode prediction is 38.8, larger by 29%. If the mass of an outer planet, such as the proposed Planet X for example, were dispersed and largely lost, the next planet inside would drift closer to the Sun. Is this what happened to Neptune, originally forming at 38.8 Astronomical Units but now on a smaller orbit, or is it just random chance making a deviation from an approximate rule?

Neptune has an atmosphere, clouds, storms, a magnetic field, rings, and moons, big and small. It was a fortunate target for a spacecraft flyby. Most of what we know in detail comes from the brief flyby through the Neptune system in 1989 by *Voyager 2*. That was 12 years, 3 planets, and many moons after its launch on August 20, 1977. Its success is a tribute to the skills and perseverance of the mission team members. *Voyager 2* gathered an amazing amount of data in little more than 10 hours as it sped by Neptune and its large moon Triton. Books have been written about this wonderful work [Cruikshank 1995, Burgess 1991, Hunt and Moore 1994], and still many questions remain unanswered.

Below the clouds, the atmosphere is thick all the way down, and blocks our view of whatever might be below it. Clouds come and go, and storms blow by, with the fastest winds in the Solar System [Kinoshita 1989, Limaye 1991, Lunine 1996]. Like the other giant planets, the atmosphere is mostly hydrogen and helium, with enough methane to color it blue.

Neptune has long been considered to be a twin to Uranus, with many remarkable similarities. In size and mass, they are closely matched. They both have offset and tilted magnetic fields. But Neptune glows with internal heat, and Uranus seems to have none. Neptune has a single large moon and several small moons; Uranus has only small moons and fragments. Both have wimpy rings.

Like Uranus, the magnetic field of Neptune is significantly tilted (47 degrees) with respect to the rotational axis (compared to the tilt of Earth's field at 11.5 degrees), which is itself tilted (by 29.6 degrees, a bit more than Earth's tilt of 23.45 degrees) compared to its orbital axis. The center of the magnetic field is offset by 13,261 kilometers from the center of the planet, more than half the planet's radius, more than the diameter of the Earth. That is well outside the central rocky core of the planet, in what is assumed to be a thick ionic (electrically conducting) shell of water, ice, and ammonia [Lodders and Fegley Jr. 1998]. The dynamo region extends to about 0.7 – 0.8 of the radius of Neptune, involving most of an isolated portion of the ionic shell. Such an offset implies that the dynamo region is quite displaced from the center, the core, of the planet. Like Uranus, this peculiarity suggests some disruption of the "normal" structure of a planet. As for Uranus, this may have been produced by the nuclear fission detonation of a large moon, possibly destabilizing the reactors that normally would power a deep, core dynamo.

The rings of Neptune, like Jupiter and Uranus and unlike Saturn, are skimpy wimpy rings. Some are not even complete, but pieces of arcs following each other around the planet. There are two broad continuous rings, named Galle and Lassell after the astronomers who discovered Neptune and Triton. Three narrow continuous rings are named after Adams, LeVerrier, and Arago. One ring is broken, and is associated with the moon Galatea. The four noticeable clumps or arcs in the Adams ring have been named after the fighting cry of the French Revolution, Fraternité, Egalité, Liberté, and an English beer, Courage. In the view of this book, the rings had to form after the capture of Triton and Nereid by an inflated atmosphere. The rings are intermeshed with the 4 innermost satellites, and probably represent debris that was not able to collect into a solid body.

There is no surface to view on Neptune, just thicker and thicker clouds until some semblance of a molten core might be reached, hidden from sight. The atmosphere of Triton, as little as there is, is crystal clear and the visual excitement of the *Voyager 2* mission came from the images of Triton's rocky-icy surface. Triton has a two-tone surface, as do many of the distant moons of the Solar System (Earth's Moon also). One side of Triton would have been blasted by the direct energy of the detonation of its parent planet, while the other side would have been resurfaced by the tidal heating produced by the capture by Neptune. Only craters produced after about 6 million years ago would be showing on its surface. Some regions of Triton are soft, some are rigid, suggesting localized internal heating. Some of its surface hasn't been explained yet.

Triton

Triton has close similarities to Pluto, and is thought to have been captured, rather than forming in place at Neptune. It may have been a large moon of the now vanished Planet X. Its strange cantaloupe surface is thought to be the oldest existing surface, and may have been melted to eruptive boiling by radiation from the fireball of the Planet X detonation, with the other side cratered by impact debris. With only about a quarter of the surface displayed by the *Voyager 2* images, it is difficult to make generalizations about the unseen surface. Features like the pit paterae and cavi "seem to require explosive activity originating at depths of tens to thousands of kilometers." [Croft *et al.* 1995]. The pit paterae are like Nevada Test Site nuclear weapon test craters, such as Sedan. The cavi bear a resemblance to the Nevada Test Site rubble domes, and are about the same size as the pancake domes on Venus. All these could have been caused by nuclear detonations deep but near the surface. The guttae and aurioles may have been formed in a similar manner.

Triton orbits in the wrong direction, in a very circular orbit, and is thought to have been captured by Neptune rather than forming in place, but the details are difficult. Triton has a very young surface, showing that it was resurfaced recently, after most of the impactors of the early Solar System were gone, lost or impacted. To explain this apparently recent acquisition of Triton, this book proposes that a planet more distant than Neptune detonated about 6 million years ago. If that planet had had large moons that survived the detonation, those moons would have been turned loose, gravitationally, as the planet was disrupted. Those moons, perhaps Triton and the Pluto-Charon pair (as an accident of the detonation), might have been captured by Neptune. Triton could have been captured into a very elliptical orbit by gas drag at Neptune, Pluto-Charon were captured into a gravitational orbital resonance still orbiting independently around the Sun, and little Nereid, in the most eccentric orbit of all, was captured by gas drag but not circularized as the inflated atmosphere shrank back to the planet too quickly.

Nitrogen freezes into a hexagonal (β) crystal structure at 63 K, which transforms into a denser cubic (α) structure at 35.6 K. This transformation has been observed to shatter the bulk material.

Perhaps the cantaloupe terrain was formed by melting of the immediate subsurface layer by radiant heat from the fireball of the detonation of Planet X. Freezing of this layer and subsequent transformation through the β to α phase transition would have caused localized slumping and buckling.

It is expected that ultraviolet radiation from the Sun would convert methane on the surface and in the atmosphere into so-called C_2 hydrocarbons, less volatile and darker in color. It has been estimated that this process should have produced 6 meters of C_2 hydrocarbons, easily detected spectroscopically. However, no trace of the C_2 hydrocarbons was found by precise near infrared spectroscopy [Cruikshank *et al.* 1993]. If Triton had originally formed as a large natural satellite of Planet X, at the predicted Titius-Bode distance, the ultraviolet radiation would have been only one-sixth as intense as at Neptune's present orbit. Further, the heavier hydrocarbons may have been submerged by mobilized nitrogen and methane following the detonation of Planet X.

Compared to their parent planets, moons are small and weak, and can be overwhelmed by the tidal effects of the planet's gravity. Close to a planet, a satellite with no internal strength would be pulled apart. The debris would crash into the planet or spread out into a ring. More importantly, a cloud of particles would not be able to accrete into a solid body. The distance within which this occurs is known as the Roche Limit. It depends on the densities of the satellite and the planet, and for Neptune it is about 71,300 kilometers from the center of the planet [Lissauer *et al.* 1995]. Four of Neptune's small regular moons orbit within this distance and probably would not have been able to form there. They are intermeshed with the wimpy rings. They have extremely small eccentricities. If the inflated atmosphere had existed for a short time, those moons could have been dragged closer to the planet than where they were able to form, with very circular orbits as a side effect. The next two moons, Larissa and Proteus, are outside the Roche Limit but were also given very circular orbits. The atmosphere shrank back to its normal size and all settled down.

The standard model for a gas-drag capture involves a protosatellite nebula surrounding the planet. This nebula must dissipate outward, away from the planet into empty space, so it continues to slow a captured satellite as the satellite moves inward, until it crashes into the planet. In the alternative view, the detonation of the distant planet would have caused the temporary inflation of the upper atmosphere of Neptune, by the absorption of ultraviolet radiation and radiant heat from the fireball. This would have facilitated the gas-drag capture of Triton and Nereid. With each pass through the edge of the atmosphere, the orbit of Triton would have become less eccentric, with the greatest distance (the "apoapse") moving in while the closest distance (the "periapse") remaining unchanged. (This method of aerobraking has been used successfully to establish close circular orbits around Venus and Mars for some of our space probes.) The temporarily overheated outer atmosphere (and only the outer part would be expanded by the radiant heat of the detonation) would cool and shrink back to Neptune. This would permit gas-drag capture without bringing Triton down to Neptune in an impact. Nereid would have been captured by its first pass through the outer atmosphere, but its orbit would not become circularized because of the shrinking of the atmosphere. It would remain in the highly eccentric and inclined orbit initially established. Nereid's orbit is far outside the orbits of all the other moons. The timing of all this is quite touchy, but the possibility seems better than the other hypotheses [Agnor and Hamilton 2006, Morbedelli 2006, McKinnon, Lunine, and Banfield 1995].

In the otherwise clear skies of Triton, two geyser-like plumes were seen, rising about 8 kilometers above the surface and streaming downwind 90 to more than 150 kilometers [Kirk *et al.* 1995]. It is estimated that the power required to create these plumes is approximately 100 million watts, concentrated in a source area of 0.02 to 2 kilometers in diameter. Since the maximum solar heating of the surface is about 1.5 watts per square meter (an average of half that for the day/night effect), heat must be concentrated from a surface area that is one hundred to one thousand times as great as the area of the source of the plume, without any losses along the way. This seems to be asking too much of a layer of ice and frozen nitrogen, and makes us ask, again, Where's the heat?

Natural nuclear fission reactors could supply this heat in a concentrated manner. Reactors can be very variable in output, which could account for the intermittent nature which is thought to occur for the plumes. On Earth, the Oklo reactors that were operating 2 billion years ago, cycled with a period of 3 hours [Meshik, Hohenberg, and Pradivtseva 2004]. On Triton, the reactors could be nitrogen moderated, like graphite-moderated (technological) reactors on Earth, or more likely, operate as fast breeder reactors with the fuel in the form of uranium nitride (UN). This material is a high temperature (refractory) material with a melting point of about 2,900 K, and an atomic density of uranium that is about 40% greater than that of uranium dioxide (UO_2), the common commercial fuel material. This greater density would result in critical masses about half that of similar uranium dioxide reactors. The high melting point could provide high heat transfer rates to the surrounding ice. Operation with a fast neutron spectrum, slightly downscattered by nitrogen from the fission spectrum, would avoid the strong neutron capture of thermal neutrons by nitrogen-14, making carbon-14. Some evolution of the reactors could diminish this parasitic capture further by the escape of methane or carbon monoxide formed from the carbon-14. This would decay later to nitrogen-14 outside the reactor region, reducing the concentration of nitrogen-14 and the associated capture. Heat from nuclear fission (or simply radioactive decay at an early stage) would have driven out the volatiles containing hydrogen, which is a significant neutron absorber.

As was discussed for Saturn's moon Enceladus, some effects may result from the extremely low temperatures. At the very low temperature observed for Triton, 37.5 to 39.3 K [Elliot *et al.* 1998], the moderated (thermal equilibrium) neutron energy is sufficiently low that the effective uranium-235 fission cross section is significantly greater than in ordinary conditions. Some possibilities are: a methane moderator would be more effective than water, allowing a lower enrichment; for reactors operating over millions of years, some of the hydrogen-1 would have been converted to deuterium (hydrogen-2) which has negligible absorption; operation as breeder reactors could have produced sufficient plutonium-239/uranium-235 fuel that the reactor material might be richer than we expect; the low temperature reduces the thermal Doppler broadening of the uranium-238 absorption resonances, thus reducing the loss of neutrons.

To paraphrase an old English proverb, "The proof of the reactors is in the heating." Neptune and Triton clearly still have much to tell us.

The Outer Limits

Beyond Neptune, we begin to lose our grasp on the Solar System. We enter a poorly defined region, a sloppy region that slips into the Solar System and extends uncertainly beyond. We must speculate as to the nature and origin of the objects we find there, much more than for the objects we can (and do) observe more closely. We must fit these distant objects into our theories, rather than develop our theories on the basis of observational facts. The theories tell us what we see. We're not even quite sure what to call them. Pluto pretends to be a planet and it somewhat seems to function as one, but probably did not form as one. It is accompanied by a moon, Charon, almost as large, that probably did not form as most moons, either large or small. The other rocks we find, somewhere out there, we may call kuiperoids, in honor of Gerard Kuiper, who speculated on their possible existence [Norris 1995], or Edgeworth-Kuiper Belt Objects, to include Kenneth Essex Edgeworth, who speculated earlier [Edgeworth 1949], and to anticipate our finding enough of them as to constitute a belt of objects. Carrying on the early tradition of much of science, Edgeworth was an Irish gentleman-scientist not affiliated with any research institution. Otherwise, Trans-Neptunian Objects (TNOs) seems fitting for those beyond Neptune, while Centaur is applied to those that orbit generally between Saturn and Neptune.

The Oort cloud, proposed by Jan Oort in 1950 as a spherical halo of cometary objects, has never been observed. The comets, without a coma, are too small and too distant to be bright enough for our telescopes to record them. The cloud is a theoretical construction based on the behavior of long-period comets, with return periods greater than 200 years, that enter the Solar System from random directions, rather than near the ecliptic plane.

Pluto and the Kuiperoids

Pluto was discovered as a planet, as Percival Lowell's Planet X, in 1930. Clyde Tombaugh, a farm boy from Illinois, had been hired as a technician at the Lowell Observatory in Flagstaff, Arizona, to perform a detailed and systematic search for a planet near a position predicted by Lowell in 1914, based on the perturbations of the orbit of the planet Uranus [Stern 1997, Hoyt 1980, Asimov 1991]. William Pickering had independently predicted several different unknown planets that would satisfy these and other perturbations. All these predictions required planets with masses similar to or greater than the mass of the Earth, in order to produce the observed perturbations.

Perturbation calculations must by their nature include the effects of all unknowns in the parameters of the analysis. For the prediscovery calculations of Neptune, the perturbations of Uranus resulted from Neptune, and any other massive body in that region. Neptune, being near Uranus, exerted the greatest perturbations. Still, after the discovery of Neptune, and inclusion of its effects on Uranus, the orbit of Uranus continued to deviate from its best, and latest, predicted orbit. This led Percival Lowell and William Pickering, in the early years of the 20th Century, to use the perturbation method to search for another unknown planet.

Pluto was discovered while looking for Planet X, predicted by Percival Lowell from the perturbations in the orbit of Uranus. That prediction produces two locations on opposite sides of the sky, much in the same manner as the Moon's perturbation of the Earth's oceans produces high tides on opposite sides of the Earth. The general location chosen for the search was picked because it was away from the clutter of the Milky Way, on the opposite side of the sky. Pluto was discovered 6 degrees from the predicted location (12 times the apparent diameter of the full Moon) and so its discovery might be considered fortuitous, a case of simply looking hard enough. Tombaugh certainly looked hard, repeatedly making photos of selected regions of the sky and inspecting them with a blink comparator, searching for a point of light that moved the right amount between the two exposures. Pluto showed only as a star-like point of light on the photos, just as Galileo's stars appeared. It was only discovered by blink-comparison of photographic plates from different nights, when its relative motion against the background of stars gave it away.

The search at the Lowell Observatory had been planned to take sequential photographs of small sections of the sky, separated by several nights to permit the distant unknown planet to show its motion. The sky would be searched within 10 degrees of the ecliptic where the known planets orbit. The photos would be taken in the "opposition" direction (opposite from the Sun) to take advantage of the reverse, retrograde motion of a distant planet. The nightly motion is dependent on the distance and any departure from the ecliptic, and would be a bit more than 1 arcminute per night for an object at the predicted distance from the Earth of 42 Astronomical Units. These were the clues for the search: near the ecliptic and backward motion of about 1 arcminute per night, and a magnitude in the range of about 12 or brighter down to 16.5 as limited by the observational capability.

As with Uranus and Neptune before it, Pluto had been observed several times before its discovery, but not interpreted as anything other than a star. After the discovery of Pluto and the calculation of an acceptable orbit that permitted estimates of its position at times before its discovery, images of it were found on earlier astronomical photographs. Two of these were actually made at the Lowell Observatory in 1915 (March 19 and April 7) as part of an earlier search, while Lowell was still alive but were not recognized as a wandering planet at the time [Hoyt 1980]. (The only distinction that Pluto could have from an ordinary faint star was that it should move with respect to the other stars from night to night. This movement was the key to discovery.) Tombaugh also imaged Pluto in March and April 1929, almost a year before its discovery. In a search for Pickering's Planet O, Milt Humason had photographed Pluto in December 1919 from the Mount Wilson Observatory. Three other images from January 1921 and January 1929 have been found, and five more from 1914 and 1925. A total of 17 images of Pluto at observatories around the world could be identified after the fact. But without a careful and detailed determination of the apparent motion, the point of light could never be considered to be anything other than a star. Even spectroscopically, Pluto would have been identified as a "solar-type star" because of its "solar-type" spectrum [V. M. Slipher quoted by Hoyt 1980]. Only the development of infrared spectroscopy has permitted detection of water, nitrogen, carbon monoxide, and methane.

As Pluto was studied more intensely, its estimated mass began to shrink. With the discovery of its moon Charon in 1978 [Christy and Harrington 1978], the mass was found to be only 0.0022 Earth masses, far too small to produce the perturbations studied by Lowell and Pickering.

This odd couple, the most evenly matched pair known, with Charon being about half the size of Pluto (equal to about 0.08 of its mass), has been hard to explain. Together, they are in a 3:2 mean motion resonance with Neptune, an orbital condition that has been found for many other kuiperoids, now named "plutinos". None of the several formation theories has been fully satisfactory. Briefly now, just consider the fragments of the detonation of a planet (Planet X) flying apart from the center of the detonation. The scene will be crowded, for all that material once made up a massive planet. Jostling among the slower chunks on the outside, chunks from the inside will collide, but gently, and couples will form. Some of these pairs, and individuals as well, will, by chance, have trajectories that carry them into orbits that fit the 3:2 resonance, and they will be kept there for us to find and ponder.

While its discovery at first seemed to be a remarkable achievement of astronomical analysis, as its size shrank, the discovery was labeled a complete accident, "one of the most incredible sets of coincidences in the history of science" [Hoyt 1980]. In fact, considering the large number of objects similar to Pluto in the outer Solar System, it was an accident waiting to happen. The coincidences are shown below in comparisons of Lowell's favored orbit for Planet X (labeled X1) Pickering's Planet O (1919), and an empirical orbit developed by A. C. D. Crommelin [Crommelin 1931, taken from Hoyt 1980. See also [Reaves 1997].

Orbital Element	Planet X1 Lowell 1914	Planet O Pickering 1919	Pluto Crommelin 1930
mean distance	43.0	55.1	39.5
eccentricity	0.202	0.31	0.248
inclination (degrees)	10	15	17.1
period (years)	282	409.1	248
mass (Earth)	6.65	2.0	<0.7
magnitude	12-13	15	15
longitude 1930.0	102.7	102.6	108.5

While the individual elements do differ considerably for the three orbits, the bottom line, the longitude at 1930, pointed to where the planet should be found, and Lowell and Pickering were remarkably close to each other, and not far from the longitude where Tombaugh discovered it. That is what makes the discovery so incredible. While the latitude (declination) could not be predicted, Lowell expected it to lie within 10 degrees of the ecliptic, as the other planets do, but Pluto was about 20 degrees north of the ecliptic when it was discovered.

Perturbation calculations must predict two nearly diametrically opposed positions in the sky. Pickering only reported on one orbit, one predicted position. Lowell favored the one that was chosen for the search for Planet X that found Pluto instead, but he reported the complementary orbit as well. The greatest discrepancy in these predictions was in the estimated mass. Lowell predicted 6.65 Earth masses, Pickering predicted 2.0 Earth masses. But when other astronomers were able to observe the new object, the mass was estimated to be less than 0.7 Earth masses, and the current accurate measure is 0.0022 Earth masses.

It became clear, with such a small mass, Pluto could not have been the perturbing object that produced the residuals in the orbit of Uranus (and Neptune and the comets) that had been analyzed by Lowell and Pickering. Yet these predictions led to the search and discovery, within 6 degrees of the predictions, an "incredible coincidence." But what of the other, complementary orbit, that alternatively resolved the residuals, but required 2 to 7 Earth masses?

The discovery of Pluto, where it was, was an accident caused by great commitment and deliberate effort. The search had continued for 15 years beyond Lowell's death and occupied the entire career of Clyde Tombaugh. Through the entire search, extending long beyond its successful discovery, Tombaugh found a globular cluster, star clusters, a cloud of galaxies, one comet, variable stars, asteroids, and extra-galactic nebulae (galaxies) [Hoyt 1980].

While Ceres, the first asteroid, was initially expected to be a full planet, filling the space where the Titius-Bode Law predicted a planet, its small size and the rapid discovery of others like it soon led to the understanding that these objects were not the same as the other planets.

Similarly, when Pluto was discovered in 1930, it was hailed as a full size planet, about as large as the Earth or larger. But the point of light was too faint to be that large, and its size was reduced as better estimates were developed. As an isolated object, its mass could only be estimated from its observed brightness, using improved calibrations, with estimated values for albedo to get the diameter, and so its volume, with an assumed density to get its mass. This showed that Pluto was smaller, less massive than expected. Pluto could not have been the Planet X (or O) that produced the perturbations of Uranus, Neptune, and the comets. Eventually, a moon was discovered in 1978, permitting an accurate determination of its mass as 0.0022 that of the Earth. Subsequently, many similar objects have been found in the Kuiper Belt; several have been found that orbit in a 3:2 mean motion resonance with Neptune as does Pluto, and several have been found to have moons of their own.

Pluto's orbit is too eccentric, at times bringing it closer to the Sun than Neptune, and then farther away. It is tilted more than any other planet. It is in a 3:2 mean motion resonance with Neptune, making 2 complete circuits of its orbit while Neptune goes around 3 times. It has a disproportionately large moon (and three small ones).

As the special nature of Pluto as a planet faded away, it was officially transformed (considered to be done officiously, by some) into a "dwarf planet" by definition by the International Astronomical Union in August 2006. Pluto is a dwarf, Uranus and Neptune are giants, and dwarfs don't shove giants around.

Just as the discovery of the minor planet Ceres, an asteroid, did not really settle the question of the missing planet where the Titius-Bode Law predicted one, Pluto was entirely inadequate for the perturbations that had been calculated for Uranus and Neptune. Those perturbations have been dismissed as due to errors in observation and an incorrect mass for Neptune [Standish Jr. 1993]. The *Voyager 2* flyby of Neptune on August 25, 1989 improved the mass determination from 17.24 Earth masses to 17.15 Earth masses, with a current uncertainty of 3 parts per million. However, Tom Van Flandern has disagreed with Standish's elimination of the significant residuals. He considers that there are known but uncorrected errors in the data [Baum and Sheehan 1997]. The situation may be further complicated by the tendency of the method of least-squares to distribute deviations among the variables. As Lowell wrote, "We have then no guarantee that our supposed elements are the real ones, but only the best attainable under the assumption *that no unknown exists.* Every theory of a planet is thus open to doubt, seeming more perfect than it is. It has been legitimately juggled to come out correct, its seeming correctness concealing its questionable character." [Reaves 1997].

Neptune's mass had previously been based on observations of Triton orbiting Neptune. The flyby of *Voyager 2* on August 25, 1989, at a closest approach distance of 29,250 kilometers, and with a large gravitational change in direction, provided information that permitted a more accurate determination. Although *Voyager 2* spent only 6 hours inside the orbit of Triton, the closest approach was less than a tenth of the orbital distance of Triton, making the gravitational effect nearly 150 times as great. However, it should be kept in mind that the *Pioneer 10/11, Galileo,* and *Ulysses* spacecraft showed small, unmodeled accelerations that have not yet been explained [Anderson *et al.* 1998].

That deviation could not be measured for the *Voyager* spacecraft, because of their more active course adjustments. While the effect is too small to cause an error during the closest approach of *Voyager 2* at Neptune, it serves to remind us that we do not yet know everything about the Solar System.

Pluto has one large moon, named Charon, so large relative to Pluto that the pair has been termed a binary planet. There are also two very small moons, Nix and Hydra, discovered by use of the *Hubble Space Telescope* [Weaver *et al.* 2006], and a newly discovered, even smaller moon, "P4" [NASA 2011].

The eccentricity of Charon's orbit is unexpectedly large, at 0.0076. Tidal effects are considered to damp out such an orbital eccentricity within 1 to 10 million years [Tholen and Buie 1997]. This suggests that, unless the small moons are sustaining the eccentricity, the system was disturbed (or created) within the last few million years. Perhaps the system was formed in the last few million years, possibly 6 million years ago by the detonation of Planet X.

Evidence suggests that both the Pluto-Charon system and Triton (Neptune's retrograde moon) were not formed in place, not as natural planets and moons. This book proposes that in both cases, fragments jostling in the crowded debris cloud around the detonated Planet X were joined together in one case, and sent on a capture trajectory to Neptune in the other. Two fragments from different parts of the disrupted planet, with inner fragments overtaking and running into outer, crustal fragments, could have formed Pluto-Charon and headed in towards the Sun, from the Planet X orbit at 77.2 Astronomical Units. Passing by Neptune, with a multitude of other fragments, Pluto and Charon, Nix and Hydra, and several dozen other fragments could have been selected to slide into a 3:2 mean motion resonance with Neptune, stable orbits where they are found today.

A flyby spacecraft, *New Horizons*, is on its way to inspect the Pluto-Charon system, intended to arrive there July 14, 2015. That should provide new data and more surprises.

Pluto is not a planet, like the others, it is not Planet X. Where is Planet X and how did the kuiperoids form?

The study of Pluto, from its origins in planetary perturbations to its recognition as a binary planet of a different sort, has evolved so that it introduces us to a whole new study of the Solar System. We would not begin to guess how many more objects like Pluto (and Charon) could exist in the outskirts of our Solar System until the discovery of the first kuiperoid (1992 QB₁) by David Jewitt and Jane Luu [Jewitt and Luu 1993]. The kuiperoids had been found.

Kuiperoids (an obvious name, suggested at least by John Norris of Pasadena, Texas [Norris 1995]) were found by looking for the home of comets. Comets are a remarkable part of our Solar System, making a brief pass through our neighborhood and putting on more or less of a show, tiny objects that dwarf everything else in size by their displays. There seems to be two sets of comets, short-period (return times of less than 200 years) and long-period (longer than 200 years). Earliest credit for a source of comets from a belt of icy bodies, comets yet to be, has been given to F. C. Leonard in 1930 [Leonard 1930] although he was actually suggesting that Pluto would be "the *first* of a *series* of ultra-Neptunian bodies". K. E Edgeworth suggested the cometary belt in 1943 [Edgeworth 1943], but the main credit is given to Gerard Kuiper for his discussion in 1951 [Kuiper 1951]. A disk of comets beyond Pluto was proposed as the source of the short-period comets, and that disk is now generally called the Kuiper Belt. For the long-period comets, J. Oort proposed a vast and very distant spherical reservoir of icy bodies, now called the Oort cloud.

In the standard story of the Solar System, these comets in cold storage are considered to be leftovers of planet formation that failed to form planets. These bodies and their sources are considered to have existed for the life of the Solar System, and to have provided short-period comets and long-period comets at a constant rate.

In a search effort that rivaled Clyde Tombaugh's search for Planet X, David Hewitt and Jane Luu started hunting for whatever they could find at the general distance (30-50 Astronomical Units) and direction (near the ecliptic plane) of the Kuiper Belt. Searching for distant objects since 1987, first with photographic plates and finally with the sensitive and versatile CCD (charge-coupled device) detectors, the team of Jewitt and Luu discovered the first Kuiper Belt object in 1992, labeled 1992 QB₁, using a computerized blink comparator [Jewitt and Luu 1993, Weissman 1993]. However, the estimated size, 250 kilometers in diameter, was 10 times the size of a large comet, and a noticeable fraction of the diameter of Pluto. Something big had been found. Seven months later, Jewitt and Luu found another similar object, and the chase was on. Thanks to the remarkable improvements in equipment, this object, at 22.8 magnitudes, was far dimmer than anything Tombaugh could have detected with his limit of 16.5 magnitudes.

In 1992, Alan Stern had written that hundreds or thousands of Pluto-size objects were out there, somewhere [Stern 1992], and now they were being found, bigger than comets, smaller than planets. By 1995, Jewitt and Luu estimated that the Kuiper Belt held at least 35,000 objects larger than 100 kilometers between 30 and 50 Astronomical Units from the Sun. These could total 0.003 Earth mass, a bit more in total than Pluto. At the other size range, Levison and Duncan estimated 12 billion comets with a total mass of 0.08 Earth mass [Weissman1995]. By 2010, diligent astronomers had found 1,101 objects categorized as Trans-Neptunian Objects, or TNOs, often called Kuiper Belt Objects.

As much material as there might be, Alan Stern estimated that the Kuiper Belt is so large that collisional accretion could not form the largest bodies in 230 billion years, far longer than the age of the Solar System, or even the Universe [News 1995]. Stern had to postulate a much greater mass of material to form objects of a few hundred kilometer diameters, by the standard model for the formation of planets by accretion. (It should be noted that these objects are not really planets, just large planetesimals.)

The discovery of 1996 TL₆₆ with a highly eccentric (0.58) and inclined (24 degrees) orbit that takes it more than 130 Astronomical Units from the Sun, introduced the "scattered disk" [Luu *et al.* 1997]. These bodies are assumed to have been scattered from regions nearer the Sun by the gravity of Uranus and Neptune. This object was also large, as it had to be to be detectable at a distance of 34 Astronomical Units, almost to Pluto's orbit. The estimated diameter, based on an assumed albedo (the reflectance of its surface), was 490 kilometers. Based on the area of the sky that had been searched, it was estimated that there should be about 800 similar objects. That would amount to about 0.008 Earth mass. Extrapolating this by an assumed size distribution suggested a total of 0.5 Earth mass for objects larger than 1996 TL₆₆, at distances between 40 and 200 Astronomical Units. (The hypothetical Oort cloud is much farther away, at about 10,000 Astronomical Units, and is completely undetectable by current techniques.) By 1998, mass estimates for the Kuiper Belt had grown to 0.26 Earth mass in 30 to 50 Astronomical Units, limited to 0.3 Earth mass out to 65 Astronomical Units.

Kuiper Belt Objects larger than 50 kilometers in diameter can survive collisions with the more numerous smaller objects, but those larger than 60 kilometers couldn't have been formed by accretion because the collisions would be too violent, and gentle collisions would be too rare [Stern 2000]. To form the larger kuiperoids by enough gentle accretional collisions, there needed to be roughly 30 Earth masses of ice, dust, and rock. Now we can estimate only 0.3 Earth mass.

Kuiperoids (and Centaurs, escapees from the Kuiper Belt that orbit within the outer planetary system) closer to the Sun than 40 Astronomical Units are either grey (colorless in the Sun's reflected light) or extremely red. Those beyond 40 Astronomical Units in a recent survey [Tegler and Romanishin 2000] are all extremely red. It appears that some dynamical process separated those kuiperoids that were, or would be, covered with a blue-absorbing material and removed (or didn't create) the others.

Larger and larger kuiperoids were found, by Robert L. Millis, director of the Lowell Observatory. The discovery of 2001 KX_{76} gave it a diameter of 1,270 kilometers, larger than Varuna and second largest of the known objects [R. C. 2001]. By use of an Internet database, "Astrovirtel", Gerhard Hahn and colleagues were able to find pre-discovery archival images from as early as 1982 that helped to determine an accurate orbit [J. K. B. 2001, R. T. 2001].

For most asteroids and kuiperoids, size must be estimated from the brightness of the image, using an assumed albedo. Size and albedo are linked in most measurements, and the albedo of a cold comet (4%) is usually assumed. Albedo and size can be measured by the reflected light and the thermal radiation emitted by the body, at a wavelength where the solar illumination is negligible. By this approach, the albedo and diameter of Varuna (2000 WR_{106}) were measured to be 7% and 900 kilometers [Jewitt, Aussel, and Evans 2001]. Even though this is brighter than an old comet, it is a very dark surface, free of frost or ice or snow.

This technique was also applied to 2003 UB_{313} (Eris) [Bertoldi *et al.* 2006]. Some assumptions are required, but the favored result is a diameter of 3,094 kilometers and an albedo of 55%. Four months later, Eris was imaged by the *Hubble Space Telescope* so precisely that a direct measure of the diameter could be made [Brown *et al.* 2006]. From these measurements, the diameter was determined to be 2,400 kilometers and the corresponding albedo was 86%. Such a bright albedo suggests that the surface must be recoated with fresh methane frost or snow, as Enceladus resurfaces itself by its geyser jets. However, there doesn't seem to be any source of heat that could drive geysers. (A somewhat more precise diameter was reported by Mike Brown on April 11, 2006, as 2,384 kilometers [Cowen 2006c]. This value was used in the analysis shown below.) While some excuses can be made for the discrepancy between 3,094 and 2,384 kilometers, taking the results at face value leaves us with a challenge to understand; two independent methods giving two different results for the same characteristics.

The method used by Bertoldi *et al.* explicitly assumes that there is no internal heat source, while that assumption is not recognized in the *Hubble Space Telescope* measurements, and may be irrelevant. If an internal source of heat is allowed, the temperature of the surface would be elevated above that calculated from the solar illumination. The temperature calculated by Bertoldi *et al.* is 23.7 K. If internal heat raised the surface temperature to 36.3 K, the same diameter (and albedo) is produced by the thermal emission measurement as by the *Hubble Space Telescope* measurement. The albedo in this case, 86%, is remarkably bright, brighter than Pluto, and almost as bright as the whitest object yet seen, Saturn's moon Enceladus, as bright as freshly fallen snow. On Enceladus, the snow comes from active geyser plumes, erupting from as yet unidentified internal heat. (Except that this book proposes natural nuclear fission reactors as the source.) Since methane, at least, is darkened by exposure to solar ultraviolet radiation and cosmic rays, to have such a bright surface as Eris appears to have, requires that the surface be quite new or frequently resurfaced, as the water ice geysers do for Enceladus. No heat source has been suggested that could power methane jets, to spray methane ice (like a snow machine) over the surface of Eris. However, an average surface temperature of 36.3 K is close to the temperature estimated to mobilize methane plumes on Neptune's moon Triton, 48 K [Kirk *et al.* 1995]. Where's the heat?

The power needed to increase the surface temperature from 23.7 K to 36.3 K can be calculated from the Stefan-Boltzmann Law. That amounts to 1.44×10^{12} watts. Using the same estimated reactor parameters as were used for the power estimates for Enceladus (uranium-235 enrichment of 2.5%, reactor critical mass of 120 kilograms, average reactor power of 6,000 watts), it would take 240 million reactors to provide this much heat. To make those reactors would require 3×10^{10} kilograms of uranium. Compared to the total mass of Eris, 1.16×10^{22} kilograms, that is 1.9 parts per trillion. Compared to the mass of the rocky core, it is 2.5 parts per trillion. These amounts are remarkably similar to the concentrations found for Enceladus, 1 part per trillion total, or 2 parts per trillion for the rock. As was found for Enceladus, these concentrations are small compared to what has been found in the meteorites, 10 to 10,000 parts per trillion.

For Eris, natural nuclear fission reactors could answer the question "Where's the heat?" for both the resurfaced albedo and the apparently larger thermal emission size.

The Kuiper Belt appears to end abruptly at about 55 Astronomical Units [R. T. 2001, D. T. 2001], but lonesome objects still linger farther out. Are they kuiperoids [Svitel 2001]? The early discoveries had suggested that the Kuiper Belt ended at 50-55 Astronomical Units, but that may have been due to the difficulty of detecting faint, slow moving objects at such great distance. The orbit of 2000 CR_{105} (discovered at 43.5 Astronomical Units from Earth) takes it out to about 400 Astronomical Units [Beatty 2001]. It seems to be well beyond the gravitational influence of Neptune. Large objects (Mars-size) at great distances could still have escaped discovery. Significantly, the trajectories of *Pioneer 10* and *Pioneer 11* cannot rule out a massive object in a highly inclined orbit. Brett Gladman speculated that a large object might have "spent all of human astronomical history out of the ecliptic plane" where searches are conducted [Cunningham 2001].

As more kuiperoids were discovered, it became clear that the Kuiper Belt had a great resemblance to the Asteroid Belt.

Mike Brown and friends discovered an object to be named Quaoar, and estimated that it was half the size of Pluto [Anonymous 2002]. Archival images were found for this object as well, interestingly, one by Charles Kowal in 1982, in a deliberate search for Planet X. He had not identified it as an orbiting object at the time. Quaoar has a highly inclined orbit, about 44 degrees, and so had not been found in the ecliptic. Another chunk of rock mistaken for a great ball of fire, by default.

Sedna is one of the larger of these outer Solar System bodies [Brown, Trujillo, and Rabinowitz 2004, Rabinowitz *et al.* 2006]. It orbits between 76 Astronomical Units and 944 Astronomical Units, beyond the studied Kuiper Belt. It has been called the strangest of the farthest out [Cowen 2006c]. According to its discoverer, Mike Brown, it was "completely unexpected." "Sedna shouldn't be there. There's no way to put Sedna where it is. ... there's no way to form anything in an elliptical orbit like that." (Sedna has an eccentricity of 0.85.) It doesn't fit into the standard model Solar System, and Scott Kenyon at the Center for Astrophysics, Harvard, has suggested that it is actually an extrasolar planet, captured by the Sun. But it orbits as if it were a fragment of a disrupted planet that detonated at a distance of about 76 Astronomical Units from the Sun, and boosted some fragments far beyond. The Titius-Bode Law predicts an orbital distance of 77.2 Astronomical Units for the next planet beyond Neptune, the unfindable Planet X.

Clyde Tombaugh not only discovered Pluto, the first of the Kuiperoids (and a multitude of asteroids and other objects near and far), he left behind a treasure trove of sequential astronomical photographs. Repeated images of selected areas of the sky, those plates contained the stories of objects that moved on their own paths. Later analysis of those objects that moved, relative to the vast population of other points of light, permitted the development of two major catalogs from the Lowell Observatory. These are the "Lowell Proper Motion Survey: 8991 Stars with m>8, PM>0.26"/year in the Northern Hemisphere" [Giclas, Burnham Jr., and Thomas 1971] and "Lowell Proper Motion Survey – Southern Hemisphere Catalog" [Giclas, Burnham Jr., and Thomas 1978]. The northern catalog includes 8,991 stars, the southern catalog includes 2,758 stars, for a total of 11,749. Those catalogs list nearly 12,000 faint objects with barely detectable motion. Considering the millions of objects in the Asteroid Belt, and as yet unnamed in the Kuiper Belt, and those on randomly distributed orbits, is it possible, is it likely, that many of these "stars" are only stars by default, and are actually nearby (in the astronomical sense) chunks of rock and ice?

The Titius-Bode Law for a Solar System with 10 Planets

The movements of the planets through the heavens were crucial to astrology and continued to be essential in the new science of astronomy. These movements depend on the size of their orbits, the distance of each planet from the Sun.

As the distances to the planets in their orbits became known, Johannes Kepler stated in 1596 that there must be an unknown planet between Mars and Jupiter. History shows that Christian Freiherr von Wolff, in 1741, thought that the planetary spacing was too irregular and wondered if one were missing. The influential philosopher-theorist Immanuel Kant echoed this idea a few years later [Hoyt 1980].

Thus, there was a concept (firmly believed in by Kepler) that there was an underlying order, a regularity, in the orbits of the planets of the Solar System. The planets had not just been thrown in by random chance. Unfortunately, then and now, there was no theoretical basis for this concept. Eddington is quoted as saying "No observation can be accepted until it has been confirmed by theory." (That precept still carries considerable weight today, although some of our theories consist of no more than giving a name to the cause of the effect.)

In 1772, when the only planets orbiting the Sun were known to be Mercury, Venus, Mars, Jupiter, Saturn, and the Earth, Johann Daniel Titius included as a footnote in a book he was editing, an ingenious rule for calculating the orbital spacing. Basing his distance scale for the Solar System on 100 parts equaling the distance from the Sun to Saturn, Titius presented a word problem statement of how to calculate the distances for each planet. He noted that his rule presented a distance just past Mars that was empty, which he reserved for "still undiscovered satellites of Mars." This moved Jupiter to the next step in order, and similarly placed Saturn in the next farther step.

Johann Elert Bode repeated Titius' rule in a 1772 edition of a book of his own, and promoted the value of this rule, favoring an undiscovered planet in the orbital space between Mars (at 1.52 Astronomical Units) and Jupiter (at 5.20 Astronomical Units).

Apparently, no thought had then been given to extending the rule beyond Saturn, but the discovery of Uranus in 1781, in an orbit of very nearly the correct distance, gave it confirmation. It soon became "Bode's Law" and its prediction of a planet at a distance of 2.8 Astronomical Units encouraged a search, both analytical and observational. Baron Franz Xavier von Zach attempted to compute its orbit in 1785. It is not clear what data he would analyze at that time, but he commented that he could not calculate its location in its orbit, its heliocentric longitude. This may have been because the asteroids, which were the sought-for planet occupying the orbit at 2.8 Astronomical Units, are evenly distributed around the Sun, so there is no definite longitude.

A conference arranged by von Zach in 1796 resulted in a suggestion by Jerome L. Lalande that a systematic search should be conducted, and four years later von Zach had organized a group of astronomers to hunt for a planet that existed only in the form of an assumed regularity of planetary orbits [Hoyt 1980].

Unfortunately for the drama of this story, the first asteroid, Ceres, was discovered by Giuseppe Piazzi before he had learned of the search, while he was making observations for a new star catalog. He noticed a faint star that moved between observations, and determined that it must be the planet that belonged between Mars and Jupiter.

Two successful predictions, Uranus and Ceres, led to the use of Bode's Law in the calculations of the orbit of an unknown planet perturbing the orbit of Uranus. However, when Neptune was discovered, and its orbital distance was found to be only 30 Astronomical Units instead of the predicted 38.8, the rule fell into disfavor as an accidental quirk, with no theoretical basis.

The Titius-Bode Law for planetary distances was developed in 1766 by Johann Titius, following on an observation by Christian Freiherr von Wolf, who suggested that there might be a regular pattern in the spacing of the planets. There appeared to be a gap between Mars and Jupiter, pointed out by Johann Lambert in 1761, and if this gap were included, Titius' sequence showed a nicely harmonic pattern. The major obstacle to accepting the rule is that it has no developed theoretical foundation, and it is now rejected or ignored. For the Solar System proposed in this book, the predictions and the actual distances are shown in Figure 61. A shift in Jupiter's order, to make room for the asteroids (or in this book, for the primordial planet "Asteria") was required, and the prediction of an original distance for Planet X at 77.2 Astronomical Units is shown.

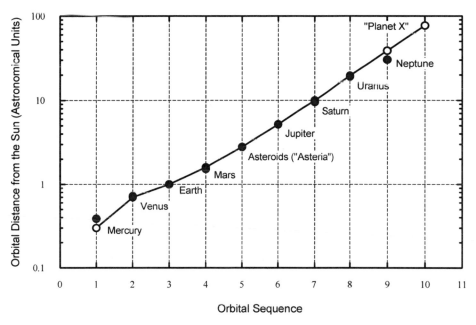

Figure 61. Titius-Bode Law for the Solar System.
A comparison of the distances of the planets from the Sun as predicted by the Titius-Bode Law, and as they actually are. The predictions are shown by the open circles; the actual values are shown by the filled circles. Only Neptune and Mercury deviate noticeably from the Titius-Bode Law prediction.

Planet X and the Milky Way

...full in the view of every visitor...
Edgar Allen Poe, *"The Purloined Letter"*, 1845

The Milky Way is an aggregation
of innumerable stars ...
Galileo, *"Siderius Nuncius"*, 1610

Planet X, planet number 10, the unknown planet, the ex-planet, what an appropriate name!

From the early times of astrology, five planets were known, named from the Greek word *planetes*, for wanderers, since they clearly wandered about the sky. These were Mercury, closest to the Sun; Venus, bright and dynamic; Mars, distinctly reddish; Jupiter, next only to Venus in brightness but slow moving; and Saturn, with yet unseen rings. Our astronomical understanding added the Earth to this list, making six small worlds orbiting a giant Sun. After the unintentional discovery of Uranus by William Herschel in March 1781, the Titius-Bode Law for planetary spacing gained greater stature, since the orbit of Uranus agreed very well with the predicted value, which had been based on the "original" six planets. This agreement was supported by the discovery of the asteroid Ceres in 1801, and others, in the otherwise empty gap between Mars and Jupiter. Although it became clear that the asteroids could not serve as a true planet, the Titius-Bode Law was considered to be valid, and perhaps there was no limit to the number of planets orbiting the Sun. When Neptune was discovered, it was found to be out of place in this scheme, its orbit was too small and seemed to be perturbed, as was the orbit of Uranus, still. The discovery of Pluto, too tiny to be a real planet, did not solve this problem, and so a search for the next planet, Planet X, continued. Improved data and calculations have reduced the significance of the residuals in the orbits of Uranus and Neptune, but as Percival Lowell wrote, orbits are fitted to the observations, and so seem to be more perfect than they are. At the great distances traversed by the outermost planets, long times pass between passings, and much of the time each planet has little effect on the others. While Myles Standish's latest orbital calculations seem to have removed the need for Planet X, Tom Van Flandern is of the opinion that true residuals still exist. The orbital message is ambiguous.

Each planet in the Solar System exerts a gravitational perturbation on every other planet, superimposed on the major gravitational attraction of the Sun, which controls the main aspects of the planetary orbits. For precise orbits, and the ephemerides that give the positions in the sky of the various planets, these perturbations must be included in the calculations. The effects can only be included for those planets that we know of, and any unknown planet may then show its presence by unexplained deviations of the observed planetary positions from the theoretical. (Actually, the mathematical form of a planet's orbit is called its "theory".)

Deviations of the observed position of Uranus from its theoretical position were seen as early as 1788, only 7 years after its discovery. Each recalculation of the elements of its orbit brought agreement, of course, but the position showed deviations again in just a few years. These deviations led to theoretical studies, the most famous being those of Urbain Le Verrier in France and John Couch Adams in England [Standage 2000]. These two calculations produced similar positions for an unknown planet, and either might have succeeded in the search, but Le Verrier was able to instigate a more effective search, at the Berlin Observatory. On the night of September 23, 1846, Johann Galle and Heinrich D'Arrest searched the sky where Le Verrier had indicated a new planet should be. Comparing stars seen with those plotted on the most recent star map, D'Arrest called out "That star is not on the map!", and Neptune had been found.

Even after correction for the perturbations of Neptune, unexplained deviations, residuals, still remained in the orbit of Uranus. Deviations began to appear in the orbit of Neptune. (Some of the deviations for Uranus have been reduced recently by use of a more accurate mass for Neptune, determined from the *Voyager 2* flyby.) These deviations suggested the presence of a more distant planet yet, which became the "Planet X" of Percival Lowell. As a result of the searches that Lowell started, Clyde Tombaugh detected the motion of a "star" that showed the correct orbital motion over several nights, and was near one of Lowell's predicted positions. This object became the planet Pluto, the first planet to be discovered by an American. Lowell's predicted mass, from his analysis of the perturbations of Uranus and Neptune, was about 6.7 times the mass of the Earth. However, Pluto was overestimated from the start, appearing to be too dim for a planet of such a size, and it has been downsized ever since. It is now classified as a dwarf planet, but perhaps should be considered to be a Kuiper Belt Object. It has become clear that Pluto is too small to cause the gravitational perturbations that produced the deviations seen in the orbits of Uranus and Neptune.

In the most recent re-calculation of planetary orbits, with Neptune nearly completing a circuit of its orbit since its discovery, and with the use of an accurate mass for Neptune, Myles Standish considers that there are no longer any significant deviations that might suggest the presence of a more distant planet [Standish Jr. 1993]. An underlying hazard in the method of analysis, although it is recognized as the best method, is that small unmodeled deviations can be hidden by the process. The method is Least Squares, and it produces the best values of the parameters that are specified in the model. The effects of unspecified parameters are accommodated by the values that are determined for the modeled forces and the uncertainties that are assigned to them [Van Flandern 1997]. Standish considers the model to be complete by inclusion of all significant known objects in the Solar System, and that the remaining deviations are within the estimated uncertainties. This is certainly shown visually by the results he has presented. Therefore, there is no longer any suggestion of Planet X, massive enough to have produced the perturbations that led to the search and discovery of Pluto.

Percival Lowell stated his judgment of the methods of mathematical analyses as: "Every theory of a planet is thus open to doubt, seeming more perfect than it is. It has been legitimately juggled to come out correct, its seeming correctness concealing its questionable character." [Lowell 1915].

The orbit of Pluto has been difficult to calculate accurately. It is seriously perturbed by the other planets, so that its period is nearly 3 years longer than it would be without these perturbations. Further, its observed positions suffer from systematic errors in the star catalogs [Malhotra and Williams 1997]. In other words, the stars in different catalogs are not in the same places. Is this because of time (proper motion) or observatory location (parallax)? In precisely determining the orbit of Uranus for the *Voyager 2* flyby, Anderson and Standish found that there were systematic differences between the star positions of the Perth (Australia) and the Smithsonian (Washington) catalogs, amounting to 2 arcsec. That is nearly three times the parallax measured for the nearest star, α (alpha) Centauri. There were also problems with positions of Uranus, because of star catalog errors.

The search for Planet X has a very erratic history. The Titius-Bode Law indicated a distance of 77.2 Astronomical Units, if Neptune at 30.1 Astronomical Units occupied the place of the planet expected to be at 38.8 Astronomical Units. Or it might have been at 38.8 Astronomical Units, if Neptune were an interloper. The perturbations of the orbits of Uranus and Neptune supported this closer distance in detail.

For over 12 years, Percival Lowell ever more meticulously calculated the orbit of an unknown perturbing planet beyond Neptune, his Planet X. These calculations, like the predictions of Neptune, were based on the deviation of Uranus from its theory, its mathematical orbit and ephemerides. By 1914, he had settled on two complementary solutions. Just as our Moon, in orbit around the Earth, raises two high tides on the opposite sides of the Earth, 180 degrees apart, perturbation calculations produce two solutions, 180 degrees apart. Sometimes only one solution has been developed, perhaps because of the extreme arithmetical labor required before electronic computers. One of Lowell's solutions showed the most likely position of Planet X at the midpoint of 1914 to be at a heliocentric longitude of 84 degrees. The other was at a heliocentric longitude of 262.8 degrees [Lowell 1915, Hoyt 1980, Russell 1930]. Lowell considered the first solution to be somewhat better, based on smaller remaining residuals, and that the latter was "one nearly inaccessible to most observatories, and therefore, preferable for planet hiding." The parameters of the two orbits that Lowell presented in 1915, referenced to July 0, 1914 [Lowell 1915], are:

	Planet X1	**Planet X2**
mean orbital longitude	22.1 degrees	205.0 degrees
mean distance	43.0 AU	44.7 AU
mass	6.66 Earths	7.59 Earths
eccentricity	0.202	0.195
longitude of periapsis	203.8 degrees	19.6 degrees
heliocentric longitude	84.0 degrees	262.8 degrees
inclination	estimated as about 10 degrees.	

By remarkable determination and perseverance, Pluto was found, but Pluto was not Planet X. The failure to find Planet X has led to the idea that the orbit must be unusual, highly elliptical and greatly inclined, or alternatively that the planet truly does not exist. T. C. Van Flandern and R. S. Herrington calculated a probable mass of 2 to 5 Earths and an orbit that ranged from 50 to 100 Astronomical Units. Their planet should have been as bright as 13 magnitudes, but since it had not been seen, it must have had a brightness of 16-17 magnitudes or dimmer [Littmann 1990]. D. P. Whitmire and J. J. Matese considered an orbit tilted to 90 degrees, and suggested that precession of the orbit disturbs comets in the plane of the ecliptic, twice in 56 million years, causing mass extinctions by impact [Littmann 1990]. Conley Powell calculated that there might be a Planet X orbiting at 60.8 Astronomical Units, with a period of 494 years, and a mass of 2.9 Earth masses [Littmann 1990]. Another calculation suggested an orbit at 39.8 Astronomical Units, with a period of 251 years, and a mass of 0.87 times the Earth. This (and Pickering's multiple hypothetical planets) shows the difficulty in producing a unique solution of the perturbations. Charles T. Kowal made unpublished calculations for Planet X [Littmann 1990]. Kowal found the asteroid comet 2060 Chiron in 1977, but he did not find Planet X. *IRAS* has searched, away from the plane of the Milky Way, and has not found a planet [Littmann 1990]. Current searches are finding objects down to the 20th magnitude, and Planet X still hasn't been found, which reinforces the idea that it really isn't there, and the orbital residuals are meaningless.

This book proposes that Planet X did indeed exist, but that it was destroyed in a nuclear fission detonation about 6 million years ago. The fragmentary debris from that detonation is seen as the Milky Way, as a reflection nebula, and the remaining core of the planet is the center of the Milky Way, still perturbing, in a weak way, the orbits of the outer planets.

Relative to the "other" orbit calculated by Lowell, the present (year 2000) heliocentric longitude of Sgr A*, which is thought to be the centerpoint of the Milky Way Galaxy, is 266.4 degrees. Precessed to the middle of 1914, when Lowell calculated the two possible positions of Planet X, the heliocentric longitude would have been 265.1 degrees. The heliocentric longitude that Lowell calculated (Planet X2) was 262.8 degrees, less than 3 degrees from where we think Sgr A* was at the time. Unfortunately for the further progress of this discussion, this position disagrees with the position indicated by Cornelius Easton, as will be discussed later. This position may appear to conflict with the motion I have considered for the center of the Milky Way, based on Cornelius Easton's map of 1900, and more recent values. Is this another "incredible coincidence"? Or the result of a very peculiar and contrary orbit? The explanation for this could be in the different points of reference used for the two comparisons. In the Easton comparison, the position of the center of the Milky Way Galaxy is referred to what may be true stars in identifiable constellations. The celestial coordinates are based on the intersection of the Earth's equatorial plane and its orbital plane. These are different points of reference, different methods of providing positions. The results of the four calculations, two of Lowell (based on the perturbation of Uranus) [Hoyt 1980], one of Brady (based on perturbation of comets) [Brady 1972], and one of the present book (based on apparent changes in the position of the Milky Way Center), with the actual orbit derived for Pluto at the time of discovery [Crommelin 1930], are shown below:

Orbital Element	Lowell X1 1914	Lowell X2 1914	Actual 1930	Brady 1972	Tuttle 2005
mean distance, AU	43.0	44.7	39.5	59.93575	59.699
period, years	282	299	248	464	461
mass, Earth	6.66	7.59	<0.7	300	----
eccentricity	0.202	0.195	0.248	----	----
inclination	about 10		17.1	120	115.22
mean orbital longitude	22.1	205.0	----	----	----
longitude of periapsis	203.8	19.6	223.4	----	----
heliocentric longitude	84.0	262.8	108.5	----	----
longitude 1930.0	102.7	279.0	108.5	----	----
perihelion longitude	204.9	20.7	223.4	----	----
perihelion date	1991.2	1994.9	1989.8	----	----
longitude ascending node	----	----	109.4	115.75	101.55

The present mass of Pluto and Charon, determined from their mutual orbit, is 0.00247 Earths. Brady's mass estimate (300 Earth masses) was exaggerated by the use of an incorrectly large mass for Pluto. While Lowell's X1 parameters are close to those actually found for Pluto, the very small mass that we now know for Pluto and Charon together rules out the possibility that they were responsible for the perturbations, and we must accept that Planet X had not been found. The orbits calculated by Brady, and by me, differ from Lowell, but are remarkably close to each other.

Neptune was discovered in 1846 and its orbit takes 165 years to complete. A most accurate orbit cannot be calculated until it has "closed the circuit" with modern, deliberate position measurements over an entire orbit. That occurred in 2011, and a new opportunity arises to find the location of Planet X, at the present time. The perturbations from a small, distant object on a highly inclined (and probably eccentric) orbit will take great skill and care to find. This complication in the perturbation calculations arises from a peculiarity of the orbits calculated by Brady and by me, showing a massive (perhaps 0.5 times Earth) object orbiting on a highly inclined and retrograde orbit. Most of the objects in the Solar System (all of the planets) orbit in nearly the same plane, the invariable plane (close to the ecliptic plane of the Earth's orbit), and in the same "prograde" direction. This situation leads the perturbation of one planet on another, Neptune on Uranus as an example, to be primarily in the same plane, causing deviations in the observed longitude, the position of a planet along its orbit. For example, Brunini has concentrated on the deviations in longitude of Uranus when searching for perturbations [Brunini 1992a]. The longitudes show the greatest deviations, and the latitudes offer little encouragement. A highly inclined orbit for the perturbing planet, the unmodeled force, produces deviations primarily in latitude. Further, a retrograde orbit results in interactions that are of shorter duration, causing smaller deviations.

A re-evaluation of Brady's calculations and consideration of the inclined, retrograde orbit that he calculated for a distant massive object might resolve the case in favor of Planet X.

Everyone's right. There was a tenth planet. There is no tenth planet. As sometimes happens in scientific controversies, both sides are right, although not quite in the way they think. There is no tenth planet. There was a tenth planet. Planet X is gone. All competent searches have failed to find it. What happened to Planet X? Why hasn't Planet X been discovered? Simply because it no longer exists as a planet.

So where is Planet X, what is Planet X? Just like Edgar Allen Poe's purloined letter, Planet X has been transformed into something that we can no longer recognize as the original, and hidden in plain sight: "... *full in the view of every visitor ...*". We can not find it as a planet because it is no longer a planet. It transformed itself, by a nuclear fission detonation, into a spiral nebula that we call the Milky Way Galaxy.

Consider the possibility that there had been a tenth planet, beyond Neptune, and that planet was disintegrated by a nuclear detonation approximately 6 million years ago. What evidence might remain?

I can briefly summarize what we might reasonably assume for Planet X, recognizing that there could be several variations on this theme. We can then explore the Solar neighborhood to try to find evidence of this planet's existence and disappearance.

Planet X probably formed twice as far from the center of the Solar System as did Neptune, according to the Titius-Bode Law, or some similar rule. If planets form from the outside in and have original masses according to the inverse distance from the center, Planet X would have had a mass approximately one-half that of Neptune, or about 8.5 times the mass of the

Earth at present. Its surface temperature would have been in the range of 10 to 30 K, and nearly all the gases except hydrogen and helium would have been frozen solid, forming a crust of frozen nitrogen, ammonia, methane, water, and carbon monoxide. Deep within, there would have been an icy-rocky kernel, possibly with a segregated core. On the basis of what we have seen for the other planets, the core probably contained a multitude of natural nuclear fission reactors, which contributed to the melting and segregation of the core, and that might provide some small amount of heat to the surface. Planet X probably had some large moons.

Let us speculate on what might have happened to Planet X, millions of years ago. For any event so far away, so long ago, we can only speculate, and so this will be written in a speculative tone. However, it may represent a close approximation to a complex reality.

Let us suppose that Planet X did exist, possibly with some large moons, like the other outer giant planets at present, except for poor Uranus, and that a nuclear fission detonation disrupted the planet. As the planet broke apart, exposing the fireball at its core, an intense burst of gamma rays would have spread through the Solar System at the speed of light, passing the Earth 6 to 10 hours later. A wave of antineutrinos might pass through the Earth a few seconds earlier, having easily escaped directly from the heart of the exploding planet, without having to wait for the rock to break apart. These antineutrinos would boost the power of the reactors of the Earth, warming the planet and producing short-lived radioactivity. Some of the large moons might be induced to detonate, by the combined effects of the antineutrinos and the neutrons released by the fissions. Large amounts of radioactive material would be produced. The mass of Planet X would be spread out in a disk by the energy of the detonation, with clumps and lumps leaving preferentially along the equator because of the reduced force of gravity resulting from the centrifugal force from the planet's rotation. This debris would retain the tangential velocity it had while rotating as parts of a solid body.

The tangential velocity (from rotation) combined with the radial velocity (outward) from the explosion would produce a raggedy spiral structure in the disk of debris (just as was seen in the rotation measurements of spiral galaxies by Adriaan van Maanen). As the mass of the planet spread out in space, it would lose its gravitational hold on the remaining moons and they would be free to continue in disturbed independent orbits around the Sun. One might be captured by Neptune, as Triton, while two others might form a binary planet like Pluto and Charon, perhaps aided by mutual magnetic attraction. (A possible alternative would be for Pluto and Charon to be remnant fragments that collided softly in the crowded cloud of debris, and stayed together ever after.) The surviving moons would have been roasted on the side facing Planet X, severely affecting their appearance. The detonated moons would orbit near the remainder of the detonated planet, as ragged clouds of debris.

Nearby globs of uranium hydride, remaining scattered through space from the primordial supernova cloud, would be induced to explode, if within range of sufficiently intense neutron and antineutrino fluxes. These would later appear to be globular clusters of stars.

Debris headed away from the Sun would be retarded by the Sun's gravity, and debris headed into the Solar System would be accelerated past the Sun, eventually forming arcs of material on the far side. (This formation is so much like the structure of the Milky Way sketched by Cornelius Easton, we must wonder if he had a better mental view than our current symmetric version.) Since the plane of the disk corresponded to the equatorial plane of Planet X, as the debris spread out, velocities would systematically exceed or fall behind the necessary Solar orbital velocities, and the disk would begin to warp, as is seen in the Milky Way Galaxy and many other spiral galaxies.

Fragments of the planet would be illuminated by the Sun, and would reflect a spectrum much like that of the Sun, a solar-type spectrum. Fine dust would scatter and attenuate the blue end of the spectrum, both for the incident sunlight and the returned reflected light, making many of the points of light appear to have somewhat redder spectra than the Sun. Some of the fragments that remained intensely radioactive, perhaps rich in aluminum-26 or iron-60 (making cobalt-60), would glow into the ultraviolet with the blue glow of Cerenkov radiation. Other fragments, still retaining the capability and nuclear fuel to operate as nuclear reactors, would be strong emitters in the infrared and red regions of the spectrum. These fragments are like the asteroids, named so because they look like stars. The fragments of Planet X are being discovered now as Kuiper Belt Objects, kuiperoids, and Trans-Neptunian Objects

Stars in our Galaxy (and in others) shine by a variety of means. True stars, those great balls of fire like our Sun, are heated to incandescence by nuclear fusion reactions deep in their cores. The asteroidal fragments of our Galaxy, that we have misunderstood to be great balls of fire like our Sun, shine by other means. Those that are radioactive and shine by Cerenkov radiation are blue. Because this blue light is in addition to the reflected light of the Sun, these "stars" are brighter and are labeled "blue giants". Similarly, those that are heated by ongoing nuclear fission reactions, glow with thermal radiation. With the addition of reflected sunlight, these "stars" are also brighter and are labeled "red giants". Those neutral fragments that only reflect sunlight are considered to be "solar-type dwarfs". Some stars may be "peculiar" as a result of fluorescence induced by beta and gamma radiation. A combination of all these sources of spectral light might be very difficult to interpret.

The neutrons released at the end of the detonation would escape into space, where they would decay into protons and electrons. The clouds of protons from neutron decay would initially have an apparent temperature on the order of 3 million K; the electrons (beta particles) would have an effective temperature of 2 billion degrees K, but would quickly radiate this energy and come into equilibrium with the protons. Protons and electrons would gradually recombine, making neutral hydrogen atoms. These atoms would have velocities that combined the planetary motions and the energy of the neutrons emitted in the detonation.

Detonated moons would appear as distant satellite clouds of fragments, also illuminated by the Sun, containing self-illuminated radioactive fragments and self-heating reactors. Eventually, a surviving (or rejuvenated) reactor might detonate and appear to be a supernova, like SN 1987A in the Large Magellanic Cloud. Powered by the energy of beta particles from radioactive decay, the remaining core of the planet would continue to act as a dynamo, with rapidly circling protons, iron ions, and electrons trapped by the magnetic field.

Consider the Milky Way: Via Lactea in Latin, made scientific as The Galaxy from the Greek word for milky, defined in a single statement by Galileo in 1610: the Milky Way is composed of a collection of innumerable stars. To us, that means stars like our Sun, more or less, spread through an immense region of space that dwarfs the Solar System. A region of space so great that billions of other suns and planetary systems might exist within its bounds. In our version of Galileo's Milky Way, our Sun is just one among many of the innumerable stars, the points of light. Our understanding of the nature of the Milky Way has developed gradually through the past centuries and, misdirected through the centuries by lack of knowledge, our understanding is wrong.

Originally perceived by the naked eye as a band of milky light across the sky, forming a stripe from one horizon to the other, the Milky Way was defined as an aggregation of innumerable stars by Galileo Galilei in 1610. His telescope allowed him to see that the clouds of milky light were made up of points of light, rather than a smoothly continuous glow. To Galileo, stars were points of light that could not be enlarged by his telescope, as the planets could be. To Galileo's telescope, the large moons of Jupiter were so small that they could not be enlarged, and so Galileo called them stars. Stars were soon known to be suns, so all of Galileo's innumerable points of light became distant suns. If they were suns, they must be distant to be so dim. Being so dim, they must be distant if they are like our Sun.

Seares has estimated that the Galaxy contains about 30 billion "stars", down to a brightness of 60 magnitudes [Rudaux and De Vaucouleurs 1959]. If we consider that the brightest of these might be like Pluto, roughly 15 magnitudes for a diameter of 2,700 kilometers, then the smallest objects that contribute to the estimated number, at 60 magnitudes, would be about 10 meters in diameter. It is likely that the size range goes on down to grains of dust, crystals of molecules, and single atoms.

Mapping the Milky Way was attempted by William Herschel in 1780 to 1790, by "star-gauging". The technique was improved by H. Von Kulige and Jacobus C. Kapteyn. As larger telescopes were developed, the distinctive structure of what became known as spiral nebulae was discovered. That spiral form began to influence the unseeable form of the Milky Way.

While exploring what we think we know about our Milky Way Galaxy, I noticed something quite strange. Cornelius Easton, a major American astronomer of a century ago, had carefully prepared a sketch of the Milky Way, showing the location of the Sun [Easton 1900, Trimble 1995]. Now, "everyone knows" that the center of the Milky Way, as a spiral galaxy, is located in the direction towards the constellation Sagittarius. In 1900, Easton clearly showed the center to be in the direction of the constellation Cygnus, about 60 degrees from Sagittarius, Figure 62. A later map, in the Larousse Encyclopedia of Astronomy [Rudaux and De Vaucouleurs 1959], shows the zero of galactic longitude in Scutum, 30 degrees away, with the note that the galactic prime meridian has moved another 30 degrees in the same direction, and showing the center now to be aligned with Sagittarius. In that view of the Milky Way Galaxy prepared before the conviction that the spiral nebulae were really Island Universes was accepted, and so the Milky Way Galaxy was an Island Universe of stars also, Cornelius Easton sketched his concept of the clouds of debris that reach out from the center of the detonation to envelop the Sun and the planets of the inner Solar System. By 1900, Easton had developed a model of the Milky Way Galaxy that fully incorporated the forms seen in the spiral nebulae. He published this carefully elaborated sketch with the remarkable disclaimer that it *"does not pretend to give an even approximate representation of the Milky Way."* (Easton's italics, [Easton 1900, Berendzen *et al.* 1976].) At this time, and seeing his work in comparison to current versions, we must wonder about this disclaimer and if he believed that deficiency to be true, why did he bother to publish the map? The agreement between the 1900 version and current views is remarkably close, considering all the knowledge that has been gained in the last century. Easton believed that the Milky Way was similar to the spiral nebulae in form, but that all the other spirals were simply small eddies in the Universe [North 1990]. Easton later published a revised view, showing a more symmetrical spiral structure, and strangely reversing the sense of the rotation of the spiral arms [Easton 1913].

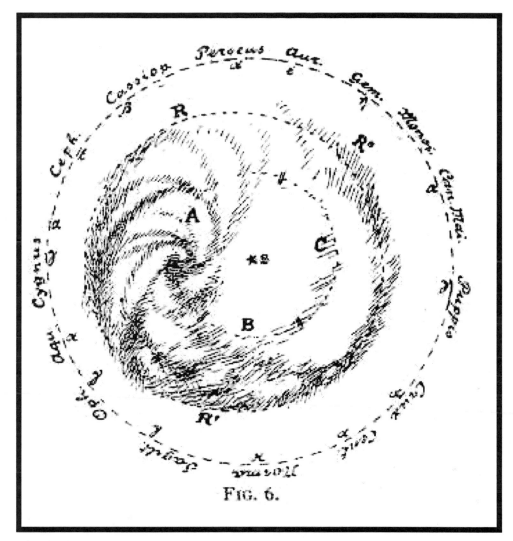

Figure 62. Cornelius Easton's concept of the Milky Way.
Conceptual view of the Milky Way, as an unknown collection of material, encircling the Sun and the Solar System, prepared 24 years before the Island Universe theory became the adopted model [Easton 1900]. The Sun is the star at the center of the diagram, at S, surrounded by the debris of the detonation of Planet X, which is seen as the Milky Way. This shows the direction through the center, from the Sun (and the Earth), to be approximately to the star γ (gamma) Cygnus.

A further, very anecdotal, indication that the Milky Way has moved during historical time comes from Aristotle (384-322 BCE). Bailey, Clube, and Napier reported that Aristotle believed the Milky Way to lie in the sublunary zone, the region between the Earth and the Moon, which the researchers interpreted to imply that the Milky Way was much nearer to the ecliptic plane then than now, "though there is no known mechanism" to cause this change [Bailey, Clube, and Napier 1990.]. The ecliptic plane was known to ancient astrologers as the Houses of the Zodiac. At present, the plane of the Milky Way makes an angle of about 63 degrees with the ecliptic. Such a change in orientation would be a natural effect of precession for the debris of a distant planet.

Because of the similarities between Easton's not-even-pretend Milky Way and our current knowledge, we must accept that he really did use objective knowledge in constructing his sketch, despite his disclaimer. Accepting that, we must recognize that Easton placed the center of the Milky Way in the direction of the star γ (gamma) Cygnus. Roughly half a century later, the zero of galactic longitude was at the constellation of Scutum [Rudaux and De Vaucoleurs 1959], and at present the center lies near the star γ (gamma) Sagittarius. How could the center of a "vast Island Universe of stars" shift so drastically?

Possibly Easton was actually mistaken as to where the center was, even though he had the details of the form right, and only now do we correctly know where the center is. Except that half-way through the century, the location of the center had been redefined as about halfway between Easton's location and the present true location. An alternative to this rather unlikely stepwise progression from error to truth may be that the center of the Milky Way Galaxy has actually moved by 78 degrees in 100 years, as it orbits our Sun, being the remnant of the detonated Planet X.

Using Easton's location and the present location in Sagittarius, and the simplifying assumption that the center of the Milky Way Galaxy is an object in a circular orbit centered on the Earth (rather than the more nearly correct "centered on the Sun" or the exact "centered on the barycenter of the Solar System"), I calculated the apparent motion of the Milky Way Center as if it were some sort of object orbiting the Sun, as if it belonged to the Solar System, rather than the reverse, being a vast Island Universe of stars. Remarkably, I later found a report by Joseph Brady on a possible cause of the perturbations in some cometary orbits, notably Halley's Comet [Brady 1972]. From the delays and advances in the returns of Halley's Comet, Brady derived an orbit for a massive object orbiting the Sun, as the cause of the observed perturbations. This also improved the calculated return times of the comets of Olbers and of Pons-Brooks. The orbit that I calculate for the Milky Way Center agrees almost exactly with the orbit calculated by Joseph Brady. Two independent calculations, done without knowledge of each other, using different methods and different types of data, gave the following results:

Orbital Element	Brady (1972)	Tuttle (2005)
mean distance	59.93575 AU	59.699 AU
inclination	120 degrees	115.22 degrees
Omega	115.75 degrees	101.55 degrees
Period	464 years	461 years

Considering the inaccuracies in the data and the calculations, the remarkable agreement might be judged to be a happy accident, or another set of incredible coincidences. (The numbers are what they are, and obviously should not be taken as indications of their precision or accuracy.)

In a study of the perturbations of Halley's orbit [Brady and Carpenter 1971], Brady had found that a secular term in the Sun's gravity was necessary to reduce the observed errors in return times. This resulted in excellent agreement with the observations. That encouraged him to proceed with a detailed perturbation calculation on the return times of Comet Halley.

Brady calculated a first approximation for the mass of the orbiting object by ratioing amplitudes of perturbations, using Pluto's then known mass as a reference. That led to a mass of 100 times the Earth. Using detailed perturbations, Brady calculated a mass of about 300 times that of Earth, based on a mass for Pluto that was thought at the time to be about equal to Earth's mass. Such a large mass made his proposal unbelievable, and his analysis unacceptable, and his results were immediately rejected [Klemola and Harlan 1972, Goldreich and Ward 1972, Seidelmann, Marsden, and Giclas 1972].

At the time that Brady was working on this, 1970-1972, the mass of Pluto was poorly known. The IAU official value [Ash *et al.* 1971] was equal to 0.92495 times the mass of the Earth. The discovery of Pluto's companion Charon in 1978 determined the mass for the pair (Pluto and Charon) as 0.00247 times the mass of the Earth. This change was never reassessed in its effect on Brady's analysis, but that leads to a much smaller estimate for the mass of the remnant of Brady's Planet X. It is not entirely clear if Brady's mass estimates can be scaled by the change in the mass of the Pluto-Charon system, but if so, the revised mass estimates would be in the range of 0.27 to 0.79 times the Earth, a "mean" value of 0.53 Earth masses. This is more compatible with our expectations, and lies well below the limit set by the *Pioneer 10/11* measurements, less than approximately 4.5 Earth masses.

Others have also made predictions based on comets. In the late 1800s, Camille Flammarion studied comets and placed them in "families" according to the distance of the aphelion of their orbits. A family was associated with Jupiter, another with Saturn. A distant family was also found, well beyond Neptune. The astronomer T. Grigull of Münster, Germany, considered the holder of that family to be about the same size as Uranus and Neptune and that it had an orbital period of 360 years [Hunt and Moore 1994].

Based on these observations, it is proposed that an additional planet may have originally formed beyond Neptune, and was disintegrated by a nuclear-fission detonation. Gravitational interaction of debris from this detonation with Neptune, and the loss of the gravitational concentration of Planet X outside Neptune's orbit, could have been the cause of the shrinking of the orbit of Neptune from the distance predicted by the Titius-Bode Law (38.8 Astronomical Units) to its present distance (30.1 Astronomical Units). It is estimated that the ejection of bodies equivalent to one-third the mass of Neptune would have produced the observed orbital reduction [Safronov 1969]. Fragments from this detonation could have bombarded Uranus, causing its tilt (although I have proposed that a lopsided detonation of Uranus, triggered by the detonation of a moon, caused the tilt), and left a remainder of objects such as Pluto and Charon, the kuiperoids, and possibly the meteorites known as the ureilites [Wasson 1985]. This detonation could have been the source of the comets that occasionally pass through the Solar System. It is estimated that these comets come from an orbit at 43,000 Astronomical Units from the Sun, and were set free by a planetary explosion that occurred 3.2 to 5 million years ago [Van Flandern 1993].

The Sun is considered to be orbiting the center of the Milky Way in a clockwise direction, so the Milky Way Galaxy (ex-Planet X) is orbiting the Sun in a retrograde orbit. That is what Brady's orbit shows, and the orbit I have calculated based on the movement from Easton's 1900 position shows that also.

The assumption that the stars are all at great distances seems to have failed us in at least three applications of data that are based on this assumption. The orbit of Uranus can not be calculated accurately because of systematic errors in the *Hubble Space Telescope* Guide Star Catalog. The perihelion of Comet Hale-Bopp can not be calculated accurately because of systematic errors in the star catalog. The orbit of Pluto can not be calculated accurately because of systematic errors in the star catalogs, Perth and Smithsonian, one from the northern hemisphere, the other from the southern. This problem shows the difference between trigonometric parallax and absolute positioning. If all the stars (the points of light) in a region are near to the Earth, at approximately the same distances, there will be no observable parallax, and the stars will be judged to be at a great distance. However, determinations of the absolute positions will differ depending on the location on Earth and the point in the Earth's orbit. Points of light, "stars", that are much closer than conventionally assumed, might be seen in slightly different positions from different locations in the Earth's orbit. These differences would show up as systematic errors in various star catalogues, based on observations at different times of the year, from different locations on the Earth. These systematic errors have plagued planetary astronomy since the discovery of Uranus to present-day calculations of the orbit of Pluto. Systematic errors in the Guide Star Catalog, used by the *Hubble Space Telescope*, were noted by Brian Marsden, in the determination of the Hale-Bopp comet orbit. [See http://cfa-www.harvard.edu/cfa/ps/HaleBopp1993.html.]

An exceptionally close and recent detonation, within the last few million years, with the Earth near the plane of debris resulting from the detonation, might look like the view we have of the encompassing lane of light and dark that we call the Milky Way.

On looking at the Milky Way through his newly built telescope and resolving the hazy glow into points of light, Galileo stated, "For the Galaxy is nothing else than a congeries of innumerable stars distributed in clusters." without any knowledge at that time of what stars are. (However, Galileo was somewhat loose in his use of the term "star". When he discovered the moons of Jupiter, he designated these as starlets, stars, and planets, using the terms quite interchangeably, as was common practice since early days or nights). If we may have the nerve to reinterpret his statement, he really meant only that the bright hard points of light that he saw in the Milky Way were the same as those clearly unresolved points of light called stars, that are seen in all directions around us. However, we have literally taken him at his word ever since, and as our understanding of our star, the Sun, grew, we extended that knowledge to the "stars" of the Milky Way. A tremendous body of knowledge has been built on this belief, that all the hard bright points of light we see are stars like the Sun, perhaps a million times brighter or ten-thousand times dimmer, a hundred times as massive, or only a tenth as much.

And so, by an accident of belief set in 1610, the detailed development of our view of the Universe was set in place. All interpretations of bright points of light in the sky would be based on this concept, and our perceptions would be adjusted and interpreted in such ways as to comply.

With apologies to Wm. Shakespeare (*'Romeo and Juliet'*) and Gertrude Stein (*'Sacred Emily'*),

> A rose by any other name
>> Would smell as sweetly.

but

> A star is a star is a star is a star,
>> Has misled us completely.

Could it be possible that the Milky Way, our Galaxy, with Galileo's innumerable points of light, is a reflection nebula of our Sun, reflecting sunlight as a solar spectrum, like Pluto does? The answer is, mostly, yes! Mostly, the Milky Way shines by reflected light from the Sun, which is why so many of the bright points of light have a solar spectrum and we label them as "normal stars". Others glow with internal heat from the continuing action of nuclear reactors, and appear to be red giant stars. Still others, paradoxically cold, shine with Cerenkov radiation into the ultraviolet, and are labeled as blue giants, superhot stars. Globular clusters, in a spherical halo about the nucleus of the Milky Way, centered on the previous location of Planet X, are the remains of globs of uranium hydride that were induced to detonate by the antineutrinos from that planetary detonation.

The spectrum of light from a planet, or any reflective object in the Solar System, strongly resembles the spectrum of the light from its source, the Sun. Only a few absorption effects alter the light, and these are not easily detected through our atmosphere, except for the gaseous planets Jupiter, Saturn, Uranus, and Neptune, where deep atmospheres give absorption by methane and other gases [Slipher 1909, Kartunnen *et al.* 1987]. Reflection from any solid surface returns a solar spectrum. That is, planet-light looks like sunlight, like starlight. Pluto was seen by Humason and Slipher to have a "solar-type spectrum" [Stern and Tholen 1997]. Color magnitudes and multiband spectra of astronomical objects have a quite different nature from the differential spectra formed with a prism or a grating, which shows spectral lines, absorption and emission. Colors may be determined for condensed objects, such as stars and planets, but spectral lines can only be formed by rarefied gas.

The color-magnitude approach has been used to study the Trojan asteroids of Neptune, those asteroidal fragments that orbit ahead of and behind Neptune at the stable Lagrangian L4 and L5 points [Sheppard and Trujillo 2006]. A plot comparing the color magnitudes of the Neptune Trojans, Kuiper Belt Objects, Centaurs, and Jupiter Trojans, all similar asteroidal fragments, shows a broad but well-defined distribution. At an extreme of the distribution but otherwise clearly belonging to it, lies the color magnitude point for the standard star G158-100 (GSC05271-00531) used for photometric calibration [Smith *et al.* 2002]. Considering the color magnitudes alone, there would be no reason to think that object was a star, a great ball of fire. Its color gives no hint that it is not a cold chunk of rock.

A reflection nebula?

In 1610, Galileo discovered that his telescope had allowed him to see that the otherwise milky Milky Way ("Via Lactea") was actually composed of tiny points of light. This led to his pronouncement that the Milky Way consisted of an aggregation of innumerable stars. In Galileo's use of the term "star", he was using an observational definition: if a point of light could not be enlarged into a disc, as the known planets could, if it remained a hard, bright point of light in his view through the telescope, it was a star. Even the Galilean moons of Jupiter were described as stars, by Galileo. In time (though I do not know exactly when), we came to understand that our Sun was a star, and stars were suns, of a wide ranging variety. This was probably from a gradual change of view from 1879 to 1921, but Richard Bentley, writing to Isaac Newton in 1693, acknowledged that most astronomers of that time considered stars to be like the Sun, and Newton used that analogy in comparison with Saturn to estimate great distances. Newton used the self-consistent reasoning that has built our view of the Milky Way in 1685 to determine the distance to stars by comparing their brightness with that of Saturn, shining by reflected sunlight, with the assumption that stars were like our Sun [Cohen 1958]. Knowing an approximate size and distance for Saturn, he determined that a star assumed to be equally as bright as our Sun but appearing only as bright as Saturn must be at a distance of 13.5 light years. This, too, set the scale of our Universe. The basis for this supposition was that stars did not show an annual parallax from the Earth's motion about the Sun, and so must be at great distances. Newton's estimate simply provided a measure of these distances. Alternatively, a possibility not previously considered, if the reference stars for parallax were as near as the object stars and moving with them, no parallax would be observed. The object stars would be judged to be at unmeasureable distances.

Most of the stars in the Milky Way do look like the Sun, which is why we call the Sun an "average" star. Other spectra are produced by the superposition of Cerenkov radiation (blue-violet-ultraviolet) and thermal radiation (up to about 3,000-4,000 K) from the heat generated by operating nuclear reactors. HI and HII clouds are the result of the decay of neutrons to protons and electrons. Gaseous nebulae are ionized by beta particles, not superhot blue stars. For an early measurement of the spectrum of the Milky Way, Vesto Slipher concluded that: "The spectra of the two regions [Sagittarius and Scutum] were similar: both were strongly the main features of the solar type spectrum. Closer inspection shows stronger hydrogen lines than belong to that type, and the spectrum would be termed 'composite' with the solar light predominant." [Slipher 1915]. The "stronger hydrogen lines" would come from the hydrogen produced by the decay of fission neutrons into protons and electrons, recombining to form hydrogen atoms.

The spectra of objects in my Milky Way are not identical, and shouldn't be. If sunlight reflects from ice, granite, limestone, basalt, iron, all the different sorts of material a planet ("Planet X") might have been made out of, the spectra will be grossly different. The Fraunhofer absorption lines, which come with the sunlight, will all be there, but the overall shape and relative intensities will be different depending on the color of the object, and those differences are what are used to classify stars into spectral types. (Look out at the world around you: in the daytime, nearly everything is reflecting the same sunlight, but the colors, the gross spectra, are all different.) Relatively few stars have had their spectra analyzed in detail, and because "interesting" stars are more likely to be studied than common ones, that tends to emphasize the differences, some of which are really real. We learn astronomy by looking at typical, but different, examples. Jaschek and Jaschek (1987) claimed 550,000 stars in the Milky Way [Jaschek and Jaschek 1987] (the *Hipparcos/Tycho* parallax survey shows about twice that number [European Space Agency *et al.* 1997]) and only 2,000 had been analyzed in enough detail to give chemical composition. Seares has estimated a total number, down to the 60th magnitude, amounting to 30 billion [Rudaux and De Vaucouleurs 1959]. In addition to reflection of sunlight, objects in the Milky Way Galaxy can glow by themselves by thermal heat (red stars), Cerenkov radiation (blue stars), clouds of HI (decayed neutrons), fluorescing gas (excited by beta particles), X-rays and gamma-rays (from radioactive decay), synchrotron radiation (electrons orbiting a magnetic body), and probably a few ways that haven't become apparent. Those fragments that provide supplemental light, by Cerenkov or internal heat or other means, will appear to be brighter than they should be and will appear to us to be giant stars.

Understanding that stars were suns established a conceptual scale for astronomy, so that the Milky Way Galaxy became a vast Island Universe of suns, with no possible alternative considered. (Indeed, until the discovery of nuclear fission, and the demonstration of the power of a nuclear detonation, no alternative was possible.)

The "Great Debate" of Shapley and Curtis in 1920 [Shapley and Curtis 1921] grew out of the question of the nature of what had been called the "spiral nebulae", as those objects had been increasingly found and defined by ever-improving optical telescopes. At one extreme, the spiral nebulae might have been exactly what the label suggested: clouds of gas and dust, formed into a spiral structure. As reflection nebulae, these spirals would glow by the reflected light of a nearby star. Otherwise, the spiral nebulae were actually distant, vast Island Universes of stars, shining with the light generated by billions and billions of stars, just like our Milky Way Galaxy. Galileo's description of the Milky Way as "stars" led to the understanding that the Milky Way Galaxy must be huge and distant, since stars like the Sun must be at great distances to appear so small and dim.

Indeed, the Milky Way is the same as the spiral nebulae, and so, the spiral nebulae are the same as the Milky Way, that is, the still incandescent radioactively hot debris of a recently detonated planet, shining in part by its own energy and in part by reflected sunlight. Both Shapley and Curtis were right, and wrong, but the Milky Way is distinctly not a vast Island Universe of stars, the stellar system of our Sun.

At the time of the debate, there was considerable observational evidence suggesting that the spiral nebulae were indeed nearby. However, the Cepheid variable distance scale, although having no absolute basis and suffering from repeatedly discovered relative errors from the start, showed that the spiral nebulae were far outside and beyond the Milky Way Galaxy. The relative distances showed that the spiral nebulae were truly outside the Milky Way Galaxy, so this was a reasonable and correct decision. Since the spiral nebulae were like the Milky Way Galaxy and the Milky Way Galaxy was like the spiral nebulae, they all became Island Universes of stars, at great distances, and therefore had to be the vast Island Universes that had been imagined for so long. Such great distances required large physical sizes, great energy output, and so the spiral nebulae were also vast Island Universes of stars, just like our Milky Way Galaxy.

However, what if the Milky Way itself were a reflection nebula, shining from the reflected light of our star, the Sun, always shining "over our shoulder"? Interesting effects may result from the fact that, at different points in the Earth's orbit, the Sun shines over a different shoulder. That makes detailed comparisons of images of the Milky Way difficult. It is a very complex structure, yet no two images of the Milky Way quite coincide. The bright clouds and dark lanes may be different illuminations by the Sun, sometimes in some areas bright, but casting shadows from the nearer clouds onto the more distant clouds. We are inside a reflection nebula, looking at the debris of a planetary detonation from the midst.

The outer planets, Jupiter, Saturn, Uranus, Neptune, and tiny Pluto, all shine by reflected light. Because of their moderate and nearly constant distance from the Sun and the Earth, these planets appear to have a nearly constant brightness. This appearance can be refined and expressed as the surface brightness of the planet, magnitudes per square arcsecond. This measure gives the brightness of the planet divided by the angular area of the planet, as viewed from Earth. Jupiter is bright both because it is near the Sun, relative to the other planets, and because it is large. If Pluto were at Jupiter's orbit, it would still be much fainter than Jupiter because of its much smaller size. However, assuming the same reflectivity, or albedo, the surface brightness of the two planets would be the same. Averaging the observational data for the four giant planets, the surface brightness at 1 Astronomical Unit from the Sun and 1 Astronomical Unit from the Earth, shows that the surface brightness for an object illuminated by our Sun should be 1.9 magnitudes per square arcsecond. For objects farther from the Sun and farther from the Earth, the surface brightness decreases in a smooth, mathematically defined manner. This leads to the curve in Figure 63, with the amplitude defined by the surface brightness values observed for the major planets. (Pluto is slightly brighter than it should be. This may be an effect of a higher albedo, or bias resulting from poor resolution of its angular size.) The surface brightness of galaxies is a well established technique of analyzing and categorizing galaxies. Anton Pannekoek reported the surface brightness of the Milky Way, in detail, in the 1920s [Pannekoek 1923]. This parameter is what categorizes "low-surface-brightness" galaxies, those that are anomalously dim. Most spiral galaxies show surface brightnesses that fall into the "Freeman Law" [Freeman 1970]. While the discovery of "low-surface-brightness" galaxies, almost invisible to our telescopes [McGaugh, Bothun, and Schombert 1995, Bothun, Impey, and McGaugh 1997] has cast doubt on this as a universal law, it is likely that these two types of galaxies, the "normal" and the "low-surface-brightness" are simply two different expressions of the spiral galaxy form. This law shows a distribution about an average surface brightness of 21.65 magnitudes per square arcsecond. That value is shown on the figure as the bar running from about 85 Astronomical Units to 102 Astronomical Units. If the average spiral galaxy (of those easily seen) were a reflection nebula at that distance from the average star, like our Sun, it would have a surface brightness of about 21.65 magnitudes per square arcsecond.

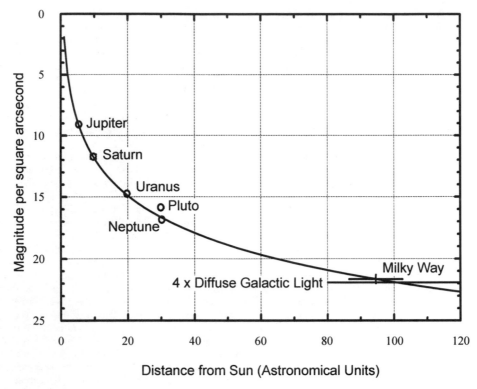

Figure 63. Reflection magnitudes of Solar System planets.
Reflection brightness of objects in the Solar System, showing the approximate location of the Milky Way, based on surface brightness estimates by Bothun [Bothun, Impey, and McGaugh 1997] and Freeman [Freeman 1970]. The equivalent brightness from the Diffuse Galactic Light is aligned with the curve, and approximates the Freeman Law [Witt 1992]. (Because larger magnitudes are dimmer, the surface brightness scale increases downward.)

At a distance of 100 Astronomical Units from the Sun, an object would have a surface brightness of only about 22 magnitudes per square arcsecond, barely brighter than the dark sky, and noticeable only if it were of a greatly extended size. This is just the surface brightness of the Milky Way.

Thus, at a distance suitable for the bulk of the Milky Way, the Freeman Law for surface brightness, the extrapolation of the planetary surface brightness curve, and a normalization of the diffuse Galactic light, all agree.

If we consider that the Milky Way is predominantly a reflection nebula of the Sun, there might be bright and dark spots, shadows and lanes. The dark lanes show well in modern photographs of the Milky Way, and give an impression of pointing away from the Sun. The clouds of debris that make up the Milky Way would be illuminated by the Sun, at the center of the Solar System, and would shine by reflected light according to the inherent brightness or albedo of the material. Illuminated by the Sun and viewed from the Earth at slightly divergent angles, some clouds might be closer than others, appear to be brighter than the background, and cast distinctive shadows.

Try to understand the Coal Sack as a shadow on the Milky Way, cast by the Sun, with the nearby bright nebula Ced 122 casting the shadow. These are a remarkable pair of nebulae in the deep southern sky, one a bright one, Ced 122 (for the Cederblad survey catalog) [Cederblad 1946], approximately 2.5 degrees in size, and the other a dark one, the Coal Sack nebula, about 6 degrees in size. Consider that the Ced 122 nebula might be a clump of fragments and dust, somewhat denser than most, somewhat closer to the Sun. It will cast a shadow (just as Charon cast a shadow on Pluto during the eclipses of 1985-1990, with a change in viewing angle of ±2 degrees over 6 months), and this shadow may show on clouds of fragments and dust that lie farther from the Sun. This shadow may be seen as the Coal Sack nebula, otherwise not any different from the material of the clouds adjacent to it, but just shadowed from the illuminating light of the Sun, and so made darker by contrast. The pair, a bright nebula and a dark nebula, are shown graphically on Chart 25 of Tirion's *"Sky Atlas 2000.0"* [Tirion 1981]. A photographic image, rather different from that in the chart, is shown on page 731 of *"Burnham's Celestial Handbook"*. Ced 122 is to the left, the Coal Sack is adjacent to the right [Burnham Jr. 1978]. A geometric comparison of the bright nebula Ced 122 with the Coal Sack as its shadow in the light of the Sun gives a distance of 20 Astronomical Units for Ced 122 and 40 Astronomical Units for the Coal Sack.

The major status of the Milky Way Galaxy in our astronomy has led to the development of a Galactic coordinate system, devised so that the Galactic Equator divides the Milky Way, as observed from Earth, into two equal halves by light and mass, and the zero of longitude passes through the line estimated for axial symmetry: equal amounts of light and mass to the left as to the right. Thus, the point on the celestial sphere, the sky, identified as 0 degrees latitude, 0 degrees longitude in this coordinate system directs our view towards the assumed center of the Milky Way Galaxy, the nucleus of an Island Universe containing the Solar System, or contrarily the remaining core of a disintegrated planet on the outskirts of the Solar System. Many active objects are seen in this region, by use of radio astronomy. One of them, designated Sgr A* ("Sagittarius A-star"), is considered to be the actual center of the Galactic nucleus, and established the most recent determination of the zero of Galactic longitude.

Since black holes have been thought to be at the center of Island Universe spiral galaxies, and Sgr A* has been thought to be at the center of our Galaxy, evidence of a black hole should be found there. However, no X-ray emission from Sgr A* has been found [Goldwurm *et al.* 1994], contrary to the expectation for a black hole. But that research showed that a nearby bipolar jet and core object emit X-radiation strongly, as would be expected for a highly radioactive fragment from a planetary detonation. (The point of mathematically infinite density in a black hole is a "singularity". The laws of physics don't work there. Around the singularity, and so having a size, is the black hole. This is the space within the event horizon, from which (almost) nothing can escape.)

The globular clusters that form a spherical halo about the Milky Way Galaxy are the radioactive remains of globs of uranium that accumulated in space, and were induced to detonate by the antineutrinos and neutrons from the detonation of Planet X. The spherical form of the distribution of the globular clusters was established by the range at which the antineutrino and neutron flux was intense enough to trigger the detonations. This effect acts in a way chillingly similar to the detonation of nuclear weapons. The antineutrino burst increases the reactivity of subcritical or just-critical globs, and then the neutron burst, moving at less than the speed of light enjoyed by the antineutrinos, pushes the globs into detonation. The radioactive fragments resulting from the detonation glow like stars, but by Cerenkov radiation rather than by intense nuclear fusion heat. The red giant stars have still operating nuclear reactors to provide internal energy, radiated as thermal light.

Thus, the Milky Way Galaxy resembles Andromeda (M31) and other spiral nebulae, as the distribution of debris from a planetary detonation, but our "sidereal system", the collection of true stars that our Sun belongs to in our neighborhood of the Universe, may more closely resemble the irregular patch of stars as William Herschel imagined many years ago [Berendzen *et al.* 1976, Paul 1993, Crowe 1994], perhaps limited only by the distance to which we can see. There are no galaxies, vast Island Universes of stars, as we have imagined and created them.

Comets and the Oort cloud

Among the most captivating (though sometimes disappointing) heavenly events in our modern skies is the appearance of a comet. Such was not always so, cultural astronomer Ed Krupp says comets were associated with disaster [Krupp 1997] (from the Greek words for a broken star). Comets have generally been considered to be warnings of disaster in many cultures, perhaps as memories of a forgotten past. As we have learned about them, and perhaps forgotten that past, we have come to see them with excitement, pleasure, wonder, awe.

Comets have comas, a fuzzy cloud around them, and cometary orbits, long ellipses. With such a simple definition, all comets are considered to be the same, primordial objects left over from the origin of the Solar System. Just as Galileo defined stars, as points of light, we define comets by their appearance. A comet must have a coma, a fuzzy cloud of gas, dust, and vapor that reflects far more sunlight than the tiny object that produced the cloud. Most comets would not be seen, or detected, if it were not for the huge size of the coma. The size of a comet is very difficult to estimate because it is hidden within the coma, but flybys of Halley's Comet showed it to be approximately 15 kilometers long, a bit smaller than the island of Manhattan in New York City. As the dust and gas are blown away from the comet by the radiation from the Sun and the Solar wind, a tail or two or more may develop. An extensive tail becomes the largest object in the Solar System (other than the remains of Planet X), spreading over a hundred million kilometers. Most comets also have extremely elongated orbits, orbits with large eccentricities, that take them from the outer bounds of the Solar System to pass near the Sun.

Visible to the naked eye, in a dark sky, sometimes appearing over the course of a year or more, a visual spectacle for weeks. A comet charges into the Inner Solar System, develops a coma, the cloud of gas, dust, and vapor that is the defining mark of a comet, and then shows a streaming tail that has become the trademark of thrilling speed. After passing the orbits of the inner planets, it whirls around the Sun, and then starts its return trip to its distant home, likely far beyond Jupiter. (Paradoxically, since the tail always streams away from the Sun, as the comet leaves, it appears to be travelling backward.) On their incoming path, comets are essentially falling towards the Sun, as all planets are, but on a much more direct path. (Some actually do fall into the Sun, never to be seen again.) Each comet temporarily becomes the fastest object in the Solar System, and then begins the long return trip to its distant home. Because its orbital speed decreases as it reaches that distant region, it spends most of its life far from the Sun, in the cold and the dark.

Several comets make their return trip on a periodic basis. The most famous is Halley's Comet, which has an orbital period of 76 years. In 1705, Edmund Halley predicted that a specific comet would return in 1759. Halley died in 1742, but the comet returned as predicted, and now bears his name. It has the longest history of repeat sightings.

A comet is seen only because of its coma, the cloud of gas, vapor, and dust that spreads out to millions of kilometers. This happens as the comet is heated by the Sun, perhaps when the comet passes within the orbit of Jupiter. Halley's Comet has an estimated albedo of 0.04, the darkest object known in the Solar System, as dark as black velvet. The black surface absorbs the solar heat very effectively and causes the ices to sublimate and outgas. The thermal velocity, and a mass velocity associated with jetting, causes the gas to expand rapidly around the comet. If the black coating were sufficiently thick and porous to act as thermal insulation, the ices would not sublimate, there would be no coma, and the comet would remain nearly invisible. Halley's Comet outgases over a small fraction of its surface, the balance either being depleted in ices or insulated by its sooty coat.

The orbital periods of comets range from 3.3 years for Comet Encke, most are less than 25 years, to more than 200 years. (Since periodic comets have been known and studied for less than 300 years, the statistics on long-period comets are not very full.) These comets are "short-period" comets. Other comets may pass through on such long orbits that we cannot tell from our observations if they will ever return. "Long-period" comets may be making their first pass through the Solar System, and have orbits that take more than 200 years. Some comets pass too close to the Sun and lose so much energy in passing through its outermost atmosphere that they become trapped and fall into the Sun itself. Others are disrupted, broken to bits by gravitational pulls or perhaps by magnetic torques. This disruption produces repeating meteor showers on Earth (and other planets) or may result in the capture and impact of the comet, as was displayed by Comet Shoemaker-Levy 9, that broke into a remarkable string-of-pearls and then sequentially blasted into Jupiter's atmosphere.

The short-period comets, with orbital periods less than 200 years, are identified by repeat apparitions, while long-period comets have calculated orbital periods of thousands of years to several million years to hundreds of millions of years [Weismann 1998], identified by analysis of their observed orbits. Orbits of the long-period comets have been analyzed by Brian Marsden and colleagues, in such a way that the original orbit can be determined, and the future orbit as well, which may indicate that it will be ejected from the Solar System or be captured into a short-period orbit [Marsden and Sekanina 1973]. That study showed a remarkable concentration of orbits with an aphelion distance near 50,000 Astronomical Units. The corresponding orbital period would be about 4 million years. These concentrated comets must have entered the Solar System for the first time, since a single orbit through the Solar System would have broadened the distribution far more than observed [Bailey, Clube, and Napier 1990]. On passing through the planetary system, and spouting jets of gas and vapor, perturbations are severe. Comet Hyakutake had a period of 18,400 years on this current pass, but the prior period is thought to have been 8,600 years before, and 17,000 years next time. The orbit of Hale-Bopp was originally 4,160 years, but next time will be 2,363 years. Comets have variable orbits because of the uncertain effects of the thermal jets produced by solar heating. The thermal jets act as small and erratic rocket engines that cause the comet to gain or lose speed in an unpredictable manner when it is near the Sun. This leads to the calculation of original and future orbits.

The Solar System is self-cleaning. Objects that get out of place are pulled by the various gravitational perturbations of the planets until the misplaced objects are either ejected into distant space or impact the Sun or one of the planets. The asteroids are small objects that have been thought to be primitive, primordial planetesimals left over from the formation of the planets. It appears more likely that they are the fragments of a fully formed planet that detonated 4.559 billion years ago. Regardless, they have survived, since they orbit in the zone of a stable planetary orbit.

Comets are also thought to be primitive, primordial planetesimals left over from the formation of the planets, perhaps even more pristine than the asteroids. However, comets travel in wild and varying orbits and are very much subject to the cleaning process. Cometary orbits vary from pass to pass, and comets will be ejected, swallowed, or dissipated in a time that is generally less than 100 million years. That is, if there were any comet-planetesimals left over at the end of the formation of the planets, they would have been subjected to the cleaning process many times over, and would have all disappeared. No comets within the planetary region could have survived for the lifetime of the Solar System. Therefore, unless we live at a special moment in time, to see comets, new comets must be created within the planetary region, or they must be brought in from outside.

If comets are the fragments of detonated moons, the source of supply is within the planetary region, among the planets. Comets will have the composition of the detonated moons; mostly ices with some differentiated core material. The surface ices will have remained cool during the detonation; the core materials will have been broken, heated, melted, or vaporized, and then recongealed.

These fragments will divide into two types according to their orbits. Most will be dispersed by the detonation in all directions and to great distances. These will gradually return at times after the detonation corresponding to their elliptical orbital periods. Fragments from a particular detonation that were ejected to lesser distances have already returned to the Solar System and been cleaned out. Fragments from the same detonation that were ejected with greater energy, to greater distances, have not yet returned. Some fragments, as the result of a combination of velocities from the satellite orbit around the planet, the planetary orbit around the Sun, and ejection by the detonation, will stay within the planetary region, in orbits that lie close to the ecliptic plane.

Tom Van Flandern has discussed how, if the long-period, new comets came from the disruption of a planet, fragments would be ejected in all directions to varying distances [Van Flandern 1993]. A study of new comets has shown that those entering the Solar System at the present time have orbits that show a mean period of 3.2 million years (±100,000 years), with a greatest distance from the Sun of 43,000 Astronomical Units [Marsden, Sekanina, and Everhart 1978]. In an independent study, Van Flandern determined that the mean period would be 5.5 million years [Van Flandern 1978]. He suggested that these comets had come from a breakup event in the asteroid belt 5.5 million years ago, "even in the absence of a suitable theory of planetary explosions." A suitable theory was developed, a nuclear fission detonation [Napier and Dodd 1973], but was rejected at the time because a mechanism for triggering the detonation could not be imagined. Jan Oort, in developing his description of what became known as the Oort cloud, considered an explosion of a planet to have created the comets. Van Flandern did, too, but placed the original planet in the Asteroid Belt, with the asteroids formed as fragments of the exploded planet.

Comets are unlikely to have come from the detonation of an outer planet such as the gas giants, because the ices on those planets are gaseous, liquid, or supercritical fluids, and would dissipate into space after disruption of the planet and the loss of its concentrated gravity. Detonation of planets produces asteroids. Detonations of icy moons, like Europa, Titan, and Triton, produce fragments of hard-frozen ice, rock, and debris. Detonations of moons produce comets. Comets may have come from a giant planet so far from the Sun that its atmosphere had frozen solid. That crust would have been shattered into fragments, and hurled in all directions.

While it has been suggested that the comets were formed by the explosion of a planet at the Asteroid Belt distance, such fragments would quickly lose their ices. For a body as small as Halley's Comet (mass of half a quadrillion kilograms), the escape velocity is less than 4 meters per second. The thermal velocity, even at 100 K, the temperature at the base of the atmosphere of Uranus, ranges about 200 meters per second. This is why comets put on such a great display as they round the Sun. Even on Pluto, ice can remain only because of the slow rate of sublimation at 38 K. On Pluto, the thermal velocity is roughly 30 times the escape velocity. On the larger planets, the ices are present as gas, liquid, or supercritical fluid and would quickly dissipate. Farther out, perhaps at 77.8 Astronomical Units, as the Titius-Bode Law calculates, all except hydrogen, helium, and neon would be frozen solid.

Calculations of the orbits of long-period comets rely on a single set of observational positions, since this is the once and only time they have been observed. That makes it difficult to calculate the observed orbit, and the original orbit, with great accuracy (compared to planetary orbits, for example). Further, choosing the bounds of the concentration of orbits is judgmental, and that choice affects the average value of the orbital period. In 1978, Van Flandern calculated a mean period, a time since detonation, of 5.5 million years. With a smaller dataset, Marsden and Sekanina gave an average distance for aphelion of 50,000 Astronomical Units, corresponding to about 4 million years. Reviewing that work, Van Flandern suggested 3.2 million years [Van Flandern 1993]. Discussing the Oort cloud in 1990, Paul Weissman suggested 8.9 million years [Weissman 1990]. I selected a set of 80 original orbits, calculated the orbital periods individually, and found an average period of 6 million years. (The grand average of these estimates is 5.5 million years.)

Here, I propose that the long-period comets that concentrate in a cluster came from the nuclear fission detonation of Planet X at 77.8 Astronomical Units, forming the Milky Way Galaxy, at about 6 million years ago. The exact time may come from information regarding evolution of life on Earth. It is interesting to note that the age of the geologic transition from the Miocene to the Pleistocene, conventionally assigned to the origin of the earliest humans, is specified at 5.2 million years. However, the numerical analysis on which that age was based shows an age of 6.0 million years as the best estimate from the data [Harland et al. 1990]. The greatest extinction of land mammals in North America occurred at that time. The mammoth split from the long elephant line at that time [Gibbons 2010b]. This could have resulted from radiation-induced genetic mutations caused by a nuclear fission detonation.

Van Woerkom proposed a diffusion process that converted a constant flux of incoming long-period comets into the short-period comets, which would then decay or be ejected from the Solar System [Van Woerkom 1948]. He found that the steady state distribution of reciprocal semi-major axes completely disagreed with the observations unless the diffusion of the long-period comets had been proceeding for only a few million years. This is consistent with the origin of the long-period comets in a planetary detonation only 6 million years ago.

Van Woerkom rejected the possibility of a planetary explosion (although this appears to have been more of a planetary eruption process [Vsekhsvyatski 1962], akin to Immanuel Velikovsky's ejection of Venus from Jupiter [Velikovsky 1950]), because he assumed that all comets had the same origin. If the comets come variously from several separate detonations of moons and planets, the objection no longer holds.

Fragments of the icy crust of a detonated planet, fractured and frothy, splattered with the vaporized rock from inside, with too little gravity to recompact, barely enough to hold together, became comets. The energy of the planetary detonation scattered these fragments in all directions, and now they return to the Solar System, as if from all directions. There is no need for a vast cloud of comets around the Sun. Occasional detonations provide the supply.

While the long-period comets appear to come from all directions of the sky, the short-period comets have generally small inclinations to the ecliptic [Bailey, Clube, and Napier 1990]. With the source of the long-period comets being a major planetary detonation that dispersed fragments in all directions, the short-period comets originated in the detonation of a large natural moon of Saturn, about 12,900 years ago, and the fragments were directed more into the planetary system.

The source of Earth's ocean water has been uncertain. Did it come from within, as the planet had been formed, or by impact, perhaps during the Late Heavy Bombardment? William Newcott wrote "A good storm of comets in a planet's formative stage could provide an ocean or two." [Newcott 1987]. Or even at some other time than its formative stage, even at several various times, we have no evidence that the ocean volume has been constant.

It has been supposed that Earth's oceans originated in their entirety during the first billion years, when we have essentially no information. This is based on the assumption that the size of the Earth has remained constant and the observation that sealevel simply fluctuates up and down. That then forms the basis for assuming that most of the extraterrestrial water came during the period of Late Heavy Bombardment, about 4.2 to 3.8 billion years ago. However, if that bombardment was by fragments from the mantle of Mercury, shocked off and ejected by a nuclear detonation about 4.25 billion years ago, essentially no water would have been brought to the Earth. Instead, Earth's oceans would be gradually supplied with water from volcanoes and from true comets. Episodic expansion of the Earth would keep the steady increase in water volume in check by intermittently enlarging the ocean basins.

Comets: beautiful portents of doom. Throughout history, until recently, the appearance of a comet in the sky has been seen as a signal of disaster. Scientific study of the nature and great distance of comets has relegated these fears to foolish folk lore. More recently, understanding the effects of a possible collision of a comet with the Earth has shown death and destruction, possibly the end of civilization, perhaps extinction. The old fears no longer seem so foolish.

New comets regularly come into the inner Solar System. These comets are brought in from the outer Solar System by the pull of the Sun and the gravitational perturbations of Jupiter, Saturn, Uranus, and Neptune. On the basis of the observed frequency and distribution of these new comets, a large, distant reservoir, the Oort cloud, was proposed [Oort 1950]. Because the long-period comets appear to come from a great distance, and from all directions in the sky, it is supposed that primordial comets exist in a great cloud, and for a great time, outside the Solar System of planets. This cloud, surrounding the Solar System beyond the farthest planets and extending halfway to our neighboring stars, seemed to satisfy the requirements for producing new comets, and has been thoroughly accepted. However, careful theoretical work has shown that the Oort cloud cannot exist [Bailey, Clube, and Napier 1990, Van Flandern 1993]. The requirements exceed its capabilities. Some other source of new comets must exist.

The Sun's sphere of gravitational influence extends out to about 2×10^{13} kilometers (133,000 Astronomical Units) where we think it is balanced by what we think are the nearest stars. The possible existence of the Oort cloud is identified by a concentration of long-period comets (greater than 200 years), which shows that a large number of cometary bodies must exist beyond 10,000 Astronomical Units [Weismann 1990].

The Oort cloud is considered to be a spherical halo surrounding the Solar System, with an inner boundary at about 1,000 Astronomical Units, roughly 30 times the distance to Neptune, and an outer bound, in the range of 10,000 - 100,000 Astronomical Units. This is assumed to be the source of long-period comets, with a different distribution of orbits than the short-period comets. The theory proposes that perturbations by distant stars disturbs the stable orbits of some of the comets, which either escape from the Sun, or drift inward and swoop around the Sun on nearly parabolic orbits. The Oort cloud is an invention, never seen, probably unobservable, to explain the details of the origin of the long-period comets. It has become part of our common knowledge, assuming its true existence [Bailey, Clube, and Napier 1990], in spite of the lack of specific evidence.

The Oort cloud is an excellent example of theory working at its outer limits. We see comets coming into the Solar System, apparently from great distances, with randomly oriented orbits. A proposed explanation for this is that the Solar System is surrounded by a cloud of comets, and perturbations send occasional comets inward. Jan Oort proposed this cloud of comets in the 1950s and it served as a successful model that continues to be used at present. None of the objects in the Oort cloud can be directly observed, and so the theory must rest on the observed distribution of a small number of comets that have been seen only once, and have very long orbits. (Oort himself had to be satisfied with only 19 well-determined orbits for the basis of his hypothesis.) The Oort cloud seems to be such a reasonable and necessary concept that it has become fully ingrained in the scientific folklore of comets. Few astronomers question its existence, few study it closely enough to have cause to doubt.

Unfortunately, the theory fails to explain how the cloud of comets formed, and is unable to explain how the cloud has survived [Bailey, Clube, and Napier 1990]. The difficulties with comets cast doubt on the standard hypothesis for the formation of the primordial Solar System. Furthermore, the theory implies that the comets come uniformly and randomly from all directions, while the observations show that the perihelia are aligned with the Galactic plane, and also with the Solar Apex, the point towards which the Sun seems to be moving, compared to neighboring stars. Or, contrariwise, the point from which the neighboring "stars" appear to be coming. That is, the debris from the detonation of Planet X, including the associated comets, is flowing past the Sun [Sharma and Khanna 1988]. The observational basis for the Oort cloud, that the long-period comets originated at locations uniformly scattered across the sky, has disappeared, but the Oort cloud remains in our science. The extreme fragility of comets and evidence that they formed at extremely low temperatures, argues against a very violent or explosive origin. However, the icy crust of a detonating moon or planet fractures without much energy, at a low temperature. The proposed detonation of Planet X took place in the core of the planet, and only mechanical energy, shockwaves and internal expansion, broke apart the surface and dispersed the pieces.

Summarizing with the words of Tom Van Flandern, "Hence we have the situation that the description of the Oort cloud and theories of its origin, both of which are intrinsically extraordinary, have to be widely accepted as necessary to explain what is observed about comets, despite numerous theoretical and observational inconsistencies." [Van Flandern 1993].

Deuterium and Hydrogen, more and again

Measurements of the ratio of deuterium (heavy hydrogen, hydrogen-2) to ordinary hydrogen have been made for most of the objects in the Solar System. We reasonably assume that the Sun has none, having rapidly consumed in fusion any deuterium that it originally had. (The average life of a deuterium nucleus in the Sun, presumably the core but probably true for anywhere, is only 6 seconds. It is gone in a flash.) Mercury probably has none near the surface, where it could be detected. The other planets, meteorites, and comets show a wide variation. A collection of observational results is shown in Figure 64. The deuterium-to-hydrogen ratio for three comets indicates that cometary water contains twice as much deuterium (309×10^{-6}) as seawater on the Earth (156×10^{-6}) does.

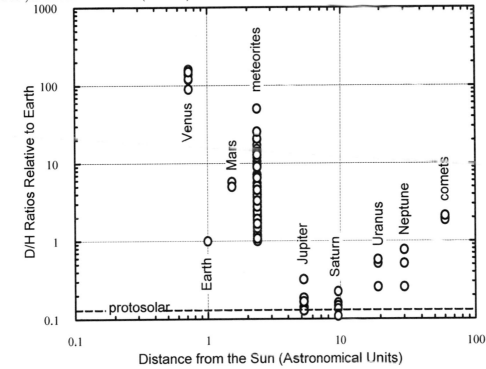

Figure 64. D/H ratios in the Solar System.
A comparison of deuterium-to-hydrogen ratios for objects in the Solar System. The value for "protosolar", shown by the dashed line, is that estimated for the nebular cloud that formed into the Solar System. The meteorites are plotted for convenience and lack of definite knowledge at the distance of the asteroid Vesta.

The estimated ratio for the protosolar nebula, the material from which the Earth (and its oceans) was made, is 20×10^{-6}. This is much less than in the comets, which are supposed to be leftovers from the formation of the Solar System. Clearly they are not, or some strong fractionation, perhaps between hydrogen gas and water ice, took place.

There do not seem to be any measurements that confidently represent unaltered mantle hydrogen, that would therefore provide information on the deep mantle D/H ratio [Kuroda *et al.* 1975, Feldstein *et al.* 1996, Kingsley *et al.* 2002]. Those measurements gave values very close to the current seawater value, so we are left with supposing what the inner Earth's water is like, to compare with the comets and our oceans.

The deuterium-to-hydrogen ratios that have been measured in the comets Halley, Yakutsk, and Hale-Bopp, are distinctly different from the other Solar System samples, but remarkably similar among themselves. Halley shows 308×10^{-6} [Balkier, Latex, and Gneiss 1995], and 316×10^{-6} [Bernhard *et al.* 1995]. Yakutsk shows 290×10^{-6} [Bockelee-Morvan *et al.* 1998]. Hale-Bopp shows 330×10^{-6} [Meier *et al.* 1998]. This strongly suggests that the water (ice) of these three comets came from much the same source, by much the same process, and further represents the neutron-scattering enhancement of bound deuterium by the loss of hydrogen-1.

The much greater ratio in cometary water than in seawater has been used to show that all the ocean water could not have been supplied by comets.

It is tempting to try to form the D/H ratio for the Earth's ocean by combining protosolar material (D/H = 20×10^{-6}), and cometary material (D/H = 311×10^{-6}) to make Earth's water (D/H = 156×10^{-6}). If deep mantle water had the D/H ratio of the protosolar material, a mixture of 47% comet water and 53% protosolar (mantle) water would match our present ocean water. However, this relies on the assumption that the comets are primordial material left over from the formation of the Solar System and stored in the yet-to-be discovered Oort cloud, beyond the reach of our research. Primitive bodies composed largely of "ices" are assumed to be stored in the far outskirts of the Solar System, in the range of 20,000 to 200,000 Astronomical Units. Occasionally, an individual comet is jostled loose from its long-term orbit by a passing star or cosmic ray, and heads into the planetary region of the Solar System. These are assumed to have the D/H ratio of the protosolar solids, as measured for three comets (Halley, Hyakutake, and Hale-Bopp) to be 311×10^{-6}. In this book, the comets, like the meteorites, come from the debris of detonated planets and satellites and do not represent primordial material any more than the meteorites do. As debris from a detonation, the comets have D/H ratios that have been severely altered by the breaking of hydrogen bonds by neutron scattering, and the subsequent loss of ordinary hydrogen (hydrogen-1).

Taking the average D/H ratio for Neptune as 80×10^{-6} and hoping that this might represent an equilibrium between solid and gaseous materials, the abundances of hydrogen and "ices" (10% and 90% respectively) leads to an estimate of the D/H ratio of the primordial solids. For a primordial gas ratio of D/H = 20×10^{-6}, the D/H ratio of the solids should be D/H = 87×10^{-6}.

With that ratio for water outgassed from the Earth's mantle, mixing with cometary water at 311×10^{-6} requires 30% comet water to produce a terrestrial ratio of D/H = 156×10^{-6}, as observed. By this interpretation, roughly one-third of Earth's surface water has come from comets such as those we have studied.

Jupiter and Saturn show ratios close to the protosolar value, lower than Uranus and Neptune. This could have resulted from the capture of the gas expelled from the inner Solar System, with a low D/H ratio, more by Jupiter and Saturn than by Uranus and Neptune.

The Local Bubble

The Sun is surrounded by a region of low-density, high-temperature gas called the Local Bubble [Abel, Morrison, and Wolfe 1991]. The gas in this region is observed by the emission of X-rays [Wynn-Williams 1992], that indicate its temperature is about 1 million degrees K. This gas has been called coronal gas, a bit misleading, since it is not thought to be related to the corona of the Sun, in the standard story. It just has a temperature similar to that of the Sun's corona. That in itself is noteworthy, suggesting a common mechanism for the production of theses gases, the Local Bubble and the Solar corona. Earlier, I described how sunspots might result from the mutual detonation (or localized conflagration) of two natural nuclear fission reactors deep in the Sun. That would produce large numbers of neutrons that would decay before being captured (in a very rarefied medium) into protons and electrons. The protons would have a random kinetic energy from the decay of 2.8 million degrees K. This could produce the million-degree gas of the Solar corona and of the Local Bubble.

The density of the gas is notably low, only 0.005 hydrogen atoms per cubic centimeter (5×10^3 hydrogen atoms in a 1-meter cube, while ordinary air contains 5×10^{25} atoms of nitrogen and oxygen). The gas is predominantly hydrogen, actually individual protons, since it is highly ionized into a plasma of protons and electrons.

In our standard view of the Universe, this region has been cleared out and the remaining gas superheated by several nearby supernova explosions. In the view of the present book, the intense neutron flux from a Solar System nuclear fission detonation (I propose the detonation of a large natural moon of Saturn at 12,900 years ago) decayed into protons and electrons (beta particles) with initial average (random) kinetic energies of 358 electron volts and 322 thousand electron volts, respectively, plus the directed kinetic energy of the neutrons released in fission. Many of the beta particles will leave the Solar System at nearly the speed of light. The random energies of the protons from neutron decay correspond to a temperature of 2.8 million degrees K, while the electrons have an initial effective temperature of 2.5 billion degrees K. The highest energy electrons, travelling at nearly the speed of light, will quickly escape from the Solar System (producing a net positive electrostatic charge on the Sun), cooling the electron component. The remaining electrons will quickly lose most of their kinetic energy, and come into equilibrium with the more massive protons.

The Solar Wind

The solar wind blows outward from the Sun
and forms a bubble of solar material
in the interstellar medium.
John D. Richardson *et al.*, *Nature* 3 July 2008

The Solar wind is the fastest material in the Solar System, with speeds up to 800 (perhaps up to 1,000) kilometers per second. It was predicted and described by Eugene Parker in 1958, but his ideas were initially rejected. Soon afterward, the Solar wind was discovered in measurements by Soviet and American spacecraft. It now forms a field of study all its own.

Measurements of its velocity and composition have been made out to 80 Astronomical Units from the Sun, as shown in Figure 65 and Figure 66. The velocity is variable with time, so we only know what we see when we see it.

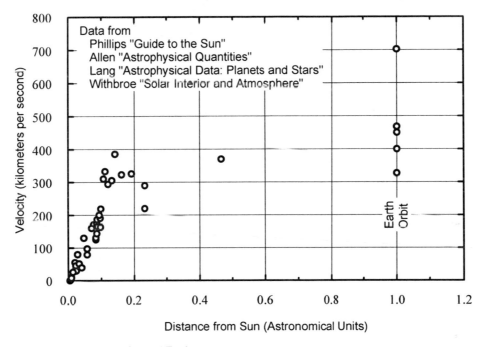

Figure 65. Solar wind between the Sun and Earth.
Solar wind velocities from the Sun to the Earth.

There appears to be no limit to the distance to which this wind can blow. If it has been blowing at 500 kilometers per second for the last 4.5 billion years, its farthest reach extends to 100,000 Astronomical Units. As it expands, the density becomes progressively less, so that the density at the Earth's orbit, about 5 protons per cubic centimeter, becomes 5×10^{-15} protons per cubic centimeter, or less than one proton in a 1,000-kilometer cube.

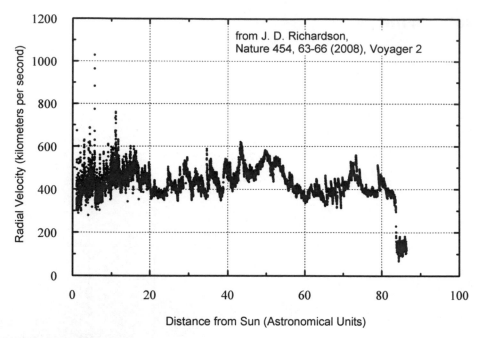

Figure 66. Solar wind radial velocity.
Solar wind velocities through the Solar System.

Summary of Part III

The development of the astronomical view of our Solar System is presented. The planets and other objects are described with emphasis on similarities and differences. The Titius-Bode Law for the orbital distances of the planets is presented.

The formation of the Solar System, with the Earth and the other inner planets formed as giant planets, is developed. This approach identifies an error in the present theory, the neglect of the effect of radioactivity in ionizing the gas of the proto-solar cloud. Uranium concentrates in solid accumulations in the cloud, and collects in the cores of the forming planets and moons, forming early natural nuclear fission reactors. The Earth was formed as a giant planet with a molten silica ocean and a massive atmosphere. Ignition of the Sun (aided by neutrons from its own natural nuclear fission reactors) blew away the atmosphere, allowing the molten silica to boil away into space. This results in an Earth with the present mass, but compressed to a greater density and a smaller diameter. The masses of the early planets are estimated and compared to present masses. Problems with why the giant extra-solar planets haven't been reduced by this process are discussed. The formation of our large Moon is shown to be a natural consequence of planetary formation, rather than from a unique impact on the Earth, and occurred for other planets as well. The current age of the Solar System is shown to have a false basis, and an alternate age of 6 billion years is proposed.

Several indications of nuclear detonations affecting the planets and satellites of our Solar System have been described. A brief, and somewhat speculative, summary is presented below.

For Mercury, Venus, and Mars, the observations show the common effects of a nuclear detonation in the loss or modification of the atmosphere, modification of the crust by mechanical, thermal, or chemical action, and destruction of most of the water. By these effects, any life or future possibility of life is eliminated, and the planet is sterilized.

Solar System Detonations

Detonated Object	Time (years)	Result
Planet "Asteria"	4.56 billion	Asteroids, setting the "clock" for dating the age of the Solar System
Planet Mercury	4.25 billion	Lost outer mantle Late Heavy Bombardment
Large moon of Jupiter	? billion	Impactors for signs of global glaciation
Large moon of Neptune	? billion	Impactors for signs of global glaciation
Planet Venus (and moons)	600 million	Radiation mutations for Ediacaran and Cambrian
Planet Mars (and moons)	180 million	Destroyed Martian oceans and glaciers
Large moon of Uranus	80 million	Four large impactors that ended the Cretaceous. Tilted and inflated Uranus, disturbed magnetic field
Planet X and moons	6 million	Milky Way Galaxy and Magellanic Clouds, radiation mutations for humans and horses
Large moon of Saturn	12,900	Rings, end of Pleistocene "Ice Age", extinction of megafauna, including large humans

Based on comparison of the disintegration of the asteroidal planet "Asteria", and the disruptive or disturbing detonations shown likely to have altered Mercury, Venus, and Mars, it appears that there are two types of detonations: those with enough energy to disintegrate and fragment the planet, and those that will only energetically destroy the surface while leaving the bulk of the planet intact. There would be a range of intensity, since the detonation energy depends on the number of critical masses that detonate, and to some extent, the degree to which the fission proceeds to superfission and ultrafission. Large moons are easier to disrupt than planets, and giant planets probably cannot be disrupted, because of the great gravitational cohesiveness.

Each of the disruptive detonations produced several large impactors, some of them hitting the Earth. Those impacts caused flood basalt eruptions, Earth expansion, mass extinctions, and initiated mantle plumes. The gamma radiation of the detonation caused mass killings and induced a wide variety of mutations, some of which were successful in the new world after the expansion.

It may be surprising to see the Magellanic Clouds as part of the Solar System, but the 30 Doradus nebula (in the Large Magellanic Cloud) is remarkably large, exceptionally large ("a truly huge HII region" resulting from protons from neutron decay), a possible consequence of our saying that it is relatively farther away than it really is, even in the standard distance scale, and that the magnitudes of the Cepheids in the Magellanic Clouds had the wrong magnitudes for use in the distance scale to other, distant galaxies. There is a peculiarity about the apparent brightness of distant objects illuminated by our Sun. The sunlight shining on the object dims by a factor of $(1/r^2)$ and then the reflected sunlight that we see is dimmed by another factor of $(1/r^2)$ as it comes to Earth, so that the brightness declines as $(1/r^4)$. This makes objects seem to be much more distant than they are (even relatively) because of their dimness, by comparison with similar objects in other galaxies, which are directly illuminated by their own starsuns.

Concepts of the Sun, as developed through our ages, are presented. The long time-span for the release of energy from the center of the Sun to its surface shows that we cannot reliably judge the nature of the Sun by surface observations. Because of its inherent density, uranium will settle to the center of the Sun. Hydrogen-moderated uranium could form individual reactors with as little as a few grams of uranium. These could act as thermal breeder reactors, converting uranium-238 to plutonium-239, which decays to uranium-235. Interaction of reactors deep in the Sun could be the source of sunspots. Protons released by the decay of neutrons have an effective temperature of nearly 3 million degrees kelvin, and could produce the unexplained high temperature of the Solar corona. Significantly fewer neutrinos are detected from the Sun than we expect from its assumed fusion power. Some of the complications are discussed. The role of fission neutrons in bypassing the slowest step in fusion, proton-proton fusion, is presented.

The remarkable division of our Solar System into the rocky inner planets and the gassy outer planets is described. Mercury may have lost half its mass by a nuclear fission detonation that shocked off the outer shell, producing the many large fragments that caused the Late Heavy Bombardment seen as impact craters on the surface of Mars and our Moon. Venus may have been overheated about 500 million years ago by a nuclear fission detonation. The amount of uranium fissioned would be a trivial fraction of the mass of the planet. Its atmosphere is nearly 100 times as massive as the Earth's but it has extremely little water. Pancake domes are seen on the surface of Venus, resembling the rubble mounds produced by underground nuclear weapon detonations on Earth. The mass of concentrated uranium needed for a pancake dome is similar to that contained in one of the ore bodies that formed the natural reactors in Gabon, Africa. The ratio of deuterium to hydrogen is different on Venus, compared to the Earth. Scattering of energetic neutrons from fission preferentially breaks hydrogen bonds in water relative to deuterium bonds, the free hydrogen is lost to space. Neutron absorption by hydrogen also produces additional deuterium.

Formation of our Moon as a natural effect of planet formation, as appears to be the case for the other large moons, avoids the need to invent a special collision of a planet with the early Earth.

Some of the features of Mars are described. Mars has pancake domes much like Venus, that were probably also formed by detonation of underground ore bodies. Mars also has very little water. Its ocean could have been destroyed by the radiation of an internal nuclear detonation. Mars has an ocean basin, with two distinct shorelines, formed before the loss of the water by the detonation, and resulting from uneven expansion of the planet in response to impacts. The major impacts produced major igneous provinces. The failure of a carbon dioxide greenhouse to keep Mars warm is discussed. Mars also has a higher ratio of deuterium to hydrogen than the Earth.

The Asteroid Belt was originally thought to be the fragments of a planet that was blown to pieces. Their appearance is starlike, hence the name. As a collection, they have orbits with randomly distributed parameters. The first asteroid was discovered at the distance from the Sun where the Titius-Bode Law predicted a planet should be, but the asteroids are too small to make up a planet. There are millions of asteroids in the Asteroid Belt. The disruption of a planet to produce the asteroids had been reviewed and rejected in several ways. However, a nuclear fission detonation is capable and possible. This detonation disturbed the isotopic systems that are measured in the meteorites and are used to determine the age of the Solar System, and so that age is incorrect. The similarity of the lightcurve of a rotating asteroid and a presumed pulsating red giant star shows that some of the millions of asteroids might be impersonating some of the millions of stars.

Our best samples of the Solar System come to us as meteorites. These show that the asteroids, as the major source of meteorites, were differentiated as if part of a planet, and had liquid water, requiring sufficient gravity and atmospheric pressure, and greater warmth than the Sun could provide. That supports the proposal that the asteroids are fragments of a detonated planet. The formation of various unique features of meteorites by a nuclear fission detonation is described. In meteorites, the deuterium-to-hydrogen ratio is very varied, and generally greater than on Earth. This is the result of neutron scattering and absorption, releasing hydrogen and producing deuterium. The distribution of helium-3 in two meteorites shows the effect of the forced detonation of small grains of uranium embedded in the iron matrix. Disturbed isotopic ratios are found in the meteorites. Some of these effects were investigated to determine the neutron fluence that might have produced them. The fluences that were determined are very similar to that near the Hiroshima atomic bomb explosion. Analysis of pairs of nuclides in the meteorites resulted in inconsistent ages for the supernova that produced the proto-solar cloud. The concentrations of uranium and uranium-235 shows the effect of a nuclear fission detonation, converting uranium-238 to plutonium-239, which decayed to uranium-235.

Jupiter, Saturn, Uranus, and Neptune are explored. Three of these show excess heat; Uranus does not. The sizes and masses differ and can be explained by the effects discussed in this book. Uranium hydride reactors in Jupiter, for the production of its excess heat, would each require less than 10 kilograms of uranium. Jupiter has four large moons, the size of our own Moon, and many fragments, in widely dispersed orbits. Jupiter's moon Io has excess heat to power its volcanoes. Io is likely to still have functioning natural nuclear fission reactors, and so poses a possible threat of detonation in the future. It has an infrared spectrum that cannot be distinguished from that of a standard star. Europa may be a possibility also, since it seems to be warmer than it should be, and may have a liquid ocean. Application of the Titius-Bode principle of orbital regularity shows that at least four moons have been lost by detonations.

It is proposed that Saturn's rings are the result of a disruptive detonation of a large icy moon. Reactors still produce heat within Saturn's moons Enceladus and Titan. This would require small concentrations of uranium to fuel the reactors that power the geysers.

Uranus, unlike its other brethren, has no excess heat and is otherwise a puzzle. It has a low mass and an inflated radius. It is severely tilted from its orbit. As for Jupiter, application of the Titius-Bode principle of orbital regularity shows that at least four moons have been lost by detonations.

Neptune's orbital distance does not agree with the Titius-Bode Law, and this might show that it drifted inward after the detonation of Planet X. Plumes of gas on Neptune's moon Triton shows that Triton may still be active, powered by natural nuclear fission reactors.

The calculational search by Percival Lowell and the observational discovery by Clyde Tombaugh of a planet beyond Neptune are described. Pluto is too small to be the planet that produced the perturbations of Uranus and Neptune that led to its discovery. It is likely that Pluto is a fragment belonging to the Kuiper Belt, along with Triton, Neptune's captured moon. There may be thousands of such large objects in the Kuiper Belt and beyond. A discrepancy in measurements of the diameter and albedo of the Kuiper Belt Object Eris can be resolved by including heat produced by natural nuclear fission reactors, requiring an extremely small bulk concentration of uranium. The Titius-Bode Law is used to estimate orbital distances for a Solar System of 10 planets, including Planet X as the outermost planet at 77.2 Astronomical Units. Planet X was not found by Tombaugh or by any subsequent searches. Recent orbital calculations have reduced the remaining perturbations in the orbits of Uranus and Neptune, but the conclusions have been disputed.

Some kuiperoids may still have functioning reactors. Any of these may detonate, with serious consequences for life on Earth. Our world is not yet safe.

A published analysis of the delays of comet returns showed the existence of an unknown object beyond Neptune. The mass and orbit were determined, but the mass was too great (300 Earths) so that the analysis was rejected. However, the mass was based on an incorrect mass of Pluto and when the present accurate mass of Pluto is used a reasonable value is produced, about one-half Earth. An analysis of the apparent changes in the location of the center of the Milky Way Galaxy produced essentially the same orbital elements as the cometary analysis. This is what remains of Planet X. Planet X no longer exists as a planet. It was destroyed in a detonation 6 million years ago, and produced what we now call the Milky Way. Our Milky Way Galaxy is a reflection nebula composed of the fragments of the detonated Planet X, making the Kuiper Belt and objects beyond. The spiral galaxies in the distant Universe are similar to the Milky Way and are the products of other planetary detonations. The surface brightness of the Milky Way is compared at its distance with the surface brightnesses of the outer planets, shining by reflected sunlight.

Comets may be periodic, returning to the inner Solar System, or new, appearing for the first time and having long calculated orbital periods. Analysis of the new comet orbits shows an origination time of 6 million years ago. That corresponds to a time of geologic change and biological extinction on Earth. The theory of the Oort cloud of comets has problems, and it is more likely that the long-period comets are fragments of the detonated Planet X.

Analysis of variations of the deuterium-to-hydrogen ratios throughout the Solar System gives an estimate that about one-third of the Earth's surface water has come from comets.

The Local Bubble, a low-density high-temperature region surrounding the Solar System at a great distance, may result from the decay of fission neutrons producing protons at a temperature of nearly 3 million degrees kelvin. High energy electrons from these decays will leave our neighborhood and produce a positive electrostatic charge on the Sun. That charge will further accelerate the positive ions of the Solar wind.

IV. LIFE (and death)

DEATH!? Seem to me that makes life a perty risky business.
Walt Kelly, *"Potluck Pogo"*, 1954

Life is filled with uncertainties. We don't know how it started, we aren't sure how it changes, we don't know the limits of the conditions it requires, and we don't know if it is everywhere or nowhere but here. At the extreme, it is difficult to define the critical character of life. Until the 1700s it was not recognized that different life forms had existed in the past that no longer exist, and never will again. We became aware of the grim finality of extinction. Origination, evolution, development, and extinction frame the major questions of life.

Life is difficult to define, but (in most cases) easy to recognize. It is what separates the living world of biology from all the other worlds of geology, astronomy, and physics. It is why we are here to ponder over why we are here. It appears to have originated on Earth almost as soon as it could, soon after the surface of the Earth cooled. Life has produced a profusion of different forms, cataloged into many types.

The simplest living organisms reproduce by simply dividing into two new individuals. One cell splits into two. Ideally, the daughter cells are identical copies of the single parent. More complicated organisms reproduce by using sex to combine genetic information from two parents to produce a new, and different, individual. In both single-celled division and in sexual reproduction, the biochemical instructions are contained in the chromosomes, composed of DNA. The DNA is composed of amino acids, known as nucleotides, and the genetic information is encoded in genes, which are composed of sequences of just four nucleotides. These nucleotides differ only by the type of molecule attached as a base, adenine (A), cytosine (C), guanine (G), or thymine (T). Since two of these bases always pair only with each other (A-T and C-G), the genes are coded in a pseudo-binary language (A-T or C-G) just as computers are programmed in binary bits (0 or 1).

While single-celled division excels at identical reproduction, sexual reproduction produces variety. At several stages in the development of the chromosomes, there are opportunities for exchanges and transformations that will make the genetic information given to the offspring slightly different from the anticipated blend of its parents' chromosomes. This gives a population genetic variation. Humans are all clearly human, but just as clearly are distinctly different individuals. It is this innate variability that aids in the survival of a species as environmental conditions change. For example, if conditions were to change so that production of successful grandchildren were favored by being tall, or having dark skin, the typical form of the species would drift towards greater height or darker skin. This is the Darwinian sense of evolution, by gradual emphasis of traits that are most effective in promoting survival, or reproductive success. Darwin needed evolution to be continuous and gradual, as was observed in the animal breeding he was familiar with, and his famous finches of the Galapagos Islands, but he rejected the fact that the fossil record says that is not the way it is. Without the existence of the genes for these traits, the simple creation of a new niche that requires height or dark skin will not bring about these traits. That theoretical view originated as the inheritance of acquired characteristics, by Jean Baptiste Lamarck, in 1809 [Postlethwait and Hopson 1989]. It became institutionalized in Soviet science under Trofim Lysenko, but has been generally rejected on the basis of many observations, but the idea still permeates evolutionary interpretations. It had been supposed that the long neck of the giraffe had developed because early giraffes stretched up to eat the leaves from taller trees. Eventually the stretched necks became hereditary. A vast amount of experimental and observational evidence refutes this. The availability of lusher leaves at the tops of trees favored giraffes with long necks, but the necessary genes had to pre-exist. Survival then selected in favor of giraffes with long necks, and with the long-necked genes. Darwinian evolution proposes that short giraffes gradually changed into tall giraffes, for better success in life.

This concept of adaptation to a newly opened niche, by the development of favorable characteristics like the giraffe's neck, survives today. But the appropriate genetic coding must be available, or the species will not succeed in the new niche. On the other hand, if changes occur in the genetic coding that produce characteristics that lead to favorable survival in new conditions, a new, successful species will abruptly appear. This is the punctuated equilibrium model of evolution. Evolution progresses in jumps.

Mutations, changes in the genetic coding, can and do occur frequently, by a variety of mechanisms. Often, these mutations are lethal, and in humans contribute to the large fraction of conceptions that are terminated early, sometimes so early that the prospective mother has no knowledge of the pregnancy. Most of the surviving mutations are quite neutral for survival and generally occur without notice. Some genetic changes result in tragic disabilities, such as Down's Syndrome.

Many of these changes happen naturally in the course of the reproduction process. Some result from the action of chemicals, diseases, or radiation. Major changes in the set of chromosomes can occur through the process of karyotypic fission [Todd 1992] or Robertsonian fusion [Redi, Garagna, and Capanna 1990]. Fission results in breakage at the centromeres of a chromosome, the weakest structural point where the two parts of a chromosome are linked. In some cases, the fragments remain separated and produce a greater number of chromosomes than were present in the parent. Otherwise, some fragments may rejoin in a novel arrangement of chromosomes. Fusion results from the joining of (usually) short-armed chromosomes, and produces a smaller number of chromosomes. Two of the chromosome pairs of the apes look like the fragments of our number 2 chromosome, broken in two. Neil Todd has described a theory of karyotypic fission [Todd 1992]. (Not nuclear fission, but a breaking of chromosomes.) Todd claims that is the cause of the major mutations that lead abruptly to speciation. Did we and the apes come from a common ancestor by karyotypic fission, and are they the result of greater chromosomal fission or less recombination, or did the apes come from us, as a result of fissioning of the chromosomes of our ancient ancestors?

The origin of man, and the difference between man and the other great apes may have resulted from the action of Neil Todd's karyotypic fission theory. At a crucial stage in development of the gamete (or the stem cell), the chromosomes may be fractured at the centromere. Surviving recombinations of these fragments make new (abruptly different) species. *Voilà!* the exclamation point in the punctuated equilibrium evolution of Stephen Jay Gould. We (humans) have 46 chromosomes (23 pairs), they (the other great apes) have 48 (24 pairs). We have a very long number 2 chromosome. The others don't have a corresponding chromosome; instead, they have two short chromosomes corresponding to the p and q arms of our chromosome number 2.

An alternative process, which is known to be induced by exposure to ionizing radiation [Schull 1995] is a failure for the chromosomes to properly separate, nondisjunction. This results in germ cells that have too few or too many chromosomes, and in the case of Chromosome 21 is the cause of Down's Syndrome. Nondisjunction results in an odd number of a particular chromosome in offspring, because the other parent is usually normal. In a catastrophic world of shattered and radiation-damaged chromosomes, both parents might occasionally match in terms of genetic disorders. This could produce offspring with the greater number of pairs of chromosomes. The "second" pair of a particular chromosome might be accommodated by suppression of its activity, as occurs for the two X chromosomes in females, and eventually by modification of its purpose. This would produce an abruptly different organism, which might or might not be well suited for survival.

In the view of this book, it is proposed that radiation from a nearby astronomical detonation induced the mutations that produced the first muticellular organisms, and changed living conditions to favor their survival. I think that brief but intense radiation is the mechanism behind karyotypic fission, producing an abundance of "hopeful monster" forms [Goldschmidt 1940]. Living conditions are abruptly changed as a result of some of the consequences of the event, and these new conditions select, in an abrupt Darwinian sense, those forms that will survive long enough for us to find evidence of their existence.

On Earth, conditions were right for life to form at least as early as 3.5 billion years ago, and their fossil remains are found as stromatolites [Postlethwait and Hopson 1989]. But other life may have existed before, deep in the Earth, thriving at high pressures and high temperatures. With the development of chlorophyll, plants used sunlight to change a carbon dioxide atmosphere into one that was rich in oxygen, and set the scene for animals.

Animals use oxygen to provide quick energy to power their activity and mobility. They eat plants, and each other, and return their waste products to the environment. One of those waste products is carbon dioxide, which the plants use. On Earth, life has established an interlocking cycle of consumption, production, and consumption, balanced by the needs of plants and animals. (The subdivisions of life on Earth are somewhat more complicated than simply plant or animal, and now break into at least five kingdoms.)

Much of the history of life has been hidden from us. Single-celled organisms leave traces of their existence and activity only by subtle chemical and isotopic changes. It is not until after 4 billion years of Earth's existence, perhaps 3 billion years or more of the presence of life, do we begin to see easily recognized traces of living organisms. These traces are indeed traces, the fossilized evidence that an animal passed by, leaving a crawling or burrowing track. Then, in rocks we estimate are 600 million years old, we begin to find the first faint impressions of complicated organisms. These are the Ediacaran fauna, named after the Ediacara Hills in Australia, where the earliest set of fossils was found. Complex life appeared on the bottom of a shallow sea.

With the faintest impressions of these organisms in the ancient rock, these fossils don't clearly tell us if they were plant or animal, or something else, flat to begin with or flattened by the rock pressure, upright on the seafloor or hugging the bottom. Much still needs to be learned before we can understand these early developments in life. Much of the difficulty in studying the Ediacaran organisms is that they had no hard parts; they were all softies and left little in the fossil rock other than faint impressions of their body forms.

While this provides a rather shaky start for so serious a topic as life, we will leave the details to the experts and only explore the major themes.

Abruptly, at about 542 million years ago, sturdy animals appeared, leaving numerous different fossils showing many New, Improved, Better Than Ever Before! animals [Gould, 1989, Conway Morris 1998, Parker 2003]. Superficially, the differences between the old, Ediacaran model of life and the new Cambrian animals are like those between a sea anemone, all soft, and a lobster, with a hard shell, or between kelp and clams. Many of these new animals had hard shells or other protective and durable parts that were preserved well. Various different animal forms were recorded, and faded away a few million years later, leaving enough evidence for scientists to continue arguments endlessly.

Abrupt appearances and abrupt disappearances have marked the major structure of the development of life on Earth. Often, the evidence is sparse, unclear, ambiguous, and our understanding has been slow to develop, and filled with errors. The development of life has led to our current world, a world that is clearly temporary, described as the blink of an eye in many efforts to explain geological time. If we are to survive into the night, beyond the next eye-blink, we must learn to understand what has happened to life, and how.

The Tree of Life, on Earth, seems to consist of three trunks, the eubacteria, archaebacteria, and eukaryotes. These trunks start so far back in the beginning that we do not really know if there is only one root. Did only one form of life originate on Earth, or two or three, somewhat independently, or perhaps from common genetic precursors?

Life forming very early in Earth's history could have developed on a surface that was at the bottom of such a thick atmosphere that no sunlight penetrated, at pressures that are now found only kilometers deep in the Earth, with no source of energy other than chemical energy from the minerals and thermal energy from volcanoes, such as the hydrothermal vents of the deep sea. As the Earth expanded and the atmosphere was depleted, these earliest microbes would have found satisfactory (and customary) living conditions only deeper in the Earth and would have effectively migrated into the rocky depths and ocean deeps, where we now find them and their descendants. Mutations might have been produced that were able to withstand the reduced pressure, eventually even sunlight, and in time developed into the surface-dwelling life (including ourselves) that we are familiar with.

We know of five catastrophic mass extinctions in the last 500 million years (plus the end of the Ediacaran, and also the end of the Pleistocene). So that's sort of one event per 100 million years. At that rate, there may have been 40 catastrophic, life-threatening (and life-changing) events in the prior development of Earth, about which we know about nothing. If any one of the five known events had happened differently, or had not happened, we quite probably would not be here. What if any of the 40 earlier events had not happened, or had no Earthly effect, or had happened more intensely?

Based on this record, even after having been formed and survived this long, animal life (and advanced plant life as well: for example, the same radiation exposure that will kill a human will kill a pine tree) faces a continuing gauntlet for survival.

Evolution by Punctuated! Equilibrium ...

The cause is hidden
but the result is well known.
Ovid, *Metamorphoses IV*, 287

Life has evolved on Earth from forms that are barely detectable in the fossil record, through jumps in complexity of unknown cause, followed by gradual refinement resulting from differential success [Gould and Eldredge 1993]. This concept may be very easily demonstrated in automobile design, as an example. New features are introduced abruptly and soon exclude the previous design. The earlier form vanishes and the new form is then slowly improved. The first autos were started manually by cranking the engine by hand. Then the electric starter was introduced and soon all new cars had electric starters and those with starter cranks disappeared. The electric starter "evolved" thereafter from a foot pedal to a dashboard-mounted pushbutton to an integral part of the ignition-security lock, and now to a smart push button. And now there are almost no cars to be found with cranks.

The difficulty with Darwin's theory of evolution is not a basic flaw in the origin of species, but the extension to the origin of new forms. As Scott Gilbert (Swarthmore College) describes it: "The modern synthesis [of Darwin's theory] is remarkably good at modeling the survival of the fittest, but not good at modeling the arrival of the fittest." [Whitfield 2009].

A major developmental change can result from the change of a single gene, and that can come from the change of a single base pair in the code of the gene.

Darwinian evolution, gradually changing one species into another, doesn't work. That's part of the problem in the conflict between Evolutionists and Creationists. One side defends a theory that doesn't explain the observations (Darwin based his description of the process on selective breeding of pigeons, and couldn't accommodate extinctions or speciation events), and the other side attacks a theory that doesn't explain the observations. The involvement of natural nuclear fission reactors in the destructiveness of extinctions, and changing the world, is balanced by the creativeness of speciation, by changing the world.

At about 500-600 million years ago, the surface of Venus was covered by the greatest eruptions of molten lava that we know of [Cattermole 1994]. This catastrophic death of the surface of a nearby planet would have occurred at just about the time of the revolutionary "Cambrian explosion" of animal life-forms on Earth [Levingdon 1992, Bowring *et al.* 1993], between 533 and 525 million years ago [Kerr 1993]. On Earth, a volcanic eruption at the start of the Cambrian Period has been dated to 543.9 ± 0.2 million years ago [Bowring *et al.* 1994]. Here, I propose that the overheating and resurfacing of

Venus resulted from an internal nuclear fission detonation, and that radiation associated with that detonation induced the great mutations that produced the great changes in the forms of life on Earth.

We are right to puzzle over mass extinctions, and the fact that what we see in "evolution" is somewhat different from what we think Darwin described. Darwinian evolution was seen as a process like selective breeding. That's what Darwin thought he saw, and he ignored the fossil record, or rather said the fossil record we had recovered was inadequate to show his way of evolution. Abrupt changes worried Darwin, and he did not manage to incorporate that effect, blaming instead the faulty fossil record, filled with gaps. We normally view extinction as a dying out of a species, with the disappearance of the last passenger pigeon, the last Neandertal, the last panda. However, if all the next generation of offspring are distinctly different from the parents, the parent species has effectively gone extinct. Some scientists prefer to call this pseudo-extinction, but I think that obscures the impact of such an extinction-speciation event. A species may go extinct because of extermination, or by mutations, or a combination of these two. Or, as we will see for the Neandertals, they may quite gently transform. The active mechanisms for these two modes of termination are likely to be different. Radiation may kill or mutate, and changes in the world may terminate and select. There is a two-sided story here: we must understand the origination of new life forms and the termination of existing life forms. We must learn of our own origin and of our own possible end.

Darwin's evolution, as it gradually gained acceptance, came to dominate the world of Science, even beyond biology. Nearly all we understand about evolution is based on Darwin's concept of gradual modification of species. This has continued even though we now recognize the mass extinctions, half-a-dozen or so, and major extinctions, numerous, that showed abrupt endings for many (in some cases most) of the life forms on Earth. These extinction events were often followed by abrupt speciation events, when new forms of life originated without gradually related ancestors, in a most un-Darwinian way. Darwin did not believe in extinctions of that sort, instead thinking that the abrupt disappearance of a line of life was simply the result of poor preservation and discovery in the fossil record. Absence of evidence was not evidence of absence, strongly for Charles Darwin. We now see those events as the most emphatic, dramatic, punctuation in the book of Life, changing an old world for a new world, filled with new creatures, from which many of the old forms had vanished.

Charles Darwin believed in speciation, the slow and gradual modification of form that eventually became sufficiently different from the ancestral form that it could, should, be called a new species. This was the foundation and theme of Darwin's Theory of Evolution, based on his observations of species in the wild, and the success of breeders of domestic animals and birds. But Darwin did not believe in extinctions, or in speciation events. He blamed the abrupt disappearances of species, genera, and orders, as simply flaws in the fossil record, where the history had not been completely preserved or discovered. Ancestral forms must gradually die out as they were replaced by the "more fit" descendent species. Now we see extinction events, and abrupt speciation events, not what Darwin imagined, as the most distinctive features of the fossil record. In an interesting inter-relation between English geology and Darwinian Theory, the present practice of labeling geological strata, era, period, epoch, stage, requires the identification of the start of the new stage by the first appearance of an identified new species. The old stage does not end with an extinction, as seems clear at present, but simply ceases to exist at some undesignated point before the start of the new stage. The new stage then represents a new life, a new world, too abrupt even so for Darwin.

In spite of the conflict between Creationists and Evolutionists, it is unlikely that there is anything mysterious or deeply significant in the gaps found in the fossil record, the "missing links". These can be excused as annoying accidents of preservation, or lack of preservation, that interfere with our knowledge and understanding of the details of the development of life.

In fact, these gaps are essential consequences of evolution by punctuated equilibrium. Evolution jumps into the next form abruptly rather than sliding slowly. The absence of transitional forms is much like a stairway. At every point along a step, one stands on a flat and level surface. Our view of past life becomes like looking down a stairway, and we see only the tops of the treads, the equilibrium, and we do not see the abrupt changes, the risers, the punctuation. And yet, by taking these abrupt steps, one eventually reaches the next floor far more easily than if climbing a smoothly continuous ramp, as Darwin supposed.

In looking for transitional traits, a partial additional neckbone, or a partial clavicle-to-become-a-wishbone, or a partial turtle shell, we are looking for what did not exist. With most lamps and light switches, we cannot see a stage between on and off, no matter how hard we look. The change has occurred in the blink of an eye. Mutations induced by radiation from a nearby detonation are intensely concentrated in time, may affect only part of a hemisphere of the Earth, and will produce drastic, not gradual, changes. A changed world then selects the winning models from this lottery of hopeful monsters.

Is this not Darwinian evolution, just because Darwin did not know of mutations produced by radiation from nuclear detonations, and so evolution is false? Or is it truly Darwinian evolution, with simply a different mechanism and cadence than he had imagined? Darwin's basic precept is correct, that life changes and that what we see now is derived from the ancient past.

Just as histories are written by the victors, the losers in evolutionary struggles for survival are not heard from nor seen. This makes evolution appear to be incredibly effective and efficient. The process seems to always produce winners, because we only see the fossils of the winners, we do not see the multitude of losers. Random genetic mutation occurs far more drastically than the slow and occasional manner imagined by Darwin, and living conditions change as drastically, and concurrently with the mutations. What works, survives, and we later find evidence of its existence. Otherwise, there are not even hints.

Consider the many ways in which an organism may fail to survive, and therefore be judged, by Nature not us, to be "unfit":

> failure of conception
> failure to survive conception
> failure to survive whatever form of gestation is required
> failure to survive birth, hatching, or sprouting
> failure to survive infancy
> failure to survive as a juvenile
> failure to survive adolescence
> failure to survive as an adult, prior to mating
> failure to mate successfully
> failure to conceive

and the cycle resumes with the next generation, with failure to survive conception. Only with success in survival at every step will this form of life have an opportunity to be recorded in the fossil record.

In between catastrophes, Uniformitarianism works quite well. In a larger view, catastrophes are part of uniformity.

Genetic mutation by radiation exposure followed by a much more radioactive environment means the rules of Life have changed. This is not just business as usual. An event occurs that kills individuals, terminates species, and creates new species. The environment itself is changed in a way of which we have little awareness. I think that change was reflected by the initial survival of the new upright dinosaur form and the protective shell of the turtle. I think similar effects are apparent in the other mass extinction events, and in other more subtle occurrences, such as the end of all existing horse species about 4 million years ago, and the creation and survival of the present horse, capable of sleeping while standing up. Those new species that have characteristics that provide survival capability in the new, temporary conditions, survive to slowly change ("improve") as the radioactive environment loses its intensity.

When mass extinction hits, I think it kills both by the "luck of the draw" and by targeting specific characteristics that had evolved during the time before the extinction event. Those organisms that were especially successful in the old world might be drastic failures in the new. Near-shore marine life dies from a major marine regression, regardless of what other characteristics the organisms might have developed. Aboveground animals directly exposed to the radiation of the detonation die, mutate, or survive intact depending on the distance and energetics of the detonation, not by any survival feature they had evolved. From that view causality is supreme. We just can't see what the effects were that produced the causes.

Extinction of a species may occur as effectively by transformation of all survivors into a new form, as it would be by the death of all existing members. The abrupt replacement of an ancestral form by its descendants is difficult to observe in the sparse fossil record that is often the only framework for study. The fossil record preserves the successful survivors, the organisms that had the right answer to the changed conditions of their new environment. Here, in our explorations, this process of replacement and change will be described as a killing event that simultaneously induces major mutations and changes the conditions in which life must survive.

The use of ionizing radiation to induce all types of mutations in laboratory animals, principally fruit flies, has been well developed [Edey and Johanson 1989]. Mutations at the extreme end of the scale are designated macromutations, producing "hopeful monsters" whose survival depends upon the environmental conditions of the time. The changed conditions select forms from the new mutations that are successful, and because these conditions are different from those before the event, the surviving forms are different from those that prospered before. Those species that survived, not only through the temporarily intense radioactivity but on into a more "normal" world, were those whose forms of radiation protection were inexpensive and provided survival value later.

Ancestors vanish and unsuccessful forms fade away. The radiation event and the temporary associated radioactive environment create a unique instigation of new forms and the selection and preservation of successful survivors. While this does not make evolution by punctuated equilibrium predictable, it does make it understandable. The path of life is shaped by causes, rather than by random chance. The causes, however, may occur by random chance.

In suggesting that a radiation burst produces sufficient surviving mutations as to produce a new viable species, I must speculate about this difference from what was observed after the atomic bombings of Hiroshima and Nagasaki. No notable mutations were observed in the offspring of survivors who had been exposed to varying levels of external radiation. I think that in those cases, the genetics of the ongoing population was dominated by the unexposed and the least exposed. Modified genetic material was diluted by unaffected genetic material to the extent that both deleterious and beneficial mutations were so rare as to be undetectable. In the case of a nearby astronomical detonation, all or nearly all of a population is severely affected, and all offspring carry mutated genetic material. Few survive; fewer still live successfully in a newly changed world. Their offspring define the new replacement species.

A mass extinction is a major shock to the system of life on Earth, and produces aftershocks in the form of scientific controversies and arguments. Specialists in one field disagree with specialists in other fields. The favorite animals of some may show no losses, so the extinction didn't occur. Other animals may have been terminated abruptly, or gradually, or at different times. Different mechanisms may be imagined and argued as if isolated effects and isolated causes could have resulted in such a disaster. In the face of confusing uncertainty, absolute explanations are offered, and lead to vitriolic professional fights. Conventional science has not managed to explain mass extinctions. It is time for an alternative view.

Let's play a game of "What If": What If the Earth in the past had had a much more massive atmosphere, like that of Venus at the present time, and What If the Earth had been smaller then, as the Earth Expansion theory proposes. What If Earth's expansion is triggered by major impacts, such as the impact at Chicxulub that is supposed to have killed the dinosaurs? How different would the world of the dinosaurs have been if the world were that different? How different would the world have been after the impacts at the end of the Cretaceous, and is that what killed the dinosaurs?

There have been proposals that neutrinos did in the dinosaurs [Collar 1996], thermal-neutron induced radioactivity killed most everything [Yayano 1983], and there is some speculation that the nitrate concentrations found in Antarctic ice cores, at levels that correspond to recent (historical) supernovae, were produced by X- and gamma-rays from the supernovae [Rood 1979]. Hints of radiation have been seen, but not yet integrated into our understanding.

The Cambrian Explosion

There'll be a change in the weather,
and a change in the sea,
so from now on there'll be a change in me.
Billy Higgins, *"There'll Be Some Changes Made"*, 1921

The fossil record shows that occasionally, major disruptions have occurred to life on Earth. In one of the most drastic disruptions, in which 96% or more of the existing species were eliminated (and replaced) [Gould 1989], the Cambrian Period opened with the development of multicelled animals with hard bodies, and those animals developed in amazing ways. It may be that the great disruption in the routine forms of life known as the "Cambrian explosion," [Levinton 1992] which occurred at about 533 million years ago [Kerr 1993, Bowring *et al.* 1993], with the amazingly odd creatures preserved in the Burgess shale [Gould 1989, Briggs, Erwin, and Collier 1994, Conway Morris 1998, Parker 2003], was the result of a nuclear fission detonation in the planet Venus, which apparently caused the resurfacing of that planet at about that time. An excess of neutrons would have been produced in the Earth's core by the antineutrino-induced positron emission from hydrogen nuclei or induced electron capture, creating free neutrons, which would have accelerated the nuclear-fission chain-reactions in Earth's reactors, and boosted volcanic activity. The increase in fission power would have been brief enough that there would have been little or no long-term climate effects, but intense enough that the ionizing radiation would have had a major effect on the living organisms, causing extinctions and speciation by radiation-induced mutations. Increased radioactivity on the sea bed surface could have continued for an extended time as volcanoes transported radioactive material from the core-mantle boundary to the surface of the Earth.

Many organisms would have been killed directly by radiation from the initial nuclear burst in Venus. Others would have been sterilized, or lethal or non-surviving mutations would have been produced. However, some mutations could have been produced that had significant survival value in the abruptly changed, highly radioactive environment. In the most extreme form of action, all existing species could have been eliminated by transmutation to new forms, but life would not have had to start all over again.

For small organisms, the major radiation dose would come from the relatively short-range beta-particle emission from the radioactive materials deposited on the ocean floor from the eruptions of submarine and above-sea volcanoes. The range of the beta particles is a few millimeters in water or sediment, and avoidance of this radiation exposure might have been crucial to the survival of individual organisms.

Features such as hard shells to protect the more sensitive body parts, particularly the gills (*Anomalocaris, Odaraia*), and the burrow shields mounted on the front and rear of an articulated halkieriid [Conway Morris and Peel 1990], and behavior that shielded the organism from the most radioactive portion of its world, the sea floor, such as swimming on its back (*Sarotrocercus*), or stand-off support (*Dinomischus*), or deep vertical burrows, would have reduced the radiation exposure to a (perhaps barely) survivable level. The legs of the "armoured lobopods" [Ramsköld and Xianguang 1991] eliminate the beta-radiation dose to the body itself from radioactive material on the surface of the sea floor.

Pikaia gracelens was a free-swimming worm with a central nervous cord, that may have been our most distant ancestor, surviving by reason of being suspended in the water, above the radioactive sea floor [Gould 1989, Matson and Troll 1994].

These body-plans and life-styles provided that the most slowly growing parts of the anatomy, and those least sensitive to radiation injury, were the parts most exposed to radiation and, by serving as shields and to exclude contact with the most radioactive layers of the sediment, protected the life of the organism. For a period of perhaps a few million years, the survival of the species was contingent upon the ability of the individuals to survive a temporarily increased radiation exposure. After time, as the induced radioactivity faded away, these features would have lost their unique survival value and the organisms, as species, would have also faded away.

Impact-induced expansion of the Earth, resulting from some fragment of a detonated moon of Venus, could have caused the change in moments of inertia considered by Kirschvink, Ripperdan, and Evans to have caused a rapid rotation of the Earth's crust at the time of the Cambrian [Kirschvink, Ripperdan, and Evans 1997]. Changes in the paleomagnetic poles as determined by remanent magnetism in rocks spanning the transition from the Vendian Period to the Cambrian Period, a time span of about 35 million years, indicated a rotation of the solid Earth (crust and mantle) relative to the core, which maintains the geomagnetic field, by approximately 90 degrees. The Earth, as an oblate spheroid, has three distinct, but similar, moments of inertia. The axis of rotation (the polar axis) is aligned with the maximum moment of inertia. So-called "inertial interchange true polar wander" results from a relative increase of the intermediate moment of inertia compared to the maximum moment of inertia. In the Cambrian, this polar wander occurred within a time period of about 16 million years.

The current relative values for the moments of inertia [Yoder 1995] are:

$$I_{minimum} = 0.99999779$$
$$I_{intermediate} = 1.00000000$$
$$I_{maximum} = 1.00330261$$

For the intermediate moment of inertia to exceed the maximum, the radius in the intermediate direction would have to increase by more than 365 kilometers on the present Earth. If the Earth moves by Earth Expansion, rather than Plate Tectonics, as is proposed in this book, by impact-induced expansion, the differential expansion would have been about 250 kilometers, on an Earth with an average radius of about 4,375 kilometers. Since the Earth must maintain a spherical figure, the average radial expansion would have been 82 kilometers. At the average rate of expansion shown in Figure 70, about one kilometer per million years, that amount of expansion would have required 82 million years. However, as the expansion events are episodic and abrupt following an impact, such an expansion could have happened in the blink of an eye. (Or a few days.)

The abrupt appearance of animal forms composed of various assortments of odd parts as if from a grab bag [Gould 1989], followed by the total disappearance of most of these forms a few million years later, suggests catastrophic mutations, making forms that were only temporarily suited to the new conditions. This would show genetics to be operating in a manner consistent with the developmental compartments of piecewise construction that is observed in modern insects [Lawrence and Morata 1993].

Perhaps, for those organisms that were the most successful adaptations to the high-radiation environment following the planetary detonation, the cost of protection became too high a price to pay to permit continued survival, after the threat of radiation injury had passed and the need for protection was gone. Some other of these inherited changes produced permanent benefits, and were retained by surviving descendants. Possessing a hard body-shell helped protect the organism inside from the hazards outside, and reduced the likelihood of being eaten. But others were more expensive than they were worth, once the radiation hazard had decayed away, and these forms were lost.

While Gould considered the survival of specific types of life-plans from this period to be a random result, relying like a lottery on the luck of the draw leading to survival [Gould 1989], perhaps the lottery was in the way the deck was stacked, in terms of life-plans that would still economically promote survival, after the extreme conditions had passed.

The isotopic composition of various elements in the sediments might show significant changes in crossing from the Precambrian deposits into the earliest Cambrian deposits.

The Cambrian, and the changes in life forms, was induced by a nuclear detonation in the planet Venus at that time. That detonation caused the global resurfacing of Venus and dissociated any carbohydrate life on the surface and carbonate rocks, making the carbon dioxide atmosphere. The strange and temporary life-forms on Earth were produced by massive radiation-induced mutations and they were selected for survival, over the next 25 million years when many disappeared, by the induced radioactivity on the surface of the Earth. (That is, I think, the bottom of shallow seas.) The radiation caused mutations that produced strange new creatures, and survival was based on possessing key characteristics to reduce the radiation dose. These creatures reduced dose by two of the standard practices in current industrial radiation protection: distance and shielding. Those new animals showed:

free swimming
stand-off stalks

hard body shells
burrowing in the seafloor bottom with shelly endcaps
walking on spines or stalks, like sea urchins today

In our world of radiation, most organisms receive little radiation dose, so those with the greatest radiation-induced mutations have the greatest dose and are most likely to perish. For the Venus detonation, nearly all the populations received lethal exposure, and the few survivors received such high doses that the mutation rate was high, producing many various new creatures. Then, only those new forms that had protective characteristics survived and propagated.

The Permian-Triassic Transformation

Tiptoe through the tulips,
Dubin and Burke
(with thanks to Tiny Tim)

A similar pattern of death and new life is shown for the end of the Permian Period and the beginning of the Triassic, at about 245 million years ago. The Permian ended with one of the greatest mass extinctions yet observed [Raup 1991, Erwin 1994], speculatively caused by the greatest continental flood basalt eruption yet discovered, the Siberian Traps [Renne and Basu 1991, Renne 1995], or at least associated with it. The record of strontium ratios suggests that this was followed by a global chilling, implying a decrease in the fission power of Earth's reactors (see Figure 24 and discussion). The Permian-Triassic extinction boundary is followed by an extensive spike of fungal spores.

At the end of the Permian Period, life came to an abrupt jolt and new forms gradually arose, standing tall or tiptoeing along with the physiques we would later recognize in the dinosaurs that are now so loved. That was an important change of form. These animals were not the dinosaurs, but their predecessors. The Triassic was a time of transition. During the Early Triassic, the dominant form of animal changed from the sprawling, close-to-the-ground stance of *Eryops* [Scarre 1993] to the nearly upright, mostly bipedal stance of *Euparkeria*, a precursor to the dinosaurs, as shown in Figure 67 [Padian 1997].

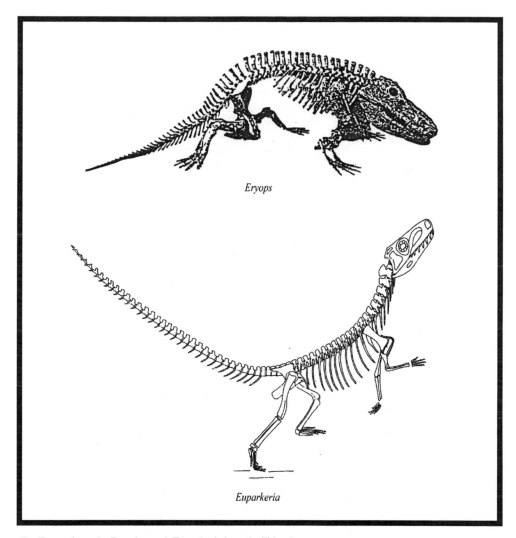

Figure 67. ***Eryops* from the Permian and *Euparkeria* from the Triassic.**
Comparison of *Eryops* from the Permian and *Euparkeria* from the Triassic, showing contrasting stance, close to the ground in the Permian, upright in the Triassic [Scarre 1993, Padian 1997].

By the end of the Triassic, culminating in a mass extinction, all the ground-hugging low-life animals died out, to be replaced by bipedal, upright dinosaurs [Benton 1997] (Figure 68). A posture away from the ground reduced the radiation dose through the bulk of the body.

Major impacts initiate flood basalt eruptions, by the shock waves that go through and around the Earth, to tear the opposite crust apart. Oakley Shields has suggested that the Siberian Traps were started by a major impact that formed the Congo Basin (Cuvette Centrale). This structure has been described [Daly *et al.* 1991] as a result of distant plate collisions, but the sketches look like an impact structure. The Congo Basin would have been nearly antipodal to the Siberian Traps at 250 million years ago.

One form of sea life that suggests a response to the special conditions of nuclear detonation and Earth Expansion is the ammonoids. These creatures, somewhat resembling the chambered nautilus, showed a remarkable recovery after the mass extinction of the end of the Permian [Brayard *et al.* 2009, Marshall and Jacobs 2009]. The ammonoids, living freely in the water column, recovered to better than ever before in less than 2 million years, but bottom-dwellers languished for 5 million years. The effects proposed here to account for the relative success of the ammonoids include a reduced sea level and ocean pressure due to impact-induced expansion, and radioactive particulates from the impactor and in lava erupted from undersea volcanoes. These effects would have been more detrimental to animals living on the sea floor than those floating in the water column, which would have been relatively free of radioactivity, and would have allowed adjustment to the optimum water depth and pressure.

Figure 68. Early upright dinosaur forms.
Early "dinosaur" forms, showing the avoidance of the ground resulting from the newly successful anatomy. Illustration by Gregory Paul [Potts 1993].

Dinosaurs originated as abrupt new life forms, from radiation mutations produced by a nearby astronomical detonation, and then succeeded in surviving, perhaps specifically as a result of their new stance. This upright structure became very advantageous, in many ways that have nothing to do with radioactivity. However, in the long run, lizards lasted to survive to the present, while dinosaurs (with the exception of their continued form in modern birds) died off.

In the ocean, thin-shelled, free-swimming or surface-sitting brachiopods were replaced by thick-shelled, burrowing mollusks. And turtles, with thick, encompassing, protective shells, were invented.

Turtles

The turtle lives tween plated decks,
which practically conceals its sex,
Ogden Nash, *"The Turtle"*, 1931

Turtles appeared so abruptly at the beginning of the Triassic that there are no known ancestors [Benton 1993]. Because of the extreme and unique specialization of the shell (carapace and plastron), it has been thought that it must have a long evolutionary development, and a gradual ancestral lineage should be apparent. However, standard Darwinian evolution, by selection through many generations, is not capable of explaining the turtle's shell [Rieppel 2009]. Perhaps the turtle was a macromutant, with major changes produced in the genetic makeup of this animal by a burst of radiation, and was then selected for survival as a result of the effective radiation protection provided by the horny shell. It would seem that not much could threaten a low-slung turtle from underneath, other than radioactive ground.

The oldest known fossil turtle, *Odontochelys semitestacea*, from 220 million years ago, had a fully formed plastron, the belly shell, or better, the belly shield, but lacked most of the upper shell, the carapace [Li *et al.* 2008, Reisz and Head 2008, Millus 2009]. This turtle is supposed to have lived at the edge of the sea, and had teeth, a retained character, and it had a hypoischium.

The hypoischium is a bone unknown in any other tetrapod, with an unknown function [Rougier, de la Fuente, and Arcucci 1995]. If form follows function, the function may be guessed to have been to provide additional radiation protection to the sensitive gonads, just as gonad shields are used by nuclear power plant workers during high exposure tasks. A most unique bone, serving a most unique function, trying to tell us about its world.

The next oldest fossil turtle, *Proganochelys quenstedti*, from 204 to 206 million years ago, had a complete shell [Gaffney 1990], providing full protection.

If this adaptation made the ancestral turtles so fit for survival, why was it not adopted by other animals? Perhaps the simplest answer is that those animals would then have been turtles also.

All these changes would reduce the radiation dose received by animals living near a newly radioactive surface. Even the increase in general body size would have been effective in reducing the dose received by internal organs. As in the Cambrian, some structural changes resulting from the radiation-induced mutations and that were selected for survival by the ensuing radioactive environment, provided significant advantages beyond radiation protection, and continued into the future.

In an interesting variation on death-and-survival, the animal forms known as bellerophonts experienced survival and death [Kerr 1994b]. These are marine animals that showed a relatively slow evolutionary process, taking hundreds of millions of years to diversify and spread around the world. This stability suggests a resistance to genetic change, that might also suggest a resistance to the destruction of species resulting from radiation-induced damage to the genetic material, and possibly some relative resistance to radiation killing, as well. The subsequent decline and disappearance of these survivors might have resulted from the fact that they were gastropods: they were snails, crawling on an intensely radioactive sea floor after a burst of nuclear-fission-produced power at about 250 million years ago. This constant high radiation exposure of sensitive tissue could easily have led to their decline.

While this suggests an initiating nearby planetary detonation, an obvious candidate for this detonation has not yet been found.

There are many possible effects of a nearby (Solar System) nuclear fission detonation, as those effects would seriously impact life. I have to emphasize the possible because the intensity may vary, and some of the effects may miss the Earth entirely:

 radiation.
 gamma-flash modification of the atmosphere, oxygen to carbon and nitrogen, nitrogen to carbon,
 disturbance of isotope ratios.
 increased surface/atmospheric radioactivity, induced by antineutrinos, neutrons, and protons from the detonation.
 marine regression from thermal expansion of Earth by induced increase in reactor power.
 volcanic eruptions from increased heating.
 release of radioactivity from eruptions.
 impact, with/without marine transgression due to additional water brought in by the impactor.
 impact induced flood basalt fissure eruptions.
 impact induced Earth expansion, causing catastrophic marine regressions and transgressions, and reducing atmospheric pressure.
These effects could be spread over a million years or so, and would have different effects on life of different forms.

Life, and Death, at the K/T

*The death of the world
is the birth of a new one.*
Icelandic saga

Extinction is forever.
World Wildlife Foundation

A more technically correct designation of this time is the Cretaceous-Paleogene boundary, or the K-Pg, which is now becoming current usage. However, K/T seems so ingrained in the current literature that I will continue to use it.

After 140 million years of success, the dinosaur form of life was abruptly terminated, in the Cretaceous/Tertiary (K/T) mass extinction. How abruptly is typically a matter of argument, and how is certainly controversial. However, the dinosaurs are certainly gone. Some claims are made that the dinosaurs live on as modern birds, but that simply clouds the question and trivializes the death of the dinosaurs. As a life form, the dinosaurs are gone. As we observe the fossil record, the end of the Cretaceous was extended and complicated, with several extinctions spread over an extended time. Attempts to explain most or all of these extinctions with a single cause has produced more argument than understanding.

It is now generally accepted that an asteroidal impact on the Yucatan Peninsula, forming the Chicxulub impact structure, mostly hidden from sight, caused the extinctions at the very end of the Cretaceous [Alvarez *et al.* 1980]. It is the change in life at that time that defines the start of the next geologic period, the Paleogene. There has been an extended battle between those who think the Chicxulub impact caused the end of the Cretaceous and those who think the Deccan Traps eruptions did it, and those few who say it was rather more complicated than either. While the impact hypothesis has received general acceptance as the cause of the extinctions [Schulte *et al.* 2010a], the controversy between the Deccan Traps and the Chicxulub impact is still active after 30 years of argumentation [Alvarez. and Asaro 1990, Courtillot 1990, Archibald *et al.* 2010, Courtillot and Fluteau 2010, Keller *et al.* 2010, Schulte *et al.* 2010b]. There seems to be enough disagreement and contradictory evidence that it is difficult for an outsider to accept either side confidently, and apparently impossible for the insiders to come to agreement. (In case you are keeping score, this most recent exchange in 2010 included 41 co-authors in favor of impact, and 37 arguing otherwise.)

Dinosaurs had been in decline for ten million years before the K/T [Sloan *et al.* 1986, Achenbach 2003], but it is generally accepted that there are dinosaur bones below a very distinctive clay layer that marks the K/T boundary and none above. That layer is dated at precisely 64.43 ± 0.05 million years ago. While some dinosaur bones have been collected from above the K/T boundary, implying an extended survival of the dinosaurs [Sloan *et al.* 1986] this has been argued against, as an accident of preservation and an artifact of collection. This is similar to the problem created by Gerta Keller, who has shown evidence in some marine deposits that the K/T ended 300,000 years after Chicxulub [Keller *et al.* 2004]. The clay layer does mark an abrupt change in life, and it is marked with evidence of a major impact. The clay contains relatively high concentrations of the rare element iridium, which is found in meteorites and at identified impact craters. (The iridium is determined by neutron activation analysis, by irradiating a sample of the clay in a nuclear reactor. The activation of the natural iridum-191 to radioactive iridium-192 is measured by gamma-spectroscopy. The absorption cross section of iridium-191 is 954 barns, so iridium present near an astronomical nuclear detonation might be depleted in iridium-191 by neutron absorption, and so not show the presence of excess iridium when tested by neutron activation analysis by researchers on Earth.) Microtektites, small spherules of glass formed from droplets of molten rock, are found immediately above the clay. Quartz that has been deformed by an intense shock has been found. Tidal wave (tsunami) deposits have been found. Most importantly, a hidden impact crater was found near Chicxulub, Yucatan, Mexico. Because the rocks of the impact site consisted largely of limestone (calcium carbonate) and anhydrite (calcium sulfate), a tremendous amount of acidic gases would be released to the atmosphere. Along with wildfires started by the impact, dust propelled high into the atmosphere, the acid rain, and the greenhouse effect from the gases, were predicted to produce an "impact winter", like the "nuclear winter" that had been claimed for a global war with nuclear weapons. This impact winter, and extended period of darkness, would have killed vegetation, so the herbivores died, so the carnivores died, so the scavengers died, and that produced the mass extinction.

The timing might be right for the final termination of the dinosaurs, but it does not explain the sudden success of the mammals immediately after the K/T, nor the protracted and earlier dying of organisms before the K/T, before the impact. It does not explain the complete extinction of the dinosaurs and the selective survival of real reptiles.

The extinction of other organisms, as events and as extended declines, brought attention to the massive eruptions of lava at the Deccan Traps in Northwest India. These have been proposed as sources of the greenhouse gas carbon dioxide, and that may have acidified ocean water and overheated the climate. These eruptions extended over 10 million years, across the K/T, which would explain the earlier extinctions and the extended declines, but there seems to be nothing special about the eruptions near 65 million years ago, to finally terminate the dinosaurs.

Life is not simple, death is not simple, and mass extinctions are certainly not simple.

In fact, both the eruptions of the Deccan Traps and major impacts seem to be recurring events near the end of the Cretaceous, and they seem to be related. There is evidence for four major impacts, starting with one near Manson, Iowa, then one directly at the Deccan Traps, one apparently in the Pacific Ocean, and the final impact at Chicxulub, dated at 64.98 ± 0.05 million years [Grieve 2000]. The Manson impact has been dated at 73.8 ± 0.3 million years ago. [Grieve 2000]. This impact site was covered by an inland sea, an epicontinental sea at that time, and tsunami deposits have been found between 73.7 and 72.4 ($\pm$ 0.4) million years ago, a few hundred miles west [Kerr 1993e, Izett 1993], in the Upper Campanion Stage, before the K/T. Geomagnetic polarity of the rocks also rules out K/T age. The Pacific Ocean impact is only hinted at by impact debris at the K/T boundary and has not been dated independently. This debris appears to be distinctly different from other impact fields at about the same time [Robin *et al.* 1993]. Remarkably, a meteorite was recovered from K/T sediments nearby, that appears to have the composition expected for the Chicxulub impactor [Kyte 1998].

The Shiva impact crater, at the Deccan Traps, has been found by Sankar Chatterjee as two split halves on opposite sides of the spreading ridge in the northwest Indian Ocean [Chatterjee 1997]. This remains as the semicircular Amirante Basin, south west of the Seychelles, and the other half has been buried by the later eruptions of the Deccan traps, but shows its existence as the Panvel Flexure. Combining these two structures produces an elongate crater, typical of an oblique impact [Chatterjee and Rudra 1996].

Chatterjee developed a related theory explaining the flood basalt eruptions of the Deccan Traps as the result of shock waves from a distant impact fracturing and heating the crustal rocks. This effect is greatest for eruptions diametrically opposite the impact, eruption at the antipode. According to Plate Tectonic reconstructions of the Earth, the Chicxulub impact site was diametrically opposite the Deccan Traps at the time of the K/T boundary. Somewhat earlier, and with a smaller expanding Earth, the Manson impact site might have been opposite the start of the Deccan Traps.

Rudists, a marine bivalve, were in decline for a few million years before the Chicxulub impact, apparently from a decrease in the area of shallow marine water [Kerr 1992c]. This would happen from expansion of the Earth, induced by the several impacts before Chicxulub.

Plesiosaurs and ichthyosaurs went into decline about 75 million years ago (about the time of the Manson impact, at 73.8 million years ago). The last stage of the Cretaceous, the Maastrichtian, began at 74.0 million years ago, the same time as the Manson impact, suggesting that the impact was the cause of the changes in life forms that defined the beginning of that stage.

Ichthyosaurs had asymmetrical tail fins, opposite in the sense compared to modern sharks. The asymmetry of the modern shark tail gives the fish an upward thrust along with forward propulsion, reducing the tendency of the shark to sink. Therefore, the ichthyosaur gained some downward thrust from the tail strokes, giving it a tendency towards a subsurface life. Plesiosaurs swallowed stones, which may have acted as ballast, as modern alligators do. Thus, an expansion of the Earth, induced by the Manson impact may have reduced the surface water density, by loss of dissolved atmospheric gases, which made the plesiosaurs and ichthyosaurs less successful.

The Cretaceous chalk deposits, so notable that the period was named for them, may have resulted from the increase in atmospheric and oceanic carbon dioxide produced by the gamma flash of the detonation, from the Giant Dipole Resonance reaction on oxygen and nitrogen in the atmosphere. Abrupt precipitation of these deposits would have been caused by the reduction in atmospheric pressure resulting from impact-induced expansion from the four major impacts. This causes the ocean to lose carbon dioxide, and the water becomes oversaturated in calcium carbonate.

Much of the trouble over the K/T extinctions seems to arise from the effort to find a single cause for what appears to have been a series of extinctions. The search for a single cause drives the efforts to show that it was indeed a single event. A nearby astronomical detonation, such as the detonation of the moon of Saturn that I propose for the end of the Pleistocene, can cause the whole suite of effects that we see in most of the mass extinctions. For the K/T extinctions, the initial, intense burst of gamma- and X-radiation would kill nearly all large animals above ground on the side of the Earth facing the detonation. This is seen, in the deaths of the vast herds of dinosaurs found in the Gobi by Norell and Novacek, and in Montana by Horner. Mike Novacek proposed that a dust storm killed them in the Gobi, but I think radiation was the cause. The deaths in both places were followed considerably later by a major flood. Novacek estimates his mass death at about 80 million years ago, which I will take as the time of the detonation. Horner dates his flood at 76.7 million years ago, which is close to the time of the Manson impact at 73.8 million years ago, and the apparently related tsunami deposits in South Dakota dated to 73.7 to 72.4 million years ago. A flood in the Gobi desert came many years later and emplaced the flood deposits about 30 meters above the Gobi skeletons. A similar flood has been found by Luis Chiappe, dated at 80 million years ago, in Patagonia [Achenbach 2003]. Given the difficulties in dating such deposits, the floods might have been at the same time.

As the atmospheric pressure decreased from impact-induced expansion, higher elevations would have become uninhabitable (for dinosaurs with inefficient lungs), so localized dyings may have occurred. Novacek had suggested that the up-and-back arching of their necks indicated a struggle to escape from the surrounding sands from a major sandstorm [Novacek 1996]. Later research showed that such drowning was not likely [Loope *et al.* 1998]. Perhaps instead it was the effect seen as the opisthotonic posture, which tightens the muscles along the spine and bends a corpse backwards, following death by damage to the central nervous system [Faux and Padian 2007]. That is the cause of death from extreme radiation exposure.

This posture is illustrated quite well in the two *Sinosauropteryx* skeletons found in China [Chen, Dong, and Zhen 1998]. While these two deaths seem to have occurred 70 million years ago earlier than the mass mortality in Gobi and Montana, similarities suggest the possibility of a similar cause. Two animals (not quite a "mass"!), a pre-adult and an adult, were dead without apparent damage, drowned in a flood and covered with mud, and the bodies survived to be fossilized without disturbance by scavengers. That suggests an abrupt, locally complete killing event, including local scavengers. I think that is what happened in Gobi and Montana.

This also happened with a hadrosaur, found with fossilized soft tissue [Science News Staff 2002]. According to Michael J. Everhart of Fort Hays State University, "Something had to shut down the normal process of decomposition within just a few days. It's difficult to explain." And there was no evidence of damage by scavengers; everything would have been killed. Intense radiation sterilizes and kills. The fossil find was dated to about 77 million years ago, which is close to the time that I propose, 80 million years ago as a guess, for the nuclear detonation that started the end of the Cretaceous Period. (Radiation sterilization does not kill bacteria. Intense radiation damages DNA so that the bacteria can no longer reproduce successfully. Thus, there is no bacterial growth and no decay.)

Impact dust in the atmosphere seems to have been inadequate to shut down photosynthesis, and selectively starve the victims [Kerr 2002]. In this book, the major killing effect was the expansion of the Earth on impact by the Chicxulub impact, decreasing the atmospheric pressure and causing the dinosaurs to collapse or to asphyxiate. Turtles, frogs, and crocodiles survived because they live in water and are supported by the nearly constant density of the water. If the dinosaurs had thrived by limiting their respiratory efforts to skin absorption of oxygen (and possibly pulmonary elimination of carbon dioxide), a reduction in atmospheric pressure would have caused death of this entire group of animals (excluding the ancestors of modern birds, who appear to have been blessed with a somewhat different respiratory system).

The ages measured for the Deccan Traps [Courtillot 1988] were deconvolved into four distinct stages of eruption, three pulses and a squirt, and these stages seem to match impacts around that time rather well: Manson, Chicxulub, one in the Pacific, and one actually at the Deccan Traps, that Sankar Chatterjee calls Shiva. Three impacts were distant, almost diametrically opposite the Deccan Traps, and one was right on the target. If these impacts caused the Earth to expand by the release of primordial compressive strain, the atmosphere would have been thinned, and the partial pressure of oxygen in the surface atmosphere would have been reduced. The dinosaurs, with inefficient lungs and relying on skin absorption, would have died of respiratory insufficiency. The small mammals, kept small by the high oxygen pressure and their efficient lungs, would have survived and grown larger, no longer threatened by oxygen poisoning. I think the whole sequence of events began about 80 million years ago with the detonation of a moon of Uranus, and Chicxulub was just the *coup de grace*.

That leads to an interesting consideration for life forms in the past. If there is an optimum size for an organism, and if that size is related to environmental conditions (almost by definition), how would expansion of the Earth affect that? It suggests that reduced air density led to the collapse of the large dinosaurs, but reduced oxygen partial pressure led to an increase in size for mammals. The reduction in air pressure resulting from expansion of the Earth triggered by these impacts killed the nonavian dinosaurs, those that had inefficient lungs, and the reduction in air density caused the collapse of the larger forms. The reduced oxygen partial pressure freed the mammals from the threat of oxygen toxicity.

Perhaps some of the opposition to impact-induced volcanism has come about because the flood basalt center is not diametrically opposite the impact site, in today's world. Asymmetric expansion can solve that, as an alternative to our standard theory of Plate Tectonics. If the Earth expands more on one side than the other, diametrically opposite points will be closer (in an angular sense) after the expansion.

It is likely that the Cretaceous ended this way:

80 million years ago: nuclear detonation of a large natural moon of Uranus. The gamma-flash caused several mass dyings, killing predator, prey, and scavengers. Some oxygen would have been converted to carbon by the Giant Dipole Resonance reaction, forming carbon dioxide, which dissolved into the ocean.

77 million years ago: a cometary stream (such as I propose later for the end of the Pleistocene) produced by the detonation flooded large areas of the Earth, as shown in Gobi and Montana.

74 million years ago: a fragment from the disrupted moon hit the Earth at Manson, causing some loss of atmosphere, inducing some global expansion, marine regression (with tsunami), thinning the air, reducing livability at upper elevations, developmentally stunting some dinosaurs, allowing some mammals to thrive in less-rich air. This impact started the Deccan Traps flood basalt eruptions. Some cap carbonates were formed as carbon dioxide was released from the oceans by the reduction in atmospheric pressure.

68 million years ago: a fragment hit in the Pacific Ocean, ablating the atmosphere some more, but not causing much other harm. The Deccan Traps eruptions were boosted.

67 million years ago: a fragment hit near the Deccan Traps, causing an abrupt outflow of lava.

65 million years ago: a large fragment hit at Chicxulub, ablating the atmosphere, causing severe global expansion with major tsunamis, depositing more cap carbonates, thinning the air to the point that ordinary dinosaurs could no longer survive, and the entire Superorder was extinguished, except for the few true birds, with efficient respiratory systems.

Latent radiation mutations succeeded, newly suited to the new world, producing the great "radiation" of bird species in the Paleocene (start of the Tertiary Era). Mammals could breathe more easily in the thinner, poorer air. Oceans sloshed around the globe, causing giant tsunamis.

It must take several effects to accomplish such a major catastrophe as a mass extinction. A world passed and a new world was born, a world fit for new birds and mammals, but no longer suitable for the dinosaurs.

If the Earth expanded stepwise in response to the apparent four major impacts, Manson, Shiva, Pacific, and Chicxulub, the atmosphere would have gotten thinner with each jolt, until there was not enough oxygen to keep the dinosaurs alive, and the air was too thin to support them buoyantly. Concurrently, the oxygen became dilute enough for mammals to grow and thrive. (The critical parameter is oxygen partial pressure rather than fractional concentration, and I haven't seen that addressed.) The expansions would have also caused marine regressions, with related extinctions, spread out over about 10 million years, which is shown. Asymmetric expansion would cause a great rush of seawater from one side of the world to the other, with apparent marine transgressions, which might have left evidence that can still be found.

Noting that measurements of oxygen trapped in air bubbles in ancient amber showed a fraction of 35%, compared to the modern amount of 21%, Keith Rigby proposed the Pele Theory, that dinosaurs died because of their less efficient respiration. He has also proposed that the extinction was very protracted, from 600 genera at 100 million years ago, to 37 genera at 65 million years ago, with some 17 genera lasting beyond that, and found in strata that lie above the K/T boundary layer. That argument is much like what Gerta Keller of Princeton has found in the marine deposits, which appear to show 300,000 years of deposition above the Chicxulub impact layer to the K/T boundary.

The Pele effect, the reduction in partial pressure of oxygen in the surface atmosphere, helps explain the extinction of the dinosaurs, the survival of the birds and the mammals, and the pre-K/T small size of the mammals.

From trackway analyses, it appears that dinosaurs were very active animals, as active as modern large mammals, perhaps possessing great stamina as well. Yet they seem to not have been warm-blooded (their nasal cavities show no turbinate structures as in birds and mammals) and seem to have had very inefficient lungs, arguing against stamina. An atmosphere rich in oxygen, as the dinosaurs developed, might explain this apparent contradiction, especially if they used skin absorption for oxygen and lungs to eliminate carbon dioxide.

A reduction in partial pressure of oxygen at or about the K/T, either by a loss of oxygen or a reduction in atmospheric pressure, might have been intolerable for the dinosaurs. A drop of one-quarter in the oxygen content of our present atmosphere is immediately fatal to humans; smaller changes take longer to kill. On the other hand, large warm-blooded animals are not likely to do well in an oxygen-rich atmosphere: they overheat too easily. Therefore, there would have been a selective pressure to keep mammals small before the K/T.

The standard description of an extinction seems to be that "for a species to go extinct, every individual must die." But all individuals of every species do die. For extinction, this must be qualified as dying "without surviving offspring representative of the species". This is somewhat of a semantic point, but it forces consideration of the important process of speciation, which is a process that has been referred to as pseudoextinction. In fact, speciation could occur in such an extreme manner as to cause extinction of a species (by total replacement) without a total dying off, and yet the extinction is just as real. Consider three possible radiation-exposure situations: in one the radiation exposure is so intense that all individuals die quickly; in another, the exposure is less intense but all individuals die sterile, without descendents; at a lower exposure, all individuals remaining fertile have major mutations in the germ cells, and their offspring are definitively different. In all three cases, the parent species ceases to exist, so they should all be considered extinctions.

The evidence for the radiation-deaths is circumstantial: the dinosaurs in Gobi died in place, *en masse*, and everything else did too. At least no scavengers were around to mess up the carcasses and so the skeletons rested in peace. Then there was a flood. The Montana deposit looks similar, with a herd (or so) dying, all about the same time (or so), and maybe the same time as in Gobi. Then there was a flood. Extinctions appear to have occurred in stages, with different categories of animal types dying or fading away. This is at the root of the volcanism/impact argument for the K-T extinction: neither one seems adequate or at quite the right time, by itself.

Juan Collar has suggested neutrinos for extinction, and Yayanos suggested radioactivity induced by thermal neutrons. I think these play a part, but slowly. The quick and abrupt deaths must be from the gamma-ray/X-ray flash from the detonation and fireball. For the major effects at the K/T, however, 15 million years after the detonation, I prefer a stepwise reduction in surface air density associated with the four major impacts (Manson, Pacific, Shiva, and Chicxulub) for both the extinctions and the speciation that followed. The dinosaurs died, gasping for breath; the ancient birds were either grounded, unable to fly in thin air, or were downsized and grew proportionately larger wings, or died; and the mammals were liberated, no longer limited in size by an overly rich (oxygen) atmosphere. So, Cretaceous birds could have flown easily with smaller wings than our birds. The current extreme examples of the effect of differences in flight-medium density on wing size are the penguins (in water) and the Andean Condor (in air at 14,000 feet).

The start of the end of the Cretaceous can be seen in the mass deaths of dinosaurs and others in Gobi and Montana about 80 million years ago, and then continuing to the final nail in the coffin, at Chicxulub at 65 million years ago. There appear to have been four major impacts during that time, Manson (74 million years ago), Pacific (68 million years ago), Shiva (68 million years ago), Chicxulub (65 million years ago), with the dates derived from the ages of the Deccan Traps lava flows.

To understand the cause of death, we must consider the mode of living.

Size Comparisons

When skeletons of dinosaurs were first discovered (scientifically), it was thought that they had lived in swamps, to support their immense weight. (Such a picture lingers on in a children's coloring book, and many others.) Eventually, it was clear that they had lived on dry land, so the question of support remained unanswered. How could the long-necked sauropods have raised their heads, when the modern giraffe seems to have the highest head possible, relative to its heart and feet, to manage its blood flow?

A brief report [Sander and Clauss 2008] has discussed how the giant sauropods could have grown so large by growing fast, but it fails to solve the problem of how they could have lived so large. Their sizes greatly exceed all other animals, except those now living in the sea, while there are unexplained obstacles to such sizes. How could the giant dinosaurs have supported themselves, how could they have run fast, or the pterodactyls flown?

These are severe problems for animals so large.

A larger territory permits the development of larger animals. This has been shown to provide reasonable correlation for the largest animals in the last 65,000 years [Burness, Diamond, and Flannery 2001]. However, the top carnivorous dinosaurs were 12 times heavier (larger) than predicted in this study, and the top herbivorous dinosaurs were 1.5 to 3 times heavier. Those great sizes remain unexplained.

The biggest animals, the biggest of the big, are extreme, but not too extreme. More would not only not be better, it would lead to death by starvation, overheating, overexertion, respiratory failure, or collapse. Nature sets upper limits for each form, animal, tree or plant, fish or bird, at whatever is appropriate for the time and scene. Deciphering the meaning of the biggest of each form can give us clues about the time and scene, while the smaller types are interesting, giving variety and counterpoint. Here, for our search, we focus only on the largest of the large.

The maximum size of the largest animals (and plants also, but we have little fossil information to work with) is subject to limitations for each type of animal. For animals living on dry land, the ability to withstand the force of gravity due to their own weight might be the most restrictive limit. The African Elephant, the largest land animal now alive, might be limited by weight and by mobility, but also by its ability to find and consume food, and to discharge excess body heat. The Blue Whale, the largest animal to ever have lived, might be limited only by food and its growth rate in a finite lifetime. Comparisons of the Mesozoic animals, the dinosaurs and their comrades, shows that the largest of the large far exceeded those animals of today, except for those who live in the ocean, buoyed up by water at nearly the same density as the animal itself.

Dividing animals into types according to where they live and how they get their food allows us to make meaningful comparisons. The easiest comparisons are shown in Figure 69, divided into Mesozoic and Modern, and into ocean, land, and air, and as grazing plant-eaters and hunters. The largest of the large is the Blue Whale, living at present in the ocean and grazing on plankton and krill. The comparable animal in the Mesozoic (*Bonnerichthys gladius*) has just recently been described, but these fishes were only (!) 12 meters long (39 feet), and so are dwarfed by the modern whales [Cavin 2010, Friedman *et al.* 2010]. The modern Sperm Whale and the Colossal Squid are somewhat larger than their counterpart in the Mesozoic ocean, *Tylosaurus proriger*. On land, however, the giant sauropod plant-eaters completely outclassed our largest modern animal. *Argentinosaurus huiculensis* was approximately 10 times as massive as the modern African Elephant. Similarly, *Tyrannosaurus rex* dwarfs all the modern land carnivores. Shown for comparison are the Elephant Seal, which spends much of its active life in the water; the Polar Bear, which spends some of its active life in the water; and the usual suspects, tigers, and lions, and bears. Even in the air, this great mismatch in sizes is shown. The pterodactyl *Quetzalcoatlus northropi* far exceeds our largest similar flying animal (with similar membrane wings supported by extended fingers and "hunting" for prey, fruit in the case of the modern fruit bat).

The most directly comparable pairs of Mesozoic and Modern animals are shown below, with their weights and ratios:

Argentinosaurus	<u>73 tons</u>	10.4 times more mass
African Elephant	7 tons	
Tyrannosaurus	<u>14,000 pounds</u>	21.2 times more mass
Tiger	660 pounds	
Quetzalcoatlus	<u>400 pounds</u>	181.8 times more mass
Flying Fox fruit bat	2.2 pounds	

Such great relative sizes demand some deliberate explanation, they are simply too extreme to excuse on the basis of better food (which it wasn't) or a more even climate (which it might have been). Throughout the ancient past, most animals (and plants) were greatly larger than their counterparts in the Modern world. Even later, in the Miocene, about 23 million years ago, the elephant-like beast *Indricotherium* was 1.6 times the mass of a modern elephant. Most recently, mammoths, mastodons, giant ground sloths, and similar animals were much larger than the animals we have in the present world.

If the density of the Mesozoic air were greater than that at present, from a more massive atmosphere (like Venus) and a smaller Earth (because of subsequent expansion), the immense bulk of *Argentinosaurus* (and the other large plant eaters) could have been buoyed up. (A smaller Earth causes an increase in surface density by two means: the area over which the mass of the atmosphere is spread is smaller, and the surface gravity is greater, pulling the mass closer to the surface.) With some weight supported by buoyancy, extra support would be available from larger (thicker) leg bones.

Figure 69. Size comparisons of Mesozoic and Modern animals.
Comparative sizes of Mesozoic animals, on the left, with their modern counterparts, on the right. The largest dinosaurs were as large as our whales, living in the ocean, and *Tyrannosaurus* dwarfs all the modern carnivores. The Mesozoic sea creatures are actually smaller than the modern forms living today.

James Maxlow produced a remarkable series of reconstructions of the Earth through time by taking the current world geological map and sequentially removing the oldest slices of ocean floor [Maxlow 2005]. (Most work on the expansion of the Earth has assumed that the Earth expands uniformly, like a well-formed balloon. In the present work, it expands unevenly, primarily as the result of the impact-induced relaxation of local regions of supercompressed minerals, remaining from the formation of the Earth as a giant planet.) Maxlow's work showed how the present continents could fit together properly on a much smaller Earth. Upon reaching the point where all the ocean floors were removed (about 245 million years ago), Maxlow continued to shrink the Earth by adjusting continental basins and rift zones, to the limit of our knowledge at 3,800 million years ago. That put the early Earth radius at a tiny 1,700 kilometers. Unfortunately, for an Earth of constant mass, that results in an unacceptably high average (or bulk) density, about 290 grams per cubic centimeter, nearly twice the density at the center of the Sun. I may suggest that two effects might have led Maxlow to overestimate the amount of crust to remove in reducing the Earth's size. This is done with some ignorance of the details of his work, and therefore some trepidation. Maxlow may have neglected the stretching of the seafloor through time as the Earth expanded and transform faults were developed. Secondly, he may have ignored the shrinkage of the continental segments due to the flattening of the general curvature on a larger Earth.

By comparing the sizes and densities of the Sun and the four large planets, I have estimated a reasonable (I think) density for the early Earth, at a final formation time of 4,500 million years ago, of 19.61 grams per cubic centimeter. (It should be recognized that the accuracy implied by figures of this sort is strictly illusionary. The estimate can only hope to be usefully close.)

Herndon has also discussed the formation of the Earth as a supercompressed gas-giant planet, "like Jupiter" [Herndon 2005, Herndon 2008]. While his process is somewhat different from what I have proposed here, Herndon estimates that the density of the kernel was in the range of 17 to 21 times the density of water. This is remarkably close to the value that I have estimated here, 19.61.

The current continental crust would require a radius of about 4,174 kilometers at that time, to fully cover the Earth. (Here, I also neglect any changes in the area of the continental crust. In passing, it should be noted that continental crust and oceanic crust differ distinctly by one being primarily granite and the other entirely basalt.) I adjusted this beginning point to Maxlow's expansion curve to produce a curve that I will use to represent the size of the Earth through time. This is shown, along with Maxlow's curve, in Figure 70.

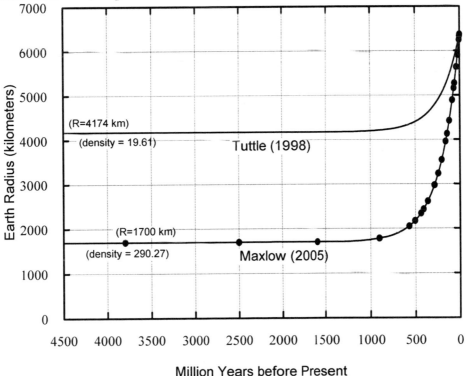

Figure 70. Radius of an expanding Earth.
Comparison of estimated radius of the Earth through time, resulting from Earth Expansion. (Density is shown in grams per cubic centimeter. The present bulk density of the Earth is 5.515.)

Using this curve for the size of the Earth, in the Middle Cretaceous, and adjusting for the extra support provided by larger leg bones, I calculate that *Argentinosaurus* would have been buoyantly self-supporting if the air density had been 0.60 grams per cubic centimeter, 60% as dense as water. Doing the same calculation for *Tyrannosaurus* at 67.5 million years ago, the air density needed to be 0.65 grams per cubic centimeter, 65% as dense as water, almost the same. To produce such a high density, the Earth's atmosphere must have been 278 to 352 times as massive as it is at present. The atmosphere of Venus is now 94 times as massive as Earth's. It might be reasonable to accept that the atmosphere of a more massive Earth (1.2 times the mass of Venus), that is farther from the Sun, and has always had a magnetic field, could be more massive than that of Venus.

Further evidence for a massive atmosphere in the past comes from a study of the accumulation of "cometary" material remaining after formation of the Earth [Delsemme 1997]. It was calculated that the early atmosphere of the Earth was 2,100 times as massive as at present. This excess must then be disposed of by subsequent impacts that dissipate the atmosphere without adding to it. By the end of the Cretaceous, an atmosphere only 278 to 352 times as massive as it is at present was required to support the giant dinosaurs and help the pterodactyloids to fly.

Many dinosaur fossils have been found with the same sorts of feathers as modern birds. Among the most interesting, *Anchiornis huxleyi* has been found at 155 million years ago, in the Late Jurassic [Hu *et al.* 2009, Witmer 2009]. It is older than the first-found dinobird, *Archaeopteryx*, by about 5 million years. This fossil clearly shows bird-like feathers, not only on the arms (forelimbs) and tail, as would be expected for an early bird, but on the legs and feet as well. *Archaeopteryx* had only wing and tail feathers. The wing areas (fore and hind) are stronger near the body, and lesser at the extremities.

This new oldest dinobird follows the functional form of a four-finned fish, sharing the forms of arm/pectoral and leg/pelvic wings/fins, differing in the alignment of the tail. Considering the possibility of a smaller Earth with a massive atmosphere providing thicker air at that time, even more so than in the Cretaceous, it appears that this form may have been adapted to swimming in the air. A horizontal tail for the dinobird would have provided flight control for flying close to the ground and over the trees, a predominantly up-and-down activity. For the fish, the vertically aligned tail would have accommodated swimming in the sunlight zone of surface waters, a more back-and-forth activity. The concentration of wing area close to the body would have been suitable for a time when the air was thicker, and wing-strokes moved a greater mass of air in producing lift and propulsion. As the air thinned in later times, this reaction mass diminished, the leg wings lost their benefit, and the forewings extended outward to gain efficiency.

The earliest ancient birds, such as *Confuciusornis* and *Archaeopteryx*, about 120 to 145 million years ago, show feathered wings in their fossils. Their ability to fly has remained uncertain. Flight feathers must provide lift without failing, and withstand additional forces for maneuvering. Overstressed flight feathers may fail by buckling, and this can sometimes be seen for modern birds flying in gusty winds, when sudden bursts "ruffle their feathers".

The stiffness of those feathers is provided by the rachis, the central shaft of the feather. In modern birds, this shaft is a thin-walled cylinder, tapering toward the ends. It is not known if the shafts of the ancient birds were hollow or solid, but they are much smaller in diameter than would be expected for the body size and feather length [Nudds and Dyke 2010]. It is predicted that these feathers would buckle at forces about 10 to 100 times less than if the rachises matched those in modern birds. That is, the wing feathers would buckle and collapse on takeoff, or any attempt at flying. Even if the shafts were solid, which would increase the weight, they would be much more likely to fail than the shafts of modern birds, in our current atmosphere.

That weakness would have made it impossible for the ancient birds to fly vigorously in our current thin air, although gliding is considered to have been possible.

This problem is eliminated by the thick air proposed here to provide support for the large dinosaurs. The same buoyancy provided to the land animals would have reduced the flight weight of the ancient birds. Flying in thick air would have been much like modern birds that actively fly under water, such as shearwaters, puffins, and penguins.

Attempts to apply this analysis to the later megafauna mammals of the Miocene gave similar results. Comparing the African Elephant with the *Indricotherium*, the largest land mammal known, required an atmospheric mass 125 times the present. These results are critically dependent on the mass assumed for the ancient beast, and the level of the mass for the modern animal, whether average or "maximum". This problem was not so severe for the Cretaceous Period, where the ratio of sizes was much greater, and the smaller Earth provided an inherently denser atmosphere. Comparing the Columbian Mammoth to the African Elephant, at 8,000 kilograms and 6,364 kilograms, respectively, the atmosphere should have been 60 times that at present. Making the comparison for masses of 6,000 kilograms and 3,900 kilograms, requires an atmospheric mass of 100 times the present. These atmospheric masses seem excessive, so either the mass estimates are in error, or this approach doesn't work for times close to the present, or a conceptual judgment is inapplicable.

Proceeding forward with the presumption that there is some truth in the analytical method, and in the great mass of air required, we must consider if that amount of air could be eliminated [Zahnle and Catling 2009]. Having determined that the Earth should have had an atmosphere roughly 300 times as massive in the Cretaceous as at present, in order to support the huge sauropods, help the large tyrannosaurs to prey on them, and make it easy for pterodactyls as large as a light plane to take off in a gentle breeze, the next problem is how to get rid of so much air. We certainly don't have it now, but we should keep in mind the possible support that thick air could have provided to the megafauna that went extinct at the end of the Pleistocene. Much of the earliest atmosphere, possibly 2,100 times as massive as at present, would have been lost in the Late Heavy Bombardment, around 4 billion years ago. Such a thick atmosphere would have absorbed much or all of the incoming impact energy, resulting in very few small impact craters from that early time.

Taking Venus as an example, some loss of air may be by conversion of some of the atmosphere to carbon dioxide, dissolving in the oceans, which Venus lacked. We think that carbon dioxide was 6 to 10 times as abundant in the Cretaceous atmosphere as at present. Carbon can be produced by the gamma-ray Giant Dipole Resonance reaction on oxygen (and on nitrogen), and the oxygen can be produced by photolytic dissociation of water high in the atmosphere. Thus, though there may be mechanisms that remove atmospheric mass into space, the mass of the atmosphere, and the oceans, should grow over geologic time. Since the Earth has about as much carbon dioxide sequestered as calcium carbonate as Venus has in its atmosphere, this process could eliminate about 94 times the present atmosphere by conversion to carbonates.

With significant expansion of the Earth, the atmospheric carbon dioxide pressure would be reduced, some oceanic carbon dioxide would be released, making the ocean water less acidic. Calcium carbonate would be deposited, perhaps non-biologically precipitated. The carbon dioxide concentration would be further reduced, so more would be absorbed from the atmosphere until essentially all of the carbon dioxide had been precipitated as calcium carbonate.

The Cretaceous Period was named for the remarkably great deposits of chalk (consider the White Cliffs of Dover, and Normandy, extending far inland, and around the world), calcium carbonate, carbon dioxide extracted from the atmosphere.

The thick air and tall atmosphere would have affected the living conditions of the Polar Regions. Greater heat carrying capacity would have kept the Polar Regions warmer, making those regions temperate and conducive to varied life. The tall atmosphere could have carried clouds much higher and provided scattered light so that even in the depth of winter, some light remained.

The fossil remains of a young and rapidly growing forest from the Late Permian (about 290 to 250 million years ago), in a polar region of ancient Antarctica, has been found and has been interpreted to show a notably warm climate [Taylor *et al.* 1992]. Growth was rapid during periods of sunlight, and there is no evidence of frost damage.

The remains of dinosaurs have been found in Australia, at Dinosaur Cove, suggesting adaptation to long dark periods [Vickers-Rich and Rich 1993]. At the time that these dinosaurs lived, Australia was close to Antarctica and Dinosaur Cove lay well within the Antarctic Circle, providing several months of darkness each winter. However, these winters may not have been bitterly cold. The Dinosaur Cove region shows evidence of a volcanic hotspot, and at the time of the polar dinosaurs, might have been associated with the current Mt. Erebus hotspot in Antarctica. This additional subterranean heat might have kept Dinosaur Cove temperate all year long, and thus the climate was controlled by supplemental heat rather than by latitude. At present, the Mt. Erebus hotspot may be providing the little extra heat that makes ice sublime in the three "inexplicably dry" valleys of Antarctica [Bond and Siegfreid 1990].

A world so different from ours should have left some evidence of its strangeness. Certainly the animals were strange, in shape and size and feature.

Higher pressure air, by a factor of 300 or so, would have facilitated absorption of oxygen by the skin. Modern frogs have been found without lungs, receiving 100% of their respired oxygen through their skin [Roach 2008]. According to David Wake of the University of California, Berkeley, for most amphibians, the majority of gas (oxygen and carbon dioxide) exchange is through the skin. (Salamanders have gill-like structures rather than lungs.) Joeys (baby kangaroos in the pouch) of a certain species obtain 50% of their oxygen through their skin [reference lost]. Even humans, in our present thin air, absorb 4% of our oxygen through our skin [Petrun 1965].

For the largest dinosaurs and other animals relying on the buoyancy of the thick air, a reduction in the density of the atmosphere (and the partial pressure of oxygen) would have led to their collapse and death, never to rise again.

The skin absorption of the joeys is significant in that the marsupials suffered many extinctions, along with the dinosaurs, at the end of the Cretaceous. Marsupials were the only mammal group to have major extinctions at the K/T.

Many ancient animals, most notably *Dimetrodon*, the fin-backed mammal-like reptile from the Permian Period, showed features that would have promoted beneficial skin absorption of oxygen [Fastovsky. and Weishampel 1996]. A similar but slightly different bony structure has been found in the related herbivore, *Edaphosaurus*. For these pelycosaurs, the area of the sail has been found to be proportional to the body volume of the animal, suggesting that the fin functioned as a temperature control [Benton 1990]. The same reasoning leads to the alternative, that the skin on the sail provided absorption of oxygen, the need being in proportion to the size of the animal. The large theropod *Spinosaurus*, from the Late Cretaceous Period, and *Ouranosaurus*, a herbivore, had very similar fins along their backs [Norman 1985]. An early tyrannosaurid, *Guanglong wucaii*, from the Late Jurassic in China, shows a remarkable nasal crest [Holtz Jr. 2010, Xu *et al.* 2010]. This has been considered to have been for ornamentation. (See Figure 71.)

Large areas of skin with little underlying tissue bring in oxygen with little demand for the use of the oxygen locally. This supplements the oxygen that is respired through the lungs, which are usually considered the only organs of respiration. Many dinosaurs and pterosaurs had crests or plates, such as *Stegosaurus* [Galton 1990], that would have also served this purpose. The fossil plates from the back of *Stegosaurus* show a great number of parallel grooves, suggesting that the outer surface of the bone was covered by blood vessels [Fastovsky and Weishampel 1996]. *Parasaurolophus* and *Corythosaurus*, from the Late Cretaceous, showed various hollow horny crests, connected to the nasal cavity [Norman 1985]. These were well lined with blood vessels, promoting absorption of oxygen. *Corythosaurus casuarius* had a hollow helmet formed by a double pair of nasal bones [Norell, Gaffney, and Dingus 1995]. It has been suggested that these structures may have functioned to produce sounds, or provided an enhanced sense of smell. Instead, they could have provided more surface area for absorption of oxygen.

If the oxygen content of the atmosphere were reduced (by conversion of oxygen by the Giant Dipole Resonance reaction, from the gamma flash of a nearby detonation), and the atmospheric pressure were reduced (by impact-induced expansion of the Earth), animals relying on such special structures to supplement respiration might die gasping for air. That would account for the extinction of the dinosaurs and so many marsupials.

The crests, collars, frills, and fins on the various dinosaurs provided extended skin surface for absorption of oxygen. While carbon dioxide may also have been released through the skin, this discharge of waste respiratory gas may have been the primary function of the lungs.

For fish in the sea, the greater atmospheric pressure (partial pressure of oxygen) would have forced more oxygen into solution, by Henry's Law. This would have made active life possible at lower atmospheric concentrations of oxygen than we would predict from present conditions [Dahl *et al.* 2010].

The pressure of the atmosphere and its fraction of oxygen (or density) were (and are) crucial parameters in respiration. Expansion of an Earth that was smaller during the Age of Dinosaurs fits into the death of the dinosaurs and the emergence of the mammals. According to an article in *Earth*, Gary P. Landis of the U. S. Geological Survey in Denver, Colorado, determined the fraction of oxygen in ancient air from bubbles trapped in amber [Yulsman 199?]. Those bubbles showed that the oxygen fraction dropped from 35% to 28% at about 66 million years ago, almost the end of the Cretaceous. A one-fifth drop in the oxygen content of our air at present would kill us. Contrariwise, if we (or mammals our size) attempted to breathe air with 35% oxygen, that would also kill us.

Unfortunately, we have no measure of the pressure of ancient air. Respiratory structure is also important, and Richard A. Hengst at Purdue University considers the respiratory systems of the dinosaurs were inefficient because they did not have active diaphragms and their nostrils were relatively tiny [Yulsman 199?]. In an oxygen-rich atmosphere, small size and efficient lungs (as mammals have) are successful survival traits. Likewise, large size and inefficient lungs (as dinosaurs appear to have had) are successful. But an animal in air cannot be large *and* have efficient lungs, in a high-oxygen atmosphere, with animal metabolism as we understand it. Absorption of oxygen through the specialized structures with no other clear purpose could have supplemented lung absorption, particularly while the pressure and fraction of oxygen were high. Loss of oxygen by its conversion into carbon by the Giant Dipole Resonance reaction and the gamma flash of a planetary satellite detonation, and the further reduction of pressure by impact-induced expansion, would have extinguished many dinosaur species during the Late Cretaceous, and the remaining survivors at the time of the Chicxulub impact. Those dinosaurs with efficient respiratory systems, yes, the birds, could have barely managed to survive into the new thinner air.

Figure 71. Ancient animals with specialized features that may have aided respiration.
Animals from a broad span of time had specialized features that would have promoted absorption of oxygen from the air.

Some suggestive evidence can be seen in the modified stance of *Tyrannosaurus rex*. Early reconstructions of the *Tyrannosaurus rex* skeleton resembled a kangaroo, with the tail resting on the ground, providing a tripod stance. Later, the tail was elevated and the spine aligned with the horizontal. However, this reconstruction forced an unnatural double kink in the neck, to raise the head. A stance that I propose, considering *Tyrannosaurus rex* to be like a shark on two legs cruising through air almost as dense as water, with feathered forelimbs to give control, straightens out the neck and spine into a nearly straight line. These various stances are shown in Figure 72 [Weishampel, Dodson, and Osmólska

1990].   The straightened-out version is close to the reconstruction of *Tyrannosaurus rex* STAN, for the Black Hills Museum of Natural History by the Black Hills Institute of Geological Research [http://www.bhigr.com/pages/info/info_stan.htm].   In that reconstruction, the neck still has two unnatural kinks.

Figure 72. Reconstructions of *Tyrannosaurus rex*.
Proposed stances for *Tyrannosaurus rex*. Top is the original (Osborn 1916), middle is more recent (Newman 1970), and the bottom is proposed here as a "shark on two legs", cruising through thick air, and guided by feathered forelimbs acting as canard fins.

Major impacts, of which there seem to have been four in the time period 74 million years ago to 65 million years ago, plus other, earlier impacts, would have dissipated some of the atmosphere into space. A thicker atmosphere may suffer greater loss from an impact because of absorbing more energy.

In passing, it should be noted that major extinctions are topped off stratigraphically by a so-called "cap carbonate". These deposits are rather poorly studied and discussed, but I think they are produced by precipitation of carbonate minerals as the result of depressurization of a carbonated ocean by the impact-induced expansion of the Earth.

In the original formation of the Earth, as a giant planet roughly 30 times as massive as Neptune is now, with a molten silica magma ocean and a massive insulating atmosphere, the kernel of the Earth, what was to remain as our world, was highly compressed, to a bulk density on the order of 20 grams per cubic centimeter. That is roughly four times the present density. When the Sun ignited, after the complete formation of the planets, the burst of energy blew away the atmosphere and that allowed the magma ocean to vaporize into space, leaving just the compressed kernel. (This ejection of the gases from the inner Solar System is much like the scenario in the standard story, where the early Sun blows away the gas leaving only rocky planetesimals to form the rocky planets.)

The compressed phases of minerals in the Earth were locked into their great densities, just as diamond is, a high pressure/high density form of carbon, and "a diamond is forever". Impacts on the surface of the Earth pass energy through its entire body, with shock waves and compression/rarefaction waves, to jolt the compressed phases and assist in relaxation and expansion. This will happen in limited regions and cause localized expansion. That regional expansion will be accommodated by the Earth always retaining a generally spherical shape due to its gravity. Such expansions will be accompanied by catastrophic surface effects.

With the general decline in the density and richness of the atmosphere during the Late Cretaceous, and the steep gradient of density with elevation, different forms of dinosaur, and other life, would have been more or less successful. Giant quadrupeds would have been less supported as the atmosphere thinned, and would have been less successful at higher elevations. Pterosaurs might have gotten larger to benefit from larger wing areas. Mammals might have grown larger at higher elevations, where the oxygen pressure was less. Dinosaurs might have been smaller there.

In a reversal of roles between dinosaurs and mammals in the Early Cretaceous, with larger mammals eating smaller dinosaurs, a mammal (*Repenomamus giganticus*) was found with the remains of a juvenile ceratopsian dinosaur (*Psittacosaurus*) in its stomach [Hu *et al.* 2005]. A terrestrial elevation sufficient to reduce the atmospheric density and the partial pressure of oxygen so that dinosaurs were not favored, and mammals were not repressed, might have produced stunted dinosaurs and overgrown mammals. This suggests that the present site of Lujiatun village in China, where the fossils were found, might have been an elevated plateau about 130 million years ago.

While mammals were generally small (tiny) during the Mesozoic, there was a great increase in the general sizes of mammals at the end of the Cretaceous. This is shown by Alroy's study of the evolution of body mass in fossil North American mammals [Alroy 1998, Kerr 1997]. As shown in Figure 73, mammal sizes were in the range of 5 to 1,000 grams in the Late Cretaceous, but increased in size immediately after the mass extinction at 65 million years ago. This growth led to the largest mammals that have ever lived on dry land. The largest late mammals were 10,000 times more massive than the largest early mammals. This rapid growth to a maximum size has been found to be a global effect [Smith *et al.* 2010], as it would be from a global change in living conditions, such as a reduced partial pressure of oxygen.

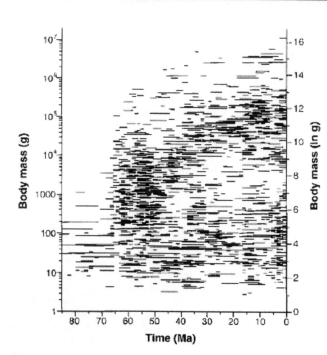

Fig. 1. Temporal distribution of Cenozoic mammalian species across the body mass spectrum. Age ranges were based on a multivariate ordination of faunal lists (*18–21*). Mass estimates were computed with the use of published regression coefficients for mass against m_1 length × width [Carnivora, Insectivora, Primates, and Rodentia (*13*)] or against m_1 length [Artiodactyla and Perissodactyla (*14*)]. Coefficients for Primates were also used for Plesiadapiformes (*15*); coefficients for Carnivora were also used for Mesonychia (*16*). Proboscidean m_1's are rarely described, and their lower cheek teeth all are relatively large; mass estimates based on m_2 area measurements and the all-mammal regression for combined p_4-m_2 area agreed with earlier literature (*17*). The all-mammal m_1 area regression was used for all remaining mammals.

Figure 73. Sizes of mammals from the Late Cretaceous to the Late Pleistocene.
Body mass of 1,534 different species during their lifetimes, showing small mammal sizes during the Mesozoic, before 65 million years ago, and the marked increase in size after the K/T mass extinction [Alroy 1998].

One of the effects that can be traced through geologic time is the transgression and regression of the oceans on the continents, the rises and falls of sealevel. These have been studied and reported by the Exxon Production Research Company, in their hunt for petroleum. Bilal Haq and associates, in work initiated by Peter Vail, have produced charts of the relative onlap of ocean on land, and the accompanying sealevel [Haq, Hardenbol, and Vail 1987, Haq and Schutter 2008]. These charts show episodes of a gradual rise in sealevel, punctuated by abrupt sealevel drops, on the order of a hundred meters. (Similar but much smaller changes in local sealevel can be seen in an earthquake zone, west Sumatra [Sieh *et al.* 2008].)

Since the discovery and final acceptance of the impact origin of Meteor Crater in Arizona (also known as Barringer Crater), the list of identified craters on the surface of the Earth has grown to nearly 200. Some of these are under the sea, but the ocean makes exploration difficult and quickly covers the scars. In Figure 74, I have plotted the cumulative number of known impacts since 550 million years ago, the beginning of the Cambrian Period, adjusted perforce, in an attempt to show the actual number that we expect have occurred, whether on land or sea. That was done simply by multiplying the actual count by the present ratio of the global area to the land area. (Admittedly, that is a bit awkward, because I am trying to show the progression of impacts from a time when there were no ocean basins, and impacts should have been recorded on land. Perhaps the ocean water was sufficient to cover many parts of the early continents, thus eroding and covering the craters as happened after formation of the ocean basins.) I have also plotted the cumulative number of sealevel drops shown by Haq and colleagues. This shows that, in any given interval of time, the number of impacts closely matched the number of sealevel drops, supporting the idea that impacts induce expansion of the Earth and expansion causes sealevel drops.

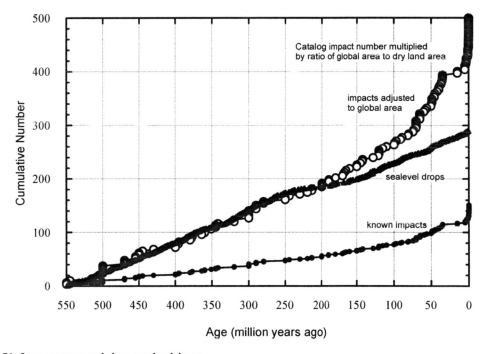

Figure 74. Impact craters and abrupt sealevel drops.
Comparison of cumulative impacts and sealevel drops, showing excellent correlation from the Cambrian (550 million years ago) to the Jurassic (206 million years ago). After the end of the Triassic (208 million years ago), impacts in deep ocean basins were cushioned by the water and may have been less effective in causing expansion, or alternatively, it may be that the Earth had significantly lost its supercompressed regions.

A further indication of impact-induced expansion comes from measurements of the physical strain rates shown by fibrous pyrite crystals from 85 to 50 million years ago [Müller, Aerden, and Halliday 2000, Ramsay 2000]. The strain rates shown by pyrite crystals from the Northern Pyrenees, representative of the Earth's crust, show an abrupt increase by a factor of 7, over the time period of 66 to 62 million years ago. This corresponds to the moment of the Chicxulub impact, at 64.4 million years ago. After this impact, the strain rates returned to normal for the balance of the measured interval, to 50 million years ago.

Pterosaur Ptakeoff

The flying takeoff of large and giant pterosaurs, such as *Quetzalcoatlus northropi*, has been a difficult problem to explain. Michael Habib (Johns Hopkins University) studied the leg bones and wing bones and concluded that *Quetzalcoatlus* jumped into the air from a four-legged stance [Habib 2008]. (*Quetzalcoatlus northropi* "weighed" 250-550 pounds, and had a wingspan of 36 feet [Fox 2009].) This has been argued against by David Unwin (University of Leicester, England) because such pterosaurs require an air speed of 30-40 miles per hour for takeoff, and that speed would be hard to achieve from a standing jump. Kevin Padian (UC-Berkeley) points out that the smaller pterosaurs were bipedal and took off from two legs.

Adjusting the atmospheric density by comparison with the Flying Fox fruit bat doesn't seem to apply well to *Quetzalcoatlus*, requiring 406 times as massive an atmosphere, which would have made *Argentinosaurus* so buoyant as to float. (It is worth noting that some of the large dinosaurs have been found with stones in their stomachs, sometimes explained as "gizzard stones", like chickens use to grind their food, but also found in alligators, for ballast, to provide a low floating level.) However, it reduces the estimated minimum flying speed from 30-40 miles per hour to about 1.4 miles per hour, making take off and flying much easier [Chatterjee and Templin 2004]. A mass of 70 kilograms was calculated for *Quetzalcoatlus*, based on volumetric estimates for several pterosaurs and a multiple regression to extend the estimates. This gave a mass of 62 to 77 kilograms, while other estimates of 85 kilograms and 200 kilograms are quoted. Michael Habib refers to estimates of 70, 200, and 250 kilograms, favoring the largest as representative of large pterosaurs with long legs and long necks. Habib considers 70 kilograms to be unreasonably small. Chatterjee and Templin concluded that it would be unlikely for a 200-kilogram *Quetzalcoatlus* to take off from the ground or to fly, difficult for an 85-kilogram *Quetzalcoatlus* to take off with a run or to maintain powered flight, and considered that 70 kilograms was an optimum limit for gliding, takeoff and landing. They restricted *Quetzalcoatlus* to gliding and formation flying (like migrating geese) to conserve energy.

Thick air would have facilitated takeoff. The drag associated with producing aerodynamic lift is inversely proportional to the density of the air, and so thick air would have reduced this component of drag to a negligible amount.

The Earth expands in spurts, asymmetrically, not like a balloon, but by continental blocks being forced up, to crumble and collapse at the margins, making "subduction zones" by overthrusting onto oceanic crust, and making mountains by flattening or flanging, or by being pushed directly up from below. This makes the geological evidence difficult to interpret and easy to overlook.

If Earth expansion occurs asymmetrically, some continental areas would be elevated faster, farther (higher) than others, causing regional extinctions and survivals of the dinosaurs.

Is there evidence for such drastic changes in the Earth? Yes, and like much of the observations discussed in this book, it is clearly ambiguous. Observations have been interpreted in a rigid uniformitarian framework that insists that only those processes we see currently at work have been effective through all geologic ages, and that the Earth has been pretty much the same for all its life.

The lesson of the dinosaurs is this: at each mass extinction, each great transition, the character of the Earth has changed permanently. Each time, old life was no longer on the new Earth. Old life, in its old form, could not exist on the new Earth. The dinosaurs could not have continued simply by being sufficiently foresighted as to have laid in supplies for a few years of "impact winter". They were not capable of living on the new Earth because they were no longer suited to the new, changed living conditions. The next time the Earth changes, our life will no longer be capable of existing in nature on the new Earth. We are only temporary inhabitants of this world.

The Human Event

Paleoanthropology,
... a discipline where ego so often triumphs over evidence.
Henry Gee, 1993

While paleoanthropology deals with the very dead, the science is very much alive, with controversy, disagreement, and conflict, found in the books and views and news, perhaps more so than most fields of study. Confrontations over personal style and approach accompany objective contradiction of alternative interpretations. Because of this, no two books are likely to give the same old same old, but each writer and each researcher presents new, other, and alternative facts and ideas. This produces rather greater uncertainty in what the observations really are, that we must try to interpret for ourselves. Comparison with what we have seen for the earlier past will help to guide our ideas of the possible.

Suppose that the Cambrian explosion was brought in with a bang, due to the direct radiation and a possible power surge in Earth's reactors induced by the nuclear detonation on Venus at about 500 (or 540) million years ago, and that the great outflowing of lava at the end of the Permian producing the Siberian Traps resulted from some other detonation, and that other lava floods occurred at about 177 million years ago as Earth's reaction to a nuclear detonation in Mars. While some effects can be produced by direct radiation, and others by radioactivity generated within the Earth, a flood basalt eruption needs a large impact about diametrically opposite. Did anything of interest occur on the Earth in response to the apparent detonation of Planet X about 6 million years ago? And does it appear to be related to the types of effects discussed so far?

As well as we can tell at present, human beings are derived from one or another of the small mammals that survived the extinctions at the end of the Age of the Dinosaurs. Various mammals survived and thrived in the changed conditions, and a new type of animal, the primate, evolved. This order would eventually produce and include the apes, monkeys, and humans. Narrowing our focus to a very small group of these primates, we need to explore the orangutan, the gorilla, the chimpanzee, the bonobo, and the human. There are important similarities and important differences. Our chemistry is nearly identical with the others, down to the genetic level. Alone, we write and read books, and to a large extent have within our power control over the fate of most life on Earth.

Our origin in time has been hazy, slowly pushed back by new fossil discoveries. The latest earliest (that is, the most recent most ancient) representation is a fragment of a child's jaw, assigned to *Australopithecus ramidus* [White *et al.* 1994] and dated to 4.4 million years ago [WoldeGabriel *et al.* 1994]. This time is close to a time that appears to have been crucial in the history of the horse, 3.7 million years ago, in North America.

The gamma flash produced by the detonation of Planet X projected sufficiently intense radiation on the surface of the Earth to kill, sterilize, or mutate the creatures living on the nearside. These new creatures could have come about by surviving genetic mutations, and the changed conditions, particularly high surface radiation exposure rates, could have marked the end of their ancestor. Severe, barely survivable mutations could have given to the chimpanzee, bonobo, and gorilla the hand structure to walk on knuckles (as never before), and to the humans the foot (so drastic a change from a hind hand) and pelvic structure to walk upright, and the potential for a vastly new and larger brain.

The histories of these four, the gorilla, the chimpanzee, the bonobo, and the earliest humans, show features that are suggestive of evidence of severe radiation effects. Unusual new forms of locomotion were adopted by these creatures, without precedent in their ancestors; knuckle walking by the gorilla, the chimpanzee, and the bonobo, and obligatory bipedalism by the humans. These provide some small measure of radiation avoidance by moving the bulk of the body, and its most sensitive organs, including the reproductive organs, away from the radioactive ground. Knuckle walking also provides, explicitly, protection of the palms of the hands from radiation damage. A similar effect is achieved by bipedalism simply by moving the hands away from the ground. This resembles the unusual life-styles adopted by the new life of the Cambrian, and also at the Triassic, apparently serving then as at this later time as a means of reducing radiation exposure.

In this time period, following the proposed detonation of Planet X at 6 million years ago, there was a surge in the island-building activity of the Hawaiian hotspot volcano [Francis 1993]: all the existing Hawaiian islands were built in the last 5.5 million years [Wood and Kienle 1990], the climate warmed [Feder and Park 1989], and the east Antarctic icesheet nearly melted away. Volcanic activity increased greatly in the Great Rift Valley in northeast Africa [Hay 1976], and the region appears to sit atop a dome created by a broad mantle plume [Strahler 1981]. Volcanic activity was common, as shown by the bipedal footprints in the Laetoli ash at 3.7 million years ago [Feder and Park 1989], the eruptions of Ngorongoro from 2.45 to 2.0 million years ago and Olmoti at 1.85 to 1.65 million years ago [Hay 1976]. These could have been effects induced by the radiation from the planetary detonation. Life on the surface of our planet would have been exposed to direct high-energy gamma radiation.

And at about this time, in that area, new forms of primates appeared, and their last common ancestor disappeared [Edey and Johanson 1989]. It is thought that the early ancestors of humans and African apes had a chromosome set of 42. Humans have 46, and the gorillas, chimpanzees, and bonobos all have 48. The orangutans also have 48 chromosomes, which would seem to argue against a much earlier origin than the other apes and us. Was our chromosome set formed by karyotypic fission from a set of 42? Were the apes created by further fission of our Chromosome 2?

If we suppose that a major astronomical detonation occurred near the Earth about 6 million years ago, the radiation from that detonation could have shattered the chromosomes of our primate ancestors. The consequences of this disruption would have varied greatly, perhaps producing similar sets of 48 chromosomes for the gorillas, chimpanzees, and bonobos, and a rather different set of 48 for the orangutan. These new sets of chromosomes would have included the genetic blueprint for a quadrupedal knuckle-walker, with the capability of temporarily walking and standing on two legs. Recombined in another way, and with special changes in individual genes, a set of 46 chromosomes could have carried the instructions for a new primate, an obligatory biped with the most powerful brain yet seen. Humans were created on Earth by an intense flash of radiation from space.

At about this same time, the orangutan (*Pongo pygmaeus*) appeared, but survived only out of Africa, in nearby Eurasia. Perhaps this animal retained more of the ancestral features than any others and so, lacking any new capability for coping with the new condition of radioactivity, vanished from Africa. Survivors from Sumatra and from Borneo now show a distinct difference in Chromosome 2 [Raloff 1995]. Following this genetic disruption, our ancestors started on a path that led to society, culture, and civilization. But was this a path of genetic evolution through a multitude of different species, or a path of achieving full development of our created genetic potential as a single continuous species?

Comparison of the banding patterns on human and ape (chimpanzee and gorilla) chromosomes has suggested that the chromosomes of one of the human pairs was formed by attachment of two of the apes' chromosomes [Feder and Park 1989], in fusion. As an extreme form of chromosome crossover, which has been observed to be a major process of genetic change [Edey and Johanson 1989], parts of two different chromosomes may join to make one. However, as an alternative to chromosomal fusion, chromosomal or karyotypic fission may have been the effective agent in producing these new forms of primates [Todd 1992]. During chromosome segregation in mitosis, the chromosomal pairs are joined by a tiny junction, the kinetochore, in the centromere. Certain conditions at that stage can lead to the splitting of these chromosomes at the centromere. This chromosomal fission, as an alternate possibility, is also suggested in Figure 75, where the extra pair of chromosomes (p and q) of the gorilla, chimpanzee, and orangutan is shown associated with our Chromosome 2 (II), as if the two were the result of chromosomal fission of our II [http://www.selu.com/~bio/gorilla/genetics/cyto.html].

Chromosomal pairs must match to permit interbreeding. Hybrids with mismatched chromosomal sets are usually sterile or non-viable, and so do not propagate [Postlethwaite and Hopson 1989]. The genes from the chromosomal pair that is missing in humans appear to be attached entire to one end of our Chromosome 2 [Nature 1994]. This form of chromosomal aberration is shown in Figure 76, where the chromosome pairs of a survivor of an atomic bombing show the disconnection of a part of Chromosome 2 and the attachment of the fragment to the lower end of Chromosome 14 [Shigematsu *et al.* 1993]. Subsequent interbreeding and the production of hybrid "chimpamans" and "humpanzees" would have been prevented by the fact that chimpanzees have 48 chromosomes (24 pairs) and humans have 46 chromosomes (23 pairs) [Dawkins 1987]. Gorillas also have 48 chromosomes (24 pairs) [Feder and Park 1989]. Chromosomal differences might be considered the strongest and most basic evidence for different species.

Figure 76 shows the effect of a splitting off of chromosomal fragments, from Chromosome 2, observed in the study of the atomic bomb survivors in Japan [Shigematsu *et al.* 1993]. These fragments broke at some distance from the centromere. Had the break come at the centromere, the two fragments would have become isolated, independent chromosomes, p and q, as seen in the karyotypes of the chimpanzee, gorilla, and orangutan (Figure 75).

These sets of fissioned chromosomes are subsequently selected for and against by the usual mechanisms of reproductive, functional, and survival fitness, and a new species stabilizes. The new species retains some of the unfissioned ancestral chromosomes and replaces others with fissioned chromosomes. The theory of karyotypic fissioning has been applied to explaining the origin of the modern apes (gorilla, chimpanzee, bonobo, and orangutan) and humans [Giusto and Margulis 1981]. In this model, an ancestor to all of us, ape and humans, with perhaps 42 chromosomes (2n = 42 in the geneticists' code, to assure understanding that it is chromosomes in pairs that are being counted), experienced a significant chromosomal fissioning event at 6 million years ago. Of the several possible sets of viable karyotypes, five continued into the present, producing humans with 2n = 46, and gorillas, chimpanzees, bonobos, and orangutans, with 2n = 48, in slightly different arrangements. The karyotypic fissioning model explains the similarities and differences in the genetic makeup of these creatures as the result of different selections of the fissioned karyotype, rather than differential and gradual evolution at different stages from a common ancestor.

One of the major objections to the karyotypic fissioning theory has been the failure to identify a mechanism by which all chromosomes would fission [Stanyon 1983] and therefore, it is claimed that a miracle must be invoked. However, chromosomal aberrations of several types are produced by exposure to ionizing radiation [Shigematsu *et al.* 1993]. This effect is proportional to dose at low doses and has been used to determine radiation doses as low as 3 rem (a somewhat obsolete unit of effective radiation dose, the "roentgen-equivalent-man". In modern units, this is equal to 30 millisieverts; natural radiation doses range from about 1 to 10 millisieverts per year). It seems possible that an intense exposure of radiation to all members of a population could lead to a significant amount of simultaneous fissioning of chromosomes in the manner suggested by the karyotypic fissioning theory.

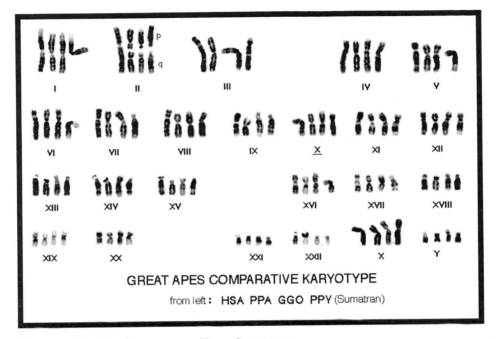

Figure 75. Karyotypes of humans, chimpanzees, gorillas, and orangutans.
Comparison of chromosomal karyotypes of humans, chimpanzees, gorillas, and orangutans, showing the apparent break in our Chromosome 2 (II) represented by the p and q segments of the great apes [http://www.selu.com/~bio/gorilla/genetics/cyto.html].

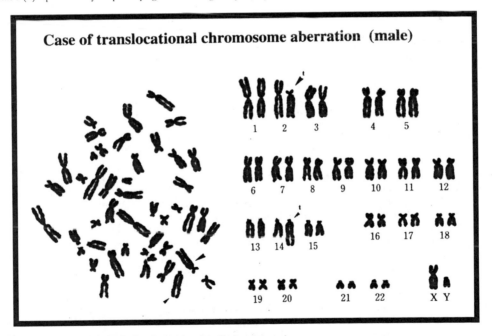

Figure 76. Aberration caused by radiation from a nuclear fission detonation.
Example of chromosomal aberration caused by radiation from a nuclear fission detonation, in the form of fragments from Chromosome 2 attaching to Chromosome 14, exhibited in an atomic bomb survivor. In this case, the Chromosome 2 fragments were broken at some distance from the centromere and so were able to join as extensions of an existing chromosome. The normal forms are shown to the left in each pairing of chromosomes [Shigematsu *et al.* 1993]. The arrowheads indicate the translocations.

Since much of my radiation career has been spent trying to soothe excessive fears, most of what I have known is related to how poor a mutagen radiation really is. Radiation-induced mutation was found in the fruit flies ("lab animals" selected and bred for the ease with which they expressed mutation of all sorts), but not in the mega-mouse experiments, and has not yet been seen in the third generation of the atomic-bomb survivors in Japan. In Neil Todd's theory of karyotypic fission, the chromosomes are broken in two and recombine, however they might. The survivors survive, the others don't. This produces what Stephen Jay Gould called "hopeful monsters" and I think he really didn't believe in them, because he didn't know of a mechanism to produce them. In the life-crisis brought on by the radiation from a nearby astronomical detonation, such as that of Venus or Planet X, the immediate survivors are those with a just barely less-than-lethal radiation dose, and I think it is likely that their chromosomes are very fractured *à la* Neil Todd.

Banding of chromosomes can indicate similarities and differences somewhat coarsely (Figure 77). The female sex-linked X chromosome in humans, chimpanzees, gorillas, and orangutans shows remarkable similarity over its entire length, essentially identical for all the great apes, us included.

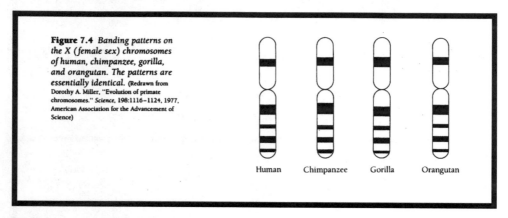

Figure 7.4 *Banding patterns on the X (female sex) chromosomes of human, chimpanzee, gorilla, and orangutan. The patterns are essentially identical.* (Redrawn from Dorothy A. Miller, "Evolution of primate chromosomes." *Science*, 198:1116–1124, 1977, American Association for the Advancement of Science)

Human   Chimpanzee   Gorilla   Orangutan

Figure 77. X chromosome banding of humans, chimpanzees, gorillas, orangutans.
Banding comparison of the X chromosome, the female sex-determining chromosome, of humans, chimpanzees, gorillas, and orangutans, showing the great similarity [from Feder and Park 1989]

The X chromosome is duplicated in human females, each one of the pair contributing active genes. In each cell, one of the pair is switched off, so that a female consists of a mosaic of the genes expressed by the X chromosomes. This makes the genes on each X chromosome subject to natural selection, and might account for the identical nature of the chromosomes in all the great apes.

While the X chromosome similarities may show that we all make good mothers (or bad), genetically, there must be differences. These show up at least (and surely more) in the autosomal chromosomes 9 and 12 [Gibbons 1998] in Figure 78.

The tectonic forces created by the power surge in the local reactor cells of the Earth started a separation in the crustal environment, like that seen between South America and Africa, on a smaller scale, with dry savannah to the east of the Great Rift Valley in Africa, wet forest to the west, and wooded uplands to the south. The chimpanzee and the bonobo, well suited to moving in the trees, with hands and feet both able to grasp, survived by living in the trees, away from the most intense radioactivity of the ground. The gorilla, a ground dweller, vanished from the area and survived in the (presumably) less radioactive areas to the south. The humans, able to spend most of the time upright on two feet, thus moving the vital organs (and the reproductive organs) farther from the ground, and perhaps pulling themselves into the sparse trees to rest and sleep, survived in the plains [Coppens 1994]. The ancestral species, not being equipped for survival in a radioactive world, perished.

In a radioactive world, bipedality was a survival-selective feature that helped our ancestors endure an Earth surface that was temporarily much more radioactive than before or since. (This affected the horse also, showing some effects in animals other than man. The many species of horse died out at the time we were formed, and a new horse arose, with radiation-resistant feet, and legs that could lock so that the horse could sleep standing up and not lying on the ground.) The crucible of human evolution was not climate change, but the gradual decline in terrestrial radioactivity, punctuated by other nuclear detonations, perhaps at 2.7 million years [Prueher and Rea 1998], followed by a related impact at 2.15 million years [Gersonde *et al.* 1997]; a detonation at 370,000 years ago that produced Neandertals and the cave bear (and the pulsar Geminga [Bignami and Caraveo 1992]); and one at (probably) 12,900 years ago that ended the Pleistocene and nearly killed off humans, destroying the existing civilization in its course.

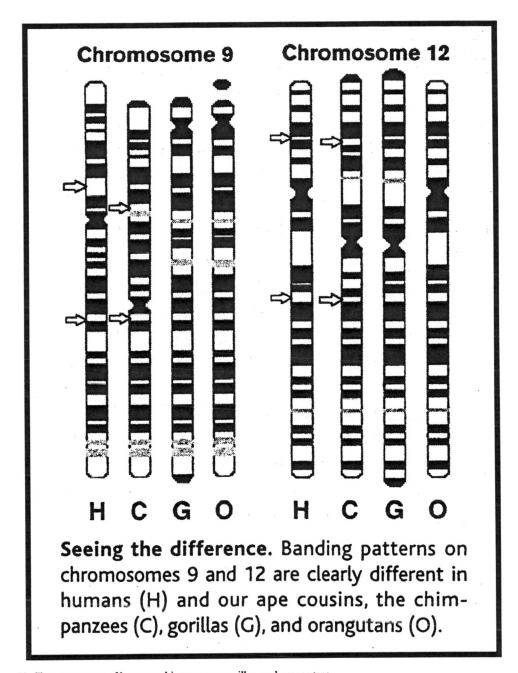

Seeing the difference. Banding patterns on chromosomes 9 and 12 are clearly different in humans (H) and our ape cousins, the chimpanzees (C), gorillas (G), and orangutans (O).

Figure 78. Two autosomes of humans, chimpanzees, gorillas, and orangutans.
Comparison of autosomal chromosomes, the ones that help determine what kind of animal we are, showing different banding patterns between humans, chimpanzees, gorillas, and orangutans [Gibbons 1998].

The origins of these five primates (including the orangutan, which may have branched off earlier), called (somewhat pridefully?) the great apes, ourselves included, are rather poorly known. There are some possible fossil ancestors, but these fossils are so incomplete and distant in time, it is not at all clear just who they are ancestral to. There is some biochemical evidence that orangutans split off from this ancestral lineage approximately 20 million years ago. Similar evidence shows that gorillas, chimpanzees, and humans all split at about 6.5 million years ago, with some researchers thinking that the gorilla split first. The evidence is all weak and uncertain. There is a possibility that the bonobo split from the chimpanzee about 1.8 million years ago. This situation is summarized roughly in Figure 79, Fossil Record for Primate Development. Identified fossils are shown by small circles, and the assumed extent of the species is shown by the dashed line. There are some remarkable points about this chart. There are many recent fossils for orangutans, from Southeast Asia. There may be more at earlier times that I have not gathered from the literature. Fossils for humans have been found nicely spread through time from the earliest immediate (bipedal) ancestor, *Orrorin turgensis*, at 5.9 million years ago. There is no fossil record for the gorilla, the chimpanzee, and the bonobo. These are the African apes, evolving, growing, living, dying, on the same continent as early (and later) humans, but there are no records of ape fossils, that I could find. The fossil record of the development of humans is a tribute to the diligence and determination of the dedicated fossil hunters. Perhaps "look and ye shall find" has been the rule, and the drive to find human fossils has forced a concentration of the looking at prospective human fossil sites. Many of the finds have been somewhat different from each other. This has led to a multitude of names, some have been changed, some are unchangeable.

Attempts to draw a clear family tree for humans, a genealogy of ancient man, resemble efforts to provide the greatest explanation with limited information. Each new discovery, and each discoverer, provides a somewhat different picture. One way of displaying the sequences of these various different named fossil names is shown in Figure 80, Fossil Record for Human Development. This starts with what has arguably (very arguably!) been termed either the first of the humans, or the last common ancestor of humans and chimpanzees, *Sahelanthropus tchadensis*. The fossil named Toumaï is covered with uncertainty, controversy, and argument, but serves the purpose of a placekeeper here very well. (There has been some confusion over the use of "hominid" and "hominin" in referring to the string of our ancestors (and us) back to (but not including) our last common ancestor with the chimpanzee. Here, I will use somewhat sloppily the term "human" to refer to all of us, back to (but not including) the Last Common Ancestor.)

An interesting part of the collection and cataloging of our ancestral bones has been in the individual statements that a certain collection of bones (or a single fragment) appears to be more "robust" than other, otherwise related bones. The others are labeled the gracile form, "gracefully slender". In the chart, the gracile forms are shown as open dots; the robust are shown as filled dots. The distinction is easily seen between the Neandertals and the Cro-Magnon, as robust and gracile, respectively. These forms seem to cluster in time.

This duality, gracile or robust, is seen repeatedly, starting with *Australopithecus afarensis* and *Kenyanthropus platyops*. Then *Australopithecus africanus* and the *Paranthropus species* showed the similarities and the differences. Cro-Magnon was the gracile form to the robust Neandertals. In Australia, a robust form dated to 15,000 years ago found at Cohuna matched a gracile form dated to 13,000 years ago, found at nearby Keilor [Gore 1997]. Also in Australia, the remains of "Mungo Lady" were found near Lake Mungo, and dated originally to about 24,000 years ago. Subsequent dating of the sediment where the bones were found gave dates of about 62,000 years ago [Gore 2000]. Some of the dating methods, such as Electron Spin Resonance and Optically Stimulated Luminescence require an assumption of the rate at which the sediment accumulates radiation excitation. This is usually measured or calculated for the present time and considered to have been constant. A brief but intense flash of gamma radiation from a detonating satellite would give such a large amount of radiation as to make the sediment appear to be much older than it actually was.

Meave Leakey has commented that the East African habitats changed dramatically from 7 million to 5 million years ago, with widespread extinctions and speciations [B. B. 2001a]. An example is that it appears that the mammoth originated about 6 million years ago [Gibbons 2010b]. Such disruption of life would be a result of the radiation from the detonation of Planet X. According to René Bobe, a similar major turnover of animal species (and our own) occurred around 2.5 million years ago [B. B. 2001b]. This was the time when *Australopithecus afarensis* vanished, and five new species of humans appeared. This could indicate the occurrence of another significant planetary/satellite detonation, with its associated radiation blast.

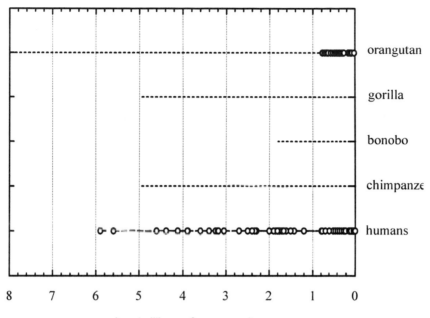

Figure 79. Fossil record for primate development.
No fossil gorillas, bonobos, or chimpanzees have been found, and the orangutan fossils are relatively recent. Human fossils may span the entire time of our existence.

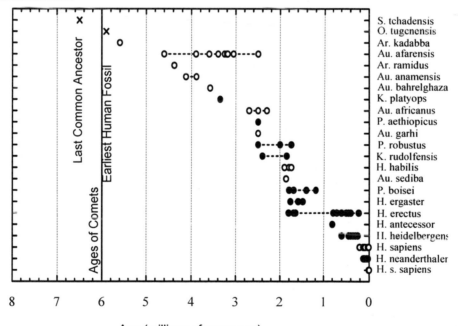

Figure 80. Fossil record for human development.
The multiple names, based on slight differences between often uncomparable skeletal fragments, presents a much more fragmented appearance than is useful. Ignoring the multiplicity, it is clear that in some times, in some areas, humans have been gracile (open dots) or robust (filled dots). In some cases, sexual dimorphism may be interpreted as two different species, or the reverse.

There is some consideration that the end of the Pliocene, the boundary between the Pliocene and the Pleistocene, should be at about 2.6 million years ago [Morrison and Kukla 1998]. Evidence for the occurrence of a nearby astronomical detonation, a supernova, has been seen in deposits in manganese nodules from the deep sea floor [Knie, *et al.* 2004]. This deposit was dated to 2.8 million years ago. At about 2.67 million years ago, an abrupt change occurred in the North Pacific sediments, with rapid swings in magnetic susceptibility, the presence of dropstones and ice-rafted debris [Prueher and Rea 1998]. (Dropstones and ice-rafted debris can be a misinterpretation of impact breccias and debris.) The researchers considered that the changes required for glaciation were too rapid for either tectonic or orbital forcing. Ash layers increased by a factor of ten, so volcanic effects were thought to be the cause of apparent cooling. The individual ash layers are associated with large increases in magnetic susceptibility. Asteroidal impacts can produce ash layers and bring increased magnetic susceptibility. Instead of volcanoes, major impacts may have altered living conditions on Earth.

A cranium dated to about 1 million years ago and eventually assigned to the African *Homo erectus* (*Homo ergaster?*) showed a combination of primitive (*ergaster*) and progressive (*sapiens*) features, a mosaic of old and evolving [Abbate *et al.* 1998]. This led the researchers to caution about the assignment to a particular species. This is one of several examples of an uncertain blend of features that are typical of the robust form and the gracile form. Such a blend could be produced by varying intensities of a variant of Paget's Disease of Bone resulting from differences in exposure to radioactivity.

In our present world of most modern humans, most of the population is gracile, while those few who spend their lives under the auroras, such as the Inuit, appear to have the robust form [Gore 1996, Ponce de León and Zollikofer 2001]. That difference may reflect exposure to beryllium-7, a radioactive isotope that comes to the radiation belts of the Earth from the Sun. It can cause much the same effects as calcium-41. Like the Inuit, the Australopithecines had barrel-shaped chests, as did the Neandertals. This will be explored further in connection with the Neandertals and Paget's Disease of Bone.

The differences used to distinguish what are called separate species of humanity may not reflect different genetic species, but rather the different variations in the effect and intensity of radiation-induced pathologies. The generally progressive development of humans over the past 6 million years reflects the gradual decline in the levels of radioactivity in the Solar System and on Earth, and not the evolutionary march to perfection. These pathologies, seen now in fossil bones only as distortions of the bones, would vary as a result of variations in exposure to external radiation (such as gamma rays from cobalt-60 from the long-lived iron-60, and aluminum-26), and to internally deposited radioactivity (such as calcium-41), due to time of exposure, intensity of exposure, duration of exposure, and other variabilities. A difference from modern Paget's Disease of Bone, here proposed as the result of accumulated naturally radioactive lead-210, is that the radioactive calcium-41 would be deposited in the growing skeleton from the first formation of the bones during gestation, whereas the lead-210 is excluded during gestation and only accumulates during the lifetime of the individual. That makes the lead-210 (modern) form a pathology of old age, while the calcium-41 (ancient) form is expressed from before birth. The generally exponential decay of radioactivity since the detonation of Planet X and the mutational beginning of humans would produce what has been interpreted as a gradual genetic evolution, passing through many different species, and bypassing some.

The unity of humanity is further suggested by the results of genetic studies by Robert Eckhardt [B. B. 2001a]. These studies showed that, in 14 fossil human species, the genetic differences were not greater than those observed for interbreeding modern wild apes.

Two skulls found in Dmanisi show an apparent transition in form between the *Homo ergaster* of Africa and the *Homo erectus* of Asia [Gabunia *et al.* 2000]. This has been interpreted as showing that these were evolutionary developments along the migration from Africa to Asia.

That work also revealed the hazards of dating. The original dating of the basalt immediately underlying the fossils showed 0.52 ± 0.02 million years ago. Subsequent, more comprehensive dating showed 1.8, 2.0, 1.85, 1.76, and between 1.77 and 1.95 million years ago. Without the diligent attention to dating, these specimens would have been considered to be much younger.

Dating difficulties were also experienced with Nanjing Man. Various methods of dating gave distinct results ranging from 130.1 thousand years ago to 620 thousand years ago [Zhao *et al.* 2001]. The researchers considered the value of 620 thousand years ago, from an oxygen isotope stage, to be best supported by the conditions of the other dating methods.

Similarly, dating the Boxgrove site in England gave geochronological ages ranging from 175.3 thousand years ago to 524 thousand years ago [Roberts 1994a, Roberts 1994b]. Boxgrove Man, a *Homo heidelbergensis* estimated to have been 1.80 meters tall (5 feet 11 inches) and weighing 80 kilograms (176 pounds), is known from a single tibia (shin bone) and some teeth. The current age estimate is between 478,000 and 524,000 years ago. The tibia shows very thick cortical bone, as is seen in Paget's Disease of Bone [Parsons 1980].

Altamura Man, considered to be a transitional form of *Homo heidelbergensis* between *Homo erectus* and pre-Neandertal, with some aspects of *Homo sapiens*, was found in the limestone Lamaluna Cave in 1993. Its age is quite uncertain, depending somewhat on interpretation of its form, and ranges from 100,000 to 400,000 years ago, with a likely age of 150,000 to 250,000 years ago.

The skeleton of a 4-year-old child was found at the base of a limestone cliff at Abrigo do Lagar Velho, with what were judged to be a combination of Neandertal and modern human features [Duarte *et al.* 1999]. Dating by radiocarbon analysis of associated material gave a date of about 24,500 years ago, the latest evidence of Neandertals. The researchers proposed that this mix had been the result of cross-breeding between the Neandertals and modern humans. Tattersall and Schwartz disagreed, arguing that such a genetic blend should have faded away over the course of 200 generations [Tattersall and Schwartz 1999]. But such a blend could result from the weak appearance of Paget's Disease of Bone, caused by a mild exposure to calcium-41.

At Krapina, a child's skull was found and dated to 130,000 years ago (but with measured ages ranging from 99,000 to 132,000), immediately above layers containing clearly Neandertal fossils [Rink *et al.* 1995]. "Krapina 1" was judged to have some aspects that grouped it with other juvenile Neandertals, but that some other characteristics approached the condition of early modern humans (Cro-Magnon). Another fossil, "Krapina 11", lacked a particular feature characteristic of those Neandertal fossils directly below it. The researchers interpreted this as showing that these semi-Neandertals might be transitional to anatomically modern humans. That was meant in an evolutionary sense, but might be true in the ecological-radiological-pathological sense, as resulting from a lesser exposure to calcium-41 and a reduced severity of Paget's Disease of Bone.

Homo antecessor was defined from adult fossils found in the Gran Dolina limestone cave in Spain [Kunzig 1997], but the most interesting of the associated fossils was that of an 11-year-old child. The fossils are dated to 800,000 years ago and show a mixture of *Homo ergaster* (of Africa) and *Homo heidelbergensis* (of Europe), even Neandertals, in between gracile and robust. But the face of the child "was a totally modern face", according to Juan Arsuaga. Did this child reach forward, through 800,000 years of evolution, to gain a modern human face? Or did it simply escape the most severe effects of environmental radioactivity, and express its inherent modern human form, 800,000 years ahead of us?

Fossils found in the limestone cave of Tabun, in the Levant (Mount Carmel, Israel), were Neandertal, while those found from the same time in the nearby rock shelter of Skhul, although more robust than modern man, were more "modern" than the Tabun fossils [Jelinek 1982]. The differences in living quarters (limestone cave or rock shelter) would have provided significantly different exposure to radioactive calcium-41, the proposed cause of the Neandertal condition as a form of Paget's Disease of Bone. In another limestone cave near Mount Carmel, Qafzeh, Neandertal fossils were found, from about 100,000 years ago [Bar-Yosef and Vandermeersch 1993]. The researchers note that anatomically modern humans had preceded the Neandertals, but that there was no evidence of local forebears of the Neandertals. Considering the Neandertals as showing a pathological condition due to intake of radioactive calcium-41, their form would have arisen abruptly upon moving into the cave and drinking the cave water, while the anatomically modern humans might have lived there during a dry spell when there was no cave water.

The development of modern human beings can be most simply seen as an origination event at 6 million years ago, corresponding to the time that the ages of the comets show as a disruption in the Solar System, and a time of great extinctions and speciations. This was followed by a nearly continuous increase in body size and inflation of the brain case, with occasional temporary "side-steps" to more robust forms. The robust forms gradually return to the "gracefully slender", or gracile form, but the overall size and form have grown during this side-step, so that the new form appears different from, is different from, the earlier. The great inflation of the brain case flattens the front of the skull, the face, and expands it.

If we consider the fossil record to represent just several changing versions of a single humanity, shaped by environmental factors (primarily radiation), rather than a sequence of 20 (or so) different species with no clear connections, the picture of human development becomes clearer. Two basic forms stand out: the gracile and the robust. Taking the gracile form, as we ourselves are, as the standard form, the robust forms appear to be offshoots that trend back towards the gracile line. That is, the gracile forms lie along the direct line of ancestry, but so do the robust forms, just set off to the side by some other effect that fades away with time, with no need to assign this to genetic evolution.

As time has passed since the earliest of our line, one of the clearest and most quantifiable measures of humanness has been the estimated brain size. Our brain was scarcely larger than that of a modern chimpanzee's brain, in the beginning. It is now larger by more than a factor of three. This increase can be fit by a mathematical function, the exponential function, and that curve is used as an indicator of progress toward our current state of humanity, in Figure 81, Fossil Record of Human Brain Size.

The smooth curve shows that the excess brain size, compared to a modern chimpanzee, doubled every 739,000 years. If we assume that this gradual increase in brain size was in response to a gradual decrease in the external radiation, it is remarkable that this doubling time is so close to the half-life of aluminum-26, 740,000 years. This radionuclide emits very penetrating, high-energy gamma rays, is produced in astronomical detonations, and can still be detected in the Milky Way. In a nuclear fission detonation it is produced from naturally occurring, stable aluminum-27, a very plentiful nuclide, by high-energy neutrons.

Anne Weaver studied the relative sizes of the cerebrum and the cerebellum in ancient and present humans [Weaver 2005]. She found that the Neandertals and Cro-Magnon (with the largest brains of the human family) had the largest cerebral hemispheres relative to cerebellum volume of any primates. In the most modern humans, the recent humans, us, the cerebral hemispheres are significantly smaller with respect to the cerebellum. Her results suggest that there is a greater difference between us and the Cro-Magnon (our presumed immediate ancestors) than any of the other fossil humans.

Radiation damage causes dividing cells to not survive to divide again, thus stunting growth and retarding development [Postlethwait and Hopson 1989]. More significantly in this case, penetrating (high-energy gamma-rays) radiation exposure of the fetus causes small heads and small brains [Schull 1995, Otake and Schull 1993, United Nations Scientific Committee on the Effects of Atomic Radiation 1986, United Nations Scientific Committee on the Effects of Atomic Radiation 1993]. Among the Japanese women exposed at Hiroshima to severe radiation doses while pregnant, infants were born with small heads: at doses above 100 rads, all infants suffered from this condition [BEIR 1980]. This was a developmental failure, not a mutation. High radiation exposure during growth also produces developmental abnormalities and mental retardation [BEIR 1980]. The Australopithecines seem to be more short in the leg than long in the arm, as are their primate relatives [White *et al.* 1993, Jungers 1994, White 1994]. Mental retardation and small body size were also observed as results of the atomic bombings [Wood *et al.* 1967a, Wood *et al.* 1967b].

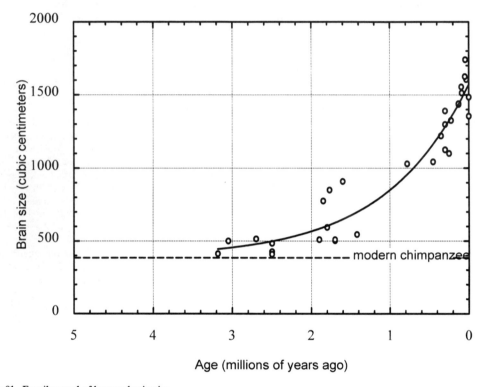

Figure 81. Fossil record of human brain size.
The fossil record of human brain size, showing the exponential growth of brain size as the postulated radioactivity from aluminum-26 declined.

In this view, the increase in human brain size with time reflected the decrease in external radiation, while the sidesteps to the robust forms were in response to temporary increases in a somewhat shorter-lived radionuclide such as calcium-41, which became an important internal radionuclide. Aluminum-26 provided penetrating whole-body irradiation, while calcium-41 caused only local cell-wide irradiation. The biological consequences were considerably different.

Planning and language are thought to have been missing in our earliest ancestors. The brain size of the first bipeds in our line is so small that modern humans with such small brains have similar lacks, being unable to care for themselves or participate in a conversation.

Our original ancestors were stunted in all ways by the effects of a level of radioactivity much greater than at present. As this radioactivity gradually decayed, the stunting effect became less, and each generation, slowly and slightly, approached the inherent human form. Nearby detonations boosted radioactivity and temporarily diverted the humans of the time into the more robust forms. As this new radioactivity decayed, the heavier growth of bone, that produced the robust form, decreased. Each of these stages has been labeled as a new species, often with confusing transitional forms.

In addition to the prompt radiation released at the instant of the detonation, a large amount of radioactivity is produced by the fission reactions. Some of this has very short half-lives and quickly decays away. A few long-lived radionuclides, such as aluminum-26, which can be detected in the Milky Way, linger on for millions of years. Evidence for the long-lived radionuclide iron-60, which decays to cobalt-60, and then to nickel-60, with high-energy gamma rays, has been found on Earth [Knie *et al.* 2004]. This deposit was dated to 2.8 million years ago.

Severe radiation exposure resulting from the atomic bombings of Hiroshima and Nagasaki caused a reduction in normal growth, roughly in proportion to the dose [Schull 1995]. These exposures were essentially instantaneous and a severe shock to the system. In most aspects, the same dose received over an extended time is less acutely damaging. However, when all cells are exposed to radiation throughout their development, some effects, such as stunted growth, may be increased. Those damaged by prenatal exposure during their mothers' pregnancy were small in stature and had small heads and small eyes. This is what we see in our early ancestors.

Continued exposure to radiation from the long-lived radionuclides produced by the detonation would have produced small-stature humans with disproportionately small skulls, containing abnormally small brains. As the radiation intensity diminished with time, the growth potential would be suppressed less, and the forms would approach the modern human.

With the seashore far lower than at present and more connected around the world, genetic transfer around the world might have been much easier than we think at present. That might have promoted a global village. Along the seashore, the land is fragmented, divided into coves and bays by sea cliffs. This might have led to the proliferation of languages that were then (at the end of the Pleistocene) transferred to higher ground. Then, with the disaster at 12,900 years ago, most of the humans on the continents were wiped out. Some, like Clovis, Kennewick Man, Spirit Cave Man, Pedras Pintada, Pedra Furada, and Monte Verde (perhaps Cheddar Man, in England), made it onto the high plateau, but perished in the dismal wasteland left by the Flood that followed the impact of cometary swarms, about 12,500 to 11,600 years ago. Africa was the least affected of all the lands. But most of the world was repopulated by immigrants, escapees, from the seaside.

By extracting and analyzing mitochondrial DNA (mtDNA) from a small fragment of a little finger bone found in the Denisova Cave in southern Siberia, Johannes Krause and colleagues have shown the likely presence of another sort of people than had been expected [Krause *et al.* 2010, Brown 2010, Balter 2010a, see also Calaway 2010, Reich *et al.* 2010, Bustamante and Henn 2010]. The results of this analysis are shown in Figure 82, with the Denisova mtDNA clearly offset from Neandertal mtDNA which is itself offset from modern humans.

Of particular interest here is the double peak with the least differences. This shows the distribution of present-day Eurasians (0-70 Pairwise nucleotide distance) and present day Africans (70-110 Pairwise nucleotide distance). The African distribution shows the greater variation in the people who stayed home in Africa and have been there all the time. The Eurasian distribution shows, according to the story of this book, the lesser variation in the descendents of "Mother Eve" and "Father Adam" and their small band of little people, who left high Africa for the seaside, 100,000 – 200,000 years ago. At the end of the Younger Dryas cold episode, otherwise called the end of the Great Ice Age, these descendents returned to the continental plateaus from seasides around the world, repopulating the devastated land simultaneously. A seaside migration out of Africa has been proposed [Stringer 2000] and is supported by findings of near-seaside occupation at about 125,000 years ago [Walter *et al.* 2000]. That is about the time that a small person left footprints by what is now Langebaan Lagoon in South Africa, 117,000 years ago [Bower 1997, Newspaper 1997].

In the comparison of Figure 82, a single specimen labeled as an Early Modern Human (EMH), from Kostenki, was analyzed. Early Modern Human usually means, in Europe, a Cro-Magnon type. However, the Kostenki skeleton, Markina Gora, was small, although fully grown at an age estimated as 20-25 years [kuntskamera 2009]. His estimated height was 1.60 meters, but Cro-Magnon males were normally about 1.80 to 2.00 meters in height. His height is the same as the supposed woman who walked in the dune sands near Langebaan Lagoon, 117,000 years ago [Bower 1997, Newspaper 1997], which is about the time of Mother Eve and Father Adam. The mtDNA differences observed for Markina Gora fall in the present-day Eurasian distribution, not near the Neandertal peak, as might be expected, which suggests that this man, estimated at 32,000 years ago, was not a Cro-Magnon but might have been an immigrant seasider, the same as us.

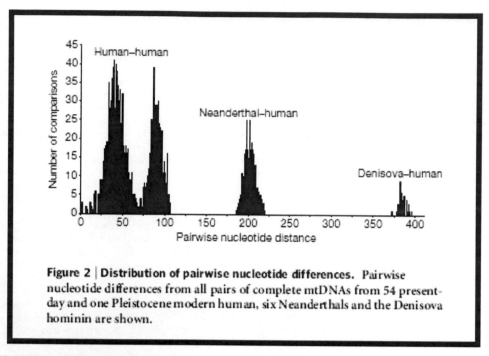

Figure 2 | Distribution of pairwise nucleotide differences. Pairwise nucleotide differences from all pairs of complete mtDNAs from 54 present-day and one Pleistocene modern human, six Neanderthals and the Denisova hominin are shown.

Figure 82. mtDNA differences among modern humans, Neandertals, and Denisova.
Distributions of mtDNA differences, showing the groups of Eurasian present-day humans, African present-day humans, Neandertals, and the new Denisova sample [Krause *et al.* 2010].

Radiocarbon dating at Kostenki failed to produce reliable results, and differed significantly from Optically Stimulated Luminescence, by about 7,600 years. A tephra layer (presumed volcanic ash) covers an Upper Paleolithic collection of artifacts. This layer is associated with Heinrich Event 4, the Laschamps geomagnetic excursion (at 39,700 to 35,200 years ago), and a related cosmogenic nuclide peak [Anikovich *et al.* 2007]. Heinrich Events are identified by so-called "ice-rafted debris" thought to have been carried by icebergs from glaciers, and by disturbances in the deuterium (hydrogen-2) and oxygen-18 isotope ratios. Cosmogenic nuclides can be produced by various types of radiation from a nearby nuclear fission detonation. The peak is indicative of a brief but intense increase in production, as would have resulted from the nuclear detonation of a moon of Saturn.

Similarly, the deuterium and oxygen isotope excursions can result from gamma-ray interactions, with the high-energy radiation from the detonation of a moon of Saturn. The electromagnetic pulse (EMP) resulting from such a large nuclear fission detonation could temporarily reverse the geomagnetic field, producing a brief excursion.

There are two competing theories for the origin of the most modern humans, us. One provides that early populations throughout the world developed the features that make us who we are, and that these populations interbred to the extent that a genetic uniformity was maintained. This is the Multiregional Theory. The other provides that our ancestors moved out of Africa about 200,000 – 100,000 years ago, and replaced all the previous peoples around the world. That is the Out-of-Africa Theory. In the view proposed in this book, an early African group migrated to the seaside, much lower in elevation at that time, and spread around the edges of the continents, essentially walking around the world to occupy seasides widely dispersed. The seaside is a one-dimensional living space, not two-dimensional as the continents are, and that forces its population to spread more rapidly. These people developed distinct cultures and languages, but were related genetically by the small number of parents in the original group. At the time of the megafaunal extinctions, most or all of the humans living on the continental plateaus were also extinguished, as should be expected by the large humans fitting the definition of "megafaunal", being heavier than 44 kilograms. At the end of the Younger Dryas, when sealevel rose abruptly and greatly due to the influx of water from the fragments of the detonated moon of Saturn, all these seaside people, all around the world, escaped to the continental plateaus, and repopulated the devastated Earth. Thus, we, the Holocene Humans, were initially from "Out-of-Africa", but occupied the world in a "Multiregional" manner.

In Catalonia, northern Spain, stone tool technology changed abruptly from the Mousterian, associated with Neandertals, to the Aurignacian, usually associated with Cro-Magnon [Straus 1989]. At one site, there is a layer of sediment with no archaeological materials, half a meter thick, indicating a considerable absence of occupation or a rapid deposit of sediment. In another, the later technology is found immediately above the older, with no apparent delay or interruption. The Neandertal technology terminated at about 40,000 (radiocarbon) years ago. Much happened at about 40,000 years ago. The stone tool technology changed abruptly, rather than transitionally, using stone from 100-150 kilometers away, and shells were found more than 500 kilometers from a likely source. This suggests that new people arrived from a distance in a deserted region.

At about that time, it is supposed that the early modern humans who had developed in Africa migrated into Europe, competed with the Neandertals as a separate population, and started to replace the Neandertals. The modern humans in Africa did not have some of the objects typical of the Châtelperronian technology developed by the late Neandertals. Once they enter Europe (and are identified anatomically as Cro-Magnon), they had the new technology [Zilhão 2010]. However, consider that the Neandertals were shaped anatomically by a form of Paget's Disease of Bone, but near the end of their time morphed into "normal" early modern humans, the Cro-Magnon, when they lived in rock shelters free of the radioactive calcium-41 of the limestone caves. Then, the culture would have continued without need for a transfer to a new people, just retained by the newly formed Cro-Magnon ex-Neandertals.

The operational assumption has been that the anatomically modern humans, finally represented by the Cro-Magnon, were the ancestors to the present human population. But the Cro-Magnon were large, and the early (post Great Ice Age, post Younger Dryas) Europeans were small.

From the bones of an adult male, it was estimated that his weight would have been at least between 93.1 and 95.4 kilograms (205-210 pounds) and his height was between 1.733 and 1.795 meters (5 feet 8 inches to 5 feet 11 inches). A female skeleton was found in a limestone cave in China dated to 260,000 years ago, and labeled as an early, archaic *Homo sapiens* [Rosenberg, Zuné, and Ruff 2006]. This individual was calculated to have been 1.6878 meters in height (5 feet 6 inches) in height and to have weighed 78.6 kilograms (173 pounds).

I propose that it was a path of development, and that the earliest ancestors that were produced at the start of this path were nearly complete *Homo sapiens*, genetically, and that the confusion of types of fossil man resulted from the various effects of continuing but declining radioactivity.

In the fossil skeletons of our early ancestors, we find that the feet are human feet, the hands are human hands, and the brain, in structure, is a human brain, but smaller [Gribben and Gribben 1993] The major differences between modern humans and those early fossil (partial) skeletons we know as *Australopithecus afarensis* are in smaller stature, apparent distortions of bone size, a narrow birth canal in the ancient pelvis, and a much smaller and misshapen skull [Johanson and Johanson 1994].

Modern humans originated in Africa, genetically, but were only able to developmentally express their humanness as they moved out of Africa, into less radioactive regions. This progression into modern human form is so subtle that similar comparisons can lead to diametrically opposed conclusions, when the either/or choice of "Out-of-Africa" or "Multiregional Development" is faced [Konigsberg *et al.* 1994, Waddle 1994a, Waddle 1994b, Bowcock *et al.* 1994].

A similar effect is seen in the progression of *Australopithecus* to larger individuals [Johanson and Johanson 1994]. The several identified later species of *Paranthropus* (*africanus, robustus, bosei*) may have been the result of living in areas, and in times, of somewhat different levels of radioactivity. Individual differences in exposure may have produced what we see, from a limited selection of specimens, as different species.

And, as time passed, and the radiation levels decreased from radioactive decay, at about 2.5 million years ago, the climate started to cool [Feder and Park 1989]. At about 2.5 million years ago, the magnetic field reversed polarity [Wicander and Moore 1989]. The reversed magnetic polarity was followed by intervals of normal and reversed polarity, and by an extended period of reversed polarity beginning about 1.6 million years ago. And throughout this time, our ancestors gradually became more human.

A more important characteristic of the brain than absolute size, is the relation of brain size to body size, and the Australopithecines showed about the same ratio as modern chimpanzees. The "quality" of the brains also shapes capability. Differences in structure, surface area, convolutions, and organization of the brain might have served to give more intelligence, or a different, more human intelligence to our earliest ancestors, with brains the size of chimpanzees.

A distinctive feature of *Homo erectus*, a robust form which was found to coexist with Australopithecines in some places for some times, was an abnormally thick and dense bone structure, the tabular bone of the skull and the cortical bone of the femur [Kennedy 1985] and possibly all the long bones. (It is sometimes difficult to generalize from a small sample of fossil remains.) The thickening appears to occur predominantly in the inner portion of the bone, next to the blood-forming marrow region. This increased thickness of the bones is not generally recognized, and results from a process that is not clear, so the mechanism is uncertain. It is unclear whether more bone grew into the central cavity of the bones, or if resorption occurred more slowly than usual. An excess amount of bone accumulated. This occurs in Paget's Disease of Bone.

Several cases of bone pathology have been found in *Homo erectus*. These cases have suggested hypervitaminosis A [Walker, Zimmerman, and Leakey 1982] or yaws [Rothschild, Hershkovitch, and Rothschild 1995]. A similar condition has been noted [M. R. 1996] on a *Homo erectus* femur dated to about 500,000 years ago from Venosa, Italy.

In these cases, the diaphyses (shafts of the long bones) were greatly thickened (as is seen in Paget's Disease of Bone), and in the Vitamin A study the bony deposit was described as "coarse-woven bone" (as is also seen in Paget's Disease of Bone [Parsons 1980]).

A condition diagnosed as calcium pyrophosphate deposition disease has been reported for Neandertal skeletons from La Chapelle-aux-Saints and Shanidar [Rothschild and Thillaud 1991]. This is a secondary form of osteoarthritis that has been found to result from Paget's Disease of Bone. Three Neandertals of Sangiran, Gibraltar, and Shanidar were found to have excessively thickened bone at the front of their skulls [Antón 1998]. That is a condition found in Paget's Disease of Bone, and examples have been shown in an article on a Viking from an Icelandic saga [Byock 1995].

In the Sima de los Huesos ("Cave of Bones") in Spain, fossils were found from at least 32 individuals, labeled as *Homo heidelbergensis* (a robust form) and dated to about 300,000 years ago [Gibbons 1997] or 33 individuals at least 200,000 years old [Arsuaga *et al.* 1999]. One skull was found to be scarred with osteitis, another showed that the person had been deaf, from growths that blocked his ear canals. Paget's Disease of Bone appears as an osteitis, and can cause deafness. Arsuaga *et al.* considered these fossils to represent the ancestors of Neandertals.

Further evidence of serious radiation effects is found in the common occurrence of symptoms of bone cancer and similar diseases in early humans. The Johansons comment [Johanson and Johanson 1994] that some *Homo erectus* limb bones were "diseased and coated with extra bone that resembled lumps of melted wax." This prevalence of disease is also seen in the large number of healed bone fractures among the Neandertals [Calvin 1990]. Healed bone fractures occur in 75% of the Neandertals over the age of 25 years. Healing of a broken bone would require immediate survival and extended care, which would seem more likely to have occurred in peaceful surroundings than in attacks on large animal prey as suggested [Calvin 1990]. These breaks occurred in bones that were larger and thicker than ours, and because of this, have been estimated to be 2 to 3 times as strong as the comparable bones of modern humans [Feder and Park 1989, Johanson and Johanson 1994].

In modern humans, bone cancer is a very rare disease [NIH 1985], showing roughly 1 case per year per 100,000 persons [Rowland *et al.* 1983]. It has been produced abundantly by excessive exposure to radioactivity, specifically internal deposition of radium [Rowland *et al.* 1983], and external exposure by X-rays [BEIR 1980]. The apparent frequent occurrence among our ancient ancestors suggests a high radiation exposure rate to the bone. Paget's Disease of Bone increases the occurrence of bone cancers by a factor of twenty [Parsons 1980].

Spontaneous fracture was seen in the radium dial painters early in the 20th Century, at radiation doses lower than those that produced evident cancers [Eisenbud 1987].

The Neandertal archetype, the "Old Man of La Chapelle-aux-Saints," showed evidence of bone disease and a tumorous growth in his jaw [Feder and Park 1989], and was probably less than 40 years old at death. Few Neandertals lived beyond 35 years of age [Calvin 1990]. It appears that Neandertals died "old" at an early age [Johanson and Johanson 1994].

While the number of reported cases of ancient bone pathology is small, the occurrence of these cases in the relatively meager sample of fossil bones that can be adequately studied, implies a surprisingly high frequency of what should be rare diseases. If a form of Paget's Disease of Bone resulted from widespread environmental factors (such as ingested radioactivity, suggested here), that in fact formed the robust nature that has been a defining feature, extreme cases might have been more common.

In considering possible radioactive causes for the condition, a particular type of radioactive decay and a specific radionuclide hold promise for an explanation. While most radionuclides produced by neutron activation decay by beta emission, some few decay by electron capture. In this decay, a proton in the nucleus captures an orbital electron and becomes a nuclear neutron. Importantly, a "hole" is left in the electronic shells around the nucleus, and the atom becomes a different chemical element. The electron shell vacancy left by the decay capture is de-energized by the ejection of orbital electrons with relatively low energy. (These are specifically called Auger electrons, for the Frenchman who first described the effect.) Filling the electron vacancy results in, paradoxically, the emission of many low-energy electrons, and these are very disruptive to biological cells. These low-energy electrons produce very dense ionization and are significantly more effective in causing biological damage than would be expected from the amount of energy available [Goorley and Nikjoo 2000]. This damage may take the form of clustered breaks in the DNA, which makes double strand breaks and chromosomal fragmentation [Sutherland *et al.* 2000]. Further, the decay leaves the decay product atom in a highly ionized state as a multiply charged ion [Persson 1994]. This may take the form of a free radical and do additional biological harm.

Radionuclides with sufficiently long half-lives that they might have been effective in creating high radiation exposure rates during the time of human development and should be investigated are aluminum-26, which requires high-energy neutrons; calcium-41, which can be produced only in material with low competing neutron absorption cross-sections, such as high-purity limestone; iron-60 (and its daughter cobalt-60), which requires two neutron absorptions within a short time (or superfission); and chlorine-36. This last isotope would be produced by neutron irradiation of common salt, sodium chloride, and might have been significant to the australopithecines living along the shores of salt lakes in ancient Africa [Hay 1976]. One of the modes of decay of chlorine-36 is electron capture.

Volcanic eruptions in the eastern part of the continent of Africa at that time may have brought large amounts of radioactive iron-55 to the surface, radioactivity that had been produced by a burst of fission reactor power. This radioactivity may have been introduced into the diet of some of the humans. The concentration of disruptive radioactivity in the red blood-cell tissue of the marrow might have produced the observed thickening of the bones. This radionuclide decays by electron capture and emits no other ionizing radiation, except short-range X-rays and Auger electrons, and would not have caused high radiation exposures, except by the internal pathway. These doses would have been concentrated in those organs where iron is concentrated, such as the liver, the spleen, the blood, and the marrow, irradiating the inner surface of the bones, including the skull [DeWitt 1989].

In the course of human development, certain uses of stone tools of particular shape and style have been noted. A peculiar characteristic of the development of early human technology is the rare appearance of a new technique or tool, followed by its long-continued use without change. Tools were produced that were ingenious in their creation at the time, in a culture that could continue their production and use, as if by rote, but was incapable of making the slightest improvement or change.

This may have been the result of the accidental and rare birth of an individual with a more nearly normal brain, one less stunted by radiation exposure during its gestational development. Perhaps his/her mother had similarly been less exposed to radiation during childhood, and so had developed a birth canal that was sufficiently large to permit birth of an "abnormally" large-skull infant. This combination could have been so rare as to produce only a few relative geniuses in hundreds of thousands of years. The inventions of these special individuals could have been passed on, unchanged, generation after generation, by tradition, training, and habit.

Approaching our own time, when human brains were as big as or bigger than ours, culture, in the forms of tools, technology, and symbolism, became an important subject for consideration as to who we were then. Towards the end of the Paleolithic, the Old Stone Age in Europe, towards the end of the Pleistocene, a major change occurred in these bits of archaeological evidence. The great change in technique, variety, and use of symbolism that occurred between the Late Middle Paleolithic and the Early Upper Paleolithic, about 40,000 years ago (standard chronology) has been confusing [Clark 1999]. We do not know who the people were who made this transformation, whether Neandertal (not likely!), or Cro-Magnon (but of course), or someone else (who?).

It is notable that in the residential limestone caves and the rock shelters, occupied by Neandertals and Cro-Magnon respectively, there are no indications of wall decorations, no paintings, not even graffiti. In the painted caves, Lascaux, Chauvet, Cosquer, Nivaux, and others, there are no indications of occupancy, no signs of who were the people responsible for the art. Except for footprints and handprints.

In Chauvet Cave, four small footprints have been found. These have been interpreted as being the footprints of a young boy, because of their small size, compared to the footprints of a Cro-Magnon adult, the presumed artist. While the art has been dated to about 31,000 years ago, the footprints are thought to be about 26,000 years old. Instead of a small child, seemingly out of place in such a cave, they might have been from a small person, a seasider from the people who would inherit the devastated Earth. Interestingly, while Chauvet may be dated at 31,000 years old, the most recent art in Nivaux is dated to only 14,000 years ago. This may show the error in radiocarbon dating due to an abrupt production of carbon-14 at the time of the detonation of a moon of Saturn.

Also most notably in the Gargas Cave, "Often they are the hands of women or adolescents, and some of children aged about two or three."... "which look like our own," [Ruspoli 1986].

In the Chamber with Heelprints at Le Tuc-d'Audoubert, "three young children walked on their heels;" "Footprints and handprints ... that could only have been made by children have been found in a number of caves." [Clottes and Lewis-Williams 1996].

As an interesting aside, small footprints were found in solidified sand near Langebaan Lagoon, South Africa [Bower 1997, Newspaper 1997]. They were found by David Roberts and interpreted by Lee Berger, who described them as made by a person who "looked just like us." They were interpreted as having been made by a small person, about 5 feet 3 inches tall (1.60 meters), and therefore likely to be a woman. These were dated to about 117,000 years ago, about the time that "Mother Eve" and "Father Adam" and their people were isolated from the rest of the human race, making a population breeding bottleneck that shows in the distribution of mtDNA and Y chromosomes. These may have been the seasiders, the small people who later repopulated a devastated Earth.

At present, there are two basic theories for the development of modern humans from our earlier ancestors. In one, it is supposed that our ancestors developed into humans in Africa and spread throughout the world. In the other, our ancestors left Africa and independently evolved into similar humans in their new distant homes. Here, it is proposed that as our ancestors moved away from the most radioactive areas of the Great Rift Valley, and as the intensity of the radioactivity decreased by decay, the humanness universally present in their genes was able to be expressed in more nearly normal development. The fossils that were left progressively resemble normal modern humans.

What has been interpreted as progressive evolution in humans was instead progressive achievement of potential development as the intensity of the radioactivity diminished. Instead of being the genetically primitive *Australopithecus*, only our ancestor, they were us, perhaps as genetically human as ourselves, our parents and children, our grandparents and our grandchildren, regardless of variations.

The tangled lineage of humans shows a multiplicity of poorly defined species. But just because a bone looks different, we shouldn't assume that its owner was different. It may be that most, or all, of the differences we see in fossil humans, through time and around the world, are the result of radiation-induced bone pathology. Different types of radioactivity produce different effects, and are present at different intensities according to the mineral content of the particular locale.

The apparent coexistence of *Homo erectus* and *Paranthropus boisei* at Konso, with Erectus thought to be associated with tools and Boisei not, is similar to the situation with Moderns and Neandertals. Both cases suggest the transition as the pathological effects of the particular radionuclides were becoming significantly less strong, and the robust character faded away.

Molecular biologists attempt to determine the genealogy and origins of humans by studying the variations in the genetic material of our present population and, to the extent they can, mummies and fossils. Finding the differences in chromosomes, genes, and markers, leads to a determination of variations, and hopefully, to a line of development. Unfortunately, it seems that the interpretation of the laboratory results attempt to grasp what is still beyond reach.

Genetic material almost seems designed to provide a means for exploring our relations and ancestors. It consists of several distinct forms that permit a variety of tracing to be done. In animal cells, tiny organelles called mitochondria provide the energy for the cell. These mitochondria have their own genetic material, called mitochondrial DNA (mtDNA), that is completely separate from the rest of the genetic material which determines the characteristics of the animal, its genus, species, sex, height, weight, on down to color of skin, hair, and eyes. Mitochondria are contained within the female germ cells, the egg, and at the base of the tail, the flagellum, in the male germ cell, the sperm. At fertilization, when the sperm joins the egg, only the nucleus of the sperm is admitted into the egg, excluding the mitochondria of the sperm, the father's mtDNA. Thus, each offspring, specifically each of us, carries mtDNA only from our mothers, and their mothers. This permits, in some way, tracing our maternal line of descent.

Human genetic material consists of 22 pairs of so-called autosomal chromosomes, that determine our major characteristics as humans, and a pair of sex chromosomes that direct our sexual development, male or female. These are labeled X and Y. A female carries a similar pair, XX, while a male carries a different pair, XY. (Other creatures have different arrangements. This discussion applies to mammals and humans.) All chromosomes except the Y undergo a complete recombination during the reproductive cycle, merging sequences of parental genes. A major part of the Y does not recombine, staying unchanged except for random mutations. The Y chromosome is inherited only from the father and so permits tracing our paternal line of descent.

Genetic material can be analyzed to determine differences between one set and another. Differences indicate changes from an earlier form, and if we knew which was the earlier form we could decide which came later, which was a descendent from an earlier population. Unfortunately, this cannot be done directly but must be done by computer analysis of large sets of data from well-defined populations. While computers do only what they are told to do, sometimes our understanding of what our instructions mean is not entirely complete.

There is a multitude of ways in which the many parts of a chromosome can be changed and rearranged. (Try rearranging the alphabet, one letter at a time, perhaps inverting or reversing a letter in the process. There is a multitude of arrangements, and many different ways to make each new sequence.)

The data are used by computer programs to produce branching trees, with the oldest population set serving as a base, and branching branches developing with each next change in the genetic pattern. Then, just on the principle that simpler is better, and the trees with the fewest changes are closest to actuality, the shortest trees are selected. At this point, problems creep in. A large number of data provide the most precise evaluation, but the number of trees increases drastically. In one of the largest studies yet, which includes 136 mtDNA sequences, with 117 identified positions, there are 10^{267} possible trees (a one followed by 267 zeroes) [Gee 1992]. This is a physically impossible number of alternatives to consider. A further flaw, in at least one of the computer programs, resulting from the tremendous number of numbers, is that the actual patterns of the trees made available for examination is affected by the order in which the data sets are entered into the computer and sorted by the program. The larger the number of data sets, the more precise the results, but the stronger this confounding problem becomes [Gee 1992].

Choices made by the researchers affect the outcome of this analysis. The first successful study of mtDNA [Cann, Stoneking, and Wilson 1987], that resulted in the mitochondrial Eve, showed that a single "mother of us all" had lived in Africa, and our current mitochondrial genetics had been determined by the population size about 200,000 years ago [Brown 1990]. This result was determined by the choice of Africa as a homeland, which was not at all unreasonable since the greatest variability was shown by African mtDNA. This is also supported by the fact that all the oldest human fossils have been found in Africa. The age depends upon the population size and the mutation rate that is assumed for mtDNA. (In one study, it has been shown that variations in population size determine everything, but that is just another stone to throw [Brookfield 1992].) The mutation rate is estimated from variations found in the native populations of New Guinea, Australia, and the New World, based on dates that are assumed for initial immigration, that are almost certainly wrong.

The existence of a single female ancestor for all our modern mtDNA is an example of what has been presented as the family name paradox. Even though the number of people increases with time, if the family name is inherited only from the father, after a long enough time, only one family name will exist. This happens because, if a male (carrying a specific family name) has no children or only daughters, his line of the family name ends. Without creation of new names (like genetic mutation), eventually the lines of every family name except one remaining will have been terminated. The same process occurs with mtDNA, passed only from mother to daughter (and son). If a mother has no daughters, her line of mtDNA ends. After a sufficient time, only one form of mtDNA will exist, except for random mutations. The problem is in determining how long that time is. The time is crucially dependent on the smallest size of the population at various times, and on the assumed mutation rate. The mutation rate is assumed to be a constant average value, but the evidence shows that it is highly variable.

So, the results of all the genetic research simply provide a vehicle for expressing the understandings of the researchers. In the end, if that is the best that can be done, it is certainly not all that bad. Africa shows the greatest genetic variability, seems to have been home to the first humans, and the ages proposed are in reasonable agreement with the fossil record. (Otherwise, they wouldn't have been proposed!) The results are reasonable, and represent the judgment, ideas, and understanding of many people who, presumably, have developed a "feel for the material" that may make their intuition far more valuable than all the computer programs and laboratory analyses. Sometimes, the process is the product.

The Hoof and Habit of Horses

A horse is a horse,
of course, of course,
"Mr. Ed", television show, 1961

The modern horse, *Equus* species, was created, genetically, at the same time as our humanoid ancestors. Most of the other forms of the horse were terminated at that time [MacFadden 1992]. The two remainders of old-style horses, with three-toed feet, unprotected by full hooves, faded away, perhaps like the bellerophonts of the Triassic Period, unable to survive successfully because of radiation damage to their feet. Modern horses survived because of two protective features that would have reduced radiation exposure: single-toed, hoofed feet that are resistant to radiation damage; and the ability to rest and sleep standing up, keeping vital organs (especially reproductive organs) away from radioactive ground. The only horse that sleeps standing up is the only horse that has survived.

A special arrangement of muscles, ligaments, and joints forms the "passive stay apparatus", to lock the legs in a standing position [MacFadden 1992]. This permits the modern horse to stand for a major part of the day, and afforded horses at the time of the human event, dated for the horse by use of the North American Land Mammal Ages (NALMA) time scale as 3.7 million years ago, soon after the beginning of the Blancan, with protection from protracted radiation exposure from lying on the ground. When *Australopithecus* went up a tree and out on a limb to rest, *Equus* just stood on four limbs and slept.

So the horse that survived this period was a new horse, a horse with protected feet (protected with the same material as protected the underside of the turtle, keratin, like your fingernails), a horse that did not need to lie on the ground to rest.

If survival is a test ...

... only the fit survive.
Robert William Service,
"The Law of the Yukon"

If survival is a test, and the turtle was the answer, what was the question? If the hard-bodied animals of the Burgess Shale survived, what was the question 540 million years ago? If dinosaurs and clams got it right 245 million years ago with high hips or bipedalism, and thick burrowing shells, what was the question then? And again, if knuckle walking, bipedalism, tree-climbing, protected feet and sleeping upright won passing grades 6 million years ago, what was the question then?

In all of these novel and abruptly successful forms, the function appears to be to reduce exposure to radioactivity at ground (or seafloor) level, and the question was: how to survive in a newly radioactive world.

The fossil record preserves the successful survivors, the organisms that had the right answer to the changed conditions of their new environment.

Punctuated equilibrium in evolution works by providing the abrupt destruction of existing species and simultaneous creation of new forms, by the action of the intense burst of radiation associated with a nuclear detonation; the selection of surviving species through the action of high levels of radioactivity; and then the gradual refinement of the details of these species, and the decline of overly protective forms, in the long period of low radioactivity.

The Neandertal Condition

... ape-like creatures who walked
hunched over with a shuffling gait
Marcellin Boule 1908

... a deaf old man with a bent back, bowed legs, curved tibia, a large skull and wobbling gait.
The arms appear relatively long and the patient ape-like.
Victor Parsons, (Paget's Disease of Bone) 1980
"A Colour Atlas of Bone Disease"

Neandertals lived in limestone caves in Europe, from about 140,000 to 28,000 years ago, and then vanished. They seemed to be distinctly different from the Cro-Magnon people who showed up in Europe at about the end of the Neandertal time, overlapping their predecessors by as much as 10,000 years in Europe, but perhaps longer elsewhere. The Cro-Magnon appeared to resemble modern humans closely, but there doesn't seem to be a certain separation, and there doesn't seem to be any certain connection. Earliest modern Europeans, after the Younger Dryas ended, were much smaller in stature.

There is something distinctive about the Neandertal form. Some later Cro-Magnon (from the Aurignacian period, about 40,000 to 26,000 years ago) have been considered to be robust, but not with the Neandertal form [Hublin 2000]. (The dating of the Aurignacian is at the limits of most radiocarbon analyses, and may be in error because of the exceptional production, and non-production, of carbon-14 by the Giant Dipole Resonance reaction from the gamma flash of the detonation of a moon of Saturn.) This similarity in form, the same but different, might result from a lesser and variable intake of calcium-41, and perhaps other radionuclides, in producing a variation of Paget's Disease of Bone. This might also account for the many transitional fossils that have been found, poorly defined for the assigned species. Paget's Disease of Bone can be very variable in aspect.

Fossils with mixed features of Neandertal and early modern human have been found from Lagar Velho in Portugal to Vogelhard in Germany to Mladeč in Moravia to Vindija in Croatia [Wong 2000]. Some Neandertal features seem to have no functional significance and are proposed to be the result of genetic drift in isolated populations. Yet these populations are distributed over the entire extent of Europe, but show the same distinctive, non-functional features. This could be the result of the variability of Paget's Disease of Bone.

Unfortunately, the debates over the Neandertals and modern humans have more argument than evidence. Even the mtDNA analyses are disputed on the grounds that the sample is too small (one, or later six) and that chimpanzees in the wild show greater genetic differences.

Neandertals have smaller, compressed, bony ear cavities. This part of the skull is formed after six months of embryological development [Hublin 2000], and could show distortion from the continuous deposition of calcium-41 beginning at conception. The alignment of the canals of the bony labyrinth has been determined to be a diagnostic feature, distinguishing between Neandertals and all others studied [Hublin *et al.* 1996]. This slight but distinct distortion could result from a form of Paget's Disease of Bone, thickening and distorting the skull.

Towards the end of the Upper Paleolithic (about 40,000 years ago to the end of the Younger Dryas), the modern humans in Europe, the supposed replacers of Neandertals and ancestors of current Europeans, lost all the features reminiscent of their tropical origins (in Africa). That is, they took on a different form than previously.

Kate Wong states that "After 40,000 years ago, Neandertal culture was static" but that modern Europeans showed many new features [Wong 2000, Hublin 2000]. Here it is suggested that at the time measured as 40,000 years ago (which is likely to be 12,900 years ago), the Neandertals were dead, extinguished like all the other megafauna of the time, by the radiation from the nuclear fission detonation of a moon of Saturn. The younger dates, after 40,000 years ago, were the result of varying production of carbon-14 in biological material, by the gamma flash of the detonation. The modern European cultures were brought by the new Europeans, up from the seaside by many different paths.

In the social climate of 19th-Century Europe, when Europeans were the pinnacle of human superiority, Neandertal was not socially acceptable as an ancestor. That curse has stuck ever since. Neandertal was too "beastly" to be an ancestor, and to not be an ancestor, Neandertal had to be beastly indeed. Even now, there are doubts as to the ability of Neandertal to talk, to have a language. All culture from that time, with the exception of obvious Neandertal burials, is assigned to the Cro-Magnon and other "anatomically modern humans" who, like us, had language, social organization, planning, cooperation, culture, even religion.

The physical appearance of the first Neandertal skeleton was initially explained away as the result of various selected diseases. The Neander Thal (or Feldhofer) skeleton was different from ours because of pathology. As a great number (20) of Neandertal skeletons were found, and identified to be the same species because of their similar distorted characteristics, it was decided that this form was normal for the Neandertal, and marked their distinctive differences from us [Lewin 1993]. Our evolutionary path becomes simpler if we accept that the Neandertal characteristics were indeed the result of an unidentified pathology that was endemic in the population and was the only cause of these differences, which became the defining differences for the species.

Boule firmly declared that Neandertal could not be ancestral to us in his extensive monograph of 1911-1915. Although much of Boule's work has been thoroughly rejected, that judgment has remained to this day as a foundation for understanding our development. Boule's opinion as to the "structural inferiority" of the Neandertal brain lingers on in the debate over a possible (or impossible) Neandertal language and the insignificance of burials without plentiful grave goods, and in judging that their gradual disappearance showed their failure to successfully compete with our ancestors, the superior Cro-Magnon. We teach what we were taught.

There were "eastern" Neandertals (earlier) (Swanscombe, Steinheim, Fontechevade, Kaprina, Teshik Tesh, Tabun, Shkul) and "western" (classic) Neandertals (later).

Another case of two forms, robust and gracile, living together like the Neandertals and the Cro-Magnon, has been described from skulls from the Willandra Lakes of New South Wales, Australia [Holden 2009]. The researchers studied 26 skulls from the end of the Pleistocene (15,000 to 40,000 years ago) and found that there were both gracile and robust skulls, like Neandertals and Cro-Magnon, and like the earlier gracile and robust species found in Africa.

Erik Trinkaus has found that morphological features in European earlier modern humans indicate an assimilation of Neandertals, and these features are "ubiquitous and variable" [Trinkaus 2007, see also Rougier *et al.* 2007, Shang *et al.* 2007].

Wolpoff has argued for an evolutionary process in the development of Neandertals and early modern Europeans (earlier than Cro-Magnon) [Wolpoff 1981]. Looking backward in time, Wolpoff saw that the modern Europeans became distinctly more like the Neandertals earlier in time, and tracing the Neandertals forward in time, they became distinctly more like the modern Europeans later in time. Rearranging this somewhat, both types became less like Neandertals, more like modern Europeans, as time progressed. This is consistent with the possibility that the differences between the Neandertals and the modern Europeans is the result of calcium-41 Paget's Disease of Bone. The Neandertals received, by their use of limestone caves as living quarters, a saturated dose of calcium-41, which declined with time, while the modern European branch of our tree, by avoiding limestone caves, received a lesser dose of calcium-41, which also declined with time.

Here, it is proposed, as has been done unsuccessfully in the past, that the form of the Neandertals was not an evolutionary effect, but the result of a very specific pathology of their bones. It appears that the Neandertal skeletons have been found predominantly in large limestone caves, while the Cro-Magnon skeletons are found in shallower rock shelters, *abri*. In Israel, neighboring caves, Tabun and Skhul, and Qafzeh and Kebara, show the contemporaneous occupation by Neandertals, Proto-Cro-Magnon, and nearly modern humans, in a confusing transitional manner [Johanson and Johanson 1994].

An ancestor to the Neandertal has been tentatively identified, in a limestone cave, with an uncertain age that may be as old as 400,000 years ago [Dorozynski 1993]. The intense neutron radiation proposed here to produce the calcium-41 that caused the Neandertal condition could have resulted from antineutrino reactions produced by the antineutrinos from a nearby planetary detonation. The pulsar identified as Geminga, which has been associated with a postulated very nearby supernova detonation [Gehrels and Chen 1993], is thought to be 340,000 to 370,000 years old [Teske 1993, Bignami and Caraveo 1992]. This is a suggestive coincidence. Using Neandertal nuclear DNA, it has been found that the apparent genetic split between us modern humans and the Neandertal ancestral population occurred 370,000 years ago [Noonan *et al.* 2006], just like the time of Geminga. Differences in the Neandertal genome may represent radiation mutations from the radiation from a nearby nuclear detonation. These mutations may or may not have been expressed in the Neandertal form, which can be well explained completely as the result of a form of Paget's Disease of Bone.

The location of the most recent surviving true Neandertal appears to have been dated to Gorham's Cave, Gibraltar [Finlayson *et al.* 2006]. However, the dating of various layers of this site has raised questions [Delson and Harvati 2006]. In a normal geological deposit, that which is under is older, and what is higher is younger. Yet, in reviewing the important dating of Gorham's Cave, Eric Delson and Katerina Harvati commented that there are "several cases where samples lower down in their [Finlayson *et al.* 2006] dig have produced ages that are younger than those above them." Further, "there are just too many instances of data younger than 28,000 years that are out of order." Variation in the nitrogen and oxygen content of the sample (at the time of exposure) and the intensity and energy of the gamma radiation from the detonation fireball could produce a range of carbon-14 concentrations, confusing the most careful excavation and analysis.

In another case of detailed dating measurements that don't quite follow the rules, the Grotte du Renne ("reindeer cave") at Arcy-sur-Cure in France, was carefully redated using accelerator mass spectrometry to measure the remaining carbon-14 atoms [Balter 2010c, Higham *et al.* 2010]. The researchers reported that the results showed "an unexpected degree of variation", suggesting that "some mixing of material may have occurred". According to the summary by Balter, this site had been identified as having 15 distinct archaeological levels. The lowest were considered to be Neandertal, and the upper levels were considered to be occupied by modern humans. The middle levels, attributed to the Châtelperronian culture, have been considered to be Neandertal, and about 30 Neandertal teeth had been found in those levels. However, the upper levels showed a date as old as 35,000 years ago, and the Châtelperronian levels showed dates ranging from 49,000 years ago, before the start of the Châtelperronian, to as recent as 21,000 years ago. That most recent age violates the rule of stratigraphy, where older is lower and younger is higher in a stratigraphic sequence. "About one-third of the dates were outside of the expected range." That forced Higham's suggestion that some materials from older and younger levels had been mixed, confusing the dating. We are faced with the appearance that as more technically precise and sensitive dating measurements are made, more sites are found to have confused stratigraphy. The radiocarbon dates do not follow the rules that are based on the assumption of a reasonably constant production and incorporation rate of radiocarbon in the archaeological deposits. This is seen most notably in Gorham's Cave, Grotte du Renne, Sandia Cave, and the Franchthi and Beldibi Caves.

These dating problems are somewhat reminiscent of the controversy over Sandia Cave in New Mexico [Hibben 1968]. Hibben claimed to have found unusual spear points dated as 25,000 to 35,000 years ago. The spear points were unusual, but his dating and the structure of the deposits are thought to be faulty. His flamboyant claims were rejected, but there might be some truth there.

The Neandertal form was initially explained as an effective adaptation to severe cold, as was thought to have existed at times in glacial Europe, and some small part of this idea lingers on. Yet, Neandertal-type skeletons have been found in areas that would appear to be balmy in comparison, poor locales for a rugged ice-man from the North, such as in the Middle East, and in north, east, and south Africa [Feder and Park 1989, Jurmain and Nelson 1994], with slight variations. Usually in limestone caves. Gorham's Cave on Gibraltar has been considered as a refugium for the last Neandertals [Finlayson *et al.* 2006]. A Mediterranean climate for cold-adapted people, while the supposed newcomers, from tropical Africa, occupied their cold European homeland?

The theory of ice age adaptation fell apart with the finding of Neandertals in the Middle East. Instead, the Neandertal could have been a pathological condition (like Paget's Disease of Bone) caused by the radioactive effects of calcium-41, which decays by electron capture, in the limestone caves where the Neandertal skeletons have been found (as opposed to the rock shelters in which the non-Neandertals have been found). The calcium-41 was produced by neutron absorption in calcium-40. For the Neandertals, the neutrons were produced by antineutrinos from a nearby astronomical detonation, like a supernova, that produced the pulsar Geminga, 370,000 years ago [Gehrels and Chen 1993, Bignami and Caraveo 1992, Teske 1993].

The concept that Neandertal was an evolutionary adaptation to glacial environments has lingered on, as the Caveman of the Ice Age. The iceman theory of Neandertal adaptation focused on the apparent similarities with the modern Eskimos, the Inuit [Jurmain and Nelson 1994], and the Eskimos and Lapps [Hublin 1996]. Neandertals have been compared to the Eskimos and Inuit, Icelanders and Laplanders, and the Tierra del Fuegians. Similarities have also been noted between the skulls of Eskimos and Icelanders, and Norwegians, and Laplanders [Hooten 1918]. Genetically, these peoples are far apart, but all drink water from the Northern Lights. A similar similarity in the structure of the pelvis has also been found between the Neandertals and the Inuit [Stringer and Gamble 1993].

In the early work on Neandertals, they were explained as adaptations to the glacial condition of northern Europe, because they resembled the modern Eskimo. In particular, the large brain of the Neandertal was essentially the same size as that of the modern Inuit, living under the Northern Lights, in the auroral zone. The aurora is formed by the energy of ions from the Sun, trapped by the Earth's magnetic field, and driven into the upper atmosphere. There these ions energize atoms of nitrogen and oxygen which then release the lovely glow of the Aurora.

The Northern Lights, the aurora borealis, and its southern counterpart, the aurora australis, the Southern Lights, result from the ionization of the upper atmosphere by charged particles that have been trapped in the extended magnetosphere of the Earth, the extension into space of the geomagnetic field. These charged particles are mainly part of the Solar wind, the stream of ions and electrons, and neutrons that decay near the Earth. The magnetosphere corrals the charged particles and channels them into the upper reaches of the atmosphere near the north and south geomagnetic poles.

All these modern people live under the auroral zones (north and south), where charged particles from space converge on the Earth's upper atmosphere. The skeletal resemblances in these people might be the result of radioactive elements ingested as part of the diet. That might be beryllium-7, a radioactive element produced in the Sun and brought to Earth in the Solar wind, captured in the radiation belts as a result of the Earth's magnetic field. It comes down to Earth in snow in the auroral regions, and freshly fallen snow supplies the drinking and cooking water of these people. One type of these ions is beryllium-7, formed by fusion reactions in the Sun. Beryllium-7 is interesting in this regard. It decays by electron capture, so it can't decay as long as it is ionized, travelling from the Sun, collected by the Earth's magnetic field, and stored in the Van Allen belts. Otherwise, it has a short half-life, 53 days, so very little material produces a considerable activity. The beryllium-7 comes down with the aurora and is taken into the bodies and bones of the "auroral people". Beryllium is a biochemical analog for calcium (as radium is, also), and so concentrates in the growing bones (and nerves) and would create the distinctive "arctic native" features that were used to explain that Neandertal was an adaptation to the Ice Age.

As charged particles, beryllium-7 atoms are collected by the Earth's extensive magnetosphere, and channeled towards the geomagnetic poles. There, in the auroral zones, the ionized beryllium-7 atoms are neutralized by recombination with electrons in the upper atmosphere and, no longer charged particles, are freed from the magnetic field. The radioactive beryllium migrates to the troposphere and falls to Earth in snow. The Inuit and Eskimos, perhaps all the auroral peoples, drink water made from the snow and year-old sea ice, and so ingest a significant activity of beryllium-7. Beryllium-7 decays by electron capture, which is considered to be as biologically damaging as alpha particles, from radium for instance.

The beryllium-7 accumulates in the plants, and in the caribou that eat the plants, but the main source to people is the use of the freshest snow for drinking water. Snow is not very drinkable water. At least one community of northern hunters solves this problem by packing snow in a pouch and putting the pouch inside their parkas, next to their body, where the excess body heat will melt the snow [Kingston 2007]. Still, the water is precipitation, with beryllium-7 as fresh as the new-fallen snow. The electron-capture process creates as much cell killing, per unit energy, as does alpha decay, such as by radium.

The human skull consists of several plates of bone that expand and grow during development, as needed to accommodate the growing brain. The brain fits tightly inside the skull, pressing against the inner surface. The brain determines the size, and to some extent the shape, of the skull. As radiation from the calcium-41 in the developing brains of the Neandertals (and beryllium-7 in the brains of the auroral people) damaged growing synaptic patterns, brains were forced to grow larger to provide the predetermined neuronal capability. In an anecdotal discussion, Anholt states that a particular enzyme is controlled by calcium, and "altered activity of this enzyme as a result of changes in the calcium concentration" causes abnormal bone structure [Anholt 1994].

Again we note that the decay of radionuclides by electron capture, as both beryllium-7 and calcium-41 decay, and also as in the decay of iron-55, is somewhat unusual in a biological sense. What had been a calcium ion mediating a biological process (or a beryllium ion acting out the role) becomes an intense explosion of electrical charge, and the wrong ion in the wrong place: wrong size, wrong charge, and wrong chemistry. This could cause a forming neural synapse in the developing brain to fail, requiring the growth of a replacement, leading to the growth of a larger brain to satisfy the functional demands of intelligence. Similar disturbances would affect growth of the skeletal bones as well.

Thus, it is likely that the large brain and sturdy stature of the Inuit has resulted from the damage done by radiation from beryllium-7, and the similar effect of the radioactive calcium-41 from the limestone caves was the cause of the Neandertal condition.

By viewing the Neandertal progression as a series of events and conditions, rather than an evolutionary process, and discarding the Glacial Theory of Louis Agassiz, a different and (I think) more consistent picture can be developed.

At the time of the earliest appearance of the Neandertals, there was a nearby astronomical detonation, of the sort that we call a supernova. The evidence for this event is seen in the pulsar known as Geminga (a pulsar is thought to be a rapidly rotating neutron star, which is thought to be the remnant core of a star that detonated as what is thought to be a supernova). Geminga formed as a pulsar and left our neighborhood 370,000 years ago [Gehrels and Chen 1993, Bignami and Caraveo 1992, Teske 1993].

Such a detonation produces an intense burst of antineutrinos, subnuclear particles, presumably with zero mass and travelling at the speed of light, which scarcely interact with other matter at all. One possible interaction of antineutrinos is the induced electron capture by protons in the nuclei of atoms, as described in Appendix A. This electron capture can release neutrons, uncharged nuclear particles that are readily absorbed by atomic nuclei, usually producing radioactive isotopes as a result. Different types of nuclei have different probabilities for absorbing neutrons, and different types of radioactive isotopes are produced. In pure limestone (calcium carbonate), calcium-40 has the highest probability of all the atoms present. Neutron capture in calcium-40 produces calcium-41 which decays by electron capture (like the beryllium-7) with a half-life of 103,000 years.

Dwellers in a limestone cave, drinking cave water saturated in calcium that would deposit in their bones, might have received a major continuous radiation dose to the skeleton, resulting in growth abnormalities that have been interpreted as a species difference, and bone cancer. Calcium in the limestone (with the calcium-41) dissolves in the seepage water in the cave, which would have formed the major source of drinking and cooking water for the Neandertals. Calcium, of course, is deposited in the bones. The electron capture decay of calcium-41 would produce the disturbed bone growth we see in Neandertals, as a form of Paget's Disease of Bone. As the calcium-41 decayed away, the intensity of Neandertalism decreased. (The 310,000 year existence of the Neandertals is just 3 half-lives of calcium-41, which leads to a reduction of the activity by a factor of 8. The radioactivity of calcium-41 is difficult to detect, and little research has been done [Moore 2004, Henning et al. 1987]). As people moved out of the limestone caves into rock shelters and the open air, ingestion of the cave water with calcium-41 and the resultant radiation exposure ended, the Neandertals disappeared, and Cro-Magnon man appeared. Neandertals faded away as they transitioned to anatomically modern humans.

In Spain and Italy, the snow-water would contain much less beryllium-7 than does the snow-water under the auroral zone where the boreal people live, at both ends of the Earth, because the beryllium-7 from the Sun comes into the atmosphere with the auroras. In a land of running water, streams and rivers, snow will not often be chosen as a direct source of drinking water. As the snow melts and seeps into the streams, the beryllium-7 will be filtered out of the water by the soil. In the case of the beryllium-7 in the modern auroral zone people, and perhaps the calcium-41 in the cave water of the Neandertals, "the dose makes the poison" is very true. Small amounts of the radioactivity do not promote the rapid resorption and growth of the bone that keeps the radioactivity concentration high and so increases the dose, which produces more pathology, and so on. If the calcium-41 idea is right, Neandertals lived in radioactive limestone caves and became Neandertals because of that; those who did not live in the caves remained anatomically modern humans. Did they become us? I don't think so.

Marcellin Boule was a French paleontologist who provided some of the earliest interpretations of Neandertals. He eventually came to see both the similarities and differences between Neandertals and modern humans [Boule and Vallois 1957]. He demonstrated these similarities and differences by comparing a reconstructed Neandertal skeleton with the skeleton of an Australian Aborigine, as shown in Figure 83.

Boule's description of the Neandertal, "... ape-like creatures who walked hunched over with a shuffling gait" is reflected by Victor Parsons' description of a classic case of Paget's Disease of Bone: "... a deaf old man with a bent back, bowed legs, curved tibia, a large skull and wobbling gait. The arms appear relatively long and the patient ape-like." [Parsons 1980]. With the exception of deafness, this is the classic description of the typical mature Neandertal skeleton, such as the "Old Man" of La Chapelle-aux-Saints, as initially described by Marcellin Boule, setting the Neandertals as primitive apemen. This fossil had lost most of his chewing teeth, and showed gross deforming osteoarthritis. Paget's Disease of Bone, when it involves the jaw, causes the teeth to loosen. By producing bone damage in joints, this disease causes osteoarthritis.

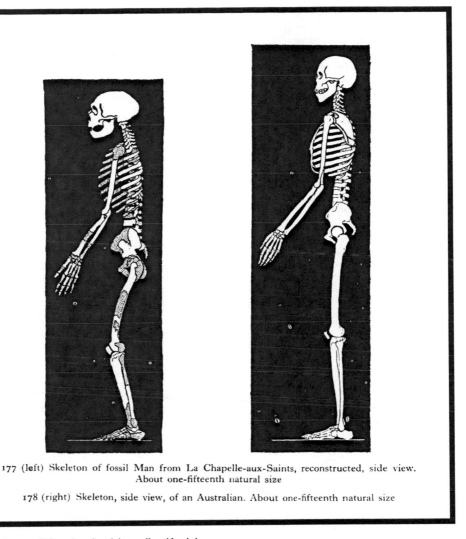

177 (left) Skeleton of fossil Man from La Chapelle-aux-Saints, reconstructed, side view. About one-fifteenth natural size

178 (right) Skeleton, side view, of an Australian. About one-fifteenth natural size

Figure 83. Skeletons of Neandertal and Australian Aborigine.
Comparison of a reconstructed Neandertal skeleton (left) and that of an Australian Aborigine (right), by Marcellin Boule [Boule and Vallois 1957].

In 1877, Sir James Paget described, as *osteitis deformans*, the disease of bone that took on his name, Paget's Disease of Bone. It is discussed in Victor Parson's *"A Colour Atlas of Bone Disease"* [Parsons 1980, also Tapley *et al.* 1989, Schwartz 1995, Paget.org, and several other sites on the Internet]. It appears as a thickening of bones, with unusual texture, weak and soft, fracturing easily and spontaneously. Microfractures occur and cause the weight-bearing bones to deform and bend. Enlargement of the skull occurs and can cause deafness by compressing the bony structure of the ear. There are some treatments but no cure. Its cause is currently unknown.

It may be that Paget's Disease of Bone is caused by the lifelong accumulation of naturally occurring lead-210, with a half-life of 22.3 years. Lead-210 is a radioactive isotope that becomes concentrated from the radon-222 that is dissolved in groundwater. The radon-222 is a decay product of radium, ultimately coming from natural uranium. Radon-222 dissolves in groundwater in proportion to the pressure (the depth of the groundwater), according to Henry's Law, and so, high concentrations of its daughter, lead-210 can accumulate. After intake, lead is deposited in the bone, continuously after birth, and accumulates there. Lead-210 decays by beta emission, with most beta particles having energies less than 17 thousand electron volts (keV). Such low-energy electrons deposit their energy within a single cell and do disproportionate damage to the cell, resulting in cell death [Goorley and Nikjoo 2000]. Low energy electrons, as these beta particles are, are also produced in other radioactive decays. Lead-210 decays to bismuth-210, which decays by a moderate energy beta particle to polonium-210. That nuclide decays by alpha emission (to stable lead-206). Alpha particles are notoriously damaging to biological cells. The radium dial painters of the early 20th Century were harmed by the alpha particles from radium and its radioactive daughters [Rowland 1994]. The accumulation of lead-210 produces bone cell damage at an increasing rate as an individual ages, making Paget's Disease of Bone, if caused by this mechanism, a disease of old age, as it is. The increasing rate of bone damage causes the bone remodeling process to speed up, essentially going out of control and destroying healthy bone and replacing it with the weak and oddly structured Paget's bone tissue.

The current form of Paget's Disease of Bone may be a disease induced by radioactivity in the diet, food or water. This may be caused by the accumulation of lead-210 from radon gas dissolved in groundwater. The dynamics of the radioactivity and bone resorption and regrowth would work together so that individual, isolated sites might express the Paget's syndrome, while other parts of the skeleton remained normal. This spottiness seems to have occurred in some of the earlier, robust ancestors also. Familial connections might arise from a tradition of cooking habits or preferred water sources, and might cause Paget's only in selected geological regions.

Treatment for Paget's Disease of Bone at present utilizes bisphosphonates, which contain two phosphorus atoms per molecule. The medicines are pharmacologically directed for adsorption onto the bone and are thought to work by inhibiting the break-down of bone tissue. Since phosphorus may be a biochemical analog to bismuth, the increase in phosphorus concentration in the active cells of the bone may interfere with the storage of bismuth-210 , newly produced from the reservoir of lead -210. Reduction of the bismuth-210 concentration will reduce the concentration of polonium-210, an alpha emitter. The reduction in the rate of radiation damage may help or may be fully responsible for the observed remission in the progress of the disease. Treatments specifically designed to remove lead from the bones, or to interfere with the retention of bismuth, might be effective in curing this disease.

Lead-210 is taken into the body in water, food, and air, and is deposited in the bone where it is firmly fixed. Lead (as the element) is stored in the bone nearly for life. In post-menopausal women with osteoporosis, it is released from the deteriorating bone tissue and can be detected in excess in the blood [Johns Hopkins Hospital 2003]. (According to a Johns Hopkins website, Paget's Disease of Bone is the second most common bone disease, after osteoporosis.) The bones of the skeleton are the single most highly radiation-exposed organ in the human body, and most of that dose comes from lead-210 and its daughters [Harley 2000, Committee on the Biological Effects of Ionizing Radiation 1980].

There is a slight possibility that some calcium-41 may remain in some limestone deposits, if it had been produced by the detonation that released the pulsar Geminga about 370,000 years ago, perhaps in the Carboniferous limestones of Northern England, where there is considerable occurrence of Paget's Disease of Bone. Individual calcium-41 atoms can be detected even in an abundance of normal calcium [Moore 2004, Henning *et al.* 1987].

Calcium-41 decays by electron capture with only low energy X-rays (4 keV, thousand electron volts) and Auger electrons of about 3 keV. These low energy radiations are particularly effective in killing cells. Similar radiation is produced in the decay of beryllium-7, also by electron capture.

In the study of the radium dial painters, Rowland quotes Aub *et al.* (1952) as stating that "The first abnormality of bone noted in roentgenograms [X-ray images] is a coarsening of the trabeculae." [Rowland 1994]. Parsons states that many cases of Paget's Disease of Bone are asymptomatic and the disease is often noted as an incidental radiographic finding and describes "coarsening of the trabecular pattern shown in the femoral heads on radiograph [X-ray images]." For radium, "Pathologic fracture is another form of injury which results from the bone destruction produced by bone-fixed radioactive materials." For Paget's Disease of Bone, Parsons states that "Pathological fractures are common." From a study by Evans (1966), Rowland reported that 26 individuals (out of a sample of 270 with radium) showed 29 bone sarcomas [cancer]. Parsons states that Paget's Disease of Bone shows an increase of twenty-fold in malignant neoplasms [cancer].

Here, it is suggested that the several different human species described as "robust" were not genetic species but variations on the basic human theme brought about by varying degrees and forms of radioactivity in the environment. In particular, the Neandertals may have suffered from a form of Paget's Disease of Bone caused by radioactive calcium-41 in the water of limestone caves. Neandertals have generally been found in limestone caves, the Cro-Magnon in rock shelters. Anatomically modern humans in the Levant had similar living quarters as the Neandertals but it may have been that the calcium-41 was not readily available, for example in cave water that would have been consumed.

A variety of bone diseases may be seen as the additional causes of the characteristics that have been interpreted as typically, and genetically, Neandertal. Osteoma is a benign, slow growing neoplasm affecting the membranous bones of the face and skull [Parsons 1980], and may have been the cause of the enlarged base of the nose and the heavy brow ridges. Neandertals have been thought to be extremely rugged and muscular, due to the heavy bone structure seen in the fossilized remains, particularly in bone attachment areas. These areas are particularly susceptible to the occurrence of bone cancers, such as osteocartilaginous exostosis, osteosarcoma, chondrosarcoma, giant cell tumor of bone, lymphomas; Ewing's tumor of bone leads to spontaneous fractures of the long bones [Parsons 1980]. The multiple fractures found in Neandertal skeletons have been compared to that of "modern rodeo bull riders in terms of damage their bodies suffered." in an otherwise healthy group of 75 individuals [LP 1999].

Metastasized cancer, which has been seen in an ancient skull [Jurmain and Nelson 1994], can be induced by excessive radiation exposure [Shigematsu *et al.* 1993, United Nations Scientific Committee on the Effects of Atomic Radiation 1994]. Neandertal skeletons from the Shanidar Cave in Iraq showed healed fractures on the shoulders and arms [Gamble 1994] as is common in Paget's Disease of Bone [Parsons 1980].

In Jean Auel's *"The Clan of the Cave Bear"* [Auel 1980], there is a scene where the elders of the clan are breaking open the base of a departed comrade's skull to extract the brain in a ritualistic cannibalism. This appears to be based on an early report of a Neandertal skull found in Monte Cicero, Italy [Bahn 1990]. Many Neandertal skulls have been found with the base of the skull opened, with the speculative interpretation that this may be evidence of cannibalism [Jurmain and Nelson 1994, Lambert 1987]. Parsons remarks that the cervical vertebrae can telescope into the soft base of the skull [Parsons 1980]. In severe cases of Paget's Disease of Bone, the base of the skull can become so soft and weak, and the skull so heavy, that the spinal column breaks through, making a distinctive opening at the foramen magnum, as seen in the Neandertal skulls.

As the limestone caves filled in with outside debris that was not significantly radioactive, (because of the diversion of neutrons into shorter lived radionuclides by elements with higher absorption cross-sections than calcium-40), the water pools would have become unusable, and outside sources of water were used. The reduced radiation exposure could have permitted the "Neandertals" to develop progressively more normally, show transitional features, and appear to be replaced by people with "Cro-Magnon" features. Not by warfare, not by competition nor from a greater mortality rate, but as a result of a natural effect due to the decay and displacement of the radioactivity. Thus, "Neandertal" was not a species, but instead, a condition.

In the Mount Carmel area of Israel, five limestone caves have housed early humans, Neandertals and anatomically early humans (like Cro-Magnon). The occupancies extended over a period from 40,000 to 75,000 or possibly 100,000 years ago for the Neandertals, in the Kebara, Amud, and Tabun caves. The anatomically modern humans were in the Qafzeh and Skhul caves at least from 80,000 to possibly 150,000 years ago. The two types of people, short, stocky Neandertals and tall, slender early modern humans, "hunted the same prey, used similar tools, and buried their dead in the same manner." [Gibbons 1996].

Erik Trinkaus and Steven Churchill found that the upper arm bones of the Neandertals were much thicker than those of the early modern humans. Trinkaus had also found earlier that the upper knob of the thigh bone, that transfers the weight of the body onto the legs, was bent more sharply in Neandertals than in early modern humans [Gibbons 1996a]. These are signs of Paget's Disease of Bone. The weakening of the bone tissue results in microfractures that allow the bones to deform almost plastically.

Further indication of a severe radiation event at the time of the Neandertal origin comes from studies of the current genetic diversity of modern humans [Gibbons 1995]. The current genetic makeup of modern humans is remarkably similar, regardless of the point of origin of the humans. That is, we are a truly uniform family of Humanity. Considering how this uniformity came about shows that the reproductive population of humans before about 400,000 years ago, when we think the Neandertals appeared, (or at 370,000 years ago with the detonation that made Geminga) abruptly dropped from roughly 100,000 individuals to 10,000. Recovery from this low level was slow, apparently not gaining momentum until after about 10,000 years ago, in association with the development of agriculture. This abrupt drop in reproductive capability, by about 90%, is suggestive of an intense burst of radiation that killed, sterilized, and produced non-surviving mutations, induced by a nearby astronomical nuclear detonation.

Interestingly, the limestone caves were shared with the cave bear, which had skeletal features that differ from its modern descendant, the brown bear, in much the same way that Neandertals differ from us. This hints that this developmental condition, induced in the Neandertals by radioactivity in the calcium of the limestone caves, may have also affected the cave bears that lived in the same caves as did Neandertal [Breakthroughs 1997, Colbert 1991, Hotton III 1968, Kurtén 1976, Ward and Kynaston 1995]. Why are there so many bear skeletons found in limestone caves? Perhaps because they are preserved there, but also because they hibernated, got overexposed to radiation, and died in their sleep, where they lay. A compounding effect in the residential limestone caves, for both Neandertals and cave bears, could be the presence of the radioactive gas radon in the air of the living spaces. Limestone deposits often contain uranium, somewhat more than the crustal average. Radon, a radioactive noble gas, is produced by the decay of uranium (and as thoron from the decay of thorium) and diffuses into the air. There, it decays to radioactive atoms that adhere to dust particles and are deposited in the nose and lungs when inhaled. (In conjunction with tobacco smoking, which reduces the ability of the lungs to clean themselves, this is the major cause of modern lung cancer.) In the nasal cavities of the Neandertals and the cave bears, this additional radioactivity might contribute to the distinctive character of the face of these two.

The cave bear, in Europe, went extinct at about the same time as the Neandertals, in Europe. In North America, the similar giant short-faced bear went extinct along with the other megafauna, about 12,000 to 10,000 years ago. It may be that the cave bear, in Europe, was the robust version of that bear, afflicted with its version of Paget's Disease of Bone, while the giant short-faced bear, in North America, was the gracile version, like the Cro-Magnon, unaffected by radioactive calcium-41.

Dating is crucial to understanding the fossil record of human development, and yet, the methods in use may be confounded by the very effects that we need to understand. Absolute dating has been provided by measurement of uranium-lead isotopic ratios [Burenhult 1993], potassium/argon [Burenhult 1993, Ninkovich and Burckle 1978, Coltorti *et al.* 1982], argon/argon [Walter *et al.* 1991, Asfaw *et al.* 1992, Clark *et al.* 1994], radiocarbon (^{14}C) [Allen *et al.* 1988], thermoluminescence (TL) [Valladas *et al.* 1988], and electron spin resonance (ESR) [Hennig *et al.* 1981, Stringer *et al.* 1989]. Use of different methods produces discrepant results and controversy [Liritzis 1982, Ikeya 1982, Hennig *et al.* 1982].

All these methods are subject to possible errors (biases) as a result of neutron irradiation or from induced radioactivity that has not been taken account of in the interpretation of the analytical data. In particular, TL dating of burnt flints in a limestone cave [Valladas *et al.* 1988] is likely to yield a greater age than geostratigraphic methods, as a result of radiation from the calcium-41, and ESR dating of teeth [Stringer *et al.* 1989] is likely to yield a still greater age as a result of the additional internal radiation dose caused by the calcium-41 contained in the teeth. Comparative dating should be directed at these crucial sites.

The Neandertals may have been fading into the anatomically modern human form as the activity of calcium-41 (produced by the detonation that produced Geminga) decayed. Some may have become the Cro-Magnon fossils we find associated with rock shelters, not in limestone caves like the apparent Neandertals. This is not at all clear, for genetic analyses [Caramelli *et al.* 2003, Serre *et al.* 2004, Krause *et al.* 2010] show that fossils identified as Cro-Magnon are close to us, genetically, and Neandertals are relatively distant. It may be that the Cro-Magnon fossils that have been genetically analyzed are actually of seasiders who migrated to the continental plateaus about 40,000 years ago (by our standard dating, but actually about 12,900 to 11,600 years ago in true time). Were there true "Cro-Magnon" people? Did the Neandertals morph into anatomically modern humans, Cro-Magnon, as the calcium-41 decayed and people moved out of limestone caves into rock shelters? Were the Cro-Magnon the first wave of seasider immigrants? Or were they all of the above? There is still much to learn, and so much more to understand.

Modern humans were created as a time capsule of humanness in Africa, sent out into the world, and did not realize their modern humanity till many generations later. But the anatomically modern humans were probably not us.

Modern Humans

There were giants in the Earth in those days.
Genesis 6:4

... how surprisingly difficult it is to define
... what is meant by "modern humans.
Roger Lewin,
"The Origin of Modern Humans", 1993

Eventually, the Neandertals, the Cro-Magnon, and the anatomically modern humans disappeared; we simply find them no more. Where did they go, where did we come from?

As we have followed the path of human development, in an attempt to understand ourselves by knowing our past, we have been forced into accepting assumptions. We assume that we are not descended from Neandertals (that savage brute!), and there is some genetic analyses that support that. However, see the work of Richard Green and colleagues [Green *et al.* 2010, Gibbons 2010a, Dalton 2010]. This sequencing of the Neandertal (nuclear DNA) genome shows that there was gene flow from the Neandertals into present-day Eurasian humans, and more so than into sub-Saharan humans. Like much good research, this work raises questions as much as answers. The affinity with current humans was found not only in Europe and West Asia, but in distant Papua New Guinea. That could trace the migration and expansion of the seasiders around the world. While interbreeding was found, "For some reason, they didn't interbreed a lot – something was preventing them." (Sarah Tishkoff, University of Pennsylvania [Gibbons 2010a].) Interactions of any sort would have been hindered by the great separation between the seaside and the continental plateau. The few resulting offspring would have remained in the seaside population, to continue in their spread around the world. There would have been no gene flow back into the Neandertal population, and so no gene flow into the stay-at-home Africans, as is observed. Perhaps only male Neandertals ventured down the great cliffs to the seaside, which would explain why the mtDNA, inherited from mothers only, shows such a clear disconnection between current humans and Neandertals, and has been interpreted as showing that there is no Neandertal ancestry in current humans. Speculatively, this interbreeding could have occurred in the Levant, near the Sinai and Mount Carmel, perhaps 100,000 years ago, as the seasiders continued their trek around the world. After meeting with a few Neandertal scouts, the seasiders could have proceeded north to Europe or south into the Red Sea, on to India, Indonesia, and Papua New Guinea. (And then farther to the New World of North and South America.) The end of the Pleistocene marked a drastic change in lifestyle for the survivors: from seacoast hunting/fishing/ harvesting, with marginal terraced agriculture, to life on flat lands that stretched for hundreds of miles (kilometers) from the sea with long-flowing rivers of fresh water.

Lewin defines the Upper Paleolithic as the time of modern humans, and the transition occurred about 40,000 years ago in Europe. Lewin proposed that as tools replaced teeth (in working leather and other materials), the face relaxed and became human. Technological change was the only possible mechanism seen behind the evolution of modern humans. The increasing use of technology released anatomical features from biological constraints. Contrary to popular scientific opinion, there was considerable anatomical variability among the available Neandertal specimens, and this was seen by Aleš Hrdlička as evolutionary progression, on the way to modern *Homo sapiens* [Lewin 1993]. Hrdlička considered that there had not been a population replacement or a cultural replacement, but that the change in form and culture had transpired along a continuous tradition. Still, there were widespread convictions against Neandertal as our ancestor.

Until the genome results are settled, our view is anchored in the past and we must consider that (maybe) we have descended from Cro-Magnon, or other anatomically modern humans as we know them, who we assume displaced the Neandertals. But it appears that all those humans disappeared in Europe about 24,000-33,000 years ago, or 11,600-12,900 true years ago. The next Europeans we find (both in Europe and in the New World), at about 11,000-9,000 years ago, were about 30 centimeters (1 foot) shorter than their predecessors. Something drastic happened, and we should not find it easy to say that the new modern Europeans are the same as the old modern Europeans.

Now we must explore the shattered time between the old and the new.

The Younger Dryas, The End of the Great Ice Age

Into my heart an air that kills,
from yon far country blows.
A. E. Housman,
'Land of Lost Content (A Shropshire Lad)" 1896

The Pleistocene lasted nearly 2 million years (or 2.6 million years, depending upon the geologists' decisions) and brought humans to the somewhat confused forms known as anatomically modern humans, Neandertals, and Cro-Magnons. Then there appears to be a sequence of great glaciations and thaws, with icesheets covering much of the northernmost part of the Earth, and then abruptly melting away. The last of these cycles has become fixed in our culture as the Great Ice Age. This concept replaced the idea of a Great Deluge, the Noah's Flood of Genesis, when Louis Agassiz seized on the similarities between geologic effects of existing and historical glaciers, and the widespread disruption that had previously been explained by a deluge and flood. From mountain and valley glaciers, Agassiz and other geologists were able to extrapolate to great, continent-wide deposits of "drift", "till", moraines, and other markers of the end of the Great Ice Age. This explanation became known as the Glacial Theory.

Before the Great Ice Age ended, there were mastodons and mammoths, giant cave bears and aurochs, huge bison, elk, armadillos, sloths, and kangaroos. There were saber-tooth cats and larger lions. There were Neandertals and Cro-Magnon. After the Great Ice Age ended, these were all gone. In this confused period of time, the world changed so that no giants reappeared, and the world became occupied by thoroughly modern humans, us.

About 12,900 years ago, a catastrophe happened in the Solar System. That was the end of the geologic time called the Pleistocene, a confusing time, a time when our measures of time itself may have become confused. Over what we think was a broad span of time, perhaps 1.6 million years, we think that great glaciers formed over the northern part of the world, and melted, only to form again, and melt again. The accepted interpretations suggest that this happened four times, with four surges of glacial ice across the land, followed by retreats as the ice melted back into the ocean. We are now living in the time following the fourth retreat.

The start of the disaster could have been just that, a "dis-aster", the breaking of a star, the detonation of a moon of Saturn. The detonation would have been far brighter than the Sun, and visible over half the Earth. In a few hours the thermal brightness of the detonation (like a gamma-ray burst) would decrease, and where Saturn had been would be a huge ragged star, then shining only by reflected sunlight, but brightly because of its large size and the white fragments of ice forming the debris cloud.

The end of the Great Ice Age, the end of the Pleistocene Epoch, brought about the final extinction of the large land animals that lived at that time, and is shown in a series of peculiar and poorly understood events, culminating in the Younger Dryas.

The Younger Dryas is considered to be a climate event of abrupt cold, most specifically determined by changes in the oxygen isotope ratios found in layers of ice in the Greenland ice cores. A decrease in the oxygen-18 concentration (relative to oxygen-16) is considered to represent a cooling, as ocean water is (presumably) sequestered in ice sheets. However, it should be recognized that this might not be the case, if the ice sheets did not exist. In the story of a detonated moon, this disturbance in the oxygen isotopes results from the influx of water from the icy fragments of the detonated moon, with a relative deficiency in oxygen-18. The isotope indication is real, but the interpretation in terms of temperature is false and misleading.

The Younger Dryas, named for a northern wildflower that flourished during that time, is shown most dramatically in the oxygen isotope curve (the ratio of oxygen-18 to oxygen-16) from the ice cores of the Greenland icesheet. Oxygen (O_2) is distributed and well mixed in the atmosphere, but may show some distinctive differences in the ocean, as water (H_2O). As water evaporates from the ocean surface, molecules of water containing the heavier oxygen-18 atoms are preferentially left behind, compared to those molecules of water containing the lighter oxygen-16 atoms. As air containing the water vapor is transported from its source to fall as snow on Greenland (and Antarctica in the South), some of the heavier water molecules (containing oxygen-18) are lost as precipitation along the way, further depleting the oxygen-18. Variations in the ratio of oxygen-18 to oxygen-16 in the ice cores from Greenland and Antarctica have been interpreted as showing changes in temperature and in the volume of water contained in large icesheets. Oxygen isotope curves are shown for Greenland (GRIP) and Antarctica (Byrd) in Figure 84 [Blunier and. Brook 2001].

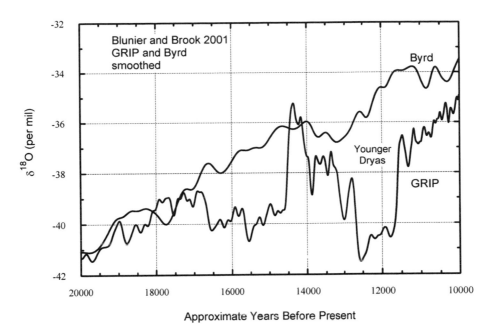

Figure 84. Oxygen isotope curves from Greenland and Antarctica.
Comparison of the oxygen isotope curves from Greenland (GRIP) and Antarctica (Byrd), showing the strong indication of a Younger Dryas event in Greenland, but none in Antarctica [Blunier and Brook 2001]. The archived data have been smoothed to emphasize the differences between the two locations. Data retrieved from the National Oceanic and Atmospheric Administration Geophysical Data Center (www.ngdc.noaa.gov/palco/paleo.html). (The time scale is approximate rather than absolute, intended to show the correlations between the two locations.)

The GRIP curve has been interpreted as showing an abrupt cooling, followed by an abrupt warming. With this approach, the variations in the oxygen isotope ratio during the last 10,000 years of the Pleistocene in Greenland are seen to show abrupt changes that are interpreted as temperature swings of the order of 10-20 Celsius degrees. The Byrd, Antarctica, ice core shows little evidence of the Younger Dryas. Dating the depth of the ice is done by meticulously counting apparent annual layers identified by a variety of characteristic markers, but the time scale here has been developed to be a consistent approximate scale used for the two sites.

Methane, as an atmospheric gas, is well mixed in the atmosphere, and is retained in small bubbles of air in the ice of Greenland and Antarctica. The concentration of methane in the GRIP and Byrd ice cores is shown in Figure 85 [Blunier and Brook 2001].

The Younger Dryas cooling appears to have been a predominantly northern effect, since the signal is loud and clear in the ice oxygen isotope curve from Greenland, but the signal is nearly absent in the curves from Antarctica. However, the well-mixed atmospheric gases, oxygen as well as methane, trapped in air bubbles in the ice, show very similar changes, North and South. Clearly, the Younger Dryas "event" was recorded equally in the North and the South, by well-mixed methane in the atmosphere. But the Younger Dryas "cooling", as interpreted from the oxygen isotopes in precipitation, was recorded only in the North.

Other studies have concluded that there was no *cooling* associated with the Younger Dryas time in the southern Pacific [Bennett, Haberle, and Lumley 2000, Rodbell 2000].

This is significant. The isotope, chemical, paleontological, and sedimentological disturbances found in the Northern Hemisphere, interpreted as incredibly abrupt changes in temperature, may have been caused by the impacts of icy fragments in the North and had no relation to cooling or warming, or any effect in the South, out of range of the impacts, except for long-distance transportation of atmospheric gases.

Alley and colleagues have shown that the end of the Younger Dryas in the North was remarkably abrupt [Alley *et al.* 1993]. The oxygen curves from Greenland suggest a transition time of only 50 years, dust concentrations show less than about 20 years, and the snow accumulation was found to double in 1 to 3 years. Such short times are difficult to accommodate in climate models or ocean circulation [Broecker 2006]. Using the snow accumulation rate variations, Alley *et al.* estimated that the Younger Dryas lasted 1,300 ± 70 years.

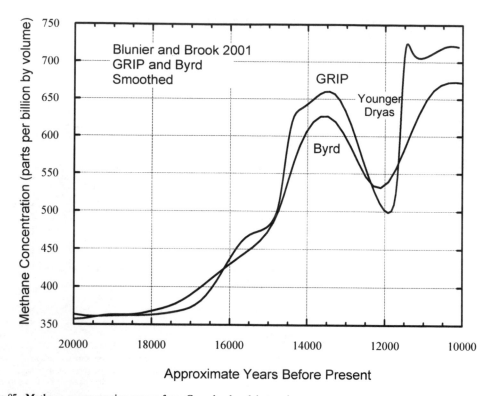

Figure 85. Methane concentration curves from Greenland and Antarctica.
Comparison of the methane concentration curves from Greenland (GRIP) and Antarctica (Byrd), showing the strong indication of a Younger Dryas event in both Greenland and Antarctica. [Blunier and Brook 2001]. The archived data have been smoothed to emphasize the similarities between the two locations. Data retrieved from the National Oceanic and Atmospheric Administration Geophysical Data Center (www.ngdc.noaa.gov/paleo/paleo.html). (The time scale is approximate rather than absolute, intended to show the correlations between the two locations.)

In discussing the evidence that the Younger Dryas was caused by an extraterrestrial impact, Richard Firestone and colleagues adopted a calendar age for the beginning of the Younger Dryas (and the end of the Clovis occupation, extinction of the megafauna in North America, and the beginning of the Black Mat) of 12,900 calendar years ago [Firestone *et al.* 2007].

This shows that the Younger Dryas extended from 12,900 years ago to 11,600 years ago.

Firestone *et al.* suggested that multiple airbursts (such as the Tunguska event) and surface impacts so disrupted the environment and climate as to cause and maintain the cooling imagined for the Younger Dryas for over a thousand years. It seems that recovery from such a massive change would be gradual and protracted, rather than as abrupt as 1-3 years. This book suggests that the evidence in the ice cores and other geologic and paleontologic materials resulted from an extended series of icy fragment impacts. When the influx of anomalous (extraterrestrial) material ceased, the Younger Dryas indicators returned to normal and the Younger Dryas ended. An attractive alternative, which is hard to justify in the face of multiple age determinations, is that the age determinations are incorrect, flawed by the assumption that the ice core layers from Greenland are the result of annual effects. Instead, they might have been produced by sequential impacts of icy fragments, each producing a separate sequence of layers. The Younger Dryas could have been much briefer.

Firestone *et al.* note the large ammonium and nitrate spikes found in the Greenland GISP2 ice core. If ammonia were present in the icy fragments of the detonated moon of Saturn, as it likely was, it would be deposited in snow as ammonium ions and, oxidized, as nitrate ions. They further note that some of the megafauna bones at the beginning of the Younger Dryas, designated the Younger Dryas Boundary, are highly radioactive compared to other stratigraphic intervals, and that this was also seen for some bones at the K/T boundary.

A mammoth bone found at Blackwater Draw was stained yellow and was notably radioactive with uranium on the upper surface only [Firestone *et al.* 2007SI. Supporting Information]. (Many uranium compounds are yellow.) This suggests that the uranium came from the skies, after the bone had been bleached dry by the Sun.

In the six Clovis-age sites that were analyzed and in the four Carolina Bays with paleosols, all had high concentrations of uranium and thorium, as was also found in the impact layers of the Chesapeake Bay and K/T impacts. The Younger Dryas Boundary has uranium and thorium up to 25 times the average crustal abundances, marking it as distinctly different from the deposits below and above. This could result from the dispersal of un-fissioned uranium and thorium from the detonating core of the moon of Saturn, sprayed into the fragmenting icy crust and eventually carried to Earth. (It is clear that these impactors, "Younger Dryas", Chesapeake Bay, and (presumably) Chicxulub, were not from the same source of the usual meteorites, which are quite depleted in uranium.)

Since no large impact crater of the right age has been identified, Firestone *et al.* rely on the shielding effect of a 2-kilometer thick Laurentide Ice Sheet (and the low acoustic impedance of ice compared to an asteroidal impactor) to minimize the shock effects of the impact. (The acoustic impedance of a material is the product of its density and the speed of sound in the material, and governs the transfer of energy from an impactor to the target.) The "ring moons" studied by the Cassini spacecraft in orbit around Saturn have exceptionally low densities for solid objects, about 0.4 to 0.7 times the density of water [Burns 2010]. Assuming a composition primarily of water ice, these are loosely agglomerated snowballs, with considerable porosity and void spaces [Charnoz, Salmon, and Crida 2010]. Such material would have a much lower acoustic impedance than solid ice, and would produce even less impact effects on hitting the Earth. Some of these snowballs might disperse so high in the atmosphere as to avoid an airburst at lower altitude or a direct impact with the surface.

In the present proposal, it is the impactor that has low impedance (even lower than solid ice because of extensive fracturing), the ice sheet was absent, and Hudson Bay (and probably Foxe Basin and Ungava Bay) was formed by impact of a major mass of ices, and was washed out by the meltwater of the icy impactor. The Canadian Shield was washed bare, as if a giant ice sheet had gradually progressed across the north of North America. Similar impacts occurred at the Gulf of Bothnia forming the gulf in Scandinavia, washing away the accumulated soil over the Siberian Traps, 250 million years old.

Earth expansion might have been induced by some more substantial impactor, as suggested by the tearalong faults that break the circular outline of Hudson Bay. Earthquake faults have been found near New York City [television program 1998] that were activated about 12,000 years ago. The faults trend to the northwest, directly toward Hudson Bay, one of the major impact sites, and the center of what we have thought was the great North American Ice Sheet. These faults may have been produced by the extreme stresses from the impact or might show the effects of impact-induced expansion. Since this book considers the dating of this time to be very suspect, a candidate crater might be Meteor Crater in Arizona. This has been dated to 49,000 years ago, but that date could be considerably in error.

In a study of lake sediments in central Poland, Goslar and colleagues found an abnormally high atmospheric concentration of carbon-14 starting in the Younger Dryas [Goslar *et al.* 1995]. Similarly, using lake sediments from Kodiak Island, Hadjas and colleagues found a significant increase in the atmospheric carbon-14 content [Hadjas *et al.* 1998]. Edwards and colleagues remarked on the large drop in the atmospheric ratio of carbon-14 to carbon-12 towards the end and after the Younger Dryas [Edwards *et al.* 1993]. Carbon-14 would be produced abundantly in ammonia (NH_3) in the ice of the detonating moon of Saturn exposed to neutrons from the nuclear fission detonation, by the $^{14}N(n,p)^{14}C$ reaction.

Along the Norwegian coast, at the same latitude as Hudson Bay, and across the Scandinavian Peninsula from the Gulf of Bothnia, two major suspected impact sites for the icy fragments, two sites were stratigraphically dated by radiocarbon analysis. In the middle of a gravel bed stratum (possibly flood-transported impact breccia) representing the Younger Dryas, lies the Vedde Ash Bed. This might be an impact debris ash bed, rather than the conventionally assumed volcanic ash bed [Bondevik *et al.* 2006]. No other ash beds are noted over the entire section above and below the Younger Dryas.

Gamma radiation from the detonation fireball would produce carbon-14 by the Giant Dipole Resonance reaction on oxygen-18 and nitrogen-15. The amount of carbon-14 produced in an object, wood or charcoal or mammoth ivory, or human bones and teeth, would depend on the oxygen and nitrogen content, and the intensity of the gamma radiation. Charcoal, with most of the oxygen gone from the original wood, would have less carbon-14 (and therefore give older dates) than wood that was exposed as unburned wood. Stringer and Davies have commented that "Where both bone and charcoal from the same level have been dated, some bone dates statistically match those on charcoal, but others are much younger and often make little stratigraphic sense." O. Jöris found that "The charcoal dates [for Middle and Upper Paleolithic sites in France and Iberia] are consistently older than those for bone, and the pattern was even clearer after calibration." [Stringer and Davies 2001]. Objects that were shielded and shadowed by massive amounts of mountainside would have less carbon-14 (and appear to be very old) than similar objects that were fully exposed. The amount of carbon-14 produced would decline from the maximum produced directly "beneath" the detonation ("ground zero") as the world turned and as the energy in the fading fireball decayed. Asia would see less energy than the immediately exposed North America, and event dates would be older. Europe would see still less energy, and events would be even older. Australia was almost over the horizon and therefore received the gamma radiation more obliquely, producing exceptionally older dates.

According to Dennis Stanford, stone blades from two southern Siberian sites, Makarova-4 ("at least 39,000 years old") and Varvarova ("more than 35,000 years"), resemble Clovis blades from the New World , dated to 11,000 years ago [Morell 1995]. That would be expected for contemporaneous cultures, one exposed in New Mexico to the early fireball of the detonation, the other exposed to the decaying and cooling fireball half a day later in Siberia. The cooler fireball, later, produced the older dates.

After the detonation fireball had cooled and faded, some explorers from the seaside, perhaps drawn to investigate the sources of corpses and carcasses washing down through their canyons, might have ventured up on the high plateau of the continents. There they would have found death and destruction, and caves that glowed in the dark from radioactive fluorescence. The sight of dead giant animals, and dead giant humans, flash burned on one side, or grotesquely distorted in the opisthotonic posture of central nervous system death, would have been recorded on walls of canyons and caves. Some of these paintings would record the detonation event itself, as a supernova beside the Moon in Baja California, a flash blocked by the Moon in Amazonia, and an expanding, brilliant debris cloud in the Chauvet Cave. In a few hours the thermal brightness (like a gamma-ray burst) would decrease, and where Saturn had been would be a huge ragged star, now shining only by reflected sunlight, but brightly because of its large size and the white fragments of ice forming the debris cloud.

Soon after the detonation, at about 12,900 true years ago, the first of a swarm of icy comets, fragments from the detonated moon, would have begun impacting the Earth. Major impacts occurred at Hudson Bay in Canada and the Gulf of Bothnia in Scandinavia. After the flood resulting from the impacts of the icy fragments from the detonated moon, which has been interpreted as the Great Ice Age, but which caused the event officially known as the Younger Dryas, dated from 12,900 to 11,600 years ago, the seasiders were forced to flee to high ground. This history is preserved in hundreds of myths and legends around the world [Vitaliano 1973]. The rise in sealevel also brought "sealevel air" to the new shores of the continental plateaus, making the territory livable for the refugee seasiders. These immigrants brought new technology, such as Clovis and Folsom points, and the unusual points found by Frank Hibben in Sandia Cave, and abruptly, civilization in new cities, such as Çatal Höyük.

I became interested in the subject of Sandia Cave when I had found that the action of natural nuclear fission reactors could have caused an event that resulted in the end of the Great Ice Age and the extinction of the large land mammals in northeast Asia and North America. The finds of frozen mammoth carcasses in Siberia, and the herds of shattered skeletons in Alaska, from about 12,000 years ago, are of tremendous importance. Hibben wrote about these in his book *"The Lost Americans"*. In that book, he briefly describes Sandia Cave. (One of the difficulties in reading Hibben's work is that he so obviously adds "color" to his story that it is hard to tell how much is just "story".) What I knew of Sandia Cave was that it had "confused stratigraphy". Settegast used similar phrases in referring to Franchthi and Beldibi Caves [Settegast 1986] and so that caught my attention. I had not found that phrase, as I remembered it, in any of the publications on Sandia Cave, so I may have heard it from Vance Haynes (at University of Arizona), one of the later investigators [Haynes Jr. and Agogino 1986, Steves and Agogino undated]. Unfortunately, it did not stick in my notes as well as it stuck in my mind.

The cave, opening onto a deep canyon in the mountains of New Mexico, may have been flooded as a result of the inability of the canyon to adequately carry off the tremendous amount of rain in the Deluge, at about 12,000 years ago, in the Spring. Soon afterward, days or weeks, the cave was briefly occupied by a small number of people that Hibben identified as "Sandia". The cave might have still been partially flooded at that time. These people left and were followed perhaps a month later by "Folsom" man, also few in number. They also left as the cave began to fill with limestone solution, from the roof of the cave. All these people were fleeing from the rising ocean, where they had lived along the seashore, and in this new land they found and scavenged dead animals.

The Younger Dryas, a rare event, was caused by a rare event. As a natural effect of nuclear fission reactors, what we see as the Younger Dryas was caused by the nuclear detonation of an icy moon of Saturn, followed by a great infall of icy fragments with anomalous isotopic ratios. Some isotopic ratios on the Earth were changed instantly by the gamma flash, the megafauna were extinguished by radiation, the permafrost was produced by the freezing cold ammonium hydroxide, the infall of water raised sealevel and flooded the homeland of the seasiders, driving them to higher ground for survival.

The Flood

The Younger Dryas falls at the time of the end of the Pleistocene and showed an abrupt drop to extremely cold temperatures (as judged from the isotopic ratios), followed by a gradual rise and then an equally abrupt return to the originally established warming trend. The basic measurement is the change in isotopic ratios of oxygen and hydrogen, and may not reflect a change in global or regional temperatures. (While the age and duration of the Younger Dryas appear to have been accurately measured, I think there may be problems with the interpretation of the data. It may have been much shorter than the about 1,000 years that we think, as a result of counting daily ice layers as yearly. Annual precipitation rates in Greenland dropped by about a factor of ten during that time.)

That period of time, the Younger Dryas, however long or short, left a landscape that was scraped and battered, covered with unusual deposits and features, and for the European Christians of the 18th and 19th Centuries, this was taken as evidence of a great flood, the Genesis Flood of Noah (Genesis, Chapters 6 through 9). This was described as a divine act intended to destroy evil mankind, and unfortunately, all other creatures living on dry land.

Evidence of a flood was seen in the great deposits of gravel, sand, and clay scattered over northern lands. These deposits often held bones of extinct animals, showing the destructive effectiveness of the Flood.

In the inland valleys of western Washington and British Columbia, the glacial till is as much as thousands of feet thick [Ward 1997]. What a tremendous amount of gravel production, transport, and deposition, compared to glacier deposits as thin veneers laid down on hard rock. These deposits of gravel seem to be better suited to a tremendous flood.

We think, from evidence that is interpreted as glacial deposits, that sometimes the Earth has been covered with glaciers, even sometimes, in spite of the White Earth Scenario (which says that a Snowball Earth would never thaw), it has been a Snowball Earth. There are great icesheets on Greenland and Antarctica, and many mountain glaciers exist. It was these mountain glaciers that gave an ambitious young Swiss geologist the opportunity to help geology escape from the bounds of the Biblical catastrophes [Gribben and Gribben 2001, Bolles 1999]. The northern shoulders of our world carry a most confused, and most confusing, collection of clues to a drastic time 12,900 years ago. The clues have been interpreted, re-interpreted, rejected, misunderstood, and still stand with their secrets hidden. In the glaciers of the Swiss Alps, Louis Agassiz found evidence to beat these clues into shape. What had been explained as the results of the great Biblical Flood of Noah, the Flood of Genesis (but also of Deucalion of the Greeks, and the Sumerian Gilgamesh, and nearly countless others), could now be explained as the results of icesheets greater than ever known, without any requirement for divine intervention.

One of the benchmarks of glacial erosion is the formation of U-shaped valleys. Most flowing rivers gradually cut downward in their beds, forming V-shaped valleys. Valley glaciers fill their valleys with erosive, abrasive ice, and gouge out U-shaped valleys. This has become a distinctive difference between the two forms of erosion. The difference is due to the filling of the valley by the ice of the glacier, acting on the full elevation of the sides of the valley, while the river acts only at the bottom of its valley.

However, when the flow of water overwhelms the capacity of the valley to carry the water away, the valley fills up with water as if it held a glacier and it takes on a U-shape. Examples have been found in the case of the massive flood that re-filled the Mediterranean basin 5 million years ago [Garcia-Castellanos et al. 2009], and in the English Channel from the end of the Pleistocene [Gupta et al. 2007]. While there has been a general attitude that ice, as glaciers, is more effective at erosion than flowing water is, recent research has shown that they are equally effective, depending mainly on the topography of the eroding land [Koppes and Montgomery 2009].

As science grew, and people began to look for more details in our explanations, practical questions about the Flood arose. How could Noah have kept all the animals of the world in an ark of limited size? How could he have preserved those animals that now exist but did not live in his land? Most importantly, where had the waters of the Deluge come from, to overtop the world's highest mountains? And where had all the water gone?

Those questions became a part of the growing conflict between a literal interpretation of the Bible, and scientifically derived theories. This was not a battle between Science and Religion, for many of the scientists were deeply religious men (indeed, most were men). This was only a disagreement over acceptance of the history of the Bible as literally and exactly true, when it seemed to be contradicted by physical evidence; or recognition of the Biblical history only as a background and vehicle for the spiritual messages of religion.

The 1800s were a time of active conflict in England against the Bible as revealed literal truth. (Remarkably, this was against the English Bible, the King James Version, created by James I of England in 1611 so his subjects could read an approved version of the Holy Writ.)

A careful chronology of the Bible by Bishop James Ussher in 1650 had shown the Earth to be about 6,000 years old. Ussher presented an exact time and date for Creation: evening, October 22, 4004 BC (with a minor correction of 4 years for an error in the date of the birth of Christ) [Dalrymple 1991]. Such a short age disagreed greatly with later estimates of geologists, who saw "no beginning" to the Earth [Wicander and Monroe 1989], and developed the theory of Uniformitarianism. This geologic theory proposed that all the features of the Earth had been developed by the action of processes that could be observed at work in the present time. This rejected the effects of occasional catastrophic events, and specifically rejected divine intervention. Charles Darwin, to support the great amount of time required by his theory of slow and gradual evolution, calculated that cutting the Weald, a valley in southern England, had required 306,662,400 years [Eldredge 1999]. We now know that most of the sediments that this valley cuts through were laid down in the Cretaceous, about 65 million years ago, rather shorter than the 300 million years estimated by Darwin. William Thompson (later Lord Kelvin) argued against such a long, or unlimited, time, but even he calculated that the Earth must be at least 20 million years old [Dalrymple 1991].

When an educated Englishman was faced with only one flood story, the Genesis story of a global flood, in that troublesome book, the Bible, rejection as a religious myth was easy. For some reason, rejection was even easier when it was determined that the Hebrews had co-opted the flood story while in Babylonian captivity, with its earlier source in the Mesopotamian and Sumerian mythology. It is perplexing why this rejection did not turn to scholarly study and acceptance as nearly 400 similar and different flood stories were found world-wide, in various cultures. This was a missed opportunity for semi-historical research.

William Whiston and Thomas Burnet, 300 years ago, saw that there was likely to be a factual basis for the story of Noah's Deluge and Flood, by way of a collision with a comet [Yeomans 1991, Cohn 1996]. If all the various, similar and different myths, legends, and religious declarations are considered, and the religification is boiled away, that view is inevitable. The objections often raised against this is that, if "that" is true, then all of the Bible (or any other religious book) is also true, absolutely, totally, and literally. That's a major obstacle, but it is not good logic, and I don't trust the lack of self-interest in the writers of the books to accept everything as unvarnished truth. We can use the physics of the situation to explain the effects related in the stories.

This religious-social conflict resulted in English Science, departing greatly from the Catholic views of the Continent. English Geology, English Paleontology, Darwin's Evolution, and finally, Agassiz's glacier-driven, global icesheet theory, offering a non-divine explanation for the tills, giant skeletons, and other geologic evidence of major changes. Divine Creation and Catastrophism were rejected, everything became Uniformitarian. Considering what little was really known at the time, this was largely based on a non-religious exercise in faith of belief.

With evolution, geology, and physics against a literal interpretation of the Bible, the time was ripe for rejection of Noah's Flood. All that was needed was a physically reasonable process to produce the jumbled mounds of gravel, silt, sand, and clay, and bones.

That process was found by Louis Agassiz at the foot of mountain glaciers in the Swiss Alps. In 1837, Agassiz described his Glacial Theory, connecting the idea of great continental icesheets to the effects of mountain glaciers by the similarities in the moraine deposits (drift and till) and erratic boulders. Hardly a generation later (new geologic theories seem to be accepted by the next generation of geologists) his Glacial Theory was completely accepted. I have reviewed the Glacial Theory and the evidence for it, as differing from a flood, and I was unable to find any true distinctions between the effects of slowly moving glacial ice and rapidly moving flood water. They are both nearly irresistible forces.

In 1837, Agassiz was 30 years old, the ambitious president of the Swiss National Geological Society. After the idea had been raised by several Swiss mountain naturalists, Agassiz became convinced that the actions of alpine glaciers, that could be seen directly and historically, were the same actions that had created the Great Flood deposits, the "diluvium", the "drift". This recognition immediately solved several perplexing problems, removed this most recent ecological disaster from the realm of religion and placed it firmly in the science of geology, or more properly, glaciology. It also launched Agassiz on a long, productive, and successful career.

Before Louis Agassiz, the Earth had great deposits of gravel, silt, sand, and clay, and the skeletal remains of giant extinct animals, all explained as the consequences of the Genesis Flood of Noah. After Louis Agassiz, great icesheets became the foundation for the disaster, and the time became known as the Great Ice Age. When we come to the end of the Pleistocene, the end of the so-called Great Ice Age, we must recognize that there may be an alternative to the story told, and sold, by the ambitious young Swiss geologist over 170 years ago. In the words of the Imbries, "With his daring imagination, bold assertions, and vigorous prose style, Agassiz had little difficulty capturing the attention of a wide audience." [Imbrie and Imbrie 1979].

The Glacial Theory is now more than 170 years old and it has shaped all the following investigations and research. A self-consistent theory has been built up that is quite resistant to alternatives. It is necessary to consider the alternative, a massive impact of water from space. If this alternative is correct, this becomes the end of the Glacial Theory and a true end of the "Great Ice Age". Our exploration of the end of the Great Ice Age must consider both the physical effects on the Earth, before and after, and the drastic changes in the forms of life.

His story found a warm welcome among the geologists who wanted an end to the literal interpretation of the Biblical stories of a 6,000-year-old Earth, and divinely directed disasters, such as the Flood of Noah. They already knew that the literal interpretations of the Genesis Flood were wrong, and rightly so, but they needed an alternative. Agassiz provided that alternative: giant ice sheets. Now we can explore an alternative to that alternative, one that brings a monstrous flood back into our view.

Replacing the Great Flood with the Great Ice Age freed geology from a rigid, literal interpretation of the Biblical story of the disaster of the Flood. The questions of where did the water come from, and where did it go, were simply answered by the Glacial Theory. The water came from the oceans, fell as snow, froze into ever-greater icesheets over the North, and then melted and returned to the ocean. No new water was required, no old hidden water need be released, and no extra water need be disposed. In the meantime, the ice slowly ground the rock of the North into bits and pieces, gravel, silt, sand, and clay.

As the science of geology developed during the 19th Century, a conflict arose between the Catastrophists and the Uniformitarians. The Catastrophists argued that the present conditions of the Earth were the results of a variety of catastrophes, such as Noah's Flood, as described in the Book of Genesis. The Uniformitarians argued instead that the Earth had been shaped entirely by those processes known today, acting uniformly through long periods of time [Hallam 1989]. As the Catastrophists became discredited, with much of their foundation resting on Scripture, Catastrophism became an anchor for Creationists. Now, even though we know that catastrophes happened (such as the extinction of the dinosaurs by the Chicxulub impact), Uniformitarinism remains as the basic guiding philosophy, and generally excludes consideration of great and abrupt changes in the Earth.

All modern observations are interpreted in a framework of huge continental icesheets over northern North America, Europe, and Asia. The Glacial Theory has even been applied to more ancient times, suggesting the possibility of a "Snowball Earth", with total glaciation, about 700 million years ago. However, Michael Rampino has found that the debris left by the outwash of an ocean impact by an asteroid can be easily misinterpreted as glacial debris, and warns against uncritical acceptance of that debris as evidence for icesheets as the agent [Rampino 1996]. Rampino specifically describes observations in the Chicxulub impact debris field in Belize that are traditionally used as indicators of glacial action: "strange boulders and cobbles … smooth facets, striations, and a mirror-like polish." This provides a key to an alternative process for the Great Ice Age (and other global glaciations), impact from space.

The evidence of the icesheets of Greenland and Antarctica, and mountain glaciers around the world, appear to so clearly match the damage left that doubting the glacial cause of the physical effects of the Great Ice Age would seem to be unreasonable. The Glacial Theory and the Milankovitch Theory have provided an effective explanation for the physical effects that can still be seen today. However, observations of floods, such as in the Channeled Scablands of Washington, and on the surface of Mars, are ignored in the interpretation of these effects. Explanations for the extinction of the largest land mammals, of every kind, have not been so successful. This will be a grand exploration. We will find certain problems, and uncertain possibilities. Someday we will understand.

The success of the Glacial Theory was assured by Agassiz's initial actions. His expansion of the theory went far beyond the available evidence, yet his personal enthusiasm promoted its acceptance. Later, evidence would be interpreted according to this theory, therefore appearing to confirm it. However, it seems that most of the effects that can be achieved by great masses of solid water (ice) can also be produced by great masses of liquid ice (water). The major difference between these two materials may just be the time scale. Geology was welcoming tremendously long time scales, rejecting religion and catastrophes, so the Glacial Theory came upon opportune times and is the sole basis for our understanding of the end of the Pleistocene.

The Glacial Theory echoes past glaciations that we surmise from similar signs at times deep in the past billion years. The astronomical theory of Milutin Milankovitch led to an understanding of how relatively small changes in the Earth's motion around the Sun might produce cold glacial periods, and warm interglacial periods. However, a detailed review [Denton et al. 2010] has shown that increasing solar heating (as predicted by the Milankovitch Theory) failed to explain two of the last four glacial terminations. Batting 0.500 in baseball would be great, but we expect better from scientific theories. The researchers suggest instead that the icesheets grew excessively and finally collapsed, disrupting oceanic and atmospheric circulation, leading to the warming that terminated the glacial stage.

Changes in relative sealevel, implying growth and decay of the icesheets, are derived from changes in the fraction of oxygen-18, in deep-sea foraminifera or ice cores. Unfortunately, the curve for the oxygen-18 data is conventionally plotted upside down and backwards. That is, the oxygen-18 fraction is increasing downwards on the plot, while time runs from right to left, in reverse of our normal (Western European) sense. This tends to obscure the appearance of the curve as a repetition of abrupt drops in the oxygen-18 fraction followed by an asymptotic exponential return towards what would be the normal (long-term) fraction. This could be caused by repetitive gamma flashes that preferentially destroyed oxygen-18 (compared to oxygen-16) by the Giant Dipole Resonance reaction.

The Biblical Flood of Genesis had been taken on faith as the obvious explanation for the various peculiar geologic deposits that appear to have been left by a flood, that are found in northern Europe and Britain. For years, as people thought about the drift deposits and other strange geologic forms, the cause was assumed to be the Biblical Flood of Genesis. However, religious writings do better at teaching us to understand God than the physical world, and this explanation fell an easy victim to the Glacial Theory, which proposed clear, specific physical processes that can still be seen in action today. When science turned away from religion, and the Ice Age Theory attacked with apparently scientific observation and analysis, The Flood had no effective defense, with little or no scientific study or support. In the battle between Uniformitarianism and Catastrophism, between Science and Religion, the Genesis Flood lost. At a time when religious dogma was beginning to become an embarrassment to science, the geologists were ready to accept observable explanations for their observations. The Glacial Theory became locked in place, succeeding in much the same manner as Plate Tectonics did a century later, in explaining the observations of Continental Drift. A slow "geologic" process is preferable to catastrophe, there was no competing theory, it fit well with cultural changes in action at the time, and it may have succeeded on the strength of the ego of its promoter.

This acceptance is described in a book by John and Katherine Imbrie, "Ice Ages, Solving the Mystery" [Imbrie and Imbrie 1979]. It is ironic that the Imbries recognize the adoption of the Glacial Theory as more of a religious process than a scientific one. We wait for them to drop the other shoe and describe how this dogma is as flawed as the previous one, but the Glacial Theory comes away unscathed. There is no indication of any detailed evaluation of the evidence, comparing a physical flood and the features produced by mountain glaciers. The geologists simply walked around and pointed, and were converted.

Still, there are lingering bits of suggestive evidence that cry out FLOOD! The edges of the continents are grooved by submarine canyons that by their form and texture show that they were cut by river water running on dry land, not by turbidity currents under the sea [Shepard and Dill 1966]. The continental shoreline is fringed by loose sediment, sands that were washed into place and filled the midsections of the canyons, disconnecting the seaward section from that on land, as is readily seen in the Monterey Canyon off California. On the east coast of North America, this sediment is most clear. As floodwaters flowed southward off the eastern plains into the Atlantic Basin, progressively more sediment was picked up and deposited into the rising sea. As the sealevel rose in response to the influx of water, the continental shelf became larger, fuller, and smoother towards the south.

Roches moutonées are supposed by the glaciologists to be smoothed and abraded on the upstream side by rock-filled ice flowing across rock outcrops. On the downstream side, chunks are plucked out by the ice and carried away. This might be done by adhesion, except that ice is slippery and has little adhesive strength, certainly far less than the tensile strength of the rock. It might be done by pressure-melted ice water entering cracks, refreezing and expanding, and forcing chunks off the major boulder. A similar effect is seen in the frost weathering of mountain rocks at high altitudes. However, at the base of an icesheet, temperatures are stable, meltwater exists because of the pressure, and there is little refreezing.

Rapidly flowing floodwater might cause the same effects by abrading the upstream side with sand, rocks, cobbles, and boulders. The water would cause the downstream plucking by two effects. High-energy impact on the upstream surface of the rock outcrop would cause shockwaves to pass through the rock. At the far surface, the shock waves are reflected, and the tension produced is doubled over what the initial wave carries. This will tend to fracture the downstream face. Turbulence in the flow shadow of the outcrop may produce cavitation that will pull fragments loose. These fragments are carried away and add to the downstream debris load.

Central Canada is stripped bare of all soil and sediment. It should be heaped high in the center where the ice sheets melted. The flow of material appears to have been outward from Hudson Bay, the lowest part of all Canada. Similarly, the surface of the Siberian Traps, a vast flow of lava that has existed for 250 million years, is now clear of any soil cover.

The elevation of snowlines around the world rose nearly uniformly by 1,100 meters. Elevation of the vegetation life zone on the Colorado Plateau in northern Arizona rose by about 4,000 feet, or 1,200 meters. Reptiles in the Sonoran Desert now live at elevations that are physically higher than before the end of the Pleistocene. The rise in sea level effectively lowered the atmospheric elevation of all land ("meters above sea level") so that animals that had adapted to vast continental plateaus at 10,000-20,000 feet elevation, suddenly found themselves nearly at sea level. Many, such as the La Brea condors in southern California, did not survive. The coral of Eniwetak and Bikini atolls is 1,200 meters thick. The standard estimate of sealevel rise at the end of the Great Ice Age is 127 to 163 meters. The thickness of the icesheets, and therefore the drop and subsequent rise in sealevel, has been estimated by the elevation of the highest levels of erosion on tall mountains [Imbrie and Imbrie 1979], which led to estimates of roughly 127 to 163 meters of sealevel rise. Estimates have also been made by measurements of rebound of the surfaces that were presumed to be under great icesheets, and by use of beach terraces. A study of the rebound of the land that was under the ice sheets suggests there was only enough ice for a sealevel rise of 105 meters, as measured by this method. The sea level measurements of the Barbados and New Guinea are relative measurements (relative to ground landmarks) and subject to uncertainty by unknown amounts of local subsidence or uprising. In my proposal, the rebound results from recovery from the force of the impacts, rather than from the dead weight of ice.

The limits of the supposed ice sheets, as mapped by the moraines ("outwash deposits" in this impact model), show flow of "ice" (or water) outward from the proposed areas of impact. Evidence of a massive, but "soft", impact at Hudson Bay, and also at the Gulf of Bothnia, is suggested by several geologic investigations.

Measurements of crustal rebound, the gradual rising of the land, show the greatest recovery over Hudson Bay, and in a similar manner, over the Gulf of Bothnia, in Scandinavia [Wicander and Monroe 1989, Strahler and Strahler 1992]. "Glacial" lineations are generally directed away from Hudson Bay, yet it is difficult to imagine how the bottom of the ice sheet could flow uphill from the lowest area [Boulton and Clark 1990].

Negative gravitational anomalies, showing where the land has been pushed down and away, show what we may interpret as impact zones centered on Hudson Bay and on the Gulf of Bothnia [Simons and Hager 1997].

Measurements of the stiffness of the mantle under North America show a central region that we might interpret as "work-hardened" rock, from the impact [Lee 2001].

Various indicators of the impact size can be used to estimate the mass of the impactors. Melosh [Melosh 1989] provides scaling information on the size of craters, crater fields, and fully overlapped crater fields, as a function of impactor mass.

As the smallest estimate of the impact in Canada, Hudson Bay itself is about 1,200 kilometers in diameter. This size crater would be formed by the impact of about 5×10^{19} kilograms of "ice", or about 2.5×10^{19} kilograms of water for the oceans, if a 50% estimate for water ice in the impactor is correct. This would raise global sealevel by about 83 meters, which seems to be too little. Other impacts, such as the Gulf of Bothnia, would increase this, but it appears to be an underestimate too small to provide a 1,000-meter rise in the oceans.

At the other extreme, the stiffened rocks of the mantle provide the largest estimate of the impact size. This area is approximately 5,000 kilometers in diameter, and would require an impactor mass of almost 6×10^{21} kilograms, or about 3×10^{21} kilograms of water. This is roughly twice the amount of water in our present oceans and is clearly far too much. However, the pattern of "work-hardening" is sparse in some places, so that considering the entire area would surely provide an overestimate. A rise in sealevel of 1,000 meters would require about 0.36×10^{21} kilograms of water, and so these estimates provide bounds that include the amount suggested by other evidence:

Hudson Bay	0.025×10^{21} kilograms of water, underestimate
1,000-meter rise	0.36×10^{21} kilograms of water
Mantle stiffening	2.9×10^{21} kilograms of water, overestimate

If we assume that 45% of the "ice" impacting the Earth is ammonia (NH_3), then even the smaller estimate provides far more nitrogen (N_2) than is contained in our present atmosphere, 3.5×10^{18} kilograms. Therefore, it may be that the 45% is an overestimate, that much ammonia was lost on the trip from Saturn to the Earth, or that most of the nitrogen was sequestered within the Earth's crust and oceans. If all the nitrogen in the present atmosphere arrived with the water required to raise sealevel by 1,000 meters, the fraction of ammonia in the impacting "ice" would be about 1.2%. Ammonia has not been found abundantly in the icy moons.

The amount of water delivered to the Earth for a 1,000-meter rise in sealevel represents about 1.3% of the water on a moon the size of Titan.

All these estimates are consistent with the delivery of water and nitrogen from space, so as to raise sealevel by 1,000 meters and greatly increase the amount of nitrogen in the atmosphere. This increase in nitrogen would result in a greater production of atmospheric carbon-14 by the standard method of cosmic ray interactions.

Other evidence seems to show that the oceans were much smaller, shallower, long ago. Remains of life deep (now) in the Atlantic show much greater sensitivity to climatic change (sea surface temperature) while they were living, than we would expect for the bottom of a deep sea, isolated from surface climate. The deep-sea drilling off New Jersey shows long dry spans, dry land, deep under the present sea. Sedimentary debris in the deep Atlantic, where there should only be fine settling detritus, is land-surface debris.

The best estimate of melted icesheet volume misses the best "standard" estimate of sealevel rise by 30-some meters. There is some evidence for a rise of 1,000 meters, and possibly 3,500 meters. Consider the difference between a world with the oceans nearly 12,000 feet lower than at present. Our present sealevel cities would be at the same elevation (above sealevel) as the tops of the Rockies at present, as high as the Altiplano in South America, where the giant condors soar. (The La Brea giant condor had lived throughout the North American Southwest, but perished at the end of the Pleistocene.) "Off eastern North America, the surface of the continental shelf shows typical features of a past land surface." [Pernetta 1994]. The shoreline would have traced around the continents at the base of the continental slope, where it looks like the seashore should have been. Rivers would flow in their dry-land canyons all the way to the sea. It would have been possible to walk along the shoreline, all around the world, perhaps even to Australia, before the Flood, if the sea were that much lower. There could have been a well-developed civilization along the seaside, limited in numbers by the relatively narrow edge of land, and generally excluded from the high continental plateau by the extreme elevations.

The standard view that it took sea level several millennia to rise is based, I think, on the time it takes to melt the ice-sheets, with the assumption that all the ocean water came from the ice-sheets. But there is some evidence of a more abrupt nature to some of the flooding [Emiliani 1975]. The USGS Map I-2206, "Landforms of the Conterminous United States", shows the Coteau des Prairies, a large structure at the northeast corner of South Dakota. This structure splits a flood channel from the north into two nearly equal parts. The sum of the stream widths downstream is equal to that upstream, showing the effect of a rapid flood.

The violent outwash of water from the impact zones left the drift and till deposits and the moraines that resemble present-day glacial moraines [Rampino 1996]. Sequential impacts left different "age" deposits, some eroding away earlier deposits, others laying down upper layers of material. Large boulders, erratics, would be carried great distances from their source. (Perhaps the stones of Stonehenge, thought to have been carried by people from Wales, might have been carried by flood waters instead of by men.) Smaller cobbles would be fractured and smoothed but not equally rounded, as is found in glacial deposits [Strahler and Strahler 1992] and in flood deposits.

In terms of drift deposits, a glacial advance implies a water flood, a glacial retreat implies the drainage of the flood, with the deposition of silt (as in the Nile floods). The Earth may have been expanding in response to an impact, perhaps Meteor Crater in Arizona. Meteor Crater has been dated by chlorine-36 at about 49,000 years ago, but all dates for this time that rely on radioactivity that might have been introduced instead by fragments of the detonation must be suspect. Chlorine-36 may be particularly susceptible to error. It is produced from natural chlorine-35 by neutron absorption, with an unusually large probability, with a cross-section of 43.6 barns and has a half-life of 300,000 years, so it could have come from outer space. Abundant salt (sodium chloride) has been found in the geysers of Saturn's moon Enceladus, and the indications have been interpreted as showing that there are liquid salt-water reservoirs that supply the geysers (which would require extra heat, so, where's the heat?) [Postberg 2011].

Consider the following sequence of events, all of them possible and likely. We will then explore the physical effects and determine if they fit with the observations.

A large icy moon of Saturn, similar to its present moon Titan, detonated about 12,900 years ago, and caused a wide variety of effects that have not been clearly recognized. It is reasonable to speculate that a moon of Saturn might have still had a functioning population of nuclear reactors, as evidenced by the currently active moon, Enceladus. If the power (fission rate) of these reactors increases too much, the plutonium-239 produced by conversion of uranium-238 does not have time to decay to uranium-235, and the dynamic behavior of the reactors becomes dominated by the plutonium-239. (Which is why nuclear weapons are built with a combination of uranium-235 and plutonium-239; they go boom better that way.) Confined by the mass of the moon, the reactors cannot expand easily and so run faster and hotter, and hotter and faster, as more positive feedback pushes them on, until they detonate.

Consider the consequences of the explosive disintegration of a large moon of Saturn. This disintegration would have scattered fragments of ice (solid water, ammonia, and methane) all around, producing the distinctive asymmetries (the "front face/back face" differences) that we see in Saturn's existing moons, and Saturn's glorious rings. If the detonation had occurred at a fortuitous point in the moon's orbit around Saturn, a vast stream of this "cometary" debris would have reached the Earth, possibly as soon as 6-10 years after the detonation. "Cometary" because the fragments would have developed the coma and tail that we associate with the comets. This stream of fragments would have developed the appearance of the broken-up Comet Shoemaker-Levy 9 just before its collision with Jupiter in July 1994.

The effects of this nuclear detonation explain the many strange observations found for the end of the Great Ice Age. This detonation produced the beautiful icy rings of Saturn. All of the other outer planets have meager rings. Saturn's rings alone are spectacular. The detonation was recorded in cave art, in Baja California and Amazonia and the Chauvet Cave, at the time, and became embedded in tradition. (This record in the Middle East and Meso-America may have been misinterpreted as signifying the planet Venus, with the resulting confusion in the ephemeris of Venus shown in the Venus Tablets.) The flash burns from the thermal radiation of this detonation, similar to those seen at Hiroshima and Nagasaki after the A-bomb explosions, are recorded in the bi-colored human images of Baja California, Barrier Canyon, and possibly Çatal Höyük [Settegast 1986], and death by central nervous system damage was recorded in the images of the "swimmers" in the Sahara, showing the opisthotonic posture typical of such a death. This posture is also shown in the engravings in Addaura Cave, Sicily [Settegast 1986].

The fireball of a nuclear detonation radiates so much heat in a brief period of time, fractions of a second to a few seconds, that exposed skin can suffer burns, from first-degree at a distance to third-degree closer, to vaporization at sufficiently close range. The Nuclear Bomb Effects Computer (a circular slide rule calculator) included in the 1977 edition of *"The Effects of Nuclear Weapons"* shows that a 9-megaton airburst will produce third-degree burns at a distance of 9 miles (15 kilometers). Second-degree burns would result from a 5.5-megaton airburst at the same range. Using the third-degree burn range as a standard value, and recognizing that the results are quite variable with explosive yield and atmospheric conditions, the energy released by a detonation at Saturn's distance, for third-degree (and second-degree) burns on the Earth, can be calculated.

At Saturn's closest approach to Earth, the distance is about 1.3 billion kilometers (770 million miles). The heat flux decreases as the inverse square of the distance, and this comparison shows that 6.6×10^{16} megatons (66 quadrillion megatons) at Saturn's distance would produce a heat flux at the Earth sufficient to cause third-degree burns. In technical units that is about 2.8×10^{32} joules, more than enough energy to disrupt a moon the size of Titan. A moon the size of Titan would be disrupted by a detonation that yielded about 3×10^{29} joules. The flash-burn paintings indicate an energy release about a thousand times greater. This energy would be released by the ordinary fission of 3.6×10^{18} kilograms of uranium. Continuing to use the existing moon of Saturn, Titan, as a reference, this much uranium would represent an average concentration in the rocky core of about 54 parts per million, or 27 parts per million for the entire mass. (For second-degree burns, 33 parts per million in the core, 16 parts per million for the entire mass, would be sufficient.)

The Earth's crust, the only part of the solid Earth that we really know, averages 4 parts per million of uranium, with an additional 12 parts per million of thorium, which could also serve as fuel for a detonation. Thus, the Earth's crust contains about 16 parts per million of nuclear fission fuel. With all the fuzzy factors, an agreement to a factor of three or so is not too bad. In fact, the agreement may be exact if the paintings show second-degree burns. It seems quite reasonable to suppose that a possible nuclear detonation that disrupted a moon of Saturn could have caused the flash burns on humans depicted on the rocks of Baja California.

As the water vapor deposited in the atmosphere by the entry of the fragments (and the complete "burnup" of many) condensed, a heavy rain would begin to fall, acid rain from oxidation of the ammonia (NH_3) to nitrate (NO_3) and water (H_2O), and rich in ammonium nitrate (NH_4NO_3), the richest nitrogen fertilizer.

Those large land animals that had survived the intense radiation of the detonation would have been killed by the impacts, the cold, the ammonia gas, and the tremendous floods resulting from the deposition of the cometary water. The ammonia gas would have been especially effective on mammals and birds, with nasal turbinate structures that swell and block the airway, causing asphyxiation. Reptiles, without turbinate structures, would be relatively immune to the ammonia gas and perhaps would also suffer less from the cold, to a small extent. Smaller mammals that live in burrows or dens would have been protected from the impacts, the cold, and the ammonia gas, and so would survive this catastrophe, to reconstitute their populations when the world returned to normal.

What caused the frozen permafrost of the North, hundreds of thousands of square kilometers frozen colder than glacial ice, hundreds of meters deep? It could have been the icefall from space, at -183 C, that ended the Pleistocene, killing and flash-freezing the mammoths, and causing the Younger Dryas. There is no permafrost in the South. There was no Younger Dryas cooling in Antarctica. It appears that there was no Younger Dryas cooling in New Zealand [Singer, Shulmeister, and McLea 1998], and therefore perhaps no Younger Dryas cooling in the Southern Hemisphere. If the Younger Dryas were the result of the impacts of icy fragments from a detonated moon, that impacted only in the Northern Hemisphere, there would be no evidence of a global climate event in the Southern Hemisphere. Variations in the isotope ratios of oxygen and hydrogen are used to define the Younger Dryas, based on the assumption that these variations represent mass fractionation occurring in evaporation and precipitation, and show the amount of glacial ice. Instead, the variations in the isotopes were produced by nuclear reactions from the detonation, in the icy moon and on Earth

In the outer Solar System, "ice" is a mixture of water (H_2O), ammonia (NH_3), and methane (CH_4). The rings of Saturn are glorious, and fresh. They are made of clean ice, with chunks from centimeters to meters in size. They are different from the scraggly little dust rings that all the outer planets have. The moons of Saturn are made largely of ice, water and ammonia, and methane. The surface temperature of the moons is about 90 K; water ice melts at 273 K, ammonia ice at 195 K, methane ice at 91 K. Much of the methane might have been trapped in the clathrate form in water ice, as we find in the permafrost and in submarine deposits. The detonation of the moon of Saturn scattered debris throughout the Solar System, and much of it came to the Earth. This ice stream of cometary debris could have arrived just a few years after the detonation. The material was bitterly cold, about 90 K (-183 C) and the first material to melt on warming was an ammonia-water eutectic at -77 C. This explains the flash-frozen mammoth carcasses of northern Siberia, which were killed by asphyxiation, by the ammonia, or more immediately by radiation. This also explains the occurrence of the permafrost in that area. The permafrost areas form a rough circle around the north of our world, similar but different from the supposed glacial ice sheet area [Birkeland and Larson 1989], but are absent in the south. Strictly, permafrost is soil with a temperature below 0 C, regardless of water content. Therefore, the "Dry Valleys" of Antarctica qualify. But the northern permafrost is filled with frozen groundwater. Ammonium hydroxide seeping into the surface would have frozen the soil, and water, to depths as great as 300-450 meters in places. In contrast to the Arctic regions, the base of the Antarctic icesheet contains lakes of liquid water, indicating temperatures closer to the melting point, at approximately –3.15 C to –2.46 C [Ellis-Evans and Wynn-Williams 1996]. "Lake Vostok" is under 4,000 meters of ice, but its surface is only 500 meters below sealevel.

On Saturn's icy moons, the ice consists of water, ammonia, and methane, closely intermingled. These form ammonium hydroxide (NH_4OH) and methane hydrate ($CH_4 \cdot H_2O$). Neutrons from the nuclear detonation would be absorbed by the nitrogen, making radioactive carbon-14:

$$^{14}N(n,p)^{14}C,$$

and gamma radiation would convert nitrogen-15 to carbon-14 by the Giant Dipole Resonance reaction:

$$^{15}N(\gamma,p)^{14}C.$$

The energy of the recoiling carbon-14 is sufficient to break the chemical bonds, but the new carbon atom would have recombined to form CH_3OH, methyl alcohol or methanol, with a release of hydrogen.

Additional carbon-14 would have been produced at the instant of detonation, in the atmosphere and in biological materials of the Earth by the Giant Dipole Resonance reactions on nitrogen and oxygen.

Saturn currently has many small moons, but only one large moon, Titan, somewhat larger than the planet Mercury. If half the mass of Titan is "ice" and half of that is water ice, then Titan contains about 20 times the amount of water in Earth's present oceans. George Wetherill estimates a 10^{-2} capture by Earth of material ejected from Mars. I extended this to Saturn and estimated 10^{-4}. In supposing that the rain of icy debris from the detonation of a moon of Saturn raised sealevel on the Earth by 1,000 meters, about 1.3% of the mass of a moon the size of Saturn's large moon Titan needs to be directed so as to collide with the Earth. Ejection towards Earth from a satellite orbit around Saturn would be needed to increase this fraction substantially, in order to achieve an increase in ocean depth by more than 1,000 meters.

This detonation resulted in a burst of gamma radiation and antineutrinos, the flash of an immense fireball, fragmentary remnants of the moon scattered around Saturn, damage to its existing moons, a supply of ice fragments to form the beautiful rings, and a cloud of icy debris headed for Earth. "Ice" on a moon of Saturn would consist of a mixture of our ordinary water ice, solid ammonia, and solid methane. Ammonia bonds to water as ammonium hydroxide, and methane forms stable hydrate clathrates in water. These are all solids at the temperature near Saturn, about 90 K (-183 C). Neutrons from the nuclear fission reactions would be absorbed by nitrogen in the ammonia, producing radioactive carbon-14. (This is the same reaction that increased the carbon-14 content of our atmosphere so much during the aboveground testing of nuclear weapons.) As this comet-like debris, ice from the crust of the broken moon, fell onto the Earth, the abrupt infusion of great amounts of carbon-14 would seriously change any regular behavior of carbon-14 dating. In addition, the gamma radiation can produce carbon-14 by the Giant Dipole Resonance reaction on nitrogen-15 and oxygen-18.

The debris stream headed toward Earth might take as little as six years to arrive, if it were ejected, partly due to the orbital motion of that moon of Saturn, into a minimum energy Hohmann transfer orbit. Changing from one planetary orbit to another in the Solar System requires energy. For a single impulse at the point of departure, Saturn in this case, to a collision with the Earth, a Hohmann elliptical transfer orbit requires the minimum amount of energy, less than about 700 joules per kilogram [Armento 1979, Prussing and Conway 1993]. The detonation calculated above provides about 2 million joules per kilogram, and therefore material will be scattered into many possible orbits, to be sorted out by gravitational perturbations. The Hohmann transfer from Saturn to Earth takes about 6 years of travel time. The great amount of excess energy available could propel fragments more directly to Earth in a shorter time. Other fragments would travel in other orbits, still held captive by the Sun's gravitation, to pass through the Solar System at later times as comets, perhaps at times to impact the Earth.

When the stream of fractured icy particles entered the Earth's atmosphere, it would have been abruptly slowed by air friction, not "burning up" as meteors generally do, but continuing through the atmosphere by sheer mass and momentum. Steam, ammonia, and nitrogen oxides would have been produced during the fall.

Impact on the surface of the Earth would have formed impact structures quite different from those normally identified. The impact would have been soft, with huge amounts of water (and ammonium hydroxide and methane hydrate) released. The surface of the impact zone would have been dented, as the debris impacted over a large area. A huge flood of water and ammonium hydroxide would have rushed out from the impact zone, carrying away soil and rock, eroding the surface, creating flood structures such as sand bars and gravel beds. The bitterly cold ammonia gas would have asphyxiated mammals and birds, nitric oxide (NO) would have caused blood vessel relaxation, and ammonium nitrate, a rich fertilizer, would have been produced. At –77 C, the liquid ammonium hydroxide would have flash-frozen the carcasses. As the ammonium hydroxide seeped into the underlying soil and sediments, the ground water would have been frozen, well below normal temperatures, creating deep permafrost. The frozen permafrost of the North, hundreds of thousands of square kilometers frozen colder than glacial ice, hundreds of meters deep, could have been caused by the icefall from space, at -183 C, that ended the Pleistocene, flash-freezing the mammoths, and causing the Younger Dryas. There is no permafrost in the South. There was no Younger Dryas in Antarctica. There was no Younger Dryas in New Zealand.

As the water flowed away from the impact zone, it would have carried eroded soil, sand, gravel, and boulders. As this material reached the continental shoreline, and as sealevel began to rise from the influx of water, existing shores would have been built out to sea, and canyons would have been filled with soft sediments at the new shoreline. The flood would have also carried loose bones and artifacts.

Are the great stones of Stonehenge, Carnac, and Callinash, "erratics" that were gathered from where they were left on the countryside after having been carried by a flood from great distances, rather than brought by men from great distances?

Seen by our ancestors, there would have been a painfully bright burst of light, followed by the appearance in the sky of a new star, in the location of the previously known dim planet Saturn, a star visible through day and night. During the following years, the size of this apparition would have shrunk, as the icy fragments dissipated and the remainder coalesced into a ring around Saturn. A vast stream of comets would be seen, approaching Earth.

The Earth's gravitational field would have drawn many of the fragments into collision with our atmosphere, resulting in spectacular "fire in the sky" and global ice falls. Starting at a temperature of 90 K, and composed of three compounds that melt and vaporize at different, but low, temperatures, many of these fragments would have been cloaked in a protective shield of gas and survived the atmospheric entry, to fall to the surface with terminal velocities, perhaps 100-300 miles per hour (50-150 meters per second). Probably all large fragments would be broken up, because of the low cohesive strength of ice and the fractured nature of the debris, and the smaller fragments would have a better chance of slowing down before impact. Terminal velocity impacts do not produce meteor craters. They do make dents. Rocky parts of the detonated moon would be sheathed in ice and could survive the atmospheric passage protected from the friction and would not develop the fusion crust that we use to identify rocks as meteorites.

As the bitterly cold blocks of ice warmed to the ambient temperature of the Earth's surface, the ammonia-water eutectic ($NH_3 \cdot H_2O$, or ammonium hydroxide, NH_4OH) would melt, at a temperature of -77 C. For falls on land, further warming would produce high concentrations of ammonia gas. Blocks falling into water, such as the oceans, would have been instantly encapsulated in shells of pure water ice. Subsequent warming and melting would have resulted in a gradual release of the ammonia with dilution in the surrounding water suppressing the release of ammonia to the atmosphere.

Such an explosive disintegration of a large body can only happen as a result of a nuclear fission detonation. Nuclear absorption of the neutrons released in this detonation, by the nitrogen of the ammonia, would produce large amounts of carbon-14. Deposited in the atmosphere and on the surface in regionally varying and large concentrations, this influx of non-cosmic ray radiocarbon would confuse and distort our present attempts at dating events from that time.

The rock-laden flood flows would carve out U-shaped valleys, leaving tributary valleys entering as so-called hanging valleys, high above the major valley floor. Giant boulders would be carried for hundreds of miles from their source. The debris flows would scour and scrape, scratch and polish the bare underlying bedrock. The high-energy streams would grind and fracture pebbles.

Emiliani [Emiliani 1975] shows evidence for a great flood of fresh water into the Gulf of Mexico at 11,600 radiocarbon years before the present. There are several notable straths under the oceans, flow channels on the sea floor; along the southern tip of Norway, along the Alaskan Peninsula, and beside Novaya Zemlya, off the north coast of Russia. These had to be cut while the continental shelf was above sea level: fresh ice water floats on cold sea water, and the Great Lakes would have acted as weirs in settling out any glacial debris that might otherwise be blamed for the channels. The Sunda Shelf and the floor of the Java Sea also show river valleys on submarine terrain that is far too flat to even invite consideration of "turbidity currents" as erosional agents under the sea.

Snowlines were elevated in South and North America after the end of the Younger Dryas, as shown in Figure 86 [from Strahler and Strahler 1992].

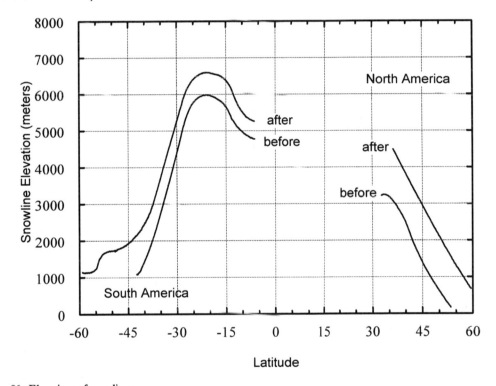

Figure 86. Elevations of snowlines.
A comparison of snowlines before and after the end of the Pleistocene, in South America and in North America [Strahler and Strahler 1992]. This shows that snowlines rose about 1,100 meters, suggesting that sealevel, as a reference, rose by that amount.

Iron-bearing sediments in the deep South Atlantic from the deserts of Patagonia were five times the present concentration, which could reflect a desert five times as large, out to the edge of the continental shelf. Existence of the Laurentide Strath, and a similar structure on the seafloor south of the Alaskan Peninsula, shows that the outwash of water from the region of the supposed Laurentide Icesheet flowed over dry land. (Freshwater is less dense than seawater, so, if there had been icesheets, the meltwater should not have impacted the ocean bottom, if the ocean had been there.) Sedimentary debris fans at the exits from the western Mediterranean, in the Atlantic, and in the eastern Mediterranean, at the Strait of Messina and the Sicilian Channel, show that the western Mediterranean overflowed, both to the west and to the east. There are submerged seamounts in the southeast Pacific, near South America, that have surfaces that were formed by surf erosion.

Many of these effects would be restricted to a large or small area around the impact, while some might be global. Freezing would be limited to the extent that ammonium hydroxide would remain sufficiently concentrated. Dispersal of the gaseous ammonia into the atmosphere might have had global effects. Would the impacts have caused the Earth to expand, as was suggested for the extinction of the dinosaurs? Probably not, as "splash" impacts may lack the energy transmission of asteroidal impacts, but impacts by fragments from the moon's core might have been intense enough to have caused expansion. One possible impact site is Meteor Crater in Arizona. This is dated well before the end of the Pleistocene, but as with all dates in this time, that date may be in error.

The span of time estimated for the Rancho La Brea tar-pit deposits, 34,000-11,000 years ago, may have really been just a short time, perhaps a few years or a thousand years. The carved bone interpreted by Alexander Marshack as a lunar record [Marshack 1991] estimated to be about 30,000 years old, may actually represent the course of a comet as it approached the Earth, about 12,900 years ago. As comets moving inward in the Solar System, as larger chunks of crustal ice headed towards Earth, the apparent position in the night sky would move back and forth through the year, tracing a zigzag curve. As the fragments moved closer to the Sun, their temperature would increase, volatile chemicals, most of the material of the debris, would outgas and form tails. These tails generally point away from the Sun and so change apparent orientation as the relative direction from Earth changes. Clouds of icy debris, propelled in part by the orbital motions of the moon of Saturn, would come streaming towards the Earth, spraying out dust and vapor, like comets. The debris would take approximately six years to approach the Earth, during which time fragments would swing back and forth in the northern night sky, as comets do at present.

By using correlation by events, rather than the analytical dating, it appears that the megafauna extinctions in Australia (46,400-30,000 years ago) happened at the same time as the megafauna extinctions in Siberia and North America (12,000-10,000 years ago). Then the Lake Mungo, Australia, geomagnetic reversal event (30,000 years ago) happened at the same time as the Gothenburg reversal event (12,500 years ago) which has been found globally (including in Australia, which would appear to be a conflict. Nothing is easy.). This is dating by event correlation, much like geological dating by stratigraphy. If similar rocks are found around the world, they may have been formed at the same time. If similar events occurred around the world, they may have occurred at the same time. The degree of similarity and uniqueness determines the strength of the correlation. Changes in technology, and the megafaunal extinction, are strong indicators.

Suppose that swarms of comets approaching the Earth from the north (as some modern comets threaten to do and as Louis Frank's Small Comets seem to) produced the unrecognized impact craters we know as Hudson Bay in Canada and the Gulf of Bothnia in Scandinavia. The floods of water scoured away all the soil and loose rock, also scouring away the cover of the Siberian Traps, 250 million years old. Careful measurements of the ongoing uplift of the Hudson Bay region and Scandinavia, and perhaps the multitudinous dents known as the Carolina Bays, can be interpreted as showing the impact patterns where momentum of the impacting material, rather than the weight of ice, depressed the surface crust, causing great dents. These effects would have been concentrated on the Northern Hemisphere.

Tremendous floods of water (and aqua ammonia) from the sky fell on Alaska and Canada and the northern United States. This killed and tumbled the herds of mammoths, mastodons, and bison, whose remains are now found in the black muck of Alaska and as scattered deposits throughout North America. The flood washed down the center of the continent, forming the flood channel that splits in two at the Coteau des Prairies in South Dakota/North Dakota. This flood of fresh water flowed down the Mississippi valley into the Gulf of Mexico, where Cesare Emiliani dates the flood at 11,600 radiocarbon years before the present [Emiliani 1975].

The sinusoidal variation of the radiocarbon age relative to tree rings may reflect the waxing and waning of Earth's interaction with the cometary debris still remaining from that detonation. This debris may be the source of the controversial "small comets" claimed by Louis Frank [Frank 1990]. The stream originally came from the North, and that is where Frank sees his comets now. The Moon has ice, at the North Pole and at the South Pole, more at the North. The southern ice comes, I think, from small comets that were diverted gravitationally by the Earth and then collided with the Moon. I recently realized that much of the ammonia could have been dissociated in the atmosphere, and increased the nitrogen content of our atmosphere. Carbon-14 is currently produced by absorption by nitrogen-14 of neutrons produced in the atmosphere by cosmic rays. A nitrogen-deficient atmosphere would have produced much less carbon-14, and everything growing in it in ancient times would look now as if it were much older, in radiocarbon years, than it should look in calendar years. Thus, everything older than about 12,000 years would look as if it was much older, perhaps 40,000 years older, by radiocarbon dating, than it really is. Dates in-between could result from contamination by water and air newly rich in carbon-14.

Such a large amount of water could have come to Earth abruptly, as the result of the detonation of a large icy moon of Saturn, a moon much like Jupiter's moon Europa. The ice from Saturn's moon would have brought very large amounts of carbon-14 with it, and also added much nitrogen to the atmosphere, where cosmic rays produce carbon-14, and a current gradual and intermittent influx of water from comets left over from that event might be the source of our continued "production" of carbon-14, which is otherwise assumed to be made by cosmic-ray spallation-produced neutrons absorbed by nitrogen-14 in the atmosphere. (Perhaps there was also less nitrogen before the icefall brought large amounts of ammonia.) If the icefall was in fact the major source of carbon-14, and little carbon-14 existed on Earth before the event, then anything that grew immediately before the icefall would date, by radiocarbon, as very much older. By that effect, I think the strange things that we date at 33,000 years ago were really from 13,000 years ago, or perhaps simultaneous with 12,000 years ago, but not exposed to the influx of carbon-14. Wood that grew up until the influx might date to 33,000 years ago; after, wood might date to 12,000, 11,400, 10,300, 9,400, perhaps even 7,400 years ago, because of the anomalously large content of carbon-14. It might take centuries for the geographic distribution of the new carbon-14 to be evened out by transport in the air and water.

Northern Eurasia and the Western Hemisphere were the most severely hit by the ice stream and flood. As the coastal plains of that time were flooded all over the world, refugees ran for high ground. In the Western Hemisphere, the Clovis people, with beautiful stone points better suited as harpoons for marine animals than as spears for land animals, killed and scavenged the last mammoths and mastodons, which were struggling in an atmosphere one-third denser than when they had thrived, due to the rise in sealevel. So, the giant Pleistocene condors perished in the new lowlands of the North American southwest, but survived in the Andes at 10,000-14,000 feet.

The Black Mat was seen as an extensive algal growth, in a wide range of locations in the Northern Hemisphere. This might be the consequence of an icefall (of water ice, ammonia ice, and methane ice) that destroyed the biotic environment but left conditions ideal for the growth of algae, with a newly wet atmosphere from the airbursts of icy impactors. Associated with this would be a jump in the surface/atmospheric carbon-14, from the $^{14}N(n,p)^{14}C$ reaction, from the neutron interactions with the nitrogen of the ammonia during the detonation, and Giant Dipole Resonance reactions on nitrogen-15 and oxygen-18, from the fireball radiation [Goslar *et al.* 1995]. A similar increase was seen for the Tunguska airburst, but was explained as the result of possible nuclear reactions during the almost-impact [Brown and Hughes 1977].

The profile of the Monterey Submarine Canyon, cut in granite and soft sediments, is similar to that of the Grand Canyon of the Colorado, cut on dry land [Stowe 1987, Birkeland and Larson 1989, Shepard 1963]. Further, the floor of the Congo (submarine) Canyon is at 3,200 meters below sealevel [Lutgens and Tarbuck 1989, credited to K. O. Emery] and the profile looks (tentatively) like the profile of the Grand Canyon. These comparisons are shown in Figure 87. It seems unlikely that turbidity currents could cut the same shape canyon underwater as a rapidly flowing dry-land river does, especially in granite rock. If the oceans had been that much lower, the coastline would have been at the base of the continental shelf, at the mouths of the submarine canyons, where it looks like a long-term coastline had existed. With asymmetric expansion of the Earth, induced by an impact, sealevel may have changed by different amounts around the world.

Flooding from the new water from the detonated moon caused the rise in sea-level, the submergence of the straths and the continental shelves, the flood structure of the Coteau des Prairies in South Dakota, and tales in the myths and legends.

The consequences are reported in many different flood legends from all over the world. A small fraction of the water from a typical large planetary satellite would have raised sealevel by a significant amount, much more than we currently estimate from the melting of the supposed ice sheets, which is about 120-150 meters. Our standard estimates of the rise in sealevel at the end of the Pleistocene are based on estimates of the amount of ice that is not in the glaciers/ice sheets at present. There is some evidence that sealevel rose abruptly by 1,000 meters. Direct measurements of sealevel seem to suffer from many problems. I think sealevel rose by about 1,000 meters at that time because that is the increase seen for snowlines around the world (up by 1,100 meters relative to present sealevel), the northern Arizona life zones went up about 1,400 meters, and the Bikini and Eniwetak corals are 1,200 meters thick. I think that sealevel may have been at the base of the continental shelf. The presence of methane hydrate deposits on the seafloor at about 3,500 meters, and perhaps not in thermal equilibrium for being formed in place, suggests a much greater rise in sealevel. (Methane would have been present in the ice from Saturn's moon and could have come from space, at 90 K.) Methane hydrate has been found at depths in the ocean of about 3,500 meters. (That evidence is complicated by the apparent fact that methane hydrate may naturally form in place, as methane gas from deeper sediments trickles upward and meets water at the right temperature and pressure. But in one location at least, the temperature distribution is not right. I think the ice brought its own temperature, 90 K, and as it warmed to the melting point of ammonium hydroxide, -77 C, the ammonia dissipated, leaving the methane hydrate.

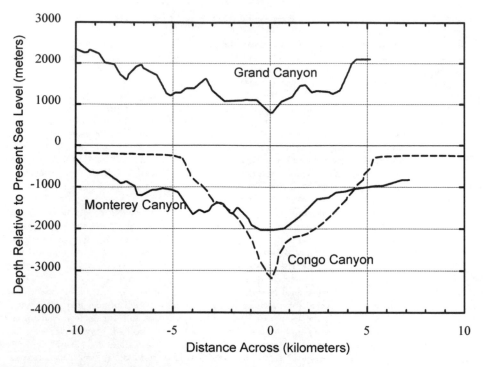

Figure 87. Profiles across submarine canyons and the dry-land Grand Canyon.
Depth profiles of the dry-land Grand Canyon and the Congo (dashed line) and Monterey submarine canyons. The profile of the
Monterey Canyon and the Grand Canyon are very similar, while the Congo Canyon is steeper and deeper.

The Flood was tremendous. The added water submerged the base of the continental shelf, and the human settlements, all around the world. An ocean so much lower than at present creates a considerable difference in the living conditions of the high continental plateaus. That is, it makes all the continents high plateaus, at a beginning elevation above sealevel starting at about 10,000 feet. So Pendejo Cave would have been at about 15,000 feet above sea level or effectively more, before the Flood. (Pendejo Cave in New Mexico was investigated by R. S. MacNeish, D. Chrisman, and G. Cunnar, and showed tools and artifacts that appeared to date from before Clovis times. The radiocarbon dating may have suffered from the nuclear reactions caused by the fireball of the detonation.) This would have been a very unattractive life-zone for most modern humans, much like the Altiplano of Bolivia, Peru, and Chile, or worse. This may explain why the findings of truly ancient humans in the New World are so sparse, and the Clovis and others who show up at about 12,000 years ago were refugees from the submerged seaside, struggling to survive in a ruined world, and failing.

The confusion of the deposits in Sandia Cave resembles the problems in two European caves. Mary Settegast discusses two caves, Franchthi in Greece (notably with evidence of seafaring) and Beldibi in Turkey, whose descriptions sound a lot like Sandia Cave. Franchthi Cave showed "a stratigraphical confusion", and Beldibi Cave showed a sterile layer ("high in iron oxide", like Sandia Cave) indicating heavy rains before reoccupation about 10,000 years ago [Settegast 1986]. Occupation, absence, and reoccupation of these two caves in southeastern Europe reflect the similar pattern shown by Clovis and Folsom-Midland occupations in the southwestern United States with a gap between. Settegast proposed that the earliest settlers of Çayönü in eastern Anatolia (about 9,500 years ago) were refugees from the drowned lands of the most ancient Greeks, following the submergence of Atlantis. In this book, these would be the people from the seaside.

As an alternative to the Glacial Theory, the proposed detonation of a moon of Saturn unifies and explains a wide variety of observations, and offers a new view for exploration.

The Megafauna Extinction

Towards the end of the Pleistocene, as we date the times, many large mammals went extinct. These giant animals, mammoths, mastodons, wooly rhinoceros, sloths, bears, and armadillos, have been labeled the megafauna. Most mammals with an adult weight greater than about 44 kilograms (100 pounds) were eliminated throughout North America and Siberia. Mass extinctions of the megafauna in Australia also occurred, dated somewhat earlier, as would be expected from the gamma-radiation production of carbon-14 perturbing the dating. Some were replaced by smaller versions. It should be noted that the known humans of that time, anatomically modern humans, Neandertals, and Cro-Magnon, were all significantly larger than 44 kilograms, and so should be considered part of the megafauna. Similarly, they were replaced by smaller humans after the Younger Dryas.

It is strange that all the small mammals seem to have survived the megafauna extinction, with the cutoff in size about 100 pounds (44 kilograms), according to Raup. Perhaps these were burrowers, or denners, and were protected from the radiation, physical impact of things falling from the sky, and from the temporary extreme cold, and possibly from the ammonia gas, because of the small entrances to relatively deep home-holes. Cave dwellers might have been more likely to have been gassed because of the larger openings to caves, and the larger openings required by larger animals. The small mammals might reproduce fast enough to recover so rapidly that we don't notice a dip in their population.

The actual duration of this event may have been only a few years, possibly only a month. We need to explore dating by correlation, like geology uses the correlations of stratigraphy. Our analytical methods may be confused and faulty. This end of the Pleistocene is a mass extinction that is rarely discussed as such.

Peter Ward agrees with the old French Catastrophist Cuvier that mass extinctions can only occur through sudden, rare, and highly unlikely catastrophes. That establishes a very important basis for studying mass extinctions. The mechanism must be catastrophic, sudden, rare, and highly unlikely. That rules out climate change, as we know it. Work on the end of the Pleistocene in northern Arizona shows that the megafauna there could have survived a climate change by just walking up the hills by 1,400 meters, and they would have stayed in the same (newly established) lifezone. Howorth remarked on this effect relative to the chamois and ibex in Europe. Where they formerly lived at low elevations, they now must live at higher elevations [Howorth 1887].

Giant condors lived all over the Southwest of North America from California to the Mississippi, then died out in North America, but survived in the Andes, 10,000-14,000 feet high. Dwarf condors now remain in a tiny mountain area of California, isolated islands of elevation, like the dwarf mammoths on their islands.

In particular, the extinction of the mammoths has been as much a fascinating problem as the earlier extinction of the dinosaurs [Hitching 1979]. It has been documented in numerous publications, some vintage, some recent, but a clear explanation has failed to develop. A presentation in 1921 by I. P. Tolmachoff in Japan provides some documentation close to the early discoveries in Siberia [Tolmachoff 1921]. Additional works include *"The Mammoth and the Flood"*, by Sir Henry H. Howorth [Howorth 1887]; *"Mammoths"*, A. Lister and P. Bahn [Lister and Bahn 1994]; *"A Book of Mammoths"*, J. Augusta and Z. Burin [Augusta and Burin 1963]; and *"Frozen Fauna of the Mammoth Steppe"*, R. D. Guthrie [Guthrie 1990]. (Guthrie's work is important in showing that fermentation of stomach contents prevented the natural freezing of a modern bison carcass in the bitter cold of an Alaskan winter. Mammoths could have been flash-frozen by a combination of the cold of liquid ammonia and the sterilization of the entire carcass by radiation.)

Skeptics of catastrophic extinction include Björn Kurtén [Kurtén 1986]; and W. R. Farrand [Farrand 1961], with response by Harold Lippman [Lippman 1962], followed by a rebuttal by Farrand [Farrand 1962]. The justification for most debunking (of flash-freezing, in this instance) appears to be "If we don't know how it could happen, it didn't really happen."

From these studies, I found that
> 1) mammoths and other large animals in Siberia were flash-frozen,
>   beyond the capability of simple winter cold.
> 2) death was accompanied by penile erection in males.
> 3) heart hemorrhage was found.
> 4) hair was loosened from the skin, or lost entirely.

I think this evidence is even stronger than it seems, because of several delicate references in the mammoth literature to carcasses with associated "tubes". The carcasses were flash frozen (in spite of the debunking of this by Björn Kurtén). Many of the skeletons of large animals in Alaska show that they were killed by impact. This is also true of mammoths in Wyoming and New Mexico. The impacts could have come from space.

While some mammoth skeletons in the southwest of North America clearly suggest a killing at the hands of human hunters, or at least the scavenging of dead carcasses, there are some massive collections of shattered bones, jumbled together in mixed deposits [Howorth 1887, Hibben 1968].

North America has shown the clearest evidence of the megafaunal extinction. At the end of the Pleistocene, in what has been labeled the Rancholabrean termination or the Terminal Pleistocene, over 35 mammal forms disappeared, mammoths, mastodons, ground sloths, horses, camels, the American lion, giant beavers and armadillos, and many others [Firestone *et al.* 2007]. All the large mammals died. Of the 35 forms that were extinguished in North America, 29 forms became globally extinct at that time, only 6 survived outside of North America [Faith and Surovell 2009]. The researchers concluded that "Results favor an extinction mechanism that is capable of wiping out up to 35 genera across a continent in a geologic instant." Of these 35 genera of large mammals, native to North America at that time, all 35 were extinguished [Faith and Surovell 2009]. Their places were filled by only 12 genera, of rather smaller animals, all immigrants. It has been assumed that these animals immigrated across the Beringia land bridge with the presumed ancestors of modern American Indians [Ward 1997]. It seems more likely that they came up from the seaside, running ahead of the rising waters.

The species extinctions occurred because of the death of all the existing individuals, rather than by evolution into a new species. Paul Martin, at University of Arizona, has tried to explain this as a result of overhunting, a blitzkrieg by modern man, just entering North America. But there are so many dead animals and so few humans that seems unlikely. Of the 35 extinguished forms, no kill sites have been found for 33. (No human skeletons from that time have been identified.) And many of the circumstances of death don't match hunting. The frozen mammoth carcasses of Siberia show that the mammoths died of asphyxiation. The male mammoths have erections. (An alternative cause of death is extreme radiation exposure, which kills by central nervous system damage, which also causes erections.) Some carcasses were flash frozen. Many of the skeletons of large animals in Alaska show that they were killed by impact, whether from space or from a flood can't be guessed. This is also true of mammoths in Wyoming and New Mexico where it is supposed that the impact was from boulders thrown by hunters.

The extinctions and the associated geology have been studied intensively, particularly in the western United States, and many explanations have been proposed. These have ranged from a blitzkrieg overkill by humans, to overwhelming disease, climate change, and extraterrestrial impact. None of these has proven successful [Haynes Jr. 2008].

Haynes reviewed 97 sites related to this extinction. These show a distinct and common stratigraphic pattern. An abrupt extinction layer (designated Stratum β1), is overlain by a "Black Mat", a deposit of highly carbonaceous soil attributed to an overgrowth of algae and wet-ground plants. This Black Mat deposit spans the time period of the Younger Dryas, from 10,900 radiocarbon years before the present to 9,800 radiocarbon years before the present, labeled Stratum β2a. Firestone *et al.* have determined the beginning of the Younger Dryas to have occurred at 12,900 ± 100 calendar years ago [Firestone *et al.* 2007]. At Stratum β1 are found abundant megafaunal fossils and Clovis artifacts, but above that stratum they are gone. In Stratum β2a are found the post-Clovis cultures, Folsom-Midland, Plainview-Goshen, and Agate Basin, but no megafauna. Above the Black Mat, ending abruptly with the end of the Younger Dryas, there are no artifacts, no megafauna fossils. Haynes reported that the earliest post-Clovis artifacts are found in the Black Mat deposit itself [Haynes Jr. 2008]. These are Folsom-Midland, Plainview-Goshen, and Agate Basin. Thus, there was a time separation between cultures, with Clovis present at the megafaunal extinction and disappearing with the extinguished megafauna, and the later cultures, which are found only in the Younger Dryas Black Mat, and then they too are gone.

A hiatus in deposition between the top of the sedimentary Stratum β1 and the bottom of Stratum β2a (labeled as a contact surface Z2) "is very short, possibly representing decades or less." [Haynes Jr. 2008]. "... Clovis sites have been found to lie on the ancient Z1 surface..." (the base of Stratum β1). The megafauna extinction and the disappearance of the Clovis culture occurred simultaneously at the time of the Z1 surface. Some short time later, the environment became excessively wet and the Black Mat plants and algae grew, burying the layer of extinction. This abnormal plant growth was terminated when the environment dried again, leaving a well-defined layer, spanning the Younger Dryas. The start of the Younger Dryas was the start of the Black Mat, and the end of the Younger Dryas was the end of the Black Mat.

This possibility of instantaneous extinction had earlier attracted the attention of Richard Firestone and a team of researchers who concluded that there had been an impact or a Tunguska-style airburst over North America that had caused the extinctions [Firestone *et al.* 2007, Firestone *et al.* 2006].

Vance Haynes had found a Black Mat, as an extensive algal growth, in a wide range of locations in the Northern Hemisphere. This might be the consequence of an icefall (of water ice, ammonia ice, and methane ice) that destroyed the biotic environment but left conditions ideal for the growth of algae, after the surface had been sterilized by radiation. Associated with this would be a jump in the surface/atmospheric carbon-14, from the $^{14}N(n,p)^{14}C$ reaction, from the neutron interactions with the nitrogen of the ammonia on the moon of Saturn, during the detonation. In addition, varying amounts (an important point) of carbon-14 would have been produced in biological materials (plants, bones, wood, tusks, and similar materials) by the Giant Dipole Resonance reactions on nitrogen and oxygen.

Directly beneath the Black Mat, Richard Firestone and colleagues found 8 different sorts of indicators of extraterrestrial impact [Firestone *et al.* 2007]. Considerable evidence was shown for an extraterrestrial impact at the Younger Dryas Boundary, but it was concluded that "the impactor was very different from well studied iron, stony, or chondritic impactors." Icy fragments from a detonated moon would indeed be very different.

Some significant research efforts followed that appeared to refute the proposal [Dalton 2009, Marlon 2009, Surovell *et al.* 2009]. These failed to find evidence for wildfires that would be expected to occur (Marlon) and were not able to confirm the presence of magnetic minerals and microspherules (Surovell) that had been found by Firestone *et al.* Continued research has also argued against the evidence presented by Firestone *et al.* [Kerr 2010c].

The impact hypothesis is consistent with the story of this book, but does not go far enough. Here, I propose that all the world-wide extinctions at that time were caused by the gamma flash radiation from the detonation of the moon of Saturn, with some hastened by the impact of ammonia-laden ice from the fragmented moon. The event occurred at the same time (within days) around the Earth, but our dating has failed us because of the disturbance of the dating methods by the radiation of the detonation. The detonation occurred just before the Younger Dryas, which represented the impact and influx of isotopically modified water. It was not a climate event, with abrupt global cooling (except near the points of impact), but water with oxygen and hydrogen isotope modification from the neutrons of the detonation. Firestone's impact hypothesis does not encompass the abrupt extinctions in Europe, Asia, and Australia [Martin and Klein 1989]. Accepting systematic errors in the dating, all these extinctions happened concurrently.

These observations can be interpreted in terms of the nuclear detonation of a moon of Saturn, followed by the impact of icy fragments, flooding the northern parts of the Earth, appearing to be a Great Ice Age.

The gamma radiation and thermal radiation from the fireball extinguished the megafauna, including the humans, as shown by the paintings on the canyon walls of Baja California [Crosby 1997]. Along with the extinction of the megafauna, it is likely that all the Clovis people were killed at the same time, by the same effects. A short time later, possibly a decade or less, a swarm of icy fragments impacted the Earth, predominantly at Hudson Bay and the Gulf of Bothnia, but also scattered in minor events over all of North America. The Carolina Bays [Firestone 2007, Allan and Delair 1995] were formed by the impacts of a cluster of fragments. (The Carolina Bays form a group of approximately 500,000 elliptical lakes, wetlands, and depressions on the Atlantic Coastal Plain, and have been suggested to be the result of impacts.) These impacts and the resultant floods (conventionally considered to be the melting of giant ice sheets) further devastated North America, killing the people represented by the post-Clovis cultures. This devastation delayed the repopulation of North (and South) America.

Radiocarbon concentrations were disturbed and dating confused by the uneven and irregular production of carbon-14 by the gamma flash of the detonation and the Giant Dipole Resonance reaction on materials with varying concentrations of oxygen and nitrogen, and local variation in the intensity of the gamma-ray flux. This was further confused by the influx of carbon-14 that had been produced in the ammonia of the icy moon by absorption of neutrons from the fission detonation. The addition of nitrogen to the atmosphere established what we now consider to be the "normal" rate of carbon-14 production by cosmic rays. (Some may still be coming to Earth in Louis Frank's "small comets" [Frank 1990].)

In Australia, 23 forms of megafauna (with masses greater than 45 kilograms) went extinct near the end of the Pleistocene. An intensive study of deposits gave a best estimate of 46,400 years ago, using Optically Stimulated Luminescence (OSL) and thorium/uranium dating [Roberts *et al.* 2001]. A previous review of 91 radiocarbon dates for this time had rejected most of them as unreliable. In the work by Roberts *et al.*, of 46 OSL dates, 7 dates were significantly younger, and 17 dates were sufficiently older so that they probably do not represent the final extinction event. Of the thorium/uranium dates on flowstone deposits, 4 of 5 were above the megafaunal unit, and showed younger ages, as expected, although the youngest were roughly 12,000 years younger. One deposit was below the megafaunal unit and was almost 14,000 years older. These problems show the difficulty in accurately and precisely determining ages for this event. Production of carbon-14 by the Giant Dipole Resonance reaction could have been very variable with Australia nearly "over the horizon" from the fireball, and exposed at a late stage in the fading of the radiation. In the Southern Hemisphere, Australia would have received little of the carbon-14 in the icy impactors. Recorded radiation energy in the sediments used for OSL could have varied greatly because of penetration of high-energy gamma radiation into sediments in the course of deposition following the killing of the animals. The extinctions in Australia could have been caused by this same event, but the dating has been disturbed by the event itself, and everything could have seemed to be older.

The disaster must have been quick, to match the many legends; it could have been quick, as the detonation of the moon of Saturn would have taken just seconds or minutes. The radiation would have quickly killed unprotected animals and plants. Later, the fragmentary debris would have blanketed the impact side of the Earth with ammonia gas and freezing cold. The "comet" (moon) debris is crucial to this. The ice would have been below 90 K (-183 C) and as the ammonia/water melted, the landscape and the exposed animals would have been treated to liquid flowing at a temperature of -77 C. It takes intense cold to flash freeze something as large as a mammoth, and a high rate of heat transfer, as could be provided by a liquid. A really cold winter might kill a mammoth (or a bison, as Guthrie showed [Guthrie 1990]), but not flash freeze it.

The mammoth, of several different species, lived in an extremely broad range [Lister and Bahn 1994]. In Europe, skeletons have been found from Northern Ireland to Spain, Italy, Poland, and even in the Middle East and Ethiopia. In

Asia, the mammoth spread across northern Siberia along the Arctic Ocean, south to China. In North America, skeletons range from Alaska to California and Florida, and the states between, and as far south as Mexico City. Yet by 10,000 years ago, at the end of the Younger Dryas, all these huge animals had disappeared.

Mammoths were about the size of the modern African elephant, but were built more heavily. Their extinction, along with the other large animals of that time, the megafauna, has remained a puzzle. Climate change in the North, to warmer and wetter, has been suggested. That would not seem to be a suitable end for mammoths living in southern California and Florida, and mammoths (and others) had survived many previous climate changes. Overkill by humans has been proposed, death by disease has also been considered. An important aspect of their extinctions is that all over the world, although less in Africa, all sorts of large mammals went extinct. Some of these were succeeded by smaller survivors, while others vanished completely. As examples, according to their home continents, some of these extinctions were:

North America	**Eurasia**
Columbian mammoth	Woolly mammoth
Giant ground sloth	Cave bear
Yesterday's camel	Woolly rhinoceros
Shasta ground sloth	Giant deer
Sabertooth cat	Irish elk
American mastodon	**Australia**
South America	Giant kangaroo
Glyptodon	Diprotodon
Litoptern	Short-faced kangaroo
Notoungulate	Giant short-faced kangaroo

Many of these animals do not seem appropriate for an overkill hypothesis and would seem to be enough different to avoid mass death by epidemics.

The dates obtained for these extinctions range from about 46,400 years ago in Australia to 11,000 years ago in North America and 10,000 years ago in Siberia. We should wonder if these dates, for what seems to have been the same kind of event all over the world, are truly different. Perhaps they have been affected by nonuniform dispersal of the carbon-14 created by the Giant Dipole Resonance reactions and also delivered with the ice from space.

In a remarkable discussion of what was even at that time contrary science, Sir Henry H. Howorth opposed the Glacial Theory, in *"The Mammoth and The Flood"* [Howorth 1887]. (His preface should make excellent reading for all who would learn an effective philosophical approach to science.) He labels the Glacial Theory, 50 years after its first proposal by Louis Agassiz, as a "glacial nightmare".

Howorth proposed that both northern and southern Siberia had similar temperate and equable climates at the time of the mammoths, but that northern Siberia is now too severe for survival. He found that mammoths had thrived there, then died, and some were abruptly frozen, and have remained frozen for 10,000 years. The present bad smell of areas containing mammoth remains shows that the carcasses were quickly frozen, to start decaying only as they were exposed in the future, our present. Climate must have changed greatly and abruptly, not over the thousands of years normally assumed. There are very few mammoth remains in Scandinavia, which was washed as bare as central Canada and the Siberian Traps. In areas where the ground is not frozen, finding a full, connected (articulated) skeleton is the same as finding a frozen carcass in the permafrost. The causes of death and preservation are likely to be the same. Some skeletons are found in an upright position, as if they had been killed in their tracks, several facing towards the North. Mammoth remains are often found with the remains of other animals, from rhinoceros to dormice, all killed at once and quickly buried. The bones and the covering drift were laid down together, at the same time. Howorth finds the vast number of mammoth skeletons remarkable, in comparison to the small number of recently dead animals found today, suggesting a mass killing.

Howorth also describes the evidence of a single flood, filling caves with greatly disturbed deposits. That may have been the effect that so confused the deposits in Sandia Cave, that Hibben described, and that were further investigated by Haynes and Agogino, and in the Franchthi and Beldibi caves [Settegast 1986].

Howorth observes that there is a great gap in quality between the "Old Stone Age" people (Paleolithic) and the "New Stone Age" people (Neolithic), and this agrees with my suggestion that the old continental plateau dwellers, those we know as Neandertals and Cro-Magnon, perished with the rest of the continental megafauna (as we should expect from their large size), and were replaced by the smaller seasiders, with a different culture.

The faunas of that time were the same on both sides of the Bering Strait. With much lower sealevel, Beringia would have been extensive dry land, and with no ice sheets, contrary to the Glacial Theory, there would have been no distinction between Siberia and Alaska. Animals would have roamed freely, forming a single, widespread fauna.

Howorth specifically proposes that the megafauna perished by some widespread catastrophe, and cites the following characteristics:

The force killed but did not break the bodies to pieces, although sometimes the skeletons are disintegrated, the bones are not weathered.

The force acted quickly to bury the bodies.

The force took up gravel and clay and laid it down.

The force swept together different types of animals, and trees, and plants, regardless of age.

He concludes that the force must have been rushing water, on a great scale. A Great Flood, without searching for a source.

He missed the cause of freezing, which I propose resulted from the ammonium hydroxide, a liquid at −77 C, and he missed the symptoms of asphyxiation, while recording a pertinent observation. In discussing a rhinoceros discovered by Schrenck, he states, "Speaking of its nostrils, he says 'They were wide open, and in the case of the one on the right side, which was uninjured, a number of horizontal folds were ranged in rows about it'." These are the turbinate structures, which in mammals serve to condition inhaled air, and are caused to swell greatly by ammonia gas. Frozen mammoths have been found well preserved in the north of Siberia, in the upper surface of the permafrost, and some of the male carcasses have penile erections.

The erections indicate death by asphyxiation, which was caused by ammonia gas, which froze the carcasses. The frozen carcass of a bison, "Blue Babe", dated to 36,425+2,575/-1,974 radiocarbon years before the present [Guthrie 1990], occurred at the beginning of the catastrophe that froze the mammoths in Siberia, and killed the mammoths and bison in Alaska and western North America, about 12,900 years ago.

The ice of the moon was bitterly cold, 90 K (about -183 C), and much of it would survive the entry through the atmosphere to fall at terminal velocity only, so no "true" impact craters would be produced. (Recall the ablation heat shields for re-entry of the early space capsules?) After the mammoths and other large animals had been killed by the gamma radiation from the detonation, an ice fall of ammonia flash-froze their carcasses. Death may have been from the direct radiation or by freezing or by asphyxiation by ammonia gas. Asphyxiation is shown by the penile erections of the male animals. Burrowing and denning animals escaped from the gas and cold, and survived. (The baby mammoth Lyuba died from radiation, but survived long enough to have lost her hair first [Mueller 2009]. Losing hair is one of the sad side–effects of whole body irradiation for the treatment of cancer.) The first material to melt would be an ammonia-water eutectic at -77C. This flash-froze the carcasses, and deep-froze the soil, producing permafrost hundreds of meters deep, even out from the present shoreline, before the oceans rose.

Large numbers of mammoth tusks, some in excellent condition, as good as a modern elephant, have been retrieved from Russia and Siberia [Tolmachoff 1929]. Tolmachoff estimates more than 46,750 mammoths had been discovered in the prior 250 years. Good ivory supplied piano keys and art objects, bad ivory was burned to make India ink. Some skeletons may have been discovered without their tusks, and so were not counted, and other animals were generally ignored by the ivory hunters, as of no economic value. Nearly all carcasses appeared to be of animals that were well fed and often fat, not suffering from deprivation, and of robust health. Plants from the stomachs of the Bolshoï Lyakhov Island mammoth and the Berezovca mammoth indicated that the animals died during late summer or early fall. (Dale Guthrie found that Blue Babe, a bison thought to have died about 36,425 radiocarbon years before the present, died in the fall or early winter. Glen Doran estimated that the Windover site was occupied in summer and fall.)

Tolmachoff reported that dogs and wild scavengers often fed off the carcasses. (Stories of human consumption seem to only be stories.) In conclusion, Tolmachoff states, regarding the variety of animals found with the mammoths,

"..., all these animals died out more or less simultaneously and probably from the same cause."

This is in agreement with the statistical study done for extinctions in North America by Faith and Surovell [Faith and Surovell 2009].

Much controversy surrounds the finding of frozen mammoth carcasses in Siberia. While nearly 50,000 mammoths have been found in the last 250 years [Tolmachoff 1929], only 39 (or 40, with a recent television expedition to "Raise the Mammoth"), have been recovered as frozen, preserved carcasses [Farrand 1961, Lippman 1962, Farrand 1962]. Mammoth carcasses are discovered as streams undercut high cliffs, the frozen soil thaws, and the carcass falls to the riverside. This seems to have led to the idea that most animals were victims of falling into a stream, freezing, and being covered with mud. However, there is general agreement that the carcasses and skeletons are usually in place on the higher parts of the tundra, "where there have been no rivers, and no rivers could ever have been." The remains are found on floodplains in Siberia. The Murray Springs skeleton, in Arizona, was found on an eroded surface, suggestive of a flood, and covered by the Black Mat deposit, a thousand years younger (apparently) than the underlying surface. This is the common occurrence in the southwestern United States.

There is considerable disagreement about the state of preservation of the frozen carcasses. As the remains thaw out of the frozen soil, decay starts, and often spoils the outside of the body. Some appear to have been recovered with the internal parts well preserved, others not.

Ivan T. Sanderson [Sanderson 1960] proposed that some mammoth carcasses had been frozen at extremely low temperatures, on the understanding that they were perfectly preserved, even to the interior. This has not been completely supported, but the carcasses are only discovered as they thaw out of the permafrost, and so decay begins. His proposal is consistent with my suggestion that the frozen conditions in Siberia and Alaska, of carcasses, soil, and ice wedges, resulted from the ammonium hydroxide ice, melting into a penetrating liquid at −77 C. The wet permafrost was formed as the ammonium hydroxide penetrated deep into the ground.

The controversy and confusion associated with the discoveries of the frozen carcasses in Siberia is shown by comparison of two writers:

Björn Kurtén [Kurtén 1986]:

> "The point here is this: Herz definitely states that it was only the superficial part of the cadaver that had been preserved. The internal organs had rotted away before the animal had become frozen."
>
> > (Kurtén may be quoting from "*Das Mammut*", by the Soviet scientist W. E. Garutt, published in 1964 by A. Ziesem.)

and I. P. Tolmachoff [Tolmachoff 1929]:

> "The mammoth was found in the best imaginable condition and comparatively little spoiled by wild animals. It has been exhibited in the Zoological Museum of the Academy as a stuffed animal with the skeleton exhibited nearby separately. The pose given to the specimen corresponds to that in which the animal was found, as if trying with its last strength to go out of some trap into which it had happened to fall. Perhaps the animal had broken through into a crevice, as thought Herz, or plunged into soft ground, as suggested by the writer, while on its pasture-ground, and died of injuries received (the pelvis, a forefoot and a few ribs were found broken, as well as the indication of a strong hemorrhage), and also of suffocation in mud. The death by suffocation is proved by the erected male genital, a condition inexplicable in any other way. However, the carcass was found, not on the very spot where the animal had perished, but within the landslide which, along with the carcass, slid down from the upper border of the high terrace of river Beresovca, these slides caused by the thawing of rock ice underlying the tundra. The flesh was so fresh and appealing that dogs devoured every piece thrown to them. Such investigations as those on the histology of stomach tissues were accomplished later with great ease. Blood, collected in great masses, owing to hemorrhage, was found to be in such a good state of preservation that it could be examined about as easily as the blood of recent animals."
>
> > (Tolmachoff references the separate reports by the expedition leaders, O. F. Herz, "*Frozen Mammoth in Siberia*", and E. W. Pfizermayer, "*Mammutleichen*".)

Can these two writers be describing the same thing, or was something lost (or gained) in translation?

These reports show that frozen carcasses of mammoths and other large animals have been found in Siberia, and appear to have been flash-frozen. Male carcasses are often found with erections. Evidence of heart hemorrhage was found in some cases. Hair was loosened from the hide.

A baby mammoth carcass, subsequently named "Lyuba", was discovered in Siberia, dead from no apparent injuries [Mueller 2009]. She was estimated to be about 1-6 months old at the time of her death, which has been put at about 42,000 years ago. Notably, she had lost most of her fur. The carcass had not decomposed, nor had it been damaged by scavengers, because of the complete killing, from mammoths to dormice. While lack of decay was proposed by the researchers to have resulted from the carcass being naturally pickled, as suggested by a distinctive aroma, decay soon after death could have been prevented by the radiation sterilization of the body and its surroundings. There were no scavengers left alive to attack the body. It was in good condition when it was frozen and buried. Although the skin was in good condition, the body was nearly hairless. (A tragedy in itself for a Woolly Mammoth.) One of the well-known side effects of whole-body radiation therapy for cancer is the loss of hair. The hairless condition of Lyuba strongly suggests a somewhat protracted dying from excessive radiation exposure, with consequent loss of hair. In spite of the excellent preservation of the carcass, Lyuba's DNA could not be recovered, suggesting that it had been damaged before her death. Destruction of DNA by radiation is the mechanism that causes so-called "radiation sickness", and loss of hair. Lyuba's DNA was damaged beyond recovery before it could be preserved by freezing. The story of Lyuba is important. She died without a certain cause; her body was not scavenged; she had lost her hair but had not decayed before being frozen; she was perfectly preserved but her DNA was damaged so badly it could not be recovered. These are signs of an intense radiation exposure.

Following the nuclear detonation of a large moon of Saturn (producing orbiting fragmental moons and the rings), as the Earth turned eastward, the radioactive debris cloud rose above the horizon in Siberia. As the cloud continued to rise, the radiation exposure on the ground became more and more intense, finally reaching lethal levels. Since most of the higher energy radiation had decayed away, little carbon-14 was produced by the Giant Dipole Resonance reaction on nitrogen-15 and oxygen-18. This resulted in there being only a small amount of carbon-14 in the body of the baby mammoth, giving it the technical appearance of great age, about 42,000 years ago.

In describing the mammoths and other animals found in Alaska, Frank Hibben produced some colorful writing [Hibben 1968]. Unfortunately, Hibben is reputed to have combined, arranged, and presented observations in his own manner, so much so that he is criticized to the point of suggestions of fraud [Preston 1995]. The black muck of the Alaskan gold fields resulted from the decay of the animal flesh and organs in place, after burial. That is, these bones were not stripped and bleached by the Sun, and then washed by a river to a point of accumulation. The animals were killed and buried in a flood, and then decayed.

Hibben states:

"In many places the Alaskan muck blanket is packed with animal bones and debris in trainload lots. Bones of mammoths, mastodons, several kinds of bison, horses, wolves, bears, and lions tell a story of a faunal population, which is the type of background we would expect in our search for early hunters. After all, if the animals were there, and the fluted points were present, could the hunters be far away?

[While flint points were collected, human remains have never been found. Did the Clovis people visit, explore, and then return to the seaside, leaving no other evidence of their existence than some lost spear points?]

"Within this mass, frozen solid, lie the twisted parts of animals and trees intermingled with lenses of ice and layers of peat and mosses. It looks as though in the middle of some cataclysmic catastrophe of ten thousand years ago the whole Alaskan world of living animals and plants was suddenly frozen in mid-motion in a grim charade."

"In summer, beneath the Alaskan sun, the frozen muck masses dripped and fell away in sludgy masses. Within these oozing piles, the bones of mammoths, camels, horses, moose, and carnivores were everywhere in abundance."

"The frozen muck had preserved, in a remarkable manner, tendons, ligaments, fragments of skin and hair, hooves, and even, in some cases, portions of the flesh of these dead animals."

"As the sun melted the black ooze in and around the bones, the stench could be smelled for miles around, the stench of some hundreds of tons of rotting mammoth meat twenty thousand years old. Apparently, a whole herd of mammoths had died in this place and fallen together in a jumbled mass of leg bones, tusks, and mighty skulls, to be frozen solid and preserved until this day."

"Mammals there were in abundance, dumped in all attitudes of death. Most of them were pulled apart by some unexplained prehistoric catastrophic disturbance. Legs and torsos and heads and fragments were found together in piles or scattered separately. But nowhere could we find any definite evidence that humans had ever walked among these trumpeting herds or had ever seen their final end."

These descriptions show an indiscriminant mass destruction of animals and plants, as if by a chaotically violent but short-lived flood, accompanied by immediate freezing.

What caused the giant/dwarf transformation at the end of the Pleistocene? At the end of the Pleistocene extinction, dwarf mammoths appear to have survived temporarily, ancient bison became (relatively) small American bison, cave bears were replaced by brown bears, giant armadillos were followed by football-sized armadillos. There is some suggestion that this affected humans as well (starting about 12,000 years ago) and we may still be recovering.

A comparative list of the victims and the survivors at the end of the Younger Dryas shows that all survivors are downsized, perhaps the lions and tigers were also?

Irish elk	moose
giant sloth	tree sloth
giant armadillo	nine-banded armadillo
giant kangaroo	red kangaroo
ancient bison	American bison
giant beaver	American beaver
giant condors	California condors
cave bear	brown bear
mastodon	Indian elephant
mammoth	African elephant
aurochs	modern cattle

and

Neandertal	current humans
Cro-Magnon	current humans

These dwarfings were quite significant. The Pleistocene giant armadillo was seven feet long, weighed 500 pounds, while the modern "giant" armadillo (*Priodontes giganteus*) is four feet long. The nine-banded armadillo in the United States weighs 15 pounds.

Thus, the question of the end of the Pleistocene is not so simple as why did so many large animals die, but also why did the recovery produce only smaller versions.

If the largest animals of a time grow as large as they can be according to their living conditions, we should compare the present-day African elephant with the mammoth of the Pleistocene. The African elephant is the largest animal currently living on land. The mammoth was the largest land animal at the end of the Pleistocene. Presumably, each animal is limited in size by how much weight it can support, how much it can breathe and eat, and how much heat it can eliminate. Our current world conditions limit the African elephant to a weight of about 7,000 kilograms. The mammoth attained weights of 10,000 kilograms. How did our world change so that the giants died and were replaced by the dwarfs?

Expansion of the Earth following a significant impact, perhaps such as made Meteor Crater in Arizona, would reduce oxygen partial pressure by spreading the atmosphere over a larger surface area and by a slightly reduced gravitational force through the atmosphere. This could contribute to the failure of the megafauna to recover as megafauna, but only as smaller versions. Oxygen in the atmosphere will be increased over geologic time by the action of photosynthesis, and by the outgassing of water from volcanoes and its eventual dissociation near the top of the atmosphere. Hydrogen escapes to space, and oxygen is left behind. At present, this is minimized by a "freeze trap" in the troposphere. Little water rises above this level. Some water from space may contribute to the growth of oxygen.

While all the megafauna of North America died off at the end of the Pleistocene, the small mammals and nearly all amphibians and reptiles survived [Holman 1995]. Small animals seem to have survived the megafauna extinction, with the cutoff in size about 100 pounds (44-45 kilograms), according to Raup.

Large land animals would have been killed by the radiation, the impacts, the cold, the ammonia gas, and the tremendous floods resulting from the deposition of the cometary water. The ammonia gas would have been especially effective on mammals and birds, with nasal turbinate structures that swell and block the airway, causing asphyxiation. In male mammals (including humans), asphyxiation would have been shown by penile erections, as discussed by Tolmachoff. Reptiles, without turbinate structures, would be relatively immune to the ammonia gas and perhaps would also suffer less from the cold, to a small extent.

However, the major cause of death was likely the intense radiation exposure from the fireball of the detonation. That would have also killed most (or all) of the potential scavengers and would have bacteriologically sterilized the carcasses, preventing immediate decay. A serious question for this proposal is the time between the detonation and extinction by radiation, and the first impacts of icy debris. The least-energy path from Saturn to Earth, assuming the orbital positions are suitable, is the Hohmann transfer orbit. That would take about 6 years. A more direct path to Earth resulting from the greater amount of energy released in the detonation could shorten the elapsed time.

Magnetic Reversals

The geomagnetic field, the Earth's magnetism, generally exists in either of two polarities, "normal" as at present, and "reversed", aligned in the opposite direction. Every million years or so, the Earth's magnetic field reverses on itself, and then remains in the same polarity for another million years or so. During these relatively stable times, occasional "events" are found, in which evidence shows an extremely brief excursion of the opposite polarity [Guyodo and Valet 1999, Langereis 1999].

One of these reversal events has been found at about 12,500 years ago, in a sediment core from Sweden, and has been named the Gothenburg event [News and Views 1971]. This excursion appears to have been found in northern and central Europe, eastern Canada, the Gulf of Mexico, and New Zealand, but has not been found in many other sediments of the same assumed age, a possible problem with dating. It may be the same as the Laschamp event, which has been variously dated between 8,000 and 20,000 years ago, and at about 43,000 or 45,400 years ago [Verosub 1975, Jacobs 1986]. The Laschamp event has not been seen in sediments that are expected to be of the same age in southern Europe, the Mediterranean Sea, and western North America.

Another reversal event has been found in Australia, at Lake Mungo [Barbetti and McElhinny 1972, Jacobs 1994]. This reversal was found in sediment baked in prehistoric fireplaces. When the sediment is hot, it takes on the magnetic polarization induced by the Earth's field. When the baked sediment cools, the polarization is locked in place. This event seems to have consisted of two excursions, at 30,780 radiocarbon years before the present and 28,310 radiocarbon years before the present. Again, dating by carbon-14 may have been confused.

A brief reversal has been found in sediment from Mono Lake in California [Jacobs 1994]. It was dated to 24,500 years ago, but was not found in sediments thought to be of the same age in nearby Clear Lake.

Other recent excursions appear to have occurred between 15,000 and 20,000 years BP; 24,000 to 25,000; 28,000 to 30,000; 32,000; and 38,000 to 40,000. This is just the time period in which we must be most suspicious of irregular deposition of carbon-14, and therefore, greatly erroneous ages. That is, these several scattered excursions might all represent a single event, confused because of erroneous dating. Excursions of this sort are not uniformly found in global investigations, and so might be suspected to be relatively isolated local or regional events, or the dates are wrong where they are found, and where they are not found.

A possible explanation for these recent events, and generally all such excursions, can be found in the electromagnetic pulse (EMP) generated when an intense burst of gamma radiation strikes the atmosphere. This has been observed with nuclear weapon detonations in near space. Tests over Johnston Island in 1962 caused an electrical blackout in the Hawaiian Islands [Glasstone and Dolan 1977].

The electromagnetic pulse is generated by the displacement of vast numbers of electrons from their parent atoms in the atmosphere, by gamma radiation, and the subsequent recombination of the electrons with the ions. The intense current flows generate a strong and rapidly changing magnetic field. Such an electromagnetic pulse produced by the 66 quadrillion-megaton detonation of a moon of Saturn could have produced the many observations of a brief reversal, apparently at different times because of the perturbations in dating caused by radiation from the detonation. An electromagnetic pulse of this intensity would damage or destroy much of the satellite electronics in orbit around the Earth, and much of our electrical systems on Earth. Could our civilization survive such a shock?

Saturn's Rings and Comets

The planet Saturn holds a special place in our planetary astronomy, worthy of attention and exploration. Its glorious rings puzzled Galileo, and continue to fascinate us all. Even though those rings are so special as to be unique, the image has become the symbol of distant planets, strange worlds. It has had important roles in our mythologies. In this book, it plays a major part in the development of our world and humanity. One of its moons was the source of the water from space that caused the Biblical Flood, and changed our world forever.

The rings of Saturn are glorious, and fresh. They are made of clean ice, with chunks from centimeters to meters in size. They are different from the scraggly little dust rings that all the outer planets have. The moons of Saturn are made largely of ice, water and ammonia, and possibly methane. The surface temperature of the moons is about 90 K; water ice melts at 273 K, ammonia ice at 195 K, methane at 90 K. If a moon of Saturn had exploded, by a nuclear fission detonation, at about 12,900 years ago, the following things would have happened:

a brilliant "supernova" in the sky, visible from Earth

intense, high-energy radiation on the hemisphere of the Earth facing Saturn

continuing radiation from the cooling fireball, as the Earth turned

production of an electromagnetic pulse, EMP, affecting surface magnetization by electric discharge of the atmosphere

formation of Saturn's rings, by capture of some of the debris

scattering of a stream of icy debris as comets into the inner Solar System

the debris could have reached Earth a few years after the detonation, creating a fire storm in the atmosphere, as chunks of ice vaporized as meteors, and the ammonia and methane released from them ignited and burned, putting the sky on fire, producing CO_2 and NO, among others

the ice chunks that fell to the ground would have killed animals by impact, freezing cold, and asphyxiation by the ammonia gas

the aqua ammonia thawing from the ice at 196 K (-77C) would have flash frozen the mammoths

rock fragments in the comets from the detonated moon would kill mammoths and bison by impact, and be ignored by future researchers, as rocks carried from far off by human hunters to kill their prey. (The distinctive fusion crust of meteorites would have been prevented from forming by the icy coating on the rocks before they entered the atmosphere.)

the rain of aqua ammonia and nitric acid from the sky would have been perceived by humans as a burning "firewater", and would have had a severe effect on vegetation, leading to the covering of the affected land by a growth of algae, discovered by Vance Haynes as the Black Mat, and noted at the conclusion of the Greek legend of Deucalion as a world covered in slime.

the water added to the Earth's oceans could have been enough to raise sealevel by about 1,000 meters.

neutrons produced by the fission detonation would react with the nitrogen in the ammonia (during the detonation) producing a great amount of carbon-14, causing an upset to the radiocarbon scale from about 11,300 radiocarbon years before the present (the date of the mammoth tusks) to 10,600 radiocarbon years before the present (the date of the Black Mat), when the material was dispersed in our atmosphere. (Adding the standard 1,000 year correction to the RC date gives 12,300 years ago, pretty close to Plato's time-setting for the submergence of Atlantis.)

the additional nitrogen in the atmosphere would result in a continuing "high" rate of carbon-14 production, which we interpret as "normal" and assume has gone on for as long as we want to date by radiocarbon.

On Saturn's icy moons, the ice consists of water, ammonia, and methane, closely intermingled. These form ammonium hydroxide (NH_4OH) and methane hydrate ($CH_4 \cdot H_2O$). Neutrons from the nuclear detonation are absorbed by the nitrogen, making radioactive carbon-14:

$$^{14}N(n,p)^{14}C$$

Carbon-14 is also made by the Giant Dipole Resonance reactions on nitrogen-15 and oxygen-18:

$$^{15}N(\gamma,p)^{14}C$$
$$^{18}O(\gamma,\alpha)^{14}C$$

The energy of the recoiling carbon-14 is sufficient to break the chemical bonds, but the new carbon atom recombines to form CH_3OH, methyl alcohol or methanol, and releases hydrogen gas.

In 1696, Whiston didn't have a good enough measure of single comets, Halley's Comet in his particular case, to accurately estimate the volume of water that could be deposited on Earth by passing through its tail. It is estimated that a single comet, Hale-Bopp, contains more water than all the Great Lakes combined, but I think much more water came to Earth at the end of the Great Ice Age. The Great Lakes gained (and lost) tremendous volumes of water around the time of this catastrophe, around 12,000 years ago. That is blamed on meltwater from the retreating ice sheet, followed by a drought [Perkins 2009b].

How much water does Comet Hale-Bopp have? We can usually see only the cloud of vapor and dust released by a comet (its "coma") and so the size of its nucleus is hard to estimate. Hale-Bopp has been estimated as having a diameter of about 40 kilometers. Typical comets are said to have a diameter of about 2-4 kilometers. Louis Frank (University of Iowa) thinks small comets, about 10 meters in diameter, are frequently captured by the Earth's atmosphere and brought most of our water, oceans, lakes, and all. Chris Chyba (Cornell University) thinks that between 3% and 8.2 times the amount of water we have at present came from very large comets early in the life of the Earth. Assuming that half the mass of Comet Hale-Bopp is ice and half of that is H_2O, the amount of water is about 8.4×10^{15} kilograms. This is about 6 millionths of the water in the oceans. Applying the same numbers to Saturn's moon Titan, that satellite has 20 times as much water as all of Earth's oceans. Supposing that the rain of icy debris from the detonation of a moon of Saturn raised sealevel on the Earth by 1,000 meters, requires about 1.3% of the mass of a moon like Titan to be directed so as to collide with the Earth.

The detonation of a moon results from the action of natural nuclear fission reactors deep within the moon. During formation of the moons, the densest material in the neighborhood, uranium hydride, will form the cores of the developing planetesimals, and the energy of radioactive decay will begin to dissociate the uranium hydride, forming metallic uranium, which is even denser. As sufficient material accumulates, in sizes from a softball to a Volkswagen, nuclear chain reactions will begin, and the glob will become a functioning nuclear reactor. So long as these reactors are free to expand as they heat up, the process is very stable, and more fuel is bred by conversion of uranium-238 to plutonium-239, and the decay of the plutonium-239 to uranium-235, so the process can go on for a very long time. Every body in the Solar System, from the Sun to the smallest (maybe) moons (as seen by the continuing activity in Enceladus), will have nuclear reactor material at its heart. The giant planets, Jupiter, Saturn, Neptune, are powered by nuclear reactors. They emit more heat than they receive from the Sun. The reactors in Uranus are dead. It is no warmer than the weak sunshine it receives. Io, a moon of Jupiter, with wonderful volcanoes, is powered by fission. It releases six times as much energy as we can figure from Sun, wind, and tide.

So, it is reasonable that a moon of Saturn might have still had a functioning population of nuclear reactors. If the power (fission rate) of these reactors increases too much, the plutonium-239 does not have time to decay to uranium-235, and the dynamic behavior of the reactors becomes dominated by the plutonium-239. (Which is why nuclear weapons are built with a combination of uranium-235 and plutonium-239; they go boom better that way.) Confined by the mass of the moon, the reactors cannot expand easily and so run faster and hotter, and hotter and faster, as more positive feedback pushes them on, until they detonate. The energy release may be enough to totally shatter the body, as happened to form the asteroids, or just shock off the crust, as happened to Mercury, or overheat the planet, as happened to Venus, half-a-billion years ago, or melt the glaciers of Mars, 177 million years ago, or shock off the icy crust, of water, ammonia, and methane, of a moon of Saturn, 12,900 years ago, dispersing the detonated core, making the rings of Saturn, and nearly killing the human race.

Art, Myths and Legends, and Plato's Atlantis

If what I have described, from the detonation of a moon of Saturn to the repopulation of the Earth from the seaside, truly occurred, surely someone would have attempted to record that, to pass on to whoever came next, what had happened to a fine and civil world. Thus, we should expect to find, in any social evidence from that time, some record of the disaster, a record fitting with the events as we can now imagine them. The evidence might have been recorded in art, in oral traditions that became embedded in myth, legend, religion, and literature. We might even find evidence in death itself.

Art

All art is but imitation of nature.
Lucius Anneus Seneca,
"Epistles,1,3"

Throughout southwestern France and along the northern edge of Spain, are many caves with paintings on their walls [Settegast 1986.] This art comes from the end of the Pleistocene, with various and uncertain dates. Some have become quite famous as art collections. They may also be serious records of a disastrous time.

Taking events in their actual sequence leads directly to some cave paintings and bone art. These were attempts of our ancestors (or others) to record and study the disaster. There seem to be three paintings that directly depict the detonation and its aftermath. These three paintings depict the same event: the detonation of a moon of Saturn as that planet passed "beneath" (to the south of) our constellation Orion. There is an undated painting near the small town of San Ignacio in Baja California [Crosby 1997], a painting in the cave at Serra da Lua, near Monte Alegre in Brazil, Amazonia [Roosevelt *et al.* 1996, Gibbons 1996b, Unknown 1996, and the Field Museum website], and a painting in the Chauvet Cave, near Bayol in southern France [Chauvet, Deschamps, and Hillaire 1996]. These three paintings portray a sequence in time and place, revealing that the detonation occurred while Saturn was over the Western Hemisphere. In Baja California, the painting shows the detonation near our Moon at the third (or first) quarter phase. In Amazonia, the detonation itself is blocked by our Moon. (Remarkably, this shadowing effect by the Moon has been used recently in a short science fiction story [Benford 2010]. The story speculatively uses a gamma-ray burster, and a small elliptical area on the illuminated side of the Earth is protected by the Moon.) In Chauvet, the image is of the debris cloud as it passed over France, after it had cooled, but reflected sunlight brilliantly. These differences are due to the passage of time, and different viewing angles.

This moment of detonation, as Saturn passed south of Orion, and was nearly in line-of-sight of our Moon, appears to have been painted in Baja California, Figure 88 [Crosby 1997].

Figure 88. Supernova and Moon, Baja California.
A rock painting in Baja California [Crosby 1997, original in color], showing an apparent supernova explosion near the Moon.

Assuming that the Baja painter did a good job of rendering relative scale, moving the viewpoint from Baja to Pedra Pintada (Serra da Lua) in Amazonia would visually move the Moon from near the explosion to directly in front of it, giving the appearance of an explosion expanding from behind an opaque disk, as the "rayed head" appears to be. Both the Caverna da Pedra Pintada painting (near Monte Alegre) and the Chauvet Cave painting show a structure above the explosion that I interpret as our constellation Orion. Saturn (and so its moons) passes "below" Orion.

In Serra da Lua in the Amazon, Figure 89 [Roosevelt *et al.* 1996]:

Figure 89. Cave painting from Amazonia, showing a detonation hidden by the Moon.
A cave painting from Pedra Pintada in Amazonia [Roosevelt *et al.* 1996], showing a detonation hidden by the Moon. The background figure may represent our constellation Orion. (Original in color.)

where our Moon blocks the direct view of the detonating moon. This relocation of our Moon is just what would be expected for the change in location from Baja California to Amazonia, if the scale of the Baja painting is correct. An outline that resembles the star pattern we call Orion appears above the detonation.

The painting from Baja California shows a "supernova" near the Moon. The painting in Serra da Lua shows what appears to be the detonation, blocked from sight by the Moon, in what may be a depiction of the constellation Orion. At the instant of detonation, as these two paintings appear to show, changing the viewing location from Baja California to Amazonia moves the Moon in front of Saturn, shielding the fireball from direct sight. Thus, the detonation was seen to be northwest of the Moon from Baja California, but was directly behind the Moon when viewed from Amazonia.

If the detonation occurred while Saturn was directly above the Western Hemisphere (midway between Baja California and Amazonia), Saturn would have been overhead in France 18 hours later. The fireball would have faded, but the expanding debris cloud would have shone brightly by reflected sunlight. The detonation appears to have been recorded in Chauvet Cave in France, far to the east of the New World (or west as the world turns), and the Moon is out of the picture, Figure 90 [Chauvet *et al.* 1996]. The Chauvet Cave painting shows the expanding debris cloud produced by the detonation, at the bottom of what may be the constellation Orion.

In the Chauvet Cave:

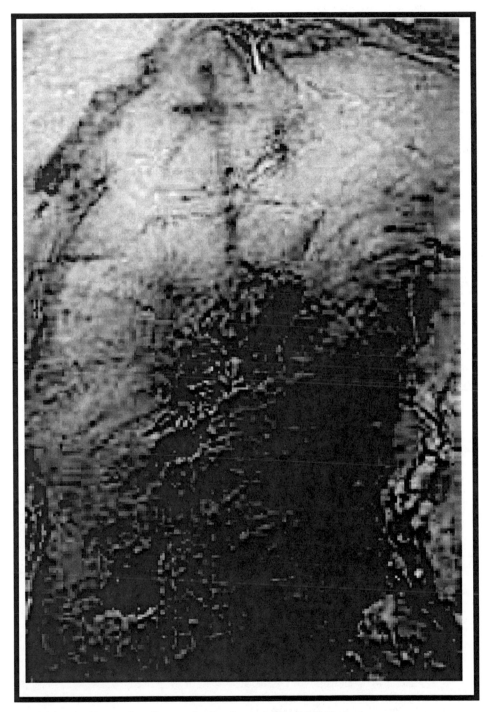

Figure 90. Painting in Chauvet Cave, France, showing the expanding debris cloud.
Cave painting in Chauvet Cave, France [Chauvet *et al.* 1996], showing the expanding debris cloud. (Original in color.)

The detonation may have been preserved, from memory or tradition, as a painting of "strange circles" in Çatal Höyük, one of the first modern cities.

Some of the Baja California paintings show apparent humans, often two-toned, red and black, nearly half-and-half, Figure 91 [Crosby 1997]. This is reminiscent of the flash burns suffered by some Japanese victims of the atomic bombings of Hiroshima and Nagasaki [Glasstone and Dolan 1977]. The intense heat of the detonation (as infrared radiation) would have caused flash burns on the exposed side of people out in the open, as happened in the atomic-bombings in Japan. This is shown by paintings in Baja California,

Remarkably, the Skidi Pawnee American Indians of Nebraska practiced human sacrifice, as recently as 1906, in a manner that reflects the rock art of Baja California [Krupp 1991]. Dr. Krupp describes the ritual as a captured neighbor maiden being painted red on the right side of her body and black on the left side, exactly as shown in the Baja painting, and then killed in celebration of Mars as the Morning Star. While art may imitate nature, it appears that for the Pawnee, life (and death) imitated art. The Pawnee lived half a continent away from Baja California.

Figure 91. Two-toned (flash burned) humans in Baja California.
Humans in Baja California [Crosby 1997], flash burned by the radiation from a detonation. (Original in color.)

Lascaux Cave also shows a scene of swimming stags in violent water [Bahn. and Vertut 1988]. The recently discovered Chauvet Cave [Chauvet, Deschamps, and Hillaire 1996] shows paintings in charcoal and in red ochre. The charcoal is well dated (by radiocarbon), to times around and after 30,000 years ago. The red ochre paintings are less securely dated. (To find red ochre splashes on 30,000 year old charcoal gives no accurate measure of the age of the red ochre, just that it is not older than the charcoal.) By the content of some of the red ochre paintings, I suspect that those paintings were done to record the event that ended the Pleistocene. For example, a bison is shown snorting blood, in what may be a later overpainting of an earlier charcoal painting.

Among the prehistoric bones stored in a French museum, Alexander Marshack found one small piece with a peculiar back-and-forth trail of engraved marks [Marshack 1991]. Marshack interpreted the markings to represent a sequence of phases of the Moon. Marshack's rendering of the marks show some marks that look more like a comet with a tail than phases of the Moon, Figure 92. The back-and-forth pattern of the marks resembles the retrograde/prograde movement of an object outside Earth's orbit, as seen from Earth. Therefore, a possible alternative to the lunar phase explanation is that the bone was held aligned with some marker stars in Orion, and the position of a comet was marked in a sequence over time, as the comet approached the Earth. Clouds of icy debris, propelled in part by the orbital motions of the moon of Saturn, would come streaming towards the Earth, spraying out dust and vapor, like comets. The debris would take approximately six years or less to approach the Earth, during which time fragments would swing back and forth in the northern night sky, as comets do at present. An astronomer of those ancient times might record the path in the sky on a bone, as studied in our time by Alexander Marshack [Marshack 1991].

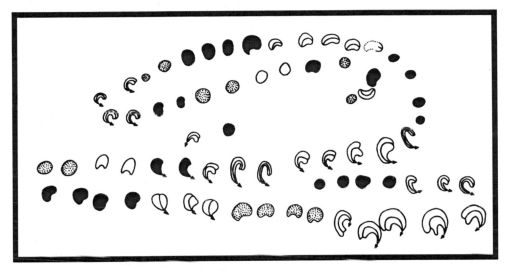

Figure 92. Carvings on an ancient bone.
Representation of carvings on an ancient bone, by Alexander Marshack [Marshack 1991]. This carving has been interpreted as a lunar calendar, but I suggest it shows a sequence of views of a comet from the detonated moon of Saturn. It shows the alternate prograde and retrograde motion of an object in the outer Solar System, as seen from the Earth, and the shapes more closely resemble a comet than phases of the Moon.

In Paul Bahn's recent book, *"The Cambridge Illustrated History of Prehistoric Art"*, he briefly discusses humanoid figures painted on the Great Gallery of Barrier Canyon, Utah [Bahn 1998]. The date is estimated to be 700-1200 CE, but he doesn't explain how. It is stated that the painting *"seems* to have been enhanced for photography by application of kerosene." (My emphasis on the 'seems'.) The radiocarbon date is 32,900 before present, which is too old to be "acceptable", and so it was concluded that kerosene "used for enhancement" perturbed the carbon isotope content, as it would, and so the date was false. It might have been painted before new carbon-14 arrived at that location, and with material that was shielded from the direct radiation of the fireball. The measured date of these paintings fits at the beginning of the problem period of 34,000-11,000 before present. They might date 32,900 by radiocarbon, but actually be only 12,900 years old.

The Barrier Canyon paintings also seem to connect to the Baja California paintings. The Barrier Canyon paintings show notably large humans, 2 meters tall, the same as the Cro-Magnon people of Europe. The Baja California paintings also show exceptionally tall humans. Those paintings are also thought to date from around 1000 CE, partly because of the idea that the supernova and Moon painting is the same as those in the southwest United States that seem to depict the Crab supernova and Moon, from 1054 CE. This is probably not true, as the phase of the Moon is distinctly different. It is recognized that the condition of the Barrier Canyon paintings looks older than that.

In the "Shaft Scene" in the Lascaux Cave, next to a typical beautifully painted bison, is the crude stick-figure of a dead man [Ruspoli 1986, Settegast 1986], with a distorted face and an erection. The notable features are the distorted face and the graffiti style of the painting, clearly not done by one of the traditional artists of the cave. An engraving on the ceiling of the Polychrome Chamber in the cave of Altamira, in Spain [Saura Ramos 1999], 300 miles from Lascaux, shows the same distorted face as the Shaft Scene stick man. An engraved bone, from Mas-d'Azil [Putman 1988], about 130 miles from Lascaux, shows a man with a distorted face and an erection. A rock carving in La Marche [Putman 1988], 100 miles from Lascaux, shows a distorted face. Either we must imagine that this was a culturally shared symbology, over distances of hundreds of miles, or that the artists were independently depicting the same observed phenomenon, death by asphyxiation by ammonia gas (or alternatively, by radiation).

The rock paintings of "The Swimmers" in the Saharan Desert, Figure 93 [Bray 2006], show humans in opisthotonic postures, with the neck bent backwards, which is a sign of central nervous system death [Faux and Padian 2007]. Such a death can be caused by extreme radiation exposure.

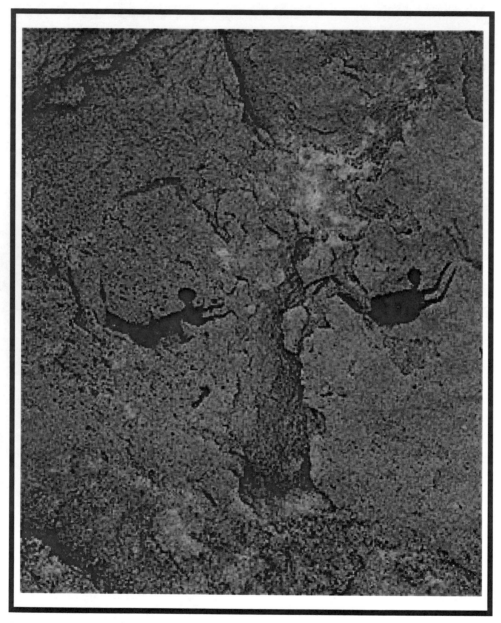

Figure 93. Rock painting labeled as "The Swimmers", in the Sahara desert.
Painting of "The Swimmers", in the Sahara Desert [Bray 2006]. These depictions probably show the opisthotonic posture resulting from central nervous system death, as caused by extreme radiation exposure.

Myths and Legends

In our skeptical, science-based world, filled with detailed reports, reviews, analyses of all events, we look with doubt on any claim that knowledge of the past might have been recorded by those people living in the past. Our judgment grows harsh as we pass through oral histories, oral tradition, legends, myths, and religious dogma. Passing back in time, we begin to discount the information contained in cultures, and reject apparent culture as artifacts. We even doubt that our not-so-long-ago ancestors could talk like humans, argue that they had no language, and little understanding of the bewildering world around them.

Yet, are we so different? The ancients used myths and legends to explain their world, we use theories and models to explain ours. Perhaps we understand our world better than we understand theirs. We should be careful not to assume that we understand their world better than they did.

In dealing with the myths, I am trying to keep on the safe side of a very thin line: I try to use the physical processes of the event (as I deduce them to be, from knowledge of the effects of nuclear fission) to explain the basis for the myths, as opposed to letting the myths guide the invention of the physical processes, as happened with Velikovsky. The difference is subtle but important.

When a world of life is wiped out, a civilization washed away and the future destroyed, how can history be written, how can tales be told? Consider yourself in that predicament, how would you preserve knowledge of an extended event that could be observed as unrecognized doom until it struck? How would you record the first disaster, when it killed in a flash, all those about you?

Could such a disaster as the death of nearly all the large mammals in the world have not had an equally severe effect on the human population? Would this disaster have gone unnoticed and unrecorded? If the anthropologists could look with this in mind, they would see the efforts that were made to preserve a record of the disaster.

Can we rely on oral traditions to hold true to the facts for several thousands of years, passed from generation to generation, scarcely modified? Homer's *Iliad* and *Odyssey*, for example, had survived for several centuries as oral tradition, before being written into history, with at least some historical accuracy. But could oral tradition survive for several millennia?

Perhaps so, if the epic relates the tale of a disaster so calamitous that it killed all but the few survivors. That is, this event exerted a life or death effect on nearly all humans. Desperate efforts might be made to preserve a record of this cataclysm, but this knowledge would become obscure or lost as time passed, peoples died, and the brightness of Saturn dimmed to that of the other planets. Ed Krupp wrote "Comets are traditionally linked to disaster." In this regard, it is intriguing that our word "disaster" comes from the Greek roots for "broken star". As much of the evidence disappeared in the cataclysm or faded from sight over time, future attempts to understand what happened, with insufficient foundations, would be ineffective and would only produce controversy.

These legends of our past, the history of our pre-history, have been saved by the strength and power and continuity of our religions. In spite of ridicule and rejection, and simple straightforward criticism, religious peoples have clung to their various beliefs, whatever creed, as important to them, containing important truths. These truths are different, as suits different peoples, but they are all important. By oral traditions, stories of that time later became myths, legends, and religious writings. Natural errors and exaggerations may have crept in, we don't know. In the conversion of history to religious fiction, the stories were distorted to produce power for the priesthood, as representatives of the ultimate cause of such a disaster, or as potential future defenders against divine wrath.

The physical observations and interpretations should provide a basis for understanding the myths. But, because we consider the myths to be ambiguous, the myths themselves can not provide direct evidence for the events that we should find in the observations.

Francis Hitching's book [Hitching 1978] discusses the many varied myths and legends regarding the Deluge and Flood. This subject is a dangerous battlefield occupied almost exclusively by fanatics, true believers on one side, debunkers on the other. Both may be equally gullible, believing absolutely their own stories and rejecting absolutely the opposition's. There appear to be enough factual truths on both sides to keep this war going forever.

Dying in place, or carefully buried by his survivors, a Lake Mungo Australian laid in rest with "his fingers intertwined around his penis." [Zimmer 1999b]. Even in death, the message was recorded.

The oldest known flood story is familiar as Noah's Flood, which traces back to Sumeria. Noah's Flood is the combination of two different stories, as if the editor wanted to make sure he had included all the recognized facts. It appears to come from the Epic of Gilgamesh, which is more than 2,600 years old, and this epic derives from an earlier Sumerian story which is 5,400 years old. The Hebrews may have learned the story in Mesopotamia, and changed the hero, and the God, and the mechanics of the flood. After all, a good story is a good story. In *"Genesis"*, it rained for forty days and forty nights, and overtopped all the tall mountains. The Greek Flood of Deucalion lasted only nine days.

Creationist researchers have diligently searched out observational evidence in support of the Biblical Flood, the Flood of Noah, the Deluge of Genesis (*"Genesis"*, Chapters 6 through 9). While many of these observations were made scientifically, rather than religiously, scientific researchers have generally avoided this subject, worrying that the proposed mechanisms require divine intervention. However, there appear to have been events that can be connected to make a pattern that matches many of the religious stories and present an amazing description of a most tremendous happening, within the cultural memory of humans.

There are several Greek flood traditions. The legend of Deucalion makes the flood worldwide. In a divine act of destruction, it rained for nine days and nights, but the top of Mount Parnassus was not covered [Vitaliano 1973].

There are many legends and myths from around the world that relate to this event, a great flood, with and without a deluge. An American Indian legend tells of a great wall of water that came down from the North and flooded the land. The people ran to the highest hills, but the water rose and washed them all away. (Which raises the question, if all the people perished, who was left to tell the tale? Perhaps the virgin girl who repopulated the Earth? I don't know the answer to that and further, I think the American Indians who repeat the story came much later and were a different people.) Those hills are what is called the Coteau des Prairies, in South Dakota and North Dakota. The Coteau clearly bifurcates a flood channel, not a flood plain, that I think was caused by a tremendous flood of water that then swept down into the Gulf of Mexico [Emiliani 1975].

The Knisteneaux and Choctaw American Indians have legends: the Choctaw say that a great mountain of water came out of the north, drowning nearly all; the Knisteneaux say the water rose to cover the Coteau des Prairies, drowning all but one virgin girl, who repopulated the Earth. (There seems to be a conceptual flaw there.) The Coteau des Prairies is shown on a marvelous USGS map (*"Landforms of the Conterminous United States - A Digital Shaded-Relief Portrayal"*, Map I-2206). It is right in the middle of the northern plains and shows how a flood came from the north, split into two channels at the Coteau des Prairies, but also overtopped the Coteau. From there it flowed to the Gulf of Mexico, as Cesare Emiliani found in 1975. Integrate the flood myths, from Deucalion of the Greeks, who landed on ground covered with algae and slime, the start of the Black Mat soil layer, to the legend of the Coteau des Prairies, where the people ran to high ground as a wall of water rushed down from the North, and Cesare Emiliani's discovery of a tremendous fresh-water flood into the Gulf of Mexico, 11,600 years ago [Emiliani 1975], and the Incas, who ran for the hills (the Andes, of course) as the sea rose, and the South Pacific people, who were left with only the tops of mountains as their island homes.

The regional variations in the legends are important. In the Near/Middle East (Greece to Babylon) the water comes as a torrential downpour, lasting for different times and taking different times to drain away. In the Greek legend of Deucalion, the hero debarks from his boat to find a world "covered with silt, slime, and algae" [Bierlein 1994] which seems like a description of the algal growth that produced the organic-rich layer found as the Black Mat.

Along the west coast of South America, the ocean rose (the legends say) and the people ran to the high hills (the Andes). The Incas may have been one of the few surviving derivative civilizations from that time. In the south Pacific, the ocean rose, covering the land to the tops of the mountains, leaving those mountaintops as islands. Here, the flood did not drain away. The Aborigines of Australia are the longest surviving civilization.

These stories have been so muddied by repeated telling, self-serving "spinning", poor translation (certainly Plato's work suffered from that), and misinterpretation that I don't think they can be used to prove anything. Their widespread existence supports the reality of a widespread disaster, and understanding the disaster can explain the legends. Here, I try to use the physics of the processes to explain the story behind the myths.

Perhaps the best summary of legends of this disaster is provided by Dorothy Vitaliano, *"Legends of the Earth"* [Vitaliano 1973], which attempts to explain all such legends as tales of geology. The summary is useful even though Vitaliano sets out to debunk the legends and discredit the storytellers. She notes the special quality of these traditions, they appear in almost every part of the world, in clearly related similarity. She contrasts the impossibility of a global flood with the possibility of a local Biblical flood, and considers local floods as the basis for the stories.

The legends vary with location, and Vitaliano uses these differences as objections to any basis in fact for a global flood. She bases many of her arguments on absolute, but unsupported, assumptions. She also requires the Ancients to have known Earth's geography completely, and argues that a flood covering "all the mountains of the world" would have had to cover Mt. Everest, and that was physically impossible. Where did the water come from, where did the water go? In my view, the flood (in those areas) covered the mountains the Ancients knew, the water came from the fragments from Saturn's moon, and it went into the ocean basins, raising sea level by about 1,000 meters or more.

Vitaliano makes much of the absence of an Egyptian flood legend [Vitaliano 1973] although it appears that there is one: in this, the gods again judged mankind as too wicked to survive, and sent a destroyer, the goddess Hathor. Blood from her killings caused the Nile to overflow and the land was flooded with blood and water. When nearly all humans had been killed, a rescuer was sent, the goddess Sektet, and the remaining humans repopulated the Earth [Bierlein 1994].

In rejecting the possibility of "exceptionally heavy rains" producing an extensive flood, Vitaliano bases her rebuttal on the assumption that "the only place that water can come from is the sea", so that if water was transferred into the clouds to fall as rain, sealevel must have been lowered, to expose more land. The rainwater would then run back into the ocean "as fast as it could", eliminating any chance of an extended flood. More significantly, while she is aware of proposals that a great amount of water came to the Earth from space [Kelly and Dachille 1953, Velikovsky 1950], she rejects this possibility. Her argument is that those theories are based on folklore (such as this section deals with) and on incorrect, farfetched interpretations of geologic features. She rejects these as simply "highly ingenious". She explains the multitude of flood stories as resulting from a multitude of floods. The widespread occurrences of a multitude of different types of floods, river floods, flash floods, ice dam floods, tsunamis, leads to the reasonable assumption that a set of flood traditions should develop. This is borne out by the down-home stories of the old folks, telling about "the great flood of ought-five", for instance. However, the reasonable floods bear little if any resemblance to the legendary floods. The legendary floods are catastrophically larger than the historic floods, involve divine generation, and often require a nearly miraculous re-population of the world. Is this simply exaggeration, amplified by the need to tell an ever-more-exciting tale? This might be done, quite sincerely, in the manner of emphasis and providing better explanations. Like original paintings that are painted over, old stories may be given a new focus, by more recent events, and like repainted paintings, seem to be newer than they are.

In the book, *"Noah's Flood, The Genesis Story in Western Thought"*, [Cohn1996] Norman Cohn also writes as a debunker, a skeptic. Cohn traces the Genesis story back to a Sumerian tale. I think if we could, the track would go back further yet, to about 12,000 years ago. That the story was modified to suit the purposes of the working authors should be no surprise, either: we all have attitudes. But I think stretching the occasional overflows of a river into a global deluge goes beyond the bounds of credibility. By focusing on Noah's Flood, Cohn ignores the vast body of "water-disaster" literature, records of a past cataclysm. He recognizes the Greek flood story, of Deucalion and Pyrrha, but does not pick up on the key physical observation, that the Earth was covered in slime and algae, the precursor to the Black Mat [Brakenridge 1981, Quade *et al.* 1998, Firestone *et al.* 2007, Firestone, West, and Warwick-Smith 2006, Haynes Jr. 2008]. Dorothy Vitaliano looked at a major fraction of that genre, but saw only fragmentation, rather than unity with local differences, and explained away all the significance of the stories.

The constellation Orion figures in the ancient Egyptian myth of Isis (the star Sirius), Osiris (the constellation Orion), and Set (I could not find agreement, but I think Set was represented by the planet Saturn). In ancient Egypt, Isis and Osiris were considered to be among the chief gods (anatomically modern humans?) who inhabited the land before it was occupied by the small people (seasiders?). Egyptian myths tell of Osiris (Orion) being killed (asphyxiated) by his brother Set (Saturn?). Isis (Sirius) sat on his erect penis to conceive their son, Horus, who then defeated Set to avenge his father's death (the magnitude of Saturn faded to less than that of Venus as the debris cloud dissipated and the rings coalesced). The dead-Osiris phallus-sitting scene is shown on page 112 of *"Beyond the Blue Horizon"* [Krupp 1991]. (Dr. Krupp and I disagreed on this, but I think the lady is Isis, not Nut (goddess of the sky), because of the absence of Nut's emblematic stars, and I think the man is the dead Osiris, because his face is turned away. He obviously has an erection. Is that artistic exaggeration, or was this an effect of abundant nitric oxide, produced by the burning of ammonia in the atmosphere?) Human males, dying of asphyxiation from overexposure to ammonia and nitric oxide, would expire with swollen faces and heroic erections. His awkward arms and legs suggest the opisthotonic posture indicative of death by central nervous system damage, as by asphyxiation or extreme radiation exposure. I think the face-toward/face-away convention for life/death is shown by contrast with the Isis (Sirius)-Osiris (Orion) scene shown on page 219 of *"Beyond the Blue Horizon"* [Krupp 1991], where the living Osiris gazes at his consort, Isis, with a smiling face. (See Figure 94.)

Sirius itself might have been affected by this detonation (accepting the possibility of a much shorter distance scale and asteroidal fragments masquerading as stars), explaining the recorded appearance of a Red Sirius by the ancients [Schlosser and Bergmann 1985, Tang 1986, Gry and Bonnet-Bidaud 1990, Tang 1991, Whittet 1999, and many others]. The record is inconsistent, the explanations are many, and the situation is confused. However, an example of how quickly a star can change its appearance was seen in the outburst of the star V838 Monocerotis [Bond, Henden, and Levay *et al.* 2003, Bond *et al.* 2003]. Changes in brightness, temperature, and size were seen in a time of less than 200 days. The cloud of debris surrounding the star was interpreted to be light echoes, but in a small scale Universe, that would be material ejected by a detonating asteroidal fragment. This event is analyzed in detail in Appendix D. That analysis, based on the supposition that V838 Monocerotis revealed the progressive detonation of natural nuclear fission reactors in an asteroidal fragment, typifying the choo-choo chug-chug huff'n'puff BOOM process, with little boom at the end, showed physically reasonable variations in the reactivity of such a system. Positive reactivities approached or exceeded the level of "prompt critical", which gives drastically fast growth in fission power, ending with negative reactivities of a few percent. These reactivities are commonly developed in "burst" reactors, experimental reactors designed and operated to produce brief but intense power excursions [Keepin 1965, Hetrick 1971].

Figure 94. Egyptian tomb paintings of Osiris and Isis.
Egyptian tomb painting of Osiris and Isis. My interpretation is that Osiris is alive in the lower painting, and dead of asphyxiation in the upper painting [Krupp 1991].

As a further depiction of this detonation, embedded in tradition, Ed Krupp shows, in *"Skywatchers, Shamans & Kings"* (page 52), a drawing of a Turkic Altaic shaman tradition, Figure 95, that bears great resemblance to the process I am describing [Krupp 1997]. At the top of the pictogram, two human-like figures stand side-by-side. The larger one to the west, if these were in the sky, is surrounded by a spray of lines, just like the figure in the Chauvet Cave painting. This drawing could be interpreted as showing the results of a detonation of a moon of Saturn (the larger rayed figure) with Saturn (the smaller figure) nearby. Saturn is then seen obscured by a cloud, which turns into a ring, which then becomes a smaller ring. I am reluctant to try to interpret what may be local, personal symbology past that point.

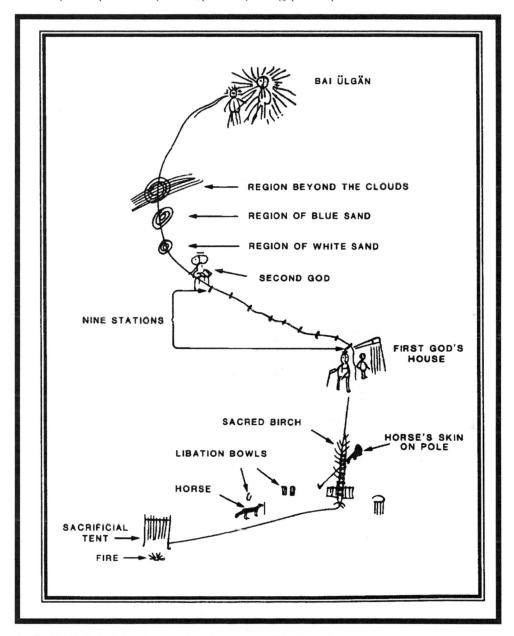

Figure 95. Turkic Altaic depiction that may show the sequence of the detonation.
Turkic Altaic depiction that may show the detonation of a moon of Saturn, and the subsequent development of the debris cloud and Saturn's rings. [Krupp 1997].

Is it possible for later generations to misinterpret the symbols of their past? I think so, in several of the situations relating to the end of the Pleistocene. The most dramatic is the re-interpretation and confusion of the detonation of Saturn's moon as the planet Venus, as the Evening Star and the Morning Star. On ancient stone pillars, stelae from Babylon (about 3,200 years ago), three astronomical emblems are shown [Aveni 1992, Reichen 1968] (Figure 96). These have been interpreted as showing the "Star of Ishtar" (Venus), the Moon, and the Sun. I think the symbol of "Star of Ishtar" originally symbolized the appearance of Saturn as "Inanna", [see E. C. Krupp 1991 page 193] at the time of the detonation of one of its satellites. As Saturn had, by later times, returned to only a normally bright planet, the Venus cult took over and assigned most important aspects of our world to the Sun, the Moon, and Venus. The goddess Inanna became Ishtar.

Figure 96. Stela of Melishipak II, with symbols assigned to Venus, Moon, and Sun.
Stela of Melishipak II showing symbols assigned to Venus, the Moon, and the Sun [Aveni 1992, Reichen 1968]. The symbol for Venus is out of scale for the other two, and probably represented the original view of the detonation of a moon of Saturn.

In a few minutes following the detonation of Saturn's moon, the thermal brightness of the detonation (like a gamma-ray burst) would decrease, and where Saturn had been would be a huge ragged star, now shining only by reflected sunlight, but brightly because of its large size and the white fragments of ice forming the debris cloud. This is shown on a stela of Melishipak [Aveni 1992], where the new star, Inanna, is placed alongside the Sun and the Moon to demonstrate its size and brightness. The three images are the same size, which is correct for the Sun and Moon, but greatly exaggerates the size of Venus, as we now see her.

For successful communication between artist and viewer, or more pertinently, between writer and reader, it is necessary for symbology to be recognizable and as directly related to reality and the message as possible. There must be agreement between the intent of the image and the perception. It is not reasonable to ask readers of the stone text to recognize an emblem as large as those for the Moon and the Sun to represent the planet Venus. As the debris cloud dissipated, it appears that Saturn and Venus became interchanged, and the symbol for Inanna was transferred to Ishtar, as representing Venus. Comparison of the Sun, Moon, and Venus in our present sky makes it clear that the relative sizes of the Sun and the Moon were truly represented, but it is not at all reasonable that the symbol for Inanna should represent Venus, even at its brightest. This is shown by comparison of Venus, the Moon, and the Sun, photographed to the same scale, in Figure 97.

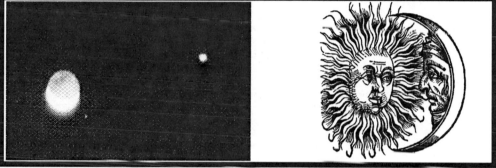

Figure 97. Images of Venus, the Moon, and the Sun to the same scale.
Images of Venus, the Moon, and the Sun to the same scale (photos by the author). The great contrast in size between Venus and the other two is clear. (The sky around the Sun is dark because of the photographic filter used to block most of the Sun's intense light.) The lower images show Venus near the Moon, for another comparison of size, and a more recent artistic rendition of the equal sized Sun and Moon, from 1493 [Krupp 1991].

Incidentally, I think that is what trapped Velikovsky into his explanations of the events from the end of the Pleistocene into Biblical times, as involving the planet Venus roving through the Solar System. This interchange between Inanna as Saturn, and Ishtar as Venus, may be the cause of the confused period in the Babylonian Venus Tablets, where the ancient appearances do not match those calculated from the present.

Plato's Atlantis

> *The houses are all gone under the sea.*
> T. S. Eliot,
> *"Four Quarters, East Coke, 1940, I"*

> *... and the island of Atlantis in like manner*
> *was swallowed up by the sea and vanished.*
> Plato, *"Timaeus"*, translation by R. G. Bury, 1929

The Greek philosopher Plato wrote two particularly interesting "dialogues", about 2,370 years ago, in which the characters exchange views on the nature of the world and civilization. In the *"Timaeus"* and the *"Critias"*, he describes an unknown land and civilization of Atlantis. He recounts a story from a distant grandfather of Critias, Solon, as told by a learned Egyptian priest. That story relates the fate of an advanced civilization at war with the most ancient Greeks, and the subsequent destruction by earthquakes and a flood that submerged Atlantis and destroyed the Greek nation [Bury 1929, Lee 1971]. This event occurred about 9,000 years before Plato (11,370 years ago), and corresponds to the flow into the Gulf of Mexico studied by Emiliani. In spite of diligent searches, Atlantis has not been found and is generally considered to be part of the earliest science fiction writing.

Searchers for the fabled land of Atlantis have claimed that evidence for "Atlantis" has been found in many cultures. These observations may reflect an ancient global seaside civilization, while Plato's references appear to be specific to a single location.

Plato's works have been translated into modern languages, from the beginning, as literature, philosophy, and social studies; a tale told to impress, rather than to inform. Little effort was made to capture the literal accuracy of Plato's statements. Bury's translation, in the Loeb Classical Library, provides the original Greek and the English translation on facing pages. Simple transliteration makes many of the Greek words recognizable, thanks to the deep roots of English. Two apparent errors were that Bury translated "Sea of Atlantis" into "Atlantic Ocean", a *non sequitur* in the Greek culture, and that he translated "your (Greek) Heraklean boundary marker" into "the Pillars of Heracles", which we now call the Pillars of Hercules, the points across the Strait of Gibraltar. However, the conversion of Heracles to Hercules is not correct. The Romans adapted the Greek hero Engonasin as the model for their Hercules. These errors change the geography greatly and are the cause of much silly searching for the lost city of Atlantis.

Atlantis may have been a real place, much as Plato related the Egyptian description, with a technology that could only be imagined, and was exaggerated for effect. Difficulties with identifying its location are the result of misdirected scholarship and the failure to recognize the geographical consequences of a major rise of sealevel in what is now the Mediterranean. People have been directed to search for Atlantis in the Atlantic Ocean, beyond the Pillars of Hercules (Straits of Gibraltar). This may have been the fault of Ignatius Donnelly, in 1882 [Donnelly 1882], although it may have been in even earlier translations. Consider that Bury intended to present *Timaeus* and *Critias* as readable literature, and viewed the story of Atlantis as "a fine piece of literary fiction" rather than an objective description based in historical fact. He therefore made a literary translation (perhaps those before him had done so as well), with little respect for Greek language and tradition.

So where was Atlantis? Considering Plato's descriptions, and recognizing that Bury and Lee tried to make their translations modernly readable, the Greeks would not have said "in the Atlantic, beyond the Pillars of Hercules". They would have said "in the Sea of Atlantis, beyond the boundary marker of Herakles". The change from Greek to Roman, in Herakles to Heracles to Hercules, is incorrect and also our geography is not necessarily their geography. To the Greeks, "Asia" was our Anatolia, to the Romans, our Tunisia was their "Africa", and the "Herakles" of the Greeks was not the "Hercules" of the Romans. Our Atlantic Ocean was a part of the Greek Okeanus, the all-encircling world ocean. The Loeb Classical Library presents the "Zurich edition" of *Timaeus* and *Critias*, in Greek text, facing the English translation of Bury. Bury places Atlantis in the Atlantic Ocean, while the Greek says " Ἀτλαντικοῦ πελάγους", Atlanticou pelagos, or Sea of Atlantis. A sea like the Aegean or Adriatic, not the world ocean, Okean, Ωκεαν. Bury and Lee and Donnelly misrepresent the Greek Ἀτλαντικοῦ πελάγους (Sea of Atlantis) as the Atlantic Ocean, terminology that the Greeks would never have used, and a modern construction that refers to a very different body of water. Bury translates Ἡρακλείας στήλας (stone marker of Herakles) as "Pillars of Heracles", which has become our modern Pillars of Hercules as we see the Greek mythology adopted by the Romans. Herakles became Hercules only in modern English literature. As John North explains, "The Romans chose to associate Hercules with the character the Greeks had called Engonasin, 'the kneeling man'" [North 1996]. I think the Pillar of Herakles is actually identified as a location on the western Greek mainland, but I can't find the reference.

Thus, we must imagine that before the Flood there were two separate seas, the eastern Mediterranean, the Ionian, containing the home of the ancestors of the Greeks, and the western Mediterranean, Atlantikou pelagos, containing the island of Atlantis. This island could have been the combined islands of the Balearics made into one larger isle by the much lower sea level. This location would have been in keeping with Plato's report of the Egyptian's stories, as told to Solon, that the Atlanteans controlled Europe to Tyrrhenia (present northern Italy) and Libya to the border of Egypt. This location is appropriate to the war of conquest started by the Atlanteans in an attempt to expand their control, and opposed by the most ancient Greeks.

For Atlantis, we must relocate the "Pillar of Herakles" to somewhere between Tunisia (the Roman "Africa"), Sicily, and the toe of Italy. Imagine a Mediterranean Sea 1,000 meters lower. The Atlantic Ocean, our Atlantic Ocean, also lower, would be walled off by a high isthmus between Spain and our Africa, and the western Mediterranean would be separated from the eastern Mediterranean at the pinch-point of the Strait of Sicily, the Malta Channel, and the Straits of Messina. There are several possible locations for the original Greek Pillar in that area, and the ancients did not describe their world for our benefit and understanding. With sealevel 3,300 feet lower than today, every coastline would have "rose precipitously" from the sea, and the mountains would have indeed been "more numerous, higher and more beautiful", as Plato said Solon said the Egyptian priest said. If sealevel had been 1,000 meters lower at the time of the Atlantean Empire, the present Balearic Islands would have been the tops of the mountains of a single island, with a well protected, nearly circular harbor in the south, sheltered from the north wind, and well-situated to control the lands to Tyrrhenia and Egypt. The present islands of Sardinia and Corsica would have been joined and connected to the mainland. This would have made Tyrrhenia much more substantial. The flooding of the western Mediterranean, the Sea of Atlantis, by the cometary deluge water flowing down the Rhône valley, would have drowned Atlantis. The overflow into the eastern Mediterranean would have drowned the Prehistoric Athens, and the Deluge of acidic rain from the entry of the cometary debris from the detonated moon of Saturn would have washed bare the mountains of the Greek highlands.

In that ancient disaster, as the sea rose greatly, rather than Atlantis sinking by itself into the sea, the stela of Herakles, representing the boundary between the Greek sphere of influence and the Atlantean, was also submerged and lost from sight. The name of a place then became a name looking for a place and as the ocean filled up with water from the skies, that place became the Strait of Gibraltar. Destruction of the existing seaside civilizations disrupted the continuity of knowledge, as was explained by Plato's Egyptian priest to Solon. Several small seas flooded together to become the Mediterranean, and the stela of Herakles became, in our world, the Pillars of Hercules.

There is physical evidence of a flood of fresh water into the Gulf of Mexico, at 11,600 radiocarbon years before the present, and the authors of that research acknowledge Plato's story [Emiliani 1975].

Thoroughly Modern Humans

... how surprisingly difficult it is to define
... what is meant by "modern humans.
Roger Lewin, *"The Origin of Modern Humans"* 1993

'Tis not too late to seek a newer world.
Alfred, Lord Tennyson
"Ulysses" 1842

The question of defining the "modern humans" raises problems. It is easy to assume that we, as the most modern humans yet, arose from the Cro-Magnon type in Europe, who is assumed to have descended from the anatomically modern humans, whose ancestors were the archaic modern humans. However, the Cro-Magnon people were big, and Europeans after the end of the Pleistocene were small. Our earliest example of a modern European is Ötzi, the 5,200-year-old Tyrolean Iceman. While Cro-Magnon males grew to a height of 2 meters (80 inches, nearly 7 feet tall), Ötzi was only 1.65 meters tall, and may have weighed only 50 kilograms (110 pounds), about half the weight of a Neandertal or Cro-Magnon. In North America, Spirit Cave Man from Nevada was only 1.57 meters tall [Tuohy and Dansie 1996]. This mummy was radiocarbon dated to 9,415 radiocarbon years before the present, or a calibrated date of 10,630 years ago. The moccasins and woven fabric indicated a remarkably advanced technology. He did not resemble current Native Americans. Kennewick Man was 1.70 meters tall (5 feet 7 inches) Gibbons 1996c], and did not resemble modern Native Americans. This is a general condition; all humans after the end of the Younger Dryas were small.

We, the truly thoroughly modern humans, attempt to discover our origins, our ancestors, our roots. In the standard story, these roots reach back to the first "anatomically modern humans", found in the Middle East, Europe, and Asia, who resembled us in many ways. The anatomically modern humans reach toward us in the distorted form of Neandertals in Europe, and the more suitable Cro-Magnon, but there and then, the trail becomes fragmented, and at the break, the so-called End of the Great Ice Age (more neutrally known as the end of the Pleistocene, and most properly, the end of the Younger Dryas), a new set of researchers comes in and continues with the peoples who might have been us. Paleontology gives way to archaeology, and the studies are disconnected. There is truly a break between the Neandertals and the Cro-Magnon, and the humans found after the end of the Great Ice Age. This is shown by molecular genetics, which finds the Neandertals too different from us to be our distant recent ancestors, and by the great change in body size. Neandertals and Cro-Magnon were 1.80-2.00 meters tall and weighed 95 kilograms. Not large by comparison with current athletes, but dwarfing the new Europeans at 1.60 meters tall and 44 kilograms. At the Windover site in Florida, adult females ranged in height from 1.57 to 1.60 meters, and the adult males ranged from 1.68 to 1.75 meters. Where did the new Europeans, and the Ainu and the Japanese, and we, come from, and when?

Modern genetic studies show a "mitochondrial Eve", identified by the uniformity in the DNA of current human mitochondria, inherited entirely from the mother, and a "Y-chromosome Adam", based on a similar result for the male-only Y-chromosome. The points in time differ somewhat, 100,000 to 200,000 years ago for Eve and 35,000 to 89,000 for Adam, but the results suggest a time in which the breeding population leading to us was extremely small and isolated.

Very few, if any, humans were living at the time of the Younger Dryas in Siberia and the North American Plateau, because, with sea-level 1,000 to 3,500 meters lower, the continental plateaus were equivalently about 3,000 to 4,000 meters (9,000 to 12,000 feet) in elevation. I am not sure that I yet understand how the seaside humans survived, except possibly those that did were shielded from the radiation and later were far enough away from the impact zone that the ammonia and methane gas had dissipated before it had any effect. That many did die of asphyxiation by the ammonia is shown by the scattered cave art showing distorted faces, dead men, and erections.

If humans came to the New World over a narrow land bridge from a limited region in Siberia, as the standard story tells, why are the New World languages so diverse? Why are the Indo-European languages so homogeneous and widespread? If the diverse languages in the New World came from diverse groups along the seaside, how (after surviving the radiation of the detonation) did the survivors survive the effects of the impacts of icy fragments? Partly because the ocean water blunted the effect of the ammonia ice, partly because they were able to move to higher ground fast enough to keep ahead of the waves. Many perished, regional populations would have been wiped out. New Guinea also has many diverse languages, as survivors occupied the island from many different areas. I think the study of Kennewick Man and Spirit Cave Man would benefit from consideration of the Jomon of Japan and the current Ainu, who are rather more "Caucasoid" than most Orientals. The Jomon seem to have come to Japan about 12,000-10,000 years ago. There appears to be some evidence of habitation in Japan from 30,000 years ago. Those two times may not be really separate. (My story of the Flood, for Japan, would fit the legends of the peoples of Pacifica who said that the ocean rose and the tops of mountains became islands.)

The work described by Maria Victoria Monsalve on Ancient Amerindian mtDNA is interesting: there are 4 main types of modern Native American mtDNA, with some few occurrences of foreigners.

The people in North/South America immediately after (and before!) the end of the Pleistocene are not the same as the American Indians. The earlier people were living mostly along the seaside, all around continents that were too high above sealevel at that time to be very hospitable. Those people ran to high ground when the oceans rose and found a very desolate ground, and died out. There are several finds of people around "9,000" years ago, such as Glen Doran's mass burial at Windover in Florida, and Kennewick Man is now the most notable, with Caucasian morphological affinities. The few artifacts we find, fabric, stone points, stone structures, indicate a very well developed culture, with no known antecedents. We find them briefly, then they are gone.

Grouping the dates related to old humans shows:

Arlington Springs Woman	13,000
Marmes Man	10,000-13,000
Kennewick Man	9,300
Spirit Cave Man	9,000
Nevada Boy	9,000
Cheddar Man	9,000
La Brea Woman	9,000±80
Wizards Beach Man	9,000
Windover Burial	7,360±80 corrected to 8,160-8,250, but there is an abrupt change in sediment age from 10,160 to 7,950.

Glen Doran and colleagues consider the Windover archaeological deposit to be a cemetery, with burials made over a thousand-year period [Doran 1991]. I think it is a tribe that died abruptly and was buried in place by a local flood. Florida, near "ground zero" of the detonation, could have been severely affected by the radiation effects, and this would produce significant errors in radiocarbon dating. A slight date inversion was shown. The base of the human bone zone was dated at 8,430 radiocarbon years before the present, while the zone immediately below that was dated at 7,950 radiocarbon years before the present [Hauswirth *et al.* 1991].

Excavations at Windover have produced at least 168 human remains, with an estimated 100-150 still in the ground, unexcavated. It appears that they were buried in water about 1-2 meters deep, as suggested by stakes that were thought to have been used to hold the corpses in place. The ages at death ranged from infants to over 65 years, with 52% under 18. The grave goods were not notably different from what might be supposed to be ordinary personal possessions for daily life. Some potentially edible plant materials were found and suggested that the site was used primarily from Summer to Fall. Because of the excellent preservation, especially of brain tissue, it was concluded that burial (immersion in peat bog water) took place in less than 48 hours after death. This preservation could have resulted from radiation sterilization.

Based on analysis of serum albumin polymorphisms, the Windover population was unlikely to be closely related to prehistoric ancestors of any group yet studied [1991] in North America.

A similar situation was discovered in Portugal, in the Muge and Sado Valley shell middens, where several hundred human skeletons have been found buried [Cunha, Umbelino, and Cardoso 2002, Cunha and Cardosa 2001]. These have been dated to about 7,000 to 8,000 years ago, similar to Windover.

Different regions and fragments of the detonated moon would carry different amounts of newly produced carbon-14 so that there would be variations in its concentration at various locations on the Earth and at different times. (Our current model for carbon-14 production proposes that cosmic ray reactions in the upper atmosphere produce neutrons that are captured by nitrogen-14, making carbon-14. Variations in the magnetic field cause variations in the cosmic ray flux, and therefore variations in carbon-14 production.) This would cause radiocarbon dating to give very different results for what might have been a simultaneous event, such as the global extinction of large mammals, the megafauna. The concentration of carbon-14 derived from calibration curves shows an irregular, surge-type pattern. The radiocarbon dating method may have been severely disturbed at the end of the Pleistocene by an influx of carbon-14 with the icy debris from the detonated moon. The nuclear detonation would have released a great burst of neutrons. Neutron irradiation of nitrogen (as in ammonia, NH_3) produces carbon-14. As different parts of the ice stream collided with different parts of the Earth, the local carbon-14 abundances were changed non-uniformly to produce rather different ages for stages that might have been at the same time. The gamma radiation from the fireball also varied from place to place and over time, and produced varying amounts of carbon-14 in the biological materials that were present at that time.

The dating of all this is crucial, and I think our present methods of dating were seriously perturbed by events that have not yet been fully described. The Giant Dipole Resonance reaction instantly created carbon-14 in biological material, bones, ivory, wood, charcoal, plants, that were exposed to the detonation fireball. More carbon-14 was produced in material that was fully exposed to the initial gamma flash, less in materials that were shielded, and less in those materials that were exposed later as the fireball cooled and lost energy. Then, the influx of carbon-14 in the icy comets, produced by the well-understood neutron absorption by nitrogen-14 in the nuclear detonation of Saturn's moon, abruptly changed the carbon-14 content of the atmosphere. This has confused our dating by carbon-14.

As sea level rose abruptly during the Younger Dryas, human settlements along the seashore were submerged, and the refugees fled to higher ground. In the Americas, the refugees, Folsom, Meadowcroft, Pedras Pintada, Monte Verde, found a land damaged so badly that they could not survive. The last Early Americans killed and scavenged the last mammoths and then perished. The land was then overgrown by algae living on the excess nitrogen compounds. We may find the remains of the last of those peoples as Kennewick Man, Spirit Cave Man, Cheddar Man, and Glen Doran's mass burial at Windover, Florida. The analytical dates may be wrong; radiocarbon was severely disturbed by radiation and input in the ice stream.

The end of the Pleistocene marked a drastic change in lifestyle for the survivors: from seaside hunting/fishing/harvesting, with marginal terraced agriculture, to life on flat lands that stretched for hundreds of miles (kilometers) from the sea with long-flowing rivers of fresh water. There, at the seaside before the "Great Flood", human civilization may have been generally restricted to a relatively narrow coastal plain, extending around the continents, around the world. It could have been possible to walk all around the world on dry, and civilized, land.

The floods from the incoming water would cause the oceans to rise rapidly and drastically, all over the world. This would have devastated (and submerged) all human settlements along the seaside, where the living was easy. The interior of the continents was dry, and higher in atmospheric elevation than at present, and would have been unattractive compared to the seaside. The seaside inhabitants would have prospered from abundant sea life, harpooning sea mammals, gathering shellfish, and harvesting the plants growing in the fog zone. That easy life ended with the rise in sealevel.

Human refugees from the rising sea would have run for the high ground, or set themselves adrift on rafts and boats. Few would survive, and those in the northern hemisphere would find a desolate landscape, covered with frozen carcasses of large animals, and developing a vast growth of algae, nourished by the plentiful water, nitrogenous fertilizer, and sunshine. Many of the refugees would perish, unable to apply their marine life-style to such an inhospitable, dry-land terrain. Human civilizations all over the world would collapse, with only vestiges of development carried to the new land, which had once been the cliffs over their heads and the mountains at their backs. Seafarers were lucky, and made it to the tops of their mountains that are now only islands. As hunters of sea mammals, these seacoast dwellers used harpoons, with points (such as the Clovis points) that were poorly suited to attacking animals on land, where every miss meant a broken point, shattered against a rock or hard dirt. But most of the animals they found were dead, and served at best as scavenger's prey. Few in number, these people did not prosper in the new New World, and all their old settlements were below the sea, hidden from modern archaeologists. The Clovis people may have developed their beautiful points for use as harpoons on marine mammals at sea, where a missed throw did not result in a shattered point. The other peoples, to the extent that we have cultural knowledge, may have been fishermen, with no need for weapons. (For example, the stomach of Spirit Cave Man yielded fish bones.)

Clovis has long been considered as the migratory forerunners of all native North and South Americans. But in this story of the Flood, Clovis was one of the few refugees from rapidly and greatly rising ocean-flooding of their seaside homelands, seen also at Pedras Pintada and Monte Verde. These peoples perished because of the disastrous condition of the land, and the forced conversion from a seaside civilization to fending for themselves in a wasteland. Later immigrants from eastern Asia found an uninhabited New World. No human skeletons representing the earliest known inhabitants, Clovis, Folsom, or Meadowcroft, have been found that we recognize.

The age of the Monte Verde settlement in Chile has been accepted as 12,500 years ago, far to the south in the New World, 1,000 years ahead of Clovis, "who came over Beringia" at the end of the Great Ice Age, and had been supposed to have spread through the New World. That, plus Pedras Pintada in Brazil, plus Kennewick Man and Spirit Cave Man, forces some new thinking about the peopling of the Americas. Monte Verde is only 30 miles from the present seaside, and only 90 miles from the pre-Flood seaside, if the ocean rose 3,500 meters. (An Incan legend says that the people of the Andes had lived beside the sea, but the sea rose and the people fled to the hills.) Similarly, in South America, refugees made it to Pedras Pintada in Brazil. However, conditions on land were so bad in the Western Hemisphere that all these peoples perished, leaving an uninhabited land for the later immigrants from eastern Asia.

It could have been possible to walk along the seaside, all around the world, perhaps even to Australia, before the Flood, if the sea were that much lower. There may have been a well-developed civilization along the seaside, limited in numbers by the relatively narrow edge of land, and generally excluded from the high continental plateau by the extreme elevations.

At a time when sealevel was much lower, the world of the seaside and the world of the high continental plateaus would have seemed quite separate and isolated from each other. Differences in atmospheric pressure would have made each region uncomfortable for the inhabitants of the other region. Climate and food and culture would have produced further barriers against mixing. In many places, the seaside would have been difficult to reach because of the steep cliffs (now submerged at the edge of the continental shelf) leading from the land to the sea.

The seaside is cut by canyons and headlands, producing coves of various sizes. Occupants of larger coves could achieve a stable, long-term society that might differ significantly in culture, tradition, and language from their neighbors in adjacent coves. This seaside could have reached around the world, providing a single human environment, peopled by humans from a genetically limited origin, split into a multitude of peoples. Those peoples became us.

Michael Richards and colleagues analyzed the carbon and nitrogen isotopes in fossil bones from modern humans of 20,000 to 28,000 years ago and European Neandertals of 28,000 to more than 100,000 years ago (standard chronologies) and showed that the modern group ate significantly more fish than did the Neandertals [Richards *et al.* 2001]. This would be natural for new immigrants from the seaside.

Samples of clay from two Gravettian settlements in the Czech Republic, Pavlov and Dolni Vestonice, dated to between 26,980 and 24,870 years ago (radiocarbon) show the impressions of fiber nets [Pringle 1997, Pringle 1998]. It was suggested that the nets had been used to capture small game. This could have been a continuation of catching small fish at the seaside. One of the researchers commented that "… at the beginning of the Upper Paleolithic, the men are really hunky, but after 20,000 years ago, they get smaller and weaker." These were the new people from the seaside, people like us.

In the Old World, conditions were more survivable after the Younger Dryas, and civilization was able to take hold again, along the shores of the enlarged Mediterranean Sea. In some locations, different peoples with different cultures were thrown together. Agriculture was boosted by the fertilization of the land by the ammonia (and the nitrate from its oxidation on entry into our atmosphere). The cometary debris that remained in space may have a return period to the Earth of about 3,000 years, and brought about "discovery" of agriculture at various different locations at various times. Sustainable agriculture requires nitrogenous fertilizer.

Humans have two modes of breathing, costal (rib or chest breathing) and diaphragmatic (abdominal or belly breathing). Costal breathing is done by expanding the rib cage for inhalation and contracting it to expel the inhaled air. This mode is shallow, and results in relatively inefficient respiration of oxygen. Diaphragmatic breathing is done by moving the diaphragm downward for inhalation and upward to expel the inhaled air. This mode is deep, and results in relatively efficient respiration of oxygen. (Try this and you will feel the difference.)

The people of the High Plateau of Tibet seem to be of a distinctly different sort than those of the Altiplano of the Andes. In Tibet, mothers give birth to normal size babies who seem comfortable in the thin air. In the Andes, however, the babies are small and struggle to survive [Moore 2005, Storz 2010, Simonson *et al.* 2010, Yi *et al.* 2010, Beall *et al.* 2010]. This suggests that the Tibetans had lived there for a long time, from before the Younger Dryas, and were acclimated to air that might have been even thinner before the rise of sealevel at the end of the Younger Dryas, after the Flood. The Andean people might have been recent escapees from the rising water along the seaside of South America, and run to the highest ground to be found to survive.

If a small group of humans moved out of Africa about 200,000 to 150,000 years ago, moving down to the seaside, and if sealevel were 1,000-3,500 meters lower then than it is now, as shown by the mouths of the submerged submarine canyons, this bimodal respiration could have been an important survival adaptation. The deep breathing would have been valuable on the high plateau of Africa, to inhale the low density air. The shallow breathing would have reduced the hazard of inhaling too much oxygen in the denser air by the seaside. The effective difference in oxygen content of the air would have been like 21% (our current normal value) on Africa, and 28% at sealevel. Such a high concentration of oxygen is detrimental to humans over a long term [Shanklin 1969], but would be ameliorated by shallow breathing. The United States Department of Labor – Occupational Safety and Health Administration recommends that the oxygen concentration of breathing air not exceed 23.5%. Over generations, a smaller human would have been favored to help protect against oxygen toxicity. This is the effect seen during the Mesozoic Era, when all mammal sizes were reduced, perhaps because of the prevailing high atmospheric pressure apparently required to support the large dinosaurs.

The early new modern Europeans, those people after the Younger Dryas at about 10,000 years ago, were smaller than the last earlier Europeans, the Cro-Magnon [Hermanusses 2003]. Hermanusses lists the Upper Paleolithic (before the Younger Dryas) males as tall and slim, 1.79 meters by 67 kilograms, and the females as small and robust, 1.58 meters by 54 kilograms. Hermanusses describes the Upper Paleolithic people as "taller and had more robust bones than the Linear Band Pottery Culture Neolithic people" (about 7,000 years ago) with "longer limbs, a shorter trunk, and similar to modern African people, very long forearms and crural [lower limb] segments." Those were different people, direct immigrants from Africa, nearly a million years ago, while the new modern Europeans were immigrants from the seaside, where they had gone 200,000 years ago.

It should be noted that after the megafaunal extinction, occurring in a poorly determined time period depending upon location, from 60,000 years ago to about 10,000 years ago (just after the end of the Younger Dryas), most animal immigrants to continental regions were significantly smaller than their predecessor relatives. Here, it is interesting to note that the megafaunal extinction extinguished nearly all animals with an adult body size exceeding 44 kilograms (100 pounds). In particular, the Neandertals, Cro-Magnon, and other known fossil humans of the time, approached adult body sizes of nearly 100 kilograms, clearly greater than the cutoff limit of 44 kilograms that permitted entry into the modern (Recent) world.

It is not clear how the people and animals at the seaside survived this extinction, in order to reoccupy the devastated continental plateaus. They may have been protected from the most severe effects of the gamma radiation from the detonation of a moon of Saturn by the additional thickness of air, attenuating the radiation.

Agriculture, the deliberate growing of food for consumption, developed abruptly in human cultures soon after the end of the Younger Dryas.

Growing plants have a variety of requirements that must be satisfied on a regular basis. The abundance or lack of these essentials determines the lushness of growth, the productivity of an area. Plants of various sorts are adapted to life over a broad range of availability of these essentials, and these adaptations, in concert with the abundances of the essentials determines the amount of animal life that can be supported by plant life, as the base of the food chain for the animals. While each of the essentials is essential, one of the most important, in terms of limitation of growth and health of the plants, is nitrogen.

With abundant nitrogen as nitrate, an overabundance of food can be produced. With an overabundance of food, wild ruminant animals can be domesticated. Ruminant animals produce nitrogen-rich manure for fertilization of the fields, making it possible to sustain an agricultural civilization.

It would appear that the initial gift of fertilizer, in the form of fixed nitrogen from the heavens, from ammonia from the detonated moon of Saturn, gave rise to an excess production of food. An excess of consumable plant crops, some not edible by humans, would be produced in a startup agricultural community. More food would be produced than was required by the founding population, and this excess would attract, and could be used to attract, various wild animals. This would lead to the domestication of a variety of suitable animals such as pigs and cattle, sheep and goats, water buffalo and chickens. These domesticated animals then served as fertilizer factories, processing plant material that was not edible by humans into fertile manure to continue the agricultural revolution. Unfertilized land losses its productivity in less than a century. In the natural state, nitrogen is fixed and converted in forms useable by plants by several processes: lightning, composting of plant and animal remains, growing legumes, and animal manure, particularly from ruminants, such as sheep and cattle.

In the natural state, equilibrium exists between plant growth and consumption by wildlife (human hunter-gatherers included) that is determined by the availability of the plant essentials. Local animals contribute to the growth of the plant community by returning fixed nitrogen.

With humans acting as hunter-gatherers, there is little disturbance of the natural equilibrium. Atmospheric nitrogen is fixed by soil organisms and lightning, and animal wastes (including that from humans) returns the nitrogen to the soil in a useable form. In a tightly closed system, the animal wastes would return all the nitrogen that was contained in the plants consumed. In the real world, there are losses by various mechanisms, and the new fixation of atmospheric nitrogen compensates for the losses. The amount of compensation determines the productivity of a region, and therefore its carrying capacity, in terms of animal life, including humans.

Agriculture, by the repeated cropping and removal of plant growth, and its consumption at other locations, disturbs this equilibrium. Continued cropping, without replenishment of nitrogen, will lead to failure of the field in times ranging from 6 years for tropical forest land to 65 years for temperate, fertile soil. In untreated soils at equilibrium, removal of crop material leads to a decline in productivity that is noticeable in the next season's crops, and rapidly leads to desolation. Without adequate nitrogen, agriculture is neither desirable nor successful, nor even suggested.

Some of the earliest cities, such as Çatal Höyük and Çayönü, seem to have been founded before the development of agriculture. Agriculture was attempted in Egypt as early as 12,000 years ago, but failed within 1,500 years [Hobson 1987]. The return of nitrogen to cropped fields is the fundamental requirement of agriculture, and failure in performing this return may have led to the failure of the earliest agriculture in Egypt. Failure to discover, or to implement this return process, or perhaps the absence of animals suited to domestication, may have led to the collapse of the earliest Egyptian agriculture.

Agriculture has been developed at various times, at different places, around the world. In some places, such as the New World and in southern Africa, agriculture remained a minor component of the human economy. North American Indians, for example, were primarily hunter-gatherers at the time of the European invasion. Small plots of corn (maize) and beans were supplied with nitrogen by the inclusion of fish bodies at planting time. However, some evidence shows that, at about the same time as in the Middle East, about 9,000 years ago, domesticated agriculture was beginning in Mesoamerica, with the domestication of corn (maize) [Balter 2010b, Piperno *et al.* 2009]. Flooded-paddy rice-growing developed in southeast Asia, with fish living in the paddies providing the nitrogen, supplemented by human waste in some cultures.

Nitrogen exists as a gas in our present atmosphere, N_2, and makes up approximately 78% of the volume of air. While the primordial atmosphere, after formation of our Earth, is thought to have lacked nitrogen gas, we suppose that it was released from abundant ammonia by bacteria. Nitrogen is moved into the soil by a variety of processes, and less is returned to the atmosphere than is extracted. The atmosphere should become depleted in nitrogen over geologic time. The residence time for nitrogen in our atmosphere is estimated to be in the range of 10 years to 1 million years, so it must be somehow replenished [Allaby 1992]. Plants cannot use this form of nitrogen directly. The nitrogen must be "fixed" by chemical reaction to form compounds such as the ammonium ion, NH_4^+ or the nitrate ion NO_3^-. The nitrate ion is more mobile in the soil and is the predominant form of nitrogen used by plants. The nitrogen in ammonia is made available by oxidation in the soil to the nitrate form. (One of the richest nitrogen fertilizers is ammonium nitrate, NH_4NO_3.) If conditions in the soil interfere with the use of the nitrate, as in waterlogged soil, the chemical reactions can proceed to releasing nitrogen as gases such as nitrous oxide, NO_2, nitric oxide, NO, and nitrogen, N_2. This removes the nitrate from the soil and returns nitrogen to the atmosphere. Fixed nitrogen in the soil may be lost, as far as plant use is concerned, by leaching past the root zone.

In our modern world, agriculture has become an industry, and industrial means are used to provide the fixed nitrogen required for modern farming. Atmospheric nitrogen is chemically processed to produce ammonia (NH_3). The ammonia can be used directly, injected into the damp soil as a pressurized liquid, or may be further processed to form ammonium nitrate (NH_4NO_3) or ammonium sulfate (($NH_4)_2SO_4$) or other compounds. A cropped field, orchard, or vineyard requires an annual resupply of nitrogen amounting to roughly one hundred to a few hundred pounds per acre (grams per square meter).

The great success of modern agriculture has been in the production of great amounts of plant crop, food, with the use of reduced amounts of land. This has been achieved by intensive care of the soil, and industrial fertilization by use of chemically fixed nitrogen.

The naturally fixed nitrogen depends directly on the nitrogen content of the atmosphere. At present that is 78%, amounting to 355,000 tons of nitrogen in the air above each acre. Lightning in thunderstorms fixes roughly 2-20 pounds of nitrogen per acre per year (2-22 kilograms per hectare). Soil bacteria associated with some plants may fix as much as 50-200 pounds of nitrogen per acre per year.

Approximately 10-20% of the soil nitrogen is returned to the atmosphere per year. A similar amount is leached below the living levels of the soil and becomes more or less permanently sequestered, removed from the atmosphere. At those rates, we expect the total nitrogen content of the atmosphere to be depleted in less than a million years. Return of nitrogen gases to the atmosphere by soil organisms slows this depletion, yet it seems that a nitrogen-rich atmosphere must be short-lived on a living, dynamic planet like the Earth [Allaby 1992]. Why is our atmosphere (still) so full of nitrogen?

This picture of the development of thoroughly modern humans, all of us, shows the following sequence:

Humans, as humans, originated genetically about 6 million years ago in a major mutation event that may have also produced the gorillas, the chimpanzees and bonobos, and possibly the orangutans. This mutation event was caused by the intense flash of radiation produced by the detonation of Planet X, resulting in what we now call the Milky Way Galaxy.

As the radioactivity from the Planet X detonation declined, human brain size increased in direct proportion. Occasionally, another nearby detonation would occur and induce radioactivity on the surface of the Earth, most often as calcium-41 in limestone. This produced a pathology like Paget's Disease of Bone in those humans that were most exposed, such as cave dwellers. We now interpret those pathological forms as the robust species, as a result of evolution rather than environment. As less radioactivity produced a less severe form of pathology, these forms transitioned into gracile forms.

As humans spread through Africa, some migrated into the Near East and Europe, some moved farther into Central Asia. Others left the familiar territory of the high continental plateau and moved to the seaside around Africa. At a time when the oceans were much lower, the seaside was 3,000 meters (or so) below the edge of the continents. Atmospheric pressure was greater, the air was richer, the climate more temperate, the living easier. Around Africa, there continued to be some exchange, of culture, language, genes, between the seasiders and the highlanders. As the seasiders began to expand along the shore, from cove to cove, from one isolated strand to the next, groups became separated, and disconnected from the original homeland.

The seasiders continued to spread along the one-dimensional world of the seashore, moving north to beyond what would eventually become the British Isles, moving east past India, Australia, and New Guinea, surrounding the high mountains of Japan, through Beringia, and then south along the west coast of the New World. At some point where the intervening land was low and not too wide, intrepid explorers crossed the New World to the east coast and spread north and south, to complete the colonization.

As time passed, the extended civilization of the seasiders diverged more from the highlanders, genetically, in culture, and in language. By the end of the Pleistocene, there were two human populations, one living on the high continental plateau, similar to the High Plateau of Tibet and the Altiplano of the Andes at present, and the others living along the seaside, in conditions similar to the shore of the Dead Sea but watered well by continental runoff and bounded by a bountiful ocean.

The disaster at the end of the Pleistocene changed all this. Most of the humans living on the high continents were killed, just as the mammoths and mastodons, cave bears and ground sloths, giant armadillos, beavers, and kangaroos came to an end, never to return. Just as the giant mammals of the Pleistocene, the mammoths, mastodons, bison, elk, sloths, armadillos, beavers, died out, so did the large version of humans. They were replaced by a smaller form from the seaside, who repopulated the Earth. As the giant animals were replaced by smaller versions, giant humans, the Neandertals and Cro-Magnon, were replaced by smaller humans, our forebears. We have grown larger during the past centuries, and may yet hope to attain the former stature of our departed relatives.

The Neandertals and Cro-Magnon were large people. A pelvis from what is thought to be a male ancestral Neandertal was discovered at Sima de los Huesos, Sierra Atapuerca, Spain [Arsuaga 1999]. This fossil is estimated to be approximately 1.764 meters tall and weighed 94.3 kilograms (207 pounds); this man was large compared to the earliest modern Europeans. Ötzi, the Tyrolean Iceman from 5,200 years ago, was only 1.65 meters tall and may have weighed 50 kilograms.

The Bog People, bodies found in Europe and England from 2,100 years ago, were small, perhaps as a result of malnutrition. However, the mummified pharaohs and other nobility of early Egypt were certainly unlikely to be undernourished. Seti I, at about 3,300 years ago, was 1.7 meters tall, not as tall and probably not as husky as the Neandertals.

As sealevel rapidly rose, seasiders rushed for high ground. This forced exodus from their seaside homelands would have resulted in the abrupt appearance of multiple advanced settlements above the new high-water mark, as has been noted along the shore of the Persian Gulf [Rose 2010]. Where the continental shelf was broad, the seashore had been a great distance from the remaining dry land, and survivors on foot failed to reach safety. Where the shelf was narrow, a brisk climb and a quick run could bring people to land above the waves. Some seafarers would have reached safety in boats, which then gained tremendous religious symbology. Many of the survivors found a desolate land, destroyed by ice, ammonia, nitric acid, and far too much water. Few lasted to found the next generation. Each brought a different dialect or language but generally, a limited set of genes, determined by the ancestors that left the plateau of Africa for the seaside a hundred thousand years earlier.

In time, the world recovered. Freshly fertilized by the new nitrogen from space, the land could support agriculture, agriculture could support domestic animals, domestic animals could support continued agriculture by returning nitrogen to the soil, and a new way of life appeared. Cities developed and mineral resources could be found and exploited. Religion attempted to preserve the knowledge of the past, but the sense and details of the stories became garbled, distorted, and lost in the promotion of priestly powers. Technology developed, in farming and metalworking, science was born, and we now strive for an objective understanding of our lost world, without yet understanding what we have lost.

Human genetic studies show that Africans have very varied gene patterns, but "all the rest of the world" have just two types [Harris and Hey 1999, Pennisi 1999]. Africa survived the end of the Pleistocene with few extinctions, and so the human population continued unaffected, preserving all the variability that had developed over 6 million years. At the seaside, the world is one-dimensional in coves isolated by seacliffs and canyons, with the nearest neighbor villages, left and right only two in number. The seaside population, derived from a limited founding group, was nearly wiped out and a limited gene pool remained to repopulate the rest of the world.

Genetic studies for the New World have been interpreted as showing that the present Native Americans are descendents of the original and only human immigrants to the New World [Lawler 2010]. However, an analysis of morphological differences of early American (about 11,000 years ago, the end of the Younger Dryas) skulls from South and Central America, compared with Amerindians (about 1,000 years ago), and skulls from East Asia showed clear differences in shape and size between the Paleoamericans and the Amerindians [Hubbs, Neves, and Harvati 2010]. These two population groups are separate populations.

Here, it is suggested that the Paleoamerican population represented the refugees from the rising ocean, the seasiders, found throughout the New World. These people failed to survive, and so left no genetic markers for the future. The Amerindians migrated from Siberia through Beringia, as conventionally proposed, and represent the full pre-Columbian inhabitants of the New World.

As genomics expands our knowledge, with the genome showing us who we are, and mitochondrial DNA showing who our mothers were, and Y chromosomes showing who our fathers were, we will eventually come to know ourselves. Until then, these questions remain very much, questions.

Digression to *Homo floresiensis*

Contrary to all expectations, a small human fossil was discovered on the island of Flores, in Indonesia, with an apparent age of 18,000 years [Dalton 2004, Mirazón Lahr and Foley 2004, Brown *et al.* 2004, Morwood *et al.* 2004]. For that age, the skull was too small, the skeleton was too small, an apparent throwback to the African fossil "Lucy". And the discovery has created controversy and antagonism ever since. Was the "Hobbit" a true laggard in human evolution out of place in time, or a modern human with a pathological condition? The arguments rage on.

As an extremely speculative possibility, consider that these people, several fossil individuals have been found, were affected by long-term exposure to radiation. Extended but non-lethal radiation exposure stunts the growth of an organism, whether a human or an elephant (or a hippopotamus). The radiation does not kill cells, but it damages the genetic material so that some of the daughter cells produced by cell division do not survive to reproduce the next generation. Thus, an abnormally small individual grows, not achieving the size and form that had been her proper legacy. Different parts are affected differently, and as in *Homo floresiensis*, various parts do not scale properly. To match the height of *Homo floresiensis*, the brain should have been 750 cubic centimeters, but was only 400 cubic centimeters. The obvious problem for this suggestion is, where and when and how did the radiation arise to affect such a small population? Could there have been a Tunguska-style airburst that scattered radioactive debris over a small but extended area? I have no answer.

We thoroughly modern humans are striving for understanding still.

Summary of Part IV

Life has developed by punctuated equilibrium evolution, with abrupt jolts nearly killing all the complex life on Earth. Natural nuclear detonations have provided the driving force for these changes, by intense radiation for killing and for genetic mutations, and breaking apart planetary satellites to provide the projectiles that have frequently impacted the Earth, changing the living conditions, permitting new forms to develop and thrive.

Questions regarding our limited knowledge of life are raised. The genetic basis for life forms is discussed. Darwin's theory of evolution and its limitations are described. The requirement for appropriate genes for evolutionary changes is discussed. A variety of changed genes for mutations is produced by radiation by several processes, including karyotypic fission and Robertsonian fusion. Mutations may be as simple as the change of a single base pair in the code for a gene. The critical difference between the chromosomes of humans and the other great apes is presented. We are distinctly different. An origin of atmospheric oxygen by photosynthesis in plants is described, and oxygen rich air (and seawater) made animal life possible.

The development of complex life occurred in the Ediacaran Period, followed by clearly complicated animal life in the Cambrian. Many of those new forms disappeared completely.

Evolution by punctuated equilibrium is described. Mass extinctions, speciation events, and a changed world form the punctuation in the story of life. Transitional forms are not found because the changes occur abruptly, produced by changes in genetic coding (mutations) and favored for survival by a new world of living conditions. Our Earth has not always been as we know it today. The extinctions, mutations, and the changes in the world are consequences of nuclear fission detonations in the Solar System.

The Cambrian "explosion" of animal life was followed three hundred million years later (with many other changes in between) by the Permian-Triassic transformation, the greatest extinction yet known and the creation of species that would lead to the dinosaurs, birds, and mammals. Speciation was slow, but eventually produced the ancestors of the dinosaurs. The standard body style changed from sprawling to upright. The strange and unique occurrence of the turtle is shown to be a form of protection against radiation.

The mass extinction that ended the dinosaurs is described as a result of the detonation of a moon of Uranus, with killing by radiation and subsequent impacts causing expansion of the Earth, changing the living conditions. Expansion reduced the density of the atmosphere, essential for buoying up the dinosaurs that were as large as modern whales, and letting pterosaurs fly. The reduction in oxygen pressure allowed mammals to grow large.

Comparisons of the sizes of ancient animals with modern counterparts show that a much denser atmosphere was required to support the early giants. Other evidence is provided for a dense atmosphere. Features promoting absorption of oxygen by skin are described.

Radiation from the detonation of Planet X at 6 million years ago is proposed as the cause of mutations that produced humans and other modern animals. Radiation effects on chromosomes, and the differences and similarities between human and the other great ape chromosomes are described. Bipedality in humans, and the ability of the modern horse to sleep while standing upright on radiation-resistant hooves, are suggested as survival features in a newly radioactive world. The dichotomy between robust human forms and gracile forms (such as us) is presented and it is suggested that we have always been humans, from our beginning. The robust forms resulted from bone damage and regrowth due to internally deposited radioactivity. Small size was the result of external radiation exposure that stunts growth. The growth of the human brain found over the past three million years is shown to follow the decay of aluminum-26, a high-energy gamma-ray emitter produced by the detonation of Planet X. It is proposed that a small group of African humans left the high plateau to live along the seaside when sealevel was much lower. These emigrants colonized the seaside around the world, isolated from the humans living and migrating on the continents.

Paget's Disease of Bone is introduced as a modern form of bone disease caused by radioactivity deposited in the bones. In present-day humans, it is the second most common bone disease, after osteoporosis. In ancient humans, the early robust forms and the Neandertals, it was caused by radioactive calcium-41. A similar effect is caused in people who live in the "auroral zones", both northern and southern. Neandertals were first thought to have been adapted to a glacial climate because of resemblances to those people, but were later found to have lived in limestone caves in the Middle East and southern Spain, warmer climate zones. Comparison of the characteristics of Paget's Disease of Bone with those seen in cases of radium poisoning (the radium dial painters of the early 20th Century) supports this proposal. Similarities can be seen in many Neandertal skeletons. Neandertals originated from the radioactivity in limestone caves induced by a nearby detonation that produced the pulsar Geminga about 370,000 years ago. Neandertals and the Cro-Magnon are examined closely, raising the possibility that the Neandertal form transitioned into the Cro-Magnon form as the radioactivity declined and as people left the limestone caves for rock shelters lacking the radioactive calcium-41. Throughout our researches into the development of humans, absolute dating has been a confused and confusing problem.

Neandertals, Cro-Magnon, and anatomically modern humans, the inhabitants of the continents, disappeared at about the time of the megafauna extinction. The Younger Dryas, considered to have been a brief cold period near the end of the supposed Great Ice Age, is suggested to be the effects of a nuclear detonation of a moon of Saturn. The Great Ice Age (and all other global glaciations) did not happen, the evidence for glaciation resulted from a massive influx of cometary ice, and the impacting fragments of the detonated moon. This influx of water caused Noah's Flood of Genesis, and the several hundred similar catastrophes described in myths and legends around the world. The Flood destroyed Atlantis. The bitterly cold ammonium hydroxide froze the soil of the North, producing ice-filled permafrost. Radiation and the thermal flash of the detonating fireball killed humans and megafauna alike. A nuclear reaction in biological materials by gamma radiation from the detonation produced differing amounts of carbon-14 (radiocarbon) locally and around the world, producing a variety of dates for a single event, the megafauna extinction. A great rise in sealevel at that time is shown by the elevation of snowlines and other evidence. Submarine canyons are shown to resemble the profile of the Grand Canyon, and were cut by rivers while sealevel was low. The continental heads of the canyons were filled with loose sediment by the Flood.

Most continental animals weighing more than 100 pounds (44 kilograms), including large humans, were extinguished. These extinctions were replaced by immigrant smaller forms. The cold ice of the impactors flash froze mammoths and other large animals of the North, which is not possible by simple severe winter cold. One mammoth carcass was found to have lost her hair, a consequence of severe radiation exposure. Her DNA could not be recovered, a result of the chromosomal damage that caused the loss of hair.

Brief magnetic reversals, events, were caused by the electromagnetic pulse (EMP) produced by the gamma radiation of the detonation.

The detailed effects of a nuclear detonation of a large icy moon of Saturn, including formation of its rings, are described.

Evidence of this catastrophe is recorded in cave paintings and bone art. Some paintings show humans flash burned on one side; others show humans in the opisthotonic posture indicating death by central nervous system damage, as caused by severe radiation exposure. The detonation is shown in three ancient rock paintings. The flow of comets to the Earth is depicted on an ancient carved bone.

Some myths and legends, preserved oral traditions, are discussed, showing the memory of a disaster and flood. Plato's writing of Atlantis shows a nearly historical account of the rise in sealevel produced by the cometary flood.

Our most recent ancestors are the surviving humans that escaped from the rising seas and repopulated the Earth. We are not direct descendants of the Cro-Magnon or the other anatomically modern humans whose remains are found on the continents, where they died with the other megafauna. Hints of our ancestry are found in the many human remains from 11,000-8,000 years ago, just after the Younger Dryas, just after the Flood.

Agriculture and the domestication of animals, a coupled industry for food production, were made possible by the influx of nitrogen fertilizer, as ammonia in the cometary impactors.

The problem of a group of small humans, found out of place and out of time on the Indonesian island of Flores, is briefly explored.

We have become, by default, the Caretaker of Life in this corner of the Universe, not knowing how big this corner is, or even how big the whole Universe is. We are the only organisms at present, perhaps here or anywhere, that are capable of insuring the continued existence of complex life against the threats of the mechanisms of the Universe. For humans such as we, a livable Earth may be available for a limited time only.

V. THE UNIVERSE
In Search of Understanding

*Astronomy compels the soul to look upwards
and leads us from this world to another.*
Plato

*Question authority,
doubt dogma.*
John Collins, 1990

Do I dare disturb the Universe?
T. S. Eliot

Beyond our home neighborhood, beyond our little planet and its Sun, beyond the Milky Way, extends the Universe, all that we can see, all that we can know. The Universe is, and always will be, all that we can observe and truly know, to the limits of our ability. Enclosing the Universe but extending to stretches of space and time beyond our grasp, is the Cosmos. We can tell stories about the Cosmos, myths and legends, and theories, but we can never truly know its character, its origin, nor its end. Our little planet, and your neighborhood, and indeed even you, are part of the Universe. We can now explore a Universe that, as yet, is known by few.

If the Universe is finite in space, and our tools and techniques become good enough, we may someday see to the edge of the Universe, even to the edge of the Cosmos. How we will know that, I do not know. If it is finite in time, we may only be able to see to the most distant reaches from which light has had sufficient time to reach us. That puts us, the Earth and Solar System, in an operational, philosophical sense, back at the center of the Universe. If the Universe is infinite in space and time, our observable Universe is limited only by the sensitivity of our techniques and the ability of light (or other radiation of any sort) to pass through the intervening space.

The Cosmos includes everything beyond our observable Universe, whatever is "outside" or "before", or perhaps "after". It is unknowable but, in certain ways, it might be designed by us from what we know of the observable Universe.

The Universe contains everything that we can directly study, matter, energy, electromagnetism, space, gravity, and time. These characters of the Universe seem to be reasonably scattered throughout and about, so that wherever it is that we are, we can find something of everything within the grasp of our observatories, satellites, and space probes. There is room enough for us to explore forever, one might think. The Universe offers us a sampling. What we make of it is our responsibility.

History, hard work in the dark

Over 400 years ago, Galileo Galilei of Florence pointed a small telescope of his own making at the night sky. He saw the heavens more clearly than anyone had before, and he started an exploration that continues today, and will likely continue long into our future. He discovered, and he described, and he determined the course of future explorations for many generations of astronomers and telescope makers. He discovered the four large Galilean moons of Jupiter, that the planet Venus went through phases much like our Moon does, and that Saturn was deformed by some sort of appendages. (These appendages were later found to be a glorious ring, as telescopes improved.)

He described the Milky Way, that broad, wavery river of lightness that passes through the heavens and is no longer visible to most city dwellers, as composed of a vast assembly of stars. This statement had a more fundamental effect on directing the path of astronomy than any other words, before or after. As we grew to understand that the Sun was a star, and therefore the stars were suns, greater or smaller, this single concept determined the scale of the Universe. As suns, stars had to be at a great distance to be so dim. And at such great distances, stars had to be large and bright to be visible as tiny points of light. How far, how big, how bright, became the overriding question of astronomy. To a large extent, that is still true today. It has led to problems that we attempt to solve by invoking magic, naming names, such as Dark Matter and Dark Energy.

The science of astronomy developed early in civilization, as a means for understanding our place, whatever that has meant to us as our understanding has grown. Our understanding has now grown increasingly complex.

Astronomy seems to be a very historical, traditional, conservative science, and this gives it much of its charm. There are few other modern fields of learning in which the terminology and philosophy extend back centuries, even thousands of years. Perhaps this is because there are relatively few astronomers in each generation, and so much must be done to pay attention to the entire Universe. Little new can be fully understood by each new generation, and even less that is old can be reinvestigated.

In addition to the generally accepted laws of physics, modern astronomy (the study of objects in our Universe) is founded on three fundamental concepts:

1. The Sun is a star, and the stars are suns,
2. The Universe is filled with great Island Universes of stars at great distances,
3. The Universe itself is expanding.

The laws of physics still hold, but we must explore alternatives to these three assumptions.

The stellar points of light we see in the Universe were declared to all be stars by Galileo in 1610. (He even considered the large "Galilean moons" of Jupiter to be "stars", because they were points of light through his telescope, not discs like the planets. This was a strictly observational consideration.) In 1750, Thomas Wright developed several descriptive models of the Milky Way. The Universe was imagined to contain distant collections of stars beyond the Milky Way, vast "Island Universes," by Immanuel Kant in 1755. Spiral nebulae, with only the Great Nebula in Andromeda visible to the unaided eye, were studied telescopically. Lord Rosse, with the largest telescope of its time, identified the spiral structure of many faintly visible "clouds". These fascinating concepts, based on so little observation and chosen mainly from imagination and for a lack of reasonable alternatives, shaped the future of astronomy. These concepts have defined the present form of our Universe.

In recent years, new instruments have brought much new information about our Universe, in radar and radiowaves, infrared, visible light, ultraviolet, X- and gamma rays, and neutrinos. Some of these observations seem to fit theoretical models based on the legacy of Galileo and Kant, and require only previously unimagined extremes of distance, speed, energy, mass. Other observations are totally undecidable: are they near or far, big or small, or what?

Our view of the Universe, the Cosmos, our cosmology, is overwhelmingly shaped by two not-so-simple measurements: astronomical distances and velocities. Within the planetary Solar System, Kepler's three laws guide us. If we know the distance at which a planet orbits the Sun, we can find its velocity. If we know the orbital velocity, we can calculate its distance. Beyond the planets, we have no guidance except what our ancestors told their children.

Recognizing the effects of nuclear-fission chain-reactions provides a foundation for explaining these observations, in a quite different manner.

Stellar Development: Stars or Starlets?

Twinkle, twinkle, little star,
How I wonder what you are!
Jane Taylor, 1806.

Stars are suns,
of course, ...
Terrence Dickinson, 1986
"The Universe...and Beyond"

It has been assumed that stars, as tremendously large, distant spheres of gas, like our Sun, but larger or smaller and much farther away, are generally at the temperature indicated by their spectra, and by the assumed brightness and size. This is not necessarily true for an object that is producing light as a result of the excitation of surrounding gas by high-energy electrons. (In fact, it is this process that produces light in fluorescent lamps, at a color temperature of 6,300 K, even though the lamp itself is barely above room temperature.)

A paradoxical situation exists with our perception of temperature and colors: we commonly think of blue as a "cool" color and red as a "warm" color. The fluorescent lamps with a higher color temperature, that radiate more blue light, are labeled as "cool" white, while those with a lower color temperature are labeled as "warm".

Similarly, the temperature scale for the Hertzsprung-Russell diagram may be quite wrong (except perhaps for the "coolest" stars indicated), and it runs paradoxically in the opposite direction to our common perception, left to right from hot to cool. The stars presented in the Hertzsprung-Russell diagrams might actually be, in many cases, asteroidal fragments and non-stellar blobs, far smaller than the stars we imagine. The optical radiation produced by these non-stellar blobs consists of three components: reddish thermal "blackbody" radiation, spectral light produced by atomic excitation resulting from the high energy electrons emitted as beta particles and photo- and Compton-electrons by the radioactive decay of fission products and activation products, and bluish Cerenkov radiation produced by high-energy electrons emitted by the radioactive materials. This requires transparent materials like ice or high-pressure gases.

T Tauri stars comprise a class of star, typified as young, irregularly variable, low mass (0.1 to 3.0 Solar masses), with emission lines, and generally with large mass outflow. The normally rare element lithium is found in the spectrum of these stars. It is thought that our Sun passed through this stage in its early development, and in fact many of the effects discussed in this book have been explained elsewhere by reference to this phase. The T Tauri stars have been further subdivided into the Classical and the Weak, and other subgroups as well. The Classical have associated dusty accretion disks, while the Weak do not show evidence of disks and are less energetic. Several of the effects described in this book have been ascribed to the energetic conditions associated with T Tauri stars, such as a strong stellar wind and an intense magnetic field. These effects seem to be more assuredly explained on the basis of nuclear fission effects, and further, the T Tauri stage may be best explained by the action of a nuclear-fission chain-reaction in the stellar core before the onset of any nuclear fusion reactions. The formative stage of a star would be controlled by natural nuclear fission reactors [Herndon 1994].

Since natural nuclear fission reactors can start with small but critical masses, and have no need for high temperatures to initiate or sustain the reactions, they can provide the kindling fire needed to start stars [Herndon 1994]. As discussed earlier in the section on formation of planetary systems, it is difficult for a newly forming star to develop the high temperatures necessary to start fusion reactions. The gravitational collapse that is assumed to produce high temperatures is forestalled by the opacity produced by radioactive materials, which retains the heat of compression, slowing the rate of collapse. However, clusters of nuclear fission reactors could increase the temperature to the fusion-ignition point. Further, some of the neutrons that are released in fission will be captured by protons (p + n) to form deuterium (hydrogen-2), which bypasses the slowest first step in the fusion chain, the proton-proton fusion, (p + p). Since hydrogen (protons) will absorb many of the neutrons in a dilute reactor, requiring greater uranium-235 content, mixing of helium, which does not absorb neutrons at all but still acts as a good moderator, will make reactors possible with lower concentrations of uranium-235, and could therefore utilize "older" uranium.

Herndon also suggests how failure of the natural nuclear fission reactors to ignite stars that formed of material deficient in uranium-235, would lead to stars over a broad range of masses, that would have failed to ignite, and would be unobservable as stars.

There may be several classes of stellar formulations. Stars with adequate nuclear fission fuel will ignite and shine brightly. Others with inadequate fission fuel will gradually collapse into darkness, provided the residual radioactivity is sufficient to generate the opacity needed to slow the collapse. Some stars, with essentially no fission fuel or radioactivity could conceivably play out the classical gravitational collapse, with sufficient shock energy to initiate nuclear fusion.

The irregular outbursts of energy seen in T Tauri stars are typical of a nuclear fission reactor where the control mechanism depends on thermal expansion and fission-product poisoning, where strong power excursions are rapidly shut down, followed by a lull while fresh fuel is accumulated and the fission-product poisons, mainly xenon-135, decay away. Random action of the many effects responsible for determining the reactivity of a stellar reactor could easily lead to chaotic behavior.

The accretion disks detected around some T Tauri stars may instead be debris disks composed of material from disintegrated newly formed planets. Study of a dust disk around β Pictoris [Lagage and Pantin 1994] shows that it is asymmetric about the star, and is depleted within 40 Astronomical Units of the star. A suggested explanation is that an existing planet is sweeping the inner area clear and concentrating material to one side of the star. An alternative, based on the planetary detonation model, is that this is a debris disk, like our Milky Way, with the apparent asymmetric concentration associated with the central remains of the detonated planet. In addition to the dusty disk, a planet has been observed, at an estimated distance of 8 to 15 Astronomical Units from its starsun [Lagrange 2010]. This is about the distance that Saturn orbits our Sun.

The FU Orionis stars, only a few of which have been found and studied, are considered to represent a brief stage in the development of T Tauri stars [Herbig 1978]. In these stars, a rapid and large brightening is seen, accompanied by a major ejection of gas from the star. The stars have high lithium abundances, and have "turned on" abruptly [Maffei 1989]. The initial surge in luminosity would represent the first accumulation of a critical mass of nuclear (fission) fuel in the center of the star.

Flare stars show similar behavior, but in a more clear-cut manner. A nearly instantaneous increase in brightness, on the order of several hundred to several thousand times the pre-burst brightness, is followed by a rapid drop in brightness and then by a pseudo-exponential decline [Byrne 1992, Zeilik and Smith 1987]. This action is identical to the fission power burst produced by pulsed reactors, such as KEWB, TREAT, and TRIGA [Hetrick 1971].

It has recently been calculated that, for large dilute hydrogen-moderated systems, as might occur in a star, the neutronic temperature effect on reactivity can be positive [Mather *et al.* 1994]. Thus, a power excursion in a star where plutonium dominates the fission reactions could end in an explosion. This might be limited to isolated reactors so that the star itself would survive, holding itself together and showing occasional outbursts.

Some stars have been identified as "barium stars" because of the high abundance of barium [Böhm-Vitense 1989] and of other elements that should be recognized as highly abundant fission products. These are strontium, yttrium, zirconium, lanthanum, europium, and gadolinium. Current theory suggests that these abundances were developed by the s-process (slow capture of neutrons) associated with a white dwarf as a binary companion. It might be more likely that the fission-product nuclides were produced by fission in the star (or a masquerading non-star), and the composition was modified by the absorption of neutrons resulting from the nuclear-fission chain-reaction.

First produced in the laboratory in 1937, and discovered in stars in 1952, the element technetium has no stable isotopes. All of its isotopes are radioactive, with the longest-lived having half-lives of 0.213 million years (technetium-99) and 4.2 million years (technetium-98). The isotope technetium-99 is produced abundantly in fission, approximately 6%. The production of technetium in red giant stars by gamma-ray-induced fission has been proposed [Malaney 1989]. This process, rather than a supposed s-process production, is shown to be adequate for most of the stars found to have technetium. However, some stars show such large amounts of technetium that a "more efficient" process is required. Neutron-induced fission chain-reactions are almost inevitable, and will produce technetium quite efficiently.

Cepheid variables are assumed to be stars that pulse as mechanical oscillators, with a period dependent on the density, the density dependent on the size, and the luminosity dependent on the size [Madore and Freedman 1992]. However, maximum light corresponds nearly with the maximum expansion rate, rather than maximum size [Abell *et al.* 1991]. This behavior suggests the action of a driving source of power, rather than a mechanical oscillator. A source of energy to drive this oscillation could be the relaxation oscillator mode of a nuclear fission reactor. Heating would lead to expansion of the star, and this expansion and the accumulation of large cross-section fission-product poisons would lead to the shutdown of the reactor. Reactor power production would resume after the star had cooled and contracted and the neutron absorbers had decayed or dissipated. A repetitive cycle of power-up and power-down would be established.

The fragments that make up the innumerable stars of our Galaxy are like the asteroids and the kuiperoids (the innermost fragments of the Milky Way). I think those stars that are away from the Milky Way and appear to be similar to our Sun, particularly those stars that are "like our Sun" and appear to have giant planets orbiting them, can be considered confidently to be true stars. I also think that the parallaxes of those stars may be correct because the measured distances are generally consistent with the objects being stars like our Sun. Stars that are labeled giants, supergiants, dwarfs, and subdwarfs, I am suspicious about.

The Sun's absolute magnitude (apparent brightness at an assumed distance of 10 parsecs) is 4.83, and the dimmest stars that can be seen telescopically are at magnitude 25. That leads to an estimate that we could see the Sun as an individual star at a distance of about 3 million light years. That is about the standard distance to M31, the Andromeda galaxy. Cepheid variables are red giant stars considered to be intrinsically much brighter than the Sun, and are detected individually at much greater distances. However, I think we are seeing the debris of detonated planets at much closer distances.

The details observed in many of these stellar phenomena can be better understood if the nuclear fission reactor process is considered.

The Sun and the Stars

The Sun is a star,
seen close.
Aristarchus of Samos, about 270 BCE

Stars exist. We know that from the clear observation that our Sun exists and our Sun is a star. The Sun serves as the defining example of a star, and that has gotten us into trouble.

The idea that stars were suns started early, at least by the time of Aristarchus of Samos (310-230 BCE), and was recognized by Giordano Bruno (1548-1600), and implicit in Immanuel Kant's philosophy (1755). It formed the basis of stellar distance estimates by James Gregory (1668), Isaac Newton (1685), and John Michell (1784) [Cohen 1958]. It was considered a generally accepted idea by Richard Bentley in 1692 [Cohen 1958].

In 1600, Giordano Bruno claimed that stars were distant suns, surrounded by inhabited worlds like our own. Bruno imagined an infinite Universe, filled with planetary systems similar to our own Solar System. He considered the stars to be suns, and so the Sun was a star, very close. For ideas like this, and others that more directly attacked the Catholic religion, Bruno was burned at the stake [Van Helden 1985]. In an earlier time (310 to 220 BCE), Aristarchus had been accused of impiety for his ideas that the Sun was a star.

Consider the Sun. It is a large, hot, intensely bright disk, dominating the daytime sky. And then consider the stars. They are tiny, cold, barely visible in twilight, gloriously scattered all over a dark night sky. What a remarkable intellectual leap it was to decide that stars were objects of the same character as our Sun, and that our Sun was one of the stars, one of a multitude. Aristarchus had this idea, based at that time entirely on his imagination and his understanding that the Sun was far larger than the Earth, and that the Earth orbited the Sun. These remarkably advanced ideas dropped from our culture. His thoughts were lost from the path of science until rediscovered by scholars in more recent times.

According to Ptolemy (about 140 CE), the stars were very far away. According to Copernicus (1543), they were really much farther away. Stars were clearly more distant than Saturn, the most distant planet then known. Saturn moved with respect to the stars, so Saturn must be closer to the Earth than the stars were [Kolb 1996].

In 1610, applying his new telescope to examine the sky, Galileo declared the Milky Way, and the four large satellites of Jupiter as well, to be "stars", without any relation to the Sun, with no consideration of the possibilities already suggested by others. For Galileo, the defining mark of a star was that it was simply a point of light, it did not show a disk as Jupiter, nor a crescent as Venus.

Even before we knew what our Sun was, what it was made of, its source of power, its lifetime, the stars were assumed to be the same as our Sun. This seems to have been based, from the earliest times, on the idea that if the stars were near, an annual parallax would be seen, much like the retrograde motions of planets that are outside the Earth's orbit. Since no parallax could be seen, the stars must be at a great distance. At a great distance, even though we see the stars as dim points of light, they must be huge and bright, to be seen at all. Therefore, the stars must be like the Sun, at very great distances, as Aristarchus of Samos said in 270 BCE.

With such a history, it is not surprising that this very durable idea has spent most of its life submerged from direct examination, an underground idea, occasionally rising to light and then submerged by the weight of established thought. An idea waiting for a reason. Philosophers provided the idea, astronomers would provide the reason.

In 1668, Isaac Newton invented the reflecting telescope, the workhorse of modern astronomy, and in 1672 described how light from the Sun could be spread into a rainbow of colors, a spectrum, by use of a glass prism. This was eventually turned into the technique of spectroscopy, permitting us to study the composition, temperature, and motion of otherwise unreachable objects in space [Meadows 1987, Pannekoek 1989, Harrison 1987].

Newton discovered (and published this in 1672) that the light from the Sun could be spread, by the use of a glass prism, into a multitude of colors. This explained the colors of the rainbow, and Newton concluded that the colors were inherently present in the sunlight. This formed the basis for the development of spectroscopic analysis, the determination of the composition of astronomical bodies (and other materials) by analysis of the spectrum of their light. Atoms give off light at specific wavelengths, colors, and each set of wavelengths can identify an atom, its ionization state, its isotopic mass, even the strength of any magnetic field. Temperature and velocity of the emitting material can be determined. All this depends on spreading the light out in a spectrum, recorded with sufficient intensity to give clear indications, and being in the right ballpark for the identifications, for some atoms can masquerade as others.

William Herschel (1738 to 1822), one of the greatest observational astronomers of the 18th Century, may have thought stars were suns, in 1785, but thought our Sun might be inhabited by beings like us [Berry 1898]. Herschel provided some of the earliest foundations of our concept that our Sun is a star, and stars are suns. In 1783 he assumed that the Sun might show the same sort of motion through space as some stars did [Pannekoek 1989]. He had assumed that stars were distant from us in proportion to their magnitudes, that is, that most stars were of the same intrinsic brightness, and only distance made them dimmer. Thus, like Aristarchus, "the Sun was a star seen close". Herschel concluded in 1784-1785 that the Milky Way was, like Galileo had said, composed of stars, and that our Sun was "one of the heavenly bodies belonging to it." In the words of his time (1785), Herschel stated that "We inhabit the planet of a star belonging to a Compound Nebula of the third form." Our place in the Universe, and the general form of astronomy, were defined over 200 years ago. The Sun is a star, the stars are suns. That was an idea that had long been waiting for a reason, and set the direction of astronomy into the present. With this idea in mind, the size of the Universe could be explored.

In 1778, Edmund Halley found that even some of the fixed stars moved, relative to their neighbors. While those stars remain as stars, we should note that such motion is the critical test for distant asteroids in the Edgeworth-Kuiper Belt, [Kolb 1996] which cannot otherwise be distinguished from stars like our Sun. Confounding this further, the visible spectrum of any solid object in the Solar System is very similar to that of the Sun, sunlight being reflected unchanged. Reflected sunlight has the same pattern of dark lines (Fraunhofer absorption lines) as sunlight itself. Slight shifts in color would allow a rock nearby to masquerade as a star of a slightly different type, at great distance. Only its relative motion would give it away.

As optical instruments improved, it became possible to find a multitude of dark lines in the Solar spectrum. Many of these lines were found to be related to the light that was emitted, as bright "spectral lines", by a variety of known elements. The dark lines are produced by atoms absorbing light emitted at their specific wavelengths, at the surface of the Sun. The science of spectral analysis, spectroscopy, was developed in the early 1800s by Joseph von Fraunhofer, Wilhelm Bunsen, and Gustav Kirchoff [Kolb 1996]. Fraunhofer has subsequently been recognized by naming the dark lines in the Solar spectrum, "Fraunhofer lines."

In 1868, spectroscopy began to show that the surface of the Sun is composed mostly of hydrogen and helium, an element that was unknown at that time. The Sun was no longer a burning coal, or an incandescent ball of molten iron. Helium was identified as an element in the laboratory in 1895, occurring in uranium ore, where it was produced by the radioactive emission of alpha particles, which was not known at that time. Natural radioactivity, including alpha emission, would be discovered in 1896.

With the discovery of nuclear reactions in the 1920s, it became possible in 1936 to propose that the Sun, and so the stars also, were powered by nuclear fusion, the combination of hydrogen into heavier and heavier atoms [Bethe 1939].

Since the Moon, planets, asteroids, and comets shine by reflecting light from the Sun, they show generally the same spectrum as the Sun, complete with Fraunhofer lines. (Differences exist in the infrared because of molecular absorption and thermal emission.) Thus, these non-stellar objects of the Solar System, particularly the asteroids and the Kuiper Belt Objects, could appear to be stars like our Sun, if we didn't know better.

As spectroscopy developed, it was found that the light from the Moon and the planets showed the same pattern of dark lines as did the Solar spectrum. This was not surprising, as sunlight reflected from solid objects should largely retain the original nature of the sunlight. Indeed, all reflection nebulae are identified as such by the similarity of their spectrum to that of the nearby star that illuminates the nebula. This effect is confounded in many cases of gaseous nebulae, where incoming incident light is absorbed and re-emitted by fluorescence just as in a fluorescent lamp, at different wavelengths. (Our commercial fluorescent lamps are carefully prepared so that our eyes see their light as various shades of white, "warmer" or "cooler".) There can be additional sources of light, such as synchrotron radiation or Cerenkov radiation, produced by high-energy electrons, and induced ionization, leading to "forbidden" emission lines, that further complicate our understanding of the light from astronomical objects.

Spectroscopy showed that various stars and clouds of stars shone with light of the same nature as our Sun. Other clouds, that could not be resolved into stars, either shone with Sun-like light, or with the light known from gaseous discharges. Those could be labeled as gaseous nebulae. Here was an observable difference between clouds of stars and true nebulae.

The final reason for the identity of stars and our Sun came in the form of distance measurements by the finally successful method of trigonometric parallaxes. Measurements on three close stars immediately showed that these stars were at distances of several light years, far farther than any of the planets in our Solar System. At such great distances, the stars must be intensely luminous, like the only intensely bright object we knew, our Sun.

To understand our Sun, our Solar System, our planets, our Earth, even our Universe, we must understand how stars form, live, and die. We have standard theories that explain that. We will look more closely later.

A star forms in space, gathering tremendous quantities of hydrogen and helium together, and the nuclear fusion reactions begin, slowly building heavier elements. Eventually, the heart of the star will run out of fuel, and the star will begin to collapse. This stellar collapse starts our voyage, so let us study stars, stars like our Sun, and stars much larger than our Sun.

It appears that stars form in clusters or groups, as the result of the gravitational condensation of clouds of predominantly hydrogen and helium gas [Phillips 1994]. According to our current theory of star formation, as a cloud shrinks, individual protostars begin to form, with masses that are determined by the local gas density, which will vary throughout the cloud.

We will see that there is a significant error in this theoretical process, due to the neglect of the radioactivity in the prestellar clouds.

According to the standard story, a protostar is cold and will contract, without increasing its internal temperature, for about 20,000 years. During this time, it is assumed that the gas forming the protostar is transparent to thermal radiation, and all the heat of compression from contraction escapes to space. Eventually, nearly all the molecular hydrogen will have been dissociated into neutral atomic hydrogen and then ionized, producing a plasma of electrons and protons, with helium. The temperature will be about 30,000 K [Phillips 1994], still too cool to start the nuclear fusion reactions necessary to power a star. This plasma will absorb all subsequent heat radiation generated, increasing the temperature. If the mass of the protostar is greater than about 0.08 that of our Sun, continued gravitational contraction will heat the central region sufficiently that the ignition temperature for fusion reactions is reached, and the star becomes self-supporting. Smaller condensations, less than about 0.08 of the Sun [Phillips 1994], cannot reach a sufficiently high central temperature for the nuclear fusion reactions to start, and will only glow dimly with residual heat as "brown dwarfs". It may be that still smaller objects form, with no significant heating, that we should call "black dwarfs", down to the size of planets and smaller [Rolfs and Rodney 1988]. (To avoid confusion with our nine (or eight) planets, perhaps we should call these distant objects "planetoids".) Condensation of these objects, with no steady evolution in their life-plans, might lead to a nearly uniform distribution of black dwarfs, planetoids, and globs, throughout space.

Protostars larger than about 50 to 100 Solar masses may break up during their formation, placing a practical upper limit on the size of stars.

As uranium accumulates in the center of a protostar, before the onset of any nuclear fusion reactions, the nuclear reactions that are normally thought to power stars, the intermingled hydrogen will serve as an effective moderator and a critical mass will probably develop quite early. This should occur for uranium isotopic compositions containing more than about 0.94% uranium-235. Since the uranium-235 content at the time of formation of the Sun was about 24%, only a few kilograms of uranium, sufficiently concentrated, will make a critical reactor, naturally. (We think there is about 2×10^{18} kilograms of uranium in the Sun, enough for about 10^{18} critical masses.) The fission reaction will start well before the onset of deuterium burning, because the fission reactor only requires sufficient fuel at a high enough concentration, independent of temperature, while deuterium burning must wait until gravitational contraction (or the fission chain-reaction) has increased the temperature to 540,000 to 600,000 K or more [Menzel *et al.* 1963, Taylor 1992]. High temperature is not needed for the nuclear-fission chain-reaction to proceed.

The fission reactions can even contribute further to the initiation of fusion in the star's core by producing significant amounts of deuterium and tritium, rare nuclides otherwise, by the ternary fission process. These nuclides fuse easily.

Deuterium is also produced by neutron capture by hydrogen. The fission reactors would build a stock of deuterium fusion fuel before the Sun finally ignites as a fusion furnace. The production of deuterium by hydrogen absorption of neutrons released in fission bypasses the extremely slow first step to fusion, the proton-proton fusion.

After initiating a star's fusion reactions, the nuclear-fission chain-reaction may continue to contribute significantly to the energy production, and to the star's development. As the fusion reaction converts the original hydrogen into helium, the core becomes a very efficient thermal breeder reactor. The neutron absorption cross-section of helium-4 is essentially zero, and so fewer neutrons are absorbed by the moderator/coolant in the progressively helium-rich core. This promotes the conversion of uranium-238 (and thorium-232) into thermally fissionable plutonium-239 (and uranium-233) and (by alpha decay) uranium-235. Because of the limited quantity of uranium (and thorium), this can last for perhaps 10-100 thousand years. (But there is evidence, in the neutrino deficit and sunspot formation, that the reactors continue to work.) Thus, nuclear fission may serve only as a "fire-starter" to raise the interior temperature to the ignition point for fusion. This question is very much dependent on how much uranium and thorium is in the core of the Sun. Because of the hydrogen moderation assumed for a stellar reactor, much smaller amounts of nuclear fission fuel are likely to be involved, and the characteristic time behavior will be considerably slower than in the planetary case, which has been assumed here to be unmoderated. Therefore, the progress of the reaction is likely to be more stable and much less likely to result in a detonation. Convection would probably be strong in these pre-stars, because of the great temperature difference between the hot fission reactor and the still cold outer regions. This would promote the transport of a variety of fission products to the surface.

After the internal temperature reaches fusion ignition temperature, the simplest fusion reaction will begin:

$$^2H + {}^2H \rightarrow {}^4He + \gamma.$$

This has an ignition temperature of about 540,000 to 600,000 K, but is considered to be limited by the small abundance of deuterium in the interstellar gas that forms the star. As the temperature increases, the proton-proton cycle can begin, in which sequential fusion reactions convert hydrogen into helium:

$$^1H + {}^1H \rightarrow \quad ^2H + \beta^+ + \nu$$
$$^1H + {}^2H \rightarrow \quad ^3He + \gamma$$
$$^3He + {}^3He \rightarrow {}^4He + {}^1H + {}^1H.$$

Note that a positron decay is necessary in the first reaction since two protons do not form a stable nucleus and one of the protons must be converted to a neutron. The positron decay is accompanied by a neutrino, shown by the Greek letter nu, ν.

In this process, energy is released, about 26 MeV (million electron volts), and four hydrogen atoms are converted into a helium atom. The kinetic energy of the positron heats the local region, and the energy of the gamma ray, and the gamma rays resulting from the mutual annihilation of the positron and a nearby electron, slowly makes its way to the surface of the star. The energetic radiation will scatter off the electrons of the plasma, leaving fractions of its energy to heat the core and surrounding gas. Hundreds of thousands of years after the gamma rays were produced in the heart of the star, photons of light will begin to shine forth into space. In most cases, the neutrino will escape unnoticed, in a direct line, traveling at the speed of light. In time, as the hydrogen in the central region is consumed, the star will begin to contract and its central temperature will increase so that the helium nuclei will begin to fuse:

$$^4He + {}^4He + {}^4He \rightarrow {}^{12}C + \gamma.$$

In stars that are sufficiently massive, perhaps as great as 20 times the Sun's mass, fusion reactions can form progressively heavier nuclei, until a core of iron begins to form. Iron is as far as fusion can proceed in a stable star. Building heavier nuclei consumes energy, rather than producing it. Heavier nuclei are built more easily by the successive absorption of neutrons that are produced incidentally by various minor reactions in the star. This normally proceeds by a sequence of neutron absorption, followed by beta decay, repeated and repeated until elements as heavy as bismuth are formed. At that point in the nuclear sequence, this slow growth, named the s-process [Burbidge *et al.* 1957], can not create heavier new nuclei as fast as they are changed to lighter nuclei by alpha decay.

However, in stars this massive, the iron core cools and starts to contract. The temperature is initially increased by this contraction, and this leads to production of photons of sufficient energy to dissociate iron nuclei. This results in absorption of energy, a cooling process. But a further cooling, or pressure-reduction process, also occurs, and this process results in the production of large numbers of neutrons and neutrinos. Electron capture, induced by the large number of electrons with energy above the threshold for this reaction, about 1.3 MeV for unbound protons, converts protons into neutrons, and results in loss of high-energy electrons, and the escape of much of the energy in the core by way of the neutrinos. The contraction becomes a cataclysmic collapse. A supernova is born, the detonation of a great star at a great distance. Understanding distances is the key to understanding the Universe.

William Herschel attempted to measure the distance to a star by parallax in 1780-81, repeatedly measuring the position of a bright star (which he assumed to be nearby, by its brightness) relative to an adjacent faint star (presumably at great distance). Since previous attempts to measure parallax had failed, any more distant star could be assumed to have a zero parallax, and so serve effectively as an infinitely distant reference. (This continued to be the basis of more modern attempts to measure stellar distances.) However, if it happens that the reference stars are actually nearby, the assumption that the reference stars are infinitely distant is false, and while the method produces very precise results, it will fail to produce an accurate measure of distance. The measured star will seem to be much farther away than it actually is. For most (or all) of Herschel's pairs of stars, the method failed, for further observation showed that the pairs were binary systems, both stars at the same distance [Hirshfeld 2001].

In 1838, Friedrich Wilhelm Bessel measured the parallax of 61 Cygni as 0.31 arcsecond. In 1840, he remeasured it as 0.348 arcsecond, equivalent to 590,000 Astronomical Units, or 9.339 light years. In 1835-38, Friedrich Georg Wilhelm Struve measured the parallax of Vega as 0.26 arcsecond, or about 12.500 light years. In 1839-40 Henderson and Maclear measured the parallax of α Centauri (alpha Centauri, the brightest star in the constellation of the Centaur) as 0.91 arcsecond, then as 0.76 arcsecond, 4.272 light years, the star closest in space to our Solar System [Berry 1898]. With such great distances, stars so faint must be very bright, like our Sun.

There is a basic ambiguity in astronomy in interpreting the points of light we see in the night sky. They are dim, strictly speaking, and this is due to them being either small and near, or large and far. Jupiter's large moons are small and near, but stars of the same brightness must be, if like our Sun, at very great distances to be seen so dim.

The fundamental conflict between this book and the standard story in astronomy lies in the astronomical distance scale, and the nature of objects as seen at a distance. The standard distance scale provides space for vast Island Universes of stars, at immense distances. In this book, the spiral galaxies are the debris fields of detonated planets orbiting stars that are far away (but relatively near), and most of the stars we see and study are asteroidal fragments in our Solar System. Obviously, this requires a much smaller distance scale than the standard scale, and the possibility that the standard scale is wrong must be considered.

The standard distance scale uses a series of relative distance relations to stretch out into the Universe, starting from what is "universally" perceived to be an absolute measure of stellar distances, by the method of trigonometric parallax. By confusing trigonometric parallax with civil surveying triangulation, the relative and ambiguous nature of the measurement has been overlooked.

The Astronomical Distance Scale

There has remained over the years considerable concern that
we are making a grave mistake.
H. Shapley, 1943.

If Cepheids are screwed up, ...
B. Madore, 1994.

The least initial deviation from the truth
is multiplied later a thousandfold.
Aristotle,
"On the Heavens", Book I Chapter 5

Question all assumptions.
Aristotle

Distance is fundamental to our understanding of the Universe. From the subatomic to the stars, we measure it in many ways, each with its own units, each with its own standards.

On Earth, we use different methods and different units to measure distances. These range from the millimeter and smaller to centimeter, meter, and kilometer. For these measurements, we eventually arrived at a permanent, accurately determined absolute standard. However, ...

Once upon a time, the standard of length was the King's royal foot. That length varied from king to king, and so from place to place and time to time. For long distances, the distance traveled by a Roman legion in 1,000 (milliare) paces became a standard still in use, the mile. In 1791, the French Revolution attempted to naturalize our measure of the world, and defined the meter (metre) as 1 part of 10,000,000 of the distance along a line of longitude from the North Pole to the Equator. The meter was defined on the basis of a physical survey from Dunkirk to Barcelona, that took 7 years to complete. This length was enshrined as a platinum-iridium bar at Le Bureau des Poids et Mesures in Sèvres, near Paris. It is a length that can be easily grasped, being about the distance between the tip of your nose and the tips of your outstretched fingers, slightly more than a yard. In principal, the circumference of the world could be measured by people standing next to each other, fingertip to the next nose tip. (This would take approximately 40,000,000 people, some of whom would have to walk, or at least stand, on water.) The meter is now defined more securely as the length of 1,650,763.73 waves in vacuum of the orange-red spectral line corresponding to an optical transition between the $2p^{10}$ and $5d^5$ electronic energy levels of krypton-86. The approximate scale is still best perceived as nose-tip to fingertip.

Similarly, as we reach out into the Universe from our observatories on Earth, we must use appropriate methods and suitable units. The first unit of astronomical length is named, appropriately, the Astronomical Unit. It is approximately the average distance from the center of the Earth to the center of the Sun.

The determination of that distance has been a remarkable story of progressive improvements in technology. Yet, the progress of these measurements has been plagued by a problem often seen in objective measurements in science. Each new measurement differed from the previous measurements by more than the uncertainties estimated by the observers. It is not wrong to make a measurement, the best at the time, with a large uncertainty, but usually the uncertainty was greatly underestimated, making the measurement look better than it actually was. The value of the Astronomical Unit did not stabilize until radar measurements of the distance to Venus in 1961 [Muhleman, Holdridge, and Block 1962]. Further refinements have been made by use of radio tracking of space probes. It is now known to the remarkable accuracy of about 1 part in 100 million, roughly an uncertainty of 1 mile from the Earth to the Sun.

In astronomy, the distances are so great that normal methods of measurement fail. The primary method that has been developed, trigonometric parallax, is based on the apparent movement of a distant object as the location of the viewer changes, or "parallax". This method is so fundamental that its name has been applied to many different methods of measuring distance, completely unrelated. The original method involves the apparent shift in location of an object when the viewing location changes. This effect can be easily demonstrated. Simply view your thumb held out at arm's length, with alternate eyes. (Don't move your head.) Your thumb will jump back and forth against the background as you switch eyes.

Parallax, in its simplest form, is often conceived to be the same method as triangulation, in civil surveying. Land surveys were conducted by use of accurately measured iron chains which were stretched (without stretching!) from point to point to determine distances. Angles are measured absolutely, and the triangle formed by the baseline and two adjacent angles permits trigonometric determination of the other two sides of the triangle. (See Figure 98.)

In horizontal parallax (Figure 99), the diameter (or a chord) of the Earth serves as the baseline for determination of the distance from the Earth to the Sun. This determination is the foundation of the astronomical measurement known as parallax, from the displacement of the limb (edge) of the Sun between viewers on the surface of the Earth, and a (virtual) viewer at the Earth's center. This is the Solar parallax, but it could never be measured directly with suitable accuracy. Its determination leads to a value for the Astronomical Unit, the semimajor axis of the elliptical orbit of the Earth around the Sun, the average of the nearest distance and the farthest distance. The measurements of the Solar parallax started in about 270 BCE and extended nearly to the present, with high accuracy determinations now based on radar ranging to the planets and radio tracking of spacecraft. In this case, the use of the word parallax reflects a tendency of astronomers to use this word to refer to any distance measurement. Many of the measurements were position and timing measurements meant to refine the parameters of the orbits of Mercury and Venus, and hence determine their distances from the Sun. The distance to the Sun was never measured directly. Summarizing the historical results shows how measured values can change with time (assuming the actual, true value remains constant) and how these values eventually converge to a measurement of high accuracy [most of these are from Pannekoek 1961].

Observer	date	method	Solar Parallax
Aristarchus	ca 270 BCE	ratio of distances	19.1 = 136 arcsec
Ptolemy	135	Solar eclipse	166 arcsec
Kepler	1606	thought	<60 arcsec
Vendelinus	1630	Moon	15 arcsec
Vendelinus	1672	Mars	9.5 arcsec
Lacaille	1751	Mars	10.2 arcsec
		Venus	10.6 arcsec
		Mercury	45 arcsec
Halley	1761	Venus transit	8.55 - 8.88 arcsec
Foucault		light speed	8.8 arcsec
	1862	Mars	8.93 - 8.96 (8.90)
	1864	1769 Venus transit	8.83 arcsec
	1874	Venus	8.76, 8.88, 8.81, 8.88, 8.88 ± 0.04
Laplace	1810	Moon	8.6
Encke	1835		8.57116 ± 0.0371
	1877	Mars, relative	8.78
	1888-1889	Asteroids, relative light speed	8.80 ± 0.01
Michelson	1900-1901	aberration of light	8.80 ± 0.01

Hinks	1900-1901	Eros	8.807 ± 0.003 photographic
Hinks	1900-1901	Eros	8.806 ± 0.004 visual
	1912	spectrographic	8.802
Noteboom	1921	Eros, mass of Earth	8.799
Spencer Jones	1924	Moon	8.805
de Sitter	1929	combination	8.803 ± 0.001
	1929	stellar occultation	8.790 ± 0.001
Spencer Jones	1929	Moon	8.796 ± 0.002
Spencer Jones	1930-1931	Eros	8.790 ± 0.001
Rabe	1950	Eros	8.7984 ± 0.0004
Muhleman	1961	Venus, radar	8.7940976 ± 0.000220
current value	(2009)		8.794143

The measurements reflect an interesting change in the methods. Initially, angles were measured (or estimated) absolutely, in the same manner in which a land surveyor uses triangulation. Later measurements used reference stars for relative measurements of the parallax angle. This has led to a misunderstanding of the nature of trigonometric parallax. Current values of the Solar parallax have no relation to parallactic angles, but rely entirely on timing of radio signals. It is still called the Solar parallax.

Understanding distances is the key to understanding the Universe.

Remote Distancing

Distance can be simple to measure, as easy as placing a ruler, meter stick, or tape measure along the distance to be measured. At greater distances, and particularly where mountains and canyons, lakes and rivers, interfere with direct measurements, civil surveyors use the method of triangulation. This technique uses the properties of a plane triangle to determine distance from three measurements: the length of one side of a triangle and the values of the two adjacent angles.

The astronomical distance scale is fundamental to our understanding (and misunderstanding) of the Universe. It starts of course with absolute (defined) distances on the Earth, where we can lay a yardstick or a meter and see both ends, measuring the distance between two points. It reaches out into space for the measurements of the distances to the Moon and planets, which define the scale of the Solar System. The astronomical distance scale starts on firm ground, on Earth, with measurements of distances to the Moon, the Sun, and the nearby planets, that are clearly accurate and precise, but difficulties interfere with our attempts to measure greater distances. Initially, attempts were made to apply triangulation, labeled as horizontal parallax, to measurements of distances in the Solar System. The method was not successful. Those measurements are now done very accurately by use of radar and laser ranging, and spacecraft radio ranging. Beyond the Solar System, distances to the objects that we perceive as stars have been measured by a method called trigonometric parallax, and misunderstanding that method has led us astray.

Errors will occur in parallax measurements, and these will be in the direction of overestimating the distances to the stars. If the reference stars, those chosen because they appear to be similar to the target star, are at about the same distance as the target star, the observed parallax will be too small to measure, so the target star will be judged to be at a very great distance, no matter how near. This is because it shows no measureable parallax. A more severe problem arises because of the difficulty in choosing only true stars for the reference points.

Triangulation

To determine the distance to an inaccessible point, the civil surveyor measures the length of a base line and the angles that two sight-lines to the point make with the base line. That forms a triangle, hence the name of the method. A trigonometric formula is then used to calculate the distance. The uncertainty in the distance is determined by the measurement uncertainties (length and angle) and the assumption that the distance is a "straight-line" distance, and that plane geometry applies. Triangulation gives the measures of the triangle absolutely, the lengths of its three sides and the values of the three angles. The triangle may be of any form, right triangle, isosceles, acute or obtuse, and the method works.

In civil surveying, used to lay out lot lines, boundaries of cities, counties, states and provinces, and nations, distances to remote points are determined by the method of triangulation. Those points may lie across rivers or lakes, or simply be included in the survey set for checking accuracy. In triangulation, the two ends of the baseline and the remote point define a triangle, Figure 98. The length of the baseline is measured (in early days by laying lengths of calibrated chains in a straight line) and the angle between that baseline and a sightline to the remote point is measured at one end of the baseline and at the other end of the baseline. In plane geometry, that gives the measurement of two angles and the included side of a triangle. With those three measurements, the triangle is completely and absolutely determined, subject to approximations of plane geometry on the irregularly curved surface of the Earth. In principle and practice, the accuracy of the distance to the remote point is the same as the accuracy of the baseline.

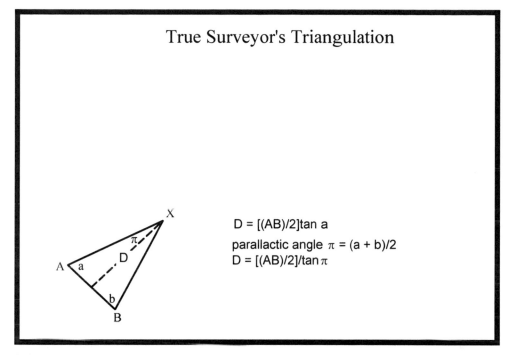

True Surveyor's Triangulation

$$D = [(AB)/2]\tan a$$

parallactic angle $\pi = (a + b)/2$

$$D = [(AB)/2]/\tan \pi$$

Figure 98. True surveyor's triangulation.
The method of civil surveying triangulation. The length of the baseline AB is measured, and the angles a and b are measured. This example of an isosceles triangle makes the trigonometry simple and easy, but it can be made to work for any triangle.

Horizontal Parallax

For angular measurements that were made directly of the Sun using graduated sectors, the surveyor's absolute triangulation method was applied. This method, in astronomy, was called horizontal parallax because it referred the parallax to the horizon.

Triangulation was applied to the measurement of the distances to the planets in a significantly similar way as in civil surveying, a way that preserved the specific absolute accuracy of civil surveying triangulation. The angle to a planet was measured relative to the zenith direction when the planet was near the horizon at the beginning of the night and again when it was near the opposite horizon at the end of the night. This is the method of horizontal parallax, Figure 99. The baseline in such a measurement was the chord through the Earth between the two positions of the observatory at the time of the measurements. The triangle was defined by the two positions of the observatory and the (moving) position of the planet. While in principle this method could produce absolute measures of distance to the planets, and thus determine the Astronomical Unit (the average distance of the Earth from the Sun), perturbations of the thick atmosphere and other systematic errors prevented its success.

Use of this method in astronomy, designated as horizontal parallax, was attempted for the measurement of planetary distances (and hence the scale of the Solar System) by measuring the angle from the local zenith and a planet (such as Mars) just after sunset and just before dawn. The length of the chord through the Earth, connecting the two positions of the observatory, served as the baseline, the measured side of the triangle, and the zenith angles gave the two adjacent angles. These three values were measured absolutely. The technique is fraught with difficulties. The observatory is moving about the (tilted) rotational axis of the Earth as the world turns, the Earth is moving in space along its orbit, and the target planet is also moving about the Sun. Atmospheric refraction displaces the image of the planet from its true position. This effect can be reduced by making the measurements closer in time to when the planet is nearest the zenith, but the shorter time difference reduces the length of the baseline and reduces the angles to be measured. Fractional errors become large. Triangulation, or horizontal parallax, was not successful in producing accurate distances to the planets. We did not accurately know the scale of the Solar System until radar ranging was applied in 1958-1961.

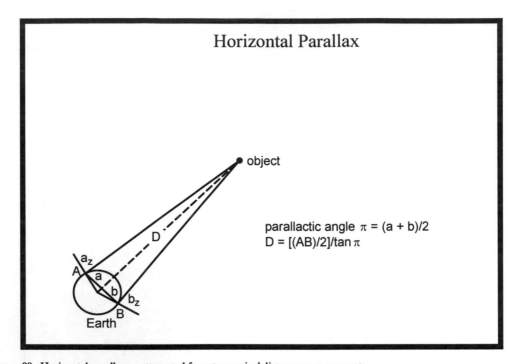

Figure 99. Horizontal parallax as attempted for astronomical distance measurements.
The method of horizontal parallax. The length of the baseline AB is calculated from measured dimensions of the Earth and the position of the observatory at the two observations, and the zenith angles a_z and b_z are measured, giving angles a and b. This method is as absolute as civil surveying triangulation, as it is exactly the same. However, the differences in the angles were too small to measure accurately.

These positional (angular) measurements have complications when very small angles must be measured. In addition to refraction in the atmosphere, the bending of light rays, two effects showed their strength in distorting the results as the measurements became increasingly precise. These were explained in 1728, as aberration, and in 1748, as nutation.

Aberration shows that the Earth's motion is not negligible compared to the speed of light. A directional instrument, such as a telescope, is pointed slightly in the direction of the Earth's orbital motion in order to properly capture light from an astronomical object. This makes the precise position of stars wander regularly back and forth, completing a full cycle in the course of a year.

The effect of nutation reflects the circular migration of the direction of the Earth's rotational axis, with a period of 18 years, superimposed on the much longer (about 25,772 years) precession. Such short-term variations in the positions of stars, caused by motions of the Earth, could amount to as much as 30 arcseconds. Even after correcting for these effects, measurements of Solar System distances by horizontal parallax were left with large deviations and uncertainties.

Trigonometric Parallax

Trigonometric parallax is the only means for measuring distances beyond the bounds of the Solar System, but this method fails us. The development of the automatically tracking telescope with a filar micrometer, and the modified version of the heliometer, permitted highly accurate measurements to be made of the angles between stars, or between a planet and a star. This method could be used to determine the relative motion of an object against the background of stars and this determination would be free of the effects of aberration and nutation, since both the observed object and the reference stars were equally affected. This technique formed the basis of trigonometric parallax.

In this method, the apparent change in position of the object, relative to the position of reference stars in the photographic field that results from the change in the observer's position due to the orbital motion of the Earth, is used to establish a triangle relating the distance of the object to the scale of the Earth's orbit and the measured angular change in the object's position, Figure 100. The success of this method relies on knowing the distance to the reference stars, or more commonly the assumption that the stars are true stars at extremely great distance compared to the object of concern. If the object is at the same distance as the reference stars, a parallax of zero (within the uncertainties) will be found. If the reference stars are nearer than the object, the parallax will be negative.

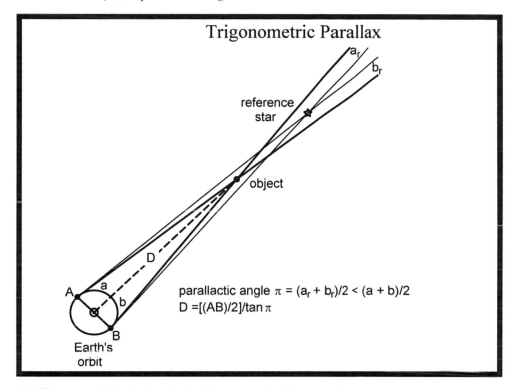

Figure 100. Trigonometric Parallax, foundation of the astronomical distance scale.
The method of trigonometric parallax. The length of the baseline AB is known as twice the Astronomical Unit, and the relative positional angles a_r and b_r are measured. This gives a parallax relative to the reference star, so it is not an absolute parallax.

The change in apparent position of a target star through the annual motion of the Earth about the Sun is used as a measure of the distance to the star. In principle, the closer the star, the greater the change in apparent position on the plane (the celestial sphere) of the sky through the course of the year. A star at "infinity" would show no apparent motion. Because of the extremely small changes in position, absolute measurements of the angles could not be made, as had been attempted for the planets. Even for the closest planets, those angles were too small to be measured accurately with the diameter of the Earth as a baseline, and extending the baseline to the diameter of the Earth's orbit for measuring stars did not improve the accuracy sufficiently to permit useful measurements. Therefore, the method was modified by measuring angular position changes relative to reference stars. The method was still perceived as one of triangulation, with absolute measurement, and that concept can be found in numerous textbooks on astronomy. This error is ingrained in the understanding of astronomers, perhaps universally so.

This modified method of triangulation was applied to measurements of the distances to nearby stars, with the baseline increased by use of the Earth's orbital motion around the Sun, but the key to absolute measurement was lost by the angular measurements being made with respect to reference stars, as shown in Figure 100. This had the significant advantage of reducing the atmospheric refraction effect, since the object being measured and the reference stars were affected essentially equally. However, the fundamental effect of this change cannot be overemphasized, for it is the fundamental cause of the error in our distance scale, and in our false perception of the Universe. The measurement had changed from an absolute measurement to a relative measurement, relative to the reference stars. This difference has not been recognized.

Trigonometric parallax differs from triangulation in two ways. First, a triangle is absolutely determined by the measurement of two angles and a side (the direction angles and the baseline). In trigonometric parallax, the baseline is accurately known, but only one angle is measured. Further, this angle is measured relative to background stars that are assumed to be so far away that their own parallactic motions have no effect on the measurement of the object. These are very small errors, but the measurements themselves are small.

If the reference stars are just barely beyond the measured object, that object's parallax will be artificially small, and it will appear to be farther away than it actually is. In the extreme case of the reference stars being nearer than the measured object, the parallax of the measured object will be negative, a clear indication of the failure of the assumption that the reference stars are essentially infinitely far away.

If the reference stars are relatively close to the distance of the target star (and the narrow field of view of parallactic telescopes limits the number of stars that can be used), the method of trigonometric parallax will give the paradoxical result that the target star is relatively far away. If the reference stars are at exactly the same distance as the target star, the distance to the target star will be rendered as "infinity", no measureable parallax. If the reference stars are closer than the target star, the distance will be "beyond infinity", a negative value. Approximately 10% of the parallaxes in the *"The General Catalogue of Trigonometric Stellar Parallaxes"* [van Altena, Lee, and Hoffleit 1995] are negative, indicating that the reference stars are actually nearer than the target star, and a distributional analysis suggests that another 10% should have been included but were edited out. These erroneous results will occur regardless of the true distance to the target star. If measurements are made in a field that is crowded with very distant asteroidal fragments, still within the scope of the Solar System, distance measurements relative to the other neighboring asteroidal fragments will give distances that are typical of our measurements of actual stars, at distances of many light years, even though the true distances are just very many Astronomical Units. This is an error of roughly a factor of 60,000, or as we will see later with regard to the extension of the distance scale to the spiral galaxies, about 100,000. The *Hipparcos* results are more precise, but are subject to the same possible error of assumption. This rung of the astronomical distance ladder, the first step beyond the Solar System, is weak and wobbly.

In most parallax measurements, made by use of a micrometer screw, the error in the position of a star relative to the reference stars increased in an unknown manner with each turn of the screw. Therefore, the highest precision work was done by using reference stars that were as near to the subject star in the photographic field as possible. Further, the field of view of a standard parallactic telescope was very small, to allow high magnification. Reference stars were chosen to have nearly the same brightness as the object star so that the images would be similar on the photographic plate. It seems likely however, that this selection of reference stars that were close in two (angular) dimensions, and in brightness, might lead to the reference stars also being close in the third dimension (distance from Earth), which affects brightness.

Measuring the parallactic shift of the object relative to a reference star has the effect of projecting the reference star to infinity, with a parallax of zero. More refined analyses correct for the measured parallax of the reference star, but that parallax was measured relative to other reference stars, and so on.

In many astronomy text books, trigonometric parallax is described as if it were triangulation, which is an absolute method. A clear example is shown in Figure 101, taken from *"Beyond the Milky Way"*, by Thornton Page and Lou Williams Page [Witherell 1944].

Unfortunately, because of the easy transition from the earlier absolute angular measurements in horizontal parallax to the new relative measurements, with greatly improved accuracy, the loss of the rigid foundation of triangulation was not noticed. Current discussions of trigonometric parallax still refer to it as if it were triangulation [Trefil 1998]. There is a perception among astronomers that this is an absolute measurement [Famous Astrophysicist 1994].

James Trefil overestimates the quality of trigonometric parallax by describing it as a form of triangulation, as used by land surveyors [Trefil 1998]. While both methods are based on triangles, the surveyor's triangulation uses direct measurements of three of the parts of a triangle (two angles and the included side, the baseline); the astronomer only measures one side and estimates the opposite angle. Subject to the assumption of plane geometry and the inaccuracies of measurement, the surveyor's result is absolute. Trigonometric parallax is a relative measurement and relies on the assumption that the reference stars used to obtain the estimate of the parallactic angle are at effectively infinite distances. It is the apparent motion (change in position) of the observed star, caused by the Earth's progress in its orbit, relative to the reference stars that provides the estimate of the parallactic angle. If the reference stars fail to be at "effectively infinite" distances, the parallactic distance for the observed star will be overestimated.

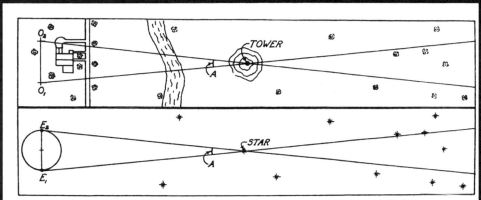

FIG. 1. The principle of triangulation is shown above as applied to a terrestrial problem and below as applied to the distance of a star from its heliocentric parallax. Observations from O_1 and O_2 or E_1 and E_2 determine the parallax angle A and hence the distance of the tower from O and the star from earth (E).

Figure 101. Conceptual presentation of trigonometric parallax as triangulation.
An example of the presentation of trigonometric parallax as if it were civil surveying triangulation. In surveying, the two angles at O_1 and O_2 are measured, rather than the parallax angle A. Triangulation is absolute, trigonometric parallax is relative to the reference stars. From Page and Page, *"Beyond the Milky Way"* [Witherell 1944].

Trigonometric parallax is not triangulation.

If this were only a simplification of the explanation, it would be innocent and little harm would be done. However, a famous astrophysicist (anonymous out of my respect) argued with me that trigonometric parallax is absolute and my complaints about the standard distance scale were foolish. It must be recognized that trigonometric parallax measurements determine the positional changes of the target object relative to reference stars whose distances are not known, except by other measurements of trigonometric parallax, and the assumption that they are suns like our Sun, perhaps bigger and brighter or smaller and dimmer, but great balls of fire nevertheless.

Is it possible for an asteroidal fragment in the Solar System to appear to be a true star? If visual evidence counts for anything, look at the discovery photos of Pluto, at about 30 Astronomical Units, or the very recent discovery by use of the *Hubble Space Telescope*, of the Kuiper Belt Object (or Trans-Neptunian Object) Quaoar, at about 43 Astronomical Units. Without the diligent and deliberate determination of the proper motion and Solar orbit of these bodies, there would be no reason to think they are not true stars, just like all the other points of light surrounding them, also presumed to be stars. Even the planet Neptune was thought to be a star by Galileo in 1610, and again by Joseph de Lalande in 1795 [Carter and Carter 2002]. Such objects would show spectra much like that of the Sun, as reflected sunlight, complete with Fraunhofer lines.

Asteroidal fragments abound in the Solar System. Unknown and unexpected until the discovery of the asteroid Ceres by Giuseppe Piazzi in 1801, asteroids (by that name) are concentrated in the Asteroid Belt between Mars and Jupiter. Some have orbits that place them near the Earth, or even inside Earth's orbit. Some are in synchronous orbits in the path of Jupiter. More recently discovered, Kuiper Belt Objects orbit beyond the orbit of Neptune and are alternatively called Trans-Neptunian Objects. Some of these have been displaced into the so-called Scattered Disk. Farther out, but not yet discovered, is the hypothesized Oort cloud, expected to be the source of comets.

Asteroids were so named because they look like stars. They shine by reflected sunlight, in some cases reddened by their own surface colors, shifting the spectrum from an apparent G type, from the Sun, to K type, or further (cooler). (It should also be noted in general, as proposed in this book, that radioactivity inside a transparent shell of ice (or similar material) will produce Cerenkov radiation, a blue light resembling the spectrum of a Type B star. To carry this idea further, a fragment with a functioning reactor will heat above the local temperature, and appear to be a red giant star.) The major distinguishing feature, and the means of identification from the very first discovery, is that careful measurement of positions over time can be analyzed to produce a Solar orbit. The difference between this orbit and that of the Earth produces proper motion that can be used to select the nearer fragments as non-stars. This effect diminishes with distance, being as slow as 1.5 arcsecond per hour at a distance of 100 Astronomical Units.

With the high resolution of modern (adaptive optics) ground-based telescopes and the *Hubble Space Telescope*, a slight enlargement of the image, from being out of focus relative to more distant objects, can be detected by careful analysis.

In numbers, it appears that asteroids and Trans-Neptunian Objects are more abundant than the points of light that we somewhat presumptively label as stars. The cataloged classical asteroids number in the hundreds of thousands and a recent estimate suggests nearly 2 million larger than 1 kilometer in diameter. Trans-Neptunian Objects are estimated to be a hundred times as plentiful as the classical asteroids, possibly amounting to 200 million objects.

The nearby stars, with parallaxes listed by the Yale University Observatory, number 8 thousand; the *Hipparcos* mission produced over 2 million parallactic stars. These points of light that we label as stars are outnumbered by asteroidal fragments by a factor of 100. We see the stars through a screen of asteroids.

Asteroids (in general) are fainter than stars that might be chosen as targets for parallax measurements. A brighter star is generally closer and would be more likely to show a measureable parallax. Similarly, a dimmer star (possibly an asteroid) would be considered as a distant star, suitable as a reference star in a parallax measurement. In the early days (nights!) of parallax measurements, the field of view of the high-magnification telescopes used was so small that often there would be very few choices of reference stars. If the target star is actually more distant than the reference points (asteroids perhaps), a negative parallax will result. If the target star is actually a bright asteroid (perhaps in the Kuiper Belt) orbiting along with slightly more distant, dimmer asteroids, the relative motion may be too small to raise concern that the object has significant proper motion from its orbit about the Sun.

In a determination of the distance to the archetype Cepheid variable star, delta Cephei (δ Cep), five stars were close enough in the *Hubble Space Telescope* field to be used as reference stars [Benedict *et al.* 2002]. While one reference star was nearly the same brightness as delta Cephei, four of the stars are significantly dimmer. The measured parallaxes were converted from relative measurements to absolute by calculating the absolute magnitudes of the reference stars, a "spectrophotometric parallax", and thereby their distances, and removing the individual parallaxes of the reference stars. The brightest reference star did not fit into this analysis and showed scattered positions, some of which fit on a plotted elliptical orbit, with a period of 1.07 ± 0.02 years (in the abstract), or 1.06 (on page 1700). This was explained as the orbit of the larger star in a binary pair, but seems close to the parallactic ellipse that would be traced by an asteroidal fragment close to the Earth. That possibility was not discussed by the researchers.

Trigonometric parallax gives a false measure if the reference stars and the target object are nearby asteroidal fragments. Since the number of cataloged asteroids, and Trans-Neptunian Objects, is comparable to the number of stars whose parallaxes have been measured, the coincidental association of asteroidal fragments must be so great as to make most of our distance measurements largely false. The night sky is much closer than we have thought, and the standard tool won't resolve the problem.

The remarkable set of parallaxes measured by the *Hipparcos* satellite has pushed the precision of the measurements to about 1 milliarcsecond, and therefore parallactic distances to 500 light years. The fundamental flaw of trigonometric parallax remains. It is not triangulation, and the distances depend on the assumption that the reference stars are infinitely far away. (*Hipparcos* used a derived frame of reference based on quasars, but the basic problem remains.)

For an astronomer, the night sky is vastly different from that seen by most city-dwellers. It is filled with stars, not just the connect-the-dots constellations that consist of only the brightest stars that can shine through the light pollution of modern cities. And to a photographic astronomer trying to measure parallax, the overabundance should be nearly overwhelming, with stellar images down to the 24th magnitude. The stars of familiar Orion, visible from the metropolis of Los Angeles, for instance, are 1st magnitude stars, brighter by a factor of nearly 2 billion times.

Cepheid Variable Stars

The astronomical distance scale was extended beyond the range of trigonometric parallax by the discovery of certain regularly variable stars, that became known as Cepheid variables, after the type star, delta Cephei, the fourth brightest star in the constellation Cepheus [Levy 1989], and discovered to be a periodic variable by John Goodricke in 1784. Because of our adherence to Galileo's declaration that all stars are stars, and for lack of any reasonable alternative, delta Cephei has always been considered to be a star like the Sun, but perhaps 3 times as massive as the Sun, 3,300 to 10,000 times as bright, and 1,031 light years away, with a diameter 25 to 30 times the Sun's. Indeed, doubting the description never came up. Until now, when it is very important to consider an alternative form for this and the other Cepheid variable stars [Burnham Jr. 1978].

In 1784, the fourth brightest star in the constellation of Cepheus (a constellation of the far north, the star itself is labeled delta Cephei, δ Cep, and is somewhat embedded in the river of light known as the Milky Way) was found to be a variable star, brightening and dimming regularly. John Goodricke of York, England, although deaf, dumb, and sadly dead at twenty-one, had nonetheless made major contributions to stellar astronomy. His study of the variations in the brightness of the star Algol had convinced him that Algol was the brighter star of a pair and when the dim star passed in front, an eclipse reduced the light. However, delta Cephei varied all by itself and its discovery would lead to identification of a distinct class of variable stars, pulsating variables. Several different types have been found, each with interesting characteristics. Since this was new territory, it was initially assumed that a pulsating star is a pulsating star, and the same period-luminosity relation would hold for them all. At this point, I think it is uncertain whether all variable stars are oscillating nuclear reactors or rotating asteroids, like 729 Watsonia, as shown in Figure 49.

It was found that stars of this type varied in a distinct manner, and that the time period, between successive peaks in brightness, or successive lows, was related to the luminosity or magnitude [Madore and Freedman 1992]. In remarkably patient work, the period-luminosity relation was determined for a large number of stars at a nearly uniform distance from Earth, in the Small Magellanic Cloud [Leavitt 1908, Berendzen *et al.* 1976, Shapley 1972].

Calibration of measuring tools must be traceable to an accepted, accurately known standard. Here, we will try to trace the calibration of the Cepheid variable distance scale, the most important link in our chain of methods for determining cosmological distances. Excellent reviews of far greater detail and scope than what is attempted here already exist and provided the material for this work. However, the history of the Cepheid variable distance scale is so full of mistakes and false discoveries [Fernie 1969] that the history of the histories is important.

These reviews have been directed at refining the Cepheid scale so as to reduce errors of a factor of 2 and 10% [Walker 1988]. Here we will conclude that the Cepheid scale has no legitimate link to a standard of length, and may be in error by a factor of about 100,000 or more. Consideration of a variety of observations related to distance will then be used to derive a better estimate of our possible error in scale.

The period-luminosity (or period-luminosity-color) relation must be calibrated, by independently determining the absolute distance to some of the Cepheid variables. The slope of the relation is easy to obtain directly from the observations, but the zero intercept, the fundamental brightness of the stars, is not and must be calibrated against some reference. Here, the history becomes strangely fuzzy, with different explanations for this calibration being provided by different authors. The calibration is complicated by the fact that during the course of development, three different types of Cepheids have been identified: Population I, Population II, and RR Lyrae stars, and since the RR Lyrae stars show no variation in period with magnitude, it is hard to accept them as Cepheids. Yet, in one of the calibrations, RR Lyrae stars were used as the base point for the period-luminosity curve [Shapley 1972]. Another calibration involves the identification of Cepheids in one open star cluster, the Hyades, in conjunction with the measurement of the distance to this cluster by the parallax method, at the extreme range of applicability, and by the moving-cluster method, with the assumption that the objects in the cluster are stars in a gravitationally bound grouping [Freedman 1992]. This one absolute distance point must then be extended to all other Cepheids with the trust that each has been properly identified as a star of a certain sort. While this is done by the use of main-sequence fitting using the Hertzsprung-Russell diagram, where intrinsic brightnesses of stars are related to spectral types, this only assures self-consistency.

The Hertzsprung-Russell diagram shows, in several variations, the distribution of stars in terms of intrinsic luminosity and temperature, color, or spectral class. Different associations of stars present different patterns in this diagram, and these patterns can provide guidance in understanding these objects.

The effective temperature is estimated from changes in various features of the spectrum, beyond the gross shape indication of a Maxwellian temperature. However, the nuclear reactors and their radioactive debris mix energies in a confusing manner. The less powerful objects appear to be hotter. Globs of molten uranium show true thermal radiation: infrared and visible light distributed like "black body" radiation. Cold, radioactive objects glow with the blue-ultraviolet light of Cerenkov radiation. Nuclear reactors and spent fuel pools have this blue glow. In addition to this intrinsic radiation is the reflected light from the adjacent starsun.

RR Lyrae stars are variable stars that differ from the Cepheid variables in having a period-luminosity plot that is nearly constant, that is, the luminosities are independent of the period and are all nearly the same. (Nevertheless, RR Lyrae stars have been used to provide a distance benchmark for the Cepheid distance scale.)

Rather than being stars that pulsate in some fashion other than that derived for the Cepheids, RR Lyrae variables are likely to be well moderated (rich in hydrogen) globs that oscillate in fission power, and therefore in thermal output, as a result of the neutron-absorbing effects of xenon-135. In the nuclear reactor field, this is known as "xenon poisoning". The RR Lyrae stars may be oscillating reactors of the sort known as xenon oscillators, the Cepheids may be oscillating reactors driven by temperature changes in reactivity, thermal oscillators [Weinberg and Wigner 1958, Hetrick 1971].

But can't a star be judged to be a star like our Sun simply because of the preponderance of hydrogen and helium? Probably not: hydrogen was bound to the uranium as it was released from the supernova and is also produced by the radioactive decay of neutrons released by the nuclear-fission chain-reactions, and helium is produced by the alpha decay of uranium, plutonium, and thorium. Objects rich in the nuclear fission fuels are likely to be surrounded by hydrogen and helium.

The Cepheid variable distance scale has served as the basis for all further extensions of estimates of distance and scale in our Universe. Each farther step into distant space is relative to distances determined by the Cepheid scale. This has resulted in the determination that spiral nebulae and elliptical galaxies are not only outside the Milky Way Galaxy, but are distant

Island Universes of stars, requiring supermassive unknowable black holes and unseeable Dark Matter to maintain their motions, and distant quasars with incredible sources of power. Faster-than-the-speed-of-light motion, superluminal motion, is remarkably common, but must be explained by uncommon geometry. It appears that this part of the distance scale is confusingly uncertain and most certainly is wrong.

Harlow Shapley was worried about errors of a factor of two or ten, that would affect the details of the description of the Universe [Shapley 1943]. Perhaps the worry of misidentification also bothered him, since identifying types of stars with certainty, and knowing just what they were, had plagued his research into the scale of the Galaxy. This underlying uncertainty still exists, though much subdued, but is honestly expressed by Barry Madore's complaint [Eicher 1994]. Could there be a grave mistake in the distance scale? Could the Cepheids be so screwed up that there is an error in the scale factor of a hundred-thousand-fold or more in distance, that produces errors of a ten-billion-fold in luminosity and so in energy, a quadrillion-fold error in mass? These are errors worth our concern.

Spiral Galaxies, Nebulae or Not?

The application of ever-larger telescopes to searching the sky led to the discovery and re-definition of many different objects. Comet-hunting became popular. Charles Messier, in France, was enthusiastically searching for comets. He discovered at least 15 and for a while was nearly the exclusive finder of comets [Mallas and Kreimer 1978]. To help in his search for comets, Messier began cataloging nebulae and star clusters that could be mistaken for a new comet, starting with the Crab Nebula in 1758. When completed, Messier's Catalog contained 100 objects of a variety of sorts that would come to be known as star clusters, diffuse and planetary nebulae, supernova remnants, and spiral nebulae. With his giant telescopes, William Herschel began the refinement, identification, and characterization of the fuzzy spots. He established the class of planetary nebulae for their appearance as a smooth disk, like a planet, without the many points of light called stars, that could be seen in star clusters. By using larger telescopes, William Herschel was able to catalog over 2,000 nebulosities. William's son John F. W. Herschel extended this to 5,000 in 1864, and Dreyer's New General Catalogue ("NGC") contained 13,000 in 1888.

Lord Rosse (William Parsons), in Ireland, with an even larger telescope, was able to discover the spiral forms that many of these nebulae showed [Berry 1898]. With the application of photography to telescopic astronomy, the forms of many spiral nebulae could be recorded. The nature of these delicate structures in space would become a major puzzle for the astronomers of the next hundred years. This puzzle was eventually resolved in the early part of the 20th Century.

The debate between Harlow Shapley and Heber Curtis in 1920 over the nature of the spiral nebulae had been framed in an either/or form, and depended on distance. Either the spiral nebulae were part of our great Galaxy of stars, the Milky Way Galaxy, or they were beyond it at great distances. If they were not a part of the Milky Way Galaxy, then they must be objects of the same sort, and were therefore Island Universes in their own right.

Immanuel Kant's philosophy had included a Cosmos filled with Island Universes, vast collections of stars, suns, and planets, and Galileo had declared that the Milky Way was composed of innumerable stars. Philosophical idea and astronomical reason had finally met.

With our explanation of the nature of (all) stars as similar to the Sun, the form and character of our Universe, the visible part of the Cosmos, was established. Our Universe was a vast collection of Island Universes, each a vast assembly of stars like our Sun.

Was the *Hubble Space Telescope* out of focus?

If the astronomical distance scale is wrong, as a result of the relative nature of trigonometric parallax, unable to distinguish between nearby asteroidal bodies and truly distant stars, telescopes focused by design at great distances would show out-of-focus images if viewing a nearby object. A telescope placed in space to eliminate the blur of the atmosphere would show a blur due to being focused too far away. (Like trying to take a portrait of your friend with the camera lens set at infinity. For some, that might be an improvement. For astronomers, it is a great disappointment.)

When the *Hubble Space Telescope* sent back its first images, they were disappointing, unsatisfactory, blurry, "out-of-focus", and not the sharp images that were expected. In this present view of a Universe much smaller than we have thought, the poor images resulted from the telescope actually being out of focus, trying to image objects that were a hundred thousand times closer than the telescope was set for. An intensive investigation for NASA concluded that the main mirror had been improperly shaped, as a result of the presumed misuse of an inspection tool, a "null corrector", miss-set by 1 millimeter. That assumed error was considered to cause the curve of the mirror to be too flat, producing an optical error in the images known as "spherical aberration". But instead, could the blurry images have been produced by focusing a hundred-thousand times farther away than the objects being imaged?

The Palomar 200-inch telescope (the "Hale") is a classic Cassegrain telescope, with a paraboloidal primary mirror and a hyperboloidal secondary mirror to reflect the light from the main mirror to the focal plane. Intensive optical testing, fraught with many difficulties, guided the fabrication of the Palomar main mirror to the goal of perfectly focusing light from infinitely distant points of light [Florence 1994]. During its "first light" testing, it produced disappointing, unsatisfactory, blurry, images. It was determined that, for the objects that were being imaged, the outside edge was too high, for objects at "infinity". No physical measurements could be made, but a high outer edge was the conclusion of the telescope testers, and the mirror was re-shaped to produce excellent images of whatever it was pointed at. (This work was poorly documented, for fear of losing funding.)

The *Hubble Space Telescope* is a Ritchey-Chrétien variation of the Cassegrain telescope, with a hyperboloidal primary mirror and a hyperboloidal secondary mirror to reflect the light from the main mirror to the focal plane. Less comprehensive testing of the *Hubble Space Telescope* main mirror, later judged by an investigation panel to have been done incorrectly, on the possibility that an inspection fixture was used improperly, was directed toward the same goal as for Palomar, focusing perfectly light from points at infinity. During its "first light" testing in orbit, it produced disappointing, unsatisfactory, blurry, images. It was determined that, for the objects being imaged, the outside edge was too low, for objects at "infinity". No physical measurements could be made, but a low outer edge was the conclusion of the telescope testers, and the mirror was later fitted with corrective lenses to produce excellent images of whatever it was pointed at. (There may have been some test indications that might have shown spherical aberration, but these were not documented nor acted upon [Anderson 1990, Allen 1990].)

Could the difference in the shapes of the primary mirrors, paraboloidal and hyperboloidal, have resulted in the two apparent errors, too high and too low, producing the same effect: blurry images from setting the focus at "infinity" for objects that were a hundred-thousand or so times closer than planned? That possibility would permit the craftsmanship of the mirror grinders to have been as close to perfect as we are allowed, but would have produced the optical failures when used to image at a great distance objects that are too near.

A similar fault was found in the mirror for the New Technology Telescope at La Silla in Chile. Eric Chaisson wrote:

"(Incidentally, the main mirror at the New Technology Telescope at La Scilla was found at its first light to have a large error in its shape, similar in form – and probably with similar cause – to *Hubble's* error, but the NTT's support actuators are sufficiently robust to strongly warp its main mirror into the desired shape, thus removing the error.)" [Chaisson 1994].

The adaptive optics actuators were strong enough to bend the mirror into focus. This supposed error was also presumed to have been caused by the misuse of a null corrector during the grinding of the NTT main mirror [Broad 1990, Wilson 1990]. Such a coincidence, the same error by the same mistake.

The Canada-France-Hawaii Telescope also suffered from "spherical aberration", eliminated in that case by bending the secondary mirror [Wilson 1990]. The MegaPrime camera, installed in 2003, gave images that were sharp in the center but blurry around the edges. The camera was "fixed" by accidently reversing one of its four correcting lenses. The cause for the blur, and the reason for the success of the fix, were never determined [Wikipedia 2009].

An excellent small telescope also suffered from the out-of-focus problem. The spacecraft *Deep Impact*, sent to drop an impactor onto Comet Tempel I in 2005, had a High Resolution Instrument, a telescope of the same sort but smaller than the *Hubble Space Telescope*, 0.3 meters diameter instead of 2.4 meters. On viewing stars while in flight, this telescope was admittedly found to be out-of-focus, for the distant stars [Dalton 2005]. The High Resolution Instrument telescope for *Deep Impact* sent back star images that were blurred. Mission scientist Lucy McFadden (University of Maryland) said "out-of-focus". The star images had point spreads of 9 pixels rather than the design of 2.5 pixels. McFadden concluded that "the flaw originated during the telescope's construction". The telescope was built by Ball Aerospace, who refused requests [by *Nature*] for an interview. The problem was overcome by computerized deconvolution. This difficulty is harder to explain, partly from lack of expert information. McFadden stated [McFadden 2005] that as a fixed focus telescope, the design "was such that its depth of field ranges from the closest approach distance of 500 km to infinity." It is hard to see how an error in distance can affect the images in that range. It may be that the focus comparison is made for the best images of point sources, and that showed poorer resolution than expected

So many telescopes with bad images:

1947	Hale	5.0 meter
1975	BTA-6 (Russia)	6.0 meters
1979	Canada-France-Hawaii	3.6 meters
1989	New Technology Telescope	3.6 meters
1990	*Hubble Space Telescope*	2.4 meters
2003	MegaPrime camera	
2005	*Deep Impact* HRI	0.3 meters

Are these remarkable coincidences that seven important astronomical telescopes showed similar problems with imaging objects that were assumed to be at great astronomical distances, hundreds to millions of light years away, truly accidental coincidences?

It is generally considered that objects at light year distances might as well be at infinity as far as optical imaging is concerned. However, even at great distances, misfocusing can produce blurry images, if the criterion of perfection is sufficiently extreme. Astronomers and telescope makers are diligent about seeking and expecting perfection. Using photos from the Subaru Suprime observations, Masanori Iye was able to identify meteor trails and artificial space objects by their different image widths [Iye *et al.* 2007]. The effect of imaging an object at a near distance *d* with a telescope focused at infinity can be calculated by:

$$w = A \sqrt{\left[f^2 \left(D^2 / d^2 \right) + \left(\delta^2 / F^2 \right) + \left(f^2 s^2 \right) \right]}$$

where D = object diameter,
    d = distance to object,
    f = focal length,
    s = typical seeing (radian),
    w = image size,
    F = focal ratio,
    δ = defocus = $f^2 / (d-f)$
A was omitted from the Iye *et al.* paper, but is needed to convert the FWHM in meters to arcseconds.

Calculations for *Hubble Space Telescope* with this equation showed that for distances less than a few hundred Astronomical Units away, the image begins to spread beyond what would be expected for a distant point. A measure of this spread is the Full Width at Half Maximum, the width (diameter) of the image at half of the maximum brightness, called FWHM. The results from the equation are shown in Figure 102. The expected FWHM for the star HD124063 was 0.052 arcseconds. The observed FWHM was actually 0.078 arcseconds. For an asteroidal object with a diameter of 500 kilometers (the size of the larger Kuiper Belt Objects) at a distance of 61.6 Astronomical Units (somewhat farther than the main Kuiper Belt), about the distance predicted by Brady from comet perturbations and in this book by the apparent motion of the Milky Way Center, the calculated FWHM is 0.078 arcseconds, in agreement with the observation. Spherical aberration is not needed to make the image blurry; focusing too far has the same effect, and is likely to be indistinguishable from the effects of spherical aberration.

That was what it appeared to be when the first images came back from space. The images were disappointing, blurry, out-of-focus, "aberrated".

A test of the focus in orbit was done by ground control. Focused farther away, the image was worse. I think that no attempt was made to focus closer because of concern that the focus mechanism might become stuck. A computer simulation showed that the blurred image could be produced by an error in the shape of the mirror, called spherical aberration. A way to cause this error by misuse of the inspection jig was found, and correction lenses were designed, made, and installed. The problem was solved. Logical consistency proved that the craftsmen had failed.

Remarkably, more distant field stars were in sharper focus in images taken before COSTAR was installed than they were after, as if the telescope had been more nearly focused for the distant stars at first.

However, if the objects (the test star and the distant galaxies) really are closer than the telescope had been designed for, the images would have been blurred, out-of-focus, and aberrated. The test star looks like an out-of-focus image. The background stars, farther away, are closer to proper focus, properly pinpoint specks of light. After the repair, the test star came in beautifully focused, but the more distant stars were blurred dots. The COSTAR lenses had moved the focus closer to the mirror, closer to the *Hubble Space Telescope*, closer to Earth. Like reading glasses, the corrective lenses have allowed us to focus on near objects, as those points of light that we have called stars actually are.

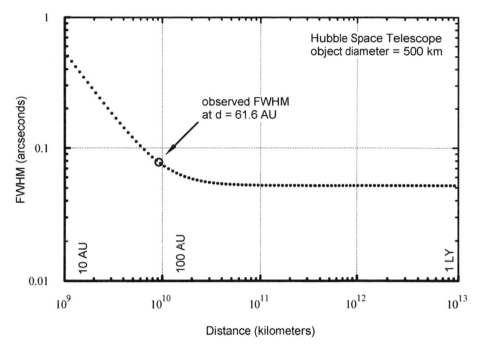

Figure 102. Defocusing effect for *Hubble Space Telescope* with an asteroidal "star".
The calculated spread of a point image in the *Hubble Space Telescope*, as a function of distance to the observed object. At distances less than 100 Astronomical Units, the image begins to spread noticeably. At a distance of 61.6 Astronomical Units from an asteroidal object 500 kilometers in diameter, the calculation matches the observed width of the *Hubble Space Telescope* stellar image.

Notice that the distant stars around the original (faulty) image of the *Hubble Space Telescope* test star are in better focus before COSTAR than after. *Hubble Space Telescope* was focused for the more distant stars originally, but the distance expected for the test star was so much greater than it actually is that the images were out of focus, and the telescope was judged to be faulty. The test images showed that the curve of the *Hubble Space Telescope* mirror was slightly flatter than it should have been to produce a sharply focused image of the test star. That is, it acted like it should be focused for a greater distance than the test star.

Both the *Spitzer Space Telescope* (in infrared) and the *Galex* telescope (in ultraviolet) show suggestions of the near-far out-of-focus effect. *Spitzer*, in imaging the object SGR 1900+14, shows blurry "foreground stars". A sharp image of the spiral galaxy M81 taken by *Galex* in ultraviolet light shows "foreground stars" with much greater spreads than the points of light in the Galaxy.

Was the *Hubble Space Telescope* truly out of focus, and the others as well?

Galaxies or Nebulae, How Far, How Near?

Adoption of the "Island Universe" model of our Universe was guided by two very strong personalities: Edwin Hubble, who was described as "Olympian" by his wife, and Harlow Shapley, who had "a complete absence of what usually passes for humility." [Whitney 1971]. (Shapley actually opposed the Island Universe concept for the spiral nebulae, but his research on the Milky Way was fundamental in forming our view.) And this was at a time when so much of what we now know was not known, not even suspected. American astronomy was growing rapidly, and nuclear fission was still unknown; induced (artificial) radioactivity was unknown; the neutron had not yet been discovered; antimatter had not yet been theoretically invented. Even the details of energy production in the Sun had not yet been worked out [Bethe 1939]. But if our description of the Universe is limited to the use of just three major sources of energy; nuclear fusion, gravitation, and electromagnetism, then all the bright points of light in the sky must be stars like our Sun, and the spiral nebulae must be Island Universes of stars at great distances. In a cycle of circular reasoning, this has led to the firm conviction that Cepheid variables are pulsating giant stars at immeasurable distances of hundreds of parsecs from the Earth, and millions of parsecs away in the Island Universes of stars. By accidents of ignorance, our observable Universe became billions of light years across.

This decision (that the spiral nebulae were distant Island Universes of stars like our Sun) was pre-determined by the form of the question, and by Galileo's declaration, 400 years ago, [Galilei 1610] that the Milky Way was made of stars. Those objects, Galileo's point-like, self-illuminated stars, became established in our astronomy as being of the same form as our Sun, some larger, some smaller. The question in conflict was "Are the spiral nebulae part of the Milky Way (Galaxy), or separate, more distant?" If separate, they must be greatly more distant. The spiral nebulae, and their associated elliptical galaxies, and the yet-to-be-discovered quasars, are indeed separate from the Milky Way Galaxy, and more distant, but the distance scale, and our understanding of the character of Galileo's stars, and thus all the stars we see, are wrong. These misconceptions were natural accidents of our state of knowledge at that time, but became so firmly established, in part by the strength of personality of the leading astronomers, that expansion of our knowledge could not overcome these errors.

It was not until 1932 that induced artificial radioactivity was discovered. This was radioactivity of the same sort as the natural radioactivity discovered and studied by Becquerel and the Curies at the turn of the century, except that it was created by nuclear absorption of neutrons. (This nuclear particle, like a proton but with no electrical charge, was discovered in 1936 by James Chadwick.) Antimatter, in the form we call the positron, was theoretically invented by Paul Dirac in 1928. But even more important for our understanding of our Universe, nuclear fission was not discovered until 1938. This process provides the source and form of energy, the radioactivity and the neutrons needed to create more radioactivity. But the Shapley-Hubble Universe had become a paralytic paradigm, preventing any re-consideration of our misunderstanding.

Galaxies, as observed, are classified as elliptical or spiral, or irregular, and "peculiar". Recent additions to this scheme are low-surface-brightness galaxies and faint blue galaxies. Often, the classification must be based on a faint diffuse image, much more poorly defined than the images of "nebulae" that were eventually decided to be galaxies, Island Universes, only ninety years ago. Often, individual stars cannot be resolved in these images, and the identification as a galaxy, a vast collection of stars, dust, and gas, is based on the diffuse nature of the image, its resemblance to known and well established galaxies, and possibly on a derived redshift or color index. The Universe is filled with galaxies.

Low-surface-brightness galaxies appear to have far fewer stars than normal galaxies, and yet are far larger [Powell 1992]. They appear to be very blue, which is taken to imply that a large fraction of the stars are young.

Faint blue galaxies are identified from the differences in brightness in the B and I color bands, indicating a blue overall color, and are found to have intermediate redshifts, about 0.4 [McGaugh 1994]. Surveys have shown an excess in the number of these galaxies that has been difficult to explain. The excess may be an artifact of the surveys, and the two similar types of galaxies, low-surface-brightness and faint blue, may be the same. Abundant blue light is produced by Cerenkov radiation.

The peculiar galaxies are those that are identified as galaxies primarily on the basis of the diffuse image, a general resemblance to standard galaxies or to what might be imagined to happen to such galaxies, and a redshift that indicates its placement at an extra-Galactic distance. If it does not fit any of the standard galaxy types, it is labeled "peculiar", but a galaxy nonetheless.

A further suggestion of an error in the distance scale is seen in the observations by *EUVE*, the *Extreme UltraViolet Explorer* satellite. Extreme ultraviolet astronomy (in space, since light of these wavelengths is completely absorbed by our atmosphere) was approached with considerable doubt as to its effectiveness, because absorption by hydrogen spread uniformly through space would limit observations to a relatively short range. Based on observations of the average density of interstellar gas, it was thought that not even the nearest stars could be seen in ultraviolet light. However, the EUVE observations show that, at least in certain directions, the opacity is a thousand times less than expected [Bowyer 1994]. This has been interpreted as a reduced density of gas, by a factor of a thousand below the average for our neighborhood along these lines-of-sight, but could equally well indicate a distance a thousand times less than supposed.

Other "peculiar galaxies" have been catalogued [Arp 1966, Kanipe and Webb 2006]. An example, NGC 4319, has been identified as a galaxy because of its resemblance to a two-armed spiral. It is labeled as peculiar because it looks as though "the arms are coming off at the roots" [Arp 1987]. Further, hydrogen-alpha emission, normally found in spiral galaxies, is nearly absent; only ionized nitrogen emission was observed. In a controversy among observers, a higher redshift (21,000 kilometers per second) object (Markarian 205) has been associated, or not, with NGC 4319, which has a much lower redshift of 1,700 kilometers per second. Gerald Cecil thinks (vehemently) the connection is an artifact.

Arp has argued that many, most, or all quasars are associated with galaxies and many, most, or all of those galaxies are peculiar [Arp 1987]. This argument places quasars at extra-Galactic, but not cosmological, distances. However, as described here, it appears likely that quasars are self-propelled planetary fragments; the associated "peculiar galaxies" are instead the debris of disruptive planetary detonations; the low-surface-brightness galaxies are debris fields illuminated from behind by their starsun.

Detonated Planets and Starsuns

If the distance scale is wrong and the nearby stars are more like asteroids than like our Sun, and the beautiful spiral galaxies are actually the spun-out debris of detonated planets, can we find their suns, their starsuns, in the many astronomical photos of galaxies? Often, images of galaxies are presented as portraits, showing just the galaxy, so that the beautiful detail may be seen best. We need family portraits to see the rest of the system.

The STScI Digital Sky Survey, operated by the Space Telescope Science Institute, has organized the photographic plates of the Palomar Observatory Sky Survey and the UK Schmidt Telescope. The images have been digitized and made available on the Internet at http://archive.stsci.edu/cgi-bin/dss_form. Named objects can be selected and viewed, in their broader environment. Images up to 60x60 arcminutes can be downloaded and saved. This broader view is important because most images of galaxies are presented as portraits that show the details of the galaxy, rather than revealing its neighbors.

In case after case, clearly photographed galaxies are found to have a "foreground star" as a nearby neighbor. With various objects scattered nearly randomly across the sky, it might be expected that some unrelated associations between foreground stars and distant galaxies (each composed of billions of stars, if the Island Universe theory were correct) might be found. In the Digitized Sky Survey, the foreground stars can be clearly recognized by the overexposed stellar disk with sharp diffraction spikes. The overexposure results from the great difference in brightness between the star and the galaxy. The diffraction spikes result when bright light from a point source is diffracted by the supports of the secondary mirror in the photographic telescope. If light from the point source is spread out, the diffraction pattern will be diluted and will lose some of its definiteness. (For readers who wish to explore on their own, object names are shown with the images I have selected for this book. Once these pairings of detonated planet and starsun are recognized, others can be found. Have fun exploring!)

The images have been selected to show the starsun/detonated planet pairs most clearly.

There are some galaxies where a starsun cannot be identified. Then it may be found that an elliptical galaxy stands in its place. Two interesting examples of this are M51 and M31. (For M31, those are probably detonated satellites.) In the pairing of M51 and NGC5195, debris from the detonated planet (M51) can clearly be seen to envelope and enshroud its starsun (NGC5195). There, the debris glows with a combination of light sources: reflected light from the starsun, thermal energy from impacts and the heat of the starsun and any significant radioactivity, and Cerenkov radiation produced by high-energy electrons from the radioactive decay.

M31, the Great Nebula of Andromeda, shows two elliptical galaxies as probable candidates for detonated satellites. There is the elongated elliptical M110 and the more spheroidal elliptical M32. A clue to identities is provided by the discovery of 8 globular clusters around M110 indicating a local nuclear detonation. M32 has no globular clusters and so may be suspected to be the starsun, enshrouded in incandescent debris, but is more likely to be a detonated satellite. The most likely candidate for the starsun is ν And (nu Andromedae), a nearby star. M110 must be suspected as a detonated satellite of the planet that became M31. Other associations similar to these can be found by exploring.

It is useful to consider some of the characteristics of planetary detonations. When a rotating planet detonates, it forms a spiral galaxy. When a rotating planet with satellites detonates, it forms a spiral galaxy with satellite galaxies, like the Magellanic Clouds in the case of the Milky Way Galaxy. The detonation of the planet induces the detonations of the satellites. These have no intrinsic rotation, only orbital rotation, and so do not develop the spiral form. They are classified as irregular galaxies.

Among the fragmentary debris, icy, rocky, or iron, the asteroidal chunks, there may be viable nuclear reactors remaining, that glow by thermal or Cerenkov radiation, or shine with reflected starsunlight. These reactors may retain the potential for detonation in the future. Such a detonation, in what is considered to be a galaxy, would appear to be a supernova. The detonation would produce a bright fireball of fractured and vaporized rock and ice, the stuff the asteroidal fragments are made of.

Interstellar dust has been found to have fosterite (Mg_2SiO_4) as a major constituent [Agladze et al. 1994]. This is the pure magnesian form of the mineral olivine, the main mineral of the Earth's mantle. While high iron concentrations have been found near the center of the Milky Way, and explained on an abundance of supernovae there, the opposite direction shows a relatively large amount of carbon and nitrogen [Wickramasinghe and Okuda 1994]. This would be consistent with our view towards the "Galactic Center" actually being a view towards the remainder of a planetary iron core, dissipated by a detonation, and looking in the other direction at the inside of streamers of debris from the crust or an atmosphere of methane and ammonia blasted past the Sun by the detonation.

Stars are predominantly made of hydrogen and helium with a small amount of heavier elements, called "metals" by astronomers. Deep in a star, in its core, the temperature and density are great enough to let nuclear fusion reactions proceed, powering the star. Those reactions convert hydrogen and helium into "metals", changing the composition of the core, but leaving the outer layers relatively unchanged, as hydrogen and helium. During the life of the star, some of the outer layers are expelled. If a star is massive enough, it begins to run out of fuel in the core and eventually collapses, to detonate as one or another type of supernova.

An extremely rare supernova, of type Ic, was observed in April 2007 [Gal-Yam et al. 2009, Langer 2009]. Designated SN 2007bi, a peculiar type Ic supernova, its spectrum was carefully measured. The spectra showed "no trace of either hydrogen or helium", but showed strong indications of "carbon, oxygen, sodium, magnesium, calcium and iron". In other words, the spectra were what would be expected from a fireball of vaporized rock.

Associations of Stars and Spiral Galaxies...

As described above, the astronomical objects we call galaxies fall into three basic forms: spirals, ellipticals, and irregulars. (There are multiple finer subdivisions, but these seem to be the fundamental types.) The various forms appear to be associated throughout the observable Universe, mingling together through space and time. There seems to be no preferred location in space, or redshift (presumed age), all three types are scattered randomly about.

Galaxies appear to be associated with stars, or with a debris-enshrouded star we interpret as an elliptical galaxy. Detonated satellites appear to be elliptical galaxies also, but may be identified by a small halo of globular clusters, the remains of detonated uranium globs still present around stars and planets. The detonation of a satellite or planet induces detonation of the globs out to a limited distance, accounting for the roughly spherical halo of globular clusters seen around our Galaxy and others.

Most galaxies are lit up by the star the planet had orbited. Some are illuminated from our side (front-lit) and appear to be bright and generally full of solar-type stars, others from the side, others from behind (back-lit). Those backlit galaxies are the otherwise unexplained "low-surface-brightness" galaxies. However, there are other sources of light inherent in the debris itself. Some fragments may contain still (or newly) operating reactors and reach temperatures of 3,000-4,000 K (maybe more) and appear to be red-giant stars, some may still retain the heat from the interior of the planet (and from the energy of the detonation) if the detonation were recent, others shine with Cerenkov radiation, the blue glow from beta particles passing through a transparent medium. The theoretical spectrum of Cerenkov radiation matches that of B-type stars, without the absorption lines. These would look like blue-giant stars. I've tried to get people interested in measuring the actual spectrum of Cerenkov radiation, at least in water to start, but haven't succeeded. Free neutrons (emitted in fission) decay in space (where they can't be absorbed by nuclei) into protons (with a few hundred electron volts of energy) and electrons (with a few hundreds of thousands of electron volts of energy). That produces clouds of hot gas, HII and HI, some of which seem to have two different temperatures. Gas is ionized and fluoresces due to radiation from the debris.

The associated star may have a redshift quite different from the detonated planet/spiral galaxy, and I haven't been able to find this. Not very many stars have had redshifts measured, and I think 21-centimeter radio, often used for galaxies, would fail for stars. This becomes confused because the redshift that is measured is, strictly, from gas expelled from the star, or galaxy, not from the objects themselves. My "associations" of stars with spiral galaxies (as detonated planets) is visual and qualitative. Since my model says there should be a star near a spiral, when I find a spiral with a nearby star, I think I have an association. I have not been able to find those stars in an ordinary catalog. When I find a star embedded in a spiral, making a kink in an arm (NGC 2997) or a star pulling material away from a spiral (Arp 188), I am pretty sure there is an association. But I have not been able to identify those stars, they are not listed in the catalog I use.

Redshift comparisons aren't always so reassuring anyway. M51 and its companion (a debris-enshrouded starsun in my model) differ considerably in velocity (467/552 kilometers per second) but are assuredly associated. Stefan's quintet has incompatible redshifts. At low velocities, the optical shift of CaII H and K lines is used. Faster away, astronomers "recognize" hydrogen lines that have been redshifted into the optical region. For large redshifts, perhaps greater than about z=0.2, the frequency shift of the 21-centimeter HI radio line can be used. Where those methods overlap, there seems to be general agreement. I worry a little about this because the HI structure of a spiral is huge compared to the visible structure. (This is also shown by planetary nebulae, which often have a halo of HI that is 10-20 times the size of the visible object.) In my model, these extended structures come from the decay of high-energy (high speed) neutrons released by fission, decaying into protons and electrons, which then recombine to form HI.

I use the Digitized Sky Survey to look at spirals that are near enough to us that they show reasonable structure, and I usually can find a star that I can say is associated with the spiral. Often that is a very judgmental call; sometimes it is clear that the star is perturbing the spiral. Is that just a chance coincidence? Perhaps, but it seems too often for that to be anything but an over-used excuse. The nearest few thousand "stars" are asteroidal fragments and have no associated galaxies. Are binary asteroids some of the extrasolar planets we find? I don't quite know.

To truly understand these objects we must try to resolve the basic astronomical ambiguity, near/small/dim or far/large/bright. This is difficult in spite of possible differences of factors of millions or trillions. This is the fundamental ambiguity that fueled the differences between Harlow Shapley's Universe and that of Heber Curtis.

For modern astronomy, the Universe was cast into the far/large/bright form by a series of self-consistent theorizing and philosophizing. Early, it was decided that because stars did not show an annual parallax (with crude observations) that they must be quite distant, and possibly all at the same great distance, on the "celestial sphere". Much later, Galileo observed that the Milky Way was composed of innumerable points of light (instead of a diffuse, milky glow), stars in his terminology. Immanuel Kant expanded on that with the idea that the Universe might include many "vast Island Universes of stars", like the Milky Way. For Shapley, the Milky Way was the whole thing, the Milky Way was the Universe, and everything we see was part of the Milky Way. Curtis (and many other astronomers in the early

20th Century) considered that the spiral nebulae were outside the Milky Way, but of the same form, and were the vast Island Universes of Immanuel Kant.

The conflict between these two views can actually be pegged to two opposing sets of observations. Adriaan van Maanen's measurements of rotation in spiral galaxies showed them to be near, or else the velocities were too great to be reasonable. Hubble's measurements of the magnitudes of Cepheid variable stars in spiral nebulae, compared to similar Cepheids in the Milky Way, showed that the spiral nebulae were much farther away and could not be part of the Milky Way. In the end, the spiral nebulae became spiral galaxies, the Milky Way, our Galaxy, became one among many vast Island Universes of stars, the stars became great balls of fire like our Sun, and a tremendously long distance scale was locked into place.

The distance scale divides naturally into three parts: planetary (the Solar System and planets around other stars), stellar (individual stars), and galactic (the Universe at large). These are generally considered to be independent realms, with no associations. Galaxies are composed of stars, on an immense scale, and are unrelated to individual stars. All subsequent observations, interpretations, and theories are compelled to adhere to this scale of distance.

Instead, this book proposes that the galaxies, like the Milky Way, are the results of nuclear fission detonations that disrupted planets and satellites orbiting distant stars, to be called starsuns because of their orbiting planets. Spiral galaxies are formed by detonated rotating planets, with the spiral structure resulting from the rotational energy of the planet. The irregular galaxies result from detonation of satellites with so little rotational motion that a spiral structure doesn't appear. (The detonated satellites are seen as satellite galaxies around the spirals.) Elliptical galaxies result from scattered debris falling into close, chaotic (but Keplerian) orbits around the starsun. The starsun may be so completely enshrouded by debris that its nature as a star is hidden.

"Rotation" of a galaxy has been hard to determine. Since the work of van Maanen (and Lundmark) no direct positional measurements have been made. Doppler shifts can directly indicate the instantaneous direction of motion (of gas) in the line-of-sight, but interpreting this motion in terms of "galactic rotation" has been confused by ambiguity.

The spirals and irregulars are illuminated by a combination of reflected starsunlight, and are seen as solar-type stars; Cerenkov radiation from radioactive fragments seen as blue stars; and by self-generated heat in fragments that have still-functioning nuclear fission chain reactions, seen as red stars. These galaxies may be illuminated from the front by their starsun or from behind, barely visible as low-surface-brightness galaxies. In this view, since the angle of illumination should be random, approximately half of the spiral and irregular galaxies in the Universe should be low-surface-brightness galaxies, from being on the near side of their starsun, towards us, and hence, backlit.

If this view is correct, spiral galaxies should be associated with individual stars, the starsuns of planetary systems in which a detonation has occurred. In some cases, the starsun may be hidden by debris and appear to be an elliptical galaxy.

Most astronomical images of galaxies are done "portrait" style, showing just the amazing beauty of the galaxy itself. Wide-field images are needed to find the associated starsun or elliptical galaxy, and any accompanying spirals or irregulars.

The STScI Digitized Sky Survey (at http://archive.stsci.edu/cgi-bin/dss_form) provides access over the Internet to collections of images displayed as wide as 60x60 arcminutes. Spirals, irregulars, and ellipticals can be found with sufficient detail to form ideas as to their nature. Bright nearby stars, "foreground" stars, can be easily identified by diffraction patterns, rings or spikes, where the abundance of light has overflowed the point image. I have chosen several images that show the features that can result from the formation of a galactic system by a planetary detonation. Out of a multitude of those images available in the Sky Survey, not all show clearly the association of a spiral galaxy with a starsun or an elliptical galaxy. In some cases, there are too many foreground stars, those with strong diffraction spikes, to make a reasonable choice. In some, the field is so empty there are no candidates. Many show clear association of a spiral galaxy and a star. Some associations show strong gravitational effects due to a clear foreground star.

A search of the STScI Digitized Sky Survey shows galaxy after galaxy in the neighborhood of a foreground star. In some of those images, it is clear that the foreground star (distinctly identifiable by the over-exposed image with diffraction spikes) is distorting the material of the spiral galaxy. In other cases, the debris of the detonated planet has enshrouded the starsun and converted it into an elliptical galaxy. Sometimes, the starsun itself can be distinguished as a sharp nucleus.

These are presented below with little comment or analysis. The reader is invited to use the Internet resources to independently discover similar associations. Unless otherwise noted, the images are from the STScI Digitized Sky Survey. (Full source information is provided in the header of the FITS file, a specialized format for astronomers.)

The Great Nebula in Andromeda (M31 or NGC 224) has been important in our (European) astronomy as the only galaxy visible to the naked eye (other than the Milky Way) in the northern hemisphere. It is shown in an image from early in the 20th Century, from the Harvard College Observatory, with two bright satellite galaxies and its proposed starsun, the star ν And, in the constellation of Andromeda. The image of NGC 151 clearly shows the starsun drawing off an arm from the spiral. This is similarly shown in NGC 613. In time, the starsuns will become enshrouded by debris and future astronomers will label them as elliptical galaxies. M51, the Whirlpool Galaxy, shows this as it is happening, with debris streaming from the spiral towards and across its companion. NGC 3184 shows an ideal spiral near its pristine starsun. (Galaxies have been cataloged, and the labels given to them reflect the catalog. M stands for Charles Messier's list of fuzzy blurs to ignore when hunting for comets. NGC stands for the New General Catalogue.) Malin 1, a low-surface-brightness galaxy, required much photographic skill to image, and stands between the starsun and us. Backlit, it is almost invisible. NGC 2997 shows how the starsun has gravitationally distorted ("broken") an arm of an otherwise ideal spiral. NGC 1097 shows a clear satellite galaxy, formed by the detonation of a planetary satellite with little rotation, and the nearby starsun. The Tadpole Galaxy, from the *Hubble Space Telescope* [at http://hubblesite.org/newscenter/archive/releases/2002/11/image/a/], shows a directed stream of material flowing from the distorted spiral towards what is seen to be a nearby starsun in an image from the STScI Digitized Sky Survey. The *Hubble Space Telescope* image is reoriented to show the streaming flow in alignment with that in the Digitized Sky Survey image.

This association of spirals and ellipticals has been confirmed by Jack Sulentic [Sulentic 1992]. Away from the crowd of clusters, individual spirals do associate with individual ellipticals. This is a natural result of a spiral being formed by the detonation of a planet orbiting a starsun, and its detonation debris enshrouding the starsun, making it appear to be what we call an elliptical galaxy. When the debris has not reached the starsun, the appearance is of a distant spiral galaxy accidentally near a foreground star. In the view of this book, they are both at the same distance, and much nearer than the cosmological distance scale provides.

I have not attempted any statistical analysis to show what probabilities there are that such associations would occur by chance. The images speak for themselves. If they are not convincing by themselves, who would believe statistics?

Examples of Associated Spiral Galaxies and Stars or Elliptical Galaxies

The Andromeda Galaxy and companions, with ν And. (from Harvard College Observatory)

NGC 1531 with elliptical galaxy

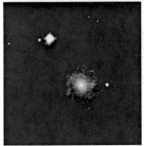

Malin 1, a low surface brightness galaxy

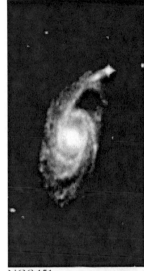

NGC 151

NGC 1566

NGC 1068

NGC 613

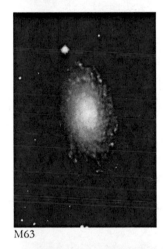

M63

M51, the Whirlpool, and companion, an enshrouded starsun

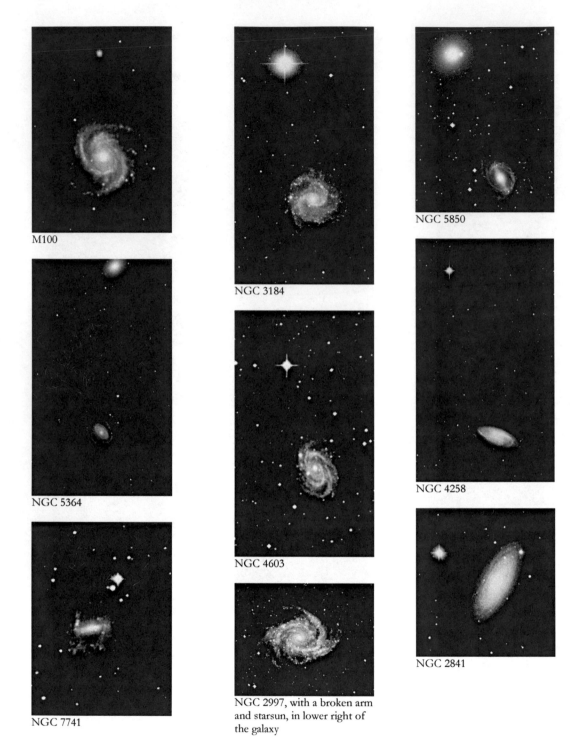

M100

NGC 5364

NGC 7741

NGC 3184

NGC 4603

NGC 2997, with a broken arm and starsun, in lower right of the galaxy

NGC 5850

NGC 4258

NGC 2841

NGC 1365

NGC 1073

NGC 1325, showing several choices

NGC 1097, showing satellite and starsun

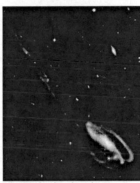

The Tadpole, showing a stream of material directed towards a foreground star, not shown. (*HST* image)

Tadpole and Starsun

The Tadpole, reoriented.

... and the Hubble Deep Field

One of the great achievements of the *Hubble Space Telescope* team was the imaging of the Hubble Deep Field [http://hubblesite.org/newscenter/archive/releases/1996/01/text/]. This image (Figure 103), captured by repeatedly targeting a small selected spot in the northern sky, accumulated light from over a thousand small (and presumably many very distant) galaxies, of all sorts. Very few stars are visible; only one possible starsun is seen in association with a spiral galaxy, the strongly diffracted foreground star. (Still, there is one, in close association of a large, and presumably not so very distant, spiral galaxy.) But if the planetary detonation model is correct, how can this absence of stars come to be?

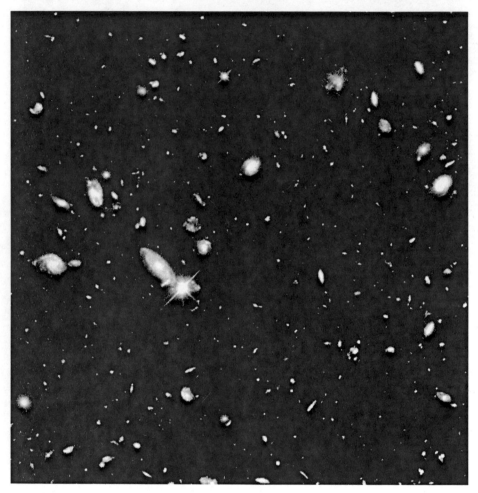

Figure 103. *Hubble Space Telescope* Deep Field, showing a multitude of galaxies.
The image of a multitude of distant galaxies taken by the *Hubble Space Telescope* in 1995, the Hubble Deep Field. (Original in color.)

This is a remarkable quirk of optics. Light from a distant object is recorded by a discrete sensor, in this case an individual pixel on a CCD (charge-coupled device). In other cases the sensor may be a grain of silver halide in a photographic emulsion, or a rod in the human eye. The quirk works the same, whatever the sensor form. Bright stars that are near show diffraction images that are spread over several sensors (pixels) and give an appearance that implies they are great balls of fire. Bright stars that are dimmed by distance produce smaller images, at greater distances covering only a single pixel. For stars of the same brightness at still greater distances, the light on the pixel and the signal out, decrease as the inverse of the square of the distance, the "inverse-square law". At sufficiently great distance, the star becomes too faint, the signal becomes too weak to be recorded, and that distant star is no longer visible. A star, a point source unresolved, can disappear.

The situation is significantly different for resolved sources, objects sufficiently extended that the true image is spread over several sensors. At greater distances, the light received still decreases as the inverse of the square of the distance, but so does the area over which the light is spread. The light per pixel remains the same, independently of distance, until the extended object is so distant that its image covers only a single pixel. Beyond that distance, the image of an extended (resolved) object will be dim and no longer resolved, just as for a point source, and it will also become undetectable in the same way.

For exposures such as the Hubble Deep Field, at the greatest distances, stars will disappear, galaxies will remain.

Most spiral galaxies show surface brightnesses that fall into the "Freeman Law" [Freeman 1970.]. While the discovery of "low-surface-brightness" galaxies, almost invisible to our telescopes [McGaugh, Bothun, and Schombert 1995, Bothun 1997, Bothun, Impey, and McGaugh 1997], has cast doubt on this as a universal law, it is likely that these two types of galaxies, the "normal" and the "low-surface-brightness" are simply two different expressions of the spiral galaxy form. This law shows a distribution about an average surface brightness of 21.65 magnitudes per square arcsecond. That value is shown on Figure 106 (relating to the Solar System) as the bar running from about 85 Astronomical Units to 102 Astronomical Units. If the average spiral galaxy (of those easily seen) were a reflection nebula at that distance from the average star, like our Sun, it would have a surface brightness of about 21.65 magnitudes per square arcsecond.

In a review of the data of McGaugh *et al.* (1995) on low surface brightness galaxies, Bothun *et al.* (1997) concluded that "up to 50% of the general population of galaxies resides in a continuous tail extending towards low [central surface brightness]." (Figure 104)

Low surface brightness galaxies were found to be otherwise indistinguishable from normal galaxies in terms of mass and size. That is, any spiral galaxy (as the debris of a detonated planet) may be fully illuminated by its starsun, or illuminated at increasingly steep angles until the orientation is such that the starsun is on the far side of the galaxy and only scattered starlight illuminates the material of the galaxy. Low surface brightness galaxies are somewhat bluer than is expected, an indication of scattered light. The galaxy itself may be detectable only as a point source, its self-illuminated nucleus, and revealed only by extreme efforts, by techniques such as Malin's method of photographic amplification.

At a distance of 100 Astronomical Units from the Sun, an object would have a surface brightness of only about 22 magnitudes per square arcsecond, barely brighter than the dark sky, and noticeable only if it were of a greatly extended size. This is just the surface brightness of the Milky Way.

The low surface brightness galaxies can be explained by considering what might happen if a spiral nebula were not well illuminated by its starsun, from our viewpoint. That is, what appearance would we see if the starsun were on the far side of the nebula? We would see the class of galaxies now labeled as low-surface-brightness galaxies. In the image of Malin 1, one of the best examples of these faint galaxies, the light of the starsun is just slightly diffused as it passes through some of the matter of the nebula.

746   BOTHUN ET AL.

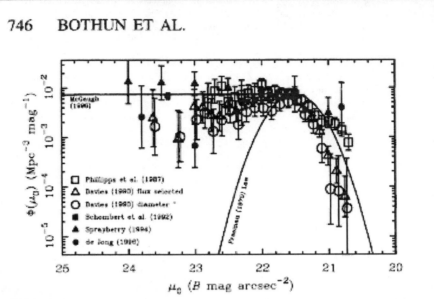

FIG. 1—The space density of galaxies as a function of central surface brightness. LSB objects appear to the left in this diagram. Raw counts from the indicated surveys have been converted to space density through the use of volumetric corrections discussed here and in more detail in McGaugh et al. (1995). The solid line shows the surface-brightness distribution which Freeman's Law suggests. The flat line fit to the data, from McGaugh (1996), has a space density which is 6 orders of magnitude higher than predicted from Freeman's Law.

Figure 104. Low-surface-brightness galaxies and the Freeman Law.
Study by Bothun *et al.* (1997), showing that approximately half of the galaxies observed show a varying surface brightness (from varying degrees of illumination by their starsuns in front) and half show a constant low surface brightness (from scattered light of their starsuns illuminating them from the rear).

If much of the light from a galaxy is reflected light from its starsun, those galaxies with their starsun on the far side, providing a "back-lit" exposure, should be quite dim, except for the self-generated light of recent detonations.

Old detonations, over a few million years old, show as the low-surface-brightness (LSB) galaxies, such as Malin 1, Figure 105. There, the starsun is barely peeking over and through an outer arm of the spiral. The diffraction spikes are only partially disrupted by the light scattered by the debris.

Figure 105. Malin 1, a low-surface-brightness galaxy.
Malin 1, the first low surface brightness galaxy to be discovered, by use of the photographic amplification technique of David Malin. This image is from the first Palomar Observatory Sky Survey, in red light. The suspected starsun, in the upper left, shows a distortion of the diffraction spike from behind the galaxy.

A similar example is seen in the Cartwheel Galaxy, as photographed by the *Hubble Space Telescope* [Hubblesite News Release Archive 1995], Figure 106 There, again, the starsun is partially obscured, at the lower left edge of the ring.

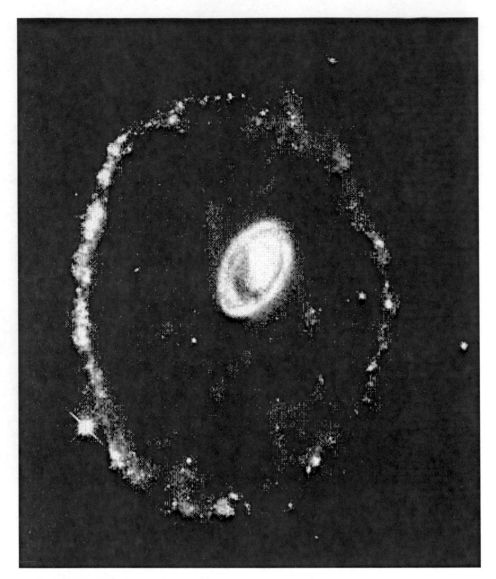

Figure 106. The Cartwheel Galaxy.
The Cartwheel Galaxy as photographed by the *Hubble Space Telescope*. The suspected starsun is seen without one of its diffraction spikes, to the lower left of the galaxy. Loss of the spike suggests that the starsun is behind the galaxy. (Original in color.)

I think we see spiral galaxies in this sort of range:

fresh & hot......................old & cold
face-lit............................back-lit

What might happen if a spiral nebula (old and cold) were not well illuminated by its starsun, from our viewpoint? That is, what appearance would we see if the starsun were on the far side of the nebula? We would see the class of galaxies now labeled as low-surface-brightness galaxies. In the image of Malin 1, one of the best examples of these faint galaxies, the light of the starsun is just slightly diffused as it passes through some of the matter of the nebula. Hot and fresh spirals that are backlit are classified as Seyfert (1 and 2) galaxies.

If a galaxy is illuminated by its starsun from the front, we see something like the full Moon illuminated by the Sun shining over our shoulder. For angles of illumination passing to the side of the galaxy, the surface brightness declines, as if we were looking at our Moon as it progresses through its phases, compounded by shadows cast by some parts of the galaxy on others. As the angle of illumination passes to the rear, from behind the galaxy, we see only scattered light, and these are the low surface brightness galaxies. With isotropic scattering of visible light, this should result in nearly constant surface brightnesses for the low surface brightness galaxies. This is exactly what is shown by Bothun *et al.* [Bothun, Impey, and McGaugh 1997]. At an extreme alignment, with the starsun directly behind the galaxy, we may see the starsun shining through the arms of the galaxy, as in Malin 1.

Spiral Nebulae

In 1610, Galileo discovered that his telescope had allowed him to see that the otherwise milky Milky Way ("Via Lactea") was actually composed of tiny points of light. This led to his pronouncement that the Milky Way consisted of innumerable stars. In Galileo's use of the term "star", he was using an observational definition: if a point of light could not be enlarged into a disc, as the known planets could, if it remained a hard, bright point of light in his view through the telescope, it was a star. Even the Galilean moons of Jupiter which he discovered were described as stars, by Galileo. In time (though I do not yet know when), we came to understand that our Sun was a star, and stars were suns, of a wide ranging variety.

Understanding that stars were suns established a conceptual scale for astronomy, so that the Milky Way became a vast Island Universe of suns.

The "Great Debate" of Shapley and Curtis grew out of the question of the nature of what had been called the "spiral nebulae", as those objects had been increasingly found and defined by ever-improving optical telescopes. At one extreme, the spiral nebulae might have been exactly what the label suggested: clouds of gas and dust, formed into a spiral structure, within our own Galaxy. As reflection nebulae, these spirals would glow by the reflected light of a nearby star. Otherwise, the spiral nebulae were actually distant, vast Island Universes of stars, shining with the light generated by billions and billions of stars, just like our Milky Way. There is an alternative to these extremes.

While there was considerable observational evidence suggesting that the spiral nebulae were indeed nearby, at the time of the debate, the question was conclusively resolved by the observation that the spiral nebulae were not in the Milky Way, and therefore had to be at great distances. Such great distances required large physical sizes, great energy output, and so they too were vast Island Universes of stars, just like our Milky Way.

However, what if the Milky Way itself were a reflection nebula, shining from the reflected light of our star, the Sun, always shining "over our shoulder"?

The outer planets, Jupiter, Saturn, Uranus, Neptune, and tiny Pluto, shine by reflected light. Because of their moderate and nearly constant distance from the Sun and the Earth, these planets always appear to have a nearly constant brightness. This appearance can be refined and expressed as the surface brightness of the planet, magnitudes per square arcsecond. This measure gives the brightness of the planet divided by the angular area of the planet, as viewed from Earth. Jupiter is bright both because it is near the Sun, relative to the other planets, and because it is large. If Pluto were at Jupiter's orbit, it would still be much fainter than Jupiter because of its much smaller size. However, assuming the same reflectivity, or albedo, the surface brightness of the two planets would be the same. Averaging the observational data for the four gas-giant planets, the surface brightness at 1 Astronomical Unit from the Sun and 1 Astronomical Unit from the Earth, shows that the surface brightness for an object illuminated by our Sun should be 1.9 magnitudes per square arcsecond.

For objects farther from the Sun and farther from the Earth, the surface brightness decreases in a smooth, mathematically defined manner. This leads to the curve in Figure 107, with the amplitude defined by the surface brightness values observed for the major planets. (Pluto is slightly brighter than it should be. This may be an effect of a higher albedo, or bias resulting from poor resolution of its angular size.)

Most spiral galaxies show surface brightnesses that fall into the "Freeman Law". This law shows a distribution about an average surface brightness of 21.65 magnitudes per square arcsecond. That value is shown on Figure 107 as the bar running from about 85 Astronomical Units to 102 Astronomical Units. If the average spiral galaxy (of those easily seen) were a reflection nebula at that distance from the average star, like our Sun, it would have a surface brightness of about 21.65 magnitudes per square arcsecond.

At a distance of 100 Astronomical Units from the Sun, an object would have a surface brightness of only about 22 magnitudes per square arcsecond, barely brighter than the dark sky, and noticeable only if it were of a greatly extended size. This is just the surface brightness of the Milky Way.

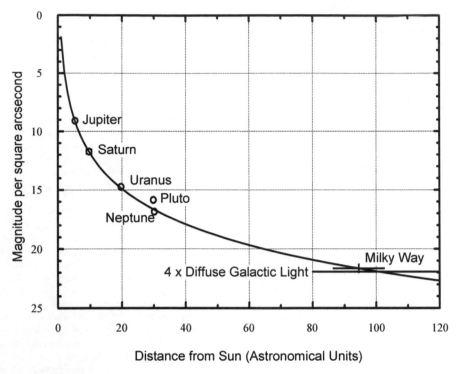

Figure 107. Reflection magnitudes of Solar System planets.
Reflection brightness of objects in the Solar System, with surface brightnesses proposed for the Milky Way.

It may be instructive to consider the likely appearance of a detonated planet. The original location of the disrupted planet will be occupied by an expanding cloud of iron vapor (including iron-60) and radioactive fission products that continues to orbit the parent star. The central core would form an expanding cloud of incandescent gas, which would be powered for several millions of years after the detonation by the radioactive decay of iron-60 and cobalt-60. Oceans would be converted into clouds of protons, streaming outward, as a result of the dissociation of the water molecules by neutron scattering. The neutrons released into space would decay into clouds of protons and electrons, eventually becoming neutral hydrogen. These decay protons and electrons have a distinctive energy spectrum (shown earlier in Figure 2 and Figure 3), mimicking extremely hot gas at temperatures of 2.77 million K for the protons, and 2.49 billion K for the electrons. The electrons quickly lose energy by X-ray emission from scattering, and come into equilibrium with the protons. The cloud will initially appear to be a HII region, composed of ionized hydrogen (and the electrons), will emit Balmer light as the protons and electrons recombine, and become an HI region, of neutral hydrogen atoms.

The ejected remains of a rotating planet would develop into lanes of debris, composed of the silicate crust, spiraling outward from the core, according to the rotational velocity associated with its previous latitudinal location on the planet and the radial (outward) velocity imparted by the force of the detonation. Detonation of nearby satellites of the exploding planet, induced by the intense flux of neutrons and antineutrinos following the nuclear fission burst of the parent planet, would produce glowing balls of radioactive debris, continuing to orbit the remains of the planet. These are interpreted as satellite galaxies, such as our Magellanic Clouds, and numerous lesser galaxies.

Various size pieces of the crust and mantle would exist as more or less radioactive blobs, depending on their characteristic composition. The most radioactive part of the innermost mantle, incandescently hot, would crowd against the less radioactive, cooler debris of the outer mantle and crust. While the expanding disk of debris would continue to orbit its sun, it will do so independently, disconnectedly, of any remaining solid core of the detonated planet. The axis of rotation of the core will continue in its steady orientation, while the plane of rotation of the disk will maintain the direction it had at the time of the detonation and in time the disk will become warped, as is seen for the Milky Way Galaxy and many other spiral galaxies. Fragments become pulsars or quasars. In those cases in which the magnetic axis of the planet was somewhat transverse to the rotational axis, the detonation may produce what we see as a barred spiral galaxy. The residual core continues to rotate with the same speed as did the whole planet. Charged particles stream out from the hot core along the magnetic field lines at the poles. These streams have the same rotational speed of the core and soon move ahead of the tangential debris. Ionized gas is trapped along the spiral arms by the magnetic field.

The globs of uranium hydride (UH_3) that condensed and crystallized from the supernova cloud, in a variety of sizes and conditions, may exist as incipient nuclear reactors, scattered through space. These globs may provide the material for the globular clusters that form spherical halos around spiral nebulae, the Milky Way, and some elliptical galaxies.

While a critical nuclear reactor, natural or not, can operate at a fission rate controlled only by its feedback mechanisms and its ability to dissipate the energy released, the fission rate in a subcritical reactor will decline to the level determined by its degree of subcriticality and the production rate of neutrons by various nuclear reactions, such as (α,n) and spontaneous fission. Thus, an incipient nuclear reactor consisting of a glob of uranium hydride formed by the negative/positive charging process of alpha/beta decay, that is barely subcritical, will have an extremely low fission rate and will continue to exist relatively unchanged long after its formation. The conversion of some of the hydrogen nuclei in such a glob into neutrons by the antineutrinos from a nearby planetary nuclear detonation, will result in the gradual production of additional, more reactive nuclear fission fuel, in the form of plutonium-239.

Thus, over a period of a few days, more or less depending upon the initial composition and size of the glob, the ingrowth of fresh plutonium would create a critical nuclear reactor, with steadily increasing reactivity due to the continuing production of highly fissile plutonium-239 by the decay of uranium-239. This could lead to a detonation that would produce bits of radioactive debris that might be interpreted from afar as spherical clusters of stars. The globs activated in this way, by the antineutrinos from a planetary detonation, would show a spherical distribution around the detonated planet, which would appear to be a spiral nebula. The interpretation based on an assumed great distance would be that this was a spiral galaxy, an Island Universe, surrounded by a spherical halo of globular clusters. Only the assumption of great distance makes it so.

The aftermath of a detonation might resemble something like the galaxy labeled NGC 5457 (M101), considered to be 15 million light years from Earth, shown in Figure 108.

The rotational velocity curve of the detonation debris, resulting from the original rotational motion of material composing a solid rotating planet, would be distinctly different from the Keplerian orbital motion expected for independent, gravitationally bound objects. The satellite debris, from a satellite that had ceased to rotate due to the tidal effects of being so close to the planet, would show only random motion, and would be identified as an elliptical galaxy, with globular clusters, or as an irregular galaxy. I think the spectroscopic rotation curves are of gas sliding off the magnetic fields of the spiral arms, not of the condensed objects that make up the arms, and so do not represent rotation of the galaxy itself.

Rotational velocity curves have been constructed from measurements of the Doppler shift of spectral lines across the spiral nebula, with a shift to the blue indicating motion towards the Earth and a redshift showing a motion away, and have led to the speculation that great amounts of invisible matter must be present in order to provide the gravitational field necessary to maintain this motion. This method of measurement, using the Doppler shift, provides information only in the line-of-sight to the galaxy, and works best for galaxies that are nearly edge-on to us, but produces absolute measures of velocity. The problem with such measurements is that the velocity that is measured is that of gas, not of identifiable objects. Magnetic fields in spiral galaxies are constrained to follow the spiral arms. The rotation measurements are of the gas sliding off the magnetic field in the arms, and not of the material in the arms themselves. This motion is contrary to the actual rotation of the arms themselves.

Direct measurements of the transverse, or proper motion, velocities of individual components of a galaxy were attempted by Adriaan van Maanen at the beginning of the 20th Century, using a stereocomparator at the Mount Wilson Observatory (actually in the Observatory office building in Pasadena). These measurements were made using telescopic photos that were taken many years apart, to emphasize the changes in positions of stars and knots of luminosity in the galaxy. The images were visually aligned, and many stars surrounding the galaxy were selected as reference points, much as for a parallax measurement. The change in position of the points within the galaxy was directly measured by using visual determinations with the stereocomparator. This machine at Mount Wilson was "owned" by van Maanen [Berendzen et al. 1976] and apparently he was quite adept in its use. The internal motions of seven galaxies were measured over a period of seven years and were remarkably consistent, one with another. An example of these measurements is shown in Figure 121, for M33 [Berendzen et al. 1976]. (Perhaps because the notations in these figures were made on the reverse of glass photographic plates, the image of M33 was reversed from the normal in the original and succeeding publications. Here, the two interpretive figures have been re-reversed, to agree with the image of M33, but the notations are backwards.)

Figure 108. Spiral galaxy NGC 5457, M101.
Spiral galaxy NGC 5457 (M101). (National Optical Astronomy Observatories, NOAO/AURA/NSF, original in color.)

van Maanen's spirals are counter-intuitive to most of us. They are not the spiral we make by stirring cream into coffee, they are not the spiral of the whirlpool down the bathtub drain. They are expanding, outward-streaming spirals.

The Form of Spiral Galaxies

The action of gravity was first described mathematically by Isaac Newton in 1679 (or 1684) [Meadows 1987]. Newton published *"Philosophiae Naturalis Principia Mathematica"* (1687, 1713, and 1726) and applied his theory of gravity to the general motion of objects in the Solar System, including comets [Cohen 1958]. Johannes Kepler, working from the detailed observations of Tycho Brahe, had shown that the planets in the Solar System orbit the Sun in elliptical paths. Motion of that sort, that can be clearly calculated for the masses of the bodies, is called "Keplerian" motion. When there are more than two bodies involved, such as in the case of the Sun-Earth-Moon, the motions can be calculated only approximately, but the approximations are very good. When there are very many objects, such as in the spiral galaxies, "representative" orbital velocities can be calculated, for an assumed amount and distribution of mass in the spiral. Reasonable assumptions lead to a Keplerian motion, with the orbital velocity increasing for short distances from the center, and then falling off strongly at greater distances. However, measurements of the orbital velocity in spirals, using the Doppler shift of the wavelengths of light, clearly show a curve that is very different from that expected.

This behavior has been found in all spirals that can be measured in this manner. Clearly, some assumption in the calculation has gone wrong. However, to correct the calculations, it is assumed that the amount and distribution of the mass are in error, so a massive spherical halo contributing roughly 90% of the total (estimated) mass of the galaxy is assumed to be present. That is, we can't see, in any way other than the assumed gravitational motions, mass amounting to 9 times the mass that we can see. This mass has been named "Dark Matter".

Dark Matter is the result of theory working on incorrect information:
1. The mass-to-luminosity ratio (M/L) for the galaxies is wrong because of the assumption that galaxies are composed of billions of stars, like our Sun, shining by self-generated fusion power.
2. The luminosity is wrong because of the error in the distance scale, so
3. The mass is wrong by a factor of roughly 10^{26}, so the material is not gravitationally bound.
4. The size of the galaxies (and clusters) is wrong because of the error in the distance scale.
5. The velocities are wrong for being measured incorrectly for the gas in the galaxies, rather than for the assembly of bright points of light, as is assumed.

If the spirals are the debris of detonated planets, particularly those of giant planets with rapid rotation, the debris would spread out with velocities that were the combination of the rotational (tangential) velocity of the planet and the radial (outward) velocity of the detonation. This leads naturally to the development of spiral arms. The tightness or looseness of the spiral winding will depend on the ratio of the detonation velocity compared to the escape velocity and the rotational velocity.

Arp states that the rotational velocity curve of the peculiar galaxies is nearly constant with distance from the center, as opposed to the orbital velocities of individual stars in a galaxy [Arp 1987]. In some cases, this has been used to show the existence of hidden mass, creating a much larger gravitational field than that due only to visible matter. Here, it is the direct result of the debris retaining the rotational velocity component of a shell of a formerly intact and rigid planet. Arp also calls attention to clouds of hydrogen associated with the same peculiar galaxies that have related quasars [Arp 1987]. These jets and clouds are likely to be the result of the disruption of the water molecules in the oceans of the detonating planet, propelled outward by the scattering collisions of neutrons, ahead of the fragments of the disrupted planet. Neutrons that escape to space and so decay rather than being absorbed, produce clouds of protons (hydrogen nuclei) and beta particles (electrons). This decay produces electrons that have an energy distribution resembling electrons in a gas at a temperature of 2.49 billion K, as shown in Figure 3, and a proton distribution resembling hydrogen gas at a temperature of 2.77 million K (Figure 2).

Three glorious objects cluster together in the northern sky, in the constellation of Andromeda. These are the Great Nebula (M31), an elliptical dwarf (M32), and a dwarf spheroid (M110 or NGC 205). The Great Nebula is a spiral, the nearest one to the Earth (other than the Milky Way), and spanning, to its utmost extent, approximately 6 degrees of arc, about 12 times the apparent diameter of the Moon. M31 has spiral arms, dust lanes, gas clouds, and a bright nucleus, just like the Milky Way. It has a spherical halo of globular clusters. M110, the neighboring dwarf spheroid, also has globular clusters, and has gas and dust, in addition to its "stars", that should be the only components of an elliptical galaxy. This is the debris of a detonated satellite of the planet that formed M31. The other major companion, M32, has no globular clusters, and appears to be a standard elliptical galaxy, but may be a smaller detonated satellite. The star ν And (nu Andromedae) appears to be the starsun of this detonated planet and satellite system, as shown in the Harvard College Observatory image presented earlier in Examples of Associated Spiral Galaxies and Stars or Elliptical Galaxies. M31 and its satellite M110 glow with reflected starsunlight, and Cerenkov radiation from the radioactivity of their nuclear detonations. These objects, one originally a giant planet, the other a large moon, detonated at the same time, one inducing the sympathetic detonation of the others, and so will appear to be of the same age.

Attempts to measure the distance to an apparent supernova in the Great Nebula of Andromeda, M31 (NGC 224), shown in Figure 109, by use of the trigonometric parallax method, yielded confusion: negative values, which indicate that the nebula was farther away than the reference stars; zero, indicating the same distance as the reference stars, whatever that was; and a notably large positive parallax that put the nebula only 7 parsecs (22 light years) away from the Sun [Bohlin 1907]. Hodge remarks, quoting Fath (1909) [Hodge 1992] that at that distance the "stars must be about the size of asteroids," that is, about the size of the fragments of a disintegrated planet. (Fath determined that the stars of the nucleus were all of the same sort, spectroscopically, and had a solar-type spectrum.) At the measured distance of 22 light years, the outer diameter of M31, given by Hodge as 240 arcminutes [Hodge 1992], corresponds to a linear diameter of 1.5 light years. That is about 100,000 Astronomical Units, roughly a thousand times larger than I estimate for the Milky Way Galaxy. That may indicate that even Bohlin's near measurement may be too far.

In the *"Carnegie Atlas of Galaxies"* [Sandage and Bedke 1994], there are images of approximately 927 spiral galaxies. Many of these are fuzzy and blurry, or are seen nearly edge-on, so that the detailed spiral structure cannot be identified. (M31, in Andromeda, is an example of this. The number and arrangement of the spiral arms cannot be identified.) However, 25% are clearly two-armed spirals, reasonably symmetrical, with the arms coming off the nucleus diametrically opposite each other. There are other spiral galaxies that clearly are not two-armed spirals. Some have several arms, others simply show a spiral-like pattern over a generally uniform disk.

The two-armed pattern shows most clearly in the barred spirals (such as NGC 1300, 1365, and 1672) and therein lies the clue to the cause. In the barred spirals, ionized atoms and electrons are streaming out of the magnetic poles, where it is easiest for them to escape. In these spirals, the rotational axis of the planet was at an angle (like Uranus at 58.6 degrees [Ness *et al.* 1991] and Neptune at 47 degrees [Ness, Acuña, and Connerney1995]) to the magnetic axis, before the detonation. The continued rotation of the remaining core of the planet, the nucleus of the spiral, then outpaces the tangential velocity of the inner ends of the spiral arms.

Figure 109. The Great Nebula in Andromeda and companions.
The Great Nebula in Andromeda, with NGC 205, a "dusty elliptical," above it. [Frommert and Kronberg 2007]. (Original in color.)

For a planet in which the rotational axis and the magnetic axis are more closely aligned, like Jupiter and the Earth, material flows would be somewhat directed by the magnetic field, and debris ejected to the "north" will form one arm, while debris ejected to the "south" will form the other arm, and these will appear to emerge from the nucleus opposite each other.

The identifying benchmarks of a stellar system, taken to prove that the nebulae were stellar associations like our Galaxy, all have counterparts on a lesser scale among the phenomena of planetary detonations. Novae and supernovae, as seen in these nearby debris fields, are the lesser and greater detonations of remnant reactor cells, not exploding stars. Blue supergiant stars are the fragments that are radioactive and glowing with Cerenkov radiation. Red supergiant stars are still active reactors heated to a red glow. Some reactors are leaking neutrons which decay into protons and electrons (beta particles) which recombine into neutral hydrogen atoms giving off Balmer light. The average star is an asteroidal fragment lit by the starsun. The low surface brightness galaxies are those barely backlit by their starsun, behind them. The globular clusters, spherically distributed about a galaxy, are not composed of stars, but the fragments of a glob of rich uranium fuel that was within range of the neutrons and antineutrinos given off by the planetary detonation. The probability of an induced detonation is related to the intensity of the antineutrino flux, which determines the spherical distribution, which is centered on the ensuing galaxy.

Galaxies are studied in visible light and by the radio emissions from neutral hydrogen atoms (HI), in the 21-centimeter line, among many other wavelengths. One of the perplexing features of these studies is the sometimes unrecognized disconnection between the two appearances, in visible light and in the HI radiation. This is shown most dramatically in a comparison of the images of NGC 628 presented in optical and 21-centimeter images of NGC 628 [van der Kruit and Shostak 1983, Gilmore, King, and van der Kruit 1990]. The HI images are up to 3 times the size of the optical image [Briggs 1982]. This comparison is shown in Figure 110. The hydrogen atoms are clearly separate from the material galaxy.

 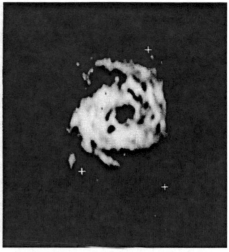

Figure 110. Spiral galaxy NGC 628 in visible light and in 21-centimeter radiation.
Comparison of the spiral galaxy NGC 628 in visible light, representing the galaxy itself, and in the 21-centimeter radio radiation from neutral hydrogen atoms, representing the decayed neutrons from the fission detonation [Gilmore, King, and van der Kruit 1990]. This shows a great difference in size, indicating a significant disconnection between the material galaxy and its cloud of hydrogen atoms.

A similar effect is shown for the M81 group of galaxies in 21-centimeter radio and in visible light [Yun, Ho, and Lo 1994]. The individual galaxies are distinctly separate in visible light, but clearly conjoined in a much larger assembly in 21-centimeter, neutral atomic hydrogen, radio radiation.

A comparison of a giant HI (21-centimeter) galaxy, NGC 262 (Markarian 348) and a normal optical NGC 266 galaxy, at the same redshift distance, is shown by Simkin *et al.* in their Figure 4 [Simkin *et al.* 1987]. The 21-centimeter image shows a galaxy that is roughly 4 times as large as the nearby NGC 266. A direct comparison of NGC 262 in 21-centimeter and optical shows a size ratio of roughly 2:1.

In a planetary detonation, the hydrogen atoms result from the decay of fission neutrons and the subsequent recombination of protons and electrons. The neutrons are emitted from the detonation with little connection to the solid matter of the detonating planet, and have velocities imparted by release from fission fragments. These neutrons decay in space into protons and electrons which form the neutral hydrogen atoms.

Random Motion in the Universe

Is the problem with our measurements of the velocities of galaxies (and all distant astronomical objects) that we are measuring redshifts of gas, not the actual semi-condensed matter of the galaxies?

In my model, all the redshift velocities are due to the Doppler effect of velocity. In standard cosmology, the large redshifts of distant galaxies and QSOs, quasars, blazars, and gamma-ray bursters, are due to the expansion of spacetime, and are not due to the Doppler shift from physical motion. For galaxies near us, about 1,000 have had their red/blue shift determined spectroscopically, some by CaII H and K lines, some by assumed hydrogen lines, others by 21-centimeter radio astronomy (from HI spin-flip of the electron). Most "farther" redshifts are measured by 21-centimeter radio.

Motion in our Solar System is guided by the Sun's gravity, and is generally well ordered. Velocities can be calculated by use of Newton's theory of gravity and Kepler's theory of orbital motion. Some random motion occurs, but it is much less than the ordered motions of the planets, satellites, and even the comets. In the Universe beyond, there may be great departures from this ordered condition, and what we see as galaxies flowing away in space may be in very random motion, instead.

Astronomical objects, in the Solar System and beyond, run at very different speeds. Those speeds are measured by various means, some more reliable than others, but usually only the "best" method is considered, so there is little cross-checking. In many cases, there is only one method that can be applied.

I have listed a collection of speeds, arranged in increasing order, in the standard astronomers' unit of kilometer per second (1 km/sec = 2,160 miles per hour, so it is clear we are dealing with a realm quite different from even the Autobahn).

Object	Speed (km/sec)
Earth, rotation	0.46
Moon orbiting Earth	1.02
Sun, rotation	2.02
Phobos, orbiting Mars	2.14
Planet X, orbiting Sun	3.85
Pluto, orbiting Sun	4.74
Titan, orbiting Saturn	5.57
Saturn, rotation	9.88
Jupiter, rotation	12.57
Jupiter, orbiting Sun	13.07
Io, orbiting Jupiter	17.33
Bright Stars, relative to Sun	18.33
Earth, orbiting Sun	29.78
Mercury, orbiting Sun	47.87
Globular Clusters of the Milky Way	109.67
Nearby galaxies	212.29
Comet, passing Sun	421.81
Solar wind (4/14/08)	438.20
Most distant galaxies	290,000.00
Speed of light	300,000.00

It is clear that most of the objects in the Solar System form one group of relatively slow (!) objects, while more distant objects (with the inclusion of Sun-grazing comets and the Solar wind) are really fast. Sun-grazing comets occasionally impact the Sun, and then they gain the full escape velocity of the Sun (in reverse), hitting its atmosphere at 618 kilometers per second (over a million miles per hour). The objects with the greatest speeds are the most distant galaxies, approaching the speed of light. No matter how fast they go, they can't exceed the speed of light.

According to current cosmology, the large redshifts of galaxies are not due to the Doppler effect, from physical motion, but are due to the expansion of spacetime (a theory that is based on the Hubble-Humason observational artifact that allowed them to only detect, or choose to study, distant galaxies that are going away. The theory may be true, but the evidence is faulty.).

Early in the past century, Einstein's field equations of General Relativity were solved in a special form that permitted the Universe to either be expanding, or contracting, or static with no change in the "size" of space. Unfortunately, it was clear that the static Universe could exist only if the net density of matter were zero. Einstein attempted to solve this problem by inserting a "Cosmological Constant", Λ, in the equations, to achieve a static Universe with matter present. Since our world is clearly composed of matter and our net density, so far as we can observe, is certainly not zero, the static Universe was quickly ruled out. That left either expansion or contraction as our possibilities.

In 1929, Edwin Hubble published measurements that carried forward the earlier measurements of Vesto Slipher, that showed that the farther away a galaxy was, based on its apparent brightness, the faster it was moving away from us, based on the redshift of its spectral lines. In current measurements, that trend appears to begin at about a few megaparsecs away, as shown in Figure 111. (A negative velocity indicates motion towards us, positive is an increase in separation, motion away from us.)

That settled the question in favor of an expanding Universe. (It should be noted that initially the redshifts were interpreted as true Doppler effect shifts. The current interpretation is that the shifts are a General Relativity effect resulting from the expansion of space and the dilation of time.) We must come back to Hubble's measurements later, but now we will focus on the easy rejection of a static Universe.

A fundamental concept of modern particle physics is that the creation of matter, by conversion of energy, must be accompanied by an exactly equal and identical creation of antimatter. (Antimatter particles are essentially the opposites of their matched normal matter particles. A positron, which is an antielectron, has a positive charge, instead of a negative charge. Similarly for an antiproton compared to a proton, an antineutron compared to a neutron. More basically, the normal nucleons are composed of quarks, and the antinucleons are composed of antiquarks.)

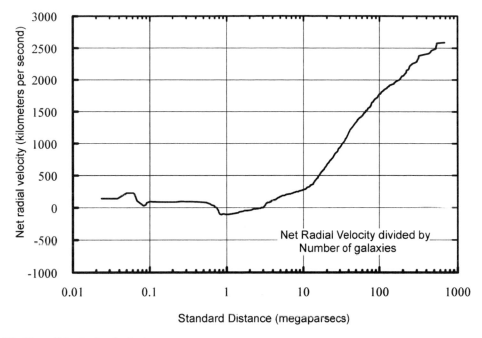

Figure 111. Net radial velocity of galaxies.
The net radial velocity of galaxies within 1,000 megaparsecs, normalized by the number of galaxies, to give an average velocity. The trend to recession begins clearly at about 3 megaparsecs and gets stronger with distance. This is the Hubble Law: the farther they are, the faster they go.

Therefore, every scientific story of creation must include the simultaneous and equal creation of antimatter particles with the normal matter particles. Acceptance of the Big Bang Theory requires some very minor differences in decay properties of K^0 mesons to show that antimatter can be totally disposed of, leaving a Universe filled with only normal matter. This seems to be a very weak argument, the differences are very small and seem to apply only in a rare case, and it is not supported by the Standard Model of Particle Physics.

While some conceptual suggestions of antimatter and its behavior had been made earlier, antimatter was predicted theoretically by P. A. M. Dirac in 1928. It did not have a serious role until the discovery of the first antiparticle, the positron, by Carl Anderson in 1932.

Particles and antiparticles energetically annihilate each other, selectively, each to only its same (opposite) kind. If equal amounts were produced, as we feel certain would be the case, anywhere particles and antiparticles mixed together, they should have annihilated each other. The Universe should be empty, and Einstein's third Universe, the static empty Universe, would be possible. But we wouldn't be here to puzzle over it.

It has been claimed that the Cosmic Microwave Background Radiation is a relic of the annihilation radiation produced by the mutual annihilation of matter and antimatter. The expansion of the Universe stretched the wavelengths from that of MeV (million electron volts) and GeV (billion electron volts) to that of the Cosmic Microwave Background Radiation, fractions of a millielectron volt, billions and trillions less energetic. Thus it has been claimed that the Cosmic Microwave Background Radiation is a relic of the initial thermal radiation of the Big Bang fireball.

Since the Cosmic Microwave Background Radiation has a clearly thermal spectrum, as well as can be measured (which is quite well), it would seem that expansion of space, if such it were, could indeed convert a thermal spectrum at very high energy (the fireball) to one at a very low energy (the Cosmic Microwave Background Radiation at 2.76 K). It is more difficult to see how a precisely thermal spectrum can result from stretching of monoenergetic photons at 938 MeV produced by mutual annihilation of protons and antiprotons, and 940 MeV produced by mutual annihilation of neutrons and antineutrons, and 0.511 MeV produced by mutual annihilation of electrons and positrons, and whatever other particle/antiparticle annihilation radiation might be considered. It would seem that the stretching of fireball thermal radiation and the annihilation line radiation originating at a later time, should produce a quite different spectrum of microwaves from the purely thermal spectrum we observe. (We must look at an alternative process for producing a pseudo-thermal microwave radiation later.)

Could the Universe be composed of equal amounts of normal matter and antimatter, without us knowing it? Joseph Leach wrote "…, astronomical observation cannot rule out large isolated concentrations of antimatter beyond the Solar System." [Leach 1993]. The key requirement is "isolation". Completely separated, there is no effect observable that distinguishes one form from the other. In fact, it is only our existence as normal matter that leads to the designation of normal matter as normal. Light from antimatter behaves the same as light from normal matter, or so we strongly think. Recent cosmic ray measurements suggest that the highest energy cosmic rays, not predicted by theory, are positrons, antielectrons. These might come from isolated concentrations of antimatter. We can speculate that in some distant part of the Universe, an antiastronomer might be making the same statements about normal matter.

What should we see, and what should we expect to see as we look at greater distances into the Universe? "Greater distances" is probably well defined in a relative sense, but it is the absolute distance scale that will give us trouble.

Here we must make a major assumption with no more basis than the statement of the "Cosmological Principle" [Narlikar 1993] that the Universe is homogeneous and isotropic. That is just a ground rule for playing our game of cosmology. It provides that wherever, whenever we look at the Universe, and in whatever direction, the Universe looks basically the same. Locally, there are obvious differences. We are very close to one star, the Sun, and distant from all the others. We think we are embedded in a spiral galaxy, the Milky Way Galaxy, while other similar objects are far away. But taken on a sufficiently large scale, the Universe looks "homogeneous and isotropic". This principal implies that whatever large part of the Universe we choose to examine, it should have roughly the same number of "objects", whichever we choose to look at. However, as we count galaxies at greater and greater distances, we find something at variance to the Cosmological Principle. Beyond a relatively short distance, of the order of 3 megaparsecs (in our standard distance scale), the number of galaxies that we observe begins to decline. This is at the same distance where the average velocity shows an increasing trend to away from us.

In the view of the Universe presented in this book, the spiral galaxies are the strewn debris of planets that have been disrupted by nuclear fission detonations, along with their satellites (which appear as satellite galaxies), illuminated from the front or the back or in between by their starsuns. These objects are roughly a hundred thousand times closer than our standard distance scale provides, based on the concept of "vast Island Universes of stars".

If the Cosmological Principle holds true, the number density of galaxies in any large regions of the Universe should be approximately the same. However, as we look at greater and greater distances (and relative distances are relatively accurate), the galaxies disappear. Some of this loss is due to geometrical dimming. The brightness of dim galaxies at greater distances is below the detection limit of our telescopes and those galaxies do not register in surveys.

The decline in the number of observable galaxies is so severe that at the greatest distance at which we can identify galaxies as galaxies, about 3,000 megaparsecs, the density of galaxies is approximately a million times less than we would expect from the local neighborhood. This is shown in Figure 112, with the data from a variety of sources. Figure 112 shows a collection of data converted to a measure of the number density of galaxies, from nearby to the furthest reach of our observations, corrected for the geometrical dimming effect. (Since the dimmest galaxies are lost from sight as the distance increases, this density has been mathematically increased by multiplying by the square of the distance to compensate for geometric dimming.)

If the density of galaxies were uniformly constant, as the Cosmological Principle requires, the data should plot along a constant horizontal band, approximately equal to the density in our local neighborhood. For galaxies within 1.32 megaparsecs of the Sun that average density is 6.76 galaxies per cubic megaparsec. (In fairness, it must be admitted that the Cosmological Principle is a philosophical attitude that we think is right, hope is right, and want to be right, but it is not a physical law.)

The number density of galaxies as a function of distance shows a great departure from our expectation that the density at great distances should be about what we see nearby. Instead, the density declines, even when corrected for our loss of galaxies with distance due to the geometrical dimming by distance alone. This is a very consistent loss with distance, possibly disturbed by changes in spectral sensitivity of different surveys. This is a major conflict with the Cosmological Principle, and we should rather look to ourselves for the cause than to discard the Cosmological Principle. It is a problem that cries out for a solution.

With the assumption that the Cosmological Principle holds, that the entire Universe is much like our part, the observed density of galaxies, when corrected for the loss of observable galaxies by dimming simply by distance, should be constant with distance and direction. That is definitely not what is observed. Figure 112 shows that beyond a standard distance of about 10 megaparsecs, the density of galaxies falls off drastically, declining by about a factor of 100 for a 10-fold increase in distance. Such small fractions seem meaningless until we consider the large number of galaxies that we might observe. The number density of galaxies starts to decrease at standard distances of only a few megaparsecs, and at the distance of the farthest identifiable galaxies, the density is only a millionth of that in our local neighborhood. Is the Cosmological Principle wrong? Is the Universe playing a trick on us? Or are we seeing a progressively smaller fraction of the galaxies that are really out there?

There seems to be only two possibilities. Either the Universe changes as we look farther away, and this might be true for an evolving Universe, but theories of galaxy evolution have not addressed this, or something happens to our observations so that we detect fewer galaxies at greater distances. The first possibility, unless the number of galaxies has grown dramatically since the early life of the Universe, violates the Cosmological Principle and brings the foundation of cosmology into question. There is probably nothing we can do to confirm what is more a statement of philosophy than a scientific fact. However, we can explore the possibility that the expected number of galaxies are really out there, but that we lose sight of most of them at great distances.

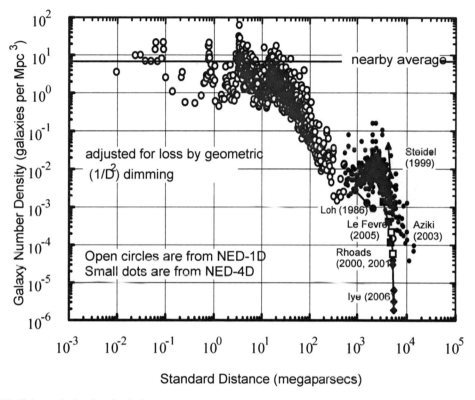

Figure 112. Volumetric density of galaxies.
The number of galaxies detected beyond about 3 megaparsecs from the Sun falls drastically below what would be expected of a uniform Universe. The Cosmological Principle would require that the density, corrected for the loss of detectable galaxies by geometrical dimming. should be similar to the nearby average at all distances, but a clear trend of loss develops at about 10^1 (=10) megaparsecs and continues progressively with distance, till it is about one-millionth of what we would expect from the nearby average density of galaxies, despite extreme efforts to find more distant galaxies. This loss is a result of looking for, and seeing, only redshifted galaxies.

If we start by assuming that the density of galaxies is in fact constant at any distance, and further assume that the galaxies are moving randomly rather than in the recessional Hubble flow, random motions like those nearby have, and not flowing away with a postulated expansion of space, we can explore the effects that might make distant galaxies disappear from sight.

There seem to be two effects in operation, and they both discriminate against blueshifted galaxies (those coming towards us), and as the distance increases, against the lesser redshifted galaxies. This leaves only the most redshifted galaxies to be observable, and observed, forming the basis for theories of the expansion of the Universe.

Thus, the loss of galaxies seen in the decline of galaxy density with distance, and the apparent Hubble flow derived from our observation of progressively redshifted galaxies at progressively greater distances, are entirely the result of observational selection against blueshifted galaxies and in favor of redshifted galaxies. This selection resulted in the early days from the observational and photographic goals and techniques of Hubble and Humason, but at greater distances it is forced on us by the attenuation of bluer light by scattering and absorption.

This removal of distant blueshifted galaxies from visibility is compounded by selection within the galaxies that we do observe by astronomers seeking a more precise, better, measure of the Hubble Constant. (See Figure 125, where the Hubble flow galaxies are compared to the NED-1D galaxies.)

Objects in random motion can be described as having a Maxwell-Boltzmann distribution, like the molecules in the air around us, defined by the value of the root-mean-square of the velocities.

The bright stars, those relatively near the Sun [Hoffleit and Warren Jr. 1991] show a distribution of velocities, towards the Sun and away, that is remarkably uniform and symmetrical, and invites a comparison with a standard distribution of velocity, the Maxwell-Boltzmann (thermal) distribution. That distribution is more often applied to collections of microscopic particles, such as air molecules, for which a temperature can be defined. The velocities of the nearby bright stars are shown in Figure 113.

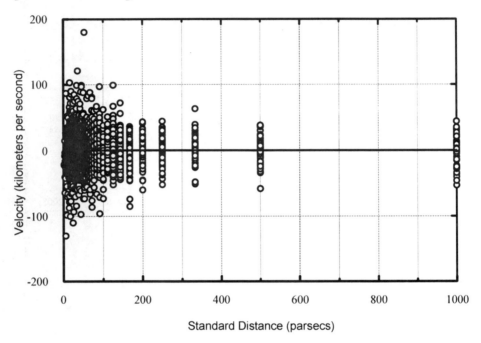

Figure 113. Radial velocities of bright stars.
Velocities of nearby, bright stars [Hoffleit and Warren Jr. 1991]. Positive velocity is away from the Sun, negative is toward the Sun. The distribution is clearly symmetric, with approximately equal numbers toward and away. Our difficulty in determining astronomical distances, even to the nearby stars, is revealed by the "quantized" distribution of the distances. These "standard" distances may be greatly in error.

The Maxwell-Boltzmann distribution can be easily calculated for this set, by calculating a "root-mean-square" of the velocities. That is 40.97 kilometers per second. A distribution with that value for the velocity parameter, and one calculated for 30 kilometers per second, are compared with the actual discrete distribution of stellar velocities in Figure 114. An intermediate velocity, 35 kilometers per second, fits remarkably well.

The specific energy (joules per kilogram) of a moving object is equal to $1/2 \, V^2$. That is 612.5 joules per kilogram for this velocity. That is a small fraction (about 0.00001) of the energy required to disrupt a planet the size of the Earth, allowing for loss of velocity (energy) as the fragments (asteroidal bodies that appear to us to be distant stars) move away from the original center of gravity of the planet (Planet X for the Milky Way Galaxy). Obviously, there is a range of disruption energy, from barely broken to dispersed to infinity. The value used for comparison here is the dispersed to infinity value.

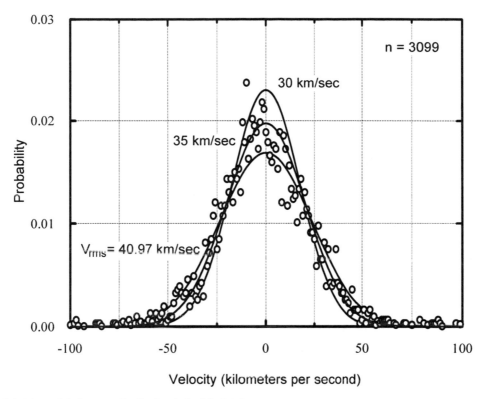

Figure 114. Maxwell-Boltzmann distribution derived for bright stars.
Probability distribution for the velocities of 3,099 nearby, bright stars. The observed stellar velocities, relative to the Sun, are fit well with a Maxwell-Boltzmann distribution with a characteristic velocity (V_{rms} is the root-mean-square velocity) of 35 kilometers per second.

Farther afield, the globular clusters of the Milky Way (most galaxies that we can observe well have these clumps of star-like objects, spherically centered on the parent galaxy) also show a symmetrical velocity distribution in their velocities relative to the Sun. In the early 1900s, Harlow Shapley studied the distribution of globular clusters (nearly spherical collections of thousands to millions of stars, having a separate identity from the clouds of stars in the Galaxy) about the Milky Way Galaxy, and considered that they were associated with our Galaxy, gravitationally bound members, even though their radial velocities were found to be quite large. In that model, the globular clusters formed a spherical distribution, and the center of that distribution would coincide with the center of the Milky Way Galaxy. That model has held up in subsequent investigations, to the extent that it has been examined.

The globular clusters show radial velocities, toward and away from the Sun, that are quite high, greater than anything else associated with the Solar System or the Milky Way Galaxy, except for the Solar wind. The measured velocities are shown relative to the standard distances, in Figure 115.

In the planetary detonation model, the globular clusters are associated with the Milky Way Galaxy (and similarly for the globular clusters detected around other galaxies) only by the accident of being nearby bystanders at the time of the detonation of the planet that formed that galaxy. (That was Planet X in our case.) These are the debris, the shrapnel, produced by the induced detonation of globs rich in uranium.

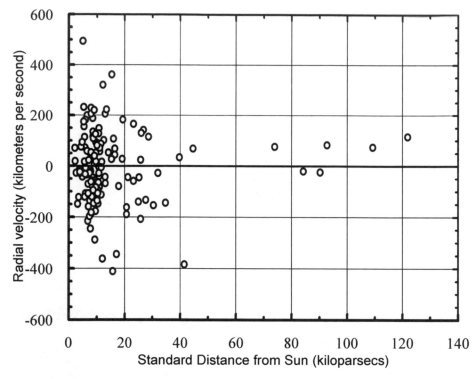

Figure 115. Spectrometric velocities of Milky Way globular clusters.
The measured velocities of Globular Clusters surrounding the Milky Way Galaxy. These velocities are determined by matching spectral emission and absorption line patterns with reference spectral templates having presumed known velocities, and interpreting the wavelength shifts as velocities, by the Doppler Theory. This shows a generally symmetric distribution, with about as many globular clusters going away from the Sun (positive velocities) about as fast, as those going toward the Sun (negative velocities).

In the same manner as for the bright stars, I have calculated a Maxwell-Boltzmann distribution for the globular clusters [Harris 1996], shown in Figure 116. In this case, the root-mean-square velocity is 142 kilometers per second. This does not provide as good a match as the bright stars, but the agreement is at least suggestive.

A nuclear fission detonation releases a large number of antineutrinos, as shown by the two pulses of antineutrinos detected from the nearby supernova in the Large Magellanic Cloud, SN 1987A (actually at a much nearer distance than our standard story claims). Those antineutrinos interact with matter by the antineutrino-induced electron capture reaction described in Appendix A. In all cases, the resultant nuclide is unstable. In some cases, a neutron is released.

In this proposal, accumulations of uranium formed from fresh supernova ejecta throughout the Universe, by the decay-charging mechanism described earlier, exist as what can only be called "globs" of uranium, forming fuel for potential, or actual, nuclear reactors, or in the case of globular clusters, fuel for nuclear detonation.

Within a certain distance, likely the same in all directions, the antineutrino flux will be sufficiently intense as to drive the masses of uranium in the globs to detonate, producing spherical clusters of fragments, spherically distributed about the detonated planet, and all within a certain distance of what takes the form of a spiral galaxy. That results in a spherical distribution of globular clusters around the center of the Milky Way Galaxy (and other galaxies, in a similar manner). The different initial sizes of the globs, and richness of the uranium, will result in the various sizes and stellar richness of the resulting globular clusters.

A major problem with this model for the globular clusters comes from their high speeds and the time since the detonation. A rough estimate of the detonation time is 5.2 million years ago, based on Van Flandern's study of long-period comets, but an alternate set gives 6 million years, and I prefer that. Using that time estimate, the fastest globular clusters, at about 400 kilometers per second, should have traveled about 7,000 light years. This is obviously not in keeping with a small scale Universe.

There are some possible explanations to eliminate this conflict.

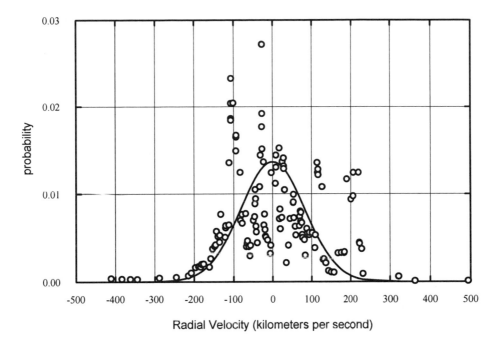

Figure 116. Maxwell-Boltzmann distribution for radial velocities of globular clusters.
Probability distribution for the velocities of the Globular Clusters of the Milky Way. The fit of the Maxwell-Boltzmann theory is not as good as for the bright stars, and this may result from attempting to make measurements at distances that are nearly a hundred times as great as for the bright stars. The actual globular cluster probabilities scatter through this distribution. The root-mean-square velocity for the measured velocities is 142 kilometers per second.

Measurements of velocity are made by recording a spectrogram averaged over the globular cluster. The absorption lines (from gas in front of the "stars" of the globular cluster) are matched to the similar lines in a template spectrum, such as one often used for 47 Tucanae, one of the brightest globular clusters. Some errors in prior measurements of radial velocities have been noted [Zinnand and West 1984]. Rather than actual observational errors, these differences might reflect varying effects over time, perhaps different flow rates of the Solar wind. The velocities may be perturbed by the Solar wind, which is the only material in the Solar System with velocities as great as 500 kilometers per second or more. It is not clear how this could result in observations that show both incoming and outgoing radial velocities. One possibility, since globular cluster velocities are usually measured by use of a spectrometric template for one cluster whose velocity is considered well known, if that velocity is in error, all the others will be also, and a major displacement could produce the symmetrical distribution that is seen.

(As an aside, it is important to recognize that astronomical Doppler shift measurements are not like the radar Doppler used by the Highway Patrol and the weathermen. Astronomical Doppler determines the shift in wavelength of emission and absorption lines in relatively dilute gas, compared to their natural wavelengths at rest. Astronomical Doppler measures the velocity of gas, not the condensed objects that we call stars in globular clusters and galaxies.)

The velocities may actually represent the flow of protons (recombining with electrons to produce hydrogen atoms) resulting from the decay of neutrons escaping from still (or newly) active reactors in the globular cluster. It is not clear how a symmetrical distribution of "toward" and "away" velocities would result from this, unless the spectroscopy happens by accident to selectively focus on protons from neutrons going toward us and away from us. One would produce negative velocities, the other would produce positive velocities. (A neutron with a velocity of 400 kilometers per second would have a kinetic energy of 826 electron volts, which is small compared to the average initial energy of a neutron released in fission, 2 million electron volts. It is a typical energy of a reactor with a high concentration of uranium, mixed with a variety of mid-mass elements, like iron.) While the distance, velocity, and time values for the globular clusters pose a significant problem for gravitationally unbound objects, it should be recognized that these parameters also pose a problem for the Island Universe concept of the Milky Way Galaxy, and force the astronomers to add great amounts of magical Dark Matter to keep the clusters in orbit. I am determined to avoid the use of magic in this book. The globular clusters are a problem yet to be solved.

Nearby galaxies also show a symmetrical velocity distribution, similar to that of the globular clusters. This can also be converted to a Maxwell-Boltzmann velocity distribution by calculating the root-mean-square velocity. For nearby galaxies, no farther than 2.0 megaparsecs (6.5 million light years) in standard distance, this velocity is 234 kilometers per second, somewhat greater than for the globular clusters. The velocities of the galaxies within 2 megaparsecs of the Sun (6.5 million light years) [Madore and Steer 2008] are shown in Figure 117. These are distributed in the same range as the globular cluster velocities, and are moderately uniform and symmetrical.

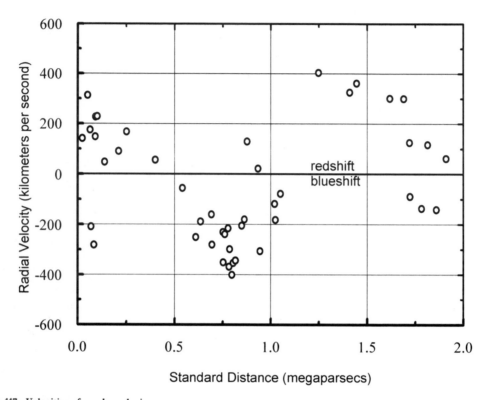

Figure 117. Velocities of nearby galaxies.
Velocities of nearby galaxies, within 2 megaparsecs (6.5 million light years) of the Sun. It is generally symmetric, but shows some clear irregularity.

While the velocities are moderately uniform and symmetrical, the small number of galaxies leads to rather poor matching to the Maxwell-Boltzmann distribution, shown in Figure 118.

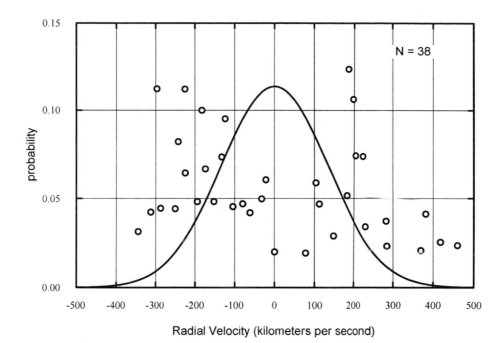

Figure 118. Maxwell-Boltzmann distribution for radial velocities of nearby galaxies.
Probability distribution for the velocities of the nearby galaxies. A rather poor fit, perhaps from problems with greater distances, or other effects may interfere. The root-mean-square velocity is 234 kilometers per second.

Replotting the volumetric density of galaxies shown in Figure 112 with the distance scale replaced by one representing conventional (Doppler) velocity gives an opportunity to compare these velocities with those we have just considered. That is shown in Figure 119.

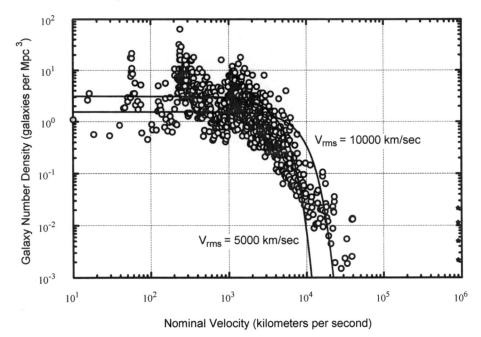

Figure 119. Galaxy number densities.
The galaxy number density, replotted with speed as a substitute for distance, limited to a distance range within which Doppler Theory can be applied. The smoothed density curve shows remarkable variations. Maxwell-Boltzmann distributions for characteristic velocities (V_{rms}) of 5,000 and 10,000 kilometers per second are shown, and come approximately close to the observations.

By "nominal velocity" I mean I am using Doppler Theory to interpret the wavelength shifts as velocities. The fastest speeds are only about one-tenth the speed of light, so this does not introduce any significant error. I have superimposed two curves representing Maxwell-Boltzmann distributions with representative velocities of 5,000 and 10,000 kilometers per second. In this plot, only velocities of recession, for galaxies going away from the Sun, are shown, as only redshifted galaxies are detected beyond a relatively short distance from the Sun.

Measurements of velocity involve determining the redshift of spectral lines, and these spectral lines are emitted by the atoms of diffuse gas that is associated with each spiral galaxy. The hydrogen atoms surrounding a spiral galaxy (see Figure 110 for example) were produced by the beta decay of neutrons to protons and electrons. The protons have essentially the same velocity distribution as the neutrons and are the source of spectral line emission as they recombine with electrons. The emissions from hydrogen atoms on the near side of the spiral galaxy (produced by the decay of neutrons escaping from the detonation in our direction along the line of sight) are blueshifted and are lost from observation. Emissions from hydrogen on the far side are redshifted and easily traverse through dust and gas, to be detected and interpreted as the recessional velocity of the spiral galaxy. In Figure 120, the redshift velocities of spiral galaxies are shown. No blueshifted galaxies are seen and the number density falls off rapidly with increasing velocity. This decline is matched well by the velocity distribution of neutrons from the VIPER reactor, a pulsed reactor operated to simulate nuclear detonations. It is the fission neutrons from the planetary detonation that produce the gas that shows the increasing redshift with distance that has been interpreted as showing that the Universe is expanding.

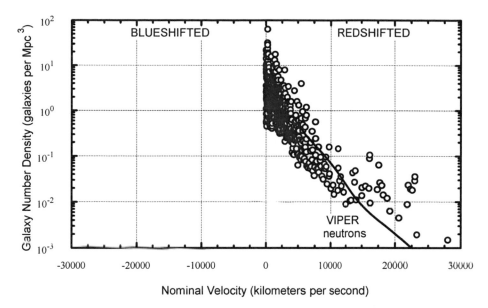

Figure 11200. Velocity distributions for distant galaxies and neutrons from the VIPER reactor.
The velocities of redshifted galaxies compared to that of neutrons from the VIPER reactor [Delafield *et al.* 1978]. This shows that we see only half of the light emitted in our direction from the gas surrounding each galaxy in the observable Universe. Blueshifted light is lost from sight, and only redshifted galaxies are recorded. The observed galaxy distribution is matched by the velocity distribution of neutrons from the reactor with a characteristic velocity of about 10,000 kilometers per second.

In the search for the Hubble Constant, farther and faster are the goals. That is shown in Figure 125, where an innocent collection of galaxies (NED-1D) shows many that depart from the Hubble flow. The determination of the Hubble Constant uses galaxies that lie along the line, and the line determines which galaxies are to be used to determine the Hubble Constant. However, not only is there deliberate selection of galaxies, in an effort to obtain the "best" value for the Hubble Constant, the Universe itself acts on the sample of galaxies that we can observe.

The Expanding Universe and the Hubble Constant

The farther they are,
the faster they go.
Hubble's Law

Edwin Hubble and Milton Humason entered astronomy at a truly remarkable time, and by strangely opposite paths. American astronomy was just entering the era of giant reflecting telescopes, photography and spectrometry were being adapted to the study of the heavens, and so much was undiscovered, undecided, and new. Hubble came by way of an elite education, where the quality of the man outweighed the content of the courses. Humason came by way of a mule path [Christianson 1995]. While Hubble was passing through Oxford and the University of Chicago, Humason had dropped out of high school and learned how to work hard, and well. Humason helped haul equipment and supplies to the new observatory being built on Mount Wilson, eventually being employed inside, and quickly learned to process photographic plates for the astronomers. Through his skillful understanding of the effects of various photographic emulsions, filters, exposures, and developing processes, Milt Humason became the observational colleague of Edwin Hubble. Humason gathered, Hubble reaped. This subtle compartmentalization of their research led to some of the most far-reaching discoveries in astronomy, in the early part of the 20th Century, discoveries that shaped our Universe and created new fields of study.

Perhaps Humason did not fully understand the character of the images that he produced, only how to produce excellent images. Perhaps Hubble did not quite understand the ramifications of how those images were produced, but could extract remarkable information from the images and the related photographic spectra.

In one of the most famous scientific careers of American science [Christianson 1995] Hubble applied himself to a broad study of the "nebulae", fuzzy-looking splotches in the sky that were being resolved into stars, dust lanes, and gas clouds, by use of the new giant telescopes. In this study, Hubble was attracted to a field that had just gone through a quiet revolution. While there were true nebulae at not very great distances (in astronomy, all is relative!), and the nebula of Orion's sword is perhaps the most easily seen example, other nebulae were at undetermined distances and were of undecided form.

From the best photographs of spiral nebulae, Hubble selected clear examples of the variety of shapes and forms, and developed a sequence showing what might be considered to be a transition or evolution from one extreme form to the other. It was not clear if this sequence truly showed evolution or development between forms, but the sequence itself and the labeling developed by Hubble has proven useful in identifying galaxies to the present day [Hubble 1936, Sandage 1961, Sandage and Bedke 1994, Binney and Merrifield 1998].

Many of the most conspicuous nebulae had been catalogued over a hundred years earlier, by Charles Messier, to help in his hunt for comets [Mallas and Kreimer 1978, Webb Society 1979, Jones 1991]. These hundred-plus objects are now known by their Messier number, M1 through M109. (Strictly, Messier himself listed only 103 nebulae, and seven more were added, to reach M110, from reviews of Messier's notes. M101 and M102 seem to be the same and there are some other conflicts and confusions.) Messier built his list so that he could quickly dismiss the known little fuzzy objects that he might otherwise mistake for comets.

Some of these, such as M1, M8, and M20, are clearly relatively near and small, clouds of gas and dust, lit by a nearby star. Many are globular clusters, densely packed clumps of stars. Others, as increasingly better telescopes revealed, were distinctly different, and many of these had a clear spiral structure.

Spiral nebulae were first identified as special astronomical objects by William Parsons, Lord Oxmantown, Earl of Rosse, who was able to see the structure and pattern of the Whirlpool Nebula (later labeled M51 by Messier and NGC 5194 by the New General Catalogue) in 1845, using a huge reflecting telescope, the largest of its time. Starting in 1845, Lord Rosse, by use of his reflecting telescope with a 6-foot diameter mirror (a giant for its day) clearly sketched the spiral structure of some of the nebulae [Berry 1898]. These became the class of spiral nebulae, different from others, gaseous, reflection, and globular clusters. As telescopes improved, it became apparent that the spiral nebulae were truly special, unique among all the diffuse objects seen in the sky. Their nature was clearly not known.

Only 90 years ago, we didn't know that the spiral galaxies were vast Island Universes of stars, they might just be nebulae within the Milky Way. It was a question of distance. Some astronomers thought them to be near, others thought they were far. (We didn't know about nuclear fusion or nuclear fission, or antimatter, or an expanding Universe, either, at that time. It was a remarkable time in astronomy.)

Throughout the last 50 years of the 19th Century and the first 20 years of the 20th Century, more and more attention was paid to determine the nature of the spiral nebulae. In 1920, Harlow Shapley and Heber Curtis formally presented a review of their conflicting understandings of spiral nebulae. Because of its historic significance in the development of the astronomy of galaxies, and our view of the Universe, this presentation has been labeled the Great Debate [Shapley and Curtis 1921]. The ideas and observations on both sides were well balanced. Shapley presented the position that the spiral nebulae were reflection nebulae, within the Milky Way, our Galaxy, which was understood to be a vast Island Universe of stars (Galileo had said so.), and perhaps constituted the entire Universe and Cosmos. Curtis proposed that the spiral nebulae were themselves vast Island Universes of stars, just like the Milky Way Galaxy, outside of it and at great distances. (Immanuel Kant might have said so.) Paradoxically, we are now able to see that they were both about half right.

Shapley had studied the globular star clusters around the Milky Way Galaxy (at that time, 1918) and found that they could be arranged in a symmetrical system centered about what was considered to be the center of the Milky Way Galaxy. This led to his view that the Milky Way Galaxy was the entire Universe and contained all the other objects that were observed in the sky. In particular, the spiral nebulae were just special forms of the various other nebulae that were being found.

Other studies were suggesting that the spiral nebulae were similar in form to the Milky Way Galaxy, and at far greater distances (at least relatively) than any thing that could be assigned with certainty to the Milky Way Galaxy itself. Thus, the Universe was vast, and the Milky Way Galaxy was just one component out of many.

These two possibilities formed the basis of a controversy that has been connected with presentations by Shapley, on his side, and Heber Curtis, on the other side, on April 26, 1920 at the National Academy of Sciences. [Shapley (Part I) and Curtis (Part II) 1921]. This became known as the "Great Debate", although most writers are clear about there being little debating and that both presentations were somewhat off-target.

Since this book proposes that spiral nebulae, spiral galaxies, (and other related objects) are of an entirely different nature than vast Island Universes of stars at great distances, and that the foundation of the distance scale is incorrect, this question became of paramount importance.

The Great Debate did not settle the question. Evidence was strong for both sides. We will postpone considering how they could both be right and wrong until after we have further explored the productivity of Adriaan van Maanen, and of Hubble and Humason.

Some of the measurements that supported the near theory were being done by an astronomer at Mount Wilson (actually his workplace was in Pasadena), Adriaan van Maanen.

During the last years of uncertainty over the nature and distance of the spiral nebulae (and other related objects), Adriaan van Maanen, an astronomer at the Carnegie Institution in Pasadena, California, made some measurements that he interpreted as roughly rotational motions in the arms of several spiral nebulae. The motions that van Maanen measured, as angular changes in position over a decade or so, were reported in angular rates of change, milliarcseconds per year. The linear velocities indicated by these angular velocities depend on the distance assumed to the particular spiral nebula. At moderate distances, the velocities had moderate values. (This does not seem to have been well documented in the published literature. Much science is done by personal discussions, at conferences, in hallways, by letter, and is often difficult to retrieve. The best I was able to find is some comparisons of velocity based on assumed distance in one of van Maanen's papers. He does not seem to be trying to quantify anything, I think he just presented a set of distances and velocities in a range that seemed appropriate.)

By use of a quirky measuring machine that "wasn't intended for his purpose" and that other astronomers didn't like, van Maanen made measurements on photographs of seven spiral galaxies (M101, M51, M33, M81, M63, M94, and NGC 2403). He was a meticulous worker and understood the problems inherent in the machine, made by Zeiss. It was very sensitive to temperature changes because it was made of different metals with different expansion coefficients, and measurements extended over 2 or 3 days. So he had his own machine made in the optical shop, which may have been at the Throop College of Technology, soon to become Caltech, which operated the Mt. Wilson Observatory for the Carnegie Institute of Washington. By very careful measurements, with specific care to eliminate asymmetric biases from the photographic plates and the exposures, and the development and the measuring machine and his own measurements, he systematically showed motions between photos taken at different times, about 10-15 years apart. Those motions were no larger than the images of stars and other small points in the galaxies. To display the motions as vectors superimposed on the image of the galaxy, van Maanen used scales that represented motions in about 1,000-1,300 years, exaggerating the length by that amount so they would be observable, but he called those motions "annual" in the figure captions. That may have set up some underlying skepticism, for to look at the figures as he presented them in journals, you would think you could just look at the images and see the flow. Hubble himself thought so and didn't see any changes when he looked, using a standard blink comparator. Those measurements were published from 1916 to 1923. The motions were about right for Shapley's near case, although they weren't really explored for that, but for the far case of Curtis they showed velocities near and beyond the speed of light, which had been set as an absolute speed limit, so they weren't physically possible.

Throughout the controversy, van Maanen's work formed one of the strong points in Shapley's arguments. If van Maanen's angular velocity measurements were to imply linear velocities that were acceptable to astronomers, the distance scale had to be short. If van Maanen's measurements were meaningless, there was no constraint on the distance scale, the spiral nebulae could be vast Island Universes of stars, and be just like the Milky Way Galaxy as galaxies in their own right.

In 1920, those two astronomers, Shapley and Curtis, on opposite sides of the near/far argument, gave presentations somewhat discussing the controversy [Shapley and Curtis 1921], and that may have started more active interest in the question. In 1926, Hubble showed that the spiral galaxies had stars that could be identified as a special type, Cepheid variables, that the Milky Way Galaxy also had, and that they were significantly farther away (by being dimmer) than anything in the Milky Way Galaxy could be. His distance measurements were about a factor of ten too close, compared to our current standard measurements, but it was clear that the near case was wrong, and so van Maanen's measurements were judged to be faulty.

Adriaan van Maanen's measurements were made by use of a stereocomparator, an instrument that he considered to be his own personal device [Berendzen, Hart, and Seeley 1976, Hart 1973], and he had developed its use to an art. This instrument superimposed photographic images of a galaxy from separated times, and van Maanen then measured the apparent offset of identifiable points in the structure of the galaxy, and interpreted these offsets as movement, rotation, from the earlier period (epoch in astronomical parlance) to the later. The stereocomparator allowed alignment of the entire image so as to achieve the best match of the background field between the two images. That differed from the usual method of measuring stellar parallax by use of a micrometer, measuring the angular distance from one star (or object) to another on the same plate and comparing that distance with a similar measurement made at a different time. That was the standard photographic method for measuring trigonometric parallaxes, the fundamental foundation of the cosmological distance scale. van Maanen's skill and art, and understanding of the instrument, allowed him to make precise and repeatable measurements of apparent changes in position of the structures in the galaxy being studied. When visiting astronomer Knut Lundmark tried to replicate van Maanen's measurements, using the unfamiliar instrument, he only produced random results. The crucial difference was that van Maanen reduced systematic errors by making eight sets of complementary measurements, but Lundmark only made three sets.

Astronomical measurements are made by use of arc lengths, measured on the celestial sphere, and expressed in terms of angular measure. This avoids the complication of converting these measurements into linear distances by use of an assumed or measured distance to the object. Interpretation of the angular motions measured for points in the spiral nebulae for large, "intergalactic distances", on the order of a million parsecs, millions of light years, resulted in velocities that were significant fractions of the speed of light and this did not seem acceptable for any reasonable amount of mass. In contrast, measurements by Knut Lundmark, a visiting astronomer from Sweden who spent a year at Mount Wilson and was provided with the same photographic plates and stereocomparator as had been used by van Maanen, showed randomly directed velocities of about the same magnitude as those of van Maanen [Berendzen, Hart, and Seeley 1976, Hart 1973]. These measurements are compared in Figure 121. Eventually, Lundmark's results were accepted as indicating the random errors inherent in the measurements. Actually, the differences were due to van Maanen measuring eight sets while Lundmark only measured three. In van Maanen's work, the biases were canceled out, but they remained in Lundmark's work.

In support of the measurements by van Maanen, and the model proposed here, it can be recognized that there are eight different modes of general motion conceivable in a spiral nebula:

1. Streaming outward along the arms, as indicated by van Maanen, according to a theory of James Jeans
2. Streaming inward along the arms, as in a whirlpool,
3. Radially outward toward the convex, outer side of an arm,
4. Radially inward toward the concave, inner side of an arm,
5. Transversely across the arms, outward, unwinding the spiral,
6. Transversely across the arms, inward, winding up the spiral,
7. Rigid rotation, like a blown pinwheel,
8. Rigid rotation in the opposite direction.

The present model of planetary detonations predicts mode 1, as a resultant of the tangential velocity of ejected material, devolving from the rotational velocity of the detonating planet, and the radial velocity outward imparted by the energy of the detonation.

van Maanen's measurements show this mode, but have been dismissed as the result of a personal bias in favor of this result, influencing measurements that were made at the limits of capability for the machine and observer. But it is not clear that this mode is the most intuitively likely form of motion, if the cause of the structure is unknown, as it was at the time of van Maanen. Midway through his measurements, van Maanen did adopt the theory proposed by James Jeans, with motion described as streaming along (outward) the arms and flowing transversely (outward) across the arms. This was also supported by a theory by B. Lindblad [Berendzen, Hart, and Seeley 1976].

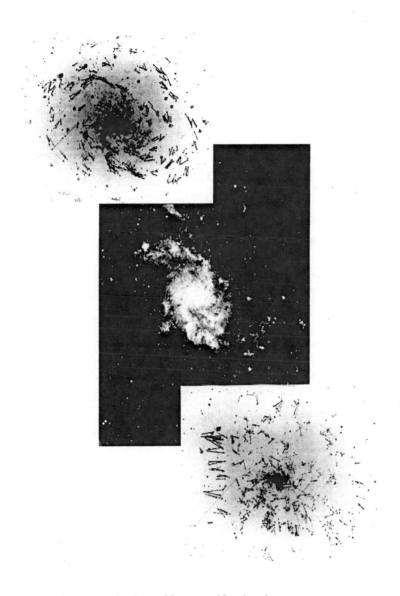

Figure 121. Measurements of galaxy rotation by van Maanen and Lundmark.
Comparison of measurements of movements in M33 made by van Maanen (in the upper left) and Lundmark (in the lower right), and a photographic image of M33. The velocities found by van Maanen agree with the planetary detonation model, while Lundmark only shows large random motion, considered to be measurement errors. (Photographic image from Carnegie Institution of Washington). The vectors showing "annual motions" are misleading for they actually represent extrapolated movement over about 1,300 years. Actual annual movements would be imperceptible, as Hubble was to observe.

The measured displacements were observed in angles of arc on the sky as recorded on the photographic plates. Those angular "annual motions" must be converted into linear velocities by multiplying by the distance assumed to the galaxy. If the distance that is assumed is millions of light years, the velocities exceed that of the speed of light, and that is a well established speed limit in our Universe. If, however, the distances are 100,000 times smaller, as shown by one lone measurement of the distance to the Great Nebula in Andromeda, M31, by Bohlin [Hodge 1992, Bohlin 1907a, Bohlin

1907b] the velocities become quite acceptable. van Maanen measured positions on images of seven nearly face-on spiral nebulae, showing rotational (angular) velocities that have been judged to be too fast. The linear velocities have changed over the years, as the distances determined by Edwin Hubble were corrected by later measurements. I have taken distances from Hubble [Hubble 1929], representing the distances that were accepted at the time of the controversy, from NED-01 [NASA/IPAC Extragalactic Database 2000] representing modern distance values, and calculated distances adjusted by Bohlin's measurement of the distance to the Great Nebula of Andromeda [Bohlin1907a, Bohlin 1907b] representing the "short" distance scale suggested in this book. (Note the change from megalight years, MLY, to light years, LY, in the change to Bohlin's distance scale.) The linear velocities implied by these distances range from too fast, to way too fast as Hubble's errors were corrected and the Universe grew by a factor of roughly 10, to quite reasonable. (The speed of light is about 300,000 kilometers per second, planetary rotational velocities in our Solar System are in the range of 0.5 to 12.6 kilometers per second.) These results are shown below.

Comparisons of Linear Velocities from van Maanen's Measurements

Galaxy	Hubble's Distance MLY	Velocity km/sec	Modern Distance MLY	Velocity km/sec	Bohlin's Distance LY	Velocity km/sec
NGC 0598 (M33)	0.86	24,929	2.8	81,886	24.2	0.7
NGC 2403	no data	no data	10.9	320,564	93.3	2.8
NGC 3031 (M81)	2.84	82,546	12.0	673,311	102.8	5.8
NGC 4736 (M94)	1.63	47,501	20.3	659,868	174.0	5.7
NGC 5457 (M101)	1.42	41,371	24.5	817,785	210.1	7.0
NGC 5055 (M63)	3.58	103,920	33.0	951,084	283.6	8.2

(MLY is million light years)

Compared to rotational speeds of planets in the Solar System (0.5 kilometers per second for the Earth to 12.6 kilometers per second for Jupiter), the "short" distance scale produces quite reasonable velocities. A graphical comparison is made in Figure 122, where the rotational speeds of the eight planets of our Solar System (open dots) are compared with the rotational components measured by van Maanen for the seven galaxies (filled dots).

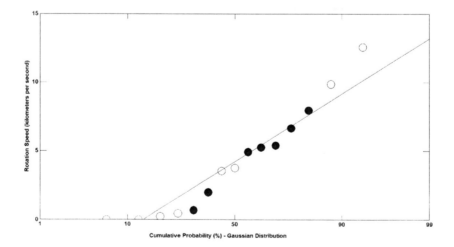

Figure 122. Planetary rotations and rotations measured by van Maanen.
Cumulative probability plot comparison of the rotational speeds of the planets of the Solar System (open dots) with the rotational components of velocity for van Maanen's seven galaxies (filled dots), calculated for the Bohlin (short) distance scale. The two sets intermesh quite well.

This shows that for the rotational velocities of seven galaxies measured by van Maanen, the entire set of galactic rotational speeds is completely consistent with the planetary rotational velocities measured for the eight planets of our Solar System, if the short distance scale based on Bohlin's measurement to the Andromeda Nebula is used. The galaxy rotational speeds intermesh with the planetary rotational speeds and there are no inconsistent outliers.

As the Island Universe theory became the accepted model of the Universe, it was decided that van Maanen's proper motion measurements were wrong, that is, they had to be wrong because of the unacceptably high velocities implied by great distances, having been affected by some unidentified "systematic error". It appears that van Maanen was working at the edge of accuracy with the instrument he was using. Later computer studies [Berendzen, Hart, and Seeley 1976, Hart 1973] with an assumed variability for the measurements showed that *if* there were a consistent bias, van Maanen's results *could* have been produced falsely, though unknowingly, by the observer. Perhaps a more objective judgment, one not invoking the Island Universe model, is that one of the observers, adept with the machine, produced measurements with greater internal consistency, less randomness, than did the other observer who was less practiced and meticulous with the machine, and unable to make a complete set of measurements. The key lay in the repetition in different alignments that eliminated biases in the material, the machine, and the measurements.

In time, the long-distance-scale model prevailed, van Maanen's measurements were rejected for a variety of reasons, and eventually his results were explained as the result of him wanting the measurements to show the spiral flow so clearly apparent in the images. That is, he subconsciously biased his multitudinous measurements to produce the rotational motions that he reported.

Acceptance of van Maanen's measurements faded and evaporated with the evidence for the far distance scale, and there were four pieces of evidence used to show that the motions he had presented were an accident of his work. No cause was discovered (unlike the cause of the out-of-focus images from the *Hubble Space Telescope*, "from improper use of an inspection jig") and a PhD thesis in 1973 had to conclude that van Maanen measured "what he expected to see" [Hart 1973].

There were four specific objections to the integrity and the reality of van Maanen's measurements. First, to provide an unambiguous basis for rejection, the disagreement (not really acknowledged by the two astronomers) between measurements by van Maanen and Knut Lundmark was taken to indicate that the measured position changes were just the result of instrumental error, but in van Maanen's work, revealed his expectations. The second and most basic objection involved the derived linear velocities, which depend on the distance scale. Close gives reasonable velocities, distant gives unacceptably great velocities. Thirdly, spectrometric measurements of the rotation of galaxies showed that van Maanen's measurements went the wrong way. Finally, Edwin Hubble used a blink comparator, the same sort of instrument for comparing different images of the same thing that was used in the discovery of the (now former) planet Pluto and saw no changes. Surprisingly, each of these objections is wrong in its own clear way, and has no strength in the rejection of van Maanen's velocities. With the objections gone, and for reasons that strengthen the reality of his work, we should accept the results and consider how they could be right, how they could lead us to a better understanding of the Universe.

The four points that justified rejection of van Maanen's measurements were:

1. Comparison measurements made by another astronomer with the same images, the same machine, and the same method, simply showed random variations, not consistent motions.
2. At the standard distances, the motions were unacceptably, physically impossibly, too fast.
3. Later spectroscopic measurements of the "rotation curve" of other, but presumably similar, galaxies showed that the galaxies were rotating very fast, but not too fast, but in the opposite direction from van Maanen's measurements.
4. By use of a blink comparator, Hubble was not able to discern any change in position of the stars and other small points.

The conflict between van Maanen's measurements and the presumed identical measurements of Lundmark is obscured in the literature by what has been interpreted as excessive courtesy toward colleagues, or what might be personal embarrassment over poor technique. Did van Maanen, who was adept with this particular measuring device, introduce incorrectly, through a very specific "personal equation", the sense of motion shown by his results, by biasing the random variability to agree with a preconceived notion as to speed and direction? Or did Lundmark, who was not an accomplished user of the instrument, introduce through his own "personal equation" enough variability that the directions became sufficiently random as to obscure the true motion?

A careful reading of the journal articles shows that reason 1 was completely invalid. van Maanen was a very meticulous measurer, making multiple measurements of each point to determine the best average value. But more importantly in these measurements, van Maanen placed the pairs of plates, early and late, in eight different complementary orientations, rotated and exchanged between the two positions of the stereocomparator, in order to eliminate as well as possible the biases introduced by imperfect plates, imperfect machine, and imperfect man. van Maanen measured each galaxy by first putting the early plate in the right side of the machine, north to the top, and the later plate in the left side, north to the top. Alignment of the two images was made by moving one from side to side and up and down and by turning it till as much of the background of the two images could be made to exactly coincide as possible. After making a full set of measurements, he then rotated the plates so that east was up for the next set, and then south, and then west. Then he swapped positions so the plates were on the opposite sides of the machine from the first set of measurements. All told, he made eight sets of measurements in these various orientations, to compensate for any asymmetries, which were known to exist in photographic images, measuring machines, and visual (through a microscope) measurements. Eight complete and complementary sets were made to compensate for errors. Lundmark made only three sets of measurements, not the full set required to eliminate the biases that were shown as random error. van Maanen got good measurements because of his skill and perseverance, Lundmark got random results because of an incomplete effort at making the measurements. Lundmark returned to Sweden after making the measurements and before preparing his report. It is likely that his time was cut short and that prevented him from doing a proper job.

Eventually, Lundmark's random results were used as indication of the errors of the machine, and that van Maanen had not been able to make the precise measurements that he claimed. But that was the fault of Lundmark. So, Point 1 is wrong.

The astronomical distance scale that is used to determine the distance of every object outside the Solar System starts out with a measurement method called "trigonometric parallax". Astronomers think that this is an absolute measurement, like triangulation in civil surveying, and perhaps they can be forgiven for that because the method was derived from the method of horizontal parallax, which was triangulation. (In triangulation, the length of a baseline is measured, and the angles of the lines of sight to a distant object are measured, and with Euclidian geometry, everything about the triangle can be calculated, specifically the distance to the distant object.) With the change to trigonometric parallax, the absolute measure of the angles was lost, and the angles were measured relative to stars that were assumed to be, mathematically, infinitely far away. Of course, if they were actually that far away we couldn't see them, but "much farther" than the star to be measured would be good enough. Those measurements of nearby stars gave a distance scale that was extended by the use of certain variable stars, Cepheid variables, first of all incorrectly because the standard stars were of one sort and the stars used for distance measurements were of a different sort. But the astronomers forged on and ignoring that experience, assumed that similar variable stars everywhere else were the same. That distance scale was extended out to the farthest reaches of the observable Universe, all of it based on a faulty assumption as to the nature of galaxies, and a belief that trigonometric parallax is absolute. (The bias introduced by the assumption that the reference stars are adequately distant, even if they aren't, gives a distance too great for the measured star.) So, Point 2 on the speed of the motion calculated from the distance has no integrity, but it is the ultimate rejection of van Maanen's measurements.

If, on the other hand, the spiral galaxies are not vast Island Universes of billions and billions of stars like our Sun, at great distances, but are instead the spiraling debris fields of detonated rotating planets (as I say, and just like the Milky Way Galaxy), they would be at distances of a few to a few hundred light years, not millions to hundreds of millions of light years, and van Maanen's measurements are quite reasonable. Reasonable velocities gave nearer distances for the spiral nebulae that were consistent with distances proposed by Harlow Shapley in his view of the Universe. Distances in the Island Universe view produced extremely great velocities, and suggested that van Maanen's measurements were faulty.

Spectroscopic measurements are thought to show rotational velocities (speed and coming or going) by use of the Doppler effect, and should be free of error, but there are a surprising number of discrepant measurements. The method can only be used with spiral galaxies that are somewhat tilted to our line of sight. Edge-on galaxies can be measured, but the direction of the spiral arms can't be guessed. That's difficult even for tilted spirals, as a view of M31 will show. Assuming those difficulties have been overcome, and I think that is generally true, the measurements showed rotation of the spirals in the opposite direction from van Maanen's measurements. van Maanen showed rotation in the direction that the arms pointed, spectroscopy showed motion crosswise to the arms, showing the galaxy was rotating in the opposite direction. But hidden in all this is that van Maanen measured condensations in the arms, points of light that might be stars or clumps of stars or HI regions or emission nebulae, and spectroscopy measured gas.

van Maanen attempted to measure displacements of condensations, tight points of light that might be stars or HI regions, or whatever contained regions might be rotating about the galaxy, whether near or far, fast or slow. The spectrometric measurements of galaxy rotational velocities, using the Doppler effect of velocity on spectral absorption lines, measures the velocity of dispersed gas, not those tight points of light [Slipher 1915]. In a spiral galaxy, a magnetic field runs along the spiral arms, developing a similar spiral arrangement. The magnetic field traps ionized gas atoms, but when the ions combine with electrons, becoming less charged ions, or neutral atoms, they slide off the magnetic field, and have a velocity directed in the opposite sense of the condensations, the arms themselves. In spectroscopic measurements of galactic rotation curves, we are seeing spectral lines caused by gas that is sliding off the magnetic fields embedded in the spiral arms, indicating a sense of rotation that is actually the opposite of the motion of the material in the arms themselves. The measurement of the gas velocity by spectroscopy is wrong for the galaxy, and van Maanen was right.

In a spiral galaxy, the magnetic field is carried by the arms, and the gas flows at high speed, sliding off the magnetic field, flowing crosswise to the arms, showing a rotation that would be opposite to that of the arms themselves. So, Point 3 is not valid.

When van Maanen plotted the vectors showing "annual motions" in the publications of his work, he was forced to exaggerate the length of the vectors in order to make the motions visible on the scale of the photographic image. That showed the motions visually as quite significant, when in fact the changes in position from one year, or even one plate to the next were nearly infinitesimal. When Edwin Hubble examined two plates of the same galaxy, using the same method that would be used successfully by Clyde Tombaugh to discover Pluto, perhaps he expected to see the jumps in position that were implied by the vectors in van Maanen's figures. However, these were exaggerated by factors of 1,100-1,300, and in fact, the displacements on the plates were no larger than the diameter of the star spots. Hubble could not have seen any changes in position, and he declared that he did not, proving that van Maanen's measurements were false, although this test was not capable of proving its point. Therefore, Point 4 is not valid.

In the meantime, Henrietta Swan Leavitt had been working with new photographic plates of the Magellanic Clouds in the Southern Hemisphere, nearby "irregular" galaxy companions to the Milky Way Galaxy. She had been able to identify many variable stars of the type known as Cepheid variables, and had discovered a remarkable regularity in the relation between the brightness of these stars and their periods [Leavitt 1908]. These stars wax and wane in brightness (presumably in size, and in color) with steady regularity, having periods on the order of a few days to a few hundred days. Many stars on many photographic plates, over an extended period of time, must be inspected to identify and characterize the variable stars. Cepheid variable stars were known in our own Galaxy, the Milky Way Galaxy. The type star, delta Cepheus (δ Cep), had long been studied. It was suspected that the periods of these variable stars, the time that passed from bright to dim to bright again, might be related to their absolute brightness, their absolute magnitude in astronomical terms. By analyzing the Cepheids in the Small Magellanic Cloud, Leavitt had selected a set of stars that were all at about the same distance, unfortunately unknown, so that this relation between period and true brightness could be explored. Her work showed clearly that the brighter the star, the longer the period, in a very simple relationship. Shapley proposed that the shortest period Cepheids should join, in absolute magnitude, the variable stars known as RR Lyrae variables, whose distances had been estimated. That provided a distance scale that could reach to the nearest of the spiral nebulae, if Cepheids could be found. Hubble seized the opportunity. With Humason as his photographic assistant, Hubble hunted for Cepheid variables and found them in M31, the Great Nebula of Andromeda, a northern constellation [Hubble 1936]. Measurement of the relative brightnesses of the Cepheids, and their associated periods, gave a distance of 897,000 light years. This was clearly far outside the Milky Way Galaxy, and therefore the spiral nebulae were galaxies on their own, just like the Milky Way Galaxy, vast Island Universes, at vast distances. This distance, quoted by Hubble as accurate to 2.5%, is far short of our current standard value of 2,365,000 light years.

Further study showed that the RR Lyrae variables and the Cepheid variables were not at all connected, and that the Cepheid variables of the Small Magellanic Cloud were so-called cluster-type Cepheids while those in M31 (and in the Milky Way Galaxy) were classical Cepheids, and so the distance scale was a bit off. Indeed, the fundamental basis of the scale was weakly determined by estimates, and the estimates were based on trigonometric parallax, which is badly flawed. The scale was recalibrated by use of the "moving cluster" method with the Pleiades. (This method assumes that the group of stars is gravitationally bound and moving together at a moderately great distance. The apparent convergence of their apparent paths, due to proper motion, provides an estimate of their distance. Then it was decided that questions of the age of the stars might affect this measurement, so the Hyades cluster was used instead. Thus, Hubble's original estimate of 897,000 light years for the Great Nebula of Andromeda, M31, grew to become our present estimate of 2,365,000 light years. Remarkably, discovery and correction of error after error only increased the confidence of the astronomers that they had developed an accurate distance scale for cosmological distances.

Hubble the astronomer was looking for Cepheid variables, to measure the distances to the spiral galaxies, and Humason the technician gave him the sharpest images possible, red-filtered photos. Naturally, Hubble found a redshift/magnitude relation, later defined as a redshift-distance law: he would not see galaxies that were not redshifted, at moderate distances. (Blueshifted galaxies are seen in the nearby neighborhood, but these disappear at greater distances because of the preferential scattering and absorption of blue light compared to red.) This effect still works, galaxies are seen (and searched for) at increasingly great distances, where only the reddest shine through. Except for the galaxies that contain the Type Ia supernovae that are said to show that expansion of the Universe has accelerated, since it was slower at the time corresponding to those distances. A key fact is that those galaxies show less redshift than is expected from the distance corresponding to the brightness of the supernovae, because there is little or no dust to dim them. They are visible even though they are not as redshifted as should be required at their distances. Those are the galaxies that fall below the Hubble line in the redshift *vs.* distance plot, the few galaxies that we can observe that fill in the broad velocity distribution I propose for cosmological objects.

The complicated details of the spectrum of the Sun were noted by Joseph Fraunhofer in 1815. It was not just a pretty rainbow discovered by Isaac Newton, but showed hundreds of dark lines, "Fraunhofer lines". Those were first noted qualitatively by William Wollaston in 1802. Most of these have been assigned to absorption of specific energies, wavelengths, of light by various chemical elements. Photographic spectrometry, the recording of astronomical spectra on photographic plates, had been developed by Henry Draper in 1843. Each chemical element, hydrogen, helium, sodium, magnesium, calcium, iron, and so on, emits or absorbs light at distinctive and identifying wavelengths. Light from an astronomical object is collected and focused by a telescope, passed through a narrow slit, and spread out into a rainbow of colors by a prism, or currently by a ruled grating that produces a spectrum like those that can be seen with the underside of a compact disc (CD). As a brightening (from emission) or a darkening (from absorption) shows as a bright or dark line across the strip of spectral color, these particular features became known as spectral lines. By comparison with light emitted by laboratory standards, the wavelength of light can be determined quite accurately. The wavelength of light emitted or absorbed by an atom is determined by its internal electronic energy levels, and those wavelengths are distinctly different for each kind of atom. A remarkable success of spectrometry was the discovery in 1868 of a new element, helium, on the Sun, before its identification in 1895 on the Earth, in a uranium mineral. It was subsequently shown that helium atoms were the same as the alpha particles emitted in the radioactive decay of uranium (and several other heavy elements).

As more was learned about spectrometry, for nothing in astronomy is simple, more details were understood, and the technique proved increasingly valuable. One of the important features of spectrometry is that it can be used as a speedometer to measure the velocity of objects and determine whether they are approaching us or moving away. This uses the Doppler effect, just as Doppler radar determines weather patterns and the highway patrol radar gun checks your driving speed. Radar waves sent out are reflected by the raindrops or car and the wavelength is shifted slightly by the motion of the raindrops or car. That shift can be measured and used to identify rain, or to measure the speed of the car. The shifts have been named in astronomy as redshifts if the wavelength is increased, or blueshifts if the wavelength is decreased. Differing in interpretation from the radar gun and the weather radar, astronomical Doppler measurements require the determination of the change in wavelength of individual spectral lines. Other effects can produce such shifts, such as gravity or the expansion of space, so some care must be exercised.

While Doppler radar can measure the speed of solid objects such as automobiles, aircraft, and raindrops, astronomical Doppler cannot. Astronomical Doppler measurements are based on the wavelength shift of spectroscopic lines emitted by, or absorbed by, low density gas. (High density gas smears out the lines too much for the measurements to be accurate, "line blanketing" becomes important, and sufficiently high density produces a continuous spectrum like a solid black body, with no observable lines.) The gas may be associated with a star or a galaxy, and we are likely to assign the velocity measured for the gas to the star or galaxy. But that should be done very carefully. The gas may simply be in our line of sight, and it may have a velocity that is very distinctly its own. The latter situation may be shown in HII/HI regions that derive from the beta decay of neutrons in space. Henry and Branch [Henry and Branch 1987] measured the spectrum of a supernova and found an oxygen emission line at a velocity of 350 kilometers per second from the supernova galaxy, but the accompanying hydrogen line had a velocity of 80 kilometers per second. This shows the difficulty in estimating astronomical velocities from the Doppler shift of wavelengths.

If the true wavelength of the light (the spectral line) can be determined, and this is assured if the chemical element (and its state of ionization) can be identified, then the speed and direction of the emitting or absorbing gas can be found.

One of the great contributions that Hubble made was his demonstration that for the spiral galaxies that he had determined distances based on the Cepheid variable distance scale, the more distant galaxies had greater redshifts [Hubble 1929], building on the spectral measurements of Vesto Slipher. Remarkably, this had been shown by Georges Lemaître in 1927 [Lemaître 1927], using published redshifts and distances based on magnitudes, but his work was not well recognized until 1931 [Lemaître 1931a, Lemaître 1931b , Way and Nussbaumer 2011, Kirshner 2009, Bartusiak, 2009, Nussbaumer and Bieri 2009, Livio 2011]. While Lemaître deserves the credit for first announcing the expanding Universe, it appears that Hubble was unaware of Lemaître's work. Considering the intensely theoretical nature of Lemaître's presentation, and Hubble's legendary aversion to theory, this may not be too surprising.

While some of the galaxies of our local group appear to be moving towards us, the redshift of the galaxies that Hubble picked from the photographic images made by Humason showed a linear relation with distance, as determined by the Cepheid variables. Because of the bias introduced by Humason's photography, Hubble was measuring the redshift of redshifted galaxies. Blueshifted galaxies, those coming towards us, are lost by the scattering and absorption of intergalactic dust, and discriminated against by the red photography. This historic discovery was published in 1929 [Hubble 1929] and astronomers have been measuring the Hubble redshift/distance relationship for redshifted galaxies ever since [Madore *et al.* 1998, Sandage *et al.* 1996].

When the first spectrometric measurements were made of shifted spectral lines, both blueshifted and redshifted, coming towards us and going away, they were all interpreted as due to the Doppler effect of velocity. Motion towards or away from an observer shortens or lengthens the wavelength of light, and causes a shift in the observed wavelength. Shorter is called a blueshift, because it is normally in the direction of the color blue in the visible spectrum. Longer is called a redshift, because it is normally in the direction of the color red in the visible spectrum. As Einstein's Theory of General Relativity came to be more fully incorporated into astronomy, it became clear that there could be a "cosmological" redshift, not caused by physical motion of the galaxies, but due to an expansion of space, changing the distance from them to us.

The expanding Universe was established in the late 1920s, in response to the astronomical observations reported by Edwin Hubble and Milt Humason, showing that galaxies were moving away from the Earth with velocities that were generally increasing in proportion to their distances from the Earth. Galaxies were identified and located by their images on photographic plates, usually taken with a red filter. Distances were determined by use of the Cepheid-variable period-magnitude relationship, and velocities in the line-of-sight were determined by the Doppler shift of identifiable absorption lines in the spectrum of the gas of each galaxy. The wavelength of light is changed by the velocity of the gas, relative to the observer. Gas moving towards us has light shifted to shorter wavelengths, blue-shifted, while gas moving away has light shifted to longer wavelengths, red-shifted. This is not just a change in color, but the change in wavelength over the spectrum results in a change in color. The relationship between the change in wavelength and the velocity is well established, even at relativistic velocities, by the Special Theory of Relativity.

To determine the Hubble Constant, H_0, thereby demonstrating the expansion of the Universe, two measurements of distant galaxies must be made. The distance to the galaxy must be found, using methods derived from the cosmological distance scale. The recession velocity must be measured, by use of the wavelength shift of spectral lines. It is desirable to use very distant galaxies to make the velocity and distance terms in the ratio defining the Hubble Constant large, increasing the precision, if not the accuracy.

The first step in distance to the great distances presumed for the galaxies is the method of trigonometric parallax. This is thought to be an absolute method, as it is similar to the civil surveyor's triangulation, which is absolute by the rules of plane geometry. In fact, however, trigonometric parallax is a relative distance measurement, based on the selection of reference stars for the determination of the small angle. This method, even if it were correct, is seen as too limited in reach to measure the distances of the galaxies. For the nearest galaxies, distance is determined by use of the Cepheid variable distance scale, which has as an underlying foundation, the trigonometric parallax.

For more distant galaxies, methods of measuring distance are used that have been calibrated by use of the Cepheid variable distance scale. Corrections are made to the recessional velocity to adjust for presumed gravitational attraction between various galaxies and clusters, and our own. The major cause of the general steady reduction in H_0 has been changes in the calibration of the Cepheid-variable distance scale.

To establish the red-shift/distance relationship, a galaxy is selected, its distance determined by any one of several accepted relative measures, its spectrum taken, and the shift in wavelength of spectral lines is used to determine the velocity of the galaxy relative to the Earth. (This process continues at present, with greatly improved instruments and modified calibrations. The results continue to produce controversy.) These measurements show that the recessional velocity is proportional to the distance. This relationship is stronger for more distant galaxies. (Nearby galaxies show considerable scatter, some approaching and some receding from the Earth.) This relationship has been explained as being due to a uniform expansion of space, and that expansion agrees with one possible solution of the equations of Einstein's General Theory of Relativity. Hubble's observations were taken as proof of the theory, and in turn, the theory was taken as confirmation of Hubble's red-shift/distance relationship.

Hubble's observations were first interpreted as a Doppler shift in wavelength, due to actual physical velocities. This was seen to support one of the three easy solutions to Einstein's field equations of General Relativity. One solution provided for an expanding Universe, one provided for a contracting Universe, and the third allowed a static, unchanging Universe, but, it was said, the Universe had to be empty of matter. (Closer examination of the equations shows that this requirement is instead that the average net density of mass must be zero. Antimatter had hardly been imagined, and not yet discovered, at that time.)

The contradictions of two theories is discussed below:

A Tale of Two Theories

Sometimes, theories that are equally well accepted for their specific purposes give contradictory results. This can be an example of "we get what we ask for" and "what we see depends on how we look". In other cases, when only one theory is recognized, the conflict may be with reality, but is not clearly revealed because we got what we asked for and it depended on how we looked.

Data from observations must be interpreted by use of a theory in order to be converted into useful information. The effect of interpretation can be clearly demonstrated by a common example in astronomy.

Astronomers use spectroscopy to measure the shift in wavelength of light from an astronomical object, tiny to large (even from blue to red and beyond), and interpret this shift in terms of velocity of the object by use of two different theories, Doppler and Special Relativity. These theories give contrary results, each consistent with its own premises.

In Doppler theory, the velocity is calculated as directly proportional to the fractional wavelength shift, by multiplying by the speed of light. This theory allows material objects to go faster than the speed of light, and velocities increase linearly without bound, as the wavelength shift increases.

In Special Relativity, the velocity is a somewhat more complicated function of the wavelength shift, to accommodate the speed limit imposed by this theory: no material object can go as fast as the speed of light.

In Figure 123, on the left, a large set of observed wavelength shifts (z) has been interpreted by the Doppler theory and, consistent with that theory, the velocities increase well above the speed of light, contradicting the theory of Special Relativity.

In Figure 123, on the right, the same set of observational data has been interpreted by use of Special Relativity and, consistent with this theory, the velocities only asymptotically approach the speed of light, contradicting the Doppler theory.

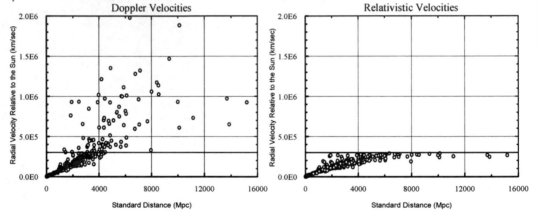

Figure 123. Comparison of the redshift/distance relation.
Comparison of the redshift/distance relation interpreted by the standard Doppler effect, and by the relativistic velocities. The horizontal line indicates the speed of light, which is the ultimate speed limit. The great scatter of the observational points is shown clearly by the Doppler interpretation

For various reasons, we have chosen to accept Special Relativity as a more nearly correct representation of reality and, in most cases, use it rather than Doppler theory. At very small velocities, such as we experience on Earth, the Special Relativity formula reduces exactly to the Doppler formula, and radar guns and Doppler weather work quite well, with no regard for the conflicts at large wavelength shifts. (As a side issue, the Doppler Velocities plot clearly shows how variable are the measurements of astronomical distances, or wavelength shift, or both. Distances can differ by a factor of nearly eight, and so can wavelength shifts. This may suggest that the measurement methods are not entirely consistent, or may be measuring different things.) This comparison shows the importance of interpretation in our understanding of reality. The Doppler theory allows things to go faster than the speed of light, and provides the evidence to prove it. Special Relativity says that things cannot go faster than the speed of light, and provides the evidence to prove it. Two different theories, each accepted in its own realm of application, produce contradictory results when used to interpret the same data.

Hubble's measurements, showing that distant galaxies were apparently moving away from us, was taken as observational support for the expanding Universe of Einstein. Acceptance of an expanding Universe led to the development of the Big Bang theory of creation, in which space, time, the Universe, and all the matter within it, were created in an instant from a point, and space has been expanding ever since.

In an expanding Universe, the redshift was identified as a shift like that due to the Doppler effect, but was named the cosmological redshift, and the recessional velocity, because it resulted from the expansion of space carrying the galaxies away from us. In this view, the galaxies do not have the actual velocities derived from the redshifts, as Doppler shifts, but the distance between us and them is getting larger because of the expansion of space in the Universe. The rate at which the recessional velocity increases with distance became known as the Hubble Constant, H_0. It is not truly constant with time, being a changing aspect of the Universe, and its measured value has hardly been constant during the time of its measurements, now almost a century long.

A plot of the values reported for this constant is shown in Figure 124 [Huchra 2008]. It is clear that for the first thirty years of its existence, astronomers were making major corrections to their methods and theories. In the past thirty years, astronomers have converged on a value of roughly 75 kilometers per second (apparent velocity) per million parsecs of distance. The best current measurements still differ within themselves more than the estimated errors suggest. This might indicate that the constant is not the significant characteristic of the Universe that it is considered to be, and that the astronomers are measuring and interpreting different aspects of the galaxies. In spite of all the efforts to produce consistent values, we must wonder if this is a Universal constant, or just a happenstance result of galaxy velocities and attenuation of distant light.

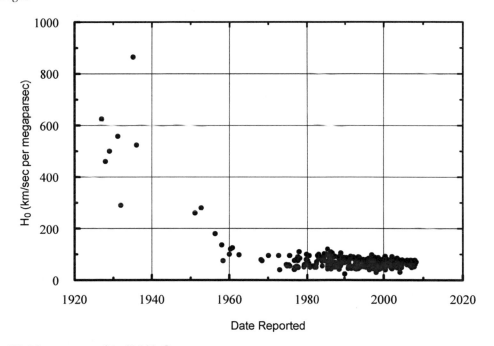

Figure 124. Measurements of the Hubble Constant.
The decline of the Hubble Constant. Values reported for the Hubble Constant, H_0, from the beginning to the major efforts of the present [Huchra 2008]. Corrections to the distance scale brought the value down to its present level, but there is still much scatter in measurements done different ways, and for different galaxies.

There are several current measurements of H_0 that disagree with each other more than the expected uncertainties should allow. At the start of this book (1993), a further problem was that the age of the Universe estimated from its expansion, originating in the Big Bang, was less than the ages estimated for what are thought to be the oldest stars. This is a problem that has been made to go away by further observations and interpretations.

More distant galaxies are sought in an effort to determine the Hubble Constant, H_0, more accurately. This search leads to rejection of data that don't perform well in the estimation of H_0. This can be seen in Figure 125, where the galaxy redshifts and distances of those used for the determination of the Hubble Constant [Huchra 2008] are plotted along with a broader collection of galaxy redshifts, from the Catalog NED-1D [Madore and Steer 2008].

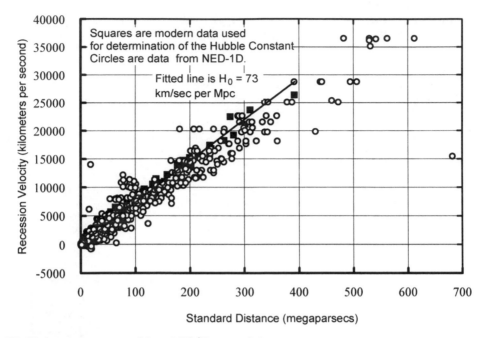

Figure 125. Modern measurements of the redshift/distance relation.
Galactic redshifts used to determine the Hubble Constant (filled squares) and those not used (open circles). Some galaxies certainly don't "go with the flow" and have been de-selected in the determination of the Hubble Constant.

This shows that there are many galaxies that have been judged to be unsuitable for the determination of one of the most important parameters in our present theory of the Universe. Including them all does not change the value much, but still must leave us wondering what to do with all the misfits.

Thus, Hubble's discovery 80 years ago, using Humason's photography, has spawned a new expanding Universe, a new Big Bang creation, and a new industry for measuring the Hubble Constant. One of the "Key Projects" identified as a reason for the *Hubble Space Telescope* was the precise determination of the extragalactic distance scale, for the sake of the Hubble Constant. While we may still be living in an expanding Universe, and the Big Bang may truly have been the start of it all, Hubble's biased measurements (and all those that follow in the same path) provide us no true reason to think so.

Some of Our Galaxies are Missing

Broad surveys have captured images of a tremendous number of galaxies. These surveys have permitted assignment of many different parameters to those galaxies, such as brightness in various colors ("passbands"), type, distance, and coupled with spectrometry, the velocity relative to the Solar System. These results have been collected into astronomical catalogs, offering easy access to the data. One of the interesting properties to explore is the distribution of galaxies in the Universe. A basic ground rule for astronomical research is the Cosmological Principle, which says that the Universe is generally the same all around us. This is a broad generalization and is more philosophical than empirical, but it is the best we can do to guide our ideas of how the Universe should act. This constraint is often coupled with an assumption that the laws of physics are the same throughout the Universe. The best that can be said is that there is little evidence to contradict these assumptions.

However, in reviewing the distribution of galaxies, we seem to have a problem. Where are all the distant galaxies? This problem was shown in Figure 112, where the number density of galaxies as a function of distance shows a great departure from our expectation that the density at great distances should be about what we see nearby. Instead, the density declines, even when corrected for our loss of galaxies with distance due to the geometrical dimming by distance alone.

The progressive decline with distance (or Doppler velocity) in the number density of galaxies shows that something is wrong, either with the Cosmological Principle, which says the Universe should be generally uniform and isotropic, or with our observations, which say that the galaxies are going away faster and faster and there are fewer and fewer of them. (Paradoxically, while that last statement might seem to explain the paradox, it is actually the reverse. We are looking back in time to when the Universe, if it is expanding, was more compact than it is at present (here and now), and therefore the number density of galaxies should have been greater (there and then).)

Optical astronomical observations record light from distant objects in selected (and now formalized) passbands. In the beginning of the great growth of astronomical photography, at the start of the 20th Century, passbands were created for each photographic image by the choices of photographic emulsion, exposure time, development, any filters used, and whatever other art the observer and darkroom technician might apply. More recently, formally defined pass bands are in use. Some of these are shown in Figure 126. There is still much art in the observations. Historically, surveys have been done using blue sensitive photographic emulsions, because of the greater sensitivity and shorter exposure times. Selected galaxies are then examined by use of filtered red-sensitive photos. The filter eliminates the blue light and provides a sharper image, allowing individual stars to be studied. Red photography also reduces one of the corrections required for distant galaxies, and became the standard technique early in galaxy research.

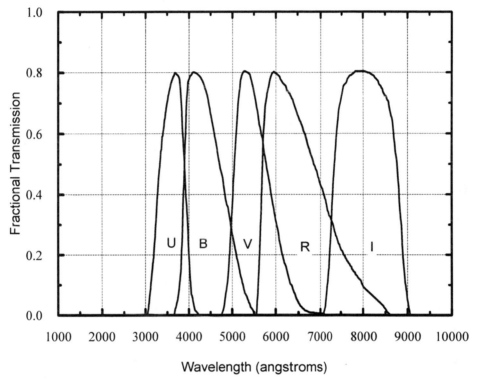

Figure 126. Astronomical passbands.
Several passbands spanning the photographic spectrum from near ultraviolet to near infrared.

The B (blue) and R (red) passbands are those most often used in astronomical photography. Red sensitive photography was favored, perhaps perfected, by Milt Humason in his efforts to provide Edwin Hubble with the sharpest possible images for the purpose of finding Cepheid variables for the determination of galaxy distances.

In order to explore the effects of Doppler shifts on the spectra and detectability of galaxies, I have collected the spectra of four galaxies, averaged them, and de-attenuated the galaxy spectra, and then extended them to shorter and longer wavelengths by attaching those parts of the Solar spectrum, as it is seen in space, above the attenuation of the Earth's atmosphere. This produces a representative incoming spectrum, and attenuating that gives a representative spectrum, a reference galaxy spectrum, as it would be seen on the ground. (Astronomical observatories choose to be at high elevations and that would reduce the absorption by our air. Thus this is only an approximation. The attenuated galaxy spectrum does match well the average of the published spectra.) This is shown in Figure 127, which also shows the BLUE (B) and RED (R) passbands.

This somewhat cumbersome process is necessary because it is the "in space" spectrum that is Doppler shifted and the shifted spectrum is then attenuated on passing through the atmosphere. While the *Hubble Space Telescope* records spectra well above the atmosphere, I am exploring the historical effects of the hunt for the Hubble Constant, and that hunt has been mostly on the ground.

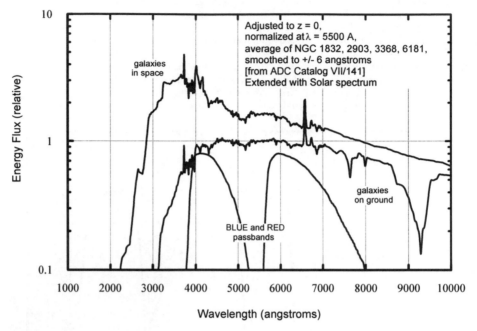

Figure 127. Galaxy spectra.
Spectra of galaxies seen in space and on the ground, with no Doppler shift, after passing through the absorption of our atmosphere. The passbands for blue and red light are well separated.

The motion of a galaxy, towards us or away, shifts the wavelengths of the spectrum by the Doppler effect or the expansion of spacetime. Motion towards us shortens the wavelengths (a blueshift), while a motion away lengthens the wavelengths (a redshift). These effects are shown in Figure 128 for motion towards us at 30,000 kilometers per second (z=-0.1) and away (z=+0.1), straddling the spectrum from a stationary galaxy (z=0). Absorption by the atmosphere reduces the effect so that all spectra are roughly the same on the ground.

At greater redshifts, the light from a receding galaxy has insufficient energy to be absorbed by a greater fraction of the intervening gas, and at a longer wavelength is less likely to be scattered by dust. The receding galaxies become more visible with increasing redshift. Galaxies that are approaching us (actually any self-luminous object), have their light shifted to shorter wavelengths, higher energies, "bluer". That light is readily absorbed by any intergalactic gas (bluer light is more energetic than redder light and so is absorbed more easily), and the bluer light is more effectively scattered in all directions. Blueshifted galaxies disappear.

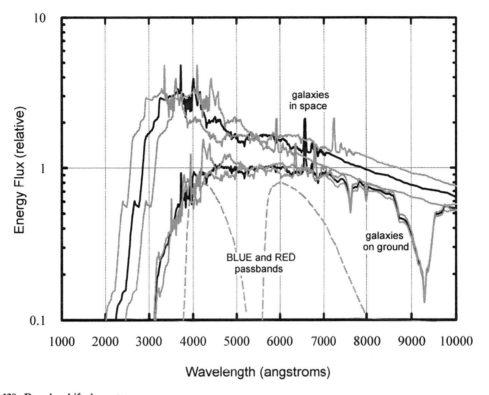

Figure 128. Doppler shifted spectra.
The effect of the Doppler shifts on the spectra of galaxies as seen in space and on the ground.

Attenuation of Light by Absorption and Scattering

Objects (in the most general sense of that word) interfere with light, they intrude on our observations. Large objects, such as planets, satellites, asteroids, rocks, simply block the light from distant sources. Astronomers refer to that as occultation or eclipse. The incident light may be reflected, scattered off into some useless other direction, or it may be absorbed, slightly warming the rock, to be reemitted later as a multitude of lower energy photons. To the observer, the light is simply gone. But a rock blocks light rather wastefully, using much more mass than is necessary, and we can't expect there to be enough rocks to interfere with our observations on a Universal scale.

For small objects, a rock finely divided into dust grains, the individual grains interact with the individual photons more intimately. Smaller grains scatter shorter wavelength photons more powerfully. Small bubbles of gas may scatter light in a similar manner to small dust grains, and individual atoms and molecules are so effective in scattering short wavelength photons that they make our sky, sky-blue.

The Universe is filled with dust and gas. Neutral hydrogen clouds (HI) drifting intermittently through space make the multitude of dips in the high energy spectra of distant sources, labeled the "Lyman-alpha forest", from the many closely spaced absorption lines, each at a different redshift. (Absorbers can be Doppler shifted just as sources are.)

Smaller objects scatter blue light (and shorter wavelengths) more strongly than they scatter red light (and longer wavelengths). All this scattering removes some of the light we wanted to see from the sight of what we are looking at, making those sources appear to be dimmer than they should be. The dimming is greatest for short wavelength light, and least for long wavelength light. For small particles, small compared to the wavelength of light, this attenuation by scattering greatly affects the spectrum. This is shown most clearly in the case of "Blue Star, Red Star", which might be the title of a children's book.

The spectral type of a star can be determined by the details of its observed spectrum. In a particular case, shown in Figure 129, a star has been observed through a cloud of dust as a dim red star, with the spectral details of a bright blue star [Bless 1996]. Because the type of star can be identified as a B2 III star, its correct spectrum can be represented by that of a nearby star of the same type, without attenuation by dust. It is clear how drastically the blue end of the spectrum has been depleted by scattering. This is the same effect that gives us blue skies and red sunsets. The blue end of sunlight is depleted by scattering, and is seen all over the sky as scattered light.

This comparison would apply to any distant source of light, depending on the size of the dust particles, their density, and the distance between the source and us, in whatever ways are peculiar to the particular source of light. Doppler shifts change the characteristic wavelengths, and that changes the effective scattering and attenuation.

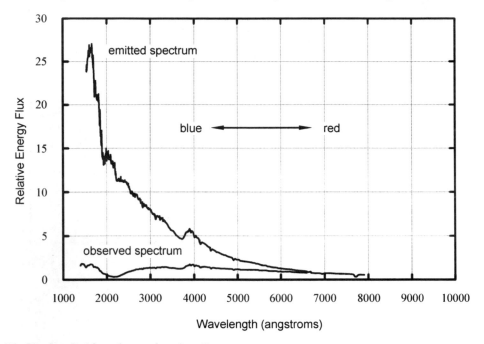

Figure 129. Blue Star, Red Star. Attenuation of a stellar spectrum.
Comparison of the energy spectra of light as observed for a B2 III blue giant star through an intervening dust cloud, and as it would be emitted according to its type, without any scattering by dust.

If there were no dust, the star would appear to be a bright blue-white, perhaps of first magnitude, nearly as bright as Sirius, to arbitrarily give it a rank. The dust, by scattering away the blue light, dims and dulls the star to a reddish hue, leaving it with a brightness of magnitude 1.73, a loss of 0.73 magnitudes, and still a fairly bright star, visually. In the B passband, the star loses 1.07 magnitudes, but in the R passband, it is practically undimmed, losing only 0.23 magnitudes. That is why examining distant objects, dimmed by dust, is done so effectively in the red. Use of red filters also reduces the size of a correction to total (bolometric) magnitude [Humason, Mayall, and Sandage 1956] and so is preferred

In detail, the scattering effect of small objects, bubbles of gas, and density fluctuations, is an intractable problem. We must consider dust grains of different shapes, materials, and sizes, gas bubbles of different shapes, materials, and sizes, and a wide range of density fluctuations, whiskers and rods spinning like a drum major's baton, always varying in "size" to the impinging photons, huge clouds of otherwise invisible hydrogen atoms, protons, electrons, diverting the light away from our telescopes.

The transmission through the dust is shown in Figure 130, in comparison with the blue (B) and red (R) passbands.

The scattering cross section of the grains of dust can be derived from this by assuming a distance of 100 parsecs (326 light years), and a grain number density of 2.56×10^{-2} grains per cubic centimeter. (That strangely precise number comes from using an average hydrogen column density of 2.66×10^{20} atoms per square centimeter from data presented in *"Extreme Ultraviolet Astronomy"* and a ratio of dust grains to hydrogen atoms of 2.97×10^{-2}, also from data in *"Extreme Ultraviolet Astronomy"* [Malina and Bowyer 1991].) The cross section calculated from the blue star attenuation is shown, with several other cross sections, in Figure 131. In spite of the wild choices of values, the comparisons are quite good. Most notable is the change of the slope of the blue star cross section from the Rayleigh cross section at long wavelengths, where the dust grains are truly small compared to the wavelength, to the constant slope of Thomson scattering as the wavelength decreases and the grains become similar in size (without changing size, of course) to the wavelength. The slope of the Blue Star cross section matches the slope of the Rayleigh cross section at longer wavelengths, where the grains are small compared to the wavelength, and approaches the zero slope of Thomson scattering at shorter wavelengths, where the grain size becomes similar to the wavelength. That is an impressive demonstration of how physics works.

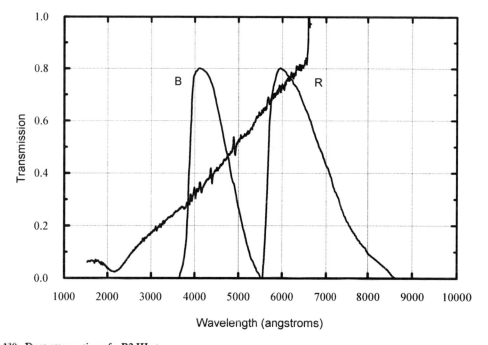

Figure 130. Dust attenuation of a B2 III star.
The fractional transmission through dust of the light from a B2 III blue giant star. The B and R passbands show how differently light would be received in these different colors after attenuation by the dust.

The Thomson cross section applies to particles that are large compared to the wavelength of light and is constant, independent of wavelength. The Rayleigh cross section applies to particles that are small, and it is a steep function of decreasing wavelength. The cross section for the particles that attenuate the blue star light matches the Rayleigh cross section (in slope) at long wavelengths, where most of the particles are smaller than the wavelength of light, and it approaches the constant form of Thomson scattering for short wavelengths, where the particles are now similar to the wavelength of light. In between, it approximates the "$1/\lambda$" scattering dependence often used as a rule of thumb [Greenberg 1968]. The transition of the particles from large to small, without changing their sizes as the wavelength increases, is quite striking.

As shown in the example of the blue-white star, the overwhelming effect of scattering by small objects, from atoms and molecules to dust grains, is to turn blue sources red. This is reddening, not a redshift, which is seen as a shift of spectral lines of specific wavelength to a different part of the wavelength scale.

Because of this effect, blueshifted galaxies, which have their light shifted into the blue region, will become dimmer, less easily detected at increasing distances where their light must pass through more scatterers, while redshifted galaxies will lose little luster.

A more drastic effect is caused by the absorption and scattering of light by tiny grains of dust, whiskers, and random variations in the hydrogen density along the line-of-sight. The standard way to calculate this effect is to find the cross section of the scatterers, multiply by their average density and by the distance, and use the result in an exponential calculation of the attenuation. It should be immediately obvious that we can make models but we really don't know the answers. Very fine particles scatter blue light far more effectively than red, and this makes galaxies disappear in blue light. I have calculated an example for the loss of blue light to make all but 0.1% of the galaxies disappear in blue at a standard distance of 400 megaparsecs. That is in agreement with the decline in the number density of galaxies previously discussed. Rather than using calculated or laboratory values for the cross section as a function of wavelength, and guessing at the average density of whatever types of particles, whiskers, and gases might lie between us and a very distant galaxy, I have used the attenuation shown by the Blue Star comparison.

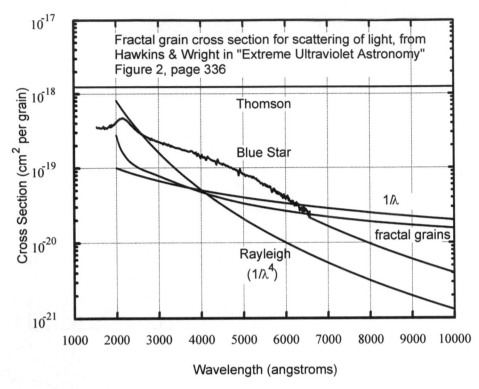

Figure 131. Attenuation of light by small grains.
Scattering cross sections derived from the blue star attenuation, and standard theoretical and practical dust grain scattering. The smooth portion of the Blue Star cross section, from 6,500 to 10,000 angstroms is an extension based on Rayleigh scattering.

In Figure 132, I show the spectrum of a galaxy in space, as if there were no attenuation, and the spectrum as we should receive it after the flight of the light through a great distance from the galaxy to the Earth, as it would appear above the atmosphere, and on the ground. It can be seen that the attenuation by dust and gas decreases the blue light (shorter wavelengths) but does not strongly affect the red light (longer wavelengths) through much of the RED passband.

This wavelength selective dimming causes the galaxy to lose 5 magnitudes of brightness in the blue, but only 0.73 magnitudes in the red. All but 0.1% of the galaxies at the standard distance of 400 megaparsecs are lost from sight in the BLUE passband, while there is little effect on galaxies seen in RED.

The question must be raised, "Is it reasonable to expect the same, or greater, amount of attenuation for a distant galaxy as for a star hidden behind a cloud of dust?" The parameter that determines the attenuation, as discussed above, is the product of cross section X number density X distance. The cross section is a function of the particles and gas pockets, the number density is poorly known, and the distance, though quite probably grossly in error, has a reasonable relation to distances used for the estimation of dimming of nearby stars. So the question devolves on density and distance. "What the Universe lacks in density, it makes up with immensity." The relative distances (in the standard scale), galaxy compared to star, are like 400 megaparsecs compared to 10 kiloparsecs, a factor of 40,000. Shouldn't there be room for enough dust particles and clouds of gas to permit a factor of 5.47 in the scattering exponent ("n") for a galaxy 40,000 times as far away as the star? It's even hard to imagine that it might be that small.

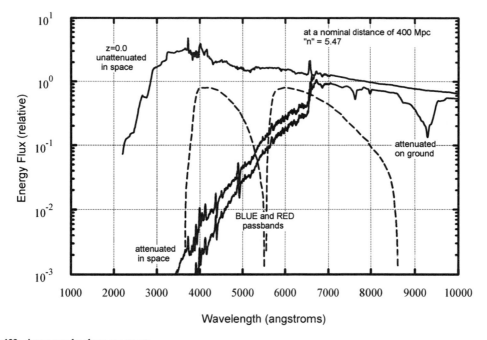

Figure 132. Attenuated galaxy spectrum.
Attenuation of light from a distant galaxy preferentially dims the galaxy in the BLUE passband, while little light is lost in the RED. This causes galaxies to disappear in blue surveys, not be detected, but allows the selected galaxies to be effectively studied in the red. That was the photographic effect discovered by Edwin Hubble's photographic assistant, Milt Humason, that led to the selection effect which produced the Hubble Law, the redshift-magnitude relation, that was interpreted as showing that the Universe is expanding.

Thus, the "observation" that the Universe is expanding can be seen to be an observational artifact stemming from the selection of galaxies in blue-sensitive surveys, and their study in red photographs. Hubble needed the sharpest possible photographic images to find and measure the Cepheid variable stars that he would use to determine distances, and which are the basis for current distance determinations in the ongoing hunt for the Hubble Constant. Humason found that the sharpest photos he could produce came from the RED. This biased Hubble's studies so that he only saw galaxies that were moving away from us, he had no chance to see those headed our way, and so it is no wonder that he found only redshifted galaxies. The Universe is not expanding, at least on the scale we have thought, so we must look to the other solution of Einstein's Equations of General Relativity, the static Universe, neither expanding nor contracting, but remaining the same size through its lifetime.

Disappearing Galaxies

What the Universe lacks in density,
it makes up with immensity.
Anonymous 1998

Galaxies that are moving towards us, emitting light that is blueshifted by the Doppler effect, are generally dimmer (all other things being equal) than galaxies that are moving away, with redshifted light. By itself, this is a small effect, amounting to only a calculated reduction in brightness of 16-20% for $z = \pm 0.1$ (30,000 kilometers per second), in the RED and BLUE passbands. However, this alone will make it more likely that the brightest, most easily studied galaxies in an observational image will be the redshifted galaxies, moving away from us.

In addition, another effect which may strongly discriminate against detection of distant blueshifted galaxies, but I have not been able to calculate it, is the absorption of emission line light from a galaxy by intervening gas. The energies of spectral line emission are increased by the blue Doppler shift and those lines are more likely to be absorbed by the same atoms or molecules in the intervening gas, making those galaxies dimmer. The energies are decreased by the red Doppler shift of galaxies going away, and are less likely to be absorbed, making those galaxies brighter. This is an effect that will uniformly result in the dimming of blueshifted galaxies, relative to similar redshifted galaxies, and will decrease their detectability in all passbands. Gas is considered to be much more plentiful in extragalactic space than dust grains are. This is clearly shown by the multitude of hydrogen absorption lines seen in distant objects, dubbed the "Lyman-alpha forest", because of line after line at different Doppler shifts, making the spectrum appear like the skyline of a forest. The emitted light from the object is severely depleted by this absorption.

The amount of "stuff" in the space between astronomical objects has always been grossly underestimated, leading for example, to Hubble's great underestimation of distances, by a factor of about ten.

The calculation of the attenuation of light passing through great distances of very dilute clouds of small particles, electrons, atoms and molecules, dust grains, is somewhat more straightforward. The calculation involves three parameters: the scattering cross section of the particles, the number density of the particles, and the distance to the galaxy. None of these can be determined with certainty, although we can reasonably use the relative distances. It is common practice to use the product of the number density and the distance, as the column density, particles per square centimeter along the line of sight in the direction of the galaxy. The product of the three parameters gives the optical depth, and this is used in an exponential function to give the attenuation. It should be clear that, even for a set distance, the other two parameters can (and will) vary, giving more or less calculated attenuation.

Using the "Blue Star" cross section, as a physically real cross section, not just theoretical, varying the column density of the intervening dust grains allows the attenuated spectrum for a galaxy to be calculated. For a reference galaxy at a standard distance of 400 megaparsecs, receding with the Hubble velocity of 30,000 kilometers per second (z = +0.1) the response (brightness) in the B and R passbands, and for the commonly used emulsions 103a-F and IN, is shown in Figure 133, as a function of column density. (Because of the relatively small effect of the Doppler shift on the received spectrum, the plot for z = -0.1 is quite similar.)

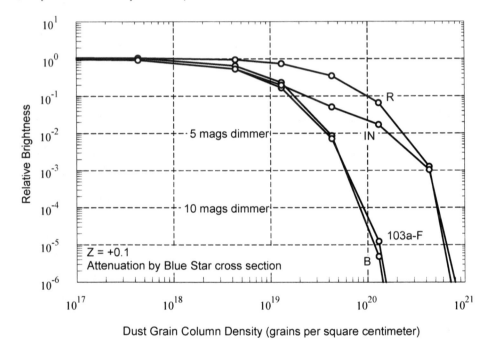

Dust Grain Column Density (grains per square centimeter)

Figure 133. Dimming by dust.
The loss of light resulting from attenuation by dust, as seen in the various passbands and emulsions. A loss of 5 magnitudes makes a galaxy invisible in most surveys.

Blue-sensitive images, in B and 103a-F, lose 5 magnitudes and more of the original brightness for reasonable column densities of about 10^{20} grains per square centimeter, while the red images, in R and IN, still show enough light to encourage the use of longer exposures. Using the most sensitive astronomical photography (blue), the distant galaxies become undetectable, they disappear. That is the reason Milt Humason used red photography for Edwin Hubble's observations of an expanding Universe, and why the only galaxies that were studied were redshifted, going away from us. (Keep in mind that the gas absorption effect will make this discrimination even stronger.)

At still greater column densities, infrared films and detectors must be used to pursue the ever more attenuated galaxies at greater distances.

With selection of the "best" red-imaged galaxies for determination of the Hubble Constant, to the neglect of the undetectable blueshifted galaxies, we see predominantly receding galaxies. We measure the redshift of redshifted galaxies.

The Hubble Constant and the Accelerating Universe

Several methods have been developed recently to extend our grasp of distance. These have been discussed by Richard Massey [Massey 2009]. One of these is the use of Type Ia supernovae as standard candles. ("Standard candles" are considered to be of a uniform brightness and so variation in the observed magnitude indicates distance, with dimmer being faerther away. This part of the distance scale must be calibrated by other methods.) When this method was first used, it discovered a new effect, accelerated expansion of the Universe, caused by Dark Energy. Further experience with this method has shown that it has some observational complications that have to be adjusted out.

A remarkable occurrence in recent cosmology happened when a new method of measuring cosmic distances was extended to the farthest reaches yet achieved for the Hubble Constant. These observations showed that the Hubble Constant had been less in the past and was greater now [Kirshner 2002, Aczel 1999, Riess, Press, and Kirshner 1995, Hamuy *et al.* 1996, Perlmutter *et al.* 1998, Perlmutter *et al.* 1999]. This is a happy example of the use of a new technique leading to new, unexpected knowledge.

Regarding the galaxies that appeared to have Supernovae Type Ia, a standard candle for determining extragalactic distances: "They aren't going as fast as they should be [Doppler effect redshift measures velocity] for their distance [by their brightness as standard candles]. They are not reddened by dust."

Estimating the focal length of presumed gravitational lenses is another new method. This has observational problems in determining the actual distortion of the imaged galaxies. This method has also had some adjustments to make the answers come out right. The presumed recession velocity is measured by a well-established method, identifying specific spectral lines and determining the shift in wavelength. However, it should be recognized that these spectral lines arise in the gas that is (presumably) associated with the galaxy. The wavelength shift measures the velocity of the observable gas. It does not necessarily measure the velocity of the galaxy. These are galaxies that fall below the Hubble line in the Velocity *vs.* Distance plot, and they fill in the velocity distribution. They are seen through no dust, so they don't need a redshift to be visible.

The development of precise techniques for using Type Ia supernovae as "standard candles" for measuring great extragalactic distances provided a greatly expanded reach for Hubble Constant hunters. But it is clear from the data shown in Figure 134 that the results depart from those measured at lesser distances. For the Hubble Constant measurements using Type Ia supernovae, the brightness of the supernova at the maximum of its light curve, a distinctive light curve with two humps which helps identify it as a Type Ia, is compared with the calibrated "standard candle" brightness to estimate distance, and the redshift of spectral lines in the host spiral galaxy is used to determine velocity.

Instead of a steadily expanding Universe, the observations suggested an acceleration in the rate of expansion, a new effect in cosmology.

The generally uniform departure of the Supernovae Type Ia measurements from the straight line extrapolation from shorter distances (the slope of the straight line is the Hubble Constant at the present time) has been interpreted as showing that the expansion of the Universe at times in the past, billions of years ago in our time frame, was slower than at present. As the straight line must bend down to pass through the Supernovae Type Ia points, its slope becomes progressively less, and so the Hubble Constant, averaged over the distance and time, becomes less.

This decrease in the Hubble Constant has been interpreted as showing that the cosmological expansion of space was less in the past, has increased to the present time (as observed at closer distances), and the expansion of the Universe is accelerating. The reason for this expansion is truly unknown. It might be related to the "Cosmological Constant", Λ, that Einstein felt he had to add to his equations of General Relativity to produce a static Universe with neither expansion nor contraction of space. The unknown cause has been named "Dark Energy" and present estimates give it 75% of the mass-energy of the Universe.

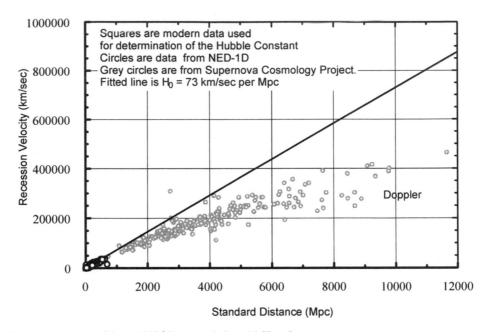

Figure 134. Measurements of the redshift/distance relation with Type Ia supernovae.
A display of the data set used to determine the decrease in the Hubble Constant with greater distance in the Universe. The small set
of values in the lower left of the plot are the recent standard set of values showing a Hubble Constant of 73 kilometers per second per
megaparsec, unchanging with distance. The grey circles are the values determined by use of the Type Ia supernovae as standard
candles. Supernova Cosmology Project data are from http://supernova.lbl.gov/Union/.

There are two extreme graphical interpretations of the departure of the Supernovae Type Ia points from the
extrapolated linear Hubble Constant line. One possibility is that, at any given recession velocity, the Supernovae Type Ia
are dimmer than they would be expected to be from the linear extrapolation, therefore farther away. This assumes that
the recession velocity (of the host galaxy) is measured exactly correctly. The other possibility is that, at any given
distance, the host spiral galaxies are moving away at a slower velocity than expected from the linear extrapolation. This
assumes that the distance is measured exactly correctly. The Dark Energy acceleration assumes that, if the expansion of
space were uniform, the Supernovae Type Ia points would lie along the linear extrapolation.

Since the Supernovae Type Ia measurements at closer distances agree well with the majority of the earlier Hubble
Constant measurements (presumably, the separate group of Supernovae Type Ia measurements that overlap the broader
set are those which provided calibrations), is there some mechanism that causes a selection effect to produce the
departure from the extrapolation?

The researchers were exceedingly careful in selecting only high-quality measurements, assuring that the supernovae
were truly Type Ia, and that the spectral redshift was accurate. (Nearly all redshifts were derived from the host spiral
galaxy spectrum; in one case the spectrum of the supernova itself was used [Kowalski *et al.* 2008].

The researchers observed that, contributing to the high quality of the data, was the unusual absence of dust. In this
case, it appears likely that the absence of dust, with the lack of severe scattering of less-red light, allowed the observation
of galaxies that were not as redshifted (lesser recession velocities) as would normally be required for detection at the
derived distance. The unattenuated galaxies were shining through clear, empty space and did not need to be so
redshifted as is the usual case, to permit detection.

Dark Energy is a brave attempt to solve what is a poorly understood problem. The observed galaxies have less
velocity, but are detectable because of the unusual absence of dust.

The Expanding Universe

Yesterday,
this day's madness
did prepare.
The Rubyait of Omar Khayam

The early years of the past century provided the greatest redefinition of our Universe since Kepler (1609) and Galileo (1610).

The expanding Universe was established in the late 1920s, in response to the astronomical observations reported by Edwin Hubble and Milt Humason, showing that galaxies were moving away from the Earth with velocities that were generally increasing in proportion to their distances from the Earth. Galaxies were identified and located by their images on photographic plates, usually taken with a red filter. Distances were determined by use of the Cepheid-variable period-magnitude relationship, and velocities in the line-of-sight were determined by the Doppler shift of identifiable absorption lines in the spectrum of each galaxy. The wavelength of light is changed by the velocity of the object, relative to the observer. Objects moving towards us have light shifted to shorter wavelengths, blue-shifted, while those moving away have light shifted to longer wavelengths, red-shifted. This is not just a change in color, but the change in wavelength results in a change in color. The relationship between the change in wavelength and the velocity is well established, even at relativistic velocities, by the Special Theory of Relativity.

To establish the redshift/distance relationship, a galaxy is selected, its distance determined, its spectrum taken, and the shift in wavelength is used to determine the velocity of the galaxy relative to the Earth. (This process continues at present, with greatly improved instruments and modified calibrations. The results continue to produce controversy.) These measurements show that the recessional velocity is proportional to the distance. This relationship is stronger for more distant galaxies. Nearby galaxies show considerable scatter, both approaching and receding from the Earth. This relationship has been explained as being due to a uniform expansion of space, and that expansion provides a possible solution of the equations of Einstein's General Theory of Relativity. Hubble's observations were taken as proof of the theory, and the theory was taken as confirmation of Hubble's red-shift/distance relationship.

In an earlier time, when the Earth occupied the center of the Universe in our stories, Hubble's Law might have been phrased just the opposite: The faster they go, the farther they are. In that Universe, everything would have been speeding away from the center, away from the Earth. That was then, this is now, and a privileged location in the Universe, such as the center, is no longer permitted in our stories of the world. However, there is a special way in which we are at the center of a Universe, and that is the Universe that we can see. We look until we can see no farther and so far as we know, that is about the same distance in any direction. That puts us at the center of *our* Universe, the Universe that we observe.

Consider that small part of the overall Cosmos, the part that we can observe, and extend our imagination to the entire space. Imagine a Universe of static space, with no expansion. Let it be filled with galaxies moving in random directions, uniformly, with velocities that are distributed like the velocities of molecules in a gas, as their velocities are measured by the wavelength shifts of spectral lines from gas. (Most moving somewhat slowly, some moving very rapidly, the Maxwell-Boltzmann distribution.) Some galaxies move towards us, a roughly equal number move away, many move more or less transversely to our line-of-sight. This random motion occupies most of the objects in our observable Universe.

Consider that the Universe is generally, perhaps nearly uniformly, filled with dust, at a very low density. (We now know this is the case. In the time of Hubble and Humason, the Universe was thought to be perfectly clear, with dust limited to dust lanes and clouds in the galaxies.) Very fine dust scatters light more strongly at shorter wavelengths. For the dust grains we think are distributed throughout distant space, the scattering cross-section is rather complicated but can be represented as proportional to the fourth power of the reciprocal wavelength, $(1/\lambda)^4$. Thus, the scattering cross-section of blue light, at a wavelength of 400 nm, is 5 times greater than for red light, at a wavelength of 600 nm. Since this cross-section determines the attenuation as an exponential function, you can see that blue light disappears in a relatively short distance, compared to red light. For example, at a distance that makes the exponential factor equal to -1 for red light and -5 for blue light, the red light would have 0.37 times its unattenuated intensity, while blue light would be reduced to 0.007 of its unattenuated value. (That is a difference in apparent brightness of over 4 magnitudes. The blue galaxy would not be seen.) Also, the wavelength shift toward the blue moves the peak of the spectrum (for an object like the Sun, or composed of many suns) away from the red-filter/film response, further dimming the image, while the red shift moves the peak towards the photographic response, enhancing the image.

At progressively greater distances, this effect becomes increasingly severe, so that the blueshifted galaxies, those moving towards us, become un-visible. Only those galaxies moving away, with increasingly great redshifts, are observable on the photographic plates (or CCDs). Only those galaxies are selected for a measurement of the spectroscopic red shift, to determine velocity, and only those are selected for a measurement of distance. Thus, only those distant galaxies that are moving away from us are plotted on the velocity *vs.* distance plot. And so, I think, the red-shift/distance relationship is entirely an artifact of the observations.

Astronomical observations in the late 1920s [Hubble 1929] were viewed in the context of the newly proposed theory of an expanding Universe [Friedmann 1922], and interpretations were caught up in the excitement of having a theory that actually described how the Universe could work. That concept determined the interpretation of the perceptions, and the theory proved the observations.

The expanding Universe was derived from a particular solution of Einstein's field equations of General Relativity. (Those equations permit a multitude of solutions, and only a few have been produced. The expanding solution caught on and has stuck with us ever since.) That solution permitted the Universe to consist of space that could expand, or contract, or remain constant, but it could be constant only if the average density of gravitational mass in the Universe were zero. That condition didn't seem likely to anyone who could pick up a rock or drink water, and Hubble had shown that the Universe was expanding so that eliminated the contracting version.

Hubble had found a correlation between the speed at which his nearby galaxies were moving away from the Earth, and their apparent brightness, which is assumed to represent their distance.

The brighter galaxies are nearer, and the dimmer galaxies are farther, and the farther galaxies are moving away faster. The plot of recession velocity, measured by the Doppler redshift of spectral lines, initially compared to the magnitude as a logarithmic surrogate for distance, became the Hubble diagram, Figure 135, and the correlation of redshift and distance became the Hubble Law.

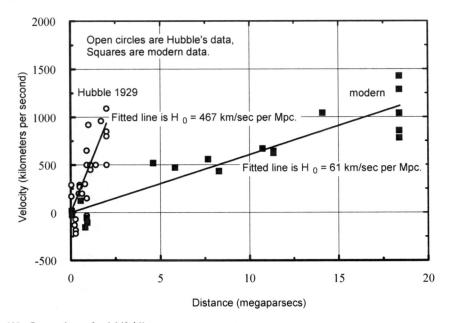

Figure 135. Comparison of redshift/distance measurements.
The Hubble plot, showing the original data and a more modern interpretation for the same galaxies, resulting mainly from corrections to the distance scale.

This showed that the galaxies were receding from the Earth at speeds that were directly proportional to their distances from the Earth. (The Doppler effect provides a secure measurement of the absolute velocity of the gas associated with the galaxies, while only the relative distances are determined from the magnitudes, with the assumption that, on average, all the galaxies have about the same intrinsic brightness.)

The expanding Universe theory provided that the recession velocity would be proportional to the distance, and the redshift data were plotted in that manner, and confirmed that the recession resulted from distance. I want to emphasize that there is nothing here that says the theory of an expanding Universe is wrong, or that the Big Bang did not happen. This book simply says that the observations were interpreted incorrectly, and there is no evidence for an expanding Universe, or the Big Bang, within the part of the Cosmos that we have been exploring. Later, it will be shown that a static Universe, one of the three easy solutions of Einstein's equations, is a possible alternative. It is time for us to re-evaluate our ideas and assumptions.

From Planet to Spiral Galaxy

The detonation of a planet forms a spiral galaxy. Variations in the detonation, and the original situation of the planet around its starsun and orbited by satellites, determine the resulting characteristics of the galaxy.

As the disintegrating planet tears apart along the crustal fracture lines, each fragment takes with it a share of the energy of the detonation, propelling it outward, and retains the rotational velocity it had as part of the rigid planet, according to its latitude. Fragments from the polar regions will have the least of this velocity, while equatorial fragments will have the greatest. The planet will break irregularly, according to its ridges and troughs, but the components of velocity, radially outward from the detonation, and tangentially from the rotation, will be determined by location on the planetary sphere and the laws of trigonometry. Thus, the symmetry seen in spiral nebulae, which is a grand symmetry of curves, not details, results from the vector addition of these two velocities.

The conventional explanation of the formation of spiral Island Universes of stars invokes density waves to initiate formation of the stars from varyingly dispersed gas and dust [Kenney 1993, Hodge 1986, Elmegreen and Elmegreen 1992, Elmegreen and Elmegreen 1993]. In this explanation, the density waves must maintain identically symmetrical action across thousands of light years distance while interacting with greatly different amounts of galactic matter. However, in detonated planetary debris, the spiral structure and the symmetric curvature is inherent in the rotation of the planet and the energy of the detonation.

Any piece of a planet has several components of motion. If the planet is rotating, each piece is moving at an instantaneous velocity that is determined by the rotation rate of the planet and the distance between the piece and the rotational axis. Each piece, as a part of the planet, is also orbiting the starsun. As the planet breaks into fragmentary debris from its detonation, a radial outward velocity, from the energy of the detonation, is added. This velocity spreads the material of the planet into a disk, possibly shaped by its magnetic field. As the disk extends far beyond the original planet's size, material that had the right orbital velocity for the size of the orbit is moved closer to or farther from the starsun. This leads to some material orbiting more slowly, and some more rapidly than it should, and a warped disk results.

After a planetary detonation, remaining uranium ore bodies in fragments of the crust could melt down and become pulsating spheres acting as nuclear reactor relaxation oscillators. Much of the uranium-238 would have been converted into the more easily fissioned plutonium-239. As such a blob cooled and contracted, a critical state would be reached and the fission power would increase until expansion made the reactor subcritical again. With power generation terminated, the blob would cool by thermal radiation and from the decay of the fission products and activation products. Cooling and contraction would lead to the critical condition again, and a repetition of the power increase and expansion, and then the cooling, contraction, and regaining criticality, in a continuing cycle. The larger blobs would oscillate more slowly and would reach greater peaks of power. These pulsing nuclear reactors would be seen as RR Lyrae stars and Cepheid variables. These misidentified objects would be small things at close distances, rather than the huge stars at immeasurably great distances on which we base our entire distance scale for the Universe.

Detonation of a planetary satellite may be induced by the neutrons and antineutrinos from the nuclear detonation of its parent planet. Those detonations in satellites that have little rotational energy will produce what we normally call irregular galaxies. Elliptical galaxies will be produced as the debris from a planetary detonation is captured at close range by the starsun and a shroud of chaotically orbiting fragments turns a star into a galaxy.

Satellite detonations can be seen with M31 (the Great Andromeda Nebula) as M110, with globular clusters, and M32. The starsun of the system is most likely to be the true star, ν And (nu Andromedae).

The multitudinous points of light seen in spiral nebulae and identified as old yellow stars may be molten blobs rich in radioactive iron-60 (and its daughter cobalt-60), which decay with relatively low energy beta emission (0.32 MeV), while the brighter stars are rich in aluminum-26, with higher energy beta emission (1.17 MeV). Higher energy beta particles will make a "star" appear to be brighter, will excite higher energy (higher temperature) spectroscopic lines, and will excite gas over a greater distance. HII regions result from the emission of fission neutrons which decay to protons and electrons (beta particles). "Young blue stars" are those radioactive fragments that have sufficient transparent material surrounding them to produce Cerenkov radiation. Another likely source of the energy of these non-stellar stars is chlorine-36, with beta emission at 0.71 MeV. (It should be noted that these energies are maximum energies, and beta particles are actually emitted over a broad range in energy, with a spectrum resembling the Maxwellian distribution used to describe thermal motion in a hot gas.) These three radionuclides have half-lives in the range of 0.3 to 1.5 million years, and would keep the stars constant, in our view of time. Mixtures of radionuclides could provide a continuous range of brightness and apparent temperature, the "hotter" the brighter, that would plot as a smooth curve on the Hertzsprung-Russell diagram.

For "front-lit" galaxies, the ones we are most familiar with, the majority of the stars will be shining by reflected light from the starsun, and will have color and spectra that resemble that star. This is seen in the Milky Way, where the majority of stars are of Solar type.

Fellow planets in the system may survive the detonation, only to undergo detonation at a later time. These later detonations could be either the lower-energy-release damaging detonations, as proposed for Mercury, Venus, and Mars, or full fledged disintegrating detonations that are seen as the supernovas. Changes in fuel composition, or perhaps the coalescence of two blobs, could also produce a system that would detonate, like a planetary detonation, only smaller, registering in our gamma ray burst detectors, producing a bright expanding speck fired by the intense radioactivity of its own production.

If these galaxies are not Island Universes but are instead the debris of detonated planets, where are the true stars that served as suns to the original planets? The star may have been converted to an elliptical galaxy by the captured debris from a detonated planet, this debris shining by reflected starlight and by its own thermal radiation. The lower temperature of these fragments would make them look like the old stars considered to be the population of elliptical galaxies. Or the star may be undisturbed and nearby, but would not be considered to be related to the "galaxy" because, of course, no star could be a companion to a galaxy. These are shown in the section on Association of Stars and Spiral Galaxies.

In the case of the Whirlpool galaxy, M51, the starsun may be its "companion galaxy" NGC 5195, now seen to be receiving a flanking blast of planetary material from a detonated planet, as shown in Figure 136. This image shows that the companion is being splashed with material from the left, that is, in a clockwise direction, in agreement with the measurements of Adriaan van Maanen. This is also shown in the pair of galaxies identified as NGC 7752/7753 [Smith 2009] (Figure 136). The companions are galaxy types known as elliptical galaxies. These are ellipsoidal bodies of light with essentially no detectable structure of their own. The largest of these are found in the center of swarms of smaller galaxies [Travis 1994b]. As would be the starsuns of families of detonated planets. These shattered planets, with many variations on the theme, would be scattered throughout the Universe.

Interpretation of elliptical galaxies as a large collection of stars formed from gas and dust causes difficulties with concepts of the age of the Universe [Flam 1994]. An "elliptical object" has been found at a distance that has been judged to represent a time of only 1 or 2 billion years after the Big Bang. This has been judged to be too short a time for the formation of an elliptical galaxy, which is thought to take about 1 billion years.

Spiral nebulae are observed to have a spherical distribution of globular clusters about them. Our Milky Way Galaxy has slightly more than a hundred identified globular clusters, tight collections of what are thought to be gravitationally bound stars, orbiting the "Galaxy." These are more likely to be the shattered remnants of the ancient uranium hydride globs that failed to accrete into planets or stars and have been drifting in space, little affected by the passage of time, other than to radioactively decay. In a delicate interaction of antineutrinos, hydrogen nuclei, uranium and plutonium, and the dynamics of a chain-reaction system, globs will be activated by the planetary detonation, out to a certain distance in all directions, and will themselves detonate with generally similar energy releases. The radioactive debris will appear to be clusters of stars, surrounding the detonated planet in a spherical halo.

Figure 136. Whirlpool Galaxy (M51) and NGC 7753/7752.
Whirlpool Galaxy, NGC 5194/5195 (M51), top, (NASA, ESA, S. Beckwith (STScI), and The Hubble Heritage Team (STScI/AURA), STScI-PRC05-12a, original in color), and interacting galaxies, NGC 7753/7752, bottom [Smith 2009].

Elliptical Galaxies, with and without Globular Clusters

If the distant galaxies are really detonated planets, and the Milky Way Galaxy is also, and the elliptical companions to the distant galaxies are individual starsuns surrounded by collected debris, what about the Magellanic Clouds, companions to our stellar association? The distance to the Large Magellanic Cloud seems to have been measured accurately and independently by a study of the illumination of the debris ring around SN 1987A [Panagia *et al.* 1991], and a variety of other techniques. However, these all suffer from preconceived ideas and assumptions that predetermine the result. Measurements that produced a result of about 60 Astronomical Units would be quickly rejected as false and foolish.

A study of infrared emission from ultraluminous infrared galaxies, thought to be powered by intense star formation, has shown that emission from rotational states of molecular hydrogen (H_2) is from the exterior of the galaxies [Zakamska 2010]. That would be consistent with the dispersal of neutrons from a nuclear fission detonation and the subsequent decay to protons, recombination with electrons, and the association of hydrogen atoms into molecular hydrogen. Because of the nearly point origin of the neutrons, each of those developments would take place at progressively greater distances from the detonation, the center of the "galaxy".

Supernovas

Supernovae (the Latin plural of the English word) are considered to be the explosive disruption of a stellar object, which shines for a short time as brightly as all the other stars in its galaxy. The occurrence of visually bright supernovae has been recorded through history, by the Chinese, Japanese, Native Americans, Kepler, and Tycho, but the first one to be scientifically studied shone in the Great Nebula of Andromeda (M31, NGC 224) in August 1885. (This contributed to some of the confusion over the cosmological distance scale and the nature of the Milky Way and spiral galaxies, by initially being considered to be a nova, much dimmer than a supernova.) Later supernovae received greater and more intense and effective study, depending on chance variations such as location on the sky, time of first detection, pre-outburst studies, and the instruments and people involved. Nearness and brightness determine how long it can be studied, as supernovae fairly quickly fade into memory, leaving at best supernova remnants and light echoes. Occurring before it was realized that there were such things, the supernova in Andromeda was given the distinctive name of S And 1885.

In August 1937, a supernova, one of the brightest observed at that time, was seen in a small, notably dim spiral galaxy, IC 4182. This supernova was also late in getting its proper name, which is now SN 1937C.

In February 1987, a supernova was seen in the Large Magellanic Cloud by astronomers in Chile. It is now the brightest ever to be studied. A multitude of instruments has been applied to its study, some asking questions that were themselves unknown to earlier researchers.

Because of this, it can be said that we know more about SN 1987A in the Large Magellanic Cloud, than we know for all the other supernovae combined. What is so remarkable about that statement is not that we know so much about SN 1987A, but how little we know about the others. It is also remarkable that so much does not agree with our expectations.

If close enough and bright enough, and noticed early enough, a naked-eye supernova can be seen to brighten from unnoticeable to attention-grabbing in a few nights, and then fading away to become undetectable in a few years. This history is presented in the form of a light curve. The quality of the light curve is subject to the vagaries of astronomical discovery and study, and the three supernovae just discussed have provided the best information yet. (While calling attention to only three supernovas over the course of a hundred years might make them seem to be extremely rare events, modern supernova watchers find roughly 60-70 each year, mostly in far distant galaxies.)

Light Curves

The light curves for these three supernovae are shown below, with the same scales to aid close comparison, converted from the astronomers' usual brightness scale of magnitudes to another measure sometimes used, luminosity, presented in the standard unit for power, watts, but far dimmer to the naked eye than a bathroom nightlight seen from the other side of town. This conversion requires knowledge , or the assumption, of the distance. The standard distances have been used here.

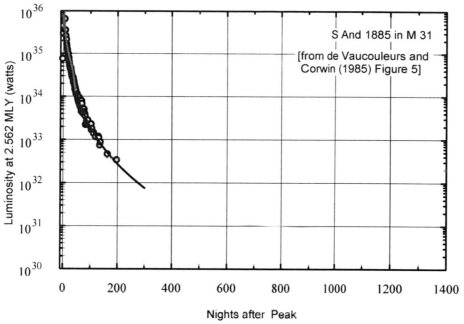

Figure 137. The observed light curve of the supernova in the Andromeda Galaxy.
This was the brightest supernova studied at its time, before supernovas had been described.

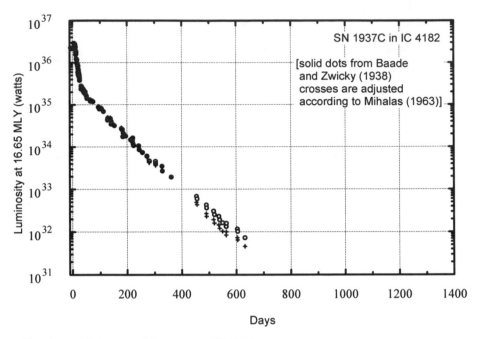

Figure 138. The observed light curve of the supernova SN 1937C.
This was another exceptionally bright supernova, studied after we had an idea as to the nature of supernovas.

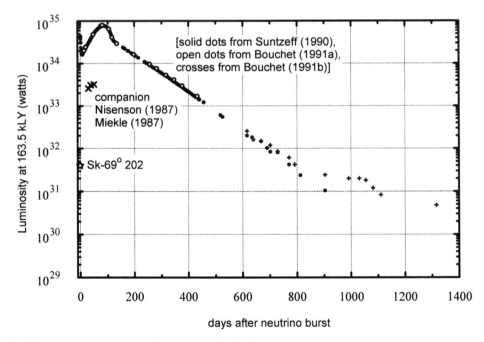

Figure 139. The observed light curve of the supernova SN 1987A .
This supernova has been the best studied, from its very beginning, but has broken the rules.

Astronomers have divided these astronomical explosions into just two types, Type I and Type II. That is, there is only Ia, Ib, Ic, I-pec (for peculiar), I½, and II-P, II-L, IIb, and III and IV (both now ignored), and V (which may be IIpec), and sometimes I? and II?. That should simplify things greatly, but the more we observe, the more closely we examine, the more differences and distinctions we find, just as it is with people, divided into men and women. Beyond that, we are all different. And, like people, supernovae change with time, from their birth to fading away, like old soldiers. For simplicity, Type I supernovas, of whatever sort, do not show spectral lines of hydrogen, and Type II do.

One of the visible distinctions from an ordinary star, and other supernovas, that SN 1987A showed, was an abrupt rise to a temporary maximum light, followed by a steep drop to a temporary minimum, then another rise and a drop, much like the light curve of a thermonuclear weapon. That roller-coaster drop, shown so well by SN 1987A, is followed by a decline that is linear on a magnitudes scale, and a logarithmic luminosity scale, inviting its interpretation as a strictly exponential decay, with energy provided at a decreasing rate by a single radioactive nuclide.

The brightness of an expanding sphere of hot gas depends on the temperature, cooling with time by radiation, and the exposed surface area, increasing with expansion, complicated by changing depth to the surface due to changing opacity. Instead of dimming rapidly to invisibility as the hot gas lost its initial heat, the supernova aftermath declines gradually, with a clearly linear logarithmic fading, indicating an exponential decay.

The differences in the logarithmically linear rate of decline, a strictly exponential decay, are clear from inspection of the light curves. Efforts have been made to assign specific radionuclides, such as californium-254 with a half-life of 60.5 days [Baade *et al.* 1956, Burbidge *et al.* 1956], when the observed half-life of SN 1937C was 55±1 days, and the half-life of californium-254 was reported as 56.2 or 61 days. A particularly appealing feature of californium-254 was its large energy release by spontaneous fission, about 200 MeV per fission decay.

The light curves of SN 1937C and SN 1987A (and others) appear to divide clearly into an exponential decay with a short half-life and one with a long half-life. That makes the nickel-56/cobalt-56/iron-56 decay sequence attractive, with half-lives of 5.9 and 77.27 days.

The standard analysis for two (or more) half-lives involves a combined or iterative least-squares solution of the data. This requires a selection of the time interval in a supernova light curve when the short half-life has become well established, after the initial transient of the detonation, ending before the long half-life becomes affected by dynamics of the expansion or observational problems, as shown by the late data for SN 1987A.

A simpler approach, which works well with good data and adequately with sparse observations, is to determine the half-life represented by adjacent pairs of observations, and then select and average those that belong, statistically, to a definite value. This provides an alternative measure of the half-life for comparison. I analyzed the light curves of S And 1885, SN 1937C, and SN 1987A in that manner, and found considerable differences in the long ("cobalt-56") half-life, as was also shown by the published values. These results are shown below.

Supernova	long decay half-life (days)	reference
S And 1885	21.7 ±1.3	present work
SN 1937C	67.8	Mihalas (1963)
	46.8	Mihalas (1963)
	51.2	Mihalas (1963)
	48.6	Mihalas (1963)
	55 ±1	Baade *et al.* (1956)
	56 ±6	Van Hise (1974)
	49.3 ±6	Barbon *et al.* (1984)
	47.4 ±1.8	present work, unadjusted
	41.5 ±1.8	present work, adjusted
SN 1987A	70.35±0.04	Suntzeff & Bouchet (1990)
	65.8 ±0.3	Suntzeff & Bouchet (1990)
	69.3 ±3.3	Suntzeff & Bouchet (1990)
	75.97 ±0.21	Bouchet *et al.* (1991)
	68.7 ±1.3	Bouchet, Danziger, & Lucy (1991)
	69.2 ±0.7	present work
cobalt-56	77.27±0.03	Firestone *et al.* (1996)

It is clear that the observed half-lives have great variability, and none match the cobalt-56 value well. At present, it appears that the nickel-56/cobalt-56 source, as a specific solution, is poorly supported by the observations, and must be considered to be a theoretical idea.

However, a supernova is complicated, and poorly suited to a single, simple solution. Supernovae show a variety of exponential decays. A variety of exponential decays calls for a variety of radioactive energy sources, unless we are willing to explain away the variety by invoking accidents of observation and interpretation. Those accidents cause enough trouble as it is, but do not seem to be a suitable overall solution. We may just be asking more of the data than they can give.

Supernova 1987A

To set the stage, imagine the Solar System as it was 6 million years ago (more or less), before humans were here on Earth to puzzle over these things. A distant planet orbited our Sun, one never seen from Earth but one searched for intensely during the past century, Planet X. Like the other giant planets, Planet X would have had several large natural moons.

Consider the detonation of one of those moons, in the choo-choo chug-chug huff'n'puff BOOM manner, forming the Large and Small Magellanic clouds, leaving a trail of surface debris and hydrogen atoms as it choo-choo chug-chugged to disaster. The antineutrinos from the detonation would have boosted the reactors within Planet X, inducing a sympathetic detonation, making what we now call the Milky Way Galaxy.

The blast of gamma radiation from that planetary detonation caused rampant mutations among the creatures living on the surface of the Earth. Different species came forth, the modern horse that sleeps standing up, and semi-apes that walked upright. The neutrons, gamma-rays, and antineutrinos created widespread radioactivity that stunted the growth of the new humans for millions of years. (Iron-60 has a half-life of 2.6 million years and decays to radioactive cobalt-60, which is a high-energy gamma-ray emitter.) Now, look at recent history and consider how the observations of SN 1987A are produced by the nuclear detonation of a pair of asteroidal fragments remaining from that extreme "disaster".

An Event Chart for SN 1987A

Observation	Alternate Interpretation
1969 Nicholas Sanduleak cataloged Sk-69°202 as a blue supergiant star in the Large Magellanic Cloud, at a distance of 162,000 light years.	The object was a small asteroidal fragment covered in ice, shining by reflected sunlight and with Cerenkov light from internal radiation, at a distance of 60 Astronomical Units.
August 10, 1985, was there a gamma-ray burst?	A nearby small fragment detonated, starting induced detonations.
February 23, 1987, 1.791 UT, a series of antineutrino counts were recorded, extending till 3.412 UT. LSD-Mont Blanc recorded 18 counts, Baksan recorded 3, and Kamiokande-II recorded 1.	Fission antineutrinos reached the Earth from the beginning of the sequential detonation of the small companion of Sk-69°202. The antineutrino spectrum suggests that this detonation was disruptive, exposing the fireball.
February 23, 1987, 2.184 UT, Ian Shelton photographed the region of the LMC containing Sk-69°202, and recorded the star to have a B magnitude of 12.12, relatively dim.	The fireball was not exposed until around 3.412 UT.
February 23, 1987, 2.750 UT, a pulse was recorded by the Rome and Maryland gravity wave detectors, interpreted as the disappearance of a mass equivalent to 2,400 solar masses.	At a distance of 60 Astronomical Units, that loss of mass is equivalent to 1.5×10^{17} kilograms, about one ten-thousandth the mass of the asteroid Ceres.
February 23, 1987, 7.593 UT, a series of antineutrino counts were recorded at Kamiokande-II (12), IMB (8), Mont Blanc (2), and Baksan (5), extending over less than a minute.	Fission antineutrinos reached the Earth from the induced detonation of the larger fragment, Sk-69°202. The antineutrino spectrum suggests that this detonation was fully contained, trapping neutrons and antineutrinos.
February 23, 1987, in correlation with the antineutrino counts, the Rome and Maryland gravity wave detectors showed an event equivalent to the loss of 8,740 solar masses.	At a distance of 60 Astronomical Units, that loss of mass is equivalent to 5.5×10^{17} kilograms, about five ten-thousandths the mass of the asteroid Ceres.
February 23, 1987, 10.632 UT, R. H. McNaught recorded the supernova at a magnitude of 6.0-6.2, roughly 7 million times as bright as the Sun.	Translating from a distance of 162,000 light years to 60 Astronomical Units, the brightness is 9×10^{23} ergs per second or 2.4×10^{-10} as bright as the Sun, one four-billionths as bright.
February 24, 4.800 UT, Oscar Duhalde noticed a brighter than usual star in the Large Magellanic Cloud.	
February 24, 5.520 UT, Ian Shelton photographed SN 1987A, at a magnitude of 5.00.	
February 24, 8.880 UT, A. Jones sighted SN 1987A, at an estimated magnitude of 6.5-7.0.	
Initial spectroscopy showed hydrogen, and it was decided that SN 1987A was a Type II supernova.	Hydrogen was formed primarily by the neutrons from the detonation of the smaller companion, little was released from the larger detonation.

The Early Spectrum of SN 1987A

Several early spectra were reported for the supernova in the Large Magellanic Cloud, SN 1987A [Branch 1990].
These three spectra, at progressively later times, are shown in Figure 140.

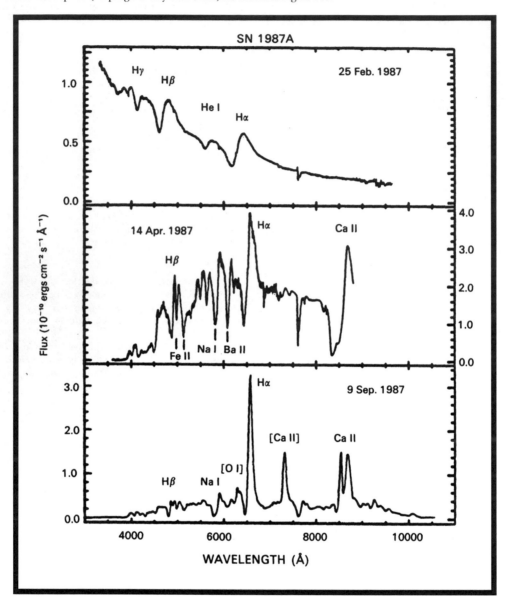

Figure 140. Spectra of Supernova 1987A.
Spectra from SN 1987A. The scale indicating an energy flux spectrum is probably incorrect. The earliest spectrum is a photon number spectrum. Immediately obvious is the fact that the earliest spectrum is remarkably different from the two later spectra. (An unfortunate complication in at least the earliest spectrum is that the intensity label implies *energy flux*, while it is actually a photon number flux. This is discussed in detail in Appendix C, "Studies of Spectra".)

The earliest spectrum was taken while the supernova was declining in brightness from its initial peak, while the second spectrum was taken during the rise to the second maximum, and the third spectrum was taken after the exponential decline was fully established. The times of the spectra are shown in Figure 141. The spectra may therefore represent significantly different situations in the progress of the supernova.

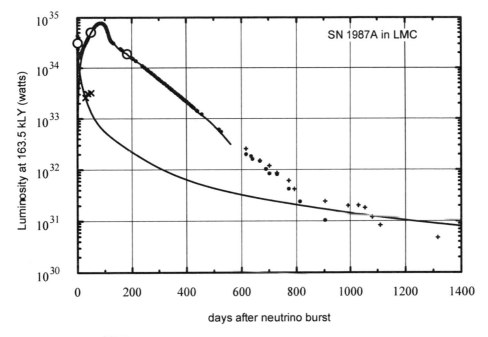

days after neutrino burst

Figure 141. Light curve of SN 1987A.
The early light curve for SN 1987A, showing the times of the three spectra (circles) shown in Figure 140. A decaying light curve is also shown for the initial burst.

The error in the scale for the earliest spectrum can easily be remedied, converting the spectrum from number flux to energy flux, by dividing by the wavelength (and renormalizing). This is shown in Figure 142, where a black body spectrum (Planck Function) at 12,000 K is shown passing nicely through the observed spectrum, and with a Cerenkov spectrum (which is independent of temperature) passing nearby.

It is difficult to judge which of the two theoretical spectra actually provides the better match. The agreement is affected graphically by the normalization, and physically by the absorption and re-emission, producing dips and peaks. It can at least be said that a black body spectrum matches the observation well, and that the Cerenkov spectrum mimics it well, also.

These two theoretical spectra represent greatly different astronomical objects. The black body spectrum indicates a blue supergiant star, as the supernova progenitor, Sk-69°202 is believed to have been, and therefore at the presumed distance to the Large Magellanic Cloud, 163,500 light years. The Cerenkov spectrum indicates a small, cold, icy, radioactive object like an asteroid at an actual distance of 60-100 Astronomical Units.

Considering a fission detonation in a relatively small object, like an asteroid, there are two possible outcomes. The detonation may be so energetic as to completely disrupt the body, exposing the fireball, terminating the chain reaction, releasing the last generation of fission neutrons to space, and allowing the antineutrinos to speed outward unimpeded. Alternatively, the body may be sufficiently large that the detonation cannot disrupt it, the chain reaction will continue to multistage fission, the neutrons will be absorbed by the surrounding material, the antineutrinos will be upscattered by the electrons in thermal equilibrium with the fissioning material, and the body will be overheated and melt. Any icy surface will be vaporized and ejected by the detonation shock wave.

Detailed analysis of the second pulse of antineutrinos from SN 1987A, presented in Appendix C, shows that the Fermi-Dirac (high-energy) distribution derived from the antineutrino energies, with a temperature (kT) of 4.2 MeV (50 billion degrees kelvin), was a consequence of attenuation during passage of the antineutrinos through the Earth, from SN 1987A to the detectors. That analysis shows that the energy distribution in the supernova was a Maxwell-Boltzmann thermal distribution with an energy (kT) of 5.3 MeV (60 billion degrees kelvin).

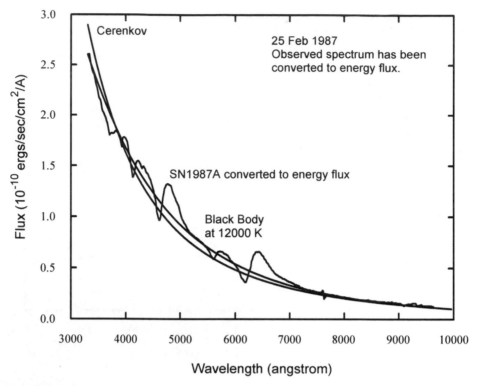

Figure 142. Energy flux spectrum for SN 1987A.
Converted spectrum of SN 1987A on 25 Feb 1987 compared with black body radiation at 12,000 K and Cerenkov radiation at any (low) temperature.

The fissioning masses would reach that temperature if 10% of the material fissioned. Nuclear fission weapons achieve a typical "efficiency" of 24% [Serber 1992].

My view of SN 1987A is that a small companion, an unseen satellite of Sk-69°202, orbiting an asteroid-sized fragment, detonated by the choo-choo chug-chug huff'n'puff BOOM process over a period of a few hours, and this caused an induced detonation in the larger body. The small body was disrupted, terminating the chain reaction and quickly releasing the antineutrinos produced in fission. The fireball was exposed and cooled quickly, giving the first spike of brightness, then cooling to be the low-temperature component seen later. A remarkable distinction in the light curve for SN 1987A is the initial drop in brightness, implying a previous peak. That peak was from the disruptive detonation of the small companion.

The detonation was fully contained in the larger body, which we have called a blue supergiant star, Sk-69°202, but overheated and melted the object, producing the thermal brightness and the high temperature component. This was the source of the major part of the supernova light curve. The antineutrinos from fission were trapped in the body by the antineutrino-induced electron capture reaction [Appendix A], to be gradually and undetectably released later in normal beta decay. While the detonation was much larger than in the companion, fewer antineutrinos were released because the body held together. The icy crust was vaporized and dispersed, so the blue Cerenkov light, typical of a blue supergiant, ceased. Sk-69°202, as a blue supergiant star, disappeared. More antineutrinos were detected from this detonation because the cross section of the positron reaction used for detection, and the detection efficiencies, are much greater for higher energy antineutrinos.

SN 1987A was a detonation of associated asteroidal fragments remaining from the detonation of a satellite of Planet X that detonated when Planet X detonated to form the Milky Way Galaxy. The satellite formed the Large Magellanic Cloud. (Other satellites formed the other satellite galaxies of the Milky Way Galaxy.)

A nuclear reactor will release neutrons to space where they will decay by beta emission. This will produce a cloud of protons (hydrogen nuclei) and electrons (beta particles). (As a side effect, this cloud may develop a large positive charge, since the emitted beta particles have velocities close to the speed of light and will quickly leave the neighborhood.) The cloud will initially be seen as an HII region, emitting light in the hydrogen spectrum, most notably in the red Balmer light of Hα.

If such a reactor detonates, a large number of neutrons will be released in a short time, decaying almost completely in less than an hour. If the reactor were moderated (neutrons slowed down by light elements, such as hydrogen, helium, carbon, and oxygen), the neutrons released in the detonation might have very little kinetic energy, and so the protons would have energy and velocity distributions that are characterized by the energetics of the beta decay. That energy distribution has a FWHM (Full Width at Half Maximum, a measure of how much spread there is in the energy distribution) equal to 600 eV. That distribution was shown in Figure 2. The apparent temperature of the protons is nearly 3 million degrees K.

Gaskell and Keel [Gaskell and Keel 1987] found strong Hα emission from the supernova 1984e in the spiral galaxy NGC 3169. Spectroscopy showed the FWHM of the hydrogen atoms emitting this light to be 326 to 603 eV, in close agreement with (and slightly degraded from) the FWHM of the protons from neutron decay, 600 eV. An image in Hα light of NGC 3169 taken on December 19, 1981, fifteen months before discovery of the supernova in that galaxy (roughly 460 days before), showed a point source of hydrogen emission, approximately a thousand times fainter than with the supernova, "at the precise position where the supernova would explode." This suggests that the supernova was triggered by a smaller detonation that occurred about 460 days before, just as SN 1987A was triggered by a nearby detonation that occurred about 559 days before.

The planetary detonation process provides insight into many peculiar observations related to the supernova in the Large Magellanic Cloud, SN 1987A. This supernova has been the best-observed stellar explosion yet, but many observations do not fit the standard model. Please see the detailed discussion of this supernova in Appendix C.

The identification of the progenitor star that became the supernova showed that it was Sk-69°202, a spectral class B3 Ib blue supergiant. Blue supergiants are not expected to become supernovas, and so a developmental path through the red giant stage had to be proposed. The instant of detonation has been determined to be indicated by a pulse of neutrinos/antineutrinos detected by the Kamiokande II detector in Japan. However, antineutrinos were detected just before the explosion, an effect that has not been accommodated by current descriptions of the supernova. These separate detections seemed to show a gap between the early detector counts and the later counts, as if there were two distinctly separate pulses, at the moment of the supernova explosion. (This gap is shown in data from three detectors: Kamiokande-II, Baksan, and IMB, of about 5 hours before the main pulse seen by Kamiokande). The first set of detections has been rejected as having no connection with the supernova.

Radioactivity, presumably created deep within the star, became observable much sooner than expected, as if the radioactive material had managed to penetrate most of the overlying stellar gas. A "mystery spot" was observed moving away from the supernova, at 0.3-0.4 times the speed of light, and a compact spot was seen in Hα (Balmer light, red), diametrically opposite it. This source later disappeared. The luminosity curve requires more cobalt-57 than is expected: the curve is too faint and too steep to begin, and too flat at late times. The explosion was abnormally asymmetric. A ring of debris was found around the star, rather than the normal nearly spherical shell, and too far from the star to have originated in this explosion. An excess of ionized iron has been detected in the vicinity. The fastest-spinning pulsar ever discovered, was found and lost, or wasn't there; thought to be seen only because of electronic interference. A radio pulsar was observed, centered on the ring, at about the opposite direction and distance as the "mystery spot" seen on the other side of the star. And probably other peculiarities that this present study did not find.

The proposed model of planetary detonations provides explanations for these observations. The initial burst of antineutrinos, which arrived 4.7 hours before the antineutrinos from the main event [Bahcall 1989] and was detected by the Mont Blanc and Baksan liquid scintillation detectors [Murdin 1990], originated in the detonation of an asteroidal satellite near the asteroidal fragment that has been labeled Sk-69°202, immediately before the supernova explosion, and represents the precipitating cause of the main supernova explosion. Because of the first detonation being "off-center", the overall detonation will be asymmetric, as observed in the subsequent explosion [Jeffery 1987, Papaliolios 1989].

In one of the most exciting images of modern astronomy, the *Hubble Space Telescope* showed the remarkable complexity of SN 1987A, in Figure 143.

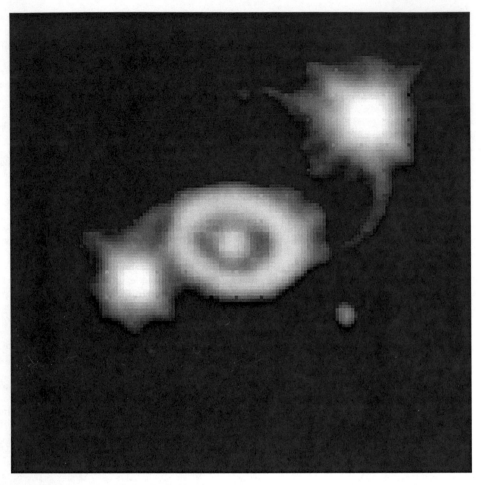

Figure 143. SN 1987A image by *Hubble Space Telescope*.
Hubble Space Telescope (improved) image of SN 1987A with ring, SE companion star, and the previously unseen "star" on the ring at the lower right. The faint outer rings are shown more clearly in the color image. (NASA, Hubble Heritage, original in color)

  Observations associated with the ring are shown schematically in Figure 144. The mystery spot sightings, at Days 30, 38, and 50, lead in reasonable progression to the "cold bright matter" found on Day 1344 [Cumming and Meikle 1993] and the "star" that coincidentally appeared in front of the ring, on Day 2511 [Mercury 1994], shown in the *Hubble Space Telescope* image in Figure 143. At a reasonable alignment on the opposite side of the ring, is the center of radio emission found on Day 1589 [Staveley-Smith *et al.* 1992].

  Two new rings were detected [Panagia 1994, Sky & Telescope 1994a] possibly showing evidence of the brief existence of the pulsar that had to be explained away, after being observed for 7 hours [Kristian *et al.* 1989]. Apparently equidistant on opposite sides of the star, these rings are mirror images of each other, and the axis of symmetry is offset from the star. (See Figure 144) The offset is a clear result of the fragment being produced from the detonation of an asteroidal satellite, orbiting at some small distance from the asteroid.

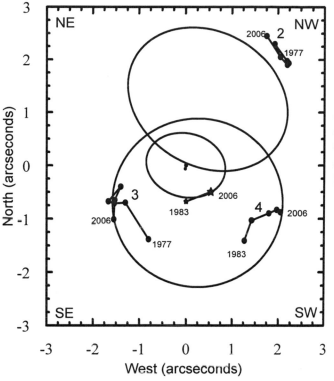

SN 1987A is at the center. Stars 2 and 3 are to the NW and SE of the supernova. Ringstar is to SW, Star 4 is farther. Small dot to SE is the companion. Circumstellar ring uses average values from Plait 1995

Figure 144. Schematic diagram of the SN 1987A system.
Schematic diagram of SN 1987A, the ring, the mystery spot(s), the "star", and early radio emission.

The "Mother of all Gravitational Waves" was detected by an instrument in Rome, but this exciting find (the first yet?) had to be discarded, since it indicated the destruction of 2,400 times the mass of the Sun at the assumed distance to SN 1987A, an obvious impossibility. The gravitational wave detector at Rome registered a signal immediately before (1.4 sec) [Amaldi et al. 1987a, Amaldi et al. 1987b] the first Mont Blanc (liquid scintillation detector, LSD) (anti)neutrino pulse [Aglietta 1987]. This series of LSD pulses is proposed here to be associated with the initiating fragment detonation, which induced the detonation of the star Sk-69°202, with its subsequent explosion. This gravitational wave signal, when adjusted to the assumed distance of 52 kiloparsecs, showed an energy corresponding to the mass of 2,400 Suns. This is such an obviously impossible energy release, in gravitational radiation, that the measurement had to be discarded.

If the energy to brighten the ring came, not from the ultraviolet flash, but from a burst of neutrons that decayed to protons and electrons before reaching the ring, the time behavior of the brightening reveals the velocity distribution of the neutrons, and therefore its energy spectrum. Higher velocity protons and electrons, from the decay of higher energy neutrons, reach the ring before slower protons and electrons, from low energy neutrons. As charged particles passing a ring of debris filled with magnetized fragments, as our asteroids appear to be, the protons and electrons would have been very much more effective in depositing energy in the ring than would ultraviolet light. The effective cross-section of the ring to charged particles would be much larger than its physical size.

Obviously, an exceptional amount of highly ionized iron [Wang et al. 1989] is dispersed throughout the region by these events. The energy of the iron-60 and other radioactive elements, such as iron-59, with a 44.5-day half-life, and short-lived fission products, contribute to the apparent energy of the supernova, over and beyond the radioactivity produced in the star itself, and make the light curve decline more rapidly than expected at first, and more slowly later on. X-radiation from the dispersed iron vapor, and the highly radioactive parts of the many fragments, that have been spread through space and are unhidden by the still massive stellar atmosphere, shows as an unexpected hard (non-thermal) X-ray emission from a region near the supernova remnant [Sunyaev et al. 1987, Dotani et al. 1987, Itoh et al. 1987]. The X-ray spectrum developed as the fragments dispersed and splintered.

Teleconnections, near and far

Teleconnections that we don't understand may seem to be Einstein's "spooky action at a distance". If we can identify both the effect and the cause, they become understandable.

There are seven possible cases of teleconnections from the supernova affecting processes in our Solar System. Because of various time responses, these are difficult to establish firmly. These are:

1. Immediately after the supernova detonation, the Sun brightened slightly. This change was so small it was overwhelmed by the daily, monthly, and 11-year variability in the Sun's irradiance [Willson 1997]. It can only be detected with determination and wishful intent.

2. In the year after the supernova, the radius of the Sun increased noticeably [Ulrich and Bertello 1995].

3. In July 1989, two and a half years after the supernova, the planet Neptune brightened noticeably [Lockwood and Thompson 1991], contrary to its normal pattern of following the solar cycle. It should be noted that the deep nuclear detonations suggested as the cause of sunspots, and therefore controlling the solar cycle, would provide pulses of antineutrinos that could keep Neptune in pace with the sunspots.

4. On September 25 1990, three and a half years after the supernova, a giant storm erupted in Saturn's atmosphere [Flasar 1991].

5. Between 1989 and 1995-1998, eight years after the supernova, the large natural satellite of Neptune, Triton, warmed noticeably [Elliot 1998, Buratti, Hicks, and Newburn Jr. 1999, Elliot 1999].

6. On September 4 and 5, 1995, Titan, the large moon of Saturn, abruptly brightened. This was apparently due to a large cloud or storm system, never seen before or after, so far [Griffith *et al.* 1998, Flasar 1998].

7. In 1997, ten years after the supernova, El Niño was among the strongest weather disturbances ever recorded. [Wolter and Timlin 1993, Wolter and Timlin 1998].

While the combination of these events may only be chance coincidence, their unusual nature calls attention to some possible effects that should be searched for in other cases. If these effects actually occur, it would seem that the only messengers capable of transmitting the signal would be antineutrinos from the supernova, acting on the responding objects by means of the antineutrino-induced electron capture reaction described in Appendix A.

Things that go bump in the night: Gamma-Ray Bursts

From ghoulies and ghosties and long-leggity beasties,
And things that go bump in the night,
Good Lord, deliver us.
Cornish prayer

Our concept of planets has changed from the earliest times, when planets were the five wandering stars in the heavens, to the eight major bodies orbiting our Sun, to the possible consideration that bodies such as these orbit many stars. Might planets form without stars? Could planetary systems exist, like Jupiter and its satellites, without an anchoring sun? Perhaps planets could exist alone or even in clusters of planets and brown dwarfs, formed when conditions broke a supernova cloud into many small condensations, and no stars formed.

So we can consider that around newly forming stars throughout the observable Universe, and perhaps all by themselves as well, planets of appropriate size and composition may have formed, with the potential for explosive disruption by a nuclear fission detonation. Planets are not exclusively a product of our Solar System, nor are the planetary detonations, and there may be a multitude of globs, rich in concentrated uranium, left over from each supernova remnant and proto-stellar cloud. It is proposed that these planetary explosions have already been observed in different forms: as gamma-ray bursts [Talcott 1991, Talcott 1992, Meegan 1994] and the explosive shrapnel [Dyson 1993] observed as spiral nebulae.

First launched in 1963, the *Vela* series of satellites was developed by Los Alamos National Laboratory in New Mexico and operated to detect nuclear detonations (nuclear weapons tests) in space, but they soon began providing evidence of intense, brief bursts of gamma radiation from scattered locations in space [Chupp 1992]. The bursts were judged to last too long to be the result of a nuclear weapon detonation and have remained an enigma. Because of the military nature of these satellites, this information was kept secret until 1973 [Klebesadel, Strong, and Olson 1973], when it was disclosed by the U. S. government [Fishman 1992]. It has been proposed that *Vela 6911* actually detected a secret test by the Union of South Africa, on September 22, 1979 [Scott 1997], possibly in cooperation with Israel. This incident is filled with ambiguity and uncertainty, and the official government position is that the evidence was not sufficiently complete and clear to reach that judgment. A scientific expert panel reviewed the data and concluded that a natural cause, although uncertain, was more likely than a nuclear detonation on the Earth [Klass 1980, Ruina *et al.* 1980]. The truth will probably remain obscured forever.

Gamma-ray bursts remain mysterious [Katz 2002], although over a hundred possible explanations have been offered. Because of their uniform distribution, they are considered to be at great cosmological distances [Paczyński 1995], and that makes them the most energetic events that we know, or they may be near [Lamb 1995]. They vary greatly in brightness, which could result from intrinsic variability or from variations in distance, or both.

Several thousand such bursts have been observed, uniformly scattered over the sky, with increasingly sophisticated equipment, providing more details on the characteristics and locations of the bursts. There appears to be little association with the Milky Way and no depletion in a "Zone of Avoidance" as is seen for the distant galaxies. Some of the identified locations have been found to be associated with repeat bursts [Sky & Telescope 1991], which might indicate the continued existence of the objects involved in those few bursts [Dermer 1992]. Until the operation of the *Beppo-SAX* satellite, which provided quick and accurate locations for gamma-ray bursts, none of the burster locations could be associated with otherwise known objects. The *Swift Gamma-Ray Burst Mission*, a multi-spectral satellite, has improved the rapid detection of these bursts [Gehrels *et al.* 2004], and other satellite observatories have joined the hunt. The goal is to locate the positions accurately and quickly so that more complete information can be gathered.

Gamma-ray bursters are postulated objects that have been observed by a variety of satellites. Attempts to explain gamma-ray bursters have focused on possible effects related to neutron stars, but have generally foundered on details of the spatial distribution of the bursts.

Gamma-ray bursters have been found, with increasing confidence, to be uniformly distributed with respect to our observation position, at the Earth [Astronomy 1993]. Thus, either the bursters must be quite distant, at least in a great halo surrounding the Galaxy (as imagined in our Island Universe theory), sufficiently far that the offset of the Solar System from the Galactic center is not noticeable, or even farther, or they must be very close, so that the limited thickness of the Galactic disk in the Sun's neighborhood does not impact on the observed distribution. While these considerations are obviously affected by our concept of the Galaxy, the significant condition is that gamma ray bursters are either near or far.

Great distances pose difficulties of placing the objects, postulated as neutron stars, at such distances, with the further complication that these bursters are not seen in association with what are considered to be nearby galaxies, and the problem of creating the requisite extreme energy corresponding to Galactic or cosmological distances. Concluding that the evidence for dimmer, redder, and longer bursts in some cases shows the effect of time dilation due to cosmological recession at half the distance to the edge of the observable Universe, results in the gamma-ray bursts being the "most powerful events ever seen, putting out the energy of a quintillion suns." [Travis 1994a].

An alternate model to the Galactic origin of the gamma ray bursts considers that comets in the Oort clouds of stars throughout the Galaxy may be captured and destroyed by neutron stars. However, this model leads to a predicted gamma-ray burst rate a hundred times greater than that observed [Astronomy 1994b]. (Since this model fails, I suspect that it indicates that our estimates of the Oort cloud and neutron stars may be in error.)

Following the neutron star model, structure in the energy spectrum has been identified as cyclotron absorption lines [Murakami 1988]. It appears instead that it is likely that these absorption lines are dips in the spectra between characteristic X-rays and mono-energetic gamma rays.

The close-in situation, such as the Oort cloud of our Sun, poses the problem of how to generate such bursts of gamma radiation without creating other observable effects at such close range [Clarke *et al.* 1994].

The gamma-ray bursts are generally similar to each other but have considerable variability. A particularly well-defined burst started with X-ray emission [Murakami *et al.* 1991] with a spectrum that is typical of a black-body source of thermal radiation at a temperature of 18×10^6 K and an apparent diameter of $(0.6 \times d)$ kilometers, where d is the assumed distance to the burster in kiloparsecs. Thus, at a distance of 30 kiloparsecs for example, or about 100,000 light years, the diameter of the radiating source is estimated to be about 20 kilometers.

Gamma-ray emission starts about 10 seconds after the X-rays and then rapidly declines in intensity. Discrete gamma lines (energies) are observed, and the spectrum peaks near 1 MeV. After the gamma-ray emission has faded away, some continued (or resumed) X-ray emission is detected in some cases.

A distribution of gamma-ray bursters will appear uniform (isotropic) if the range of detection is less than the distance to any boundary of the distribution. While the gamma-ray burster distribution appears to be isotropic in detection events, detailed evaluation of the distribution according to intensity shows that fainter sources occur more often near the plane of the Milky Way [Atteia *et al.* 1991]. This may indicate diffuse attenuation of the gamma rays, rather than a concentrated distribution of gamma ray bursters. Diffuse attenuation by iron dust may cause the fall-off of number distribution at fainter bursts.

One of the early additions to the satellites equipped to study gamma ray bursts is the *Burst And Transient Source Experiment*, BATSE [Kniffer 1993]. This instrument is part of the *Compton Gamma-Ray Observatory*, along with three other specialized systems: the *Oriented Scintillation Spectrometer Experiment* (OSSE); the *Imaging Compton Telescope* (COMPTEL); and the *Energetic Gamma Ray Experiment Telescope* (EGRET). The *Compton Observatory* was placed in orbit in April 1991 by use of the Space Shuttle *Atlantis*. BATSE observed gamma-ray bursts at a rate of about 800 per year [Meegan *et al.* 1992]. If gamma-ray bursts were produced by neutron stars in our Galaxy, each neutron star would have to produce 10,000 bursts to achieve this rate [Atteia 1993].

If a planetary detonation releases a few tenths of a percent of its energy as prompt gamma radiation, *BATSE* [Dermer 1992, Meegan *et al.* 1992, Chupp 1992] should be capable of detecting it at distances up to a few hundred parsecs away, within our neighborhood. Thus, most of the gamma-ray bursters may be local events, just as most of the stars seen at night by the unaided eye live nearby [Goldsmith 1990].

Analysis of the *BATSE* observations has shown a strong division into two groups, bright and dim [Norris *et al.* 1994]. In that research, three groupings of the 131 bursts analyzed were used, but the "dim" and "dimmest" groupings gave essentially the same results, leading to a conclusion that there are only two significant groupings. This research was directed toward showing that the dimmer bursts were cosmologically distant and that time dilation had slowed down the more distant bursts. Based on the observations, two types of bursts exist: one has a characteristic time of about 0.1 to 5 seconds, while the other group falls into a range of about 5 to 500 seconds. A similar division into two groups has been found with respect to variability during the burst [Flam 1993]. Studies of the time behavior of the bursts [Link *et al.* 1993] have lead to the characterization of the bursts as being typical of an explosive action. In the present proposal, the bursts are all nuclear fission detonations at near distances, ranging through our local neighborhood, and differ distinctly by the form of the detonating material. The detonating objects are likely to be single critical masses, which provides a basis for the uniformity of the bursts, and may differ in the retention of hydrogen, or its loss.

It appears that the "dimmer, redder, longer" bursts are likely to be from those detonations of hydrogen-moderated globs. The faster detonations are from globs that have lost their hydrogen.

Some gamma-ray bursts have been subsequently found with X-ray aftermaths, optical transients, and even an indication of a supernova [Caraveo *et al.* 1997, Fruchter *et al.* 1999, and Galama *et al.* 2000]. The study of gamma-ray bursts is difficult because of their brief intensity. Bursts may be detected from great distances, cosmological distances have been proposed, but none last longer than a few minutes. The *Beppo-SAX* satellite has successfully pinpointed the location of two events (February 28, 1997 and May 8, 1997) so that optical telescopes directed quickly towards those locations have found fading optical transients that suggestively appear to be the remnants of the burst (and more have been located since this writing). The fading of these two bursts has been measured both optically and in the X-ray range. The decline of the brightness of the optical transient follows the "after-glow" decay heat of a nuclear detonation very well. This is shown in Figure 145 for GRB970228 in the R-band (solid dots) and for X-radiation (2-10 keV, X symbols and bars), and for GRB970508 in the r-band (open dots). The smooth curves are normalized presentations of the ANSI/ANS standard decay heat curves for a fission burst. Normalization of the standard (watts of decay heat produced per joule of fission energy released in the detonation) to the optical and X-ray observations indicates that the burst energy flux amounted to about 10^{-8} joules per square meter in both cases. This corresponds to a burst energy of about 10^{32} joules at a distance of 1,000 light years, or 2.5×10^{22} joules at 1,000 Astronomical Units. (Greater energy at a greater distance, lesser at a closer distance.) The distant energy would be released in a nuclear detonation of a planet the size of Venus or Earth (roughly 6×10^{24} kilograms), containing 1 part per million of nuclear fuel, while the nearer case would require an object of only 1.5×10^{15} kilograms, smaller than the asteroid Eros.

A detailed analysis of *Hubble Space Telescope* images taken on March 26, 1997 and April 7, 1997, suggested a clear motion of GRB 970228 relative to a certain set of stars (1, 2, 3, 4) [Caraveo *et al.* 1997]. The derived proper motion, relative to four reference stars nearby, amounted to about 550 milliarcseconds per year. This was rejected, twice, based on other detailed analyses of the same images, using a different set of reference stars (2, 3, 4, 5) [Sahu *et al.* 1997, Fruchter *et al.* 1999]. If the reported proper motion were correct, it would represent a velocity (in the plane of the sky) of only 226 meters per second at the distance of 17,301 Astronomical Units, the distance where the circular orbital velocity around the Sun would be the same. (This would be a short distance into the hypothetical Oort cloud, presumably gravitationally bound as far as 200,000 Astronomical Units from the Sun.) At this greater distance, detonation of an object with twice the mass of the asteroid Eros and a uranium content of 30 parts per million would suffice for the observed energy. These are still reasonable numbers.

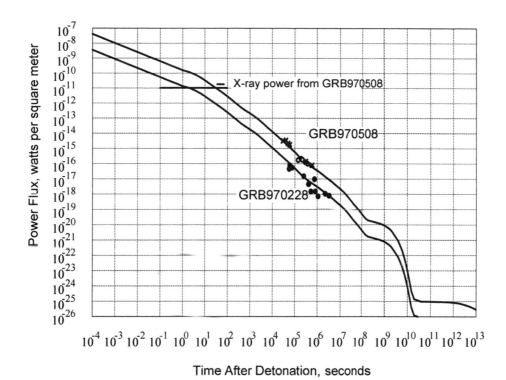

Figure 145. Power flux from GRB970228 and GRB970508.
Power flux from GRB970228 and GRB970508. The curves are calculated from the ANS Standard for Decay Heat in Light Water Reactors [Subcommittee ANS-5 1973]. The bars indicate the early X-ray power for GRB970508.

These optical transient observations often produce estimates of a power law decline in brightness, ranging about $\alpha = -1.2$. That is the value shown to apply to the energy release of a nuclear detonation in the similar time range [Glasstone and Dolan 1977]. For GRB 970228, the decline was found to become steeper at later times, with the power law index approaching $\alpha = -1.73$ [Galama *et al.* 2000]. This change in the power law index is also shown for nuclear detonations [Glasstone and Dolan 1977].

While the gamma spectrum of Gamma Ray Bursts (Figure 146) matches the fission spectrum well in the mid-range, it clearly has more intensity at low energies and at high energies. This could result from the detonation being confined in a massive object, producing higher temperatures and releasing more attenuated (scattered) photons. The fission spectrum was measured for a thin target, with essentially no photon scattering.

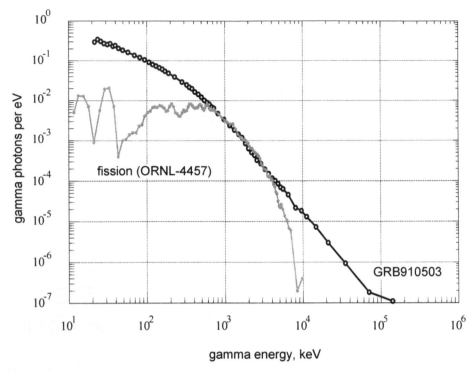

Figure 146. Gamma-ray spectrum from GRB910503 compared to a fission spectrum.
The Gamma Ray Burst gamma spectrum matches that of fission gamma-rays in the mid range, but has more high-energy and low-energy gamma-rays. This may result from a fission detonation that is strongly contained.

The gamma ray bursts, observed by our instrumented satellites at a rate of about one per day, are the result of similar detonations of "globs" of uranium spread through space. Nuclear fission detonations, converting large amounts of mass into energy, might produce gravitational waves, as Joe Weber thought he was finding at a rate of hundreds per year. Just like the gamma ray bursts.

Quasars, Blazars, and Pulsars

> *... who can tell which of her forms*
> *has shown her substance right?*
> William Butler Yeats

Quasars and blazars may be the same, going in opposite directions, the polar fragments of a detonation. Perhaps the observed redshifts are from the ion exhaust, the blueshifted ions are invisible. Pulsars are ion-propelled, like the quasars and blazars, but the tangential thrust makes them spin. The spin rate will vary from the maximum possible for equatorial fragments of a planetary detonation, to zero for the polar fragments.

Quasars belong to a strange menagerie of poorly characterized and understood objects, that also includes Seyfert galaxies, BL Lac objects, and blazars. Each has distinguishing properties, so each has a name. These seem to merge by slight differences, with no clear boundaries. Here I will concentrate on quasars and blazars, and perhaps unexpectedly, pulsars.

Quasars are considered to be intensely energetic, the most luminous objects in the Universe, hundreds to thousands of times brighter than their associated host galaxies. They are nearly point sources of radiation, with variability that shows sizes on the order of light-days. They have redshifts that range up to the greatest measured, as a consequence of their great brightness, and so are considered to be at great distances. However, the redshifts range from small to great, and the relation to distance is not at all certain. In optical images, they appear to be much like stars, and have been named quasi-stellar objects, which led to QSO and quasar.

Blazars and BL Lac objects seem to be two versions of the same thing, much like quasars without the distinctive emission lines.

Pulsars are described as rapidly rotating neutron stars, the dense remainders after a core collapse supernova. A strong magnetic field traps plasma around the star, whirling it and accelerating it to close to the speed of light. This plasma radiates energy that sweeps around the sky. Each time the beam passes an observer a pulse can be seen. While much of the radioactive decay energy is directed into spinning up the pulsar, some will be directed along the axis of the magnetic field. This will act like an ion engine, just as in the quasars. This has been seen in the Crab pulsar and the Vela pulsar. These two pulsars are "flying through space pole-on." [News Notes 2000]. This self-propulsion seemed unlikely to astronomers, with insufficient momentum in the observable jets to accelerate a neutron star with roughly the mass of the Sun (in the standard story). However, for fragments of a planetary detonation, much smaller than neutron stars, and much closer, so that the derived velocities are much smaller, ion propulsion is quite adequate. Acting over hundreds or thousands of years, the gentle push can move the fragment. Otherwise, some hypothetical "kick" must be given to the neutron star by the remains of the supernova.

The great brightness of quasars, making them the brightest steady sources in the Universe, results from the great distance assigned to the high redshift quasars, and so depends entirely on the assumption that the quasar redshifts are the same as the redshifts of ordinary galaxies and agree with the Hubble flow. There is no independent evidence for the great distances.

The assumed distances, the consequent extreme brightness, and possible connections to nearby galaxies, have been a long-term controversy among astronomers, as described by Halton Arp [Arp 1987]. Arp has shown many examples of quasars in clearly identifiable spiral galaxies, but not centered, and quasars adjacent to and apparently connected to galaxies. The redshifts of the quasars and galaxies differ greatly. The quasar redshifts are always greater than the redshifts of the galaxies. Contrary to the great brightnesses conventionally assigned to the quasars, hundreds to thousands times brighter than normal galaxies, those quasars seem to have brightnesses appropriate to their accompanying galaxies. The standard story requires that the quasars be roughly ten times farther away and be shining through the nearby galaxies.

Many apparent paired quasars associated with nearby galaxies have been described by E. M. Burbidge [Schilling 2001]. Again, the redshifts of the quasars are much greater than those of the associated galaxies, implying a great separation in distance.

Quasars are thought to result from the infall of matter into a supermassive black hole [Disney 1998]. This is usually at the center of a galaxy, but quasars are often seen away from galaxies. Jerry Kristian demonstrated in 1973 that at least some low-redshift quasars showed a faint fuzzy halo. This has been interpreted as light from the stars of the presumed host galaxy, but the galaxies themselves could not be resolved. More recent images by the *Hubble Space Telescope*, with extremely careful processing, have shown what appear to be host galaxies for about half the cases. Colliding galaxies are thought to provide the disturbance that fuels the black hole, to create the quasar. In general, the host galaxies of distant quasars cannot be seen because the quasars are so bright as to overwhelm any light from a galaxy. However, it may be that the host galaxies of quasars that are actually near cannot be seen because the quasars are not hosted, but have a significant separation from the associated galaxy, as is the case for NGC 4235, a quasar, and a BL Lac object, described below.

Similar image processing has shown a quasar to be separated from a nearby galaxy, but with a "blob" of light-emitting gas partly enshrouding the quasar [Megain *et al.* 2005]. This strongly shows that there is no host galaxy present. The associated galaxy is within about 1 arcsecond of the quasar, so this association might be significant. It may be that such a blob of light-emitting gas is an integral part of quasars and could be seen as the fuzzy halos in the other cases, rather than host galaxies of stars. The blob shows strong emission from singly ionized oxygen, but very faint hydrogen emission.

A possibly similar blob of light was studied in the radio galaxy PKS2152-69 [Aligheri *et al.* 1988, Tadhunter *et al.* 1987, Ward 1988]. Tadhunter *et al.* found emission from highly ionized atoms, requiring ionization energies in the range of 25 to 93 electronvolts, considerably above the 13.6 electronvolts required to ionize hydrogen. It was concluded that ionization by bright young stars could not produce such a spectrum.

High levels of ionization, requiring ejection of tightly bound electrons from deep within the atomic cloud of electrons, can be produced easily by the beta particles and gamma rays from fission products resulting from a nuclear fission detonation, as proposed for the formation of all spiral galaxies. These radiations have energies in the range of a few hundred kiloelectronvolts (keV) to a few million electronvolts (MeV).

Arp has proposed (insisted!) that quasars have been ejected from the associated galaxies and are nearby, regardless of the different redshifts. A very clear association of a quasar, a Seyfert galaxy, and a BL Lac object is shown by Arp [Arp 1997]. An image of this area in X-rays by the *ROSAT* Position Sensitive Proportional Counter (PSPC) is shown in Figure 147. This image, as interpreted by Arp, showed that a quasar was directly in line with the minor axis of the Seyfert galaxy NGC 4235, suggesting that it had been ejected in a direction perpendicular to the plane of the galaxy, and so in the direction of the polar axis of a detonated planet as proposed in this book. A BL Lac object is seen at a similar angular distance from the galaxy, in the opposite direction, and nearly in line with the minor axis direction.

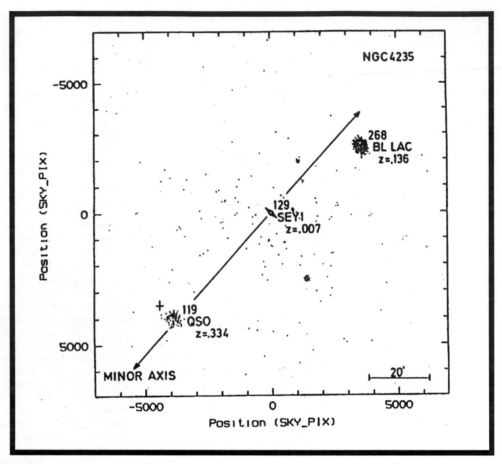

Figure 147. The field of objects around the spiral galaxy NGC 4235.
The field around the spiral galaxy NGC 4235 as imaged in X-rays by the ROSAT PSPC. Arp has inserted a replica of the optical image of NGC 4235 itself, and indicated the minor axis, perpendicular to the major axis, and so perpendicular to the plane of the galaxy. The integer numbers are the number of counts per 1,000 seconds in the PSPC; the redshifts are indicated by z [Arp 1997].

The controversy arising from these measurements comes from the very different redshifts:

quasar	z = 0.334,	mag 15.68
NGC 4235	z = 0.007,	mag 12.62
BL Lac object	z = 0.136,	mag 16.1

In this example, the much greater brightness normally stated for quasars is not shown. The quasar is actually only 6% as bright as the galaxy.

While the emphasis has been on single quasar and galaxy associations, and pairs, Arp interprets observations of the neighborhood of NGC 1097 as showing about 40 quasars associated with it [Arp 1997].

Michael Disney described the relative abundance of quasars in the nearby Universe, within 1 billion light years (standard measure), as 1 quasar per million galaxies [Disney 1998], but that at a redshift of 2, the ratio increased to 1,000 quasars per million galaxies. This has been interpreted as showing that in the past, quasars were 1,000 times as abundant as at present. However, if the distant ratio is restated to reflect a constant abundance of quasars as 1 quasar per thousand galaxies, the implication is that the number of observable galaxies at a redshift of 2 is 1,000 times less than we see nearby. This is in agreement with the decline in galaxy density with distance that is shown in Figure 112. In that figure, the density of observed galaxies declines by a factor of 1,000 at a standard distance of 8,000 megaparsecs, corresponding to a redshift of 2. The presumed overabundance of distant quasars is instead a strong confirmation, by different observations, of the disappearance of distant galaxies. Quasars were not more abundant in the past, at great distances, but instead, we just see fewer galaxies.

Hewitt and Burbidge have compiled a catalog of all quasars and BL Lac objects known at the time, 1992, with position, magnitude, and redshifts [Hewitt and Burbidge 1993]. As part of their review, a Hubble diagram was presented, showing the relation between the redshift, z, and the apparent magnitude, for brightness. (This is the way the earliest Hubble diagrams were drawn, before distances to the galaxies could be determined. Brightness substitutes for distance, with the dimmer objects assumed to be farther away.) Using the data in the catalog, I have plotted the redshifts and magnitudes in a similar way, in Figure 148, and for comparison, a not-so-similar Hubble diagram for field galaxies also in Figure 148, using data taken from Figure 5.6 of Peebles [Peebles 1993], who cites his source as T. Shanks, Private communication 1991. (Field galaxies are isolated, not associated with any clusters, and so should show the truest indication of the Hubble flow, unaffected by gravitational perturbations.) The diagram of the field galaxies shows that at progressively greater redshifts, the brighter galaxies (smaller magnitudes) are not seen. At low redshifts, galaxies disappear.

The major problem lurking in those two Hubble diagrams is that redshift has been accepted as a true measure of distance for galaxies, and by analogy, for distance to the quasars. The lower plot is for galaxies, while the upper plot is for those objects that we identify as quasars. As Peebles points out [Peebles 1993], "If the redshifts of quasars did not follow the redshift-distance relation observed for galaxies, it would show we have missed something very significant." Peebles states that the controversy over the difference in redshift distributions has been resolved, without stating the resolution, and cites the clustering of galaxies around low redshift quasars and the gravitational lensing of high redshift quasars by galaxies well in front of them, as the basis for this resolution. But the distributions definitely do not agree, the quasar redshifts do not match the redshift-magnitude relation observed for galaxies, at all. And it becomes a problem to decide what represents distance for quasars, is it redshift and the Hubble flow, or is it brightness, with dimming at great distances?

1. The quasar redshifts are roughly a factor of ten greater than for the field galaxies in this brightness range.
2. The quasar redshifts are essentially independent of brightness, and therefore has no relation to distance.
3. The quasars are "born" fast, with few slower than z = 0.1, and none negative. The missing negative z quasars may actually be invisible BL Lac objects, blazars, quasars coming toward us and blueshifted out of sight.
4. The field galaxies have a remarkable scatter in z for the same values of brightness, and in brightness for the same z.
5. The quasar distribution shows the vanishing effect at low redshifts for dim (distant) quasars, as seen for disappearing galaxies.
6. The quasars "top out" at a redshift of about 3.3, equal to a relativistic speed of 90% the speed of light, and this is independent of brightness (and distance). Three quasars have been found at a redshift of 6, making them the highest redshift objects yet known [Fan 2001]. (Some higher redshift objects have been found since this writing.)

It is clear that the Hubble diagram for quasars is thoroughly different from that for galaxies.

(There is a similar problem with the use of redshift for distance shown in comparisons of "giant" spirals, "normal" spirals, and "dwarf" spirals. The size of the spiral galaxy must be estimated from its angular size and its cosmological distance, which is determined by its redshift. Yet, NGC 309 is about 10 times as large as a very similar M81 galaxy [Block and Wainscoat 1991]. At the other end of the scale, dwarf spirals (not the dwarf galaxies seen as companions to normal spirals) average 1/7 the size of M31 (the Great Nebula of Andromeda), and are 1/50 as bright [Schombert *et al.* 1995]. Since size (and brightness) are derived from the observations by use of the distance, which is determined from the redshift, this implies that the redshift distances aren't even relatively right. The giants are closer than we think, and the dwarfs are farther.)

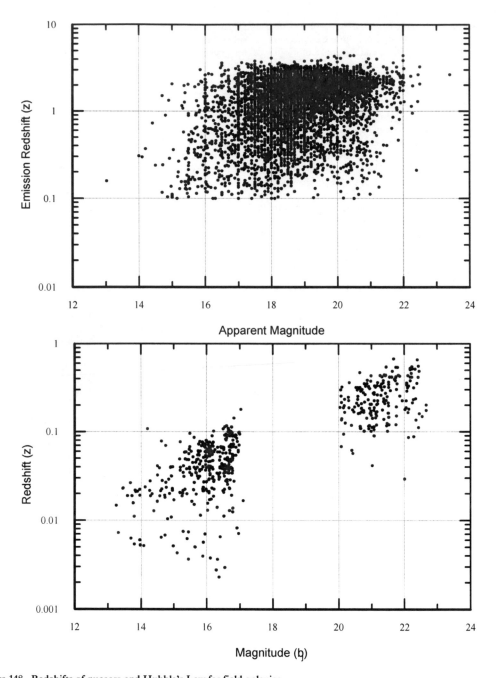

Figure 148. Redshifts of quasars and Hubble's Law for field galaxies.
The redshift of quasars plotted against brightness in magnitudes, as an approximate representation of distance. The low redshift quasars begin to disappear at about magnitude 18. The redshifts of some field galaxies plotted in a similar manner, against brightness. The two distributions are quite different.

Some few quasars have been found to show unusually strong emission from iron, indicating abundances much greater than in our neighborhood. Two in particular have extremely bright iron emission. These are IRAS 18508-7815 [Lipari, Machetto, and Golembek 1991] and IRAS 07598+6508 [Lawrence *et al.* 1988]. These two quasars are also notable in matching each other closely in optical and infrared characteristics, and having very similar redshifts, 0.1618 and 0.1483, respectively. They are also seen roughly on the opposite sides of the sky [News Notes 1991a] Actually, they are very nearly 180 degrees apart on the sphere of the sky, making them truly opposites. In the special situation of a very nearby planetary detonation, a pair of matching quasars might be seen to be moving away from the Earth, in opposite directions, in opposite locations in the sky. The quasars IRAS 18508-7815 [Lipari *et al.* 1991] and IRAS 07598+6508 [Lawrence *et al.* 1988] form just such a pair. Both show strong iron emission, have similar optical spectra and far-infrared characteristics, and similar redshift. They are close to opposite points in the sky, as viewed from Earth.

If our Milky Way Galaxy was produced by the nuclear fission detonation of Planet X, approximately 6 million years ago, these twin quasars might belong to our Galaxy, ejected by the detonation from the poles where there would be no tangential velocity from rotation to spin them into the disk.

Particularly energetic planetary detonations may occur in large planets with a generous amount of nuclear fuel. These detonations would produce fragments with extremely high initial velocities and great amounts of radioactivity. Ejection of vaporizing material from the rear of the fragment, both by radioactive heating and thermal evaporation, and directly by radiation itself, will further increase this velocity. The development of a strong electrostatic charge, from the emission of negatively charged beta particles at nearly the speed of light will lead to exceptionally efficient ion-propulsion engines. Gravitational re-shaping of a large fragment may have a significant effect in the development and enhancement of this jet. Radiation from the fragments that are moving in the direction towards the Earth will be blocked by the forward, cool region of the fragment, and we will only see the expanding "exhaust" plume. Fragments moving away from the Earth will be observed as quasars.

As a result of the emission of high-speed electrons, ejected by gamma-ray photons, the fragment will develop a positive electric charge, limited in voltage by the highest energy electrons. That is, with no electrical leakage (neutralization) the voltage on the fragment would approach 1.33 million volts, corresponding to electrons ejected by the higher energy cobalt-60 gamma ray.

This high voltage will repel positively charged iron, cobalt, and nickel ions that have been released from the rear surface of the fragment, thus acting as an ion propulsion engine. As in all well designed ion engines, electrons immediately behind the exhaust serve to neutralize the positive ion stream.

If the fragment is powered by iron-60 and its daughter cobalt-60, the ions will have an energy of about 1.33 MeV per charge, and thus will be accelerated to velocities on the order of a million kilometers per hour as they leave the fragment. The ions and the photon-ejected electrons and beta-particle electrons will recombine in a vast cloud behind the fragment. The ion propulsion system will work well, perhaps only, if the magnetic field of the fragment is parallel to the line of thrust. Thus, fragments from the two opposite magnetic pole regions of the disintegrated planet are the most likely to develop as full-powered ion engines. This is likely to produce pairs of high-speed fragments, moving directly away in opposite directions from the central, expanding core of vaporized iron.

In the rocket industry, the figure of merit of a reaction propulsion engine (chemical, nuclear, or ion) [White 1986, Nelson 1998] is the specific impulse, the thrust divided by the rate of fuel consumption. Chemically powered rockets have a specific impulse of about 400 seconds, nuclear powered rockets, about 800 seconds, and ion propelled rockets have a specific impulse of 3,000-15,000 seconds. Ion propulsion engines have been used on the *Deep Space One* and *Dawn* space probes. *Dawn* is scheduled to visit the asteroids Vesta (in 2011) and Ceres (in 2015). The specific impulse of the *Dawn* ion engine is 3,100 seconds, much less than can be provided by a natural engine powered by cobalt-60.

The cobalt-60 powered ion engine would have a specific impulse on the order of 100,000 seconds or more. With initial accelerations in the range of 10 kilometers per second squared, about the surface gravity of the Earth, it is likely that material would slide off the edges of the fragment, leaving a trail of crumbs behind. The fragment could reach a significant fraction of the speed of light. These fragments would be spin-stabilized, and their axial magnetic field would constrain and direct the rearward propulsive stream of ions. The axial field would be reinforced by the ion stream encircling the fragment, formed by emission of ions and electrons that are directed transversely to the field.

Experimental ion propulsion engines show a beautiful fuzzy blue glow from recombination of the ions with electrons. Perhaps this is the origin of the Big Blue Bump [Wills 1992] observed in the optical spectra of QSOs and quasars, and the excess ultraviolet radiation seen in objects that have been identified as high-redshift radio galaxies [di Serego Alighieri *et al.* 1989] and in blazars [Antonucci 1992].

A natural consequence of this model of quasars is the generation of large concentrations of ionized iron. This has been observed for high-redshift quasars [Elston *et al.* 1994].

Those fragments that have lost their remanent magnetization will not radiate at radio frequencies, while those that have retained a magnetic field will radiate radio energy and synchrotron radiation as a result of the stream of electrons (beta particles as well as energetic electrons ejected by the photoelectric and Compton effects) moving through this magnetic field. The radio emission would be modulated by any spin of the fragment. The nebulosity seen around some QSOs and quasars would result from the evaporation of material from the rear of the fragment. As the iron-60 decays away, the ion propulsion engines run out of fuel and the plasma surrounding the fragment dissipates. The quasar fades from sight, but now the radio emission of the rapidly moving magnetized fragment is no longer absorbed by the plasma, and the quasar becomes radio-loud. Such radio-loud quasars, not noticeable by their optical emission, have recently been discovered [Jayawardhana 1994b]. The fragment will cool, and particularly with the loss of the radioactively emitted beta particles generating Cerenkov radiation, its color will change from blue to red. Recent observations have suggested that over half of all the detectable quasars are more red than blue [Sky & Telescope 1994b].

These objects would be relatively near the Sun, and would also appear to be associated with objects that are seen as other nearby galaxies, but would not be at the cosmological distances that are indicated by their red-shifts, which are entirely due to the high speeds of the repelled gas. As the rear section of the fragment, the "engine", vaporizes and erodes, the fragment would have an erratic path, leading to the production of a twisted wake.

Many polar fragments of a detonated planet might have a sufficiently well aligned magnetic field and enough radioactive fuel to become quasars. These would appear to cluster in opposite directions from the origin of the detonation. In the case of the Milky Way, clusters of quasars would be seen toward the north and south galactic poles, as they are (Figure 5 and 6 in [Hewitt and Burbidge 1993]), shown in Figure 149. Other clusters, in twos or threes or fours, might appear from behind the parent planetary detonation, and would appear to be gravitationally lensed by what is judged to be a massive spiral galaxy at moderate distance, but is actually the spiral debris of a planetary detonation, that spawned the quasars. These quasars, born in the same detonation from the same planet, would be likely to have similar or identical spectra and redshifts.

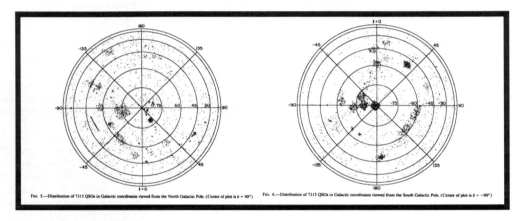

Figure 149. Clusters of quasars related to the Milky Way Galaxy.
Clusters of quasars related to the Milky Way Galaxy. (Figures 5 and 6 in [Hewitt and Burbidge 1993]). The cluster "below" the South Galactic Pole is particularly pronounced.

Propulsion of planetary fragments in this manner would result in the dispersal of large amounts of highly ionized iron. Extraordinarily high iron abundances in the space between the Earth and the Large Magellanic Cloud have been found [Malaney and Clampin 1988, Pettini *et al.* 1989] and discarded [Wang *et al.* 1989]. The identification, based on spectroscopic determination of the absorption of Fe X at 6375 Å, was rejected for lack of a mechanism to create such large abundances. The radioactive heat of these fragments will cool at a rate corresponding to the radioactive half-life of iron-60, 2.6 million years, as the radioactive iron changes to cobalt and then to nickel, and the iron/nickel vapor will condense into fine, dispersed crystals, crystalline whiskers.

The number of quasars within our Universe is estimated to be 4 million [Powell 1991]. This quantity seems more in keeping with fragments from detonated planets than with unknown objects that have been labeled the most powerful objects in the Universe.

The possibility that pulsars could be rocket-propelled, in somewhat the same manner as discussed here for quasars, has been suggested before [Tademaru and Harnison 1975]. Fragments that have a magnetic field directed across the stream of ions will produce an acceleration that acts as a torque, and spins the fragment to high speeds.

If these self-propelled fragments appear to be quasars with great redshifts as they move away from us, perhaps the blazars are the opposite fragments, moving towards us.

Here it is proposed that blazars and BL Lac objects are the same as quasars, polar fragments from a nuclear fission detonation of a planet, but traveling in the opposite direction, with a component of motion toward the Earth

These fragments are born intensely radioactive, and possibly contain still active nuclear reactors. This surrounds the fragments with ionized gas (positive ions). The emission of beta particles (negatively charged electrons) from the fission products near the surface creates a strong positive charge on the fragments. The beta particles travel at nearly the speed of light, and so reach considerable distances from the fragment in a short time. Electrons are also ejected from the surface by the high energy gamma-rays from deeper inside the fragment. The positive electrostatic charge on a fragment repels the surrounding positive ions, accelerating them to high velocities. The positive ions are neutralized by the electron cloud, at a considerable distance. Nature makes an excellent ion propulsion engine.

Using data from observations of the supernova remnant G29.7-0.3 in X-rays by the *Advanced Satellite for Cosmology and Astrophysics (ASCA)* X-ray observatory, Elizabeth Blanton and David Helfand found the remnant to consist of a hard X-ray central source surrounded by a soft X-ray thermal plasma shell [Blanton and Helfand 1996, News Notes 1996]. Spectral line X-ray emission was seen from ions that require over a hundred times as much energy (in a single photon) to ionize as does hydrogen, which can be ionized by ultraviolet light from young stars. The most distinctive X-ray lines were identified with Mg XI (the recombined +11 ion that gives an X-ray at 1.305 kiloelectronvolts), Si XIII (the recombined +13 ion that gives an X-ray at 1.8389 kiloelectronvolts), and S XV (the recombined +15 ion that gives an X-ray at 2.4720 kiloelectronvolts). The high-energy beta particles and gamma-rays from fission products, with energies of a few hundred kiloelectronvolts to a few million electron volts can provide this ionization. The continuum X-rays were found to have a characteristic temperature corresponding to an average electron energy of 0.75 million electron volts. This is very close to the maximum energy of the beta particles emitted in the beta decay of free neutrons, 0.782353 million electron volts. Space immediately surrounding a nuclear reactor, continuously operating or detonated, would be filled with free neutrons which would produce the electrons that give off the X-rays.

As the ions recombine with nearby free electrons, they emit characteristic spectral lines. (The spectral lines are emitted by the gas ions, not from the central object itself. The redshifts are therefore representative of the recombining gas ions, rather than the central object.) Ions that recombine while moving away from the Earth emit redshifted photons. Those ions moving towards the Earth emit blueshifted photons. Redshifting reduces the energy of the photons, making them less likely to be absorbed on their way out of the surrounding gas cloud. Blueshifting increases the photon energy, making them more likely to be absorbed. Furthermore, redshifting increases the photon wavelength, reducing scattering by dust, while blueshifting reduces the wavelength, increasing scattering. These effects allow redshifted emission lines to be detected, while the blueshifted lines are blocked and lost from sight. Motion of the blazar towards the Earth increases the blueshift effect, but the observed spectrum is likely to come from the "exhaust" gas, which has its own redshift, while motion of the quasar away from the Earth increases the redshift effect.

Broad emission lines have been found, with widths generally from 1,000 to 10,000 kilometers per second, and even approaching 50,000 kilometers per second [Wills 1992]. Different stages of ionization will result in the different ions being accelerated to different speeds, giving the very broad lines.

Jets are produced by the channeling of ionized material along the magnetic field lines at the opposite magnetic poles, which are aligned to permit the escape of charged particles. The material traveling in a jet pointing towards the Earth emits blueshifted spectral lines, which are lost. A jet pointing away from the Earth emits redshifted lines, which remain to be seen.

The problem of superluminal motion, and how observational selection could result in so many examples, is eliminated by avoiding the use of the redshifts observed for the repelled gas ions as redshift-distance indicators for the quasars/blazars and accepting the distances as those of the associated galaxies. (Of course, accepting the short distance scale implied by the formation of spiral galaxies by the detonation of rotating planets completely eliminates the problem.)

While only rare fragments from a detonation, as polar fragments, will have the physical and magnetic alignment necessary to produce quasars and blazars, in keeping with their "one quasar in a million galaxies" abundance, pulsars (in the Milky Way Galaxy) are found abundantly. As of November 1998, only 31 years after their discovery, 1,000 pulsars in and near our galaxy had been found. The number continues to grow.

Pulsars are considered to be rapidly rotating, highly magnetized neutron stars, with the mass of the Sun in an object 10-30 kilometers in diameter, with a density exceeding that of nuclear matter. The bulk of the star, the remnant of a Type II supernova explosion, is composed of neutrons. Theories of pulsars, based on rapidly rotating neutron stars, are well developed and generally explain the observations [Lyne and Graham-Smith 1990]. (In particular, the binary pulsar analysis is so detailed and complete, and agrees so well with the General Relativity model, it is quite a problem for this book.)

Here, I can only offer an alternative view that might be possible in the framework of a planetary detonation.

Fragments near the magnetic equator of a detonating planet will be produced in much greater numbers than the rare polar fragments. Positive ions repelled transversely to the magnetic field of the fragment will create a torque causing the fragment to spin. Continued release of energy, beta particles, and positive ions, will accelerate the fragment to some considerable rotation rate or until the fragment breaks apart from centrifugal force. Technological devices have been built that spin faster than pulsars: the Capstone turbine at 1,600 revolutions per second, automobile engine turbochargers at 2,500 revolutions per second, and the fastest known device, a microengine one millimeter in size, at 8,333 revolutions per second. These are quite small and rely on the tensile strength of the metal to hold together. Spinning fragments with strong magnetic fields would be held together by the compression of the magnetic field. Clearly, the actual physical size of these objects is a crucial question.

Trapping of the surrounding plasma (positive ions and negative electrons) by the spinning magnetic field will produce pulses of radiation in the same manner as proposed for neutron stars. In this model, however, a neutron star with a density exceeding nuclear matter is not required.

The complementary nature of quasars and blazars, and pulsars, as polar and equatorial fragments respectively, of a detonated planet, those fragments with magnetic fields parallel to the ion propulsion thrust or transverse, is shown by their spatial distributions, Figure 150. The quasars are found at the opposite poles of the Galactic coordinate system, as was shown in Figure 149 [Hewitt and Burbidge 1993], while the pulsars generally lie along the Galactic equator, as shown in Figure 150 [Lyne and Graham-Smith 1990].

Pulsars are thought to be the remnant neutron star produced in the collapse of the core of a supernova. Having magnetically trapped nearly all the angular momentum of the progenitor star, the neutron star would spin at a high rate and beam out radio and optical radiation as its misaligned magnetic axis sweeps around. The pulsar is born fast and slows down, losing energy by radiation from its magnetic field. It is theorized that the fast, millisecond, pulsars are old pulsars with weak fields that have been spun up to extremely high speeds by accretion from a companion star in a binary system. Some pulsars are found in binary systems.

In globular clusters, which are the result of an induced detonation of a glob of fission fuel in the neighborhood of a detonating planet, the residual kernel of a detonated glob could be spun up because of the radial electron emission crossing the magnetic field. This form would act more like the traditional neutron star model.

In special cases, a fragment of a planetary detonation could exhibit action that resembles a neutron-star pulsar. Fragments that were not so well aligned, thrust with center-of-mass, as to become self-propelled, but with the thrust transverse to the magnetic field, would direct their energy into a rapidly increasing rate of spin. A spinning object is subject to precession and nutation effects that cause it to appear to wobble [Julian 1987, Sekaniana 1987]. The spin rate of the fragment, driven by the decay of iron-60 with a half-life of 2.6 million years, would be observed as the pulse rate of the pulsar, and the precession and nutation would modulate the radio signal to give the appearance of a massive object in a small, fast orbit, or one surrounded by two planets. In cases where this precession obscures the hot, radioactive, radiating part of the fragment, it would give the appearance of an eclipsing binary system.

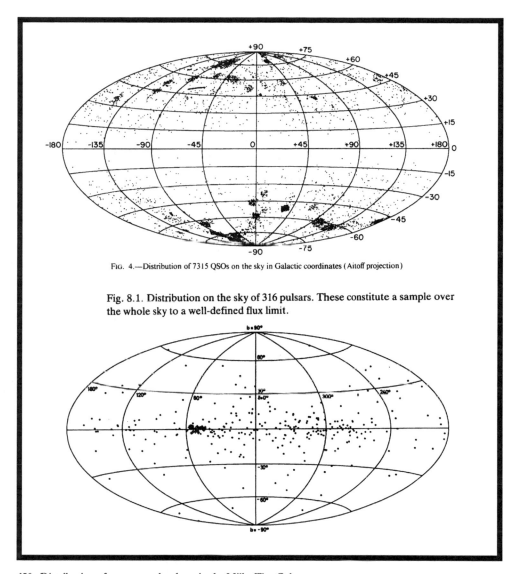

FIG. 4.—Distribution of 7315 QSOs on the sky in Galactic coordinates (Aitoff projection)

Fig. 8.1. Distribution on the sky of 316 pulsars. These constitute a sample over the whole sky to a well-defined flux limit.

Figure 150. Distribution of quasars and pulsars in the Milky Way Galaxy.
Distributions of QSOs (quasars) and pulsars in the galactic coordinate system. Clusters of quasars are seen to be concentrated at the poles [Hewitt and Burbidge 1993]. Pulsars are objects in our galaxy and show abundantly along the galactic plane [Lyne and Graham-Smith 1990].

An obscure and arcane technique of interpretive analysis takes the famous bell-shaped curve of the Gaussian distribution, the distribution called Normal because of its common occurrence, and turns it into a straight line. Straightening out the Normal distribution provides a visual means for easily identifying data values that do not belong to a data set by their readily apparent deviations from the straight line. If some of the data values do not belong to the data set, it is unlikely that the objects that these data values represent would belong to the class of objects represented by the bulk of the data.

A distributional plot [McGinnis 2009] of the rotation rates of the pulsars listed in the 1993 catalog of Taylor, Manchester, and Lyne [Taylor, Manchester, and Lyne 1993] is shown in Figure 151. (In this sort of plot, if points lie along a straight line, they belong to a Log-Normal distribution, one in which the logarithm of sample values has a Normal or Gaussian distribution.) It is clear that, although the rotation rates uninterruptedly span the range of about 0.1 revolutions per second to nearly 1,000 revolutions per second, the distribution is irregular.

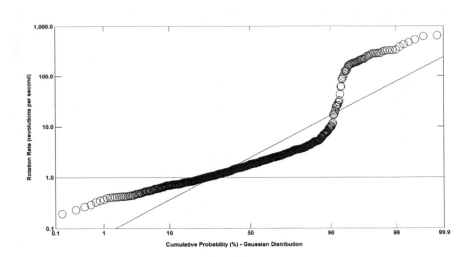

Figure 151. Rotation rates of pulsars.
A Log-Normal distribution plot of the rotation rates of the entire Taylor-Manchester-Lyne Catalog of Pulsars, showing an irregular distribution, an inhomogeneous population.

Statisticians refer to this situation as an "inhomogeneous" population, where the full data set cannot be represented by a single distribution function. This indicates that some of the data objects do not really belong with the others. The set of pulsars divides into a fast group (the millisecond pulsars), shown in Figure 152, and a slow group, shown in Figure 153. Dividing the set into two groups, as shown in Figures 152 and 153, gives good agreement with the Log-Normal distribution. These two distributions, separately, are excellent fits to different Log-Normal distributions. (A Log-Normal distribution is common in nature. It reaches a maximum at a somewhat low value of the variable, rotation rate in this case, and has a long tail extending to higher values.)

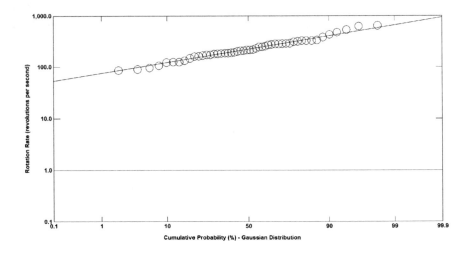

Figure 152. Rotation rates of fast pulsars.
A Log-Normal distribution plot of the rotation rates of the fast group of pulsars, showing excellent agreement with a Log-Normal distribution.

The millisecond pulsars are thought to be old pulsars whose magnetic fields have decayed away, but were then spun up to rapid rotation rates by accretion of material from a binary companion. It is just a little strange that the fast pulsars do not join with the slow pulsars and gradually depart from the slow distribution, but instead are so clearly a separate distribution. At a time (1994) when only 25 millisecond pulsars were known, the number estimated for the Milky Way Galaxy was between 25,000 and 40,000 [Thorsett 1994]. It is estimated that there are 10,000 pulsars in the Galactic globular clusters alone [Bailyn 1990].

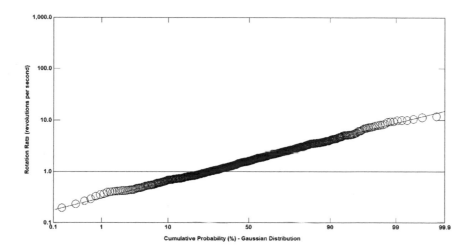

Figure 153. Rotation rates of slow pulsars.
A Log-Normal distribution plot of the rotation rates of the slow group of pulsars, showing excellent agreement with a different Log-Normal distribution.

The conclusion of this analysis is that the short-period pulsars consist of a distinct class of objects, separate from the great majority of objects that have been classified as pulsars. It is possible that this distinction is simply that the short-period pulsars form a class of pulsars that are not detectable by routine survey methods, and so are not at all special or different in kind from the rest. If the pulsars belong to distinctly different distributions (like jockeys and basketball players), it is likely that there are some different distinguishing characteristics. Even if the mechanism for producing the pulsars is the same, there must be some underlying differences to produce such different distributions.

In the formation of pulsars as ion-spun fragments from a planetary detonation, such as Planet X making our Milky Way Galaxy, two possible mechanisms may occur. The two different spin rate distributions may result from two distinct size distributions, representing different fracturing behavior of regions in a mineralogically segregated planet. An alternative could be a difference in the residual radionuclides, with different beta decay energies or half-lifes, from the detonation of a chemically segregated planet. Since we know the Earth is both mineralogically and chemically segregated, either or both of these mechanisms could be at work.

Pulsars have the fastest speeds of any objects in our local neighborhood, with an average transverse velocity of 300 kilometers per second [Lyne and Lorimar 1994], and a top observed transverse velocity of 2,300 kilometers per second [Frail and Kulkarni 1991]. Transverse velocities are determined from the observed proper motion of an object and its assigned distance. Large proper motions and great distances produce very fast velocities. These velocities often exceed the escape velocity of binary systems, globular clusters, and the Milky Way Galaxy. Yet, these objects are still in place.

Pulsars have large proper motions, presumably directed away from their points of origin, as a result of a kick received during creation. This would be a natural result of a detonation acting on an off center fragment, rather than an asymmetric effect on the centrally located neutron star remnant of a supernova. Axially-directed ion propulsion of a spinning fragment could further increase the linear velocity, but it is more likely that the actual velocities should be reduced by about a factor of 60 Astronomical Units/8 kiloparsecs = 3.6 x 10⁻⁸. Thus, the derived velocity of 2,300 kilometers per second at a distance of 8 thousand parsecs becomes a much more leisurely 8.4 centimeters per second at 60 Astronomical Units. This is an embarrassingly slow speed for a gravitationally bound object. This shows the great effect that our assumptions of great distance can have.

A spinning-up pulsar has been observed [Lyne 1989, Wolsczczan *et al.* 1989]. The acceleration in the spin has been explained as perhaps due to gravitational acceleration of the pulsar towards the Earth. Or, this could be the result of the accelerating spin of a fresh transverse ion engine. Loss of a fragment of mass would allow the ion engine to spin the pulsar faster.

Glitches (abrupt increases in the spin or pulse rate) may result from a decrease in the moment of inertia of the spinning fragment as a splinter is broken loose due to the centrifugal force overcoming both the tensile strength of the material and the compressive force of the magnetic field. While some angular momentum will be carried away by the splinter, magnetic contraction of the fragment may occur after this loss, reducing its moment of inertia and increasing its rotation rate.

Misalignment of the ion propulsion thrust (for a pulsar) with the axis of rotation will result in three separate rotational motions, one associated with each of the three moments of inertia. These periodic motions may be misinterpreted as providing evidence of gravitationally bound orbiting planets. An alternative is that the pulsar fragment is actually accompanied by one or more orbiting fragments, perhaps held in place by their joint magnetic attraction.

The X-ray binary Hercules X-1 is supposed to be a neutron star orbiting the faint star HZ Herculis [Kaufman III 1991]. Evidence that it is a highly magnetized neutron star is found in the detection of what has been interpreted as cyclotron emission at an X-ray energy of 58 ± 5 keV [Trümper *et al.* 1978]. It seems more likely that HZ Her and Her X-1 are one and the same object, a spinning, precessing, radioactive fragment emitting gamma-radiation characteristic of iron-60 at an energy of 58.6 keV.

In those cases where the inner core of the detonating planet remains intact, this core will emit large amounts of radiation: gamma rays, X-rays, blackbody (thermal) radiation, electrons, and highly ionized iron. It will act as a dynamo, with radially outward current, and the iron ions, or magnesium ions from aluminum-26, accelerated away from the electrically charged core, will be turned tangentially by the magnetic field to form a high velocity swirl, "orbiting the nucleus" as if the core were a massive black hole.

Cosmic rays have been thought to be produced by supernovas [Friedlander 1989, Gaisser 1990] and by pulsars [Beskin *et al.* 1993]. Detonating planets and the resulting fragments produce exploding magnetic fields, high-velocity magnetic fields, and rapidly rotating magnetic fields, and might also be sources of cosmic rays, by the acceleration of the ionized atoms in their neighborhood.

Most planetary detonations will be self-inflicted as a matter of the normal progress of development, but for a planet on the verge, orbiting a star that becomes a supernova or near a detonating planet, the extremely intense antineutrino flux produced by the supernova explosion could trigger a detonation by converting a large number of protons (hydrogen nuclei) in the core into neutrons, or by the antineutrino-electron capture reaction. The induced (or sympathetic) concurrent detonation of many objects in a planetary system would leave as evidence a "cluster of galaxies".

There are thus several types of astronomical "objects" that consist entirely of observed effect: no actual object has been observed, but various objects have been invented or adopted to explain the observations. These include quasars, blazars, and pulsars, and powerful radio galaxies.

The formation of spiral galaxies by the nuclear fission detonation of rotating planets produces quasars, blazars and BL Lac objects, and pulsars. Quasar redshifts do not provide even a relative measure of distance. The problems of superluminal motion and observational selection are eliminated.

Superluminal Motion

*... it takes all the running you can do
to keep in the same place.
If you want to get somewhere else,
you must run at least twice as fast as that!*
Lewis Carroll

One of the absolute laws of physics in our Universe is a speed limit. Nothing can move faster than the speed of light. At 2.99792458×10^5 kilometers per second (186,000 miles per second) that might seem fast enough, yet it takes slightly more than 8 minutes for sunlight to reach the Earth from the surface of the Sun. The transmission time for radio messages from Mission Control to Tranquility Base on the Moon and back showed a noticeable gap between statement and reply. This is due to the finite speed of light.

In the development of his theory of relativity, Albert Einstein proposed that the speed of light in a vacuum was the absolute speed limit for matter or energy in our Universe [Einstein 1961]. Yet, in spite of this fundamental speed limit, astronomers sometimes measure angular velocities for objects that imply, at the distances assigned to those objects, actual velocities that greatly exceed the speed of light [Cohen 1992]. Similar to supersonic flight, this is called "superluminal motion". Since objects cannot exceed the speed of light in our Universe, measurements that show superluminal motion must be in error, and the astronomers have decided that the error is in the assumption that the objects are moving directly apart, in the plane of the sky, where the angular motions are measured. However, if the motion actually involves very great velocities, near (but less than) the speed of light, and is directed towards us, an illusion of superluminal speeds can result.

Some quasars have been found to have a single jet extending some considerable distance from the core of the quasar. The quasar 3C 273, one of the first discovered and one of the brightest, has shown a jet that appears to move on the plane of the sky with a speed about 2 to 5 times that of the speed of light [Davis, Muxlow, and Conway 1985, Davis, Unwin, and Muxlow 1991]. Since such a speed is not allowed, special interpretation is required. For a beam of material moving nearly at the speed of light, pointing nearly in our direction allows the projected, proper motion, to exceed the speed of light.

As of about 1992, proper motions had been measured for 33 objects, and 23 were found to have superluminal motion based on the standard distances [Cohen 1992]. Since the true motion must be within a narrow angle from our line of sight, on the order of 20 degrees, this fraction of superluminal cases seems remarkable. There should be many more cases in which the special interpretation is not needed. Selection is suspected, with relativistic beaming increasing the brightness of the jet and thus making the superluminal cases more easily detected.

(In transparent materials, such as air, water, and glass, light is slowed to less than its speed in a vacuum. In these materials, electrons (and other particles) can travel faster than the speed of light *in that material*. This leads to the release of Cerenkov radiation, bleeding energy from the speeding particles and slowing them down to the local speed of light. In nuclear reactors surrounded by water, that makes a lovely blue glow.)

First, we must recognize that the estimated speeds are entirely based on the estimated distances to the objects. Outside the Solar System, astronomical "distances" are measured in terms of angular distances, arc lengths on the sphere of the sky. These objects are estimated to be at great distances from the Earth so that the motion of their parts must be great to produce an observable angular displacement, even so small as is seen. For example, the distance from the Earth to the quasar 3C 345 is estimated (from its redshift, $z = 0.595$, and the assumption of an expanding Universe) to be approximately 1,700-3,500 megaparsecs (6-11 billion light years). The motions in 3C 345 have been observed by use of Very Long Baseline Interferometry (VLBI) with radio telescopes, since 1979. These observations show that one component is moving away from the major component at a speed of about 7-14 times the speed of light. This clearly violates the speed limit established by standard physics.

However, this high speed has been explained as an optical illusion resulting from the motion of the component being directed mostly towards us, along our line-of-sight. For this to work, the angle must be less than about 17 to 34 degrees.

The superluminal motion, like the great brightness, is a result of the great distance assigned to the quasars, and would not be interpreted as such for the nearer distances that are consistent with Arp's view of quasars, or the even shorter distances proposed in this book.

Cosmology

Cosmology:
... it generates almost as much nonsense as religion ...
M. Disney

Time, not reason,
Separate real from absurd.
Edward W. Constant, 1980, quoted by Robert W. Smith,
"The Space Telescope", 1989

Like modern astronomy, our current cosmology, which attempts to explain the Universe and beyond, is founded on certain assumptions that have guided its development. Our cosmological theories have developed from four assumptions, and one single observational conclusion. The assumptions are:

1. On a large scale, matter in the Universe is homogeneously distributed. Scatterings of stars, planets, galaxies, in one place are balanced by similar scatterings at every other place.
2. The properties of the Universe are the same in all directions: the Universe is isotropic; there is no up and down, no left and right.
3. The physical laws that describe the behavior of matter, energy, space, and time, are the same everywhere, and always have been. (The special behavior of the Universe during the Inflationary Era is accommodated by inventing forces that are no longer effective, under the present conditions.)
4. There is no special place in the Universe. To the extent that we can see equally distant in all directions, we are at the center of our Universe, but any other location should produce a Universe that is essentially the same.

There are no observations to irrefutably support these assumptions, they are simply reasonable, based on our knowledge and current philosophies.

The observational conclusion that forms the single foundation stone of all current cosmology is the redshift/distance relation found by Edwin Hubble and Milt Humason in the 1920s. While this relation was quickly confirmed by theory, consideration of the effect of dust scattering the light from distant, randomly moving objects shows there is no reason to think that these observations show that our Universe or the Cosmos is expanding

In 1916 Albert Einstein published his General Theory of Relativity [Einstein 1916, Einstein 1961] which provided a mathematical description of a Universe like ours. Many experimental observations have been interpreted as verifying his predictions. Conceptually, it is a beautiful theory, integrating space, time, matter, and energy. Many, perhaps most, of the observations may truly support the theory. The theory is probably correct, and has been accepted as the foundation of our cosmology. In combination, solutions of Einstein's General Theory of Relativity that showed the Universe could expand, and Edwin Hubble's measurements showing that the Universe might be expanding, have formed the foundation of current cosmology.

A difficulty arose in the development of the theory, in terms of Einstein's personal concept of the Universe. In order to have his equations describe a constant-size (static) Universe, he found that he had to add a hypothetical parameter, the "Cosmological Constant", Λ, which seemed to have no justification other than to permit the existence of the Universe that Einstein imagined. He disliked this hypothetical term [Aczel 1999] and eventually called it his "biggest blunder", or so it has been said [Will 1986].

However, Alexander Friedmann realized that the Cosmological Constant could be set equal to zero (actually omitted), and Einstein's equations would predict either an expanding or a contracting Universe. In 1929, Hubble's measurements from Humason's photography showed a redshift/distance relation that supported the concept of an expanding Universe [Hubble 1929]. That led to the Einstein-Friedmann-de Sitter Universe that has formed the basis of our current cosmology. (A steady-state theory by Fred Hoyle and Thomas Gold was considered to be a serious alternative during the 1940s, but as proofs of the General Theory of Relativity accumulated, and as theoretical refinements were added to the Expanding Universe theory, all competition has faded away. The Expanding Universe theory, as a theory of observations, led naturally to the Big Bang theory, a theory of processes .)

Here, it is important to recognize that the much smaller observable Universe proposed in this book, and the fallacy of Hubble's redshift/distance relation, do not contradict the Expanding Universe theory, nor the Big Bang. Here, we must only face the fact that none of our observations provide support for the cosmology of those theories. Those theories may describe a Universe that is bigger than we can observe, a Cosmos containing within itself that smaller Universe that we do observe. The Cosmos may be explained by those theories. The smaller Universe, where we live, is explained (in part) by this book. We will find that the static Universe is a possible alternative.

Antimatter and Antigravity

Creation exists only in the equilibrium
of opposing forces.
Zoroastrian religious belief

I have suggested that the expanding Universe measurements of Edwin Hubble and Milt Humason, "the farther they are, the faster they go", might have been the result of an observational artifact, caused by the preferential scattering (and loss) of blue light and the greater penetrating power of red light. We see redshifted galaxies at greater distances, and more sharply, than blueshifted galaxies.

If Hubble's measurements and all those subsequently in the hunt for the Hubble Constant, were actually due to observational artifacts, to a selective bias, was Einstein wrong? Yes, and no.

Einstein's field equations of General Relativity describe how the Universe should function. The equations consist of a set of 6 independent equations with 10 unknowns [Narlikar 1993]. Obviously, there are too many unknowns for too few equations to determine an analytic solution. In principle, there are an infinity of solutions. However, by making some simplifying assumptions and hoping that some of the unknown parameters will be determined by observation, or simply remain as variables, a set of three related solutions was produced in the early stages of General Relativity [Coles 2005]. These solutions show the time rate of change of the cosmic scale constant and the average density. There are three choices for a parameter (k) that determines the topological shape of space: k may be positive, negative, or zero. For k = 0, space is flat, "Euclidian", the same form as we are directly accustomed to.

These three solutions provide that space in the Universe will expand, or contract, or remain constant, static, but in that last case, only if there is no matter in the Universe [Peebles 1993]. While "no matter" is the conventional specification, the actual condition required by the equations is that the average net density of matter must be zero. That allows some interesting interpretations.

The contracting Universe was considered to be excluded by Hubble's measurements (without any investigation of observational bias), and the expanding Universe was accepted, as shown by Hubble (without any investigation of observational bias). Hubble's observations proved Einstein's theory, and Einstein's theory proved Hubble's observations.

We must make a bit of a detour to see where this is leading, and consider a problem of our theories of creation of the Universe. No matter what the theory of the Universe, whether the Big Bang or Steady State or whatever, to have a Universe filled with matter as we observe it to be, requires that matter be created, abruptly or gradually. As well as we know the laws of physics, creation of matter requires creation of an equal and identical (in an anti sense) amount of antimatter. Antimatter (as antielectrons) was invented theoretically by Paul Dirac in 1928, and after the discovery by Carl Anderson of evidence for positrons (antielectrons) in 1932, antimatter particles have been created and studied across the range of nuclear particles. Protons are matched by antiprotons, neutrons are matched by antineutrons, and electrons are matched by antielectrons, known as positrons. Protons and electrons combine to form hydrogen atoms, and antiprotons combine with positrons to make antihydrogen atoms. At a distance, these atoms are indistinguishable.

There is one problem that the Big Bang Theory completely fails to explain. It is a problem that could not even be recognized or foreseen at the time of the invention of the Expanding Universe. At that time, following Hubble's measurements of galaxy redshifts, an Expanding Universe appeared to be inevitably required by Einstein's Theory of General Relativity. This all changed without much critical notice with the invention of antimatter by Paul Dirac in 1928, and the discovery of the first antimatter particle, the positron, by Carl Anderson in 1932. The problem is that the laws of physics require that when matter is created, which must have occurred because our Universe is filled with matter, an equal and identical amount of antimatter must be created. We see no evidence of antimatter in the entire Universe, except for minor occurrences of positrons in high energy processes. The Universe should be half-and-half, but it is all-and-none. In the Big Bang Theory, there is no explanation for this. This should be a fatal objection, but it has simply been excused on possible asymmetries, and is hardly addressed.

One of the problems with creating a Universe such as we can see around us, in a scientifically acceptable way, is that the rules of physics require that for every particle of matter that is created, normal mater as we commonly know it, an exactly equal but exactly opposite particle of antimatter must be produced, at the same time and the same place. The most common forms of antimatter that we know experimentally are the positrons that are emitted in some radioactive decays, and that can be produced as desired by various particle accelerators. These are antielectrons, identical in every way we can determine, except that they have the opposite charge, positive to the normal electron's negative charge. Enough other features are just the reverse so that these two particles, electron and positron, annihilate each other when they combine, in a flash of gamma rays. Other particles of our subatomic world also have antiparticles. For protons, there are antiprotons, and neutrons are matched by antineutrons. Antihydrogen atoms can be produced, and presumably there is no limit to the extent to which our normal world could be matched by an antiworld. However, in every case, normal mater and antimatter annihilate each other with great energy. They cannot co-exist.

So far as we know, everything about an antiparticle is opposite to the corresponding characteristics of the corresponding normal particle. Charge is opposite (if different from zero), and every other characteristic that can be considered has been found to be opposite. This mutual oppositeness becomes dramatically clear in the complete mutual annihilation of a particle/antiparticle pair. Their material existence disappears in (usually) a flash of electromagnetic energy, photons, just equivalent to the sum of the particle masses. Nothing is left over.

However, one fundamental characteristic seems to have been overlooked, and that is not only surprising, it has some surprising consequences. The mass of the particles and antiparticles has been assumed to be identical, completely the same. We ought to consider that mass has at least three ramifications: gravitational mass, inertial mass, and annihilation mass (or existence mass). We have difficulty measuring the gravitational mass, we know that the particles and antiparticles act as if their inertial mass were the same, and we know that the annihilation mass is exactly opposite.

For charged particles, such as a proton and an antiproton, or an electron and a positron, the electrostatic attraction is overwhelming, and they combine to the point of mutual annihilation. For neutral particles, such as a hydrogen atom and an antihydrogen atom, or a neutron and an antineutron, or any mix of any neutral particles, the electrostatic force is non-existent, and only the gravitational force can be involved.

Now, consider the effect of that force if the gravitational masses of particle and antiparticle are opposite. The gravitational force will be repulsive between the opposite particles, not attractive as we normally assume for normal matter. Hydrogen attracts hydrogen, but hydrogen and antihydrogen repel each other, while antihydrogen and antihydrogen attract each other. Similar attraction, repulsion, and attraction applies to the neutron and antineutron. Neutral matter clumps together gravitationally, and neutral antimatter clumps together gravitationally, but neutral matter and neutral antimatter repel each other with all the strength of their gravity. A Universe containing equal and (anti)identical amounts of normal matter and antimatter, as we think it ought to, will develop into isolated cells of normal matter and antimatter repelling each other, each holding on to its own kind so as to maintain an isolation without interaction. This proposed form of gravitation presents a nice complementary symmetry between the electromagnetic and gravitational forces. In electromagnetism, like charges repel, unlike charges attract. In the proposed gravitation, like matter attracts, unlike matter repels.

Einstein's static solution for the Universe, requiring "zero" matter, might be well represented by a Universe filled with alternating cells of normal matter and antimatter.

A Universe of that sort would have an average net density of matter equal to zero, because of the exactly equal production of normal matter and antimatter in the process of creation. With the parameter k equal to zero, space is flat, and all the time rates in the field equations are zero, and the Universe is flat and constant, a very ordinary Universe. In space that is flat, straight lines that are parallel stay parallel no matter how far, and triangles made of straight lines have 180 degrees in the interior angles. It may be boring to some, but I think it's pretty nice. No magical mysterious unequal destruction of antimatter is required, just a deeper consideration of what it means to be anti, and the ordinary rules of physics.

Is inertial mass also opposite? Does that force kinetic energy and momentum to be negative for antimatter? Energy and momentum are often presented as squared terms in quantum mechanical formulations, and that would eliminate the negative sense. This is true for the Klein-Gordon relativistic quantum theory equation for a particle (and therefore for any collection of particles), where all terms involving the inertial mass are squared [Goebel 1993]. Thus, "annihilation mass" is clearly opposite for particles and antiparticles, "inertial mass" is ambivalent, and "gravitational mass" may be opposite, which would solve the problem of matter and antimatter in the Universe, by mutual repulsion. It seems reasonable that the three forms of mass would carry the same sign, and would be opposite for matter and antimatter.

Such a Universe with zero average net density satisfies Einstein's "third" solution, a static Universe, and is consistent with Hubble's red shift-distance measurements being a result of the specific photographic imaging used, measuring the redshifts of redshifted galaxies.

Some early stories proposed that positrons (antimatter electrons, a type of particle known as leptons) might change into protons (normal matter, a different type of particle known as baryons) forming hydrogen atoms by combining with normal electrons [Penrose 2005], or that since some K mesons can decay by two different modes, the survival of normal matter, with the complete destruction of antimatter, is explained. The first story doesn't comply with the laws of physics that cause the problem in the first place, and the second is just a fringe effect that might suggest something worth exploring, but certainly is not the answer we need.

If normal matter, the sort of our world, is so contrary to the existence of antimatter, and exactly equally the opposite, why is there any matter in our Universe at all? Why has it not combined with its equal amount of antimatter and left the Universe empty except for the blaze of high energy gamma rays produced by the complete annihilation of all the normal matter and antimatter that was ever produced? Why do we not see any nearby antimatter? So far, those questions are barely touched on by hopeful answers, none of which are really any good.

Einstein (and Newton and Galileo before him) considered that gravitational mass, reflecting the attractive force of gravity of one body on another, and inertial mass, reflecting the response of a single body to an applied force, are equivalent. That has never been tested for antimatter. We might suppose, for the sake of producing a possible solution to the problem of why does our universe contain any matter, that there is a difference between gravitational mass and inertial mass that is not readily observed. Suppose that gravitational mass is equivalent to, but distinctly separate from inertial mass. Suppose further that the gravitational mass of normal matter has a positive sign mathematically associated with it, and that antimatter has a negative sign. Then the gravitational force between particles of normal matter would be attractive $[-(mGm/r^2)]$, the negative sign shows that the force tends to reduce the distance r, and I'm still trying to avoid equations, and it would be attractive for particles of antimatter as well, and exactly the same, for $[-((-m)G(-m)/r^2) = -(mGm/r^2)]$ since mathematically, the negative signs balance to a positive sign in multiplication.

Thus, antimatter and normal matter would behave exactly the same, gravitationally, so long as the particles were exposed to interactions with their own kind. However, the gravitational force becomes repulsive for a pair of normal and antimatter particles (or any similar pairing of normal and antimatter bodies) as $[-((m)G(-m)/r^2) = +(mGm/r^2)]$, where the positive sign on the result shows that the force tends to increase the distance r, as a repulsive force.

In such a Universe, if it were possible to suddenly appear on the surface of an antiEarth, without being instantly annihilated, you would immediately fall into space, quickly gaining the same speed in the "upward" direction that you would have falling "down" to the normal Earth, hitting the surface at about 11 kilometers per second (25,000 miles per hour).

Upon creation, particles of a pair of normal and antimatter would be repelled so long as the electrostatic attraction of positive proton for negative antiproton, and negative electron for positive positron, were overcome. There would be no electrostatic attraction between the electrostatically neutral atoms of normal hydrogen and antihydrogen, or between normal neutrons and antineutrons. Antimatter and normal matter would gravitationally separate from each other, gathering together with particles of their own kind, and increasingly drift apart. Clumps would develop that would repel clumps of the other sort and attract clumps of the same kind. These clumps would continue to grow, excluding the others until regions of the Universe were developed that were dominated exclusively by either normal or antimatter. There would be little or no mixing of the two types, peacefully coexisting in different parts of the Universe, alternating cells of normal and antimatter, indistinguishable from a distance.

A Universe of alternating normal and antimatter, one kind the negative of the other in terms of gravitational mass, would have an average gravitational mass density equal to zero, just as proposed by the other choice in the solution of

Einstein's equations. The Universe would be static, its space unchanging, filled with regions of normal matter and regions of antimatter, where normal astronomers and antiastronomers would ponder their otherwise identical Universes, and its average gravitational mass density would be zero. No expansion, no Big Bang. Einstein is still right.

The gravitational constant G is the most poorly known of all our physical constants. The measurements scatter about the derived mean value by more than the uncertainties estimated by the experimentalists. Could this be from an effect of our "matter" Universe being surrounded by a mixture of matter and antimatter Universes, arranged like the bubbles in soap foam?

Recent measurements of the gravitational constant have shown major discrepancies with previous measurements. These differences are in the range of hundreds of parts per million for measurements with uncertainties on the order of 20 parts per million [Reich 2010a, Davis 2010]. If the Universe is actually filled with alternating cells of positive and negative matter, as suggested here, gravitational space may be wavier or bumpier than we assume. These unexpected variations might be the cause of the differences in measurements made at different times.

The Cosmic Microwave Background Radiation seems to have been enlisted, by separate researchers, for two rather different purposes. As an explanation for, and evidence of, an expanding Universe, it is the redshifted relict of the Big Bang fireball, cooled by the expansion of space to about 3,000 K, when hydrogen atoms were able to form, and the Universe became transparent. Starting from a thermal (Maxwell-Boltzmann) energy distribution at 3,000 K, the spectrum has been downshifted to what is considered to be a precise Maxwell-Boltzmann (thermal) distribution at about 2.73 K. This has been used as the strongest indication of cosmological expansion.

Otherwise, the counts of Cosmic Microwave Background photons is used to show that only one proton of each billion that were created, survived mutual annihilation with antimatter [Peskin 2002]. However, the transition from this annihilation source spectrum to the thermal equilibrium of the Cosmic Microwave Background requires that essentially all the matter-antimatter annihilations occur before the Universe becomes transparent, and the photon energy be absorbed into the remaining fireball distribution. This is at a later time and with a distinctly different energy distribution. Lingering annihilation, or significant distortion of the thermal distribution should have left noticeable distortions in the Cosmic Microwave Background spectrum.

Assuming, for simplicity, that the annihilations take place between protons (positive charge) and antiprotons (negative charge), and electrons (negative charge) and positrons (positive charge), the annihilation source spectrum consists of two lines of monoenergetic gamma rays, one at 938 million electron volts, the other at 0.511 million electron volts. Presumably there were equal quantities of protons and electrons, although I do not know of any strong requirement for that equality. This assumption puts considerable stress on any asymmetric decay of matter and antimatter to eliminate the antimatter equally as hadrons (heavy particles, protons in this example) and leptons (light particles, electrons in this example). Both types of particles must suffer the same amount of asymmetry. If this equality holds, there would be equal numbers of gamma rays at 0.511 million electron volts and 938 million electron volts. Photons at such greatly different energies interact with matter (and other photons) very differently, but must all be absorbed into the somewhat downshifted fireball spectrum before the Universe becomes transparent.

While each of these gamma-ray lines consists of monoenergetic photons (Doppler shifted somewhat by the relative velocities of the annihilating pairs), the proton gamma rays are roughly equivalent to a thermal distribution $[E=(3/2)kT]$ with a temperature of 7.26×10^{12} K (seven trillion degrees kelvin). In a similar sense, the gamma rays from the annihilation of electrons and positrons are equivalent in energy to a thermal distribution with a temperature of 1.98 billion degrees kelvin. Can two such gamma-ray lines at such different energies and with such a different distribution from the fireball thermal distribution be accommodated in time to vanish without a trace?

Measurements of the cosmic-ray spectrum by PAMELA [Adriani *et al.* 2009] have shown an anomalously large number of high-energy particles, and these are positrons, antielectrons. If these are escaping high-energy particles from an antimatter subuniverse, where are the matching antiprotons? They are still trapped in the antimatter subuniverse because of their greater mass, 1,836 times that of the positrons, while the positrons, with sufficient energy, can escape. An excess of high-energy positrons has been seen by PAMELA and the Fermi Large Area Telescope [BMS 2009, Abdo *et al.* 2009].

Antihydrogen atoms have been made by combining antiprotons and positrons (antielectrons) [Gabrielse 2010]. These atoms are electrically neutral and, except for the magnetic moment that allows them to be trapped for study, should be free of all forces except for gravity. The observational problem is that the antihydrogen atoms escape more rapidly than expected. This might be the result of being gravitationally repelled by the Earth's normal matter mass, rather than being gravitationally attracted. Studies of gravitational effects are being attempted at CERN [Reich 2010b, Andresen *et al.* 2010].

A satellite particle detector, the Alpha Magnetic Spectrometer, was scheduled for launch in 2010 by use of the Space Shuttle [Barry 2009], but delays have postponed the launch (for installation on the *International Space Station*) until 2011. This detector will search for, among other things, antihelium nuclei in the high-energy cosmic rays. If my proposal that the gravitational mass of antimatter is opposite that of normal matter, this detector should not see any antihelium.

Those particles would be more tightly held in their antimatter subuniverse, and more forcefully repelled by our normal matter subuniverse, than positrons are, by a factor of 7,360.

If our current physical description of the Universe is wrong: if distant galaxies are nearby detonated planets; Cepheid variables are pulsing reactors or rotating asteroidal fragments; quasars are self-propelled fragments; pulsars are not neutron stars but self-spun splinters; and there are no black holes; perhaps that is why cosmologists have had such a fertile field and yet, have not yet got it right [Nature 1994a, Nature 1994b].

This book has presented the reader with a variety of evidence that the world we live in is rather different from that described by our theories. The implications of this proposal for cosmology are six-fold:

1. Quasars are not cosmological objects that are the "oldest, most distant, most powerful" objects in the Universe, but easily explained local objects, the ion-propelled fragments of frequent planetary disintegrations.

2. The nuclide abundances observed in the Universe have been significantly affected by a process that has not been considered in estimation of the primordial abundances. Therefore, the estimates of primordial abundances are likely to be in error.

3. The unfound mass needed to produce the gravitational motions of clusters of galaxies has resulted from an evolutionary process: production of nuclear fission fuel in supernovae and dispersal of iron by planetary detonations; and from errors in the estimation of masses and velocities, due to great (but incorrect) distances.

4. The Cosmic Microwave Background Radiation, presumed to represent ancient radiation that originated in the formation of the Universe and which is taken as evidence of the Big Bang, instead originates in a universal haze of microscopic iron crystals.

5. The nebulae identified as distant Island Universes filling the Cosmos to indefinite distance, and presumed to provide us with records of the early times, are instead the strewn fields of detonated planets, relatively small and nearby.

6. The nearly universal redshift effect, used to support theories of an expanding Universe, is no indication of cosmological recession at all. It is, instead, an artifact of the observations, resulting from scattering and absorption of blue light, making receding galaxies more visible and approaching galaxies invisible.

These implications will seriously alter our understanding of the Universe.

The Big Bang

The Big Bang is the currently favored theory explaining how our Universe came into being [Pagel 1992, Trefil 1988, Ferris 1988, Kolb and Turner 1990, Narlikar 1993, Maffei 1989, Gribben 1988, Peebles 1993]. In utmost brevity, it states that a point of mass/energy expanded abruptly, cooled, and produced space and time and the atoms and photons of our Universe. The redshifted echo of the early extreme temperature of the Universe exists now as microwave radiation, and the space of the Universe is still expanding. Since its earliest nurturing by George Gamow from the original theory by Georges Lemaître in 1927, over 80 years ago [Lemaître 1927, Lemaître 1931a, Lemaître 1931b], this theory has been tailored to fit the job, and nearly does so [Peebles *et al.* 1991, Peebles *et al.* 1992]. But not quite right [Arp *et al.* 1990, Arp *et al.* 1992, Oldershaw 1990], and with only a little malice, its condition might be compared to that of early models of the Solar System that used complicated epicycles on planetary orbits around the Earth to make the planets fit [Seeds 1992, Kaufman III 1991, Abell *et al.* 1991]. Only a few miracles are invoked. It has survived because it explains misunderstood observational data (incorrect information) better than any alternative.

One of the keystones in the Big Bang theory is the prediction of the elemental composition of the Universe. The nuclide abundances that were established before star formation, the primordial abundances, are important in describing the conditions of the developing stages of the expansion, and must be estimated to a certain extent from what is observed now, based on our understanding of the processes that have modified these abundances. These include primarily fusion and the s- and r-processes, operating in stars of various sorts. Nuclear fission has been overlooked and its effects have been omitted from consideration in estimating the primordial (or cosmological) abundances.

It appears that the fission process is as important in the formation of the trace nuclides (deuterium, helium-3, lithium, beryllium, and boron) as the other processes are for the major nuclides. Neglect of this fact would yield incorrect estimates of the primordial abundances of these elements.

To show these effects more specifically, fractional abundances of the stable trace nuclides as estimated for the Big Bang theory [Kawano 1992, Smith *et al.*], as observed [Cameron 1982], and as produced in ternary fission [Wagemans 1991], are shown below:

	primordial estimate (Kawano, T=10^7 K)	current estimate (Cameron)	ternary fission (Wagemans)
^{1}H	9.263×10^{-1}	9.366×10^{-1}	9.08×10^{-3}*
^{2}H	6.72×10^{-5}	1.55×10^{-5}	4.54×10^{-3}
^{3}He	1.43×10^{-5}	1.13×10^{-6}	5.72×10^{-2}
^{4}He	7.36×10^{-2}	6.34×10^{-2}	9.102×10^{-1}
^{6}Li	2.73×10^{-14}	1.57×10^{-10}	1.54×10^{-2}
^{7}Li	1.18×10^{-10}	1.96×10^{-9}	3.63×10^{-4}
^{9}Be	nil	4.23×10^{-11}	1.82×10^{-4}
^{10}B	nil	6.34×10^{-11}	2.81×10^{-3}
^{11}B	nil	2.54×10^{-10}	1.82×10^{-4}

* It should be noted that protons will also be produced by fission neutrons that decay in space.

(While it might be thought that 10^{-14} (0.00000000000001) is nearly nil, really nil is smaller still (10^{-25}, 0.0000000000000000000000001) [Kawano 1992].) Estimates for the primordial abundances are those for the final temperature in the calculation, 10^7 K. At this point in the calculation, all abundances have stabilized. These three abundance distributions are shown in Figure 154.

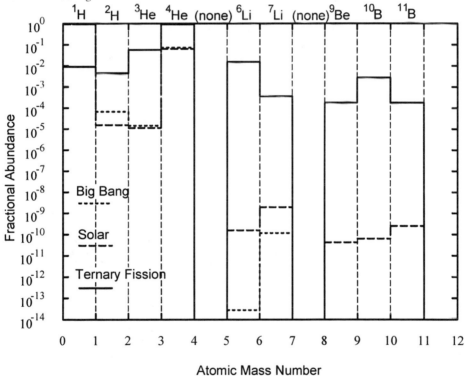

Figure 154. Abundances of nuclides produced by the Big Bang.
Fractional abundances of nuclides produced in the Big Bang, observed in the Sun, and produced in ternary fission. The rare trace elements, ^{6}Li, ^{7}Li, ^{9}Be, ^{10}B, and ^{11}B, that are barely (or not at all) produced in the Big Bang are produced amply in ternary fission.

For most nuclides, the Solar distribution clearly lies between the Big Bang abundances and the ternary fission abundances: small amounts, fractional parts per million, of the ternary fission yields would bring the supposed primordial composition to that observed for the Sun.

It is clear that radically different amounts of the trace elements are produced by ternary fission, compared to the conventional processes, particularly for lithium-6 and beyond.

Since the observed deuterium abundance has been interpreted as resulting from the Big Bang nucleosynthesis, without production by any other source [Peebles *et al.* 1991], but only destruction [Steigman and Tosi 1992], the production in stars and planets by ternary fission and by fission neutron absorption by free protons, should be considered.

The isotopes of lithium are difficult to produce by fusion reactions, and this has posed a major problem in both primordial and stellar nucleosynthesis. In the Earth's crust, lithium is a moderately abundant trace element, and the isotope lithium-6 is present with a variable abundance ranging from 6.77% to 9.28%. While small amounts of lithium have been detected spectroscopically in stars, evidence for the lithium-6 isotope is scarce. These observations are difficult to reconcile.

However, both nuclides are produced by ternary fission, at a $^7Li/^6Li$ ratio which may be as low as 0.0235:1. Since that would provide a lithium-6 content of 97.7%, there is clearly plenty of lithium-6 to spare. Most of the lithium-6 is thought to come from beta decay of helium-6 produced by fusion in stars and the Sun. At low neutron energies, the neutron absorption cross-section of lithium-6 is large. In a stellar interior, neutrons from the nuclear-fission chain-reaction process are rapidly moderated to relatively low energy by scattering off the abundant hydrogen atoms. Absorption of these thermal neutrons by the lithium-6 results in the conversion into helium-3 (via hydrogen-3) and helium-4. In a planetary core, the fission neutrons lose their energy slowly. At these energies (tens to hundreds of keV) the absorption by lithium-6 is much less. At a neutron energy of 255 keV, a strong resonance in lithium-7 leads to neutron absorption in this nuclide. The nuclear structure is highly conducive to the reaction $^7Li(n,2n)^6Li$, resulting in a reduction of the lithium-7 abundance and an increase in the lithium-6 abundance. In a planet, the lithium produced in the core will be preferentially transported to the crust. The variable isotopic composition may partly reflect differences in the neutron irradiation history.

In addition, neutron absorption will have modified most of these abundances, and dilution by large amounts of hydrogen-1 and helium-4 will affect the overall abundance ratios. It has been noted that older stars, which would be expected to have less uranium than stars that were formed of more recent material, material that has been processed through more supernovae than for the older stars, appear to have less lithium in their atmospheres [Hogan 1989, Gilmore 1992, Martin *et al.* 1992, Sky & Telescope 1993a], by about a factor of 10. This is consistent with the production of lithium predominantly by ternary fission of uranium in the more recently formed stars. On the other hand, these observations may simply be confused by our poor ability to truly distinguish stars and to guess their ages [Israelian *et al.* 2009].

For stars with heavy-element abundances that are less than a tenth of those in the Sun, the lithium abundance is about a tenth of that predicted by the standard Big Bang model [Gilmore 1992]. Stars with greater heavy-element abundances show proportionately more lithium [Martin *et al.* 1992]. Along with increased heavy-element abundance will come increased uranium content. This will lead to fission chain-reactions in the early life of the stars and production of lithium by ternary fission. The standard homogeneous Big Bang nucleosynthesis [Thomas *et al.* 1993] is not able to produce significant amounts of Be and B, nor to approach the observed ratios of $^7Li/^6Li$. In view of the ease of production of these trace elements in ternary fission, the Big Bang shouldn't be needed to produce them. Inclusion of the fission process, for the production of these trace nuclides, in the development of estimates of the primordial abundances may eliminate the need for excessive adjustment of the conditions assumed for the early Universe.

Problems arise in the production of fluorine. This element consists of only one stable nuclide, fluorine-19, and it is not easily formed in stars, by proton capture or by alpha capture. It is anomalously rare compared to its neighbors, oxygen and neon, because of this. The only practical manner of production appears to be by nuclear reactions on the surface of a white dwarf star in a binary system, where the companion star is losing matter to the white dwarf [Maddox 1992]. An alternate process, easily producing fluorine where neutrons are abundant, is by sequential neutron capture by oxygen. The major isotope, oxygen-16, which is easily produced in stars by helium burning (alpha capture), is converted into oxygen-17 and then oxygen-18, which becomes radioactive oxygen-19 with another neutron absorption, and this decays to fluorine-19.

The presence of unseen Dark Matter in our Galaxy [Sky & Telescope 1991b, Kaufman III 1991], other galaxies [Lake 1992, Tremaine 1992], and in clusters of galaxies [Ponman and Bertram 1993] has been determined by studies of the gravitational motions of the visible stars [Seeds 1992]. The nature of this hidden mass has not yet been identified [Jayawardhana 1994a]. Dark Matter disappears in the view of this book, with spiral galaxies actually being the debris fields of detonated planets, and the observations of non-Keplerian rotational velocity curves is explained as another artifact of observation. Spectroscopy measures the Doppler shift of gas streaming out of the galaxy, not the rotational velocity of "objects" that constitute the structure of the galaxy.

The need for Dark Matter in clusters of galaxies is again an artifact of our assumptions, and of the mass to light ratio used to calculate the gravitational mass at distances that are a hundred thousand times too great. If the galaxies are actually fields of debris with their light being, to a major extent, reflection of starlight from a nearby starsun, the ratio of physical mass to observable light (the M/L ratio used to calculate masses) will be greatly underestimated. Are the gravitational dynamics calculated so incorrectly as to completely mislead us? The mass of the objects themselves is that of rock and iron in asteroidal fragments, not hydrogen and helium in stars. Determination of the true mass will solve the Dark Matter problem.

In our current understanding, every time a supernova explodes, fresh uranium and thorium are produced and dispersed into interstellar space. This material is incorporated in the formation of new stars, and in new planets. Every time a planet explodes, and this seems to be a frequent fate, solid matter that will cool to the coldness of space in a few million years is also dispersed. The Universe turns into iron, as the most stable form of matter. These fragments of iron are too small to block light, and too large to scatter it, and too cold to produce discretely detectable radiation. Except perhaps for occasional intermittent glimmers at energies slightly above the cosmic background, as found by Matsumoto [Matsumoto 1987, Matsumoto *et al.* 1988].

Olbers' Paradox

Olbers' Paradox has a long history. The question was first discussed by Thomas Digges in 1576, then by Johannes Kepler in 1610, Edmund Halley in 1721, and by Wilhelm Olbers in 1826. The problem is a paradox only if we attempt to force our ideas of how the Universe ought to be: static, infinite, and eternal, uniformly filled with stars. If that were the case, the entire sky should be as bright and as hot as the Sun and we would be cooked.

Olbers' Paradox should also be explored relative to the antineutrinos from distant supernovas. If the rate of supernova explosions is constant with distance, the number of supernovas at any distance would be proportional to the square of the distance. As the antineutrino flux decreases with the square of the distance from the Earth, the flux at the Earth would be infinite for an infinite Universe, unless the antineutrinos were absorbed along their flight to the Earth. Our detectors would be swamped. This is not the case. Does this same logic apply to gravitational waves, whether from black holes or neutron stars or (more likely) the detonations of gamma-ray bursters? Are the detectors (and Joe Weber's antennas) swamped by a constant flux of gravitational waves, in and out of phase with each other, constituting an annoying and meaningless noise?

Often, a paradox is paradoxical because the question posed is cleverly wrong. Olbers' paradox about the night sky may be of that form. During the day, if we can see the Sun, we see the surface of a star, blindingly bright, burning hot. At night, why do we not see the surface of a star, no matter how far, in every direction? Why is the sky at night dark?

In any direction we look in the sky, at some distance, we will be looking at the surface of a star, so the sky should never, ever be dark. If most of the "stars" are fragments of detonated planets, and their light fades away in a few million years, does this still hold? Maybe, for infinite is awfully big, and eternity is awfully long. This would heat the Universe to the average surface temperature of the stars, about 1,000 K. But what if most of the stars are cold?

Olbers' paradox asks:

> "If the Universe is infinite, and filled with stars,
> Why isn't the sky uniformly bright?"

Several explanations, any one of which would seem to be adequate have been offered. [Hamilton 1987].

But the question could also be phrased in a more appropriate manner:

> "If the Universe is infinite, and filled with dust,
> why shouldn't the sky be dark?"

And it is dark, except at the wavelengths of radiation emitted by the dust, where it should be uniformly bright. This is exactly what we observe and call the Cosmic Microwave Background Radiation. In a Universe that is infinite in time and space, the energy released by stars, and planets, and detonations would build up to a high temperature, similar to the hottest objects, so we cannot have a Universe that goes on forever, and has gone on forever. Except that the temperature of the whiskers is low, stays low, and no amount of low-temperature radiation can heat the Universe. This energy is just distributed and redistributed among the whiskers, which absorb, scatter, and re-emit the Cosmic Microwave Background Radiation.

In olden times, the Universe we saw was simply the spherical shell of the heavens, and it could easily be accepted that the stars were scattered sparsely, so that much of the sky was empty and dark. If the Universe were infinite, the ever-more-distant stars, ever greater in number, would contribute small twinkles of light, and there would be no dark.

The problem of an infinite Universe could be solved, however, by taking away the eternal nature of the Universe and its stars. Edgar Allen Poe did this in 1848, by proposing that the Universe had been created some time ago, and light from its most distant stars had not yet reached us. That is precisely what we accept in our current cosmology.

The Cosmos may be infinite in extent but was created (by the Big Bang) only (!) 16 billion years ago. Therefore, the farthest stars that we can see can only be 16 billion light years away. In this model, it is not that there is nothing beyond the most distant stars that we see, but simply that light from more distant stars has not yet reached us. This permits us to define a Cosmos, infinite in extent but limited in time, containing the observable Universe within it, enclosing us. Our Universe is limited in extent by the finite velocity of light. (There would remain the problem of the formation of an infinite Cosmos from the expansion of a point singularity, as the Big Bang proposes.)

The conflict between our assumptions and our observations created the paradox. Once our assumptions are changed (for they are still only assumptions), the paradox disappears.

This paradox has been reviewed in impressive detail by Edward Harrison in *"Darkness at Night – A Riddle of the Universe"* [Harrison1987]. He shows that there are several solutions (more than enough!). His ultimate explanation is that a bright night sky (and equally uniformly bright during the day) requires a certain energy density, and the total mass-energy density of our Universe is not sufficient to supply that density of radiant energy. This is a solution regardless of the age of the Universe, or whether it is expanding.

The paradox is that if the Universe is infinite and filled with stars, an infinite number of them, then every where we look we should see the surface of a star, no matter how far and dim, because there will be enough stars to make up for the distance and the dimming. (A star dims with the square of the distance, but the number at that distance increases by the square of the distance, so everything evens out.) But we don't see that, there are great empty spaces between stars, whether we look with our naked eyes or the largest telescopes in the world, the night sky is mostly dark. (One rustic philosopher said that night was caused by a lack of sunshine.) What happened to all the stars, the starlight, that we should be seeing? The stars should fill the sky with starlight, perfectly uniform and blindingly bright, so that we couldn't see anything but a glow from an infinite number of stars. What happened to all the stars that we don't see? (That has been solved in the Standard Story by having the Universe expand so fast at great distances that those stars can't shine on us. That makes this a problem for my book, since I say the Universe doesn't expand, Hubble's measurements were observational accidents, due to the techniques he used.)

In 1912 or so, Einstein invented his General Relativity, with some equations that, if solved, would tell the Universe how it had to be. Some theoreticians came up with a solution, part of which said it had to expand. In the 1920s, Edwin Hubble made some measurements that seemed to show that the Universe was expanding, and that was the start of the Big Bang Theory. That led to the prediction that there would be a remnant radiation from the fireball of the Big Bang, redshifted a tremendous amount to a very low energy of about 5 degrees K. (Our Earth is about 287 K.) In 1965, that radiation, the Cosmic Microwave Background Radiation, was discovered by Penzias and Wilson, and found to have a temperature of 3 K, now measured to be 2.73 K or so, very precisely. The tremendously high energies of the fireball radiation were downgraded by the expansion of space stretching the wavelength to the microwave region. That radiation fills the sky, almost perfectly uniform at 3 K, blindingly bright since nothing can be seen past it in that energy range. In my story, the Cosmic Microwave Background Radiation comes from the scattering of present starlight by tiny iron whiskers, acting like dipole radio antennas (this is microwave energy), filling the sky so we can't see past the last scattering by the fog of whiskers.

If the Cosmic Microwave Background Radiation is produced by a distribution of iron whiskers such that some average length is similar to the wavelength of the Cosmic Microwave Background Radiation (nominally 3.5 millimeters), light in the Universe is downshifted so that the intensity of the Cosmic Microwave Background Radiation increases, but the temperature does not. The Universe does not become hotter because of the Cosmic Microwave Background Radiation. Before expansion solved the paradox, it was thought that the light from distant stars was absorbed by intervening matter, and that stopped us from seeing distant stars. The counterargument to that was that because of the conservation of energy, the intervening material would soon get to an equilibrium temperature with the stars, and the sky would still be blindingly bright, just as though we were seeing the stars. (This would also make the sky as hot as the surface of the Sun, so we need to understand why we weren't cooked and the Earth vaporized.) This is a confusion between conservation of energy and conservation of temperature. Energy is conserved, temperature is not, it is shared energy. When that energy is shared over many particles, the temperature decreases. At every stage in its life, when a photon interacts with almost anything, it loses energy. That energy isn't lost, it is just redistributed among the interacting particles. Everything goes downhill until it can't go down any more. The sky would certainly be blindingly bright, totally filled with radiation, but not at its original temperature. The sky would be filled with radiation at a microwave temperature where about all it could do is bounce off tiny iron whiskers, filling the sky with a Cosmic Microwave Background Radiation. And that is what we observe. The Cosmic Microwave Background Radiation fills the sky, almost perfectly uniformly, blinding us to anything beyond, at microwavelengths.

Since we would not be seeing, unobstructed, the surfaces of an infinite number of stars, and energy can't flow from a cold object (the iron whiskers or the Cosmic Microwave Background Radiation)) to a hotter object, there would be no transfer of energy to heat up the Universe, as considered by Olbers. (Narlikar says solving Olbers says the Universe must expand; I say it's the whiskers that does it.)

If ion-engine-propelled fragments disperse iron vapor throughout the Universe, where it condenses into fine iron crystals, and if this is the source of the 3 K microwave background radiation [Wickramasinghe 1992], perhaps this also explains the dipole moment of the background radiation distribution. This dipole asymmetry has been interpreted as showing the motion of our Galaxy (really, our Solar System) towards a Great Attractor [Lindley 1992].

Space is expanding, according to the Big Bang theory. Traditional mechanical models that show how this could be include coins on an inflating balloon, and raisins in bread, as shown in Figure 155. This raises a perplexing problem: space between the "galaxies" expands, but the space within the "galaxies" does not. As gravitationally bound objects, the "galaxies" remain the same physical size as they speed away from each other, carried by the expansion of the Universe, but not expanding themselves. And so, the observational appearance is just as if the objects were flying out into a non-expanding space. Puzzling ...

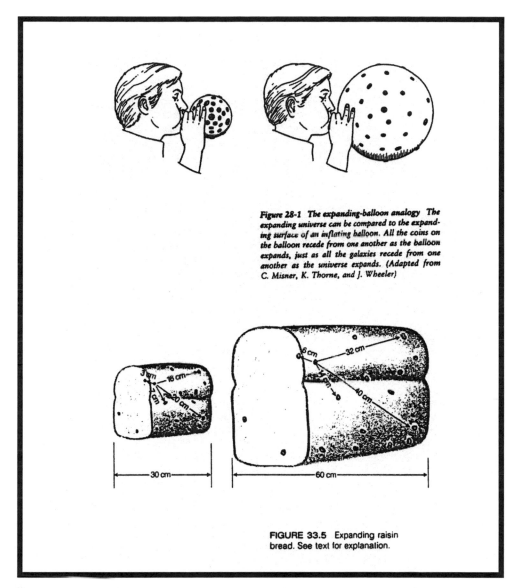

Figure 28-1 The expanding-balloon analogy The expanding universe can be compared to the expanding surface of an inflating balloon. All the coins on the balloon recede from one another as the balloon expands, just as all the galaxies recede from one another as the universe expands. (Adapted from C. Misner, K. Thorne, and J. Wheeler)

FIGURE 33.5 Expanding raisin bread. See text for explanation.

Figure 155. The expansion of space, without expansion of bound objects.
Typical mechanical demonstrations showing how space can expand, to explain the recession of galaxies from each other. The inflating balloon is from Kaufman III 1991, and the raisin bread is from Abell *et al.* 1991.

Summary of Part V

Some of the recent history of astronomy, which has shaped our view of the Universe, is explored. Our view is based on the concept that all stars are somewhat like our Sun; the Universe is filled with Island Universes of stars at great distances; and the Universe is expanding. There is an alternative to this view. The effects of nuclear-fission chain reactions form the basis of this alternative view.

The role of natural nuclear fission reactors in the development of true stars, and of supposed stars that are actually asteroidal fragments, is described.

The development of our concept of the stars as suns is described. Astronomical spectroscopy is presented. Reflection of the Sun's light by an asteroid appears to be starlight from a star. The power source of the Sun and stars was explained by nuclear fusion reactions. Development of larger stars to detonate as supernovas is described.

A basic ambiguity in astronomy, and the conflict that this book presents, is the question of small, dim, and near or large, bright, and far, for astronomical objects. Resolution of this conflict relies on measuring astronomical distances correctly. The basic technique currently used is flawed. Distance measurements are reviewed, and the fundamental technique for measuring distances to the stars is shown to be a relative measurement, not the absolute measurement it is assumed to be. This technique fails to resolve the ambiguity, but its results have always been interpreted as showing great distances. That supports the supposition that stars are suns, large and bright at great distances.

Cepheid variable stars, as uncalibrated extensions of trigonometric parallax in the distance scale, are described. (It was shown earlier that the pulsational lightcurve of a Cepheid variable star is remarkably similar to the rotational lightcurve of an asteroid.)

The conclusion of uncertainty over the nature of spiral nebulae as Island Universe galaxies is described. The spiral galaxies are like our Milky Way Galaxy, but none of these are Island Universes of stars.

There are many cases where observations of planetary fragments have been misinterpreted. Rapidly spinning or tumbling fragments may be seen to be pulsars; self-propelled fragments moving away from us may be seen to be quasars or QSOs with great redshifts indicated by repelled gas ions; cold, strongly magnetic spinning fragments may appear to be powerful radio galaxies. The tremendous energies that must be assigned to these objects due to the great apparent distances diminish as the distances shrink from cosmological to nearby, at hundreds of light years rather than millions to billions of light years. Superluminal motion, apparently observed in many objects as a result of their great assigned distances [Zensus *et al.* 1987], is no longer a problem needing a contrived explanation, but returns to a more normal realm of speeds at the closer distances provided by a new understanding.

As these discussions suggest, an astronomical object that does not provide a distinguishable image, but is judged to be at a great, cosmological, distance because of this lack of identity and because of a large redshift, may instead be a relatively nearby planetary fragment, surrounded by high velocity ions propelled by the decay of the products of the nuclear-fission detonation. Even those that provide such beautiful images that are taken to be Island Universes of stars may be small and near, simply masquerading by our misunderstanding. The composition, size, structure, and dynamics of the detonating mass, whether planet or satellite, large or small, or smallest globule, will affect the outcome of the detonation and determine the resulting form of the astronomical objects.

The poor, blurry imaging of the *Hubble Space Telescope* was blamed on spherical aberration, but a similar problem of unidentified origin has plagued a least seven major telescopes since 1947. One of the initial poor images from the *Hubble Space Telescope* is shown to result from the imaging by a perfect telescope of an object that is 500 kilometers in diameter at a distance of 61.6 Astronomical Units, while focused at "infinity". Other evidence for focusing errors in otherwise excellent telescopes is offered.

The close association of stars with spiral galaxies, often dismissed as "foreground stars in our own galaxy", is presented. Numerous astronomical images show a spiral galaxy in close proximity to a foreground star; in some cases, clear interaction is shown. The Hubble Deep Field is explored, and the reason for the disappearance of stars at greater distances is explained.

The Freeman Law for the surface brightness of spiral galaxies, and its failure to predict low-surface-brightness galaxies, as are observed, is discussed. Low-surface-brightness galaxies are shown to be reflection nebulae illuminated by their starsuns from behind. The surface brightness of the outer planets in our Solar System is shown to predict the surface brightness of the Milky Way at a distance from the Sun of about 100 Astronomical Units, as a reflection nebula.

The detonation of residual globs of uranium, induced by the detonation of a planet in forming a spiral galaxy, produces the spherical distribution of globular clusters seen around those galaxies.

The problem of Dark Matter is discussed.

The nature of blue and red supergiant stars is discussed, as the result of radioactively generated Cerenkov radiation and internally generated thermal radiation adding to the reflected starsunlight of asteroidal fragments. The much larger sizes of spiral galaxies in HI radiation (21-centimeter radio) compared to their sizes in visible light is discussed.

Random motion (most probably of the clouds of gas that are spectroscopically observed) of objects in the Universe is explored. Nearby galaxies have essentially random velocities relative to the Sun. Farther away, a predominantly receding velocity develops.

The possibility that the third solution to Einstein's equations of General Relativity, that provided for a static Universe, with space neither expanding nor contracting, is shown to be a consequence of the essential creation of equal amounts of normal matter and antimatter.

Distant galaxies, particularly those moving towards us with blueshifted spectra, disappear from sight by the scattering and absorption of their light. This leaves only redshifted galaxies, necessarily more redshifted at greater distances to remain visible, as the only distant galaxies we can see. This is the foundation of Hubble's Law. Hubble (and others) only measured the redshifts of redshifted galaxies. The decline in the number density of galaxies with distance is a key feature of this view.

The radial velocities of bright stars, globular clusters, and nearby galaxies are explored to develop the possible Maxwell-Boltzmann distribution for these objects, and it is shown that only the receding half of this distribution is observed for spiral galaxies. The origin and progress of Hubble's Law, leading to Einstein's expanding Universe and the Big Bang cosmological theory, are discussed.

Questions of the scale of the Universe are discussed. Measurements of physical rotation of spiral galaxies by Adriaan van Maanen show reasonable velocities, comparable to planetary velocities in our Solar System, if a close distance scale, based on a measurement to the Andromeda galaxy, is used. The reasons that were presented for rejecting those measurements are shown to be false.

The conflict between two accepted theories, each correct in its own realm, is illustrated by differing interpretations of redshifts.

Attenuation of light by absorption and scattering shows discrimination against blueshifted light, causing distant galaxies that are approaching us to become invisible. We can only see redshifted galaxies, and so observe greater redshifts at greater distances. This effect has led to the observation of distant galaxies that aren't receding as "fast as they should" for their distance. This is due to an absence of dust along our line of sight to those galaxies, allowing galaxies with lesser redshifts to be seen. There is no acceleration of the cosmological expansion, and so, no need for Dark Energy.

The formation of spiral galaxies, irregular satellite galaxies, and elliptical galaxies is shown to be a result of planetary detonations.

Supernovas are discussed. Problems with the lightcurve and spectrum of SN 1987A are explored. The possibility of the detection of gravitational waves from that supernova is discussed. Widespread effects that might have been caused by antineutrinos from that detonation are presented.

Gamma-ray bursts are described and it is shown that the optical transient lightcurve follows the decay heat curve of a nuclear fission detonation. Were the gravitational wave detections first reported by Joe Weber at a rate of hundreds per year, the result of nearby nuclear fission detonations converting mass to energy, seen as gamma-ray bursts, at a rate of hundreds per year?

Strange objects, quasars, blazars, pulsars, are the expression of ion propulsion formed by radioactive fragments of a detonation that have retained a magnetic field. Beta particles from radioactive decay leave the fragments at nearly the speed of light, producing a positive electrostatic charge that repels positive ions, propelling the fragments to high speed. As with all spectroscopy, the spectra we see of these objects is from the gas, not the underlying object itself.

In a key comparison, the redshift/magnitude plots for quasars and spiral galaxies are shown to be definitely different. This has been considered to show that "we have missed something very significant." Pulsars are shown to belong to two distinctly separate groups, those with fast rotation and those with slow rotation.

Motion of objects at speeds faster than the speed of light, superluminal, has been found for many objects, based on the assumed great distances. A special interpretation is required to explain this apparent violation, but it seems to be needed more often than is appropriate. A shorter distance scale eliminates this problem.

The fundamental rules for our cosmological theories are presented. The expanding Universe shown by the Hubble/Humason redshift/distance measurements became associated with one of the predictions of Einstein's General Theory of Relativity, and that led to the Big Bang theory. This has provided the framework for all subsequent interpretations of cosmological observations. However, antimatter must be created in amounts equal to normal matter, and that antimatter would balance the normal matter, providing an average net density of zero, the requirement in Einstein's equations for a static Universe, neither expanding nor contracting.

The Big Bang theory was initially formulated by Georges Lemaître in 1927, before antimatter had been seriously proposed. After antimatter was theoretically invented, and after its discovery in 1932, the theory was not re-evaluated. Mutual gravitational attraction and gravitational repulsion of opposites would establish alternating isolated cells of normal matter and antimatter that would result in an average net density of zero. The Universe may be in the simplest form possible, static and flat.

The Big Bang theory and some of its shortcomings are discussed.

Olbers' Paradox, as to why the sky is not blindingly bright in an infinite Universe, is answered to the contrary: it is blindingly bright in microwaves, the Cosmic Microwave Background Radiation.

This path of discovery has led to a confusing duality of nature: When are we looking at stars and when are we seeing the incandescent or glow-in-the-dark fragments of planetary detonations? And uncertainty in the astronomical distance scale remains a worrisome problem.

VI. CONCLUSION

All's well that ends well:
Shakespeare
but
Nothing quite new is perfect.
Cicero

The nuclear fission process has been overlooked and omitted from the interpretation and explanation of a wide variety of effects that are observed in our Solar System, our Galaxy, and our Universe. Because of this, misunderstanding of effects that are likely to be due to this process has interfered with the proper interpretation of observations and the development of accurate cosmological theories. Consideration of this process in the interpretation of trace element abundances, the description of objects that are considered to be distant, ancient, and powerful, and the calculation of the gravitational behavior of the Universe, leads to a simpler, more complete, and more unified understanding of these observations.

In planets, the nuclear-fission chain-reaction appears to proceed initially in a steady phase that contributes to planetary evolution. The reactors in the Earth's core are still functioning at present, in the steady phase. Abrupt disturbances in the reactors of other planets and satellites, including nearby disruptive detonations, produced bursts of radiation that provided the exclamation points in the punctuated equilibrium of life on Earth.

Natural nuclear fission reactors provided the heat necessary to keep the Earth from freezing over when the Sun was too faint. Radioactivity in the protoSolar cloud prevented the presumed gravitational collapse to form the Sun and forced the formation of the planets from the outside in, all as gas giants. The Earth was formed as a supercompressed gas giant that was downsized by the loss of its atmosphere and the evaporation of its molten silica ocean. That compression has been intermittently released causing the Earth to expand. That is the cause of Continental Drift, not Plate Tectonics. That theory fails by requiring that the Pacific margins of the continents are moving apart, and the Atlantic margins are also moving apart. That can only occur on an expanding Earth. An atmosphere more massive than on Venus and an Earth that was smaller in ancient times provided thick air which buoyed up the giant dinosaurs, otherwise too large to support themselves.

Humanity originated in a radiation event that produced unique mutations, and our form has been progressively revealed as environmental radiation and radioactivity faded away.

The nuclear chain reaction may become destructive, causing disruption of the planet or radical modification of its character. Scattered throughout space, these detonations are observed as gamma-ray bursts and supernovas, and the dispersed disks of debris of past detonations are now seen as spiral galaxies, formerly called nebulae. The consequences of planetary detonations in our Solar System are seen in the disrupted condition of the inner planets Mercury, Venus, and Mars; in the fragments of the asteroids and the kuiperoids; and in the strewn field of the Milky Way.

In the Universe, the distance scale is wrong by a factor of about 100,000, and the spiral nebulae are remnants of planetary detonations, rather than massive, distant collections of stars. Elliptical galaxies are their associated starsuns, enveloped in a chaotic cocoon of planetary debris.

Does it seem incredible that a few self-consistent mistakes of interpretation made by a few prominent astronomers nearly one hundred years ago could have formed the basis for our current explanation of the Universe? The process of self-consistency, guided by the principle that "we see what we have been trained to see" [Aveni 1992] has led to stories of an expanding Universe, well founded in theories of that time, and a Big Bang with miraculous inflation to make things come out right, and matters of faith to control the expansion. This Universe is filled with unseeable black holes, unknown quasars and blazars, gamma-ray bursters and pulsars, the "most powerful, most distant, most energetic, most dense" objects imaginable. Indeed, our Universe has become the "most" imaginable, and is mostly imagined.

How did this happen? Why was the wrong path, once taken, never abandoned? Perhaps simply because no alternatives could be seen. Knowledge of the process of nuclear fission and its effects was born in secrecy, limited among its practitioners, rejected by the public, shunned by academia, and ignored by researchers who did not recognize its possibilities, from not knowing its character. Antimatter did not become real in science until after our cosmological theories had become accepted and dominant.

All this left the landscape littered with gems for a lone voyager to discover and admire. I hope you have also enjoyed this voyage, and will do much with what you have discovered. If these suggestions succeed in producing more complete explorations by specialists in the appropriate fields, I will have succeeded in my task. And so now we may know less than we did before, but perhaps more of what we know is closer to reality than it was before.

VII. APPENDICES
Appendix A: Antineutrino-Induced Electron Capture

Early in the development of this book, I realized that there might be a nuclear reaction in which an antineutrino caused a nuclear proton to capture an electron, becoming a nuclear neutron.

In 2008, I prepared a detailed discussion of the antineutrino-induced electron capture reaction, and submitted it to the *Physical Review* for publication. Those calculations showed good agreement with experimental observations. (While I can only claim "good" agreement, and the KamLAND analysis shows better agreement, the details are important. The KamLAND interpretation of the deficit of reactor antineutrinos fit the count results to 55 independent equations (representing each individual reactor distance and power) with 2 adjustable parameters, assuming neutrino oscillation [Decowski 2009]. In this paper, the predicted attenuation is presented directly as calculated, with no adjustments.) This paper was rejected on August 15, 2008 by the *Physical Review C* editor, on the basis of reviews by two referees. I disagreed with the reviewers, but found it pointless to argue. I made some minor changes, rearranging and explaining more, but that version was also rejected. I present the original version here. The referees argued that the cross sections were "too large" (but actually they are only as large as produces good agreement with the KamLAND observations, and are much smaller than neutron reaction cross sections, on which the calculational model is based), and that "antineutrino cross sections increase with energy, while this cross section decreases" (that is inherent in the model, and this reaction is not the same as the one that increases with energy). In my view, the referees did not understand the calculational method. If the cross sections are in error, it is in the way I have formulated the calculation, not that they are too large or the wrong shape.

In addition to offering a solution to the KamLAND observations of a deficit in reactor antineutrinos, without invoking neutrino mass and oscillation, this reaction and the cross sections may apply to two other poorly explained sets of observations.

The radioactive decay rates of two radionuclides have been found to show a slight variation over year-long time scales [Jenkins *et al.* 2008, Fischbach *et al.* 2009]. This variation appears to be seasonal, and has been correlated with the distance from the Sun to the Earth. It also seems to have a correlation with the rotation of the Sun [Fischbach 2009].

An experiment intended to detect Dark Matter has shown a variation in the count rate from ultra high purity sodium iodide scintillation crystals [Bernabei *et al.* 2008, see also Bhattacharjee 2011]. The variation is seasonal, and so should also correlate with the distance between the Sun and the Earth, but has been interpreted as evidence for Dark Matter [NEWS 2009]. If there are natural nuclear fission reactors in the Sun, as has been described in this book, antineutrinos from the decay of fission products could cause the small effects seen in these two studies, with a correlation to the distance and orientation of the Sun.

In the early Reines and Cowan experiments, there may be an indication that this reaction was detected and rejected [Reines and Cowan Jr. 1959]. Excess false counts were recorded when the reactor was ON compared to OFF. I was not able to find enough information to resolve this question.

Antineutrino-induced electron capture

Robert J. Tuttle

Moorpark, California

USA

rjtuttle@earthlink.net

The reaction converting a nuclear proton into a neutron by the absorption of an antineutrino into the nucleus with subsequent electron capture is discussed. A form for the cross section is developed based on the Breit-Wigner theory for neutron absorption. Calculations of this cross section for a variety of nuclides show it to have a significant magnitude and interesting structure. Calculation of the attenuation of antineutrinos from distant nuclear power reactors shows good agreement with the KamLAND observations, eliminating the basis for proposing oscillation of massive antineutrinos. Similar attenuation that is suggested in individual reactor experiments is discussed. This reaction provides an alternative to the hypothesis of neutrino oscillation, which requires a departure from the Standard Model. An appendix explains an error in the use of the Fermi factor in calculating beta and neutrino spectra is corrected and describes some overlooked reactions that may produce significant sources of antineutrinos in nuclear reactors are described.

PACS Numbers: 25.30.Pt, 23.40.-s, 24.30.-v

I. INTRODUCTION

Recent measurements have shown a depletion in the antineutrino flux from nuclear reactors. This disappearance of antineutrinos has been interpreted as evidence for neutrino oscillations, requiring non-zero rest mass for the neutrino, which is a departure from the very successful Standard Model of Nuclear Particles. A reformulation of the cross section for antineutrino-induced electron capture, presented in this paper, shows good agreement with the observations and avoids the need for "new physics".

Access to high powered nuclear reactors and great improvements in detectors of the sort used by Reines and Cowan to first detect antineutrinos from nuclear beta decay [1,2,3,4,5,6] have led to remarkably precise measurements of a ghostly particle [7].

Detailed measurements of the antineutrino flux from a large set of power reactors have been presented by The KamLAND Collaboration, using a liquid scintillation detector in a mine in a mountain in Japan [8,9,10]. The currently favored explanation of the observations involves oscillation of massive neutrinos, related to a similar explanation for the Solar Neutrino Problem [11]. That explanation invokes "new physics", not yet elucidated. The interpretation presented here avoids this difficulty.

The possible reactions in nuclear beta decay form a complete, nearly symmetrical set, as shown below:

Lepton creation

$$n \rightarrow p + \beta^- + \overline{\nu}_e \qquad\qquad p \rightarrow n + \beta^+ + \nu_e, \qquad\qquad (1a,b)$$

Lepton conversion

$$e^- + p \rightarrow n + \nu_e \qquad\qquad e^+ + n \rightarrow p + \overline{\nu}_e, \qquad\qquad (2a,b)$$

$$\overline{\nu}_e + p \rightarrow n + \beta^+ \qquad\qquad \nu_e + n \rightarrow p + \beta^-, \qquad\qquad (3a,b)$$

Lepton destruction

$$\overline{\nu}_e + e^- + p \rightarrow n \qquad\qquad \nu_e + e^+ + n \rightarrow p. \qquad\qquad (4a,b)$$

These reactions are often referred to as "semi-leptonic" reactions, since half the particles are leptons. The first two reactions (1a,b) are commonly known from natural and artificial radioactivity as beta decay and positron decay. Reaction (2a), spontaneous electron capture is also important in nature and competes with reaction (1b) where it is energetically favored. Spontaneous positron capture (2b) should be extremely rare because of the general lack of positrons, their short lifetime in ordinary matter, a lack of attraction to the neutron, and repulsion by the positively charged nucleus.

Reactions (3a,b) are used to detect antineutrinos and neutrinos, respectively, and have in most cases led to observations of a deficit in the (anti)neutrino flux from nuclear (fission) reactors and fusion reactions in the Sun, compared to the best theoretical expectations. Reaction (4b) can reasonably be ignored for the same reasons as (2b).

I have maintained a distinction between the beta particles $(\beta, -/+)$ and the electrons $(e, -/+)$, because of the polarization of the one, proportional to v/c, and the absence of polarization of the other, regardless of velocity [12]. While helicity might be considered a minor and fleeting distinction, it should be recalled that this helicity broke the conservation of parity in 1957 [13].

Several of these reactions can occur only for complex nuclides, and may require atomic (orbital) electrons. The antineutrino-induced electron capture in particular and in general, Reaction (4a), is more properly written as:

$$\bar{v}_e + e^- + {}^A Z \rightarrow {}^A(Z-1)^*, \qquad (5)$$

where the asterisk signifies an excited level (of most significance in this reaction). With sufficient excitation energy, many exit channels are available, so the reaction can be described more completely as, for example,

$$\bar{v}_e + e^- + {}^{24}\text{Mg} \rightarrow \left({}^{24}\text{Mg}^v\right) \rightarrow {}^{24}\text{Na}^* \rightarrow {}^{24}\text{Mg} + \beta^- \text{ or } \gamma + \beta^-, \text{ or } \alpha \text{ or n or p or etc.} \qquad (6)$$

Here, an intermediate step, $\left({}^{24}\text{Mg}^v\right)$, has been shown explicitly. If the conditions are not right, the reaction does not proceed, but reverts to the initial state, and the antineutrino is released, passing by as if there had been no interaction at all.

Reaction (4a) has been studied briefly, and found to have a very small interaction probability, as calculated [14,15,16,17]. For a process that has been specifically identified as resonance orbital electron capture, the effective cross section for fission antineutrinos has been calculated to be of the order of 10^{-46} cm^2. The earlier formula was derived using the coupling constant, which was then related to the energy width of the reaction. The basic premise underlying this formulation is that antineutrino-induced electron capture only occurs in a transition from the initial nucleus to the final nucleus (ground state to ground state or excited level) for antineutrinos with energy equal to $Q_\beta + E_B + E_x$ for the opposing beta decay energy (Q_β) adjusted for atomic binding energy (E_B) and the energy of the excited level(E_x), within an infinitesimal energy increment. The energy width of the incoming antineutrino was incorrectly taken to be that of the corresponding beta decay reaction. That restricted the electron capture reaction to populating the ground state of the resultant nuclide, with a very small energy width. This naturally yields a very small result, in keeping with the other very small interaction cross sections for neutrinos.

I have re-evaluated that reaction with an alternative (though standard) approach, consistent with the earlier formula, and found the calculated cross section to be significantly large. This reaction offers an explanation for the deficits found for reactor antineutrinos in the KamLAND measurements [8,9,10] and may also be the cause of deficits found somewhat uncertainly at other reactor experiments. This explanation remains within the framework of the Standard Model and avoids the need for massive neutrinos that express flavor oscillations.

II. THE CROSS SECTION MODEL

I have taken a rather mechanistic approach to designing the form of the cross section for antineutrino-induced electron capture, and then used a standard method to calculate it.

I model the reaction based on an incoming antineutrino with a specific energy, which defines its wavelength, giving it a "size". The wavelength of a particle is a significant physical characteristic that is commonly used in the calculation of cross sections, as for example, neutrons [18,19,20] and gamma rays [21]. The greater this size, the more likely the reaction. The proton which will be induced by the antineutrino to capture an atomic electron and convert to a nuclear neutron resides in a nucleus which is immersed in a cloud of atomic electrons. Some of these electrons, as point charges, have a very definite probability of existing briefly within the nucleus, and have kinetic energies sufficient to contribute to the reaction. Imagined as point charges, these electrons pass through the nucleus many times as the antineutrino passes by. The greater the probability that an electron is within the nucleus and has energy available as needed, the more likely the reaction. The antineutrino-nucleus pair have a joint energy width that constitutes a measure of the "admittance" of the antineutrino to the combined nucleus-electron pair. The greater this energy width is, the more likely the reaction. The newly formed nucleus, having been a stable nucleus that has traded a proton for a neutron, has a variety of ways to relax to a stable form, and this establishes an outgoing energy width for the reaction. The greater this energy width is, the more likely the reaction. The various possible combinations of spin states contribute a small effect, increasing or decreasing the probability of the reaction, as this factor is generally close to unity.

The most significant difference from the earlier resonance orbital electron capture model is in the consideration of the energy width of the reaction in terms of the incoming antineutrino, and the multitude of exit channels that this makes accessible.

In beta decay, the decay reaction has an energy width generally defined by the inverse lifetime, and the combined energies of the beta particle and the antineutrino has a well defined value, subject to this decay energy width. Together, the beta particle and the antineutrino nearly exactly share the decay energy, ignoring nuclear recoil and some other small corrections. Individually, however, each particle may have an energy ranging from zero to the full energy of the decay, provided that the sum is limited to the decay energy. This gives each of the particles, the beta and the antineutrino, individual energy widths equal to the full energy of the decay.

Since this model proposes transitions to multiple excited levels in the daughter nucleus, I have used the method of Breit and Wigner [22,23] for resonance absorption of a neutron to defined excited levels in the daughter nucleus (see also [24,25,26,27]). This formulation differs from the Breit-Wigner one-level model in two major aspects. For any selected antineutrino energy, the cross section is the sum of transitions to many excited levels in the product nucleus, the multi-level scheme to the extreme [28,29]. Also, three particles, rather than the two considered for neutron capture, interact in the entry channel. It may be important to note that this reaction is unique among the semi-leptonic reactions in that no leptons are created.

III. THE CROSS SECTION

Here, as an alternative to the earlier resonant orbital electron capture calculations, I have attempted to derive the "antineutrino-induced electron capture" cross-section in the same way as neutron resonance cross-sections are derived. The reformulated cross section has a considerable resemblance to that for resonance orbital electron capture (a name which unfortunately offers no clear distinction from antineutrino-induced electron capture), but the major difference, from assigning a large energy width to the incoming antineutrino, results in the calculated cross section being relatively large, compared to the other neutrino cross sections that have been studied.

In my model of the reaction, the incoming antineutrino is treated as a plane wave from a distant source, with a reduced wavelength $\lambdabar_\nu = \hbar c / E_\nu$. The electron is treated as a point particle with a probability density function at the nucleus of $|\psi|^2 = 4(Z/a)^3 e^{-2ZR/a}$, where R is the radius of the nucleus and a is the radius of the first Bohr orbital. The nucleus is a homogeneous volume of neutrons and protons, $V = (4\pi/3)(R_0 A^{1/3})^3$.

My approach has been simply to take the geometric cross-section of the incoming antineutrino $(\pi \lambdabar_\nu^2)$, reduced by the fraction of time that an electron is in the nuclear volume and can be captured (the probability density function at the nuclear surface multiplied by the nuclear volume), and then modify that by the spin statistics factor and the resonance shape factor as in the Breit-Wigner single-level formula. The extent that the electron can contribute energy to the reaction is limited by supposing that the available energy is distributed as a Maxwellian, in the energy width of the decay channel. That is,

$$\sigma_{EC}(E_\nu) = P \qquad \text{factor of 2 for parity non-conservation,}$$

$$\times \left[\pi (\hbar c/E_\nu)^2 \right] \qquad \text{geometric cross-section of the antineutrino, } <\text{cm}^2/\nu>,$$

$$\times \left[4(Z/a)^3 e^{-2ZR/a} \right] \qquad \text{density of K electrons at the nuclear surface, } <\text{e}/\text{cm}^3>,$$

$$\times \left[(1 + L/K) \right] \qquad \text{capture of L shell electrons } <\text{dimensionless}>,$$

$$\times \left[\left(4\pi/3\right)\left(R_0 A^{1/3}\right)^3 \right] \quad \text{volume of the nucleus, } <\text{cm}^3/\text{nucleus}>,$$

$$\times \left[\left(2J+1\right)\big/\left(\left(2\ell+1\right)\left(2I+1\right)\right) \right] \quad \text{spin-statistics factor, } <\text{dimensionless}>,$$

$$\times \left[\Gamma_v \Gamma_d \big/ \left(\left(\Gamma/2\right)^2 + \left(E_v - E_0\right)^2\right) \right] \quad \text{the resonance shape factor, } <\text{MeV}^2/\text{MeV}^2>,$$

$$\times \left[\left(\left(2/\sqrt{\pi}\right)\big/kT\right)\left(|Q + E_x - E_v|/kT\right)^{1/2} \left(e^{-\left(|Q+E_x-E_v|/kT\right)}\right) \Gamma_d \right] \text{limiting electron energy.}$$

This formula for the cross section is:

$$\sigma_{EC}\left(E_v\right) = 2 \left[\pi\left(\hbar c/E_v\right)^2 \right] \left[4\left(Z/a\right)^3 e^{-2ZR/a} \right] \left[\left(1 + L/K\right) \right] \left[\left(4\pi/3\right)\left(R_0 A^{1/3}\right)^3 \right]$$

$$\times \left[\left(2J+1\right)\big/\left(\left(2\ell+1\right)\left(2I+1\right)\right) \right] \left[\Gamma_v \Gamma_d \big/ \left(\left(\Gamma/2\right)^2 + \left(E_v - E_0\right)^2\right) \right]$$

$$\times \left[\left(\left(2/\sqrt{\pi}\right)\big/kT\right)\left(|Q + E_x - E_v|/kT\right)^{1/2} \left(e^{-\left(|Q+E_x-E_v|/kT\right)}\right) \Gamma_d \right], \tag{7}$$

where

$\hbar$ = reduced Planck's constant = $6.58211889 \times 10^{-22}$ MeV-seconds,

c = speed of light = $2.99792458 \times 10^{10}$ cm/second,

E_v = antineutrino energy, MeV,

Z = atomic number of target nucleus,
a = radius of first Bohr orbital = $5.291772083 \times 10^{-9}$ cm,

R = nuclear radius = $R_0 A^{1/3}$,

R_0 = 1.43×10^{-13} cm,

A = atomic mass number of target nucleus,

J = nuclear spin of product nucleus,

ℓ = orbital angular momentum of incoming antineutrino,

I = nuclear spin of target nucleus,

Γ_v = energy width of incoming antineutrino = $Q + E_x$ MeV,

Q = Q value for beta decay from the product nucleus ground state, MeV,

E_x = energy of the excited level in the product nucleus, MeV,

Γ_d = energy width of level in product nucleus, from lifetime or energy data, MeV,

Γ = total width = $\Gamma_v + \Gamma_d$, MeV,

E_0 = energy available for beta decay from product nucleus = $Q + E_x$ MeV,

k = Boltzmann's constant = 8.617342×10^{-11} MeV per kelvin,

T = the effective temperature of the atomic electron, $kT = \left(2/3\right)\left(\hbar c/R\right)/2$,

this energy is calculated from the uncertainty relation, with
$$\Delta E_e \geq \left(\hbar c/\Delta x\right)/2, \text{ and } \Delta x = R.$$

The units are <electron (captured) per nucleus per unit antineutrino flux>, as is proper. This complete unit is usually shortened to simply <cm²>, or "barn" $\left(= 10^{-24} \text{cm}^2\right)$.

IV. THE CALCULATION

While the calculations are complicated, they are not difficult, and the major task is one of bookkeeping. Thus, a spreadsheet forms an excellent structure for the calculation.

For each level in the product nucleus (the nuclide formed from the target nucleus by electron capture), an energy-dependent cross section is calculated, by use of the formula just described. This involves knowledge of the level energy ($E_x > 0$ for levels above the ground state), the spin and parity of the level, and the energy width of the level. Since the electron capture reaction on a stable nuclide produces an unstable nuclide, often short-lived, with very sparse data, a synthetic level structure was created by supplementing the published data with level information from a variety of stable nuclides. (For an alternative approach, see [30].) Since some of the known characteristics of the unstable nuclide and its decay are firmly included in the calculation, it was hoped that the use of many excited levels (generally 200-500), albeit from foreign nuclides, would provide a statistically representative estimate of the actual cross section. In that sense, the calculated cross section for each target nuclide must be considered a representative cross section, not exactly correct in its features, but showing features that may be considered typical of the reaction. Level data were taken from the "Table of Isotopes" [31].

In supplementing the sparse data for each product nuclide, the lower of the neutron or proton separation energies was used as a benchmark. Above this energy, the levels are truly virtual states, and have markedly larger energy widths, for all nuclides considered, than most of the lower energy quasi-stable levels. The energy of an excited level from a foreign nuclide was matched to the energy of the product nuclide above its separation energy, and the foreign level energy and width were assigned to the product nuclide. Spin and parity values were limited to those values reported for the product nuclide levels, with average values of the calculated spin factor used for all those levels without that information.

For each selected antineutrino energy, the contributions from all the product nuclide levels were summed to make the cross section at that energy. (In the calculations described here, the antineutrino energy was varied from 0 to 12 MeV in steps of 0.1 MeV.)

In spite of the use of many of the same foreign levels in creating the synthetic level structure for diverse product nuclides, the results show considerable diversity, supporting the hope that this approach might be effective.

The results of this calculation for a variety of target nuclides are shown in Fig. 1. The cross sections are generally larger than the resonant orbital electron capture cross section by roughly 10^{15}. For comparison, the electron capture cross section calculated in this manner for neutral atomic hydrogen, ^{1}H, which has no excited levels, is only about 5×10^{-82} cm^2 at 0.915 MeV. The cross section increases (although not uniformly) with increasing atomic number, due to the increased density of atomic electrons in the nucleus, resulting from the greater nuclear charge.

The cross section for the heavier nuclides show observable dips, not found in neutron resonance capture. They result here, as a result of the mathematical treatment, from especially low probabilities for the electron energy to contribute effectively to the reaction. The peaks in the ^{113}Cd cross section arise in the same way as in neutron capture, from the product level having an unusually large energy width compared to the adjacent levels.

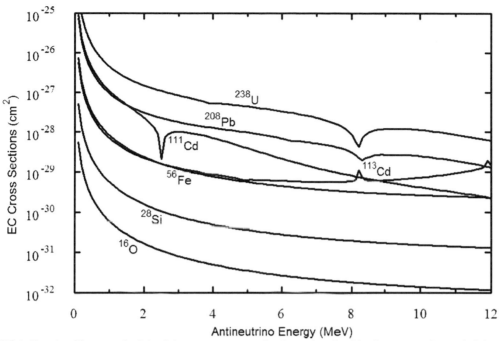

FIG. 1. Examples of the antineutrino-induced electron capture cross sections for a variety of nuclides. At any energy, the magnitude increases with increasing atomic number, and shows the normally expected resonance peaks seen in neutron capture. In addition, unusual dips are also seen, a result of the interaction of the electron, as a third particle.

V. APPLICATION

Nuclear power reactors, with high fission rates and consequently high rates of beta-decay of the neutron-rich fission fragments and their products, serve as intense sources of antineutrinos. These antineutrinos can be detected, above the threshold of 1.8 MeV, by the induced positron emission reaction, $\overline{v}_e + p \rightarrow n + \beta^+$. Unfortunately that threshold lies above approximately 67% of the antineutrinos emitted following fission, but this reaction provides effective selection and background rejection by using detection of both the positron and the neutron.

Several experiments have used this detection reaction in conjunction with high power reactors to first of all, demonstrate the existence of antineutrinos away from the site of beta decay [6] and secondly, in attempts to quantify the hypothetical flavor oscillation of massive neutrinos that has been used to explain the Solar Neutrino Problem [32,33,34]. In spite of the intense sources, and because of the very small effective cross section (on the order of 10^{-43} cm²), detection rates are low, detectors must be large, and long counting times are required, on the order of years. The detectors are comparable in size to the reactors themselves, or even larger. The most successful of these experiments, in terms of observing an energy dependent effect that might represent antineutrino oscillation, has been the KamLAND experiment, in Japan.

A. KamLAND

In the Kamioka Observatory, in the Mozumi Mine, under Mount Ikenoyama in the Japanese Alps, a large liquid scintillator was used to detect antineutrinos predominantly emitted by the many nuclear power reactors in the region. Those reactors are located at various distances, as shown in Fig. 2. Twenty reactor sites, some with multiple units, at distances ranging from 81 km to 976 km, were in operation during the measurements. Data provided by the reactor operators gave information on the actual power (fission rate) and the ongoing changes in the fuel isotopic mix. The entire detector required 1 kiloton of scintillator and occupied a space about 19 m in diameter by 26 m high. The data collection time extended from March 9, 2002 to January 11, 2004, and then further to May 12, 2007, a total of over five years. Positrons produced by the induced positron emission reaction on protons (Reaction 3a, $\overline{v}_e + p \rightarrow n + \beta^+$) in the organic scintillator were detected and their kinetic energy measured, and the detection was confirmed by the gamma radiation detected from capture of the accompanying neutron by hydrogen in the scintillator. Unwanted reactions that produced detectable scintillations were suppressed as background by various selections, the most important being the delayed coincidence of a neutron pulse after the positron pulse. Antineutrinos from beta decay of natural radioactive elements in the Earth were removed by calculation [8,9,10] (Hereafter, KamLAND 1, 2, and 3, respectively.).

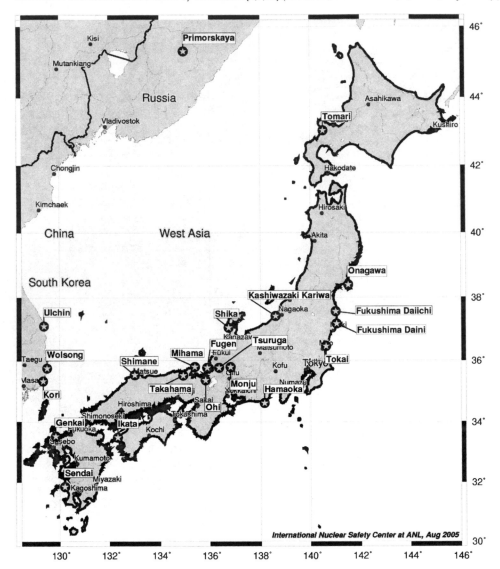

FIG. 2. Locations of nuclear power reactors involved in the KamLAND experiment. The Kamioka Observatory is located approximately at the lower right corner of the label for Fugen, at about Lat 36 N Lon 137 E. (Map of nuclear power reactors in Japan is from http://www.insc.anl.gov/.)

The results showed a clear deficit of antineutrinos at lower energies, which was interpreted as evidence of oscillation of antineutrinos. Oscillation requires that antineutrinos of different flavors must have different masses, and massive antineutrinos are not part of the current Standard Model of Nuclear Particles, nor the weak interaction theory of beta decay. Since oscillation requires "new physics", the evidence should be carefully examined and reasonable alternatives should be seriously considered. The antineutrino-induced electron capture reaction, with cross sections calculated as described above, offers a mechanism for attenuation that agrees with the observations and is consistent with the Standard Model.

In FIG. 4 of KamLAND3, the "Survival Probability" of antineutrinos is shown plotted against a ratio of distance to antineutrino energy, based on the interpretation that the deficit in antineutrino flux is due to oscillations. The antineutrino spectrum used to produce the expected spectrum used as a reference was generated according to the reactor fuel isotope estimates ("burnup") from spectra derived from isotopic laboratory data. This fundamental spectrum may be subject to the potential errors discussed in the Appendix. I have replotted those data in my Fig. 3, with the interpretation that the loss of antineutrinos is due to exponential attenuation by absorption in the induced electron capture reaction. The length/energy variable in FIG. 4 of KamLAND3 was converted to energy alone by the distance specified in their caption, 180 km. There is no manipulation other than this, and the replotting is mainly a clarification and re-naming, plotting "Transmission" against antineutrino energy. The results from KamLAND2 (their FIG. 3) are shown as ghosts, and the great improvement achieved by KamLAND3 is clear. A smooth curve, which passes reasonably well through the observations, was calculated for the attenuation from each reactor site, using the microscopic cross-sections described above, with the atom densities of the 13 most abundant isotopes in the average Earth's crust. No adjustable parameters were used.

Because of the interesting peak in the observations between 2 and 3 MeV, and the accidental occurrence of a dip in the cross section I have calculated for ^{111}Cd (my Fig. 1), I calculated the attenuation for a combination of ^{111}Cd and ^{113}Cd, which is shown as the more structured curve (dashed), with a peak remarkably in agreement with that shown by the observations. I must emphasize that this is an accident, since the level structure for these nuclides is mainly artificial. However, the comparison shows clearly that the antineutrino-induced electron capture reaction, with cross sections as calculated here, offers a promising explanation for observations that otherwise are being seen as a need to invoke "new physics".

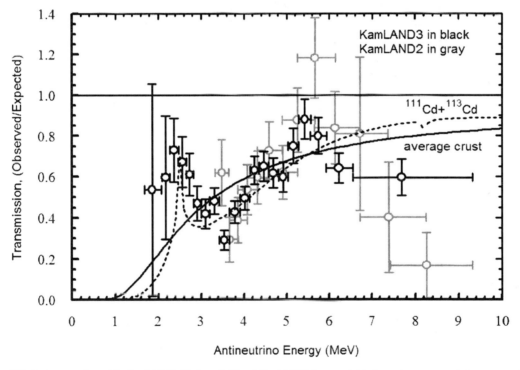

FIG. 3. Interpretation of the KamLAND3 (black symbols) and KamLAND2 (gray symbols) observations as transmission, compared with the unmodified transmission (line at 1.0) for the reference reactor spectrum, with exponential attenuation by the antineutrino-induced electron capture reaction (smooth curve) and a selected nuclide cross section (dashed line).

The Mozumi Mine, containing the KamLAND detector, exploited rich deposits of zinc, silver, and lead, and abundant amounts of related elements might be expected to also be present in the surrounding rocks of the Japanese Alps, differing considerably in composition from the average of Earth's crust.

The agreement of the predicted transmission curve with the observations is remarkably good, particularly in recognition of the difficulties in accurately defining the materials (minerals) in the various flight paths. This problem is suggested in Fig. 4, showing two hypothetical flight paths through greatly different materials, with different nuclides, and different reaction cross sections. (In that figure, the curvature of the Earth is exaggerated, the topography is emphasized, and the geology is imagined, to make some of the complications clear.)

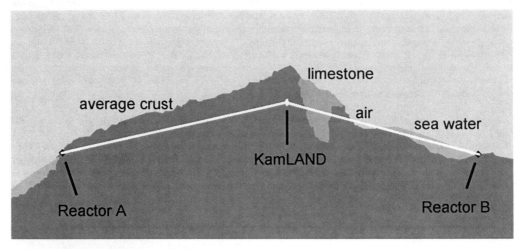

FIG. 4. A representative sketch of the KamLAND situation, relative to two distant reactors. The figure is exaggerated. Transmission from Reactor B is likely to be different from that from Reactor A.

There are many complications that I have bypassed. The set of excited states for each target nuclide in the set of "average crust" is wrong and incomplete. The mineralogy of the flight paths in general and in particular can certainly not be accurately represented as "average crust". The treatment of the energy of the electron may only be a distant approximation. However, the match with the observations is so good as to offer promise of a better understanding of antineutrino interactions.

B. Other Reactors

In addition to the KamLAND measurements with multiple distant reactors, several measurements have been done with smaller detectors near individual nuclear power sites. In some cases the detectors have been within the reactor building itself; in the most distant case, slightly over 1 kilometer away from the reactors. These are:

    Savannah River "P" reactor [34,36],
    CHOOZ [37,38],
    Rovno [39],
    Palo Verde [40,41],
    Gösgen [42,43],
    Bugey [44,45,46,47].

The published positron energy spectra have been converted to antineutrino spectra by adjusting the energy and dividing by the cross section for the antineutrino-induced positron emission, as described in the Appendix. These spectra are shown in Fig. 5, in comparison to, and normalized to, the adopted spectrum for ^{235}U thermal fission from Vogel [48], Vogel & Engel [49], and Avignone [50], also described in the Appendix.

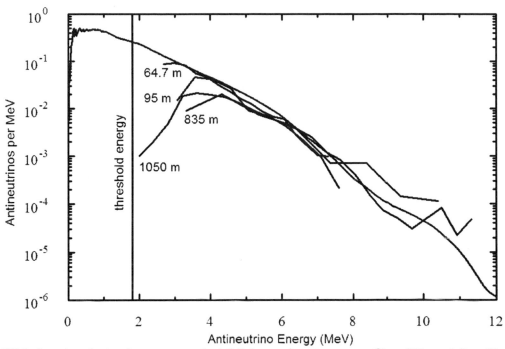

FIG. 5. Comparison of antineutrino spectra measured at individual reactor sites. Measurements at Gösgen (64.7 meters), Bugey (95 meters), Palo Verde (835 meters), and CHOOZ (1050 meters), show depletion of the antineutrino flux that generally increases with increasing distance. The experimental spectra have been arbitrarily normalized to the adopted standard spectrum.

From the view of antineutrino attenuation, the most interesting of these measurements are labeled with the distances between the source of antineutrinos and the detector. The most distant measurements show a decrease in antineutrino flux, compared to the standard spectrum and the near spectra, and that decrease becomes more significant with increasing distance. This may be an observational artifact, resulting from attempts to interpret data taken at the limits of detectability. (The cross section declines to zero at E_ν = 1.8 MeV, with decreasing energy, making detection at low energies difficult.) However, the progressive decrease with distance is at least suggestive. The decreases are somewhat greater than would be expected from the present KamLAND analysis, at a much greater distance, but may reflect intervening materials with atomic numbers greater than the average Earth's crust. This material might be stored (uranium) fuel between the source and the detector, or lead shielding, but cannot otherwise be estimated from published reports. This opens the opportunity for future explorations.

VI. CONCLUSIONS

A re-examination of the cross section for the antineutrino-induced electron capture reaction, converting a nuclear proton to a neutron, has shown that this reaction has a high probability of occurring, relative to previous calculations and to other neutrino reactions. This cross section permits calculation of the results of the KamLAND measurements on antineutrinos from distant nuclear power reactors as being caused by attenuation by the crustal rocks intervening between the sources and the detector, with good agreement. Structure in the cross sections for various other nuclides which might be present in the surrounding rock suggests the possibility of nearly exact agreement, using more detailed calculations. The far measurements at several individual reactor sites show some low energy attenuation that deserves further investigation.

The appendix discusses the use of the Fermi factor in calculations of spectra from beta decay, which is appropriate for the beta particle, but is incorrect for the uncharged antineutrino. Several reactions that will occur in nuclear power reactors have been overlooked in estimating the antineutrino source spectra are also described.

ACKNOWLEDGMENT

I wish to thank J. Marvin Herndon and L. A. Mikaélyan for essential contributions to the development of the cross section model. This work was supported entirely by personal funds.

APPENDIX:
ADJUSTMENTS TO THE REACTOR ANTINEUTRINO SPECTRUM

It is often stated that high accuracy is needed in the antineutrino spectrum, and considerable effort has been made to achieve this. However, there appear to be two aspects in which the accuracy is unnecessarily diminished. Correction of these deficiencies should improve the accuracy of the antineutrino spectra.

A collection of 23 spectra has been gleaned from the literature, produced from 1959 to 2007, and is shown with an adopted standard spectrum, in Fig. A-1. This discussion is limited to the antineutrinos produced by thermal neutron fission of ^{235}U in nuclear reactors, but also applies to antineutrino spectra in general. Some information on other nuclides and low energy antineutrinos produced by absorption and activation reactions can be found in published reports. (See for example, [39,51,52,53,54].)

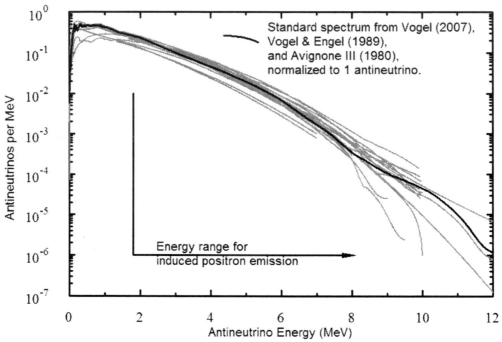

FIG. A-1. Comparison of antineutrino spectra produced for thermal fission of 235U. A standard spectrum has been adopted, from Vogel (2007) [48] [1 to 10 MeV], Vogel & Engel (1989) [49] [0 to 1 MeV], and Avignone (1980) [50] [10 to 12 MeV]. The spectra have been normalized so that the integral from 0 to 12 MeV is equivalent to 1 antineutrino. Spectra spanning only part of this range have been normalized to contain the same quantity as the standard spectrum over the corresponding energy range. The induced positron emission reaction is sensitive only to antineutrinos with energy greater than 1.806 MeV.

The intent of this comparison is simply to justify the use of a single, representative standard spectrum for later comparisons, and to suggest that, in the historical aggregate, current estimates of the accuracy of these spectra, approaching 2%, may be overly optimistic.

1. The Fermi Factor

The spectrum of reactor antineutrinos is derived from data on radioactive fission products: fission yields, beta-decay branching ratios, and end-point energies. These are used to calculate the individual beta-particle spectra and the antineutrino spectra are taken to be the complement in energy. That is,

$$N_v(E_v) = N_\beta(E_0 - E_v). \qquad \text{(A1)}$$

However, this approach imposes the form of the interaction of the charged beta-particle with the nuclear Coulomb field on the uncharged (anti)neutrino. That is physically incorrect. A better representation of the Fermi factor, as described below, actually prevents this error and inherently accommodates negative, positive, and neutral leptons.

The beta particle spectrum of each radioactive fission product is calculated [55] by use of a kinematic term modified by a nuclear Coulomb correction factor, the Fermi factor $F(Z, W)$:

$$N_\beta(E_\beta) = n\left[\left(W^2 - 1\right)^{1/2}\left(W_0 - W\right)^2 W\right]F(Z, W), \qquad \text{(A2)}$$

where n is a normalization factor, and W is related to $E_\beta = T_\beta + mc^2$ by

$$W = \left(T_\beta + mc^2\right)\Big/mc^2.$$

Physically, the Fermi factor modifies the shape of the spectrum to account for the retarding effect of the positive nuclear charge on the momentum of the negatively charged beta particle. (An alternative form accounts for the acceleration of the positively charged beta particle in positron decay.) This factor was derived by Fermi [56] from the work of Hulme [57] and was presented as a theoretical relativistic approximation [58] as

$$F(Z, \eta) = \left\{4(1+2s)\Big/\left[\Gamma(3+2s)\right]^2\right\}\left\{2R/(\hbar/mc)\right\}^{2s}\left\{\eta^{2s}e^{2\pi y}\left|\Gamma(1+s+iy)\right|^2\right\}, \qquad \text{(A3)}$$

where η is the beta particle momentum relative to mc, R is the nuclear radius, and α is the fine structure constant:

$$\eta = \left(W^2 - 1\right)^{1/2}, \qquad \text{(A4a)}$$

$$s = \left(1 - \gamma^2\right)^{1/2} - 1, \qquad \text{(A4b)}$$

$$\gamma = \alpha Z, \qquad \text{(A4c)}$$

$$y = \gamma(1+\eta)^{1/2}\Big/\eta. \qquad \text{(A4d)}$$

Evaluating the complex gamma function $\Gamma(1+s+iy)$ has been a major obstacle to the use of this form in the calculation of beta particle spectra in nuclear reactors for hundreds of energy points and many hundreds of distinct beta decays.

Avignone [59] has used an approximation that eliminates the complex gamma function:

$$F(Z, \eta) = \left\{4(1+2s)\Big/\left[\Gamma(3+2s)\right]^2\right\}\left\{2R/(\hbar/mc)\right\}^{2s}\left[2\pi y\Big/\left(1-e^{-2\pi y}\right)\right]\left\{\left[(1+\eta^2)(1+4\gamma^2)-1\right]\Big/4\right\}^s$$
$$\text{(A5)}$$

with the same definition of terms as above. This approximation was originally presented [60] in use with a form of the spectrum that used the 4th power of the (anti)neutrino energy (W_0-W), in order to better match the experimental results on hand at that time, rather than the 2nd power as is currently used. Regardless, comparisons of approximations by Feister [61] show that it is quite accurate. Feister's work led to the publication of tables of the Fermi function by the National Bureau of Standards [62]. That work is long since out of print, and modern calculations are far more easily done with analytical forms in computers.

While several other approximations and approaches have been noted (see for example Bahcall [33]), the simplest method at present seems to be that used by Goorley and Nikjoo [63]. The gamma functions are evaluated as part of the calculation by the use of a commercial mathematical software program, Mathcad, in their application. Others are available and alternatives can be found when needed by searching the Internet.

The Fermi factor is a natural correction to the observable beta-particle spectrum, recognizing the loss of energy as the beta-particle leaves the atom, or the corresponding gain by a positive beta-particle. However, as pointed out by Bethe and Bacher [60] although the beta particle and the nucleus interact through the Coulomb field, the (anti)neutrino, with no electric charge and no magnetic moment, does not.

"One must simply insert a plane wave for the neutrino wave function, while the electron wave function ψ is to be taken in the Coulomb field of the disintegrated nucleus." (page 191)

F(Z,W) should not be applied to the antineutrino spectrum, except in a rather awkward way that has been implied by several researchers; by setting the value of the nuclear charge Z equal to zero. With that change, the Fermi function becomes equal to unity. That is,

$$F(Z, W) = F(0, W) = 1. \qquad (A6)$$

With F(Z,W)=1, the calculation obviously becomes far simpler, and actually, correct. Not only is there no complex gamma function to evaluate (nor the real gamma function as well), there are no other terms to evaluate. Only the normalization factor and the kinematic terms must be calculated.

Since this aspect, the absence of a Coulomb interaction for the antineutrino, was so easily forgotten, it may be worthwhile to consider embedding the effect directly in the formula by a slight generalization. This can be done simply by replacing terms in Z by $-Zz$, where Z is still the nuclear charge number and z is the lepton [electron, positron, or (anti)neutrino] charge number.

$$F(Z,\eta) = \left\{ 4(1+2s) \big/ \left[\Gamma(3+2s) \right]^2 \right\} \left\{ 2R \big/ (\hbar/mc) \right\}^{2s} \left[2\pi y \big/ \left(e^{2\pi y} - 1 \right) \right] \left\{ \left[\left(1+\eta^2\right)\left(1+4\gamma^2\right)-1 \right] \big/ 4 \right\}^s ,$$

$$(A7)$$

with

$$\eta = \left(W^2 - 1 \right)^{1/2} , \qquad (A8a)$$

$$s = \left(1-\gamma^2\right)^{1/2} - 1 , \qquad (A8b)$$

$$\gamma = \alpha Z z , \qquad (A8c)$$

$$y = \gamma \left(1+\eta\right)^{1/2} \big/ \eta . \qquad (A8d)$$

Note that the only change to the formula itself is in the denominator of the third major bracket, $\left[2\pi y \big/ \left(1 - e^{-2\pi y}\right) \right] \rightarrow \left[2\pi y \big/ \left(e^{2\pi y} - 1\right) \right]$. As subterms, γ, s, and y are also changed by the inclusion of z. This generalization automatically implements the rules of using +Z for negative beta decay and −Z for positive beta decay [58] and enforces the rule of setting the Fermi factor to unity for the (anti)neutrino, by making Zz = 0, from the fact that z = 0 for the (anti)neutrino. The use of Zz is also more consistent with the form of the fine structure constant, involving the square of the electronic charge, $\alpha = e^2/\hbar c$. (See for example, [64].) This is actually nearly shown in the derivation of Hulme [57],which served as the basis for Fermi's application to beta decay [56]. Hulme's Eq (24) shows this form except for the implicit use of z=1 in the potential energy perturbation.

Early on, the Coulomb interaction of an electron with the charge of the atomic nucleus was written as $Ze^2/\hbar c$, with the charge number of the electron (z=1) implicitly hidden. Dimensional analysis, with the elementary charge being ±e <coulombs per unit charge>, shows that this form of the interaction leaves one <per unit charge> unaccounted for, by ignoring the electron charge number. The <per unit charge> is not truly a dimension, but rather a specification that still must be considered in dimensional analysis.

It is suggested that this modification be used (it is easier and more nearly correct) in future calculations of the beta decay (and reactor) antineutrino spectra, and the results applied in future measurements and any re-evaluations of past observations.

As a side issue, Ishimoto et al. [65] show a comparison of the theoretical and the observed beta spectrum for the decay of [80]As, and demonstrate a considerable discrepancy. I'm not sure if this represents an error in $F\left(Z = 34, E_0 = 5.641 \text{MeV}\right)$, or an unrecognized complexity in the decay scheme. The decays are I=1+ to 0+, and 1+ to 2+, with no parity change, which indicates an allowed, unfavored transition.

2. The high energy discrepancy

The reactor antineutrino spectra presented in the introduction to this appendix, and the other spectra used for interpreting measurements of antineutrinos from nuclear reactors, are based on laboratory-grade measurements, such as thin-sample beta-spectrometry, radiochemistry, and mass spectrometry. The actual production of antineutrinos in a reactor is an industrial-grade situation, rather different from the laboratory, and includes processes that are not evident in the laboratory.

There are eight sets of published measurements of the positron spectrum from induced positron emission, using reactor antineutrinos. Six of these are from commercial pressurized water reactors (PWRs) with uranium oxide fuel and ordinary "light" water as a moderator/coolant, while two are from a U. S. Government nuclear materials production reactor, the Savannah River Plant "P" reactor, which used natural uranium metal fuel and heavy water (D_2O) as moderator and coolant. These different materials are likely to show different effects in the production of antineutrinos. The PWRs use slightly enriched uranium (2-4% ^{235}U) with a significant buildup of plutonium, the Savannah River reactor used natural uranium (0.72% ^{235}U) with the buildup of plutonium strictly limited to minimize the production of ^{241}Pu. (In the most recent measurements at Savannah River, the ^{239}Pu fission fraction was estimated as less than 8% with ^{238}U contributing less than 4% [36]. This may be an increase in the ^{239}Pu fission fraction from the 1966 measurements due to a change from plutonium production to tritium production.)

The antineutrino spectrum for each of these measurements can be extracted from the observed positron spectrum by dividing by the cross section for the reaction, with an appropriate relation between positron and antineutrino energy. (A complete determination of the spectrum also involves the time interval of exposure, total number of target protons, detector efficiencies, and center-to-center distances, but these can be ignored at present by focusing solely on the shapes of the spectra.)

Extraction of the antineutrino spectra was done using the approach given by Vogel [66] and Fayans [67], and used by Zacek *et al.* in their analysis of measurements at Gösgen [43]. This provides corrections for neutron recoil, weak magnetism, and outer radiation effects. The "spec" correction of Vogel, to account for the steep decline of the positron spectrum with increasing energy, does not explicitly appear.

The basic cross section, as a function of antineutrino energy (kinetic energy E_v with $m_v=0$) and assuming that the proton and neutron are at rest and infinitely massive (no recoil, correction is made separately), is evaluated by Zacek *et al.*[43] (and others) from:

$$\sigma_\beta\left(E_v\right) = \left\{2\pi^2 c^2 \hbar^3 \ln 2 / \left[\left(mc^2\right)^5 \text{ft}\right]\right\}\left[E_v - \left(M_n - M_p\right)c^2\right]\left\{\left[E_v - \left(M_n - M_p\right)c^2\right]^2\right\}^{1/2} \text{ cm}^2. \quad (A9)$$

This is cast in the "experimentalist's" form with energy terms in MeV, length in cm, and time in seconds. Since this study focuses solely on the shapes of the spectra, the constant term is ignored, and only the energy-dependent form of the cross section is used.

The antineutrino energy is related approximately [see Vogel (1984) [66]] to the positron energy by

$$E_v = \left\{E_\beta + \left(M_n - M_p\right)c^2\right\} + 2\left\{\left[E_\beta\left(E_\beta + \left(M_n - M_p\right)c^2\right) + \left[\left(\left(M_n - M_p\right)c^2\right)^2\right]/2\right]\right\} / \left(M_n + M_p\right)$$

(A10)

At $E_\beta = 0, E_v = 1.806065$ MeV, using the current best values for the masses.

The energy of the beta particle E_β implicitly includes the rest mass energy of the positron, $E_\beta = T_\beta + mc^2$. The experimental spectra, extracted from the positron spectra, are shown in Fig. A-2, compared with the adopted standard spectrum. These have been somewhat subjectively normalized to the standard. It should be recognized that information on the antineutrino energy distribution can only be obtained from the induced positron emission reaction for antineutrino energy greater than the threshold at 1.806 MeV, which includes only about one-third of the fission antineutrinos.

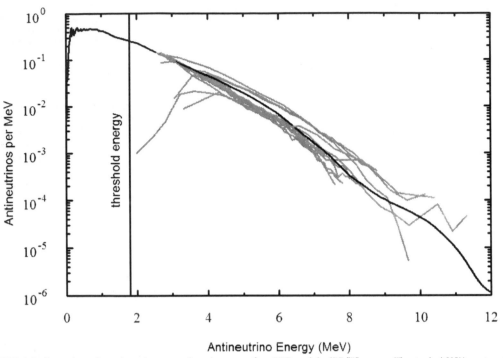

FIG. A-2. Comparison of experimental reactor antineutrino spectra from PWRs and the SRP "P" reactor. The standard 235U spectrum from Vogel (2007) [48], Vogel & Engel (1989) [49], and Avignone (1980) [50] is shown as the heavy line, with all experimental spectra normalized to produce reasonable connections. (The standard spectrum has been normalized so that the integral from 0 to 12 MeV is equivalent to 1 antineutrino.) Clear departures from this spectrum are shown in various ways by the experimental spectra, and it does not appear that these can be readily explained as due to mixed fuel fission.

It can be seen that the PWR spectra form a similar set generally below the standard spectrum between about 4 MeV and 6 MeV, and the heavy water reactor ("P") is distinctly different. [The positron measurements of Nezrick and Reines (1966) [35], were reported as a "raw" histogram, and as a "true" spectrum. The true positron spectrum yielded an unusually smooth antineutrino spectrum which differed from all others. The raw spectrum yielded a much more reasonable variation and is used in this comparison. That spectrum falls along the bulk of the PWR spectra till about 6 MeV, where it crosses over to join the much more recent measurements of Greenwood et al. (1996) [36]. Rovno [39], the highest spectrum of all, and showing the greatest decrease at high energy, may be misplaced because of the normalization.] Significantly, all the reactor spectra differ from the adopted standard spectrum (and so from all the spectra used to analyze reactor antineutrino measurements). Considerable buildup of plutonium occurs in PWRs, resulting in progressively more fissions of ^{239}Pu and ^{241}Pu during a fuel cycle, which causes the expected antineutrino spectrum to generally decrease in energy, whereas the discrepancy shows as an increase. This effect would not be significant in the "P" reactor, where production of ^{239}Pu, not utilization, is intended. The significant departure from the standard spectrum at higher antineutrino energies, here called the high energy discrepancy, may result from the neglect of several sources of antineutrinos, in some cases with notably high energies.

Some comparisons can be made between experimental antineutrino spectra that might lead to some improvements in the accuracy of the reference (expected) spectra. Two pairs of measurements seem particularly suitable: the measurements at the SRP "P" reactor reported by Greenwood et al. (1996) [36] and by Nezrick & Reines (1966) [35], and those at the Bugey reactors reported by Achkar et al. (1995) [46] and by Cavaignac et al. (1984) [44]. These are shown, with my adopted standard spectrum for reference, in Fig. A-3. The SRP measurements show remarkable agreement for being done 30 years apart by different experimentalists with different equipment with a reactor that may have had many changes itself. (SRP reactors were used to produce plutonium and tritium, and changing needs and priorities for ^{239}Pu, ^{6}Li, and ^{3}H may have caused changes in loadings and operations.)

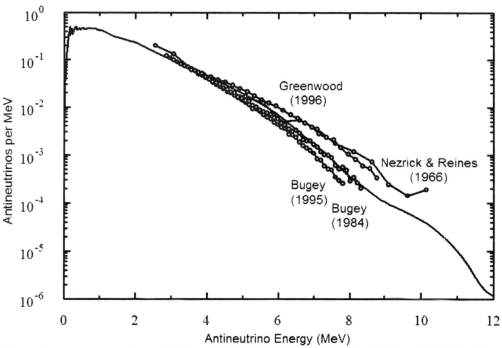

FIG. A-3. Comparison of spectra from the SRP "P" reactor [Greenwood et al. (1996) [36] and Nezrick & Reines (1966) [35]] and the Bugey reactors (1984 and 1995) [44,45,46], with the adopted standard spectrum. Normalizations were chosen to produce the best agreement about 3.5 MeV.

The two SRP spectra deviate a small amount in the mid-energy range (4-6 MeV) but appear to coincide well at higher energies. Greenwood *et al.* (1996) [36] state that the fission fractions for ^{239}Pu and ^{238}U were respectively, less than 8% and less than 4%, so that the expected spectrum should have been within 1.5% of the reference spectrum. (It is unlikely that any significant amount of ^{241}Pu would have been present.) However, at high energies, the observed spectrum falls above my chosen spectrum by nearly a factor of three. From 6 MeV on, the SRP measurements are clearly above the standard (^{235}U) spectrum. This deviation should have been in the direction of a softer spectrum, which is the opposite to what is seen. (It should be noted that this occurs where the intensity is so small that the effects are insignificant.)

Similarly, the Bugey measurements are in good agreement, but deviate significantly from each other above 5 MeV. Both the 1984 and the 1995 measurements show spectra that is clearly softer than the SRP spectra, and this would be expected for reactors with significant fission fractions from the higher nuclides.

The following reactions appear to be worth consideration as possible sources for the higher energy antineutrinos:

^{16}O(n,p)^{16}N(βν)^{16}O

This reaction requires high energy neutrons fresh from fission, at a threshold of about 11 MeV, and is more likely in the PWRs with oxide fuel, but may still occur in the oxygen in the heavy water in the "P" reactor. While moderation by hydrogen can cause a neutron to lose most of its energy in a single collision, sufficient ^{16}N is produced in the coolant of PWRs as to create a significant radiation shielding problem, due to the high energy gamma-rays emitted in its decay to ^{16}O. Greenwood *et al.* (1996) [36] remarked (p. 6056) about "… the large flux of γ radiation associated with the β decay of ^{16}N in the water of the heat exchangers above our head." High energy antineutrinos would also be associated with the β decay of ^{16}N, and may have affected the recorded positron spectrum. One branch of the beta decay has a maximum energy of 10.4 MeV. The production cross section is very small, but ^{16}O is the most abundant nuclide in PWRs.

The cross section for the ^{16}O(n,p)^{16}N reaction is compared with the fission cross section for ^{238}U (from ENDF/B-VII.0 at National Nuclear Data Center, http://www.nndc.bnl.gov) in Fig. A-4.

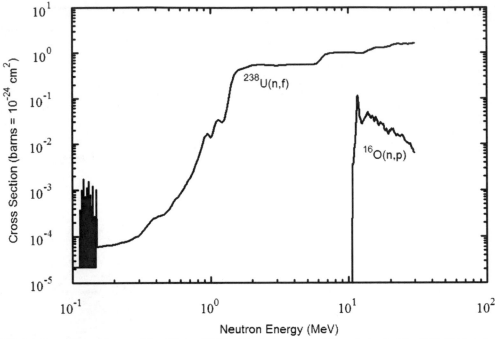

FIG. A-4. Cross sections for neutron-induced fission of 238U and the (n,p) reaction on 16O, producing short-lived 16N which beta-decays with high-energy antineutrinos. Cross sections are from ENDF/B-VII.0 at http://www.nndc.bnl.gov/.

$^7\text{Li}(n,\gamma)^8\text{Li}(\beta\nu)^8\text{Be}^*(2\alpha)$

Lithium hydroxide is used in PWRs to control the corrosion activity of hydrogen in the water. Lithium enriched in ^{7}Li (depleted in ^{6}Li) is used to avoid neutron absorption, but some small amount occurs in ^{7}Li and results in a high energy beta decay, with an endpoint of 13 MeV, by decaying to the 3 MeV level of ^{8}Be. The thermal neutron cross section for production of ^{8}Li is 0.0454 barns.

$^{11}\text{B}(n,\gamma)^{12}\text{B}(\beta\nu)^{12}\text{C}$

Natural boron, with 80% ^{11}B, is used as a burnable poison in some PWRs, to reduce the amount of reactivity that must be controlled by moveable control rods. This isotope has a small neutron absorption cross section, but the beta decay produces antineutrinos with an end-point energy of 11.5 MeV. The thermal neutron cross section for production of ^{12}B is 0.0055 barns. Gadolinium may be used as a burnable poison instead of boron, and does not produce any significant antineutrinos.

These reactions can only be well characterized by use of the reactor-specific data, such as the neutron flux spectrum and the elemental number densities, that are available to the reactor operators

The antineutrino spectra for these reactions, which can be accurately represented, are shown in Fig. A-5. The high energy beta branch for ^{16}N, the decay to the ground state of ^{16}O, is shown as calculated with a shape correction for a first-forbidden transition, which it is.

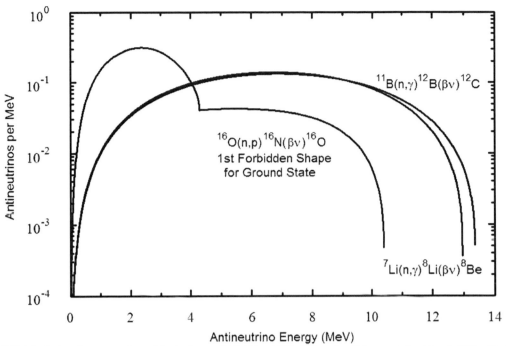

FIG. A-5. Antineutrino spectra calculated for the decays of 16N, 8Li, and 12B, which produce high-energy antineutrinos in PWRs. The high-energy beta decay from 16N to 16O is a first forbidden transition and a shape correction has been applied. The change from the usually assumed allowed shape is small.

Fusion-fission with oxygen

The higher energy fission fragments (the low-mass branch) have sufficient energy to fuse with oxygen nuclei, resulting in a secondary fission. The beta decay properties of the resulting fission products have not been explored, but might be expected to be different from those of the usual neutron-induced fission of U and Pu.

Coulomb fission

Fission fragments may cause the ^{238}U to fission by Coulomb excitation. The antineutrino spectrum produced in the decay of the ^{238}U fission might be different from the neutron-induced fission product spectrum that is calculated for this isotope. (See for example, V. Oberacker *et al.* [68].)

The effects of the last two reactions are difficult to estimate, and the reactions have not been thoroughly explored. Fusion-fission with oxygen has been studied in a somewhat backwards manner by accelerating ^{16}O ions against heavy target nuclei [69,70]. Some differences might be expected because of differences in atomic electron screening.

I attempted to correct the standard spectrum to the Greenwood *et al.* [36] spectrum by applying the ^8Li antineutrino spectrum, on the possibility that large quantities of natural lithium were being irradiated to produce tritium (^3H). I also considered the ^{16}N antineutrinos, because of the shielding problem posed by the intense gamma radiation from the ^{16}N in the heat exchangers. Neither of these was successful. I did find, as shown in Fig. A-6, that adding a small amount of a high-energy beta decay, 1.8% with $E_0=8.9$ MeV, to the standard spectrum gave a very good match over the full energy range of the experimental data. Of course, it is unlikely that a single beta decay with a specific end point energy would be responsible for the entire adjustment. Small errors in the yields and branching ratios for the highest energy, shortest lived fission fragments, most difficult to quantify, are the most likely source of this discrepancy.

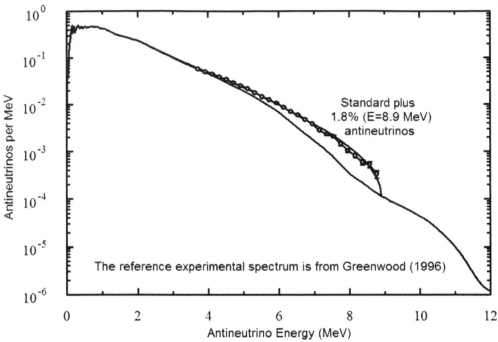

FIG. A-6. Adjustment of the standard spectrum to match the SRP measurements of Greenwood et al. (1996) [36] by the addition of a small amount of antineutrinos from a high-energy beta decay.

The later Bugey (1995) [46] measurements showed a significant decrease at higher energies compared to the (1984) [44] measurements. This might be considered to be the result of a more mature fuel mix with more fission from the higher nuclides. That may be the case (no data on this point were presented in Achkar *et al.* (1995) [46]), but comparison of the measurements in 1984 at 13.63 meters, with a ^{235}U fission fraction of 62.1%, with those at 18.60 meters and a ^{235}U fission fraction of 47.9% shows no significant difference in shape.

I attempted to account for the difference by subtraction ("removal") of ^{12}B antineutrinos from the Bugey (1984) [44] spectrum, on the possibility that the early core loading had used boron as a burnable poison, but that the later core might have used gadolinium, which produces no significant antineutrinos. This attempt was unsuccessful, making too large a change at high energies, where the intensity is low, and not enough change in the middle energy range. I have not identified any other cause of the difference between what should be very similar results, and therefore must let the question rest.

As even more powerful nuclear reactors are built, perhaps with considerations of antineutrino experimentation included in the early construction design, it can be hoped that we will see continuing refinement in our knowledge of these difficult particles.

VIII. REFERENCES

1.   F. Reines and C. L. Cowan, Jr., Phys. Rev. **90**, 492-493 (1953)
2.   F. Reines and C. L. Cowan, Jr., Phys. Rev. **92**, 830-831 (1953)
3.   F. Reines, C. L. Cowan, Jr., F. B. Harrison, H. W. Kruse, and A. D. McGuire, Science **124**, 103-124 (1956)
4.   Frederick Reines and Clyde L. Cowan, Jr., "Neutrino Physics", Phys. Today, August 1957, pp. 12-18
5.   Frederick Reines and Clyde L. Cowan, Jr., Phys. Rev. **113**, 273-279 (1959)
6.   F. Reines, C. L. Cowan, Jr., F. B. Harrison, A. D. McGuire, and H. W. Kruse, Phys. Rev. **117**, 159-173 (1960)
7.   F. Reines, Rev. Mod. Phys. **68**, 317-327 (1996)
8.   K. Eguchi *et al.* (KamLAND Collaboration), Phys. Rev. Lett. **90**, 021802-1-6 (2003)
9.   T. Araki *et al.* (The KamLAND Collaboration), Phys. Rev. Lett. **94**, 081801 (2005)
10.   The KamLAND Collaboration, arXiv:0801.4589 (5 Feb 2008)
11.   A. B. McDonald, J. R. Klein, D. L. Wark, "Solving the Solar Neutrino Problem", Sci. Am. (April 2003), pp. 40-49

12. Allan Franklin, *Are There Really Neutrinos?* (Perseus Books, Cambridge, MA, 2001)
13. T. D. Lee and C. N. Yang, Phys. Rev.**105**, 1671-1675 (1957)
14. L. A. Mikaélyan, B. G. Tsinoev, and A. A. Borovoi, Sov. J. Nucl. Phys. **6**, 254-256 (1968)
15. C. Avilez, G. Marx, and B. Fuentes, Phys. Rev. D **23**, 1116-1117 (1981)
16. L. M. Krause, S. L. Glashow, and D. N. Schramm, Nature (London) **310**, 191-198 (1984)
17. G. Domogatsky, V. Kopeikin, and L. Mikaélyan, arXiv:hep-ph/0409069, v2 (8 Sep 2004)
18. G. Breit and E. Wigner, Phys. Rev. **49**, 519-531 (1936)
19. C. W. Reich and M. S. Moore, Phys. Rev. **111**, 929-933 (1958)
20. H. Feshbach, C. E. Porter, and V. F. Weisskopf, Phys. Rev. **96**, 448-464 (1954)
21. H. A. Bethe and G. Placzek, Phys. Rev. **51**, 450-484 (1937)
22. G. Breit and E. Wigner, Phys. Rev. **49**, 519-531 (1936)
23. H. A. Bethe and G. Placzek, Phys. Rev. **51**, 450-484 (1937)
24. Alvin M. Weinberg and Eugene P. Wigner, *The Physical Theory of Neutron Chain Reactors* (The University of Chicago Press, Chicago, 1958)
25. John M. Blatt and Victor F. Weisskopf, *Theoretical Nuclear Physics* (John Wiley & Sons, New York, 1952)
26. Sergio DeBenedetti, *Nuclear Interactions* (John Wiley & Sons, New York, 1964)
27. K. H. Beckurts and K. Wirtz, *Neutron Physics* (Springer-Verlag, New York, 1964)
28. C. W. Reich and M. S. Moore, Phys. Rev. **111**, 929-933 (1958)
29. D. K. Olsen, G. de Saussure, R. B. Perez, E. G. Silver, F. C. Difilippo, R. W. Ingle, H. Weaver, Nucl. Sci. Eng. **62**, 479-501 (1977)
30. D. K. Olsen, G. de Saussure, R. B. Perez, F. C. Difilippo, R. W. Ingle, H. Weaver, Nucl. Sci. Eng. **69**, 202-222 (1979)
31. Richard B. Firestone, *Table of Isotopes*, 8th edition, edited by Virginia S. Shirley with Coral M. Baglin, S. Y. Frank Chu, and Jean Zipkin, (John Wiley & Sons, New York, 1996)
32. C. Bemporad, G. Gratta, P. Vogel, Rev. Mod. Phys. **74**, 297-328 (2002)
33. John N. Bahcall, *Neutrino Astrophysics* (Cambridge University Press, Cambridge, 1989)
34. S. N. Ahmed *et al.*, Phys. Rev. Lett. **92**, 181301 (2004)
35. F. A. Nezrick and F. Reines, Phys. Rev. **142**, 852-870 (1966)
36. Z. D. Greenwood, W. R. Kropp, M. A. Mandelkern, S. Nakamura, E. L. Pasierb-Love, L. R. Price, F. Reines, S. P. Riley, H. W. Sobel, N. Baumann, H. S. Gurr, Phys. Rev. D **53**, 6054-6064 (1996)
37. M. Apollonio *et al.*, Phys. Lett. B **466**, 415-430 (1999)
38. M. Apollonio *et al.*, EPJ C **27**, 331-374 (2003)
39. V. I. Kopeikin, L. A. Mikaélyan, and V. V. Sinev, Phys. At. Nucl. **60**, 172-176 (1997)
40. F. Boehm *et al.*, Phys. Rev. Lett. **84**, 3764-3767 (2000)
41. F. Boehm *et al.*, Phys. Rev. D **64**, 112001-1-10 (2001)
42. K. Gabathuler, F. Boehm, F. v. Feilitzsch, J. L. Gimlett, H. Kwon, R. L. Mössbauer, J.-L. Vuilleumier, G. Zacek, V Zacek, Phys. Lett. B **135**, 449-453 (1984)
43. G. Zacek *et al.*, Phys. Rev. D **34**, 2621-2636 (1986)
44. J. F. Cavaignac, A. Hoummada, D. H. Koang, B. Vignon, Y. Declais, H. de Kerret, H. Pessard, J. M. Thenard, Phys. Lett. B **148**, 387-394 (1984)
45. Y. Declais *et al.*, Phys. Lett. B **338**, 383-389 (1994)
46. B. Ackhar *et al.*, Nucl. Phys. B **434**, 503-532 (1995)
47. B. Ackhar *et al.*, Phys. Lett. B **374**, 243-248 (1996)
48. P. Vogel, Phys. Rev. C **76**, 025504 (2007)
49. P. Vogel and J. Engel, Phys. Rev. D **39**, 3378-3383 (1989)
50. F. T. Avignone and Z. D. Greenwood, Phys. Rev. C **22**, 594-605 (1980)
51. O. Tengblad, K. Alekett, R. Von Dincklage, E. Lund, G. Nyman, G. Rudstam, Nucl. Phys. A **503**, 136-160 (1989)
52. A. A. Hahn, K. Schreckenbach, W. Gelletly, F. von Feilitzsch, G. Colvin, B. Krusche, Phys. Lett. B **218**, 365-368 (1989)
53. F. von Feilitzsch, A. A. Hahn, K. Schreckenbach, Phys. Lett. B **118**, 162-166 (1982)
54. V. Kopeikin, L. Mikaélyan, V. Sinev, arXiv:hep-ph/0308186 v1 (19 Aug 2003)
55. E. J. Konopinski and G. E. Uhlenbeck, Phys. Rev. **48**, 7-12 (1935)
56. E. Fermi, Z. Phys. **88**, 161-177 (1934)
57. R. H. Hulme, Proc. R. Soc. A **133**, 381-406 (1931)
58. Robley D. Evans, *The Atomic Nucleus* (McGraw-Hill, New York, 1955), p. 552
59. F. T. Avignone and Z. D. Greenwood, Phys. Rev. D **17**, 154-158 (1978)
60. H. A. Bethe and R. F. Bacher, Rev. Mod. Phys. **8**, 82-228 (1936)
61. I. Feister, Phys. Rev. **78**, 375-377 (1950)
62. NBS Applied Mathematics Division, *Tables for the Analysis of Beta Spectra* (U. S. Government Printing Office, Washington, D. C., 1952)
63. T. Goorley and H. Nikjoo, Radiat. Res. **154**, 556-563 (2000)
64. John Dirk Walecka, *Theoretical Nuclear and Subnuclear Physics* (Oxford University Press, New York, 1995), p.535
65. S. Ishimoto, T. Omori, H. Arima, K. Ishibashi, J. Nucl. Sci. Tech. **39**, 670-672 (2002)
66. P. Vogel, Phys. Rev. D **29**, 1918-1922 (1984)
67. S. A. Fayans, Sov. J. Nucl. Phys. **42**, 590-597 (1985)
68. V. Oberacker, W. Greiner, H. Kruse, W. T. Pinkston, Phys. Rev. C **20**, 1453-1466 (1979)
69. T. Sikkeland, Phys. Rev. **135**, B669-B677 (1964)
70. M. Dasgupta, D. J. Hinde, A. Diaz-Torres, B. Bouriquet, C. I. Low, G. J. Milburn, J. O. Newton, Phys. Rev. Lett. **99**, 192701-1-4 (2007)

End of paper

Appendix B: Studies of Spectra

I have been interested in comparisons of astronomical spectra because of my suspicion that there are objects in space that are radioactive and enclosed in a transparent medium, such as ice, gas, or plasma, and would be radiating the blue glow of Cerenkov radiation. Objects such as the icy moons of Jupiter, Saturn, and Neptune, and planetary nebulae might be candidates for such a condition, and that could suggest that many more objects might qualify.

Most astronomical spectra are modified from the measurements by elimination of the so-called "continuum", the smoothly varying baseline of radiation that provides the bulk of the energy that is radiated. This is done to normalize and emphasize the spectral line emission and absorption, which is of most interest to the observer. However, that eliminates the underlying shape of the spectrum, that might provide information as to the process of the radiation. I have gathered five full spectra that were not modified in that manner, and make comparisons for each in the plots below.

In Figure 156, I show the measured spectrum of the white dwarf companion, Sirius B, of our brightest star, Sirius (A). That is the black, somewhat jagged curve. The upper smooth curve is a black body (Planck Function) spectrum calculated for the temperature (25,193 K) determined by Barstow *et al.* [Barstow *et al.* 2005].

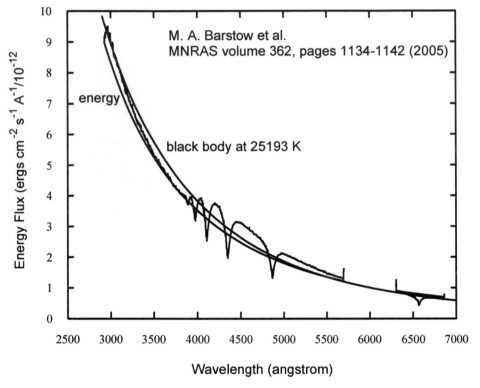

Figure 156. Spectrum of the white dwarf star Sirius B.
Optical spectrum of the white dwarf star Sirius B, compared to a black body spectrum at a temperature of 25,193 K, and a Cerenkov spectrum (by energy).

I have normalized the calculated spectrum to the observed spectrum at the long wavelength (6,800 Å) end of the spectrum. The lower smooth curve is a Cerenkov spectrum, with no adjustable parameters (what you get is what you see, for Cerenkov) and normalized to the observed spectrum at the short wavelength (2,900 Å) end of the spectrum. It is hard to judge (for me, at least) which of these curves fits the observed spectrum better, and the comparison obviously depends on the normalization.

I think this comparison clearly shows that Cerenkov radiation, emitted at any low temperature, can mimic a black body radiator at very high temperature.

In Figure 157, I show an unmodified spectrum reported for the supernova of 1987 in the Large Magellanic Cloud, SN 1987A [Branch 1990]. This spectrum was taken just two days (nights) after the supernova outburst. It is somewhat peculiar in that it differs greatly from the later spectra, as was shown in Figure 140.

Here we get into some further confusion. Disregarding the label on the graph (here and in the original) indicating that this is energy flux, I have plotted the Cerenkov spectrum as the photon "number" spectrum. That is the flux of photons counted by the spectrometer detector, a CCD (charge-coupled device). It is different from the energy spectrum previously shown in being lower at the short wavelength. I have normalized this number spectrum to the observed spectrum at the long wavelength (9,800 Å) end of the spectrum. The Cerenkov "number" spectrum matches the reported spectrum remarkably well.

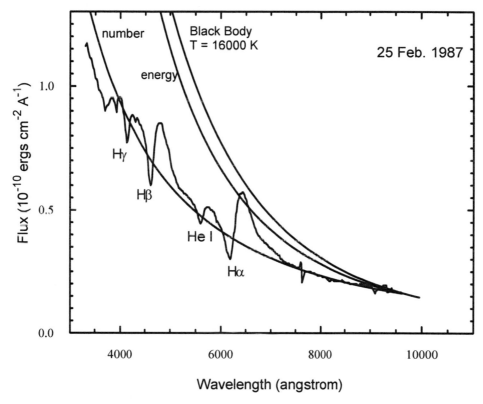

Figure 157. Critique of earliest spectrum of SN 1987A.
Comparison of reported visible light spectrum for SN 1987A within three days of the antineutrino pulses. While the reported spectrum had been labeled as an energy flux, it appears more likely that the plot was for a number flux, as would come from a CCD spectrograph, without the necessary energy modification to the data. Curves are shown for a black body at 16,000 K, and for the number flux of Cerenkov radiation. The Cerenkov number flux matches this version of the SN 1987A spectrum.

A black body spectrum is plotted for a temperature of 16,000 K (the typical temperature of a blue supergiant star, as the progenitor, SK-69°202, is considered to have been). It is clear that the black body spectrum does not match the observed spectrum.

This leaves me with a problem. The good match of the Cerenkov number spectrum (and the poor match of the black body spectrum) suggests that the observer incorrectly presented the spectrum as energy flux, while it was actually photon number flux. Since the initial data from the spectrometer are in photon counts at each wavelength position (number flux), that data must be converted to energy flux by the computer program. Could it be possible that this step was accidently omitted? That leaves me in the uncomfortable position of having to assume that an error was made.

I have converted the observed spectrum to energy flux, based on that assumption. That is shown in the next plot, Figure 158. The black body curve for 16,000 K matches fairly well, and the Cerenkov energy spectrum matches even better. I think that is good evidence that an error was made in processing the spectrometer data, and the published spectrum is a photon count (number) spectrum.

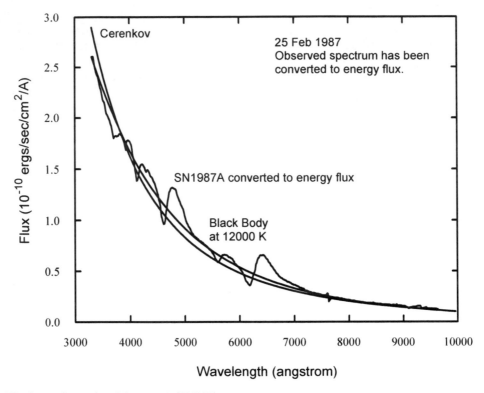

Figure 158. Comparisons of model spectra to SN 1987A.
A black body spectrum compared to the earliest spectrum of SN 1987A, and the Cerenkov energy spectrum. The Cerenkov spectrum matches the reported spectrum as well as does the black body spectrum with a temperature of 16,000 K.

The next plot (Figure 159), of WX Ceti, a repeating nova, shows the contrast between the Cerenkov energy spectrum, and the Cerenkov number spectrum. Howell *et al.* determined that two parts of the nova WX Ceti had distinctly different temperatures, 21,000 K and 72,500 K [Howell *et al.* 2002], and I have plotted black body spectra for those temperatures. Neither spectrum matches very well. All calculated spectra are normalized to the observed spectrum at the long wavelength (8,000 Å) end. The Cerenkov spectrum isn't a great match, drifting away at longer wavelengths, but it is clearly better than the two black body spectra.

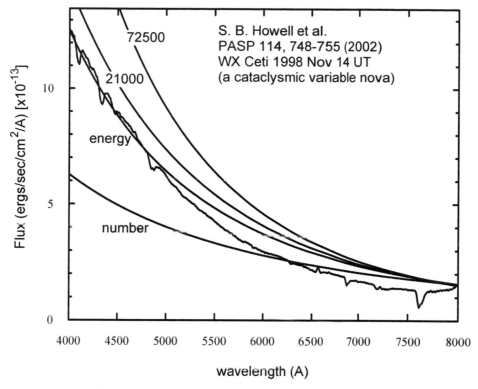

Figure 159. The spectrum of WX Ceti.
The observed spectrum of the nova WX Ceti, compared to the two black body spectra derived from the reported temperatures, 21,000 K and 72,500 K, and the Cerenkov energy spectrum.

The "Typical Blue Horizontal Branch Star Spectrum" was developed by Xue *et al.* (2008) from a collection of spectra for stars in the temperature range of 7,000 K to 10,000 K [Xue *et al.* 2008] and is shown in Figure 160. The black body spectrum for 10,000 K fits reasonably well, although it looks like a slightly hotter spectrum would do better. Perversely, the observed spectrum lies equidistant between the Cerenkov spectrum for number and the spectrum for energy, as if the reference spectrum had been produced from a collection with equal numbers of photon count spectra and photon energy spectra. The stars were selected from the Sloan Digital Sky Survey, subject to the availability of existing spectra. It may be that the combination of spectra produced the in-between result shown here.

(Note that the label for the flux in the original is in error, showing a cm^{-1} instead of the correct cm^{-2}. That label, with its error, is used here.)

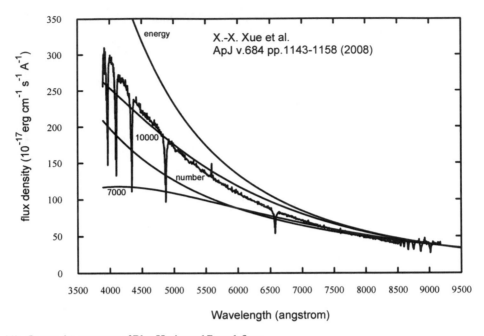

Figure 160. Composite spectrum of Blue Horizontal Branch Stars.
A typical spectrum for Blue Horizontal Branch Stars, derived from a collection of spectra. (The units in the flux density label are incorrect in the original.)

The next plot (Figure 161) shows the long wavelength spectrum of NGC 7027, a planetary nebula. (Note the extremely long wavelength scale, in angstroms like all the others.) Liu *et al.* determined the effective temperature to be 13,000 K [Liu. *et al.* 1996], but the black body spectrum for that temperature does not fit well. (It should be recognized that a planetary nebula compares to a black body like a fluorescent light compares to an incandescent light.) However, the Cerenkov (energy flux) spectrum matches almost exactly, except for drifting a bit above at the longer wavelengths. The model I would propose here is a radioactive central star radiating beta particles into the gas that makes up the planetary nebula, thereby generating Cerenkov radiation and exciting the line emission as well. Both calculated spectra were normalized to the observed spectrum at the long wavelength (1.9×10^6 Å) end of the spectrum.

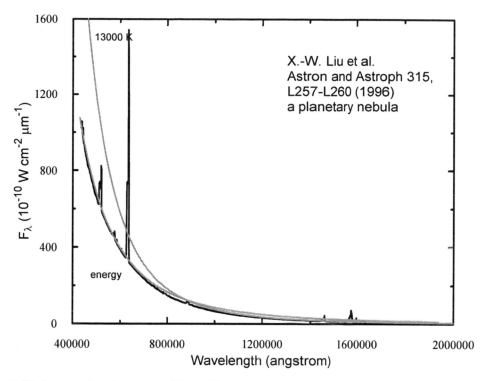

Figure 161. The long-wavelength spectrum of NGC 7027.
The spectrum of the planetary nebula NGC 7027, in the long wavelength range. The Cerenkov energy spectrum matches the observed spectrum well (the two nearly obscure each other), but the derived 13,000 K black body spectrum (the grey curve) does not.

I found an additional full supernova spectrum, for SN 2005cs, at
http://cfa-www.harvard.edu/cfa/oir/Research/supernova/spectra/sn2005cs-20050630.flm.gif.
That is shown in Figure 162, with the Cerenkov energy spectrum and a black body spectrum at a temperature of 14,000 K. The black body spectrum matches the observation generally better than the Cerenkov spectrum does, but I think this does not account for the line absorption, which would reduce the flux at the shorter wavelengths.

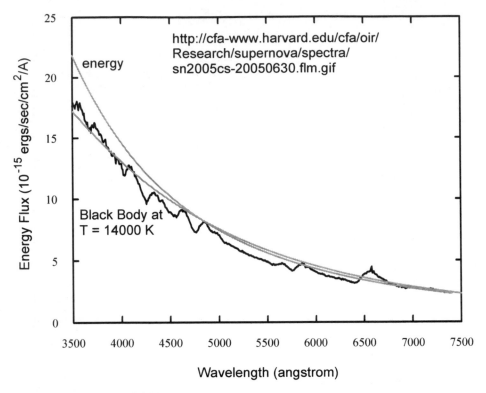

Figure 162. Spectrum of supernova SN 2005cs.
The observed spectrum shows good agreement with a black-body spectrum, as energy flux, and fair agreement with the Cerenkov (energy) spectrum, also as energy flux.

To further support the idea that the SN 1987A spectrum was incorrectly reported, and was actually a photon number spectrum, I have converted that spectrum to energy by dividing by the wavelength (and renormalizing) and comparing those spectra with the spectrum for SN 2005cs. That comparison is shown in Figure 163. I think the comparison is conclusive.

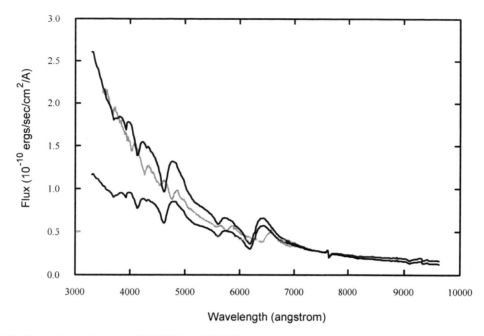

Figure 163. Comparison of spectra of SN 1987A and SN 2005cs.
This shows good agreement between the two spectra, with SN 1987A converted from number flux to energy flux. The original published spectrum of SN 1987A (lower curve) differs greatly from the energy flux spectra.

Errors, or at least unrecognized deviations, do occur in complicated software. Fairall found that a new software program for converting spectral data to velocity gave "totally different velocities" [Fairall 1992]. Similar problems might be lurking, unrecognized, in other programs.

So, I am left with some perplexions. Are the spectra sufficiently accurate over the energy ranges to permit distinguishing between alternative theoretical spectra? Can processing errors, such as photon counts instead of energy flux, occur? Can spectra be so blended as to confound what the underlying form might be?

Appendix C: Detailed Analysis of SN 1987A

The 1987 supernova in the Large Magellanic Cloud, SN 1987A, was the brightest, nearest supernova yet, studied from its beginning with the broadest arsenal of instruments. These observations have raised more unanswered questions. This supernova was not "right".

Analysis of Antineutrino Data from SN 1987A

SN 1987A was the first supernova to be observed by antineutrino detectors.

The interpretation of the various antineutrino observations, and the gravity wave measurements, is too involved to easily put into the text of this book. Here, I present a thorough discussion of the path of analysis for these results, to show that SN 1987A was not the explosion of a blue supergiant star a hundred thousand light years from Earth, but an asteroidal detonation by nuclear fission, at a distance placing the Large Magellanic Cloud just at the outer edge of the Solar System. This is not a thoroughly convincing story, given the data, but it is a consistent story. It is not intended to criticize the work of the supernova theoreticians. They are doing an excellent job at trying to explain something that isn't what they are trying to explain.

The neutrinos associated with electrons may be either "neutrinos" (normal matter) or "antineutrinos" (antimatter). While neutrinos and antineutrinos are created and released by the theoretically modeled standard stellar supernova in roughly equal numbers, the sensitivity of our detectors is such that they are most likely to detect the antineutrinos. It is generally accepted that all the counts were from antineutrinos. If SN 1987A was a nuclear fission detonation, or a pair of detonations, then all the neutrinos were antineutrinos.

Immediately before the first sighting of the brave new star that became SN 1987A, four neutrino observatories detected a flurry of antineutrino counts, enough in short periods of time to be noticeable. The detectors involved were

Kamiokande-II (a water Cerenkov detector in Japan), Baksan (a liquid scintillation detector in the Caucasus in Russia), IMB (a water Cerenkov detector in Ohio, USA), and Mont Blanc (a liquid scintillation detector under Mont Blanc at the border of France and Switzerland). The antineutrino detections occurred in two pulses clearly separated in time by nearly 5 hours.

Early information was reported at several conferences. A consensus developed, for various reasons, that the first pulse of detections was probably spurious, and should be ignored as it was at least "not associated with SN 1987A" [Bahcall 1989]. I have always been reluctant to throw anything away, and this is no exception.

I think it is not possible for an outsider (or even most insiders) to gather all the information that exists for a situation such as the unprecedented detection of antineutrinos from a star (or other distant astronomical object, other than routinely from our Sun). However, I may have a nearly complete chronology of SN 1987A's entry into our science. That is presented below, showing the time relative to midnight beginning February 23, 1987, the supernova's birthday. The time is relative to Greenwich or the Prime Meridian, variously labeled as GMT, UT, and UTC. Some minor difficulties crept into the data with regard to times.

Chronology of SN 1987A

Time relative to 2/23.00 hour	location	(seconds)	character
1.797	LSD-Mont Blanc	6468.800	antineutrino count
2.178	LSD-Mont Blanc	7840.100	antineutrino count
2.178	LSD-Mont Blanc	7840.310	antineutrino count
2.181	LSD-Mont Blanc	7851.410	antineutrino count
2.184	Shelton	7862.400	nothing notable
2.194	LSD-Mont Blanc	7897.040	antineutrino count
2.285	LSD-Mont Blanc	8225.050	antineutrino count
2.640	LSD-Mont Blanc	9504.890	antineutrino count
2.733	LSD-Mont Blanc	9838.500	antineutrino count
2.750	Rome & MD	9899.971	mass of 2400 Suns
2.761	LSD-Mont Blanc	9938.840	antineutrino count
2.872	Baksan	10337.826	antineutrino count
2.873	Kamiokande-II	10342.000	unreported antineutrino count
2.876	Baksan	10353.913	antineutrino count
2.877	LSD-Mont Blanc	10356.790	antineutrino count
2.877	LSD-Mont Blanc	10356.792	antineutrino count
2.878	LSD-Mont Blanc	10360.649	antineutrino count
2.878	LSD-Mont Blanc	10361.007	antineutrino count
2.879	LSD-Mont Blanc	10362.696	antineutrino count
2.879	LSD-Mont Blanc	10363.800	antineutrino count
2.886	Baksan	10391.087	antineutrino count
3.052	LSD-Mont Blanc	10987.300	antineutrino count
3.078	LSD-Mont Blanc	11082.080	antineutrino count
3.412	LSD-Mont Blanc	12284.040	antineutrino count
7.593	Kamiokande-II	27335.000	antineutrino count
7.593	Kamiokande-II	27335.107	antineutrino count
7.593	Kamiokande-II	27335.303	antineutrino count
7.593	Kamiokande-II	27335.324	antineutrino count
7.593	Kamiokande-II	27335.507	antineutrino count
7.593	Kamiokande-II	27335.686	antineutrino count
7.593	Kamiokande-II	27336.541	antineutrino count
7.594	Kamiokande-II	27336.728	antineutrino count
7.594	Kamiokande-II	27336.915	antineutrino count
7.595	IMB	27341.370	antineutrino count

7.595	IMB	27341.790	antineutrino count
7.595	IMB	27342.020	antineutrino count
7.595	IMB	27342.520	antineutrino count
7.595	IMB	27342.940	antineutrino count
7.596	IMB	27344.060	antineutrino count
7.596	Kamiokande-II	27344.219	antineutrino count
7.596	Kamiokande-II	27345.433	antineutrino count
7.596	IMB	27346.380	antineutrino count
7.596	IMB	27346.960	antineutrino count
7.597	Kamiokande-II	27347.439	antineutrino count
7.600	LSD-Mont Blanc	27360.000	antineutrino count
7.603	Baksan	27371.000	antineutrino count
7.603	Baksan	27371.450	antineutrino count
7.604	Baksan	27372.730	antineutrino count
7.605	Baksan	27378.750	antineutrino count
7.605	LSD-Mont Blanc	27379.589	antineutrino count
7.606	Baksan	27380.120	antineutrino count
10.632	McNaught	38275.200	6.0 magnitudes
10.632	McNaught	38275.200	6.2 magnitudes
28.800	Duhalde	103680.000	supernova
29.520	Shelton	106272.000	5 magnitudes
32.880	Jones	118368.000	6.8 magnitudes

It should be noted that this listing includes "at least one" count from Kamiokande-II, with unspecified energy, in the early pulse [Simone 1998, citing de Rújula 1987 and Schramm 1988].

The detected antineutrino pulses can be clearly displayed in a plot of energy and time. The early pulse, the rejected one, is shown in Figure 164, and the later pulse, the only counts eventually accepted, in Figure 165. Because of the large separation in time compared to the duration of the pulse, this must be done in two separate plots, but I have arranged them to have the same energy scale. Note that the first pulse is quite extended in time, occupying nearly an hour and a half, while the second pulse is almost instantaneous, existing for less than a minute.

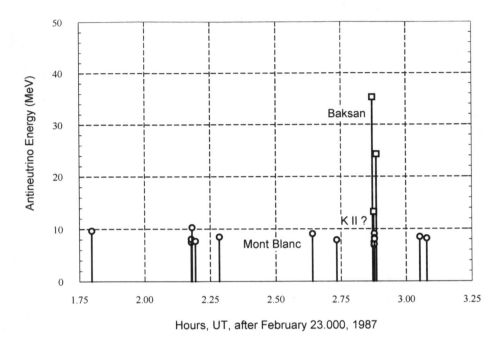

Figure 164. First pulse of antineutrinos from SN 1987A.
Antineutrinos were detected by the Mont Blanc and Baksan detectors, and possibly by Kamiokande II.

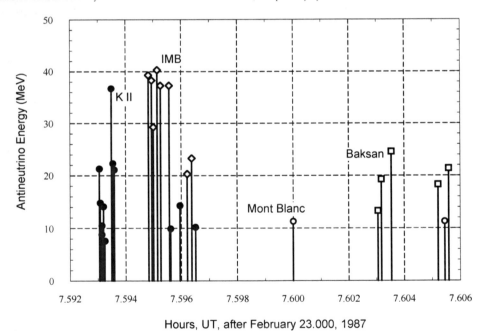

Figure 165. Second pulse of antineutrinos from SN 1987A.
Antineutrinos were detected by the Mont Blanc, Baksan, IMB, and Kamiokande II detectors.

The counts are clearly distributed over a broad energy range, with little indication of what that distribution might be. The theoretical distribution for a standard supernova model is one labeled as a Fermi-Dirac distribution, with a characteristic energy of 4.2 MeV [Bahcall, Spergel, and Press 1987, Mayle, Wilson, and Schramm 1987, Myra, Lattimer, and Yahil 1987]. That is based on analysis of the Kamiokande-II and IMB counts in the later pulse only. The Fermi-Dirac distribution is similar to the Maxwell-Boltzmann distribution, which represents the energies of particles in full thermal equilibrium, such as the molecules in the air around us. This distribution might be proposed for the antineutrinos from a fully confined nuclear fission detonation, with the antineutrinos being thermalized by multiple scattering in an extremely hot detonation. A third possibility, considering the termination of a nuclear fission detonation by the disruption of the surrounding body, is the distribution of antineutrinos from nuclear fission.

By use of a technique called cumulative probability plotting [McGinnis 2009] it is possible to analyze the energy distribution for each of the observatories separately. By taking advantage of the similarity of all these energy distributions (Fermi-Dirac, Maxwell-Boltzmann, and a fission distribution) to a log-normal probability distribution, it is possible to derive a representative shape for each distribution. The results of this analysis are shown in Figure 166, along with the theoretical Fermi-Dirac distribution as dashed lines.

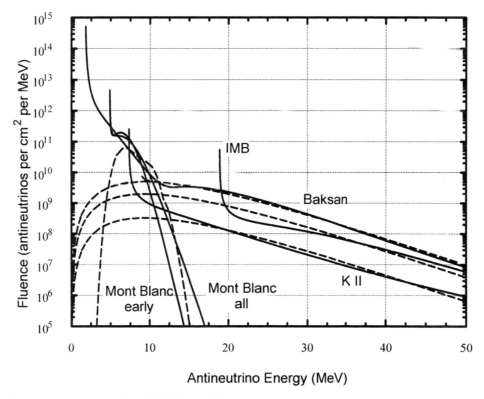

Figure 166. Antineutrino fluences from SN 1987A for different detectors.
The antineutrino fluences (time-integrated fluxes) for four detectors, showing the differences in energy distributions. Fermi-Dirac spectra are shown as dashed curves for the IMB, Baksan, and Kamiokande II detection spectra, while a fission spectrum is shown for Mont Blanc.

It is clear that the Kamiokande-II, Baksan, and IMB shapes match the Fermi-Dirac distribution well, as should be the case, since the effective temperature (kT=4.2 MeV) was derived from the observations. It is equally clear that the Mont Blanc shapes do not fit the Fermi-Dirac distribution.

Some excuses and explanations are called for. It can be seen that each curve has an unreasonably steep rise at lower energies. This is the result of dividing a slowly varying response function by the detection efficiency for each detector. At low energies, the efficiency factors (shown in Figure 167) are zero, and then climb steeply to a nearly constant value. The steep rise in efficiency causes the peculiar upturn in the curves.

Further, if the full set of Mont Blanc counts is used (in Figure 166), the representative shape lies generally above the fission spectrum, while limiting the Mont Blanc analysis to the early pulse only, produces good agreement. Similarly, because of the small number of detections at Baksan, the combined set gives a shape which has an awkward hump, apparently corresponding to a small component of the fission spectrum, represented in the counts in the early pulse. The derived shapes are distinctly different and are consistent with the proposed underlying spectra.

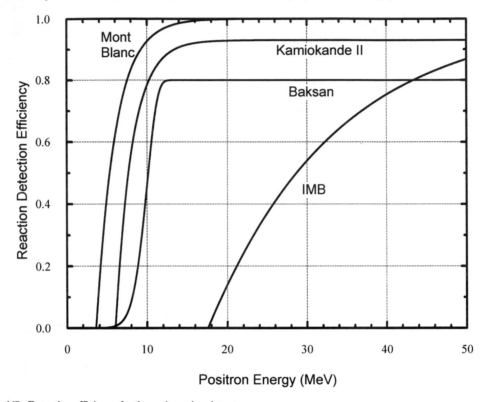

Figure 167. Detection efficiency for the antineutrino detectors.
Detection efficiency for the antineutrino detectors. Below a certain threshold energy, different for each system, the detection efficiency is zero.

The fluence (time-integrated flux) at each detector can be calculated and compared with model spectra that assume the incoming antineutrino spectra (before passing through the Earth to impinge on the various detectors) are a Maxwell-Boltzmann(thermal) distribution with kT = 5.3 MeV for Kamiokande-II, Baksan, and IMB, and a fission spectrum for Mont Blanc, representing the early pulse. The results of those calculations are shown in Figure 168, and it can be seen that the attenuated Maxwell-Boltzmann spectra match the observations just as well as the unmodified Fermi-Dirac distribution does. The attenuated fission spectrum is a somewhat poorer match to the Mont Blanc data, which show more lower energy antineutrinos. That may reflect an error in the values used for the attenuation calculation.

The count sequence shown in Figures 164 and 165, and the spectral distribution derived for the unattenuated fluence (incident on the farside of the Earth), Figure 168, lead to the following description of the events and conditions of the supernova. It must be recognized that this is based on the proposition that our present distance scale is greatly in error, and that our concept that (almost) all the bright little points of light in the night sky are stars of the same sort as our Sun is grossly misdirected. This will be borne out in the next section, discussing the gravitational waves from SN 1987A.

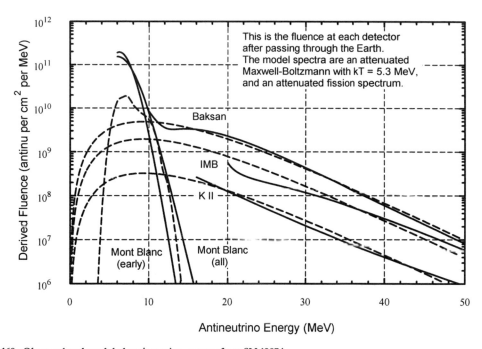

Figure 168. Observed and modeled antineutrino spectra from SN 1987A.
Antineutrino fluences adjusted for attenuation in passing through the Earth. The observed spectra are shown as solid lines, the model spectra are dashed lines.

Consider an asteroid and its satellite, such as Ida and Dactyl (combined mass = 4.2 x 10^{16} kilograms) or Pluto and Charon (combined mass = 1.5 x 10^{22} kilograms). Imagine that the main body is radioactive, as a result of ongoing nuclear fission (at a very low rate, so as to not significantly heat it) or from activation resulting from a nuclear fission detonation that produced the Large (and Small) Magellanic Cloud, within which these two fragments reside. If that larger body is coated with ice, of whatever sort, Cerenkov radiation will be produced, with a spectrum that is indistinguishable from the spectrum measured immediately (February 25, 1987) after the detonation of SN 1987A [Branch 1990], indistinguishable from the spectra measured for blue supergiant stars. A small, nearby, cold, icy object can appear to be, in our current astronomy, a huge, distant, tremendously hot ball of gas, due to its self-emitted light and our errors in distance and character.

Because we believe the stars to be brilliant objects like our Sun, they must be distant to be so dim, and since they are so distant they must be brilliant to be like our Sun.

Consider that the asteroidal fragment that we knew as the blue supergiant star Sk-69°202 contained embedded natural nuclear fission reactors, as would be likely from its origin as part of a planetary satellite that suffered a nuclear fission detonation, millions of years ago. The fragment's fragmentary companion (perhaps 10^{-6} to 10% as large, as we see for Ida and Dactyl, and Pluto and Charon) might have such reactors as well, since both bodies might have been produced from material near the central core of the detonated satellite.

If some of the reactors in the smaller companion were critical, or nearly so, a simple disturbance, such as a high velocity impact by another fragment, could, by briefly compressing the reactors by the impact shock wave, increase the reactivity of the reactor (the multiplication constant or criticality constant k), and cause the power level to increase above its prior equilibrium level. (The reactivity of a nuclear reactor is proportional to the square of the density, so an increase of 0.5% in density would cause an increase in reactivity of 1%, which is a very significant increase, pushing a critical reactor into the "prompt critical" regime, when a detonation can occur.) An alternative mechanism would be the gradual production of an accumulation of uranium-235 and plutonium-239 from the radioactive decay of curium-247 (half-life = 15.6 million years) that was produced by rapid neutron capture by uranium-238 during the disruptive detonation of the planetary satellite. (It is interesting that in the extreme, this produces an accumulation of nearly pure uranium-235 and plutonium-239 in a mixture that is used in modern nuclear fission weapons and triggers of thermonuclear bombs (H-bombs).)

While the initial surge of increased power might be easily tempered, it would transmute some of the associated uranium-238 to plutonium-239, with a time scale of roughly 2½ days. The loss of uranium-238 and the gain of plutonium-239 has a significant effect on the reactivity of the reactor, amounting to a positive effect more than twice as great as the negative effect due to the consumption of some uranium-235. By whichever initiating mechanism, the reactor is set on what I will call the path of "choo-choo chug-chug huff'n'puff BOOM!". More appropriate to a children's story, I admit, but a very clear description of the process. Each succeeding power excursion may be shut down by the inherent character of the reactor, whether it is thermal expansion of the material and its surroundings, or by a negative Doppler effect on reactivity in a predominantly uranium-235 reactor. But each succeeding surge will transmute more uranium-238 to plutonium-239 and combined with the plutonium-239 resulting from the long-ago promised decay of curium-247, each next surge will be increasingly more powerful. Eventually, as the power of the reactor and its temperature increase more rapidly than negative feedback can compensate, a nuclear detonation will result. A significant fraction of the nuclear fuel will have been consumed, sending out pulses of fission antineutrinos in the process, and at the endpoint, converting so much mass to energy as to send gravitational waves rippling out through Einstein's space. (Gravitational waves, radiation from abrupt changes in mass, are not to be confused with gravity waves, which are made by clouds, the surface of the sea, and football fans in a stadium.)

But for SN 1987A, the show is not yet over. If the main fragment and its companion are close, neutrons escaping from the detonation may penetrate to reactors inside the main fragment, boosting its reactors. Antineutrinos from the detonation will release neutrons throughout the fragment by the induced electron capture reaction, with neutron emission. A nuclear detonation will be initiated in the larger fragment, which we may suppose from the evidence is sufficiently large, massive, and rigid, incompressible deep inside, so that no expansion will temper the initial increase in reactor power, and the detonation will be fully confined. A detonation will be fully confined by the inertia of its surrounding material if that material is incompressible, and if the time taken by the shock wave, at a velocity of about 10 kilometers per second, to reach the surface, be reflected, and return to the detonating reactor, is longer than the time scale of the detonation, perhaps a few millionths of a second. That will be the case in a remarkably small body, provided sufficient pressure exists and the material is incompressible. Pressure takes kilometers of depth, and the surrounding material, no matter how solid to begin with, will quickly be vaporized and become a compressible gas. However, sufficient confinement can be provided by an asteroid-sized fragment to significantly affect the progress of the fission detonation.

Whatever happens after that time, the fission detonation has already saturated its region with energy, achieving tremendously high temperatures. A nuclear weapon produces a central temperature of 10 billion degrees kelvin [Serber 1992]. (That corresponds to a particle energy (kT) of 0.86 MeV.) In a confined detonation, the temperature can rise much higher as the fission fuel is more completely fissioned. As the temperature increases, the fission neutrons will be up-scattered and "thermalized" to high energies, increasing the reactivity (higher energy neutrons produce more neutrons from fission), and increasing their effectiveness in causing fission in almost any nuclide they may collide with. Some of this energy is shared with the antineutrinos by electron scattering, boosting their average energy to the kT = 5.3 MeV as seen in the analysis of the SN 1987A antineutrino spectra. The relatively low density of material in the fission detonation, compared to that in a supergiant star, results in a high-temperature Maxwell-Boltzmann distribution, rather than in the Fermi-Dirac distribution postulated for a stellar supernova. As shown above, attenuation through the Earth's crust changes the Maxwell-Boltzmann distribution into one resembling the Fermi-Dirac distribution. Multistage fission will occur, with fission products fissioned, and nuclear statistical equilibrium will be approached, so as to mimic our standard nickel-56/cobalt-56 fueled model supernova.

Since astronomical brightness is measured in magnitudes, and those are logarithmic, in a sort of backwards sort of way, the linear decline in brightness often seen after the initial burst was naturally suspected to be the result of the exponential decline associated with short-half-life radionuclides. An early suggestion [Burbidge *et al.* 1956] was californium-254 (half-life 60.5 days), formed by neutron capture in the r-process occurring in the final stage of the supernova collapse. Of considerable significance in this choice was the great amount of energy released in the spontaneous fission of this nuclide. This was later rejected, in favor of nickel-56/cobalt-56 (half life 70.3 days), because it was thought that the r-process would have to produce far more of the intermediate mass nuclides, on the way up from iron, than are seen in supernovae. However, in the fission detonation, californium-254 is easily produced by 16 sequential neutron captures on uranium-238, and the intermediate mass nuclides that are only moderately seen would result in multistage fission from high energy fission fragments colliding with high energy neutrons (or even with each other) and splitting into smaller fragments.

The outer coat of ice is lost to space, Cerenkov radiation ceases, and the blue supergiant disappears from our sight, even though a slightly disrupted asteroidal fragment may remain. We are left with a puzzle of observations which defy clear explanation in the current astronomical models, there are pieces we don't understand, and we can only reject some of the more peculiar data.

Mass Estimates for SN 1987A

Fission Fuel

A very significant result of this analysis is the estimate of the quantities of uranium required for such a detonation as SN 1987A. The mass that was converted to energy, from the integrated light curve at a revised distance of 60 Astronomical Units, is 3.1×10^8 kilograms. Assuming 200 MeV per fission, that required the fissioning of 3.4×10^{14} kilograms of uranium. If the detonation was 10% efficient in fissioning its fuel material, as indicated by the Maxwell-Boltzmann distribution at $kT = 5.3$ MeV underlying the Fermi-Dirac distribution (after attenuation through the Earth's crust) at $kT = 4.2$ MeV, the fuel mass was 3.4×10^{15} kilograms of uranium. In a body of the size of our largest asteroid Ceres, with a mass of 1.2×10^{21} kilograms, that would have an average concentration of 2.8 parts per million. In a smaller body, the concentration would be proportionately larger. The average concentration of uranium in the Earth's crust, the only part of our planet we can truly sample, is 4 parts per million [Weast 1980]. The point of that comparison is not so much the remarkable agreement between a proposed body that detonated 60 Astronomical Units away to make SN 1987A and what we know with reasonable certainty here on Earth, but that this comparison does not need 4 million parts per million, and that the crust is not limited to something so small as 4 parts per trillion. We are truly in the ball park. (Standing on the same base, so to speak.)

Temperature Increase

Astronomers consider that roughly 99.9% of the energy released in a supernova is emitted in the form of neutrinos and antineutrinos, scarcely detectable by our largest devices, vanishing into space. The heat energy released in the detonation is converted to kinetic energy of the material and of the atoms, raising the temperature. (In the stellar supernova, the material starts out at an extremely high temperature. For a small object glowing with the blue light of Cerenkov radiation, the temperature may be that of open space, close to the 3 K temperature of the cosmic microwave radiation.)

In a nuclear fission detonation, only about 5% of the energy released is dissipated by antineutrinos, the rest becomes energy that can be observed and measured. In an extensive and detailed series of measurements [Suntzeff and Bouchet 1990, Bouchet *et al.* 1991, Bouchet, Danziger, and Lucy 1991], the astronomers have carefully recorded the photon energy radiated by SN 1987A. These measurements are shown in Figure 169, converted to conventional energy units (watts of power) as if the supernova were at a distance of 163,500 light years from Earth. (Here, the term "bolometric" means that measurements across the spectrum were combined and adjusted to provide a true measure of the energy radiated by the supernova.)

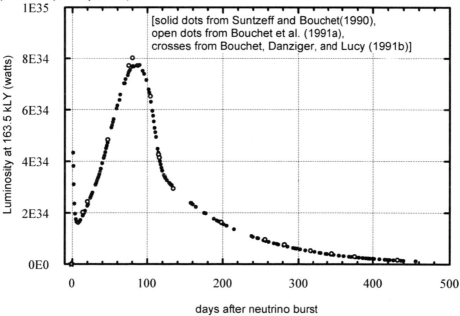

Figure 169. Bolometric light curve of SN 1987A.
Measurements of the luminosity of SN 1987A. A unique feature is the early decline from an apparent initial spike of energy.

The integral of that light curve is 8.23 x 10⁴¹ joules (equal to watt-seconds; a hairdryer operated for 20 minutes releases roughly a million, 10⁶, watt-seconds). At my assumed distance of 60 Astronomical Units, the energy released is equivalent to 2.77 x 10²⁵ joules. That energy is available to heat and melt the main object. In this story, the main detonating object remained intact, so that kinetic energy of any fragments can be ignored.

Measurements in 14 spectral bands at day 202, ranging from the near ultraviolet (U) to the far infrared (Q0) have been analyzed into two distinct temperature distributions [Danziger *et al.* 1987]. This showed a bright high temperature component, at 4,800 K, and a dimmer low temperature component, at 675 K. That component cooled rapidly, as shown in Figure 170. I interpret these results to show the large, undisrupted body at 4,800 K, and the smaller, disrupted body at 675 K, rapidly cooling to the local environment, probably below 100 K.

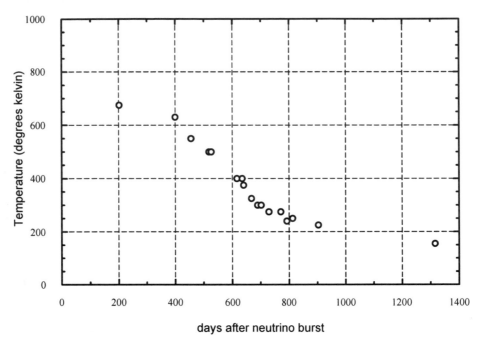

Figure 170. Temperature of the cooler object in SN 1987A.
Cooling of the cooler object detected in the SN 1987A.

Assuming an average heat capacity for "rock" of 969 joules per kilogram per degree K, and ignoring heat going into melting the material, that energy, the energy calculated for 60 Astronomical Units, could heat an object consisting of 5.96 x 10¹⁸ kilograms of rock, or about 0.006 the mass of Ceres, our largest asteroid. Of course this changes the estimate of the concentration of uranium presented above, from 3.4 parts per million for a body the size of Ceres to 567 parts per million, or approximately 0.06%. The uranium concentration in the ores of Oklo ranged from 0.1% (1,000 ppm) to 64%.

Gravitational Waves from SN 1987A

General Relativity predicts that local changes in mass, such as a conversion of a fraction of a star's mass to energy in a supernova explosion, should send out "ripples" of gravitational waves into the Universe as space adjusts to a new gravitational curvature. That also should happen in a nuclear fission detonation, as the fission process releases great amounts of energy by conversion of mass. A small nuclear (fission) weapon converts approximately 24% of its plutonium/uranium mass into energy. Fully confined detonations in asteroids and planets would convert a higher fraction. Attempts at detecting gravitational waves have been a long on-going (and controversial) program of significant proportions [Collins 2004, Blair and McNamara 1997].

Usually, new scientific information is first delivered at meetings, more or less informally. It may then be included in published proceedings, which are more or less accessible. If the information is generally accepted and supported by the scientific community (colleagues, journal editors, referees) it may be published formally and become readily retrievable.

Unfortunately, that process often makes unaccepted frontier observations difficult or impossible to find. (It's very hard to find something if you don't know to look for it.)

That has been the case with the gravitational wave measurements for SN 1987A.

While light, neutrons, and antineutrinos may be trapped and retarded in the bulk of the supernova, I think that gravitational waves should be almost immediately observable, at the speed of light. The conversion of matter to energy should certainly produce gravitational waves, if General Relativity is correct. However, the interpretation of the detection of gravitational waves from SN 1987A in terms of mass is fraught with difficulties. Several layers of assumptions go into the interpretation, but all the mass losses are much greater than that calculated above and below by other means, which requires a great violation of astronomical knowledge in putting the Large Magellanic Cloud at a distance of 60 Astronomical Units, as part of our Solar System.

Gravitational wave antennas were in operation at the University of Maryland and the Gran Sasso National Laboratory in Italy during the detonation of SN 1987A. These antennas showed significant signals, "temperature innovations" in the Italian terminology, that coincided with each other and with the antineutrino counts [Amaldi *et al.* 1987a, Amaldi *et al.* 1988a, Amaldi *et al.* 1987b]. Interpretations as to the meaning in terms of a change in gravity or a change in mass somewhere in the Universe were rather obscure. A loss of mass was associated with the detonation of SN 1987A, and that was at a supposed distance of 50-52 kiloparsecs (about 160,000 light years). From the energy of the gravitational wave signals, this mass loss, for the early pulse, was reported as 100,000 solar masses [Guido Pizzella quoted by David Blair and Geoff McNamara [Blair and McNamara 1997]], or 20,000 solar masses (Giuliano Preparata) or 1/3 solar mass (Joe Weber) [Collins 2004] or 2,400 solar masses [Amaldi *et al.* 1987a]. I think the reader can see some of the difficulty in dealing with numbers such as these. I have chosen to use the last value, 2,400 solar masses, primarily because I read it first, but also because it was presented in a formal proceedings and it lies somewhere midway between the others.

The correlation of gravitational wave antenna signals with the Mont Blanc and Kamiokande-II antineutrino signals was reported at the 14th Texas Symposium on Relativistic Astrophysics [Amaldi *et al.* 1988]. My analysis of those results gives a value of 386 K for the temperature innovations accumulated during that time, the time of the second pulse of antineutrino signals. This compares with the reported value of 106 K for the first pulse, equaling the loss of 2,400 solar masses [Amaldi *et al.* 1987a]. Therefore, I calculate that, in the same sense as the 2,400 solar masses, the temperature innovation of 386 K is equivalent to a loss of 8,740 solar masses.

From the published light curves [Suntzeff 1990, Bouchet *et al.* 1991a, Bouchet, Danziger, and Lucy 1991], the visible energy released by the supernova was 8.2×10^{41} joules (watt-seconds) at a distance of 163,500 light years. Scaling this energy to a distance of 60 Astronomical Units, as an approximate distance for a detonated satellite of Planet X, it becomes 2.8×10^{25} joules. Assuming an energy release of 200 MeV per fission, a fission detonation would have required the fissioning of 3.4×10^{14} kilograms of uranium, with a conversion to energy of 3.1×10^8 kilograms of matter. The mass of uranium represents approximately 0.3 parts per million of the mass of the asteroid Ceres, but would be about 5% of the asteroid Eros. The uranium concentrations reported for the Oklo ores ranged from 0.056% to 65%. The richest values are about ten times what would be required for a fragment the size of Eros. Scaling the estimated detected mass loss of 2,400 Suns at 163,500 light years to a smaller mass loss at 60 Astronomical Units gives 1.6×10^{17} kilograms. This is far greater than the mass loss estimated from the energy release, 3.1×10^8 kilograms.

There is clearly something going on with the gravitational wave detectors that I (and everyone else) do not understand. Either gravitational waves are stronger (per unit mass converted to energy), or the detectors are far more sensitive than the grandest hopes, or gravitational waves are focused and concentrated by local disturbances in space, such as the Earth and the Sun. (That possibility might be considered to be more reasonable in a Universe with a distance scale that puts the Milky Way Center (*née* Planet X) at a distance of only 60 Astronomical Units, on the outskirts of the Solar System. This has been a trip through the gravitational battleground, and trying to make sense of it in a real world has not been possible. All of this depends on uncertain calibrations and ingenious calculations. Who knows how strong gravitational waves should be, or how sensitive the detectors are? I think no-one knows. A violent controversy surrounded the gravitational wave research, almost from the beginning. Measurements were being done without knowing what was important in the detectors or the data processing, or even with any certainty, what was being seen. Different experimenters got different results. At the root of the controversy, most likely, was the difficulty in understanding the equipment, the presumed distant sources of gravitational waves, and how best to extract a signal from the noise. Assumptions were made that once a signal was seen, it should be seen in the same way, over and over. That is obviously not true for supernovae, in any radiation, at any point in the spectrum. These come and go, rarely in the same place again. Detonations converting nuclear mass into energy by fission may occur at various locations in space, making understanding the signal more difficult, especially if there is a firm assumption as to the mechanism and neighborhood of these pulses. Gravitational waves need not be produced only by a black hole devouring a neutron star in a crowded part of the Milky Way Galaxy.

The pioneer of gravitational wave research, Joe Weber [Collins 2004], saw signals above the noise, that few others saw. By 1974 he had become marginalized, rejected and ignored, largely because he saw hundreds of signals per year, best described as gravitational waves, perhaps showing conversion of mass to energy. The observation of gamma ray bursts had been announced in 1973 [Klebesadel 1973], and with improved satellite detectors, hundreds of gamma ray bursts were being detected each year, and were labeled as the most energetic events in the Universe. Presumably these bursts were the result of conversion of huge amounts of mass to energy, with the associated emission of gravitational waves, hundreds of times a year. Joe Weber found some correlations, but died on September 30, 2000, and there is probably no-one who can now come to his defense.

Is my story one of certain reality? Certainly not, but I think it is one of possibility.

The Companion

A month after the explosion, a bright source was discovered near the supernova, using speckle interferometry [Papaliolios *et al.* 1987]. It was found again during the following weeks, but then faded from sight. In this present view of the supernova being the detonation of an asteroidal fragment, detonation of this companion was the initiating event, producing the first set of antineutrinos previously discussed, being the source of the initial burst of light, which quickly declined below detection, and inducing the detonation of the primary object, Sk-69°202. The light curve shows an early declining brightness that can be separated into a component that represents light from the companion [Nisenson *et al.* 1987, Meikle, Matcher, and Morgan 1987].

Here, it is proposed that the first burst of light came from the nuclear fission detonation of a satellite of Sk-69°202, as an asteroidal fragment. That detonation of the satellite induced the detonation of its primary (Sk-69°202) that became the supernova. The detonation of the satellite may have been induced by the much earlier detonation of what is now seen as the bright spot on the ring, considered to be an accidental star, just coincidently exactly placed on the ring. This sequence is clear when the light echoes are interpreted as spherical shells of excitation from fission neutrons escaping from the primary magnetic field before decaying to protons, like the mysterious "Domes of Light" over Scandinavia in 1989-1990 [Reed 2008], and the U. S. high altitude tests of nuclear weapons in 1958 and 1962 [Glasstone and Dolan 1977]. The outermost halo is off-center, and that one should be from the ringstar.

The detonated companion would shine in its own light, powered by the decay of radioactive fission products. The American Nuclear Society developed a safety standard for calculating the heat released by fission products in a reactor, to provide design guidance to mitigate any possible accidents [Subcommittee ANS-5 1973]. The decay heat from an instantaneous pulse of fission of uranium-235, as calculated from that standard, is shown in comparison to the light curve of SN 1987A (Figure 171). The zero time of the calculation has been shifted to 1.2 days before the antineutrino pulses, and the curve was normalized to the third brightness observation. A good fit was obtained, not only to the initial burst of light, but to the later speckle measurements of the companion, and to the residual light from the supernova, which has not been explained by the conventional radioactivity of nickel-56.

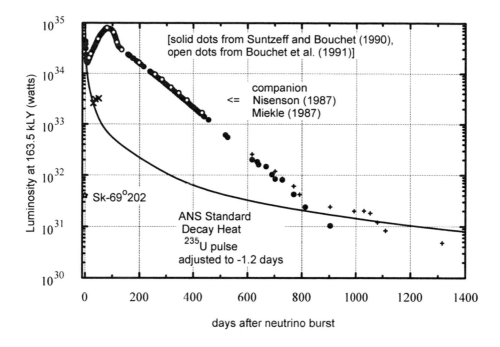

Figure 171. Comparison of fission decay heat with SN 1987A light curve.
Fission decay heat calculated for the nuclear detonation of SN 1987A, compared to the observed energy release light curve.

The goodness of the early fit is shown in Figure 172. (The zero time of the ANS decay heat curve was offset by 1.2 days to improve the fit. This may be justified as accounting for fission products accumulated during the approach to the detonation.)

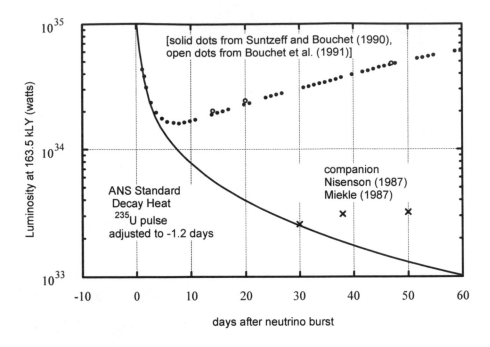

Figure 172. Early portion of SN 1987A light curve.
Early portion of SN 1987A light curve, showing fit of ANS Standard to the initial energy release.

Using the observed match of the decay heat curve to the supernova light curve, at three different "epochs" gives a total number of fissions in the companion alone as 1×10^{36}. This is equal to a mass of 4×10^{12} kilograms, or about 555 parts per million of the mass of the asteroid Eros [Yeomans *et al.* 1999]. The gravitational wave observation of the Maryland-Rome detectors was equal to 1.5×10^{17} kilograms, far more than this analysis produces.

Did Joe Weber actually measure gravitational waves? That question is so deeply mired in scientific socio-political troubles [Collins 2004] that we shall probably never know the answer. But, at one moment in time, it looks like he came close.

Further Discussions of Supernova 1987A in the Large Magellanic Cloud

I have collected all of the positional measurements I can find, as of December 2008, for objects around SN 1987A, composing what I call the SN 1987A system. It is surprisingly complex, resulting from excellent and intensive observations, made possible only because of its relatively close location. The system is shown in Figure 173. The system consists of the Supernova itself, the dot in the center of the sketch, with a "Companion", shown as the smaller adjacent dot. Surrounding these is the Circumstellar Ring, with a "superposed star" which I will call the Ringstar. Its position in 1983 is my interpretation of a suggestion by S. R. Heap and D. J. Lindler [Heap and Lindler 1987] that Sk-69°202 was a double star with a north-south separation of 0.68 arcsec (my estimate from their Figure 1b). Plait has insisted that the superposed star has no relation to the circumstellar ring [Plait *et al.* 1995] but that it just happens to be on our line of sight to the ring. Farther out are three other stars, 2, 3, and 4. (Sk-69°202, the progenitor of the supernova, has been labeled Star 1 from the early days of observations.) Later discussions here will present more of the observed characteristics of these stars.

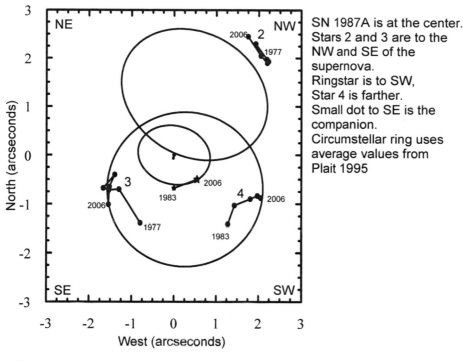

SN 1987A is at the center. Stars 2 and 3 are to the NW and SE of the supernova. Ringstar is to SW, Star 4 is farther. Small dot to SE is the companion. Circumstellar ring uses average values from Plait 1995

Figure 173. Schematic diagram of the SN 1987A system.
The SN 1987A system, composed of the Supernova, its Companion, the Circumstellar Ring and its Ringstar, the Northern Outer Ring, the Southern Outer Ring, and Stars 2, 3, and 4. Observations show that the Southern Outer Ring is significantly distorted and bent, while the other two rings are good ellipses. The geometry in this sketch was unable to show that distortion and bend, probably resulting from gravitational perturbation by Star 3 and control of the rest of the ring by a general, and uniform, magnetic field. In this figure, the observational uncertainties are generally about the same size as the symbols.

Star 4 has been on the so-called Southern Outer Ring since its discovery (as the ring has slowly expanded) but Burrows considers this star to also be an accidental coincidence and dismisses any significance [Burrows *et al.* 1995]. The fact that SN 1987A itself lies on the Northern Outer Ring is generally ignored. I think that 3 out of 3 "accidental" occurrences stretches credibility to the breaking point. (In my sketch, Star 3 appears to be on the Southern Outer Ring also, but in most *Hubble Space Telescope* images it appears to be inside. This ring is shown by Burrows *et al.* (their Figure 3) to be distorted and bent in the neighborhood of Star 3, an effect I was not able to produce with my simple geometry. Star 2 has always been shown clearly outside the Northern Outer Ring.

I suggest that the system consists of asteroidal fragments orbiting in a combined weak gravitational and magnetic field, with the rings formed of the debris ejected from the surfaces by ongoing nuclear fission reactions. This material is primarily composed of protons and hydrogen atoms resulting from the beta decay of neutrons leaking from the embedded nuclear reactors, but would also include ions of helium (from the alpha decay of the nuclear fuel isotopes), carbon (from methane and carbon monoxide/dioxide), nitrogen (from ammonia), and oxygen (from water), and other elements that might be vaporized or otherwise ejected from the surface. The ions are constrained in circular orbits by the magnetic field. (A uniform magnetic field produces circular orbits, as observed, while a point gravitational attraction produces an off-center elliptical orbit, as has also been tentatively reported.) A similar effect, a belt of hydrogen, oxygen, and sodium, encircles Jupiter, emitted from Jupiter's large moon Io [Judge *et al.* 1976]. It has been difficult to explain what is considered to be "a huge escape rate of hydrogen", needed to keep the orbital cloud filled with atomic hydrogen [Brown and Yung 1976]. As we have seen, atomic hydrogen results from the recombination of protons and electrons, which are produced by the beta decay of neutrons in space. ("Recombination" of protons and electrons from the beta decay of neutrons is particularly apt. Originally, the particles were combined as nuclear particles, neutrons, now "recombined" as hydrogen atoms.) The "huge escape rate" requires 10^{11} hydrogen atoms (neutrons) per square centimeter per second. A light water reactor, used for the production of electrical power, has a neutron flux (at a power of 3,000 megawatts) of 5×10^{13} neutrons per square centimeter per second [Hu and Kohse 2008]. The plasma torus surrounding Jupiter in Io's orbit is confined by Jupiter's magnetic field [Rothery 1992].

The current conditions of the system resulted from the "choo-choo chug-chug huff'n'puff BOOM" detonation of the Ringstar, as shown from an analysis of the halos that have been (mis)interpreted as light echoes from dust clouds near our line of sight to the supernova. There are four distinct halos, one off-center. The outermost halo is off-center, and represents neutrons/protons from the detonation of the Ringstar, which induced the detonation of the Companion, forming the middle halo, which in turn induced the detonation of Sk-69°202, forming the inner halo. The innermost halo may be from the mystery spot. The protons have been trapped by the magnetic field and have slowly moved outward, forming the halos/echoes, as the protons and electrons recombine, and as the material impacts dust along the way. The astronomical objects known as Soft Gamma-ray Repeaters (SGRs) may be objects in the choo-choo stage of this process.

As shown in Figure 173, the published positions of the stars have had large changes, year to year, and these changes in position are the most significant part of this discussion. These changes give clear indication that the standard distance scale cannot be correct for the Large Magellanic Cloud, and that forces a re-evaluation of the nature of such astronomical objects.

Taking Star 4 (a "foreground star") as the clearest example, the observations show that it has had a displacement from 1/26/1983 to 12/5/2006 (my estimate from a *Hubble Space Telescope* image) of 0.943 arcsecond, in a time interval of 7.529×10^8 seconds. Those are two strictly observational values.

At the assumed standard distance of 50,000 parsecs for SN 1987A, that displacement requires a velocity of at least 9,369 kilometers per second (3% of the speed of light). That is 20 to 10,000 times the gravitational speeds generally encountered in the Solar System. (The Moon goes around the Earth at about 1 kilometer per second, comets about to impact the Sun are travelling at about 450 kilometers per second.) This might be accommodated by assuming that Star 4 is actually only a few tens of light years away, but that would lead to a great mismatch in apparent brightness of the several stars in the system, some assumed to be near, some far.

On the other hand, at my proposed distance of (roughly) 60 Astronomical Units (twice the distance to Neptune) the velocity is only 5.451×10^{-5} kilometers per second, or about 5 centimeters per second. For comparison, a binary asteroidal pair (2001 QW$_{322}$) was recently discovered in the Kuiper Belt of our Solar System [Petit *et al.* 2008]. The mutual orbital velocity of the two parts is 4.26 centimeters per second at a separation of 11.4×10^4 kilometers. With the assumption that the two components of the binary asteroid are equal, the researchers estimated diameters of 54 kilometers and masses of about 1.4×10^{18} kilograms. Using my distance of 60 Astronomical Units to SN 1987A, Star 4 is 9.7×10^4 kilometers from the remains of Sk-69°202, and Star 4 shows a velocity of 5 centimeters per second. These similarities are remarkable.

I think this shows that the standard distance is grossly in error, and that my shorter distance, 60 Astronomical Units, based on the proposal that the Magellanic Clouds, and the Milky Way Galaxy, are the debris fields of the detonation of Planet X, is more likely to be correct. My model agrees with the observations.

The Light Halos of SN 1987A

Around the exploded star Sanduleak-69°202 (Sk-69°202), that became SN 1987A, there are at least four observable rings of light [Gouiffes 1988, Suntzeff *et al.* 1988, Chevalier and Emmering 1988, Crotts 1988, Bond *et al.* 1989, Chevalier and Emmering 1989, Sparks, Paresce, and Macchetto 1989, Bond *et al.* 1990, Couch, Allen, and Malin 1990, Xu, Crotts, and Kunkel 1994, Spyromilio *et al.* 1995, Xu and Crotts 1995, Rest *et al.* 2005]. I call them halos, the astronomers call them light echoes and explain them as reflections of the first flash of light from the supernova off dust clouds between the supernova and our telescopes. My halos are the expanding shells of protons resulting from the beta-decay of neutrons released in the nuclear fission detonations that occurred in several bodies in the Sk-69°202 system, as asteroidal fragments. The protons have average energies from the beta decay of about 358 eV, up to 750 eV, plus any significant energy from the neutrons before decay, such as 2 MeV for prompt fission neutrons. However, the protons, as charged particles, are trapped by the magnetic field surrounding the supernova, and are forced to travel in spiral paths, slowly winding their way from their source to the outskirts of the system. The observations, the radius of each ring at progressive times, are shown in Figure 174 as dots, and a fit of a simple function is shown as a curved line.

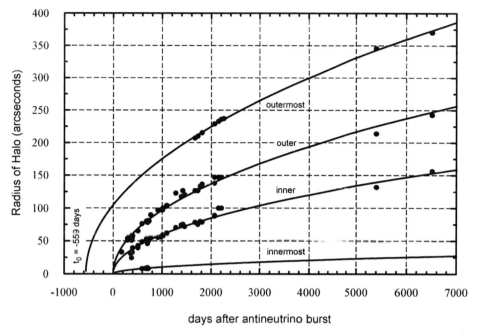

Figure 174. Expansion of halos around SN 1987A.
Expansion of the rings of light around the SN 1987A system, with a fit of an expansion curve to the observations that shows the early origination of the off center outermost halo/light echo.

The fits are very good, but what is so remarkable about this plot is that the curve through the dots of the outermost ring clearly shows that one object (probably the Ringstar, the bright "superposed star" on the circumstellar ring, since the outermost halo is clearly off center to the south, in the direction of the Ringstar at that time), exploded 559 days before everything else (about August 10, 1985). Was there a Gamma Ray Burst detected near Sk-69°202, about 559 days before the supernova, to produce the outermost halo? This detonation destabilized the nuclear fission reactors in the other objects so that they were subject to nuclear fission detonations of their own (the Companion and Sk-69°202), leading to the two pulses of antineutrinos that were detected on February 23, 1987, as a prelude to the supernova. This is an interpretation that cannot occur in conventional astronomy.

The radial expansion velocities of the halos can be calculated from the analytical fit, and are shown in Figure 175.

Considering that these halos may represent excitation of gas and dust in the supernova environment by protons resulting from the decay of neutrons released by the nuclear fission detonations, the velocities near the bends in the curves, in the range corresponding to velocities of 10^2 to 10^4 kilometers per second, represent neutron energies of 52 eV to 0.52 MeV. These energies are in the range of nuclear fission (prompt neutrons have average energies of 2 MeV, those that have been scattered, slightly moderated, are in the electronvolt range), and the halos would be the result of neutrons escaping from the magnetic field before decaying to protons. Therefore the halos and Panagia's Ring show a separation between escaped protons and trapped protons. Using a proposed distance of 60 Astronomical Units, and with the early velocities shown in Figure 172, most of the (uncharged) neutrons would pass the circumstellar ring before decaying into (charged) protons. Only the charged protons would be confined by the magnetic field. After decaying into protons and electrons, the neutrons would be effective in energizing the environmental medium.

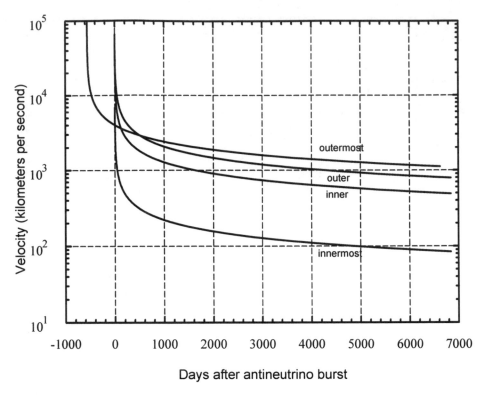

Figure 175. Expansion velocities of SN 1987A halos.
Expansion velocities of the halos. The steep portion near time zero may be an analytical artifact, or may show the high initial energies before slowing down in the magnetic field.

Measurements of neutron activation from the atomic bomb at Hiroshima showed a significant flux of high energy neutrons, with energies above 1.5 MeV [Straume *et al.* 2003].

Panagia's Ring

Astronomers have supposed that the first brightening of the circumstellar ring around SN 1987A was caused by the initial flash of light from the supernova explosion, arriving at the ring about 90 days after the antineutrino burst [Panagia *et al.* 1991, Panagia 1999]. The observations have been interpreted to determine the distance from the supernova to Earth, with the assumption that the energizing flux traveled at the speed of light. That assumption produced the right answer, the standard distance, and so it was accepted as an absolute assumption. However, the great release of neutrons from my proposed nuclear fission detonation produced an equally great burst of protons and electrons that would energize the ring even more effectively than would light. The charged particles are constrained by the magnetic field (shown by the synchrotron radiation [McCray1997] and the excess polarization of light, a factor of two stronger than expected) to spiral slowly outward, and it was the arrival and passage of this pulse of protons that energized the ring.

Protons from the decay of neutrons, free neutrons in the vacuum of space, have energies up to about 750 electronvolts, as was shown in Figure 2. This gives them an average energy of 358 electronvolts, and a FWHM (full width at half maximum) of 600 electronvolts. The average energy corresponds to an effective temperature of 2.77 million degrees kelvin, which could account for the million degree gas clouds seen within galaxy clusters.

The electrons, beta particles, have the energy distribution that is typical of beta decay, and this was shown in Figure 3. (Note that the energy scale, while numerically the same for the two figures, is in kiloelectronvolts for the electrons, a factor of 1,000 larger.) The average energy is 322 kiloelectronvolts, giving an effective temperature of 2.49 billion degrees kelvin. Electrons lose energy rapidly and will come into equilibrium with the environment.

Detailed analysis of the energy release, the light curve, the objects in the system, antineutrinos, gravitational waves, and the light halos, supports the proposal that SN 1987A was a complicated explosion of at least three components, consisting of asteroidal size bodies at a distance of 60 to 100 Astronomical Units.

Appendix D: Nuclear Reactor Analysis of V838 Monocerotis

A remarkable outburst of a star, V838 Monocerotis, has been described [Bond, Henden, and Levay *et al.* 2003, Bond *et al.* 2003]. Changes in brightness, temperature, and size were seen in a time of less than 200 days. A trace of the power of V838 Monocerotis is shown in Figure 176 [http://www.stsci.edu/~bond/lightcurve.dat]. The cloud of debris surrounding the star was interpreted to be light echoes, but in a small scale Universe, that would be material ejected by a detonating asteroidal fragment. A nuclear reactor analysis of the power trace, based on the supposition that V838 Monocerotis revealed the progressive detonation of natural nuclear fission reactors in an asteroidal fragment, typifying the choo-choo chug-chug huff'n'puff BOOM process, with little boom at the end, showed very reasonable variations in the reactivity of such a system. Positive reactivities approached or exceeded the level of "prompt critical", which gives drastically fast growth in fission power, ending with negative reactivities of a few percent (Figure 177). These reactivities are commonly developed in "burst" reactors, experimental reactors designed and operated to produce brief but intense power excursions [Keepin 1965, Hetrick 1971].

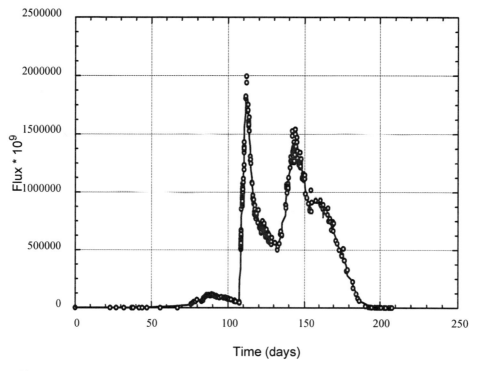

Figure 176. The power trace of V838 Monocerotis.
Relative power trace for the outburst of V838 Monocerotis [Bond *et al.* 2003].

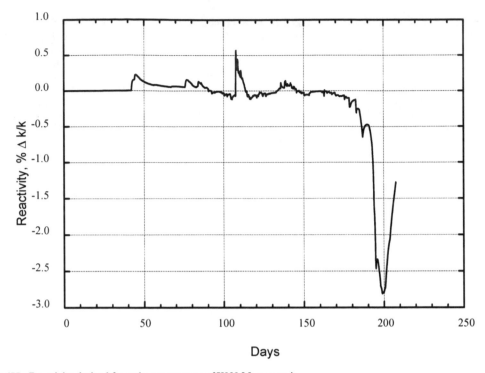

Figure 177. Reactivity derived from the power trace of V838 Monocerotis.
Calculated reactivity for the power trace shown for V838 Monocerotis. Small (0.5 percent in this case) departures from zero result in rapid increases in power (for positive reactivity) and shutdown of the reactor for negative departures.

Is this a reactor in the sky? Would behavior like this explain the labeling of Sirius, one of the brightest, whitest stars we see, as a red star in ancient times, and we reject that character because we don't understand how it could be?

A Red Sirius was recorded by the ancients [Schlosser and Bergmann 1985, Tang 1986, Gry and Bonnet-Bidaud 1990, Tang 1991, Whittet 1999, and many others]. This has created a controversy of contradictions. The star has appeared in colors other than the blue-white we recognize today, but at various times has had its modern color. Homer added an embellishment: golden plumes [Homer 850BCEa] or flaming hair [Homer 850BCEb]. It is listed as a Type AIV star, with a spectrum that is somewhat less blue-white than its dwarf companion Sirius B (see Figure 156) which matches a Cerenkov spectrum quite well. Could Sirius, or its dwarf companion, have gone through a stage like that we observed as the outburst of V838 Monocerotis, with ejected material showing as light echoes, or golden plumes, or flaming hair? The spectra of both stars are rich in hydrogen (Sirius B shows only hydrogen), which would result from the release and decay of fission neutrons, recombining as protons and electrons to make neutral hydrogen. The outburst of V838 Monocerotis shows that changes can happen quite quickly.

References

Abbate, E. *et al.* 1998. A one-million-year-old *Homo* cranium from the Danakil (Afar) Depression of Eritrea. *Nature* **393**, 458-460 (1998).

Abdo, A. A. *et al.* 2009. Measurement of the Cosmic Ray e⁺+e⁻ Spectrum from 20GeV to 1TeV with the Fermi Large Area Telescope. *Physical Review Letters* **102**, 181101-06 (2009).

Abel, G. O., D. Morrison, and S. C. Wolfe 1991. *"Exploration of the Universe".* 6th Edition. Saunders College Publishers, Philadelphia (1991).

Achenbach, J. 2003. Dinosaurs Come Alive. *National Geographic* **March**, 2-33 (2003).

Aczel, A. D. 1999. *"God's Equation".* Dell Publishing, New York (1999).

Adams, J. B., M. E. Mann, and C. M. Ammann 2003. Proxy evidence for an El Niño-like response to volcanic forcing. *Nature* **426**, 274-278 (2003).

Adriani, O. *et al.* 2009. An anomalous positron abundance in cosmic rays with energies 1.5-100 GeV. *Nature* **458**, 607-609 (2009).

Agladze, N. I. *et al.* 1994. Reassessment of millimetre-wave absorption coefficients in interstellar silicate grains. *Nature* **372**, 243-245 (1994).

Aglietta, M. *et al.* 1987. On the Event Observed in the Mont Blanc Underground Neutrino Observatory During the Occurrence of Supernova 1987a. *Europhysics Letters* **3**, 1315-1320 (1987).

Agnor, C. B. and D. P. Hamilton 2006. Neptune's capture of its moon Triton in a binary-planet gravitational encounter. *Nature* **441**, 192-194 (2006).

Agueda, N. *et al.* 2011. On The Near-Earth Observation of Protons and Electrons from the Decay of Low-Energy Solar Flare Neutrons. *Astrophysical Journal* **737** 53 doi: 10.1088/0004-637X/737/2/53 (2011).

Alexander, A. F. O'D. 1965. *"The Planet Uranus".* Faber and Faber, London (1965).

Aligheri, S. *et al.* 1988. A blue, polarized continuum source near radio galaxy PKS2152-69, *Nature* **334**, 591-593 (1988).

Allaby, M. 1992. *"Air".* Facts on File Limited, Oxford (1992).

Allamandola, L. J. *et al.* 1993. Diamonds in Dense Molecular Clouds: A Challenge to the Standard Interstellar Medium Paradigm. *Science* **260**, 64-65 (1993).

Allègre, C. J., P. Sarda, and T. Staudacher 1993. Speculations about the cosmic origin of He and Ne in the interior of the Earth. *Earth and Planetary Science Letters* **117**, 229-233 (1993)

Allen, C. W. 1976. *"Astrophysical Quantities"*, Third Edition. The Athlone Press, London (1976).

Allen, D. A. and M. G. Burton 1993. Explosive ejection of matter associated with star formation in the Orion nebula. *Nature* **363**, 21-22 (1993).

Allen, J. *et al.* 1988. Pleistocene dates for the human occupation of New Ireland, northern Melanesia. *Nature* **331**, 707-709 (1988).

Allen, L. 1990. *"The Hubble Space Telescope Optical Systems Failure Report".* NASA, November 1990 (1990).

Alley, R. B. *et al.* 1993. Abrupt increase in Greenland snow accumulation at the end of the Younger Dryas Event. *Nature* **362**, 527-529 (1993).

Allison, M. *et al.* 1991. Uranus Atmospheric Dynamics and Circulation. Pages 253-295 in *"Uranus"*, J. T. Bergstrahl, E. D. Miner, and M. S. Mathews, Editors. University of Arizona Press, Tucson (1991).

Alroy, J. 1998. Cope's Rule and the Dynamics of Body Mass Evolution in North American Fossil Mammals. *Science* **280**, 731-734 (1998).

Alvarez, L. W. 1987. Mass extinctions caused by large bolide impacts. *Physics Today* **July**, 24-33 (1987).

Alvarez, L. W. *et al.* 1980. Extraterrestrial Cause for the Cretaceous-Tertiary Extinctions. *Science* **208**, 1095-1108 (1980).

Alvarez, W. 1997. *"T. rex and the Crater of Doom"*. Princeton University Press, Princeton (1997).

Alvarez, W. *et al.* 1984. Impact Theory of Mass Extinction and the Invertebrate Fossil Record. *Science* **223**, 1135-1141 (1984).

Alvarez, W. and F. Asaro 1990. (What Caused the Extinction?) An Extraterrestrial Impact. *Scientific American* **October,** 78-84 (1990).

Amaldi, E. *et al.* 1987a. Data Recorded by the Rome Room Temperature Gravitational Wave Antenna during the Supernova 1987a in the Large Magellanic Cloud. *Europhysics Letters* **3**, 1325-1330 (1987).

Amaldi, E. *et al.* 1987b. Analysis of the data recorded by the Maryland and Rome room temperature gravitational wave antennas in the period of the SN 1987A. Pages 453-462 in *"Supernova 1987A in the Large Magellanic Cloud"*. Edited by Minas Kafatos and Andrew Michalitsianos. Cambridge University Press, Cambridge (1988).

Amaldi, E. *et al.* 1988. Coincidences among the Maryland and Rome Gravitational Wave Detector Data and the Mont Blanc and Kamioka Neutrino Detector Data in the Period of SN 1987A. *Annals of the New York Academy of Sciences* **571**, 561-576 (1988).

Anders, E. 1963. Meteorite Ages. Pages 402-495 in *"The Moon, Meteorites, and Comet"*. B. M. Middlehurst and G. P. Kuiper, Editors. University of Chicago Press, Chicago (1963).

Anders, E. and N. Grevesse 1989. Abundances of the elements: meteoritic and solar. *Geochimica et Cosmochimica Acta* **53**, 197-214 (1989).

Anderson, C. 1990. No end of blame. *Nature* **348**, 471 (1990).

Anderson, D. L. 1990. Planet Earth. Pages 65-76 in *"The New Solar System"*, 3rd Edition. J. K. Beatty and A. Chaikin, Editors. Sky Publishing Co., Cambridge (1990).

Anderson, D. L. 1993. Helium-3 from the Mantle: Primordial Signal or Cosmic Dust? *Science* **261**, 170-176 (1993).

Anderson, D. L., T. Tanimoto, and Y. Zhang 1992. Plate Tectonics and Hotspots: The Third Dimension. *Science* **256**, 1645-1651 (1992).

Anderson, G. C. 1990. "Spinning pulsar" claims retracted. *Nature* **343,** 679 (1990).

Anderson, J. D. *et al.* 1998. Indication, from Pioneer 10/11, Galileo, and Ulysses Data, of an Apparent Anomalous, Weak, Long-Range Acceleration. *Physical Review Letters* **81**, 2858-2863 (1998).

Anderson Jr., A. H. 1971. *"The Drifting Continents"*. G. P. Putnam's Sons, New York (1971).

Andresen, G. B. *et al.* 2010. Trapped antihydrogen. *Nature* **468**, 673-676 (2010).

Andrews, P. and C. Stringer 1993. The primates' progress. Pages 218-251 in *"The Book of Life"*. S. J. Gould, Editor. W. W. Norton & Co., New York (1993).

Andrews-Hanna, J. C., Maria T. Zuber, and W. B. Banerdt 2008. The Borealis basin and the origin of the martian crustal dichotomy. *Nature* **453**, 1212-1215 (2008).

Angela, P. and A. Angela 1993. *"The Extraordinary Story of Human Origins"*. Translated by G. Tonne. Prometheus Books, Buffalo (1993).

Anholt, R. R. H. 1994. *"Dazzle 'em with style: the art of oral scientific presentation"*. W. H. Freeman and Company, New York (1994).

Anikovich, M.V. *et al.* 2007. Early Upper Paleolithic in Eastern Europe and Implications for the Dispersal of Modern Humans. *Science* **315**, 223-226 ((2007).

Anonymous 2002. *Caltech News* **36**, 6 (2002).

Antón, S. C. 1998. Endocranial hyperostosis in Sangiran 2, Gibraltar 1, and Shanidar 5. *American Journal of Physical Anthropology* **101**, 111-122 (1998).

Antonucci, R. 1992. Active Galaxies and Quasistellar Objects, Blazars. Pages 6-7 in *"The Astronomy and Astrophysics Encyclopedia"*. S. P. Maran, Editor. Van Nostrand Reinhold, New York (1992).

Apt, K. E. 1976. Investigations of the Oklo Natural Fission Reactor, July 1975-June 1976. LA-6575-PR, Los Alamos Scientific Laboratory, Los Alamos (1976).

Archibald, J. D. *et al.* 2010. Cretaceous Extinctions: Multiple Causes. *Science* **328**, 973 (2010).

Arden, J. W. 1977. Isotopic composition of uranium in chondritic meteorites. *Nature* **269**, 788-789 (1977).

Armento, W. J. 1979. Hohmann Ellipse Transfer Data (Minimum Energy of Transfer) pages F-188-F-191 in *"Handbook of Chemistry and Physics"*, R. C. Weast and M. J. Astle, editors, CRC Press, Boca Raton (1979).

Armitage, M. H. 1992. The implications of pleiochroic radiohalos in biotite. *American Laboratory* **November**, 25-28 (1992).

Armitage, M. H. 1993. Internal radiohalos in a diamond. *American Laboratory* **December**, 28-30 (1993).

Arp, H. C. 1966. Atlas of Peculiar Galaxies. *Astrophysics Journal Supplement* **123**, 1-20 (1966).

Arp, H. C. 1987. *"Redshifts, Quasars and Controversies"*. Interstellar Media, Berkeley (1987).

Arp, H. 1997. Pairs of X-ray sources across Seyferts: the NGC4235 field. *Astronomy and Astrophysics* **328**, L17-L20 (1997).

Arp, H. C. *et al.* 1990. The extragalactic Universe: an alternative view. *Nature* **346**, 807-812 (1990).

Arp, H. C. *et al.* 1992. Big Bang contd... *Nature* **357**, 287-288 (1992).

Arsuaga, J.-L. *et al.* 1999. A complete human pelvis from the Middle Pleistocene of Spain *Nature* **399**, 255-258 (1999).

Asfaw, B. *et al.* 1992. The earliest Acheulian from Konso-Gardula. *Nature* **360**, 732-735 (1992).

Ash, M. E. *et al.* 1971. The System of Planetary Masses. *Science* **174**, 551-556 (1971).

Ash, R. 1994. Small spheres of influence. *Nature* **372**, 219-220 (1994).

Asimov, I. 1991. *"How Did We Find Out About Pluto?"*. Walker and Company, New York (1991).

Assistant Secretary for Nuclear Energy 1983. *"Atoms to Electricity"*. DOE/NE-0053. U. S. Department of Energy, Oak Ridge (1983).

Astronews 1997. Ganymede Loses an Ocean, Gains a Core. *Astronomy* **April**, 26 (1997).

Astronews 1998. How to make Earth's Moon. *Astronomy* **January**, 24 (1998).

Astronomy 1993a. Asteroid Gaspra Surprises Astronomers. *Astronomy* **April**, 20-21 (1993).

Astronomy 1993b. Brightest Gamma-Ray Burst Seen. *Astronomy* **August**, 19 (1993).

Astronomy 1993c. Andromeda Galaxy Has A Double Nucleus. *Astronomy* **November**, 20 (1993).

Astronomy 1994a1. Looking Below Mercury's Surface. *Astronomy* **February**, 22 (1994).

Astronomy 1994a2. Miranda: Seeing Its History Afresh. *Astronomy* **February**, 20 (1994).

Astronomy 1994b. Gamma-Ray Bursts from Comets? *Astronomy* **March**, 20 (1994).

Astronomy 1994c. Little Dark Matter in M81. *Astronomy* **May**, 18 (1994).

Astronomy 1994d. Two Shells of Glowing Gas. *Astronomy* **December**, 18 (1994).

Atteia, J.-L. 1993. Gamma-ray burst observations. *Astronomy and Astrophysics Supplement Series* **97**, 35-38 (1993).

Atteia, J.-L. *et al.* 1991. Statistical evidence for a galactic origin of gamma-ray bursts. *Nature* **351**, 296-298 (1991).

Auel, J. M. 1980. *"The Clan of the Cave Bear"*. Crown Publishers, New York (1980, 2001).

Augusta, J. and Z. Burin 1963. *"A Book of Mammoths"*. Paul Hamlyn, London (1963).

Aveni, A. 1992. *"Conversing with the Planets -- How Science and Myth Invented the Cosmos"*. Times Books (Random House), New York (1992).

B. B. 2001a. Our family tree does the splits... *Science News* **159**, 232 (2001).

B. B. 2001b. ...and then takes some lumps. *Science News* **159**, 232 (2001).

Baade, W. *et al.* 1956. Supernovae and Californium 254. *Publications of the Astronomical Society of the Pacific* **68**, 296-300 (1956).

Bahcall, J. N. 1989. *"Neutrino Astrophysics"*. Cambridge University Press, Cambridge (1989).

Bahcall, J. N. 1990. The Solar-Neutrino Problem. *Scientific American* **May**, 54-61 (1990).

Bahcall, J. N., D. N. Spergel, and W. H. Press 1987. Phenomenological Analysis of Neutrino Emission from SN 1987A. Pages 172-184 in *"Supernova 1987A in the Large Magellanic Cloud"*. Edited by Minas Kafatos and Andrew Michalitsianos. Cambridge University Press, Cambridge (1988).

Bahcall, J. N. and R. K. Ulrich 1988. Solar models, neutrino experiments, and helioseismology. *Reviews of Modern Physics* **60**, 297-372 (1988).

Bahcall, J. N. and M. H. Pinsonneault 1992. Standard Solar Models, with and without helium diffusion, and the Solar Neutrino Problem. *Reviews of Modern Physics* **64**, 885-926 (1992).

Bahn, P. G. 1990. Eating people is wrong. *Nature* **348**, 395 (1990).

Bahn, P. G. 1998. *"The Cambridge Illustrated History of Prehistoric Art"*. Cambridge University Press, Cambridge (1998).

Bahn, P. and J. Vertut 1988. *"Images of the Ice Age"*. Facts On File, New York (1988).

Bailey, M. E., S. V. M. Clube, and W. M. Napier 1990. *"The Origin of Comets"*. Pergamon Press, Oxford (1990).

Bailyn, C. 1990. Good things come in threes. *Nature* **346**, 12-13 (1990).

Baker, V. R. *et al.* 1991. Ancient oceans, ice sheets, and the hydrological cycle on Mars. *Nature* **352**, 589-594 (1991).

Balachandran, S. C. and R. A. Bell 1998. Shallow mixing in the solar photosphere inferred from revised beryllium abundance. *Nature* **392**, 791-793 (1998).

Balsiger, H., K. Altwegg, and J. Geiss 1995. D/H and O-18/O-16 ratio in the hydronium ion and in neutral water from in situ ion measurements in comet Halley. *Journal of Geophysical Research* **100**, (A4) 5827-5834 (1995).

Balter, M. 2010a. Ancient DNA from Siberia Fingers a Possible New Human Lineage. *Science* **327**, 1566-1567 (2010).

Balter, M. 2010b. In Archaeologist's Hands, Tiny Fossil s Yield Big Answers. *Science* **329**, 28-29 (2010).

Balter, M. 2010c. Reanalysis of French Cave Could Deal Setback to Neandertal Smarts. *Science* **330**, 439 (2010).

Barbetti, M. and M. McElhinny 1972. Evidence of a Geomagnetic Excursion 30,000 yr BP. *Nature* **239**, 327-330 (1972).

Bar-Yosef, O. and B. Vandermeersch 1993. Modern Humans in the Levant. *Scientific American* **April**, 94-100 (1993).

Barrett, P. J. *et al.* 1992. Geochronological evidence supporting Antarctic deglaciation three million years ago. *Nature* **359**, 816-818 (1992).

Barron E. J. 1987. Explaining glacial periods. *Nature* **329**, 764-765 (1987).

Barron E. J., S. L. Thompson, and S. H. Schneider 1981. An Ice-free Cretaceous? Results from Climate Model Simulations. *Science* **212**, 501-508 (1981).

Barry, P. 2009. In Search of Antimatter Galaxies. SCIENCE@NASA **08.14.2009** (2009).

Barstow *et al.* 2005. *Hubble Space Telescope* spectroscopy of the Balmer lines in Sirius B. *Monthly Notices of the Royal Astronomical Society* **362**, 1134-1142 (2005).

Barszak, T. 1998. email to R. J. Tuttle 12/22/98 (1998).

Barth, C. A. *et al.* 1997. Galileo ultraviolet spectrometer observations of atomic hydrogen in the atmosphere of Ganymede. *Geophysical Research Letters* **24**, 2147-2150 (1997).

Bartusiak, M. 1998. Outsmarting the Early Universe. *Astronomy* **October**, 54-59 (1998).

Bartusiak, M. 2009. *"The Day We Found the Universe"*. Pantheon Books, New York (2009).

Basu, A. R. *et al.* 1993. Early and Late Alkali Igneous Pulses and a High-³He Plume Origin for the Deccan Flood Basalts. *Science* **261**, 902-906 (1993).

Baum, R. and W. Sheehan 1997. *"In Search of Planet Vulcan"*, Epilogue note 11. Plenum Trade, New York (1997).

Beall, C. M. *et al.* 2010. Natural selection on *EPAS1* (*HIF2a*) associated with low hemoglobin concentrations in Tibetan highlanders. *Proceedings of the National Academy of Sciences USA* **107**, 11459-11464 (2010).

Beasley, A. 1997. Chondrites and the Solar System. *Science* **278**, 76-77 (1997).

Beatty, J. K. 1995. Ida & Company. *Sky & Telescope* **January**, 20-23 (1995).

Beatty, J. K. 1996. Into the Giant. *Sky & Telescope*, **April**, 20-22 (1996).

Beatty, J. K. 2001. Far-Out Kuiper-Belt Object Confounds Dynamicists. *Sky & Telescope* **September**, 24 (2001).

Becker, S. 1993. Approximating the r-Process on Earth with Thermonuclear Explosions. Pages 453-456 in *"Origin and Evolution of the Elements"*. N. Prantzos, E. Vangioni-Flam, and M. Cassé, Editors. Cambridge University Press, Cambridge (1993).

Beckwith 1994. Getting off to a rocky start. *Nature* **368**, 190-191 (1994).

BEIR 1980. *"The Effects on Populations of Exposure to Low Levels of Ionizing Radiation"*. National Academy of Sciences, Washington (1980).

Bell, J. F. 1989. The Big Picture. Pages 921-945 in *"Asteroids II"*. R. P. Binzel, T. Gehrels, and M. S. Matthews, Editors. University of Arizona Press, Tucson (1989).

Belton, M. J. S. *et al.* 1994. Galileo Multispectral Imaging of the North Polar and Eastern Limb Regions of the Moon. *Science* **264**, 1112-1115 (1994).

Benedict, G. F. *et al.* 2002. Astrometry with the *Hubble Space Telescope*: a Parallax of the Fundamental Distance Calibrator δ Cephei. *Astronomical Journal* **124**, 1695-1705 (2002).

Benford, G. 2010. Penumbra. *Nature* **465**, 836 (2010).

Bennett, K. A. 2001. Uncool Callisto. *Nature* **412**, 395-396 (2001).

Bennett, K. D., S. G. Haberle, and S. H. Lumley 2000. The Last Glacial-Holocene Transition in Southern Chile. *Science* **290**, 325-328 (2000).

Benton, M. J. 1990. *"The Reign of the Reptiles"*. Crescent Books, New York (1990).

Benton, M. 1993. Four feet on the ground. Pages 78-125 in *"The Book of Life"*. S. J. Gould, Editor. W. W. Norton & Co., New York (1993).

Benton, M. J. 1997. Extinction, Triassic. Pages 230-236 in *"Encyclopedia of Dinosaurs"*. Edited by P. J. Currie and K. Padian. Academic Press, San Diego (1997).

Benz, W., W. L. Slattery, and A. G. W. Cameron 1988. Collisional Stripping of Mercury's Mantle. *Icarus* **74**, 516-528 (1988).

Berendzen, R., R. Hart, and D. Seeley 1976. *"Man Discovers the Universe"*. Science History Publications, New York (1976).

Bernabei, R. *et al.* 2008. First results from DAMA/LIBRA and the combined results with DAMA/NaI. *European Physical Journal C* **56**, 1434-6044 (2008).

Berner, R. A. 1990. Atmospheric Carbon Dioxide Levels Over Phanerozoic Time. *Science* **249**, 1382-1386 (1990).

Berry, A. 1898. *"A Short History of Astronomy"*. Dover Publications, New York (1961), originally published by John Murray (1898).

Bertoldi, F. *et al.* 2006. The trans-neptunian object UB₃₁₃ is larger than Pluto. *Nature* **439**, 563-564 (2006).

Beskin *et al.* 1993. *"Physics of the Pulsar Magnetosphere"*. Cambridge University Press, New York (1993).

Bethe, H. A. 1939. Energy production in stars. *Physical Review* **55**, 434-456 (1939).

Bhattacharjee, Y. 2011. Possible Sighting of Dark Matter Fires Up Search and Tempers. *Science* **332**, 1144-1147 (2011).

Bierlein, J. F. 1994. *"Parallel Myths"*. Ballantine Books, New York (1994).

Bignami, G. F. and P. A Caraveo 1992. Geminga: new period, old γ-rays. *Nature* **357**, 287 (1992).

Binney, J. and M. Merrifield 1998. *"Galactic Astronomy"*. Princeton University Press, Princeton (1998).

Binzel, R. P. and T. C. Van Flandern 1979. Minor Planets: The Discovery of Minor Satellites. *Science* **203**, 903-905 (1979).

Binzel, R. P., M. A. Barucci, M. Fulchignoni 1991. The Origins of the Asteroids. *Scientific American* **October 1991**, 88-94 (1991).

Birck, J-L. and C. J. Allègre 1988. Manganese-chromium isotope systematics and the development of the early Solar System. *Nature* **331**, 579-584 (1988).

Birkeland, P. W. and E. E. Larson 1989. *"Putnam's Geology"* Fifth Edition. Oxford University Press, Oxford (1989).

Blair, D. and G. McNamara 1997. *"Ripples on a Cosmic Sea"*. Helix Books, Addison-Wesley, Reading (1997).

Blanton E. and D. Helfand 1996. *ASCA* Observations of the Composite Supernova Remnant G29.7-0.3. *Astrophysical Journal* **470**, 961-966 (1996).

Bless, R. C. 1996. *"Discovering the Cosmos"*. University Science Books, Sausalito (1996).

Blichert-Toft, J. and F. Albarède 1994. Short-Lived Chemical Heterogeneities in the Archean Mantle with Implications for Mantle Convection. *Science* **263**, 1593 (1964).

Block, D. L. and R. J. Wainscoat 1991. Morphological differences between optical and infrared images of the spiral galaxy NGC 309. *Nature* **353**, 48-50 (1991).

Blum, J. D. *et al.* 1987. 'Domestic' origin of opaque assemblages in refractory inclusions in meteorites. *Nature* **331**, 405-409 (1987).

Blunier, T. and E. J. Brook 2001. Timing of Millennial-Scale Climate Change in Antarctica and Greenland During the Last Glacial Period. *Science* **291**, 109-112 (2001).

BMS 2009. Signs of dark matter? *Physics Today* **January**, 16 (2009).

Boaz, N. T. 1993. *"Quarry"*. The Free Press, New York (1993).

Bockelee-Morvan *et al.* 1998. Deuterated water in Comet C/1996 B2 (Hyakutake) and Its Implications for the Origin of Comets. *Icarus* **133**, 147-162 (1998).

Bodiselitsch, B. *et al.* 2005. Estimating Duration and Intensity of Neoproterozoic Snowball Glaciation from Ir Anomalies. *Science* **308**, 239-242 (2005).

Bogard, D. D., L. E. Nyquist, and P. H. Johnson 1984. Noble gas contents of shergottites and implications for the Martian origin of SNC meteorites. *Geochimica et Cosmochimica Acta* **48**, 1723-1740 (1984).

Bogard, D. D. *et al.* 1995. Neutron-capture ^{36}Cl, ^{41}Ca, ^{36}Ar, and ^{150}Sm in large chondrites: Evidence for high fluences of thermalized neutrons. *Journal of Geophysical Research* **100** (E5), 9401-9416 (1995).

Bohlin, K. 1907a. Versuch einer Bestimmung der Parallaxe des Andromeda-Nebels. *Astronomiska Iakttagelser och Undersökningar à Stockholms Observatorium* **8** (4), 1-70 plus Plate 1 (1907).

Bohlin, K. 1907b. Versuch einer Bestimmung der Parallaxe des Andromedanebels. *Astronomische Nachrichten* **176**, 205 (1907).

Böhm-Vitense, E. 1989. *"Introduction to stellar astrophysics"*. Vol. 1. Cambridge University Press, Cambridge (1989).

Bok, B. J. 1991. The Milky Way Galaxy. Pages 299-320 in *"The World Treasury of Physics, Astronomy, and Mathematics"*. T. Ferris, Editor. Little, Brown and Co., Boston (1991).

Bolles, E. B. 1999. *"The Ice Finders."* Counterpoint, Washington (1999).

Bonatti, E. 1994. The Earth's Mantle Below the Oceans. *Scientific American* **March**, 44-51 (1994).

Bond, C. and R. Siegfreid 1990. *"Antarctica - No Single Country, No Single Sea"*. New Holland Publishers, London (1990).

Bond, H. E. *et al.* 1989. Supernova 1987A in the Large Magellanic Cloud. *IAU Circular* No. **4733** (1989).

Bond, H. E. *et al.* 1990. Discovery of an inner light-echo ring around SN 1987A, *Astrophysical Journal* **354**, L49-L52 (1990).

Bond, H. E. *et al.* 2003. An energetic stellar outburst accompanied by circumstellar light echoes. *Nature* **422**, 405-408 (2003).

Bond, H., A. Henden, and Z. Levay *et al.* 2003. V838 Light Echo: The Movie. http://antwrp.gsfc.nasa.gov/apod/ap030402.html. (2003).

Bondevik, S. *et al.* 2006. Changes in North Atlantic Radiocarbon Reservoir Ages During the Allerod and Younger Dryas. *Science* **312**, 1514-1517 (2006).

Boothroyd, A. J., I.-J. Sackmann, and W. A. Fowler 1991. Our Sun: II: Early mass Loss of $0.1M_o$ and the Case of the Missing Lithium. *Astrophysical Journal* **377**, 318-329 (1991).

Boslough, M. B. *et al.* 1994. Axial Focusing of Impact Energy in the Earth's Interior: Proof-of-Principle Tests of a New Hypothesis. Pages 14-16 in *"New Developments Regarding the KT Event and Other Catastrophes in Earth History"*. Lunar and Planetary Institute, Houston (1994).

Boss, A. P. 1995. Timing is everything. *Nature*, **375**, 13-14 (1995).

Boss, A. P. 1995. Hot times in the solar nebula. *Nature* **377**, 578-579 (1995).

Bothun, G. D. 1997. The Ghostliest Galaxies. *Scientific American* **February**, 56-61 (1997).

Bothun, G., C. Impey, S. McGaugh 1997. Low-Surface-Brightness Galaxies: Hidden Galaxies Revealed. *Publications of the Astronomical Society of the Pacific* **109**, 745-758 (1997).

Bouchet, P., I. J. Danziger, and L. B. Lucy 1991. The Bolometric Light Curve of SN 1987A: Results from Day 616 to 1316 After Outburst. *Astronomical Journal* **102**, 1135-1146 (1991).

Bouchet, P. *et al.* 1991. The bolometric light curve of SN 1987 A II. Results from visible and infrared spectrophotometry. *Astronomy and Astrophysics* **245**, 490-498 (1991).

Boule, M. and H. V. Vallois 1957. *"Fossil Men"*. Translated into English by Michael Bullock. The Dryden Press, New York (1957).

Boulton, G. S. and C. D. Clark 1990. A highly mobile Laurentide ice sheet revealed by satellite images of glacial lineations. *Nature* **346**, 813-817 (1990).

Bouzigues, H., R. J. M. Boyer, C. Seyve, and P. Teulieres 1975. Contribution a la Solution d'une Enigme Scientifique. Pages 237-243 in *"The Oklo Phenomena"*. International Atomic Energy Agency, Vienna (1975).

Bowcock, A. M. *et al.* 1994. High resolution of human evolutionary trees with polymorphic microsatellites. *Nature* **368**, 455-457 (1994).

Bowell, E. 2003. Orbits of Minor Planets and Comets. Catalog B/astorb, Astronomical Data Center, University of Maryland (2003).

Bower, B. 1997. Ancient human saunters into limelight. *Science News* **152**, 117 (1997).

Bowring, S. A. *et al.* 1993. Calibrating Rates of Early Cambrian Evolution. *Science* **261**, 1293-1298 (1993).

Bowyer, S. 1994. Extreme Ultraviolet Astronomy. *Scientific American* **August**, 32-39 (1994).

Brady, J. L. 1972. The Effect of a Trans-Plutonian Planet on Halley's Comet. *Publications of the Astronomical Society of the Pacific* **84**, 314-322 (1972).

Brady, J. L. and E. Carpenter 1971. The Orbit of Halley's Comet and the Apparition of 1986. *Astronomical Journal* **76**, 728-739 (1971).

Brakenridge, G. R. 1981. Terrestrial Paleoenvironmental Effects of a Late Quaternary-Age Supernova. *Icarus* **46**, 81-93 (1981).

Branch, D. 1990. Spectra of Supernovae. Pages 30-58 in *"Supernovae"*. A. G. Petschek, Editor. Springer-Verlag, New York (1990). The primary reference is Phillips, M. M. 1987. *"Proceedings of the George Mason Conference on SN 1987A"*. M. Kafatos, Editor. Cambridge University Press, Cambridge (1987).

Bray, M. T. 2006. People followed the rains in ancient Sahara. *Cosmos* **July 21**, (2006) credited to *Science*.

Brayard, A. *et al.* 2009. Good Genes and Good Luck: Ammonoid Diversity and the End-Permian Mass Extinction. *Science* **325**, 1118-1121 (2009).

Brazzle, R. H. and C. M. Hohenberg 1995. Iodine-Xenon dating: Discordancy between uranium/lead and iodine/xenon ages in phosphates, and implications for the formation of secondary minerals in meteorites. *Meteoritics (Supplement)* **30**, 491(1995).

Breakthroughs 1997. Cavalcade of Carnivores, Part II – Bear Fare. *Discover* **June**, 26-27 (1997).

Brearley, A. 1997. Chondrites and the Solar System. *Science* **278**, 76-77 (1997).

Brennan, M. S. 1921. *"Familiar Astronomy"*. B. Herder Book Company, St. Louis (1921).

Brennecka, G. A. *et al.* 2010. $^{238}U/^{235}U$ Variations in Meteorites: Extant ^{247}Cm and Implications for Pb-Pb Dating. *Science* **327**, 449-451 (2010).

Bridges, N. T. 1994. Comparisons of Venusian Lava Domes and Seafloor Volcanoes: Implications for the Volcanic History of Venus. *Eos Supplement - 1994 Spring Meeting*, 214 (1994).

Briggs, D. E. G., D. H. Erwin, and F. J Collier 1994. *"The Fossils of the Burgess Shale"*. Smithsonian Institution Press, Washington (1994).

Briggs, F. H. 1982. An Outlying Ring of Neutral Hydrogen Around the ScI galaxy NGC 628. *Astrophysical Journal* **259**, 544-558 (1982).

Broad, W. J. 1990. Panel Finds Error by Manufacturer of Space Telescope. *The New York Times*, August 10, 1990 (1990).

Broecker, W. S. 2006. Was the Younger Dryas Triggered by a Flood? *Science* **312**, 1146-1148 (2006).

Broecker, W. S. and G. H. Denton 1990. What Drives Glacial Cycles? *Scientific American* **January**, 48-56 (1990).

Broecker, W. S. *et al.* 1989. Routing of Meltwater from the Laurentide Ice Sheet during the Younger Dryas cold episode. *Nature* **341**, 318-321 (1989).

Brookfield, J. F. Y. 1992. Importance of DNA ages. *Nature* **388**, 134 (1997).

Brookins, D. C. 1978. Oklo reactor reanalyzed. *Geotimes* **March**, 27-28 (1978).

Brown, J. C. and D. W. Hughes 1977. Tunguska's comet and non-thermal ^{14}C production in the atmosphere. *Nature* **268**, 512-514 (1977).

Brown, M. E. *et al.* 2006. Direct Measurement of the size of 2003 UB313 from the *Hubble Space Telescope*. *Astrophysical Journal* **643**, L61-L63 (2006).

Brown, M. E., C. Trujillo, and D. Rabinowitz 2004. Discovery of a Candidate Inner Oort cloud Planetoid. *Astrophysical Journal* **617**, 645-649 (2004).

Brown, M. H. 1990. *"The Search for Eve"*. Harper & Row, New York (1990).

Brown, P. *et al.* 2004. A new small-bodied hominin from the Late Pleistocene of Flores, Indonesia. *Nature* **431**, 1055-1061 (2004).

Brown, R. A and Y. L. Yung 1976. Io, its Atmosphere and Optical Emissions. Pages 1102-1145 in *"Jupiter"*. T. Gehrels, Editor. University of Arizona Press, Tucson (1976).

Brown, T. A. 2010. Stranger from Siberia. *Nature* **464**, 838-839 (2010).

Brunini, A. 1992a. On the ephemerides of Uranus. *Astronomy and Astrophysics* **255**, 401-404 (1992).

Brunini, A. 1992b. Would a collision on Uranus explain its systematic residuals? *Astronomy and Astrophysics* **264**, 292-295 (1992).

Bryant, E. A., G. A. Cowan, W. R. Daniels, and W. J. Maeck 1976. Oklo, an Experiment in Long-Term Geologic Storage. Chapter 6 in *"Actinides in the Environment"*. A. M. Freidman, Editor. American Chemical Society, Washington (1976).

Bugbee, B. G. and F. B. Salisbury 1978. Controlled Environmental Crop Production: Hydroponic vs. Lunar Regolith. Pages 107-129 in *"Lunar Base Agriculture"*. D. W. Ming and D. L. Henninger, Editors. American Society of Agronomy, Madison (1978).

Buratti, B. J., M. D. Hicks, and R. L. Newburn Jr. 1999. Does global warming make Triton blush? *Nature* **397**, 219 (1999).

Burbank, D. W., L. A. Derry, and C. France-Lanford 1993. Reduced Himalayan sediment produced 8 Myr ago despite an intensified monsoon. *Nature* **364**, 48-54 (1993).

Burbidge *et al.* 1957. Synthesis of the Elements in Stars. *Reviews of Modern Physics* **29**, 547-650 (1957).

Burbidge, G. R. *et al.* 1956. Californium-254 and Supernovae. *Physical Review* **103**, 1145-1149 (1956).

Burenhult, G. (general editor) 1993. *"The First Humans: the illustrated history of humankind"*. HarperCollins, New York (1993).

Burgess, E. 1991, *"Far Encounter – The Neptune System"*. Columbia University Press, New York (1991).

Burness, G. P., J. Diamond, and T. Flannery 2001. Dinosaurs, dragons, and dwarfs: The evolution of maximal body size. *Proceedings of the National Academy of Science USA* **98**, 14518-14523 (2001).

Burnham, R. 1991. Venus: Planet of Fire. *Astronomy* **September**, 32-41 (1991).

Burnham R. 1994. Here's Looking at Ida. *Astronomy* **April**, 38-39 (1994).

Burnham, R. 1996. Into the Maelstrom. *Astronomy*, **April**, 42-45 (1996).

Burnham Jr., R. 1978. *"Burnham's Celestial Handbook"*. Dover Publications, New York (1978).

Burns, J. A. 1985. The Evolution of Satellite Orbits. Pages 117-158 in *"Satellites"*. J. A. Burns and M. S. Matthews, Editors. University of Arizona Press, Tucson (1985).

Burns, J. A. 2010. The birth of Saturn's baby moons. *Nature* **465**, 701-702 (2010).

Burrows, C. J. *et al.* 1995. *Hubble Space Telescope* Observations of the SN 1987A triple ring nebula. *Astrophysical Journal* **452**, 680-684 (1995).

Burton, K. W., H-F. Ling, and R. K. O'Nions 1997. Closure of the Central American Isthmus and its effect on deep-water formation in the North Atlantic. *Nature* **386**, 382-385 (1997).

Bury, R. G. 1929. *"Plato: IX Timaeus, Critias, Cleitophon, Menexenus, Epistles"*. Harvard University Press, Cambridge (1989).

Bustamante, C. D. and B. M. Henn 2010. Shadows of early migrations. *Nature* **468**, 1044-1045 (2010).

Byock, J. L. 1995. Egil's Bones. *Scientific American* **January**, 82-87 (1995).

Byrne, P. B. 1992. Stars, Red Dwarfs and Flare Stars. Pages 783-786 in *"The Astronomy and Astrophysics Encyclopedia"*. S. P. Maran, Editor. Van Nostrand Reinhold, New York (1992).

Caird, R. 1994. *"Ape Man"*. Macmillan, New York (1994).

Calaway, E. 2010. Fossil genome reveals ancestral link. *Nature* **468**, 1012 (2010).

Caldeira, K. and J. F. Kasting 1992. Susceptibility of the early Earth to irreversible glaciation caused by carbon dioxide clouds. *Nature* **359**, 226-228 (1992).

Calvin, W. H. 1991. *"The Ascent of Mind"*. Bantam Books, New York (1991).

Cameron, A. G. W. 1982. Elemental and nuclidic abundances in the solar system. Pages 23-43 in *"Essays in Nuclear Astrophysics"*. C. A. Barnes, D. D. Clayton, D. N. Schramm, Editors. Cambridge University Press, Cambridge (1982).

Cameron, A. G. W. 1995a. The first ten million years in the solar nebula. *Meteoritics* **30**, 133-161 (1995).

Cameron, A. G. W. 1995b. Trying to make sense of aluminum-26 abundances. *Meteoritics* **30**, 363-364 (1995).

Cameron, A. G. W. *et al.* 1988. The Strange Density of Mercury: Theoretical Considerations. Pages 692-708 in *"Mercury"*. F. Vilas, C. R. Chapman, and M. S. Matthews, Editors. University of Arizona Press, Tucson (1988).

Cann, R. L., M. Stoneking, and A. C. Wilson 1987. Mitochondrial DNA and Human Evolution. *Nature* **325**, 31-36 (1987).

Canup, R. M. and E. Asphaug 2001. Origin of the Moon in a giant impact near the end of the Earth's formation. *Nature* **412** 708-712 (2001).

Canup, R. M. and K. Righter 2000, Editors. *"Origin of the Earth and Moon"*. University of Arizona Press, Tucson (2000).

Capo, R. C. and D. J. DePaolo 1990. Seawater Strontium Isotopic Variations from 2.5 Million Years Ago. *Science* **249**, 51-55 (1990).

Caramelli, D. *et al.* 2003. Evidence for a genetic discontinuity between Neandertals and 24,000-year-old anatomically modern humans. *Proceedings of the National Academy of Sciences USA* **100**, 6593-6597 (2003).

Caraveo, P.A. *et al.* 1997. *HST* Data suggest proper motion for the optical counterpart of GRB 970228. *Astronomy and Astrophysics* **326,** L13–L16 (1997).

Carey, S. W. 1976. *"The Expanding Earth"*. Elsevier, Amsterdam (1976).

Carey, S. W. 1988. *"Theories of the Earth and Universe, A History of Dogma in the Earth Sciences"*. Stanford University Press, Stanford (1988).

Carlson, R. W. and G. W. Lugmair 2000. Timescales of Planetesimal Formation and Differentiation Based on Extinct and Extant Radioisotopes. Pages 25-44 in *"Origin of the Earth and Moon"*. R. M. Canup and K. Righter, Editors. University of Arizona Press, Tucson (2000).

Carroll, A. R *et al.* 1998. Eolian-derived siltstone in the Upper Permian Phosphoria Formation: Implications for marine upwelling. *Geology* **26**, 1023-1026 (1998).

Carr, M. H. 1976. The Volcanoes of Mars. *Scientific American* **January**, 33-43 (1976).

Carr, M. H. 1981. Mars. Pages 53-64 in *"The New Solar System"*, 3rd Edition. J. K. Beatty and A. Chaikin, Editors. Sky Publishing Corporation, Cambridge (1981).

Carr, M. H. 1987. Water on Mars. *Nature* **326**, 30-35 (1987).

Carr, M. H. 1996. *"Water on Mars"*. Oxford University Press, Oxford (1996).

Carr, M. H. *et al.* 1998. Evidence for a subsurface ocean on Europa. *Nature* **391**, 363-365 (1998).

Carter, Bill and Merri Sue Carter 2002. *"Latitude"*. Naval Institute Press, Annapolis (2002).

Castro, J. M. and D. B. Dingwell 2009. Rapid ascent of rhyolitic magma at Chaitén volcano, Chile. *Nature* **461**, 780-783 (2009).

Catalano, P. 1997. On the Trail of Rogue Planets. *Astronomy*, **December**, 37-40 (1997).

Cattermole, P. 1994. *"Venus, the geological story"*. The Johns Hopkins University Press, Baltimore (1994).

Cavin, L. 2010. On Giant Filter Feeders. *Science* **327**, 968-969 (2010).

Cederblad, S. 1946. Studies of bright diffuse Galactic nebulae. *Catalogue VII/231*, Centre de Données Astronomiques de Strasbourg (1946).

Chaisson, E. J. 1994. *"The Hubble Wars"*. HarperCollins, New York (1994).

Chandler, M. 1993. Not just a lot of hot air. *Nature* **363**, 623-624 (1993).

Chang, Y. C. *et al.* 1981. Light Curves of Asteroids (IV). *Chinese Astronomy and Astrophysics* **5**, 434-437 (1981).

Chapman, C. R. 1988. Mercury: Introduction to an End-Member Planet. .Pages 1-23 in *"Mercury"*. F. Vilas, C. R. Chapman, M. S. Matthews, Editors. University of Arizona Press, Tucson (1988).

Chapman, C. R. 1994. Ida's Moon. *Mercury* **March/April**, 22-23 (1994).

Chapman, C. R. *et al.* 1995. Discovery and physical properties of Dactyl, a satellite of asteroid 243 Ida. *Nature* **374**, 783-785 (1995).

Chapman, G. A. 1992. Sun, Atmosphere, Photosphere. Pages 857-858 in *"The Astronomy and Astrophysics Encyclopedia"*. S. P. Maran, Editor. Van Nostrand Reinhold, New York (1992).

Charnoz, S., J. Salmon, and A. Crida 2010. The recent formation of Saturn's moonlets from viscous spreading of the main rings. *Nature* **465**, 752-754 (2010).

Chatterjee, S. 1995. The Last Dinosaurs of India. *The Dinosaur Report*, **Fall**, 12-17. The Dinosaur Society, Guildford (1995).

Chatterjee, S. 1997. *"The Rise of Birds"*. The Johns Hopkins University Press, Baltimore (1997).

Chatterjee, S. and R. J. Templin 2004. Posture, locomotion, and paleoecology of pterosaurs. *Special Paper 376*. The Geological Society of America, Boulder (2004).

Chauvet, J.-M., E. B. Deschamps, and C. Hillaire 1996. *"Dawn of Art: The Chauvet Cave"*. Harry N. Abrams, New York (1996).

Chen, J. H. and G. J. Wasserburg 1980a. A search for isotopic anomalies in uranium. *Geophysical Research Letters* **7**, 275-278 (1980).

Chen, J. H. and G. J. Wasserburg 1980b. U and Pb Isotopes in Allende Inclusions and Meteoritic Whitlockite. *Meteoritics* **15**, 271 (1980).

Chen, J. H. and G. J. Wasserburg 1981. The Pd-Ag Systematics in IVA and IVB Iron Meteorites and in Pallasites. *Lunar and Planetary Science Conference XX* (1981).

Chen, P.-j., Z-m. Dong, and S-n Zhen. 1998. An exceptionally well-preserved theropod dinosaur from the Yixian Formation of China. *Nature* **391**, 147-152 (1998).

Chevalier, R. A. and R. T. Emmering 1988. The interstellar light echoes around SN 1987A. *Astrophysical Journal* **331**, L105-L108 (1988).

Chevalier, R. A. and R. T. Emmering 1989. Illuminating a red supergiant wind around SN 1987A. *Astrophysical Journal* **342**, L75-L78 (1989).

Christianson, G. E. 1995. *"Edwin Hubble; Mariner of the Nebulae"*. University of Chicago Press, Chicago (1995).

Christy, J. W. and R. S. Harrington 1978. The satellite of Pluto. *Astronomical Journal* **83**, 1005-1008 (1978).

Chupp, E. L. 1992. The Gamma-Ray Cosmos. *Science* **258**, 1894-1896 (1992).

Cisowski, S. M. and L. L. Hood 1991. The Relict Magnetism of Meteorites. Pages 761-784 in *"The Sun in Time"*. C. P. Sonett, M. S. Giampapa, and M. S. Matthews, Editors. University of Arizona Press, Tucson (1991).

Clark, D. H. and F. R. Stephenson 1977. *"The Historical Supernovae"*. Pergamon Press, Oxford (1977).

Clark, G. A. 1999. Highly Visible, Curiously Intangible. *Science* **283**, 2029-2032 (1999).

Clark J. D. *et al.* 1994. African *Homo erectus*: Old Radiometric Ages and Young Oldowan Assemblages in the Middle Awash Valley, Ethiopia. *Science* **264**, 1907-1910 (1994).

Clarke, T. E. *et al.* 1994. Do Gamma-Ray Bursts Come from the Oort cloud? *Astronomical Journal* **107**, 1873-1878 (1994).

Clayton, D. D. 1994. Production of ^{26}Al and other extinct radionuclides by low-energy heavy cosmic rays in molecular clouds. *Nature* **368**, 222-224 (1994).

Clayton, R. N. 1999. Primordial Water. *Science* **285**, 1364-1365 (1999).

Clemens, S. C., J. W. Farrell, and L. P. Gromet 1993. Synchronous changes in seawater strontium isotope composition and global climate. *Nature* **363**, 607-609(1993).

Close, L. 2010. A giant surprise. *Nature* **468**, 1048-1049 (2010).

Clottes, J. and D. Lewis-Williams 1996. *"The Shamans of Prehistory"*. Harry N. Abrams, New York (1996).

Coffin, M. F. and O. Eldholm 1993. Large Igneous Provinces. *Scientific American* **October**, 42-49 (1993).

Cohen, I. B. (editor) 1958. *"Isaac Newton's Papers & Letters on Natural Philosophy"*. Harvard University Press, Cambridge (1958).

Cohen, M. H. 1961. Radiation in a Plasma. I. Cerenkov Effect. *Physical Review* **123**, 711-721 (1961).

Cohen, M. H. 1992. Active Galaxies and Quasistellar Objects, Superluminal Motion. Pages 17-19 in *"The Astronomy and Astrophysics Encyclopedia"*. S. P Maran, Editor. Van Nostrand Reinhold, New York (1992).

Cohn, N. 1996. *"Noah's Flood"*. Yale University Press, New Haven (1996).

Colbert, E. H. 1991. *"Evolution of the Vertebrates"*. Wiley-Liss, New York (1991).

Coleman, M. and K. Hodges 1995. Tibetan plateau uplift 14 million years ago. *Nature* **374**, 49-51 (1995).

Coles, P. 2005. "The state of the Universe". *Nature* **433**, 248-256 (2005).

Collar, J 1996. Biological Effects of Stellar Collapse Neutrinos. *Physical Review Letters* **76**, 999-1002 (1996).

Collins, H. 2004. *"Gravity's Shadow"*. University of Chicago Press, Chicago (2004).

Collins, L. G. and C. W. Hunt 1992. Silane Systematics: Interpretations of Granitization *in situ*. Chapter V in *"Expanding Geospheres"*. C. W. Hunt, Editor. Polar Publishing, Calgary (1992).

Coltorti, M. *et al.* 1982. Reversed magnetic polarity at an early Lower Palaeolithic site in Central Italy. *Nature* **300**, 173-176 (1982).

Committee on the Biological Effects of Ionizing Radiation 1980. *"The Effects on Populations of Exposure to Low Levels of Ionizing Radiation"*. National Academy of Sciences, Washington (1980).

Connerney, J. E. P. *et al.* 2005. Tectonic implications of Mars crustal magnetism. *Proceedings of the National Academy of Sciences USA* **102**, 14970-14975 (2005).

Conway Morris, S. 1998. *"The Crucible of Creation"*. Oxford University Press, Oxford (1998).

Coppens, Y. 1994. East Side Story: The Origin of Humankind. *Scientific American* **May**, 88-95 (1994).

H. S. F. Cooper Jr. 1993. *"The Evening Star: Venus Observed"*. Farrar Strauss Giroux, New York (1993).

Couch, W. J., D. A. Allen, and D. F. Malin 1990. Photographic imaging of light echoes from SN 1987A. *Monthly Notices of the Royal Astronomical Society* **242**, 555-559 (1990).

Courtillot, V. 1988. Deccan flood basalts and the Cretaceous/Tertiary Boundary. *Nature* **333**, 843 (1988).

Courtillot, V. E. 1990. (What Caused the Extinction?) A Volcanic Eruption. *Scientific American* **October**, 85-92 (1990).

Courtillot, V. and F. Fluteau 2010. Cretaceous Extinctions: The Volcanic Hypothesis. *Science* **328**, 973-974 (2010).

Cowan, G. A. 1976. A Natural Fission Reactor. *Scientific American* **July**, 36-47 (1976).

Cowan, G. A. and H. A. Adler 1976. The variability of the natural abundance of ^{235}U. *Geochimica et Cosmochimica Acta* **40**, 1487-1490 (1976).

Cowen, R. 1999. Found: Primordial Water. *Science News* **156**, 284-285 (1999).

Cowen, R. 2000. Ganymede May Have Vast Hidden Ocean. *Science News* **158**, 404 (2000) citing David J. Stevenson.

Cowen, R. 2006a. The Whole Enceladus. *Science News* **168**, 282-284 (2006).

Cowen, R. 2006b. Propelling Evidence. *Science News* **169**, 198 (2006).

Cowen, R. 2006c. Brilliant! *Science News* **169**, 230 (2006).

Cox, A. 1968. Lengths of Geomagnetic Polarity Intervals. *Journal of Geophysical Research* **73**, 3247-3260 (1968).

Crabb, C. 1998. Atomic Rock. *Earth* **April**, 18-19 (1998).

Craig, H. 1994. Retention of Helium in Subducted Interplanetary Dust Particles. *Science* **265**, 1892-1893 (1994).

Craik, D. 1995. *"Magnetism"*. John Wiley & Sons, Chichester (1995).

Croft, S. K. and L. A Soderblom 1991. Geology of the Uranian Satellites. Pages 561-628 in *"Uranus"*. J. T. Bergstrahl, E. D. Miner, and M. S. Mathews, Editors. University of Arizona Press, Tucson (1991).

Crommelin, A. C. D. 1931. The Discovery of Pluto. *Monthly Notices of the Royal Astronomical Society* **91**, 380-385 (1931), taken from Hoyt 1980.

Crosby, H. W. 1997. *"The Cave Paintings of Baja California"*. Sunbelt Publications, San Diego (1997).

Crotts, A. P. S. 1988. Discovery of optical echoes from supernova 1987A: new probes of the Large Magellanic Cloud. *Astrophysical Journal* **333**, L51-L54 (1988).

Crotts, A. P. S. and S. R. Heathcote 1991. Velocity structure of the ring nebula around supernova 1987A. *Nature* **350**, 683-685 (1991).

Crowe, M. J. 1994. *"Modern Theories of the Universe from Herschel to Hubble"*. Dover Publications, New York (1994).

Crowley, T. J. and S. K. Baum 1991. Toward Reconciliation of Late Ordovician (~440 Ma) Glaciation With Very High CO_2 Levels. *Journal of Geophysical Research* **96**, 22597-22610 (1991).

Crowley, T. J., J. G. Mengel, and D. A. Short 1987. Gondwanaland's seasonal cycle. *Nature* **329**, 803-807 (1987).

Cruikshank, D. P. 1995. *"Neptune and Triton"*. D. P. Cruikshank, Editor. University of Arizona Press, Tucson (1995).

Cruikshank, D. P. *et al.* 1993. Ices on the Surface of Triton. *Science* **261**, 742-744 (1993).

Cumming, R. J. and W. P. S. Meikle 1993. Cold bright matter near supernova 1987A. *Monthly Notices of the Royal Astronomical Society* **262**, 689-698 (1993).

Cummings, J. R. *et al.* 1993. SAMPEX Observations of Geomagnetically Trapped Anomalous Cosmic Rays. *Proceedings of the 23rd International Cosmic Ray Conference, Calgary 1993* (1993).

Cunha, E. and F. Cardosa 2001. The osteological series from Cabeço Da Amoreira (Muge, Portugal). *Bulletins et memoires de la Société d'Anthropologie de Paris* **13 (3-4)** (2001).

Cunha, E., C. Umbelino, and F. Cardoso 2002. New Anthropological Data on the Mesolithic Communities from Portugal: the Shell Middens from Sado. *Human Evolution* **17**, 187-197 (2002).

Cunningham, C. J. 2001. The Kuiper Belt and Beyond. *Mercury* **September-October**, 11 (2001).

Curtis, A. R. 1990. *"Space Almanac"*. Arcsoft Publishers, Woodsboro (1990).

Cuzzi, J. N. and C. M. O'D. Alexander 2006. Chondrule formation in particle-rich nebular regions at least hundreds of kilometers across. *Nature* **441**, 483-485 (2006).

Cuzzi, J. N. and P. R. Estrada 1998. Compositional Evolution of Saturn's Rings Due to Meteoroid Bombardment. *Icarus* **132**, 1-35 (1998).

Cuzzi, J. N. *et al.* 2010. An Evolving View of Saturn's Dynamic Rings. *Science* **327**, 1470-1475 (2010).

Dahl, T. W *et al.* 2010. Devonian rise in atmospheric oxygen correlated to the radiations of terrestrial plants and large predatory fish. *Proceedings of the National Academy of Sciences USA* **107**, 17911-17915 (2010).

Dalrymple, G. B. 1991. *"The Age of the Earth"*. Stanford University Press, Stanford (1991).

Dalton, R. 2004. Little lady of Flores forces rethink of human evolution. *Nature* **431**, 1029 (2004).

Dalton, R. 2005. Image problems jeopardize comet mission's impact. *Nature* **434**, 685 (2005).

Dalton, R. 2009. Impact theory under fire once more. *Nature* **461**, 861 (2009).

Dalton, R. 2010. Ancient DNA set to rewrite human history. *Nature* **465**, 148-149 (2010).

Daly, M. C. *et al.* 1991. Late Palaeozoic deformation in central Africa: a result of distant collision? *Nature* **350**, 605-607 (1991).

Daly, R. A. 1943. Meteorites and an Earth-Model. *Bulletin of the Geological Society of America* **54**, 401-456 (1943).

Danziger, I. J. *et al.* 1987. SN 1987A: Observational Results Obtained at ESO. Pages 37-50 in *"Supernova 1987A in the Large Magellanic Cloud"*. M. Kafatos and A. Michalitsianos, Editors. Cambridge University Press, Cambridge (1988).

Davis, R. 2010. Big G revisited. *Nature* **468**, 181-183 (2010).

Davis, R. and A. N. Cox 1991. Solar Neutrino Experiments. Pages 51-85 in *"Solar Interior and Atmosphere"*. A. N. Cox, W. C. Livingston, M. S. Matthews, Editors. University of Arizona Press, Tucson (1991).

Davis, R. J., T. W. B. Muxlow, and R. G. Conway 1985. Radio emission from the jet and lobe of 3C273. *Nature* **318**, 343-345 (1985).

Davis, R. J., S. C. Unwin, and T. W. B. Muxlow 1991. Large scale superluminal motion in the quasar 3C273. *Nature* **354**, 374-376 (1991).

Dawkins, R. 1987. *"The Blind Watchmaker"*. W. W. Norton & Company, New York (1987).

de Bergh, C. *et al.* 1991. Deuterium on Venus: Observations from Earth. *Science* **251**, 547-549 (1991).

Decowski, P. 2009. Best-fit to theory for KamLAND 2008. Personal communication, email to R. J. Tuttle. June 02, 2009 (2009).

De Hon, R. A. and E. A. Pani 1993. Duration and Rates of Discharge: Maja Valles, Mars. *Journal of Geophysical Research* **98**, E5, 9129-9138 (1993).

Delafield, H. J. et al. 1978. Measurements of the neutron leakage spectrum from the VIPER reactor. Health Physics 35, 471-480, (1978).

Delsemme, A. 1997. The Origin of the Atmosphere and the Oceans, Chapter 2 in *"Comets and the Origin and Evolution of Life"*, P. J. Thomas, C. F. Chyba, and C. P. McKay, Editors. Springer, New York (1997).

Delson, E. and K. Harvati 2006. Return of the last Neanderthal. *Nature* **443**, 762-763 (2006).

Denton, G. H. *et al.* 2010. The Last Glacial Termination. *Science* **328**, 1652-1656 (2010).

DePaolo, D. J. and B. L. Ingram 1985. High-Resolution Stratigraphy with Strontium Isotopes. *Science* **227**, 938-941 (1985).

Dermer, C. D. 1992. The answer is near. *Nature* **351**, 272 (1992).

de Rújula, A. 1987. May a supernova bang twice? *Physics Letters B* **193**, 514-524 (1987).

Desch, S. 2006. How to make a chondrule. *Nature* **441**, 416-417 (2006).

de Silva, S. 2003. Eruption linked to El Niño. *Nature* **426**, 239-240 (2003).

DeWitt, W. 1989. *"Human Biology -- Form, Function, and Adaptation"*. Scott, Foresman and Company, Glenview (1989).

Dia, A. N. *et al.* 1992. Seawater Sr isotope variations over the past 300 kyr and influence of global climate cycles. *Nature* **356**, 786-788 (1992).

Di Achille, G. and B. M. Hynek 2010a. The case of an Integrated Global Hydrosphere on Early Mars: Clues from the Distribution of Ancient Deltas and Valley Networks. *41st Lunar and Planetary Science Conference* 2366-2367 (2010).

Di Achille, G. and B. M. Hynek 2010b. Ancient ocean on Mars supported by global distribution of deltas and valleys. *Nature Geoscience* **3**, 459-463 (2010).

di Serego Alighieri, S. *et al.* 1989. Polarized light in high-redshift radio galaxies. *Nature* **341**, 307-309 (1989).

Disney, M. 1998. A New Look at Quasars. *Scientific American* **June**, 52-57 (1998).

Dixon, R. K. *et al.* 1994. Carbon Pools and Flux of Global Forest Ecosystems. *Science* **263**,185-190 (1994).

Dobrzhinetskaya, L., H. W. Green II, and S. Wang 1996. Alpe Arami: A Peridotite Massif from Depths of More Than 300 Kilometers. *Science* **271**, 1841-1845 (1996).

DOD 1962. *"The Effects of Nuclear Weapons"*, Revised Edition. S. Glasstone, Editor. U. S. Department of Defense, U. S. Government Printing Office, Washington (1962).

DOD and ERDA 1977. *"The Effects of Nuclear Weapons"*, Third Edition. S. Glasstone and P. J. Dolan, Editors. U. S. Department of Defense and Energy Research and Development Administration, U. S. Government Printing Office, Washington (1977).

Dodson, M. H. 1989. Diamond dating anomalies. *Nature* **337**, 207 (1989).

Donahue, T. M. 1995. Evolution of water reservoirs on Mars from D/H ratios in the atmosphere and crust. *Nature* **374**, 432-434 (1995).

Donahue, T. M. and J. B. Pollack 1983. Origin and Evolution of the Atmosphere of Venus. Pages 1003-1036 in *"Venus"*. D. M. Hunten *et al.*, Editors. University of Arizona Press, Tucson (1983).

Donahue, T. M. *et al.* 1997. Ion/Neutral Escape of Hydrogen and Deuterium: Evolution of Water. Pages 385-414 in *"Venus II"*, S. W. Bougher, D. M. Hunten, and R. J. Phillips, Editors. University of Arizona Press, Tucson (1997).

Donnelly, I. 1882. *"Atlantis: The Antediluvian World"*. Egerton Sykes, Editor. Originally published in 1882. Gramercy, New York (1949).

Dones, L. 1991. A Recent Cometary Origin for Saturn's Rings? *Icarus* **92**, 194-203 (1991).

Doody, D. 1994. Magellan Aerobrakes into Venus' Atmosphere. *The Planetary Report* **March/April**, 6-13 (1994).

Doran, G. H. 1991. Florida Archaeological Wet Sites – Windover. Pages 205-228 in *"The Art and Archaeology of Florida Wetlands"*. B. A. Purdy, Editor. CRC Press, Boca Raton (1991).

Dotani, T. *et al.* 1987. Discovery of an unusual hard X-ray source in the region of supernova 1987A. *Nature* **330**, 230-231 (1987).

Draganić, I. G. *et al.* 1993. *"Radiation and Radioactivity On Earth and Beyond"*. CRC Press, Boca Raton (1993).

Drake, M. T. 1990. Experiment confronts theory. *Nature* **347**, 128-129 (1990).

Driscoll, R. B. 1988. Nuclear Detonation of a Planet. Abstract JE5, American Physical Society Meeting, Crystal City, January 25-28, 1988. (See also, Abstract GM12, American Physical Society Meeting, Baltimore, April 18-21, 1988.) (1988).

D. T. 2001. The Solar System's Edge? *Sky & Telescope* **March**, 26 (2001).

Duarte, C. *et al.* 1999. The early Upper Paleolithic human skeleton from the Abrigo do Lagar Velho (Portugal) and modern human emergence in Iberia. *Proceedings of the National Academy of Sciences USA* **96**, 7604-7609 (1999).

Durisen, R. H. 1999. Planetary Rings: Moonlets in a Cosmic Sandblaster. *Mercury* **September/October**, 11-23 (1999).

Dyson, J. 1993. Shocking behavior of young stars in Orion. *Nature* **363**, 21-22 (1993).

Earth 1994. Solar Paradox Resolved? *Earth* **March**, 14 (1994).

Easton, C. 1900. A New Theory of the Milky Way. *Astrophysical Journal* **12**, 136-158 plus Plate XI (1900).

Easton, C. 1913. A Photographic Chart of the Milky Way and the Spiral Theory of the Galactic System. *Astrophysical Journal* **37**, 105-118 plus Plate III (1913).

Eberhardt, P. *et al.* 1995. The D/H and $^{18}O/^{16}O$ ratios in water from comet P/Halley. *Astronomy and Astrophysics* **302**, 301-316 (1995).

Eddy, J. A. 1976. The Maunder Minimum. *Science* **192**, 1189-1202 (1976).

Eddy, J. A. and A. A. Boornazian 1979. Secular decrease in the solar diameter. *Bulletin of the American Astronomical Society* **11**, 437 (abstract) (1979).

Edey, M. A. and D. C. Johanson 1989. *"Blueprints: Solving the Mystery of Evolution"*. Penguin Books, New York (1989).

Edgeworth, K. E. 1943. The evolution of our planetary system. *Journal of the British Astronomical Society* **53**, 181-189 (1943).

Edgeworth, K. E. 1949. The origin and evolution of the Solar System. *Monthly Notices of the Royal Astronomical Society* **109**, 600-609 (1949).

Edmond, J. M. 1993. Himalayan Tectonics, Weathering Processes, and the Strontium Isotope Record in Marine Limestone. *Science* **258**, 1594-1597 (1993).

Edwards, R. L. *et al.* 1993. A Large Drop in Atmospheric $^{14}C/^{12}C$ found in the corals of Papua New Guinea and Reduced Melting in the Younger Dryas, Documented with ^{230}Th Ages of Corals. *Science* **260**, 962-968 (1993).

Eicher, D. J. 1992. A New Member of the Family. *Astronomy* **December**, 38-39 (1992).

Eicher, D. J. 1994. Candles to Light the Night. *Astronomy* **September**, 32-39 (1994).

Einstein, A. 1916. Die Grundlagen der allgemeinen Relativitätstheorie. *Annalen der Physik* **49**, 769-822 (1916).

Einstein, A. 1961. *"Relativity"*. Random House, New York (1961).

Eisenbud, M. 1987. *"Environmental Radioactivity"*. Academic Press, Orlando (1987).

Elder, J. W. 1986. *"The Structure of the Planets"*. Academic Press, London (1986).

Eldredge, J., D. Walsh, and C. R. Scotese 1997. Plate Tracker. http://www.scotese.com/software.htm. (1997).

Eldredge, N. 1999. *"The Pattern of Evolution"*. W. H. Freeman and Company, New York (1999).

Eliot, J. and R. Kerr 1984. *"Rings, Discoveries from Galileo to Voyager"*. The MIT Press, Cambridge (1984).

Elliot, J. L. 1999. The Warming Wisps of Triton. *Sky & Telescope* **February**, 42-47 (1999).

Elliot, J. L. *et al.* 1998. Global warming on Triton. *Nature* **393**, 765-769 (1998).

Ellis-Evans, J. C. and D. Wynn-Williams 1996. A great lake under the ice. *Nature* **381**, 644-646 (1996).

Elston, R. *et al.* 1994. Detection of strong iron emission from quasars at redshift z > 3. *Nature* **367**, 250-251 (1994).

Elmegreen, D. M. and B. G. Elmegreen 1992. Density Wave Resonances in Galaxies. Pages 276-285 in *"Evolution of Interstellar Matter and Dynamics of Galaxies"*. J. Palouš, W. B. Burton, P. O. Lindblad, Editors. Cambridge University Press, Cambridge (1992).

Elmegreen, D. M. and B. G. Elmegreen 1993. What Puts the Spiral in Spiral Galaxies? *Astronomy* **September**, 34-39 (1993).

Emiliani, C. 1975. Paleoclimatological Analysis of Late Quaternary Cores from the Northeastern Gulf of Mexico. *Science* **189**, 1083-1088, (1975).

Emiliani, C. 1992. *"Planet Earth"*. Cambridge University Press, Cambridge (1992).

Endress, M., E. Zinner, and A. Bischoff 1996. Early aqueous activity on primitive meteorite parent bodies. *Nature* **379**, 701-703 (1996).

Epstein, S. *et al.* 1987. Unusual stable isotopic ratios in amino acid and carboxylic acid extracts from the Murchison meteorite. *Nature* **326**, 477-479 (1987).

Erba, E. *et al.* 2010. Calcareous Nanoplankton Response to Surface-Water Acidification Around Oceanic Anoxic Event 1a. *Science* **329**, 428-432 (2010).

Erickson, J. 1996. *"Marine Geology: Undersea Landforms and Life Forms"*. Facts on File, New York (1986).

Ernst, R. E. and W. R. A. Baragar 1992. Evidence from magnetic fabric for the flow pattern of magma in the Mackenzie giant radiating dyke swarm. *Nature* **356**, 511-513 (1992).

Erwin, D. H. 1994. The Permian-Triassic extinction. *Nature* **367**, 231-236 (1994).

Esposito, L. W. *et al.* 1984. Saturn's Rings: Structure, Dynamics, and Particle Properties. Pages 463-545 in *"Saturn"*. T. Gehrels and M. S. Matthews, Editors. University of Arizona Press, Tucson (1984).

European Space Agency *et al.* 1997. *"The Hipparcos and Tycho Catalogues"*. Volume 1. ESA Publications Division, Noordwijk (1997).

Fairall, A. P. 1992. "A Caution to Those Who Measure Galaxy Redshifts". *The Observatory* **112**, 286-287 (1992).

Fairbanks, R. G. 1989. A 17,000-year glacio-eustatic sea level record: influence of glacial melting rates on the Younger Dryas event and deep-ocean circulation. *Nature* **342**, 637-642 (1989).

Faith, J. T. and T. A. Surovell 2009. Synchronous extinction of North America's Pleistocene mammals. *Proceedings of the National Academy of Sciences USA* **106**, 20641-20645 (2009).

Famous Astrophysicist 1994, personal communication, November 21, 1994. (1994).

Fan, X. 2001. A Survey of z>5.8 Quasars in the Sloan Digital Sky Survey. I. Discovery of Three New Quasars and the Spatial Density of Luminous Quasars at z~6. *Astronomical Journal* **122**, 2833-2849 (2001).

Farinella, P. *et al.* 1994. Asteroids falling into the Sun. *Nature* **371**, 314-317 (1994).

Farrand, W. R. 1961. Frozen Mammoths and Modern Geology. *Science* **133**, 729-735 (1961).

Farrand, W. R. 1962. Frozen Mammoths. *Science* **137**, 449-452 (1962).

Fastovsky, D. E. and D. E. Weishampel 1996. *"The Evolution and Extinction of the Dinosaurs"*. Cambridge University Press, Cambridge (1996).

Faux, C. M. and K. Padian 2007. The opisthotonic posture of vertebrate skeletons: postmortem contraction or death throes? *Paleobiology* **33**; 201-226 (2007).

Feber, R. C., T. C. Wallace, and L. M. Libby 1984. Uranium in the Earth's Core. *Eos* **65**, 785 (1984).

Feder, K. L. and M. A. Park 1989. *"Human Antiquity"*. Mayfield Publishing Company, Mountain View (1989).

F. F. 1993. A Stellar Blast From the Past. *Science* **262**, 1372 (1993).

Feldstein, S. N. *et al.* 1996. Ferric-ferrous ratios, H_2O contents and D/H ratios of phlogopite and biotites from lavas of different tectonic regimes. *Contributions to Mineralogy and Petrology* **126**, 51-66 (1996).

Fernie, J. D. 1969. The Period-Luminosity Relation: A Historical Review. *Publications of the Astronomical Society of the Pacific* **81**, 707-731 (1969).

Ferris, T. 1988. *"Coming of Age in the Milky Way"*. William Morrow and Company, New York (1988).

Filippelli, G. M. 2010. Phosphorus and the gust of fresh air. *Nature* **467**, 1052-1053 (2010).

Fink, U., H. P. Larson, and T. N. Gautier III 1976. New upper limits for atmospheric constituents on Io. *Icarus* **27**, 439-446 (1976).

Fink, U. *et al.* 1992. The Steep Red Spectrum of *1992 AD*: An Asteroid Covered with Organic Material? *Icarus* **97**, 145-147 (1992).

Finlayson, C. *et al.* 2006. Late survival of Neanderthals at the southernmost extreme of Europe. *Nature* **443**, 850-853 (2006).

Fireman, E. L. 1958. Distribution of Helium-3 in the Carbo Meteorite. *Nature* **181**, 1725 (1958).

Firestone, R., A. West, and S. Warwick-Smith 2006. *"The Cycle of Cosmic Catastrophes*. Bear & Company, Rochester (2006).

Firestone, R. B. *et al.* 2007. Evidence for an extraterrestrial impact 12,900 years ago that contributed to the megafaunal extinctions and the Younger Dryas cooling. *Proceedings of the National Academy of Sciences USA* **104**, 16016-16021 (2007).

Fischbach, E. 2008. Decay rates and the Sun. Personal communication, email to R. J. Tuttle. September 9, 2009 (2009).

Fischbach, E. *et al.* 2009. Time-Dependent Nuclear Decay Parameters: New Evidence for New Forces? *Space Science Reviews* **145**, 285-335 (2009).

Fischer, A. G. 1996. Mineralogy: Carved in Stone. *Planetary Report* **July/August**, 8-10 (1996).

Fischer, D. 2009. Searching for New Worlds: The power of many. *The Planetary Report* **29**, 12-16 (2009).

Fishman, G. J. 1992. Gamma-Ray Astronomy, Space Missions. Pages 282-284 in *"The Astronomy and Astrophysics Encyclopedia"*. S. P. Maran, Editor. Van Nostrand Reinhold, New York (1992).

Flam, F. 1993. A Double Dose of Gamma-Ray Bursts. *Science* **260**, 756 (1993).

Flam, F. 1994. The Space Telescope Spies on Ancient Galaxy Menageries. *Science* **266**, 1806 (1994).

Flasar, F. M. 1991. At the eye of the storm. *Nature* **349**, 21 (1991).

Flasar, F. M. 1998. Titan weather report. *Nature* **395**, 541-553 (1998).

Florence, R. 1994. *"The Perfect Machine - Building the Palomar Telescope"*. HarperCollins, New York (1994).

Forget, F. and R. T. Pierrehumbert 1997. Warming Early Mars with Carbon Dioxide Clouds. *Science* **278**, 1273-1276 (1997).

Foukal, P. 1994. Stellar Luminosity Variations and Global Warming. *Science* **264**, 238-239 (1994).

Fox, S. 2009. Leapin' Lizards. *Scientific American*, **May**, 25-26 (2009).

Frail, D. A. and S. R. Kulkarni 1991. Unusual interaction of the high-velocity pulsar PSR1757-24 with the supernova remnant G5.4-1.2. *Nature* **352**, 785-787 (1991).

Francis, P. 1993. *"Volcanoes, a Planetary Perspective"*. Oxford University Press, Oxford (1993).

Frank, L. A. 1990. *"The Big Splash"*. Carol Publishing Group, New York (1990).

Frank, L. A. *et al.* 1997. Outflow of hydrogen ions from Ganymede. *Geophysical Research Letters* **24**, 2151-2154 (1997).

Frankel, C. 1996. *"Volcanoes of the Solar System"*. Cambridge University Press, Cambridge (1996).

Freeman, K. C. 1970. On the Disks of Spiral and S0 Galaxies. *Astronomical Journal* **160**, 811-830 (1970).

Frei, R. *et al.* 2009. Fluctuations in Precambrian atmospheric oxygenation recorded by chromium isotopes. *Nature* **461**, 250-253 (2009).

Friedlander, M. W. 1989. *"Cosmic Rays"*. Harvard University Press, Cambridge (1989).

Friedman, M. *et al.* 2010. 100-million-Year Dynasty of Giant Planktivorous Bony Fishes in the Mesozoic seas. *Science* **327**, 990-993 (2010).

Friedmann, A. A. 1922. On the curvature of space. *Zeitschrift für Physik* **10**, 377-386 (1922).

Froelich, P. N. 1993. Ruling in the improbable. *Nature* **363**, 585-587 (1993).

Frommert, H and C. Kronberg 2007. Messier 31. Students for the Exploration and Development of Space. http://seds.org/messier/Jpg/m31.jpg (2007).

Fruchter, A. S. *et al.* 1999. The Fading Optical Counterpart of GRB 970228, 6 Months and 1 Year Later. *Astrophysical Journal* **516**, 683-692 (1999).

Fujiwara, A. *et al.* 1989. Experiments and Scaling Laws for Catastrophic Collisions. Pages 240-265 in *"Asteroids II"*. R. P. Binzel, T. Gehrels, and M. S. Matthews, Editors. University of Arizona Press, Tucson (1989).

Fukai, Y. 1984. The iron-water reaction and the evolution of the Earth. *Nature* **308**, 174-175 (1984).

Gabrielse, G. 2010. Slow antihydrogen. *Physics Today* **March**, 68-69 (2010).

Gabunia, L. *et al.* 2000. Earliest Pleistocene Hominid Cranial Remains from Dmanisi, Republic of Georgia: Taxonomy, Geological Setting, and Age. *Science* **288**, 1019-1025 (2000).

Gaffey, M. J., J. F. Bell, and D. P. Cruikshank 1989. Reflectance Spectroscopy and Asteroid Surface Mineralogy. Pages 98-127 in *"Asteroids II"*. R. P. Binzel, T. Gehrels, and M. S. Matthews, Editors. University of Arizona Press, Tucson (1989).

Gaffney, E. S. 1990. The comparative osteology of the Triassic turtle *Proganochelys*. *Bulletin of the American Museum of Natural History* **194**, 1-236 (1990).

Gaisser, T. K. 1990. *"Cosmic Rays and Particle Physics"*. Cambridge University Press, Cambridge (1990).

Gal-Yam, A. *et al.* 2009. Supernova 2007bi as a pair-instability explosion. *Nature* **462**, 624-627 (2009).

Galama, T.J. *et al.* 2000. Evidence for a supernova in reanalyzed optical and near-infrared images of GRB970228. *Astrophysical Journal* 536:185-194 (2000), and arXiv:astro-ph/9907264v1. (1999).

Galilei, G. 1610. *"Siderius Nuncius or the Sidereal Messenger".* Translated by A. Van Helden. University of Chicago Press, Chicago (1989).

Galton, P. M. 1990. Stegosauria. Pages 435-455 in *"The Dinosauria".* D. B. Weishampel, P. Dodson, and H. Osmólska, editors. University of California Press, Berkeley (1990).

Gamble, C. 1994. The Peopling of Europe, 700,000 – 40,000 Years before the Present, Chapter 1 in *"The Oxford Illustrated Prehistory of Europe".* B. Cunliffe, Editor. Oxford University Press, Oxford (1994).

Garcia-Castellanos, D. *et al.* 2009. Catastrophic flood of the Mediterranean after the Messinic salinity crisis. *Nature* **462**, 778-781 (2009).

Gaskell, C. M. and W. C. Keel 1987. Another Supernova with a Blue Progenitor. Pages 13-15 in *"Supernova 1987A in the Large Magellanic Cloud".* M. Kafatos and A. Michalitsianos, Editors. Cambridge University Press, Cambridge (1988).

Gauthier-Lafaye, F. 1997. The last natural fission reactor. *Nature* **387**, 337 (1997).

Gee, H. 1992. Statistical cloud over African Eden. *Nature* **355**, 583 (1992).

Gehrels, N. and W. Chen 1993. The Geminga supernova as a possible cause of the local interstellar bubble. *Nature* **361**, 706-707 (1993).

Gehrels, N. *et al.* 2004. The *Swift* Gamma-ray Burst Mission. *Astrophysical Journal* **611**, 1005-1020 (2004).

Gentry, R. V. 1973. Radioactive Halos. *Annual Review of Nuclear Science* **23**, 347-362 (1973).

Gentry, R. V. 1974a. Radiohalos in a Radiochronological and Cosmological Perspective. *Science* **184**, 62-66 (1974).

Gentry, R. V. 1974b. 'Spectacle' array of ^{210}Po halo radiocentres in biotite: a nuclear geophysical enigma. *Nature* **252**, 564-566 (1974).

Gentry, R. V. 1992. *"Creation's Tiny Mystery",* 3rd Edition. Earth Science Associates, Knoxville (1992).

Gentry, R. V. *et al.* 1976. Radiohalos and Coalified Wood: New Evidence Relating to the Time of Uranium Introduction and Coalification. *Science* **194**, 315-318 (1976).

Geodigest 1992. That asteroid again? *Geology Today* **January-February**, 13 (1992).

Gersonde, R. *et al.* 1997. Geological record and reconstruction of the late Pliocene impact of the Eltanin asteroid in the Southern Ocean. *Nature* **390**, 357-363 (1997).

Gerya, T. 2010. Dynamical Instability Produces Transform Faults at Mid-Ocean Ridges. *Science* **329**, 1047-1050 (2010).

Gibbons, A. 1995. The Mystery of Humanity's Missing Mutations. *Science* **267**, 35-36 (1995).

Gibbons, A. 1996a. Did Neandertals Lose an Evolutionary "Arms" Race? *Science* **272**, 1587 (1996).

Gibbons, A. 1996b. Archaeology: First Americans: Not Mammoth Hunters, But Forest Dwellers? *Science* **272**, 346-347 (1996).

Gibbons, A. 1996c. DNA Enters Dust Up Over Bones. *Science* **274**, 172 (1996).

Gibbons, A. 1997. Into the Pit of Human History. *Science* **276**, 1332 (1997).

Gibbons, A. 1998. Which of Our Genes Make Us Human? *Science* **281**, 1432-1434 (1998).

Gibbons, A. 2010a. Close Encounters of the Prehistoric Kind. *Science* **328**, 680-684 (2010).

Gibbons, A. 2010b. Tiny Time Machines Revisit Ancient Life. *Science* **330**, 1616 (2010).

Giclas, H. L., R. Burnham Jr., and N. G. Thomas 1971. *"Lowell Proper Motion Survey: 8991 Stars with m>8, PM>0.26 "/year in the Northern Hemisphere".* Lowell Observatory, Flagstaff (1971).

Giclas, H. L., R. Burnham Jr., and N. G. Thomas 1978. *"Lowell Proper Motion Survey: – Southern Hemisphere Catalog".* Lowell Observatory, Flagstaff (1978).

Gilmore, G. 1992. Lithium given and taken away. *Nature* **358**, 108-109 (1992).

Gilmore, G., I. R. King, and P. C. van der Kruit 1990. *"The Milky Way as a Galaxy"*. University Science Books, Mill Valley (1990).

Gilmour, J. D. *et al.* 1995a. Iodine-xenon studies of Bjurböle and Parnallee using RELAX. *Meteoritics* **30**, 405-411 (1995).

Gilmour, J. D. *et al.* 1995b. Xenon isotopes in irradiated and unirradiated samples of Allan Hills 84001. *Meteoritics* **30**, 510-511 (1995).

Gilmour, J. D., J. A. Whitby, and G. Turner 2001. Negative correlation of iodine-129/iodine-127 and xenon-129/xenon-132: Product of closed-system evolution or evidence of a mixed component. *Meteoritics and Planetary Science* **36**, 1283-1286 (2001).

Ginenthal, C. 1995. *"Carl Sagan & Immanuel Velikovsky"*. New Falcon Publications, Tempe (1995).

Gladman, B. and J. Coffey 2009. Mercurian Impact Ejecta: Meteorites and Mantle. *Meteoritics and Planetary Science* **44**, 285-291 (2009).

Glanz, J. 1994. A Dusty Road for Space Physics. *Science* **264**, 28-30 (1994).

Glasstone, S. and P. J. Dolan 1977. *"The Effects of Nuclear Weapons"*. U. S. Government Printing Office, Washington (1977).

Gleick, J. 1987. *"Chaos, Making a New Science"*. Viking, New York (1987).

Goebel, C. J. 1993. Relativistic quantum theory. Pages 1193-1194 in McGraw-Hill Encyclopedia of Physics, Second Edition, S. P. Parker, Editor in Chief. McGraw-Hill, New York (1993).

Goel, P. S. 1962. Calculation of Production Rates of Specific Nuclides in Iron Meteorites. Pages 36-67 in *"Researches on Meteorites"*. C. B. Moore, Editor. John Wiley & Sons, New York (9162).

Goldberg, M. D. *et al.* 1966. *"Neutron Cross Sections"*. Volume IIA, BNL-325, Second Edition, Supplement No. 2. Sigma Center, Brookhaven National Laboratory, Upton (1966).

Goldhaber, A. S. and M. Goldhaber 2011. The neutrino's elusive helicity reversal. *Physics Today* **May 2011**, 40-43 (2011).

Goldreich, P. and W. R. Ward 1972. The Case Against Planet X. *Publications of the Astronomical Society of the Pacific* **84**, 737-742 (1972).

Goldschmidt, R. 1940. *"The Material Basis of Evolution"*. Yale University Press, New Haven (1940).

Goldsmith, D. 1990. *"Supernova, The Violent Death of a Star"*. Oxford University Press, Oxford (1990).

Gönnewein, F. 1991. Mass, Charge, and Kinetic Energy of Fission Fragments, Chapter 8 in *"The Nuclear Fission Process"*. C. Wagemans, Editor. CRC Press, Boca Raton (1991).

Gooding, J. L. 1992. Soil Mineralogy and Chemistry on Mars: Possible Clues from Salts and Clays in SNC Meteorites. *Icarus* **99**, 28-41 (1992).

Gooding, J. L. *et al.* 1991. Aqueous alteration of the Nahkla meteorite. *Meteoritics* **26**, 135-143 (1991).

Goorley, T. and H. Nikjoo 2000. Electron and Photon Spectra for Three Gadolinium-Based Cancer Therapy Approaches. *Radiation Research* **154**, 556-563 (2000).

Gore, R. 1993. Dinosaurs. *National Geographic* **183**, 2-53 (1993).

Gore, R. 1996. The Dawn of Humans – Neandertals. *National Geographic* **January**, 2-35 (1996).

Gore, R. 1997. The Dawn of Humans – Expanding Worlds. *National Geographic* **May**, 84-109 (1997).

Gore, R. 2000. The Dawn of Humans – People Like Us. *National Geographic* **July**, 84-109 (2000).

Goslar, T. *et al.* 1995. High concentration of atmospheric ^{14}C during the Younger Dryas cold episode. *Nature* **377**, 414-417 (1995).

Gostin, V. A., R. R. Keays, and M. W. Wallace 1989. Iridium anomaly from the Acraman impact ejecta horizon: impacts can produce sedimentary peaks. *Nature* **340**, 542-544 (1989).

Gostin, V. A. *et al.* 1986. Impact Ejecta Horizon Within Late Precambrian Shales, Adelaide Geosyncline, South Australia. *Science* **233**, 198-200 (1986).

Gough, D. O. 1981. Solar Interior Structure and Luminosity Variations. *Solar Physics* **74**, 21-34 (1981).

Gouiffes C. *et al.* 1988. Light echoes from SN 1987A. *Astronomy and Astrophysics* **198**, L9-L12 (1988).

Gould, S. J. 1989. *"Wonderful Life"*. W. W. Norton & Co., New York (1989).

Gould, S. J. and N. Eldredge 1993. Punctuated equilibrium comes of age. *Nature* **366**, 223-227 (1993).

Graham, D. W. *et al.* 1993. Mantle Plume Helium in Submarine Basalts from the Galápagos Platform. *Science* **262**, 2023-2026 (1993).

Graham, J. A. 1993. Emission-Line Objects Near R Coronae Australis. *Publications of the Astronomical Society of the Pacific* **105**, 561-564 (1993).

Graham-Smith, F. 1995. Personal communication to R. J. Tuttle, January 20, 1995 (1995).

Gratz, A. J., W. J. Nellis, and N. A. Hinsey 1993. Observations of high-velocity, weakly shocked ejecta from experimental impacts. *Nature* **363**, 522-524 (1993).

Green, R. E. *et al.* 2010. A Draft Sequence of the Neandertal Genome. *Science* **328**, 710-722 (2010).

Greenberg, J. M. 1968. Interstellar Grains. Chapter 6 in *"Nebulae and Interstellar Matter"*. B. M. Middlehurst and L. H. Aller, Editors. University of Chicago Press, Chicago (1968)

Greenberg, R. *et al.* 1991. Miranda. Pages 693-735 in *"Uranus"*, J. T. Bergstrahl, E. D. Miner, and M. S. Mathews, Editors. University of Arizona Press, Tucson (1991).

Gribben, J. 1988. *"The Omega Point"*. Bantam Books, Toronto (1988).

Gribben, J. 1991. *"Blinded by the Light: the secret life of the Sun"*. Harmony Books (Crown Publishers), New York (1991).

Gribben, J. and M. Gribben 2001. *"Ice Age"*. Barnes & Noble, New York (2001).

Grieve, R. A. F. 1990. Impact Cratering on the Earth. *Scientific American* **April**, 66-73 (1990).

Grieve, R. A. F. 2000. Terrestrial Impact Structures, http://gdcinfo.agg.emr.ca:80/crater/ (2000).

Griffith, C. A. *et al.* 1998. Transient clouds in Titan's lower atmosphere. *Nature* **395**, 575-578 (1998).

Grimesey, R. A., D. W. Nigg, and R. L. Curtis 1990. *"COMBINE/PC- A Portable ENDF/B Version 5 Neutron Spectrum and Cross-Section Generation Program"*. EGG-2589, EG&G Idaho, Idaho Falls (1990).

Grinspoon, D. H. 1993. Implications of the high D/H ratio for the sources of water in Venus' atmosphere. *Nature* **363**, 428-431 (1993).

Gross, M. G. 1987. *"Oceanography, A View of the Earth"*. Fourth Edition, Prentice-Hall, Englewood Cliffs (1987).

Grossman, J. N. *et al.* 2000. Bleached chondrules: Evidence for widespread aqueous processes on the parent asteroids of ordinary chondrites. *Meteoritics and Planetary Science* **35**, 467-486 (2000).

Grotzinger, J. and L. Royden 1990. Elastic strength of the Slave craton at 1.9 Gyr and the implications for the thermal evolution of the continents. *Nature* **347**, 64-66 (1990).

Gry, C. and J. M. Bonnet-Bidaud 1990. Sirius and the colour enigma. *Nature* **347**, 625 (1990).

Guest, J. E. *et al.* 1992. Small volcanic edifices and volcanism in the plains of Venus. *Journal of Geophysical Research* **97**, 15949-15966 (1992).

Gupta, S. *et al.* 2007. Catastrophic flooding origin of shelf valley systems in the English Channel. *Nature* **448**, 342-345 (2007).

Guthrie, B. N. G. and W. M. Napier 1996. Redshift periodicity in the Local Supercluster. *Astronomy and Astrophysics* **310**, 353-370 (1996).

Guthrie, R. D. 1990. *"Frozen Fauna of the Mammoth Steppe"*. University of Chicago Press, Chicago (1990).

Guyodo, Y. and J.-P. Valet 1999. Global changes in intensity of the Earth's magnetic field during the past 800 kyr. *Nature* **399**, 249-252 (1999).

Haag, R. A. 1992. *"Field Guide of Meteorites"*. 10th Anniversary Edition, Robert A. Haag, Tucson (1992).

Haag, R. 2003. *"The Robert Haag Collection of Meteorites"*. Robert Haag Meteorites, Tucson (2003).

Habib, M. 2008. Comparative evidence for quadrupedal launch in pterosaurs. *Zitteliana* **B28**, 159-166 (2008).

Hadjas, I. *et al.* 1998. Cold reversal on Kodiak Island, Alaska, correlated with the European Younger Dryas by using variation of atmospheric ^{14}C content. *Geology* **26**, 1047-1050 (1998).

Hahn, O. 1958. The Discovery of Fission. *Scientific American* **February**, 76-84 (1958).

Hahn, O. and F. Strassman 1939. Über den Nachweis und das Verhalten der bei der Bestrahlung des Urans mittels Neutronen entstehenden en Erdalkalimetalle. *Naturwissenschaften* **27**, 11-15 (1939).

Hallam, A. 1989. *"Great Geological Controversies"*. Second Edition, Oxford University Press, Oxford (1989).

Halpern, I. 1971. Three fragment fission. *Annual Review of Nuclear Science* **21**, 245-294 (1971).

Hamilton, E. 1987. *"Darkness at Night"*. Harvard University Press, Cambridge (1987).

Hamilton, D. P. 2009. Secrets that only tides will tell. *Nature* **460**, 1086-1087 (2009).

Hammer, W. R. and W. J. Hickerson 1994. A Crested Theropod Dinosaur from Antarctica. *Science* **264**, 828-830 (1994).

Hamuy, M. *et al.* 1996. The absolute luminosities of the Calán/Tololo Type Ia supernovae. *Astronomical Journal* **112**, 2391-2397 (1996).

Hansen, M. and K. Anderko 1958. *"Constitution of Binary Alloys"*. Second Edition, McGraw-Hill, New York (1958).

Haq, B. U, J. Hardenbol, P. R. Vail 1987. Chronology of Fluctuating Sea Levels Since the Triassic. *Science* **235**, 1156-1167 (1987).

Haq B. U. and S. R. Schutter 2008. A Chronology of Paleozoic Sea-Level Changes. *Science* **322**, 64-68 (2008).

Harland, W. B. *et al.* 1990. *"A geologic time scale"*. Cambridge University Press, Cambridge (1990).

Harley, N. H. 2000. The 1999 Lauriston S. Taylor Lecture – Back to Background: Natural Radiation and Radioactivity Exposed. *Health Physics* **79**, 121-128 (2000).

Harris, E. E. and J. Hey 1999. X chromosome evidence for ancient human histories. *Proceedings of the National Academy of Sciences USA*. **96**, 3320-3324 (1999).

Harris, W. E. 1996. A catalog of parameters for globular clusters on the Milky Way. *Astronomical Journal* **112**, 1487-1488 (1996), Catalogue VII/202 (1997).

Harrison, E. 1987. *"Darkness at Night"*. Harvard University Press, Cambridge (1987).

Hart, D. 1987. *"The Encyclopedia of Soviet Spacecraft"*. Exeter Books, New York (1987).

Hart, R. C. 1973. *"Adriaan van Maanen's Influence on the Island Universe Theory"*, Ph.D. thesis. Boston University (1973).

Hart, S. R. *et al.* 1992. Mantle Plumes and Entrainment: Isotopic Evidence. *Science* **256**, 517-520 (1992).

Hartmann, W. K. year, *Meteoritics and Planetary Science*, Front Cover, reference lost (year).

Hartmann, W. K. 1999. Martian Cratering VI: Crater count isochrons and evidence for recent volcanism from Mars Global Surveyor. *Meteoritics and Planetary Science* **34**, 167-177 (1999).

Hartmann, W. K. 2000. Renaissance. *Astronomy* **July**, 36-41 (2000).

Harwit, M. 1988. *"Astrophysical Concepts"*. Second Edition, Springer-Verlag, Berlin (1988).

Hatchwell, P. 1998. Letter: Seismic sea changes. *New Scientist* **10 January**, 49 (1998).

Haug, G. H. and R. Tiedemann 1998. Effect of the formation of the Isthmus of Panama on Atlantic thermohaline circulation. *Nature* **393**, 673-676 (1998)

Haug, G. H. *et al.* 2001. Role of Panama uplift on oceanic freshwater balance. *Geology* **29**, 207-210 (2001).

Hauswirth, W. W. *et al.* 1991. 8000-year-old brain tissue from the Windover site: Anatomical, cellular, and molecular analysis. Pages 60-71 in *"Human Paleopathology: Current Synthesis and Future Options"*. D. J. Ortner and A. C. Aufderheide, Editors. Smithsonian Institution Press, Washington (1991).

Hay, R. L. 1976. *"Geology of the Olduvai Gorge"*. University of California Press, Berkeley (1976).

Haynes Jr., C. V. 2008. Younger Dryas "Black Mats" and the Rancholabrean termination in North America. *Proceedings of the National Academy of Sciences USA* **105**, 6520-6525 (2008).

Haynes Jr., C. V. and G. A. Agogino 1986. Geochronology of Sandia Cave. *Smithsonian Contributions to Anthropology, Number 32.* Smithsonian Institution Press, Washington (1986).

Head, J. W. *et al.* 1991. Venus Volcanism: Initial Analysis from Magellan Data. *Science* **252**, 276-288 (1991).

Heaman, L. M. and J. Tarney 1989. U-Pb baddeleyite ages for the Scourie dyke swarm, Scotland: evidence for two distinct intrusion events. *Nature* **340**, 705-708 (1989).

Heap, S. R. and D. J. Lindler 1987. Deconvolution of a pre-outburst picture of SN 1987A. *Astronomy and Astrophysics* **185**, L10-L12 (1987).

Heide, F,. and F. Wlotzka 1995. *"Meteorites"*. Springer-Verlag, Berlin (1995).

Hein, J. R., Editor 2004. *"Life Cycle of the Phosphoria Formation"*. Elsevier, Amsterdam (2004).

Heki, K. 2011. A Tale of Two Earthquakes. *Science* **332**, 1390-1391 (2011).

Heller, C. *et al.* 2009. An orbital period of 0.94 days for the hot-Jupiter planet WASP-18b. *Nature* **460**, 1098-1100 (2009).

Henbest, N. 1992. *"The Planets"*. Viking, London (1992).

Henderson, E. P. and S. H. Perry 1958. Studies of seven siderites. *Proceedngs of the U. S. National Museum* **107**, 339-403 (1958).

Hennig, G. J. *et al.* 1981. ESR-dating of the fossil hominid cranium from Petrolana Cave, Greece. *Nature* **292**, 533-536 (1981).

Henning, W. *et al.* 1987. Calcium-41 Concentration in Terrestrial Materials: Prospects for Dating Pleistocene Samples. *Science* **236**, 725-727 (1987).

Henry, R. B. C and D. Branch 1987. The spectrum of the type II-L supernova 1984E in NGC 3169; Further evidence for a superwind. *Publications of the Astronomical Society of the Pacific* **99**, 112-115 (1987).

Herbig, G. H. 1978. Some Aspects of Early Stellar Evolution that may be Relevant to the Origin of the Solar System. Pages 219-235 in *"The Origin of the Solar System"*. S. F. Dermott, Editor. John Wiley & Sons, New York (1978).

Hermanusses, M. 2003. Stature of early Europeans. *Hormones* **2**, 175-178 (2003).

Herndon, J. M. 1992. Nuclear Fission Reactors as Energy Sources for the Giant Outer Planets. *Naturwissenschaften* **79**, 7-14 (1992).

Herndon, J. M. 1993. Feasibility of a Nuclear Fission Reactor in the Center of the Earth as the Energy Source of the Geomagnetic Field. *Journal of Geomagnetism and Geoelectricity* **45**, 423-437 (1993).

Herndon, J. M. 1994. Planetary and protostellar nuclear fission: Implications for planetary change, stellar ignition and dark matter. *Proceedings of the Royal Society of London,* **A445**, 453-461 (1994).

Herndon, J. M. 1998. Examining the Overlooked Implications of Natural Nuclear Reactors. *Eos* **79**, 451-456 (1998).

Herndon, J. M. 2005. Whole-earth decompression dynamics. *Current Science* **89**, 1937-1941 (2005).

Herndon, J. M. 2008. *"Maverick's Earth and Universe"*. Trafford Publishers, Canada (2008).

Herndon, J. M. 2010. Impact of recent discoveries on petroleum and natural gas exploration: emphasis on India. *Current Science* **98**, 772-779 (2010).

Hess, J., M. L. Bender, and J-G. Schilling 1986. Evolution of the Ratio of Strontium-87 to Strontium-86 in Seawater from Cretaceous to Present. *Science* **231**, 979-984 (1986).

Hetrick, D. L. 1971. *"Dynamics of Nuclear Reactors"*. University of Chicago Press, Chicago (1971).

Hewitt, A. and Burbidge, G. 1993. A Revised and Updated Catalog of Quasi-Stellar Objects. *Astrophysical Journal* **87**, 451-947 (1993).

Hibben, F. C. 1968. *"The Lost Americans"*. T. Y. Crowell, New York (1968).

Higham, T. *et al.* 2010. Chronology of the Grotte du Renne (France) and implications for the context of ornaments and human remains within the Châtelperronian. *Proceedings of the National Academy of Sciences USA* **October 18, 2010** (2010).

Hirose, K. 2010. The Earth's Missing Ingredient. *Scientific American* **June**, 76-83 (2010).

Hirshfeld, A. W. 2001. *"Parallax"*. Henry Holt and Company, New York (2001).

Hitching, F. 1979. *"The Mysterious World—An Atlas of the Unexplained"*. Holt, Rinehart and Winston, New York (1979).

Hiyagon, H. 1994. Retention of Solar Helium and Neon in IDPs in Deep Sea Sediment. *Science* **263**, 1257-1259 (1994).

Hobson , C. 1987. *"The World of the Pharaohs"*. Thames and Hudson, New York (1987).

Hodge, P. 1986. *"Galaxies"*. Harvard University Press, Cambridge (1986).

Hodge, P. 1992. *"The Andromeda Galaxy"*. Kluwer Academic Publishers, Dordrecht (1992).

Hodgman, C. D. *et al.*, Editors 1956. *"Handbook of Chemistry and Physics"*. Chemical Rubber Publishing Co., Cleveland (1956).

Hoffleit, D. and W. H. Warren Jr. 1991. *"The Bright Star Catalog"*, 5th Edition, preliminary. Astronomical Data Center, NSSDC/ADC (1991).

Hoffman, D. C. *et al.* 1971. Detection of Plutonium-244 in Nature. *Nature* **234**, 132-134 (1971).

Hoffman, P. F. and D. P. Schrag 2002. The snowball Earth hypothesis: testing the limits of global change. *Terra Nova* **14**, 129-155 (2002).

Hogan, C. J. 1989. Signals from the Big Bang. *Nature* **339**, 15 (1989).

Holden, C. 1997. Crater Pattern says Europa Still Active. *Science* **276**, 203-204 (1997).

Holden, C. 2009. Not Out of Java (referring to M. Westaway and C. Groves, *Archaeology in Oceania*). *Science* **326**, 649 (2009).

Hollenbach, D. F. and J. M. Herndon 2001. Deep-Earth reactor: Nuclear fission, helium, and the geomagnetic field. *Proceedings of the National Academy of Sciences USA* **98**, 11085-11090 (2001).

Holman, J. A. 1995. *"Pleistocene Amphibians and Reptiles in North America"*. Oxford University Press, New York (1995).

Holtz Jr., T. R. 2010. A Jurassic tyrant is crowned. *Nature* **439**, 665-666 (2010).

Homer 850BCEa. *The Illiad*. Translated by S. Butler. Barnes and Noble Books, New York (1995).

Homer 850BCEb. *The Illiad of Homer*. Translated by A. Pope. The Macmillian Company, New York (1965).

Homoky, W. B. *et al.* 2009. Pore-fluid Fe isotopes reflect the extent of benthic Fe redox recycling: Evidence from continental shelf and deep-sea sediments. *Geology* **37**, 751-754 (2009).

Honda, M. *et al.* 1993. Noble gases in submarine pillow basalt glasses from Loihi and Kilauea, Hawaii: A solar component in the Earth. *Geochimica et Cosmochimica Acta* **57**, 859-874 (1993).

Hooten, E. 1918. On certain eskimoid characters in Icelandic skulls. *American Journal of Physical Anthropology* **1**, 53-76 (1918).

Jenkins, G., H. Marshall, and W. Kuhn 1993. Precambrian Climate: The Effects of Land Area and Earth's Rotation Rate. *Journal of Geophysical Research* **98**, 8785-8791 (1993).

Jenkins, J. and E. Fischbach 2009. Perturbation of nuclear decay rates during the solar flare of 2006 December 13. *Astroparticle Physics* **31**, 407-411 (2009).

Jenkins, J. H. *et al.* 2008. Evidence for Correlations Between Nuclear Decay Rates and Earth-Sun Distance. *Astroparticle Physics* **32**, 42-46 (2008), arXiv:0808.3283 (2008).

Jewitt, D. 2000. Eyes wide shut. *Nature* **403**, 145 (2000).

Jewitt, D., H. Aussel, and A. Evans 2001. The size and albedo of the Kuiper-belt object (20000) Varuna. *Nature* **411**, 446-447 (2001).

Jewitt, D. and J. Luu 1993. Discovery of the candidate Kuiper belt object 1992 QB$_1$. *Nature* **362**, 730-732 (1993).

J. K. B. 2001. Kuiper Belt Object Dethrones Ceres. *Sky & Telescope* **December**, 25 (2001).

Johanson, D., L. Johanson (and B. Edgar) 1994 . *"Ancestors"*. Villard Books, New York (1994).

Johns Hopkins Hospital 2003. Television press release, reference lost.

Johnson, R. A. 1995, personal communication, January 31, 1995 (1995).

Jones, K. G. 1991. *"Messier's Nebulae and Star Clusters"*, second edition. Cambridge University Press, Cambridge (1991).

Jones, S. E. *et al.* 1989. Observation of Cold Nuclear Fusion in Condensed Matter. Preprint, March 1989, of *Nature* **338**, 737-740 (1989).

Judge, D. L. *et al.* 1976. *Pioneer 10* and *11* Ultraviolet Photometer Observations of the Jovian Satellites. Pages 1068-1101 in *"Jupiter"*. T. Gehrels, Editor. University of Arizona Press, Tucson (1976).

Judson, S. and M. E. Kaufman 1990. *"Physical Geology"*. Prentice- Hall, Inc., Englewood Cliffs (1990).

Jueneman 1998. Brown's Gas. *R&D Magazine* **August**, 11 (1998).

Julian, W. H. 1987. Free precession of the comet Halley nucleus. *Nature* **326**, 57-58 (1987).

Jungers, W. L 1994. Ape and hominid limb length. *Nature* **369**, 194 (1994).

Kaiho, K. *et al.* 2001. End-Permian catastrophe by a bolide impact: Evidence of a gigantic release of sulfur from the mantle. *Geology* **29**, 815-818 (2001).

Kaler, J. B. 1989. *"Stars and their spectra"*. Cambridge University Press, Cambridge (1989).

Kanipe, J. and D. Webb 2006. *"The Arp Atlas of Peculiar Galaxies"*. Willman-Bell, Richmond (2006).

Kargel, J. S. 1997. The Rivers of Venus. *Sky & Telescope* **August**, 32-37 (1997).

Kargel, J. S. 2006. Enceladus: Cosmic Gymnast, Volatile Miniworld. *Science* **311**, 1389-1391 (2006).

Kargel, J. S. and J. S. Lewis 1993. The Composition and Early Evolution of Earth. *Icarus* **105**, 1-25 (1993).

Kargel, J. S. and R. G. Strom 1992. The Ice Ages of Mars. *Astronomy* **December**, 41-45 (1992).

Kargel, J. S. and R. G. Strom 1996. Global Climatic Change on Mars. *Scientific American* **November**, 80-88 (1996).

Karlsson, H. R. *et al.* 1992. Water in SNC Meteorites: Evidence for a Martian Hydrosphere. *Science* **255**, 1409-1411 (1992).

Karttunen H. *et al.* 1994. *"Fundamental Astronomy"*, 2nd edition. Springer-Verlag, Berlin (1994).

Kasting, J. F. 1992. Paradox lost and paradox found. *Nature* **355**, 676-677 (1992).

Kasting, J. F. 1993. New spin on ancient climate. *Nature* **364**, 759-760 (1993).

Kasting, J. F. and T. P. Ackerman 1986. Climatic Consequences of Very High Carbon Dioxide Levels in the Earth's Early Atmosphere. *Science* **234**, 1383-1385 (1986).

Kasting, J. F. and D. H. Grinspoon 1991. The Faint Young Sun Problem. Pages 447-462 in *"The Sun in Time"*. C. P. Sonett, M. S. Giampapa, and M. S. Matthews, Editors. University of Arizona Press, Tucson. (1991).

Kasting, J. F., O. B. Toon, and J. B. Polack 1988. How Climate Evolved on the Terrestrial Planets. *Scientific American*, **February**, 90-97 (1988).

Kasting, J. F, D. P. Whitmire, and R. T. Reynolds 1993. Habitable Zones around Main Sequence Stars. *Icarus* **101**, 108-128 (1993).

Katterfel'd, G. N. 1962. *"The Face of the Earth and its Origin"*. NASA TT F-533, NASA, Washington (1969).

Katz, J. I. 2002. *"The Biggest Bangs"*. Oxford University Press, Oxford (2002).

Kaufman, A. J., A. H. Knoll, and G. M. Narbonne 1997. Isotopes, ice ages, and terminal Proterozoic earth. *Proceedings of the National Academy of Science USA* **94**, 6600-6605 (1997).

Kaufman III, W. J. 1991. *"Universe"*. W. H. Freeman & Company, New York (1991).

Kawano, L. 1992. *"Let's Go: Early Universe II, Primordial Nucleosynthesis The Computer Way"*. FERMILAB-Pub-92/04-A, NASA/Fermilab Astrophysics Center, Batavia (1992).

Keepin, G. R. 1965. *"Physics of Nuclear Kinetics"*. Addison-Wesley, Reading (1965).

Keishner, R. P. 1988. Supernova – Death of a Star. *National Geographic* **May,** 619-647 (1988).

Keller, G. *et al.* 2004. Chicxulub impact predates the K-T boundary mass extinction. *Proceedings of the National Academy of Sciences USA* **101**, 3753-3758 (2004).

Keller, G. *et al.* 2010. Cretaceous Extinctions: Evidence Overlooked. *Science* **328**, 974-975 (2010).

Kelly, A. O. and F. Dachille (1953). *"Target Earth"*, publisher unknown, Pensacola (1953).

Kelley, S. P. and J-A. Wartha 2000. Rapid Kimberlite Ascent and the Significance of Ar-Ar Ages in Xenolith Phlogopites. *Science* **289**, 609-611 (2000).

Kennedy, B. M. *et al.* 1985. Intensive sampling of noble gases in fluids at Yellowstone: I. Early overview of the data; regional patterns. *Geochimica et Cosmochimica Acta* **49**, 1251-1261 (1985).

Kennedy, G. E. 1985. Bone thickness in *Homo erectus*. *Journal of Human Evolution* **14**, 699-708 (1985).

Kenney, J. 1993. More whirls in the whirlpool. *Nature* **364**, 283-84 (1993).

Kerr, R. A. 1992a. 1991: Warmth, Chill May Follow. *Science* **255**, 281 (1992).

Kerr, R. A. 1992b. Planetesimal Found Beyond Neptune. *Science* **257**, 1865 (1992).

Kerr, R. A. 1992c. Extinction With a Whimper. *Science* **255**, 161 (1992).

Kerr, R. A. 1993a. Magnetic Ripple Hints Gaspra is Metallic. *Science* **259**, 176 (1993).

Kerr, R. A. 1993b. Did Venus Hiccup or Just Run Down? *Science* **259**, 1400-1401 (1993).

Kerr, R. A. 1993c. Fossils Tell of Mild Winters in an Ancient Hothouse. *Science* **261**, 682 (1993).

Kerr, R. A. 1993d. Evolution's Big Bang Gets Even More Explosive. *Science* **261**, 1274-1275 (1993).

Kerr, R. A. 1993e. New Crater Age Undercuts Killer Comets. *Science* **262**, 659 (1993).

Kerr, R. A. 1994a. An Asteroidal Family Adds a Little One. *Science* **264**, 35 (1994).

Kerr, R. A. 1994b. Who Profits From Ecological Disaster? *Science* **266**, 28-30 (1994).

Kerr, R. A. 1996a. Galileo Hits a Strange Spot on Jupiter. *Science* **271**, 593-594 (1996).

Kerr, R. A. 1996b. Galileo Finds Mysterious Magnetic Field at Ganymede. *Science* **273**, 311 (1996).

Kerr, R. A. 1997. For Mammals, Bigger is (Usually) Better. *Science* **278**, 1018 (1997).

Kerr, R. A. 1998a. Planetary Scientists Sample Ice, Fire, and Dust. *Science* **280**, 38 (1998).

Kerr, R. A. 1998b. Geologists Take a Trip to the Red Planet. *Science* **282**, 1807 (1998).

Kerr, R. A. 2002. No "Darkness at Noon" to Do In the Dinosaurs? *Science* **295**, 1445-1446 (2002).

Kerr, R. A. 2005. Cosmic Dust Supports a Snowball Earth. *Science* **308**, 181 (2005).

Kerr, R. A. 2007. Warped Shorelines On a Rolling Mars. *Science* **315**, 1788-1789 (2007).

Kerr, R. A. 2009. A Primal Crust Found on the Moon, While Mercury's Proves Elusive. *Science* **324** 161 (2009).

Kerr, R. A. 2010a. Did a Deep Sea Once Cover Mars? *Science* **328**, 1467 (2010).

Kerr, R. A. 2010b. Did a Deep Sea Once Cover Mars? *ScienceNOW*, http://news.science mag.org/sciencenow,2010/06/did-a-deep-sea-once-cover-mars.html (2010).

Kerr, R. A. 2010c. Mammoth-Killer Impact Flunks Out. *Science* **329**, 1140-1141 (2010).

Kerr, R. A. 2010d. First Goldilocks Exoplanet May Not Exist. *Science* **330**, 433 (2010).

Kerr, R. A. 2011. Enceladus Now Looks Wet, So It May Be ALIVE! *Science* **332**, 1259 (2011).

Keszthelyi, L. *et al.* 2007. New estimates for Io eruption temperatures: Implications for the interior. *Icarus* **192**, 491-502 (2007).

Kiel, K. *et al.* 1997. Constraints on the role of impact heating and melting in asteroids. *Meteoritics and Planetary Science* **32**, 349-363 (1997).

Kim, J. S. and K. Marti 1992. Evidence for Neutron Irradiation in the Early Solar System. *Lunar and Planetary Science* **23**, 689-690 (1992).

King Jr., D. T. 1996. Stratigraphy: Layers of Evidence. *Planetary Report* **July/August**, 10-11 (1996).

Kingsley, R. H. *et al.* 2002. D/H ratios in basalt glasses from the Salas y Gomez mantle plume interacting with the East Pacific Rise: Water from old D-rich recycled crust or primordial water from the lower mantle? *Geochemistry Geophysics Geosystems* **3**, 1-26 (2002).

Kingston, D. 2007. King Island. email to R. J. Tuttle, May 3, 2007 (2007).

Kinoshita, J. 1989. Neptune. *Scientific American* **November**, 82-91 (1989).

Kippenhahn, R. and A. Wiegert 1990. *"Stellar Structure and Evolution"*. Springer-Verlag, Berlin (1990).

Kirk, R. L. *et al.* 1995. Triton's Plumes: Discovery, Characteristics, and Models. Pages 949-989 in *"Neptune and Triton"*. D. P. Cruikshank, Editor. University of Arizona Press, Tucson (1995).

Kirko, G. E., M. T. Telichko, and A. G. Shrinkman 1983. Self-excitation of a magnetic field (MHD dynamo) in the central region of a nuclear reactor with liquid-metal cooling. *Soviet Physics Doklady* **28**, 502-503 (1983).

Kirshner, R P. 2002. *"The Extravagant Universe"*. Princeton University Press, Princeton (2002).

Kirshner, R. 2009. The genesis of modern cosmology (book review). *Physics Today* **December 2009**, 51-52 (2009).

Kirschvink, J. L, R. L. Ripperdan, and D. A. Evans 1997. Evidence for a Large-Scale Reorganization of Early Cambrian Continental Masses by Inertial Interchange True Polar Wander. *Science* **277**, 541-545 (1997).

Kirsten, T. and L. Wolfenstein 1991. Solar Neutrinos. pages 586-610 in *"Neutrino Physics"*. K. Winter, Editor. Cambridge University Press, Cambridge (1991).

Kivelson, M. G. *et al.* 1993. Magnetic Field Signatures Near Galileo's Closest Approach to Gaspra. *Science* **261**, 331-334 (1993).

Kivelson, M. G. *et al.* 1996. Discovery of Ganymede's magnetic field by the Galileo spacecraft. *Nature* **384**, 537-541 (1996).

Klass, P. J. 1980. Clandestine Nuclear Test Doubted. *Aviation Week & Space Technology* **August 11**, 67-70 (1980).

Klebesadel, R.W., I. B. Strong, and R. A. Olson 1973. Observations of Gamma-Ray Bursts of Cosmic Origin. *Astrophysical Journal* **182**, L85-L88 (1973).

Kleine, T. 2011. Earth's patchy late veneer. *Nature* **477**, 168-169 (2011).

Klemola, A. R. and E. A. Harlan 1972. Search for Brady's Hypothetical Trans-Plutonian Planet. *Publications of the Astronomical Society of the Pacific* **84**, 736 (1972).

Knauth, L. P. and S. Epstein 1976. Hydrogen and oxygen isotope ratios in nodular and bedded cherts. *Geochimica et Cosmochimica Acta* **40**, 1095-1108 (1976).

Knauth, L. P., D. M. Burt, and K. H. Wohlhetz 2005. Impact origin of sediments at the Opportunity landing site on Mars. *Nature* **438**, 1123-1128 (2005).

Knie, K. *et al.* 2004. ^{60}Fe Anomaly in a Deep-Sea manganese Crust and Implications for a Nearby Supernova Source. *Physical Review Letters* **93**, 171103-171106 (2004).

Kniffer, D. O. 1993. The Gamma-Ray Universe. *American Scientist* **81**, 342-349 (1993).

Knitter, H.-H. *et al.* 1991. Neutron and Gamma Emission in Fission. Chapter 11 in *"The Nuclear Fission Process"*. C. Wagemans, Editor. CRC Press, Boca Raton (1991).

Kolb, E. W and M. S. Turner 1990. *"The Early Universe"*. Addison-Wesley, Redwood City (1990).

Kolb, R. 1996. *"Blind Watchers of the Sky"*. Addison-Wesley, Reading (1996).

Konigsberg, L. W. *et al.* 1994. Modern human origins. *Nature* **372**, 228-229 (1994).

Koppes, M. N. and D. R Montgomery 2009. The relative efficacy of fluvial and glacial erosion over modern to orogenic timescales. *Nature Geoscience* **2**, 644-647 (2009).

Kouřimský, J. 1993. *"The Illustrated Encyclopedia of Minerals and Rocks"*, translated by V. Gissing. Chartwell Books, Secaucus (1993).

Kowal, C. T. 1988. *"Asteroids: Their Nature and Utilization"*. John Wiley & Sons, New York (1988).

Kowalski, M. *et al.* 2008. Improved Cosmological Constraints from Old, New and Combined Supernova Datasets. *Astrophysical Journal* **686**, 749-778 (2008).

Krause, J. *et al.* 2010. The complete mitochondrial DNA genome of an unknown hominin from southern Siberia. *Nature* **464**, 894-897 (2010).

Krinov, E. L. 1960. *"Principles of Meteoritics"*. Translated by I. Vidziunas, edited by H. Brown, Pergamon Press, Oxford (1960).

Kristian, J. *et al.* 1989. Submillisecond optical pulsar in supernova 1987A. *Nature* **338**, 234-236 (1989).

Krupp, E. C. 1991. *"Beyond the Blue Horizon"*. Oxford University Press, New York (1991).

Krupp, E. C. 1997. *"Skywatchers, Shamans, and Kings"*. John Wiley and Sons, New York (1997).

Kuiper, G. P 1951. On the Origin of the Solar System. *Proceedings of the National Academy of Sciences USA* **37**, 1-14, 233 (1951).

Kump, L. 1993. Oceans of change. *Nature* **361**, 592-593 (1993).

kuntskamera 2009. Early Upper Paleolithic Man from Markina Gora. Peter the Great Museum of Anthropology and Ethnology, http://www.kunstkamera.ru/en/temporary_exhibitions/virtual/gerasimov/09. (2009).

Kunzig, R. 1997. Atapuerca. *Discover* **December**, 88-101 (1997).

Kuroda, P. K. 1956. On the Nuclear Physical Stability of the Uranium Minerals. *Journal of Chemical Physics* **25**, 781-782 (1956).

Kuroda, P. K. 1960. Nuclear fission in the early History of the Earth. *Nature* **187**, 36-38 (1960).

Kuroda, P. K. 1975. Early History of the Fossil Nuclear Reactor and Plutonium-244 in the Solar System. Pages 479-487 in *"The Oklo Phenomena"*. International Atomic Energy Agency, Vienna (1975).

Kuroda, Y. *et al.* 1975. D/H ratios of the coexisting phlogopite and richterite from mica nodules and a peridotite in South African kimberlites. *Contributions to Mineralogy and Petrology* **52**, 315-318 (1975).

Kürster, M. *et al.* 2000. An extrasolar giant planet in an Earth-like orbit. *Astronomy and Astrophysics* **353**, L33-L36 (2000).

Kurtén, B. 1976. *"The Cave Bear Story"*. Columbia University Press, New York (1976).

Kurtén, B. 1986. *"How to Deep-Freeze a Mammoth"*. Columbia University Press, New York (1986).

Kurz, M. D., W. J. Jenkins, and S. R. Hart 1982. Helium isotopic systematics of oceanic islands and mantle heterogeneity. *Nature* **297**, 43 (1982).

Kyte, F. T. 1998. A meteorite from the Cretaceous/Tertiary boundary. *Nature* **396**, 237-239 (1998).

Lagage, P. O. and E. Pantin 1994. Dust depletion in the inner disk of β Pictoris as a possible indicator of planets. *Nature* **369**, 628-630 (1994).

Lagrange, A.-M. 2010. A Giant Planet Imaged in the Disk of the Young Star β Pictoris. *Science* **329**, 57-59 (2010).

Lainey, V. *et al.* 2009. Strong tidal dissipation in Io and Jupiter from astrometric observations. *Nature* **459**, 957-959 (2009).

Lake, G. 1992. Cosmology of the Local Group. *Sky & Telescope* **December**, 613-619 (1992).

Lamb, D. Q. 1995. The Distance Scale to Gamma-Ray Bursts. *Publications of the Astronomical Society of the Pacific* **107**, 1152-1166 (1995).

Lambert 1987. *"The Field Guide to Early Man"*. Facts On File, New York (1987).

Langer, E. 2009. Different stellar demise. *Nature* **462**, 579-580 (2009).

Langereis, C. G. 1999. Earth science: Excursions in geomagnetism. *Nature* **399**, 207-208 (1999).

Langmuir, C. H. 1990. Ocean ridges spring surprises. *Nature* **344**, 585-586 (1990).

LASL 1950. *"The Effects of Atomic Weapons"*. Revised September 1950, Los Alamos Scientific Laboratory, U. S. Government Printing Office, Washington (1950).

LaTourette, T. Z., A. K. Kennedy, and G. J. Wasserburg 1993. Thorium-Uranium Fractionation by Garnet: Evidence for a Deep Source and Rapid Rise of Oceanic Basalts. *Science* **261**, 739-742 (1993).

Lawler, J. E., W. Whaling, and N. Grevesse 1990. Contamination of the Th II line and the age of the Galaxy. *Nature* **346**, 635-637 (1990).

Lawler, A. 2010. Was the New World Settled Twice? *Science* **328**, 1467 (2010).

Lawrence, A. *et al.* 1988. Extreme Fe II emission from an *IRAS* quasar. *Monthly Notices of the Royal Astronomical Society*, **235**, 261-268 (1988).

Lawrence, P. A. and G. Morata 1993. A no-wing situation. *Nature*, **366**, 305-306 (1993).

Le Grand, H. E. 1988. *"Drifting Continents and Shifting Theories"*. Cambridge University Press, Cambridge (1988).

Leach, J. 1993. Antimatter. Pages 58-59 in *"McGraw-Hill Encyclopedia of Physics"*, 2nd edition. S. P. Parker, Editor-in-Chief, McGraw-Hill, New York (1993).

Leavitt, H. S. 1908. 1777 Variables in the Magellanic Clouds. *Harvard College Observatory Annals* **60**, 87-108 (1908).

Lebreton, J.-P. *et al.* 2005. An overview of the descent and landing of the *Huygens* probe on Titan. *Nature* **438**, 758-764 (2005).

Lee, D. 1971. *"Timaeus and Critias"*. Penguin Books, London (1977).

Lee, S. v. d. 2001. Deep Below North America. *Science* **294**, 1297-1298 (2001).

Lee, T. 1978. A local proton irradiation model for isotopic anomalies in the solar system. *Astrophysical Journal* **224**, 217-226 (1978).

Lehman, S. J. *et al.* 1991. Initiation of Fennoscandian ice-sheet retreat during the last deglaciation. *Nature* **349**, 513-516 (1991).

Leifer, R. *et al.* 1987. Detection of Uranium from Cosmos-1402 in the Stratosphere. *Science* **238**, 512-514 (1987).

Lemaître, G. 1927. Un Univers homogène de masse constante et de rayon croissant rendant compte de la vitesse radiale des nébuleuses extra-galactiques. *Annals de la Societe Scientifique de Bruxelles* **A47**, 49-59 (1927).

Lemaître, G. 1931a. A Homogeneous Universe of Constant Mass and Increasing Radius accounting for the Radial Velocity of Extra-Galactic Nebulae. *Monthly Notices of the Royal Astronomical Society* **91**, 483-490 (1931).

Lemaître, G. 1931b. Expansion of the universe, the expanding universe. *Monthly Notices of the Royal Astronomical Society* **91**, 490-501 (1931).

Leonard, F. C. 1930. The New Planet Pluto. *Leaflet of the Astronomical Society of the Pacific* **1**, 121-124 (1930).

Levinton, J. S. 1992. The Big Bang of Animal Evolution. *Scientific American*, **November**, 84-91 (1992).

Levy, D. H. 1989. *"Observing Variable Stars"*. Cambridge University Press, Cambridge (1989).

Lewin, R. 1993. *"The Origin of Modern Humans"*. Scientific American Library, New York (1993).

Li, C. *et al.* 2008. A new global model for *P* wave speed variations in Earth's mantle, *Geochemistry, Geophysics, Geosystems* **9**, Q05018 (2008).

Li, C. *et al.* 2008. An ancestral turtle from the late Triassic of southwestern China, *Nature* **456**, 497-501 (2008).

Limaye, S. S. 1991. Neptune's weather forecast: Cloudy, Windy and Cold. *Astronomy* **August**, 38-43 (1991).

Linde, A. T. and I. S. Sacks 1998. Triggering of volcanic eruptions. *Nature* **395**, 888-890 (1998).

Lindley, D. 1988. SN1987A pulsar slow to reveal itself. *Nature* **331,** 394 (1988).

Lindley, D. 1992. Not so Great Attractor? *Nature* **358**, 657 (1992).

Lindstrom, D. R. and D. R. MacAyeal 1993. Death of an ice sheet. *Nature* **365**, 214-215 (1993).

Link, B., R. I. Epstein, and W. C. Priedhorsky 1993. Prevalent Properties of Gamma-Ray Burst Variability. *Astrophysics Journal* **408**, L81-L84 (1993).

Lipari, S. *et al.* 1991. Southern *IRAS* quasar with extreme Fe II emission. *Astrophysics Journal* **366**, L65-L67 (1991).

Lipman, P. 1997. Chasing the Volcano. *Earth* **December**, 32-39 (1997).

Lipman, P., M. Dungan, and O. Bachmann 1997. Comagmatic granophyric granite in the Fish canyon Tuff, Colorado: Implications for magma-chamber processes during a large ash-flow eruption. *Geology* **25**, 915-918 (1997).

Lippman, H. E. 1962. Frozen Mammoths. *Science* **137**, 449-450 (1962).

Lissauer, J. J. 1997. It's not easy to make the Moon. *Nature* **389**, 327-328 (1997).

Lister, A. and P. Bahn 1994. *"Mammoths"*. Macmillan, New York (1994).

Littmann, M. 1988, *"Planets Beyond – Discovering the Outer Solar System"*, John Wiley and Sons, New York (1988).

Littmann, M. 1990. *"Planets Beyond"*. Updated and Revised. John Wiley & Sons, New York (1990).

Liu, X.-W. *et al.* 1996. The ISO LWS grating spectrum of NGC 7027. *Astronomy and Astrophysics* **315**, L257-L260 (1996).

Livio, M. 2011. Lost in translation: Mystery of the missing text solved. *Nature* **479**, 171-173 (2011).

Lockwood, G. W. and D. T Thompson 1991. Solar cycle relationship clouded by Neptune's sustained brightness maximum. *Nature* **349**, 593-594 (1991).

Lockwood, G. W. *et al.* 1992. Long-term solar brightness changes estimated from a survey of Sun-like stars. *Nature* **360**, 653-655 (1992).

Lodders, K. and B. Fegley Jr. 1998. *"The Planetary Scientist's Companion"*. Oxford University Press, New York (1998).

Longhi, J. *et al.* 1992. The Bulk Composition, Mineralogy and Internal Structure of Mars. Pages 825-834 in *"Mars"*. H. H. Kieffer, B.M. Jakosky, C. W. Snyder, and M. S. Matthews, Editors. University of Arizona Press, Tucson (1992).

Loope, D. B. *et al.* 1998. Life and death in a Late Cretaceous dune field, Nemegt basin, Mongolia. *Geology* **26**, 27-30 (1998).

Love, S. G. and K. Keil 1995. Recognizing Mercurian Meteorites. *Meteoritics* **30**, 268-278 (1995).

Lowell, P. 1915. Memoir on a Trans-Neptunian Planet. *Memoirs of the Lowell Observatory*, **1**, (1) (1915).

LP 1999. Fit Fossils. *Scientific American Discovering Archeology* **November/December**, 12 (1999).

Lucchitta, B. K. 1987. Recent Mafic Volcanism on Mars. *Science* **235**, 565-567 (9187).

Lugmair, G. W. and S. J. G. Galer 1992. Age and isotopic relationships among the angrites Lewis Cliff 86010 and Angra dos Reis. *Geochimica et Cosmochimica Acta* **56**, 1673-1694 (1992)

Lugmair, G. W. and A. Shukolyukov 2001. Early Solar System events and timescales. *Meteoritics and Planetary Science* **36**, 1017-1026 (2001).

Luhmann, J. G. *et al.* 1992. The Intrinsic Magnetic Field and Solar-Interaction of Mars. Pages 1090-1134 in *"Mars"*. H. H. Kieffer, B. M. Jakosky, C. W. Snyder, and M. S. Matthews, Editors. University of Arizona Press, Tucson (1992).

Lunine, J. I. 1996. Neptune at 150. *Sky & Telescope* **September**, 38-44 (1996).

Lupton, J. E. and H. Craig 1981. A Major Helium-3 Source at 15°S on the East Pacific Rise. *Science* **214**, 13-18 (1981).

Lutgens, F. K. and E. J. Tarbuck 1989. *"Essentials of Geology"*. Third edition. Merrill Publishing Company, Columbus (1989).

Luu, J. and D. Jewitt 1993. *IAU Circular 5730*, May 10, 1993 (1993).

Luu, J. *et al.* 1997. A new dynamical class of object in the outer Solar System. *Nature* **387**, 573-574 (1997).

Lynch, P. 2003. On the significance of the Titius-Bode law for the distribution of the planets. *Monthly Notices of the Royal Astronomical Society* **341**, 1174-1178 (2003).

Lyne, A. 1989. Spin-up more apparent than real. *Nature* **337**, 510 (1989).

Lyne, A. G. and F. Graham-Smith 1990. *"Pulsar Astronomy"*. Cambridge University Press, Cambridge (1990).

Lyne, A. G. and D. R. Lorimar 1994. High birth velocities of radio pulsars. *Nature* **369**, 127-129 (1994).

Lyons, T. W. and C. T. Reinhard 2009. Oxygen for heavy-metal fans. *Nature* **461**, 179-181 (2009).

Machetto, F. D. and M. Dickinson 1997. Galaxies in the Young Universe. *Scientific American* **May**, 92-99 (1997).

MacPherson, G. J., A. M. Davis, and E. K. Zinner 1995. The distribution of aluminum-26 in the early Solar System-A reappraisal. *Meteoritics* **30**, 365-386 (1995).

Maddox, J. 1989. Who will see a nearby supernova? *Nature* **339**, 335 (1989).

Maddox, J. 1992. The anthropic view of nucleosynthesis. *Nature* **355**, 107 (1992).

Madore, B. F. and W. L. Freedman 1992. Stars, Cepheid Variable, Period-Luminosity Relation and Distance Scale. Pages 722-725 in *"The Astronomy and Astrophysics Encyclopedia"*. S. P. Maran, Editor. Van Nostrand Reinhold, New York (1992).

Madore, B. F. and I. P Steer 2008. NASA/IPAC Extragalactic Database (NED), (http://nedwww.ipac.caltech.edu/level5/NED1D) (2008).

Madore, B. F. *et al.* 1998. A Cepheid distance to the Fornax cluster and the local expansion rate of the Universe. *Nature* **395**, 47-50 (1998).

Maffei, P. 1989. *"The Universe in Time"*, translated by M. Giacconi. The MIT Press, Cambridge (1989).

Malaney, R. A. 1989. Production of technetium in red giants by gamma-ray-induced fission. *Nature* **337**, 718-720 (1989).

Malaney, R. A. and M. Clampin 1988. A Search for Interstellar [FeX] Toward SN 1987A. *Astrophysical Journal* **335**, L57-L60 (1988).

Malcolm, G. 2000. CCD Photometry of Minor Planet 729 Watsonia. *The Minor Planet Bulletin* **27**, 45-46 (2000).

Malhotra, R. and J. G. Williams 1997. Pluto's Heliocentric Orbit. Pages 127-157 in *"Pluto and Charon"*. S. A. Stern and D. J. Tholen, Editors. University of Arizona Press, Tucson (1997)

Malin, D. 1993. *"A View of the Universe"*. Sky Publishing Corporation, Cambridge (1993).

Malina, R. F. and S. Bowyer, Editors 1991. *"Extreme Ultraviolet Astronomy"*. Pergamon Press, New York (1991).

Mallas, J. H. and E. Kreimer 1978. *"The Messier Album"*. Sky Publishing, Cambridge (1978).

Mamyrin, B. A. and L. N. Tolstikhin 1984. *"Helium Isotopes in Nature"*. Elsevier, Amsterdam (1984).

Manabe, S. and K. Bryan Jr. 1985. CO_2-induced change in a coupled ocean-atmosphere model and its paleoclimate implications. *Journal of Geophysical Research* **90**, 11689-11707 (1985).

Mandeville, C. W. *et al.* 1994. Paleomagnetic evidence for high-temperature emplacement of the 1883 subaqueous pyroclastic flows from Krakatau Volcano, Indonesia. *Journal of Geophysical Research* **99**, 9487-9504 (1994).

Mann, A. K. 1997. *"Shadow of a Star – The Neutrino Story of Supernova 1987A"*. W. H. Freeman and Company, New York (1997).

Manuel, O. K. and G. Hwaung 1983. Solar Abundances of the Elements. *Meteoritics* **18**, 209-222 (1983).

Maringer, R. E. and G. K. Manning 1962. Some observations on Deformation and Thermal Alterations in Meteoritic Iron. Pages 123-144 in *"Researches on Meteorites"*. C. B. Moore, Editor. John Wiley & Sons, New York (1962).

Marinova, M. M., O. Aharonson, and E. Asphaug 2008. Mega-impact formation of the Mars hemispheric dichotomy. *Nature* **453**, 1216-1219 (2008).

Marlon, J. R. 2009. Wildfire responses to abrupt climate change in North America. *Proceedings of the National Academy of Sciences USA* **106**, 2519-2524 (2009).

Marois, C. *et al.* 2010. Images of a fourth planet orbiting HR 8799. *Nature* **468**, 1080-1083 (2010).

Marschall, L. A. 1988. *"The Supernova Story"*. Princeton University Press, Princeton, (1988).

Marsden, B. G. and Z Sekanina 1973. On the distribution of "original" orbits of comets of large perihelion distance. *Astronomical Journal* **78**, 1118-1124 (1973).

Marsden, B. G., Z. Sekanina, and E. Everhart 1978. New osculating orbits for 110 comets and analysis of original orbits for 200 comets. *Astronomical Journal* **83**, 64-71 (1978).

Marshack, A. 1991. *"The Roots of Civilization"*. Moyer Bell, Mt. Kisco (1991).

Marshall, C. R. and D. K. Jacobs 2009. Flourishing After the End-Permian Mass Extinction. *Science* **325**, 1079-1080 (2009).

Marti, K. and H. Craig 1987. Cosmic-ray-produced neon and helium in the summit lavas of Maui. *Nature* **325**, 335-336 (1987).

Marti, K. *et al.* 1989. Xenon in chondritic metal. *Zeitschrift für Naturforschung A* **44a**, 963-967 (1989).

Martín, E. L. *et al.* 1992. High lithium abundance in the secondary of the black-hole binary system V404 Cygni. *Nature* **358**, 129-131 (1992).

Martin, P. S. and R. G. Klein 1989. *"Quaternary Extinctions"*. University of Arizona Press, Tucson (1989).

Marvin, U. B. 1994. Ernst E. F. Chladni (1756-1827) and the Beginnings of Meteoritics. *Meteoritics* **29**, 496-497 (1994).

Marzoli, A. *et al.* 1999. Extensive 200-Million-Year-Old Continental Flood Basalts of the Central Atlantic Magmatic Province. *Science* **284**, 616-618 (1999).

Mason, B. 1962. *"Meteorites"*. John Wiley & Sons, New York (1962).

Massalski, T. B. 1962. Some Metallurgical Aspects in the Study of Meteorites. Pages 107-122 in *"Researches on Meteorites"*. C. B. Moore, Editor. John Wiley & Sons, New York (1962).

Massey, R. 2009. Dark is the new black. *Nature* **461**, 740-741 (2009).

Mather, D. J. *et al.* 1994. *"CRITEX - A Code to Calculate the Fission Release Arising from Transient Criticality in Fissile Solutions"*. Report No. AEA/CS/R1007/R, AEA Technology, Risley, Warrington Cheshire (1994).

Mathews, G. J. and J. J. Cowan 1990. New insight into the astrophysical r-process. *Nature* **345**, 491-494 (1990).

Matson, B. and R. Troll 1994. *"Planet Ocean"*. Ten Speed Press, Berkeley, CA (1994).

Matsuda, J. *et al.* 1993. Noble Gas Partitioning Between Metal and Silicate Under High Pressure. *Science* **259**, 788-790 (1993).

Matsumoto, T. 1987. *"Comets to Cosmology"*. 3rd International IRAS Conference, London, 6-10 July 1987. *Nature* **328**, 291 (1987).

Matsumoto, T. *et al.* 1988. The Submillimeter Spectrum of the Cosmic Background Radiation. *Astrophysical Journal* **329**, 567-571 (1988).

Maurette, M. 1976. Fossil Nuclear Reactors. *Annual Review of Nuclear Science* **26**, 319-350 (1976).

Maxlow, J. 2005. *"Terra non Firma Earth: Plate Tectonics is a Myth"*. Terrella Press, Perth (2005).

Mayle, R., J. R. Wilson, and D. N. Schramm 1987. Neutrinos from Gravitational Collapse. *Astrophysical Journal* **318**, 288-306 (1987).

McCray, R. 1997. SN 1987A enters its second decade. *Nature* **386**, 438-439 (1997).

McDermott, F. and C. Hawkesworth 1990. The evolution of strontium isotopes in the upper continental crust. *Nature* **344**, 850-853 (1990).

McElroy, M. B. and M. J. Prather 1981. Noble gases in the terrestrial planets. *Nature* **293**, 535-539 (1981).

McFadden, L. 2005. email to R. J. Tuttle, 4/20/2005 (2005).

McGaugh, S. S. 1994. A possible local counterpart to the excess population of faint blue galaxies. *Nature* **367**, 538-541 (1994).

McGaugh, S. S., G. D. Bothun, and J. M. Schombert 1995. Galaxy Selection and the Surface Brightness Distribution. *Astronomical Journal* **110**, 573-585 (1995).

McGill, G. E. *et al.* 1983. Topography, Surface Evolution, and Tectonic Evolution. Pages 69-130 in *"Venus"*. D. M. Hunten, L. Colin, T. M. Donahue, and V. I. Moroz, Editors. University of Arizona Press, Tucson (1983).

McGinnis, E. R. 2009. ProbPlot 3.0. http://www.radprocalculator.com/Probability.aspx (2009).

McKay, C. P. 1993. Did Mars Once Have Martians? *Astronomy* **September**, 26-33 (1993).

McKay, C. P., R. D. Lorenz, and J. I. Lunine 1999. Analytic Solutions for the Antigreenhouse Effect: Titan and the Early Earth. *Icarus* **137**, 56-61 (1999).

McKinnon, W. B., C. R. Chapman, and K. R. Housen 1991. Cratering of the Uranian Satellites. Pages 629-692 in *"Uranus"*. J. T. Bergstrahl, E. D. Miner, and M. S. Matthews, Editors. University of Arizona Press, Tucson (1991).

McKinnon, W. B., J. I. Lunine, and D. Banfield 1995. Origin and Evolution of Triton, pages 807-877 in *"Neptune and Triton"*. D. P. Cruikshank, Editor. University of Arizona Press, Tucson (1995).

McPhaden, M. J. 1999. Genesis and Evolution of the 1997-98 El Niño. *Science* **283**, 950-954 (1999).

McSween Jr., H. Y., 1987. *"Meteorites and their parent bodies"*. Cambridge University Press, Cambridge (1987).

McSween Jr., H. Y., and R. P. Harvey 1993. Outgassed Water on Mars: Constraints from Melt Inclusions in SNC meteorites. *Science* **259**, 1890-1892 (1993).

Meadows, J. 1987. *"The Great Scientists"*. Oxford University Press, New York (1987).

Meegan, C. A. 1994. The Mystery That Won't Go Away. *Sky & Telescope* **August**, 28-32 (1994).

Meegan, C. A. *et al.* 1992. Spatial distribution of gamma-ray bursts. *Nature* **351**, 143-145 (1992).

Megain, P. *et al.* 2005. Discovery of a bright quasar without a massive host galaxy. *Nature* **437**, 381-384 (2005).

Meier, R. *et al.* 1998. A Determination of the HDO/H2O Ratio in Comet C/1995 O1 (Hale-Bopp). *Science* **279**, 842-844 (1998).

Meikle, W. P. S., S. J. Matcher, and B. L. Morgan 1987. Speckle interferometric observations of supernova 1987A and of a bright associated source. *Nature* **329**, 608-611 (1987).

Meisel, T. and T. Pettke 1994. Strontium isotopic composition of carbonates from the Sumbar KT boundary. *Meteoritics* **29**, 501-502 (1994).

Meitner, L. and O. R. Frisch 1939. Disintegration of Uranium by Neutrons: A New Type of Nuclear Reaction. *Nature* **143**, 239-240 (1939).

Melillo *et al.* 1993. Global climate change and terrestrial net primary production. *Nature* **363**, 234-240 (1993).

Melosh, H. J. 1989. *"Impact Cratering"*. Oxford University Press, New York (1989).

Melosh, H. J. 1995. Cratering Dynamics and the Delivery of Meteorites to the Earth. *Meteoritics* **30**, 545-546 (1995).

Melosh, H. J. 1993. Blasting rocks off planets. *Nature* **363**, 498-499 (1993).

Melosh, H. J. and W. B. McKinnon 1988. The Tectonics of Mercury. Pages 374-400 in *"Mercury"*. F. Vilos *et al.*, Editors. University of Arizona Press, Tucson (1988).

Menard, H. W. 1986. *"The Ocean of Truth"*. Princeton University Press, Princeton (1986).

Menzel, D. H. *et al.* 1963. *"Stellar Interiors"*. John Wiley & Sons, New York (1963).

Mercury 1994. "We Nailed It!" A First Look at the New and Improved Hubble Space Telescope. *Mercury* **January/February**, 6-15 (1994).

Meshik, C. M. Hohenberg, and O. V. Pravdivtseva 2004. Record of Cycling Operation of the Natural Nuclear Reactor in the Oklo/Okelobondo Area in Gabon. *Physical Review Letters* **93**, 182302-1-4 (2004).

Mewaldt *et al.* 1994. Anomalous Cosmic Rays: Interstellar Interlopers in the Heliosphere and Magnetosphere. *Eos* **75**, 185,193 (1994).

Miller, S. *et al.* 1992. Identification of features due to H_3^+ in the infrared spectrum of supernova 1987A. *Nature* **355**, 420-422 (1992).

Michaud, G. and S. Vauclair 1991. Element Separation by Atomic Diffusion. Pages 304-325 in *"Solar Interior and Atmosphere"*. A. N. Cox *et al.*, Editors. University of Arizona Press, Tucson (1991).

Michaux, C. M. 1967. *"Handbook of the Physical Properties of the Planet Mars"*. NASA SP-3030, NASA Scientific and Technical Information Division, Linthicum Heights (1967).

Millus, S. 2009. A turtle's shell isn't so terribly bizarre after all. *Science News* **August 1**, 5-6 (2009).

Mirabel, I. F. *et al.* 1992. A double-sided radio jet from the compact Galactic Centre annihilator 1E1740.7-2942. *Nature* **358**, 215-217 (1992).

Mirazón Lahr, M. and R. Foley 2004. Human evolution writ small. *Nature* **431**, 1043-1044 (2004).

Misch, A. 1999. Casting for a Horoscope. *Mercury* **28**, 21-25 (1999).

Mittlefeldt, D. 1997. (reply). *Mercury* **January-February,** 4 (1997).

Moberg, A. 2005. Highly variable Northern Hemisphere temperatures reconstructed from low- and high-resolution proxy data. *Nature* **433,** 613-617 (2005).

Moffatt, W. G. 1984. *"The Handbook of Binary Phase Diagrams"*. Genium Publishing Co., New York (1984).

Molnar, P. and P. England 1990. Late Cenozoic uplift of mountain ranges and global climate change: chicken or egg? *Nature* **346**, 29-34 (1990).

Molnar, P. and J. Stock 1987. Relative motions of hotspots in the Pacific, Atlantic and Indian Oceans since late Cretaceous time. *Nature* **327**, 587-591 (1987).

Montelli, R. *et al.* 2004. A catalogue of deep mantle plumes: New results from finite-frequency tomography. *Geochemistry, Geophysics, Geosystems* **7**, Q11007 (2004).

Moore, I. D. 2004. Counting Individual ^{41}Ca Atoms with a Magneto-Optical Trap. *Physical Review Letters* **92**, 153002-153005 (2004).

Moore, L. G. 2005. Maternal O_2 transport and fetal growth in Colorado, Peru, and Tibet high-altitude residents. *American Journal of Human Biology* **2**, 627-637 (2005).

Moore, P. 1995. The Discoveries of Neptune and Triton. Pages 15-36 in *"Neptune and Triton"*. D. P. Cruikshank, Editor. University of Arizona Press, Tucson (1995).

Morbedelli, A. 2006. Interplanetary kidnap. *Nature* **441**, 162-163 (2006).

Morell, V. 1995. Siberia: Surprising Home for Early Modern Humans. *Science* **268**, 1279 (1995).

Morrison, D. R. O. 1993. The rise and fall of the 17-keV neutrino. *Nature* **366**, 29-32 (1993).

Morrison, D. and J. A. Burns 1976. The Jovian Satellites. Pages 991-1034 in *"Jupiter"*. T. Gehrels, Editor. University of Arizona Press, Tucson (1976).

Morrison, L. V. *et al.* 1988. Diameter of the Sun in AD 1715. *Nature* **331**, 421-423 (1988).

Morrison, P. 1998. Where Fiction Became Ancient Fact. *Scientific American* **June**, 99-101 (1998).

Morrison, R. and G. Kukla 1998. The Pliocene-Pleistocene (Tertiary-Quaternary) Boundary Should Be Placed at About 2.6 Ma, Not at 1.8 Ma! *GSA TODAY* **August**, 9 (1998).

Morwood, M. J. *et al.* 2004. Archaeology and age of a new hominin from Flores in eastern Indonesia. *Nature* **431**, 1087-1091 (2004).

Mosqueira, I. 1995. Rally Around the Ring. *Mercury* **March-April**, 10-14 (1995).

M. R. 1996. Yaws Origins. *Archaeology* **May/June**, 21 (1996).

Mueller, B. E. A. *et al.* 1992. Extraordinary Colors of Asteroidal Object (5145) 1992AD. *Icarus* **97**, 150-154 (1992).

Mueller, T. 2009. Ice Baby. *National Geographic* **May**, (2009).

Muhleman, D. O., D. B. Holdridge, and N. Block 1962. The Astronomical Unit Determined by Radar Reflections from Venus. *Astronomical Journal* **67**, 191-203 (1962).

Müller, W., D. Aerden, and A. N. Halliday 2000. Isotopic Dating of Strain Fringe Increments: Duration and Rates of Deformation in Shear Zones. *Science* **288**, 2195-2198 (2000).

Mukerjee, M. 1995. Mystery of the Missing Dynamo. *Scientific American* **January**, 24-26 (1995).

Muller, R. A. 1986. Evidence for Nemesis: A Solar Companion Star. Pages 387-396 in *"The Galaxy and the Solar System"*. R. Smoluchowski, J. N. Bahcall, and M. S. Matthews, Editors. University of Arizona Press, Tucson (1986).

Murdin, P. 1990. *"End in Fire, The supernova in the Large Magellanic Cloud"*, Cambridge University Press, Cambridge (1990).

Murakami, T. 1988. Evidence for cyclotron absorption spectral features in gamma-ray bursts seen with Ginga. *Nature* **335**, 234-235 (1988).

Murakami, T. *et al.* 1991. A γ-ray burst preceded by X-ray activity. *Nature* **350**, 592-594 (1991).

Murdin, P. 1990. *"End in fire, the supernova in the Large Magellanic Cloud"*. Cambridge University Press, Cambridge (1990).

Myra, E. S., J. M. Lattimer, and A. Yahil 1987. Neutrino Emission from Cooling Neutron Stars. Pages 213-216 in *"Supernova 1987A in the Large Magellanic Cloud"*. M. Kafatos and A. Michalitsianos, Editors. Cambridge University Press, Cambridge (1988).

Naeye, R. 1993. Supernova 1987A Revisited. *Sky & Telescope* **February**, 39-43 (1993).

Naeye, R. 1995. Was Einstein Wrong? The Mystery of DI Hercules. *Astronomy* **November,** 54-59 (1995).

Nagy, B. *et al.* 1991. Organic matter and containment of uranium and fissiogenic isotopes at the Oklo natural reactors. *Nature* **354**, 472-475 (1991).

Nagy, B. 1994. Gabon's natural reactors. *Nuclear Engineering International* **February**, 30-31 (1994).

Napier, W. M. 2003. A statistical evaluation of anomalous redshift claims. *Astrophysics and Space Science* **285**, 419-427 (2003).

Napier, W. McD. and R. J. Dodd 1973. The Missing Planet. *Nature* **242**, 250-251 (1973).

Narlikar, J. V. 1993. *"Introduction to Cosmology"*. Cambridge University Press, Cambridge (1993).

NASA 2011. Hubble Discovers a New Moon Around Pluto. *News release* **July 20, 2011** (2011).

NASA/IPAC Extragalactic Database 2000. http://nedwww.ipac.caltech.edu (2000).

Nash, D. B. *et al.* 1986. Io. Pages 629-688 in *"Satellites"*. J. A. Burns and M. S. Matthews, Editors. University of Arizona Press, Tucson (1986).

Nash, J. M. 2002. *"El Niño"*. Warner Books, New York (2002).

Nataf, H.-C. and J. VanDecar 1993. Seismological detection of a mantle plume? *Nature* **364**, 115-120 (1993).

National Geographic Society 1990. Map Supplement. *National Geographic* **January** (1990).

Nature 1994a. The best cosmology there is. *Nature* **372**, 15-16 (1994).

Nature 1994b. Holes in the big bang. *Nature* **372**, 16-18 (1994).

Naudet, R. 1977. Les Reacteurs d'Oklo: Cinq Ans d'Exploration du Site. Pages 3-24 in *"Natural Fission Reactors"*. IAEA Staff, Editors. IAEA, Vienna (1978).

Naudet, R. and C. Renson 1975. Resultats des Analyses Systematiques de Teneur Isotopiques de l'Uranium. Pages 265-291 in *"The Oklo Phenomenon"*. IAEA Staff, Editors. IAEA, Vienna (1975).

Neelin, J. D. and M. Latif 1998. El Niño Dynamics. *Physics Today* **December**, 32-36 (1998).

Nellis, W. J. 2000. Making Metallic Hydrogen. *Scientific American* **May**, 84-90 (2000).

Nelson, R. M. 1997. Mercury, The Forgotten Planet. *Scientific American* **November**, 56-67 (1997).

Nelson, R. M. 1998. Deep space one: preparing for space exploration in the 21st century. *Eos, Transactions of the AGU* **79**, 493 (1998).

Ness, N. F., M. H. Acuña, and J. E. P. Connerney 1995. Neptune's Magnetic Field and Field-Geometric Properties. Pages 141-168 in *"Neptune and Triton"*. D. P. Cruikshank, Editor. University of Arizona Press, Tucson (1995).

Ness, N. F. *et al.* 1991. The Magnetic Field and Magnetospheric Configuration of Uranus. Pages 739-779 in *"Uranus"*. J. T. Bergstrahl, E. D. Miner, M. S. Matthews, Editors. University of Arizona Press, Tucson (1991).

Newcott, W. R. 1987. The Age of Comets. *National Geographic* **December,** 96-109 (1987).

Newman, M. J. and R. T. Rood 1977. Implications of Solar Evolution for the Earth's Early Atmosphere. *Science* **198**, 1035-1037 (1977).

NEWS 2009. Dark-matter test faces obstacles. *Nature* **462**, 23 (2009).

News 1995. First Drops in a Comet Reservoir. *Astronomy* **1995**, 18-19 (1995).

News and Views 1971. A New Magnetic Reversal at 12,500 Years? *Nature* **234**, 441 (1971).

News Notes 1991a. *Sky & Telescope* **March**, 246 (1991).

News Notes 1991b. The Sun's Missing Lithium. *Sky & Telescope* **December**, 586 (1991).

News Notes 1995. New Clues to Lunar Origin. *Sky & Telescope* **October**, 15-16 (1995).

News Notes 1996. Pulsar's Parent Revealed. *Sky & Telescope* **April**, 14 (1996).

News Notes 2000. Do all Pulsars Fly Pole-First? *Sky & Telescope* **September**, 20 (2000).

Newsom, H. E. and S. R. Taylor 1989. Geochemical implications of the formation of the Moon by a single giant impact. *Nature* **338**, 29-34 (1989).

Newspaper 1997. Journey through history. undated, Associated Press (1997).

Nichols Jr., R. H., C. M. Hohenberg, and K. Marti 1992. ^{129}I-derived and ^{244}Pu-fission Xe in individual neutron-irradiated phosphate crystals from the Acapulco meteorite. *Meteoritics* **27**, 268 (1992).

Nichols Jr., R. H. *et al.* 1991. Xenon and neon from acid-resistant residues of Inman and Tieschitz. *Geochimica et Cosmochimica Acta* **55**, 2921-2936 (1991).

Niemann, H. B. *et al.* 2005. The abundances of constituents of Titan's atmosphere from the GCMS instrument on the Huygens probe. *Nature* **438**, 779-784 (2005).

NIH 1985. "*Report of the National Institutes of Health Ad Hoc Working Group to Develop Radioepidemiological Tables*". National Institutes of Health, Bethesda (1985).

Nimmo, F. *et al.* 2008. Implications of an impact origin for the martian hemispheric dichotomy. *Nature* **453**, 1220-1223 (2008).

Ninkovich, D. and L. H. Burckle 1978. Absolute age of the base of the hominid-bearing beds in Eastern Java. *Nature* **275**, 306-308 (1978).

Nisenson, P *et al.* 1987. Detection of a very bright source close to the LMC supernova SN 1987A. *Astrophysical Journal* **320**, L15-L18 (1987).

Noonan, J. P. *et al.* 2006. Sequencing and Analysis of Neandertal Genomic DNA. *Science* **314**, 1113-1118 (2006).

Norell, M. A., E. S. Gaffney, and L. Dingus 1995. "*Discovering Dinosaurs*". Alfred A. Knopf, New York (1995).

Norman, D. 1985. "*The Illustrated Encyclopedia of Dinosaurs*". John Sibbick, Illustrator. Crescent Books, Avenel (1994).

Norris, J. P. *et al.* 1994. Detection of Signature Consistent with Cosmological Time Dilation in Gamma-Ray Bursts. *Astrophysical Journal* **429**, 540-545 (1994).

Norris, J. 1995. Here Come the Kuiperoids. *Astronomy*, about **April/May** (1995).

North, J. 1994. "*The Norton History of Astronomy and Cosmology*". W. W. Norton & Company, New York (1994).

North, J. 1996. "*Stonehenge*". The Free Press, New York (1996).

North, J. D. 1990. "*The Measure of the Universe*". Dover Publications, New York (1990).

Novacek, M. 1996. "*Dinosaurs of the Flaming Cliffs*". Doubleday, New York (1996).

Nudds, R. L. and G. J. Dyke 2010. Narrow Primary Feather Rachises in *Confuciusornis* and *Archaeopteryx* Suggest Poor Flight Ability. *Science* **328**, 887-889 (2010).

Nussbaumer, H. and L. Bieri 2009. "*Discovering the Expanding Universe*". Cambridge University Press, New York (2009).

Oberacker, V. *et al.* 1979. Characteristics of Coulomb fission. *Physical Review C* **20**, 1453-1466 (1979).

Oberbeck, V. R., J. R. Marshall, and H. Aggarwal 1993a. Impacts, Tillites, and the Breakup of Gondwanaland. *Journal of Geology* **101**, 1-19 (1993).

Oberbeck, V. R., J. R. Marshall, and H. Aggarwal 1993b. Impacts, Tillites, and the Breakup of Gondwanaland: a Reply. *Journal of Geology* **101**, 679-683 (1993).

Ocampo, A. 1996. Belize: Rosetta Stone of the K/T Boundary. *Planetary Report* **July/August**, 4-5 (1996).

O'Connor, J. M. and A. P. le Roex 1992. South Atlantic hot spot-plume systems: 1. Distribution of volcanism in time and space. *Earth Planetary Science Letters* **113**, 343-364 (1992).

O'Dell, C. R. and A. Van Helden 1987. How accurate were seventeenth-century measurements of solar diameter? *Nature* **330**, 629-631 (1987).

Oldershaw, R. L. 1990. Cosmology theory compromised. *Nature* **346**, 800 (1990).

Oort, J. H. 1950. The structure of the cloud of comets surrounding the solar system and a hypothesis concerning its origin. *Bulletin of the Astronomical Institute of the Netherlands* **11**, 91-110 (1950).

Oppo, D. W., Y. Rosenthal, and B. K. Linsley 2009. 2,000-year-long temperature and hydrology reconstructions from the Indo-Pacific warm pool. *Nature* **460**, 1113-1116 (2009).

Orr, J. B. 1949. Uranium^{235} in Thucholite. *Physical Review* **76**, 155 (1949).

Östlund, H. G. 1985. *"Atmospheric Tritium 1968-1984, Tritium Laboratory Data Report No. 14"*. University of Miami, Miami (1985).

Otake, M. and W. J. Schull 1993. Radiation-related small head size among prenatally exposed A-bomb survivors. *International Journal of Radiation Biology* **63**, 255-270 (1993).

Ott, U. 1993. Interstellar grains in meteorites. *Nature* **364**, 25-33 (1993).

Ott, U. 2000. Salty Old Rocks. *Science* **288**, 1761-1762 (2000).

Ovenden, M. W. 1972. Bode's Law and the Missing Planet. *Nature* **239**, 508-509 (1972).

Ovenden, M. W. 1976. The principle of least interaction action. Pages 295-305 in *"Long-Time Predictions in Dynamics"*. V. Szebehhely and B. D. Tapley, Editors. Reidel, Dordrecht (1976).

Owen, T. 1992. The Composition and Early History of the Atmosphere of Mars. Pages 825-834 in *"Mars"*. H. H. Kieffer, B. M. Jakosky, C. W. Snyder, and M. S. Matthews, Editors. University of Arizona Press, Tucson (1992).

Owen, T. and R. D. Cess 1979. Enhanced CO_2 greenhouse to compensate for reduced solar luminosity on early Earth. *Nature* **277**, 640-642 (1979).

Owen, T. *et al.* 1988. Deuterium on Mars: The Abundance of HDO and the Value of D/H. *Science* **240**, 1767-1770 (1988).

Ozima, M. 1986. Looking for Missing Xenon. *Nature* **321**, 813-814 (1986).

Ozima, M. 1987. *"Geohistory: Global Evolution of the Earth"*. Springer-Verlag, Berlin (1987).

Ozima, M. *et al.* 1989. Origin of the anomalous ^{40}Ar-^{39}Ar age of Zaire cubic diamonds: excess ^{40}Ar in pristine mantle fluids. *Nature* **337**, 226-229 (1989).

Paczyński, B. 1995. How Far Away Are Gamma-Ray Bursters? *Publications of the Astronomical Society of the Pacific* **107**, 1167-1175 (1995).

Padian, K. 1997. Origin of Dinosaurs. Pages 481-486 in *"Encyclopedia of Dinosaurs"*. P. J. Currie and K. Padian, Editors. Academic Press, San Diego (1997), after Ewer 1965. The anatomy of the thecodont reptile *Euparkeria capensis* Broom. *Philosophical Transactions of the Royal Society of London B* **248**, 379-435 (1965).

Pagel, B. E. J. 1992. Cosmology, Big Bang Theory. Pages 147-149In *"The Astronomy and Astrophysics Encyclopedia"*. S. P. Maran, Editor. Van Nostrand Reinhold, New York (1992).

Paget.org. http://www.paget.org/, http://www.paget.org.uk/info/outline.htm

Panagia, N. *et al.* 1991. Properties of the SN 1987A Circumstellar Ring and the Distance to the Large Magellanic Cloud. *Astrophysical Journal* **380**, L23-L26 (1991).

Panagia, N. 1994. Origins revealed in demise. *Nature* **369**, 354-355 (1994).

Panagia, N. 1999. Distance to SN 1987A and the LMC. Pages 549-563 in *"New Views of the Magellanic Clouds"*. Y. –H. Chu *et al.*, Editors. Astronomical Society of the Pacific, San Francisco (1999).

Pannekoek, A. 1923. Photographic photometry of the Milky Way and the colour of the Scutum cloud. *Bulletin of the Astronomical Institutes of the Netherlands* **II**, 19-46 (1923).

Pannekoek, A. 1989. *"A History of Astronomy"*, Dover, New York (1989), reprint of George Allen and Unwin, London (1961).

Papaliolios, C. *et al.* 1987. SN 1987A and Companion. Pages 225-229 in *"Supernova 1987A in the Large Magellanic Cloud"*. M. Kafatos and A. Michalitsianos, Editors. Cambridge University Press, Cambridge (1988).

Papaliolios, C. *et al.* 1989. Asymmetry of the envelope of supernova 1987A. *Nature* **338**, 565-566 (1989).

Pappalardo, R. T. *et al.* 1998. Geological evidence for solid-state convection in Europa's ice shell. *Nature* **391,** 365-368 (1998).

Parker, A. 2003. *"In the Blink of an Eye"*. Perseus Publishing, Cambridge (2003).

Parkinson, J. H. *et al.* 1980. The constancy of the solar diameter over the past 250 years. *Nature* **288**, 548-551 (1980).

Parrington, J. R. *et al.* 1996. *"Nuclides and Isotopes"*, 15th edition. General Electric Co. and Knolls Atomic Power Laboratory, Schenectady (1996).

Parsons, D. K. 1988. *"ANISN/PC Manual"*. EGG-2500, EG&G Idaho, Inc., Idaho Falls (1988).

Parsons, V. 1980. *"A Colour Atlas of Bone Disease"*. Wolfe Medical Publications, London (1980).

Paul, E. R. 1993. *"The Milky Way Galaxy and Statistical Cosmology 1890-1924"*. Cambridge University Press, Cambridge (1993).

Pauli, W. 1991. On the earlier and more recent history of the neutrino. Pages 1-25 in *"Neutrino Physics"*. K. Winter, Editor. Cambridge University Press, Cambridge (1991).

Pavri, B. *et al.* 1992. Steep-Sided Domes on Venus: Characteristics, Geologic Setting, and Eruption Conditions From Magellan Data. *Journal of Geophysical Research* **97** (E8) 13445-13478 (1992).

Paxton, H. C. and N. L. Pruvost 1987. Critical Dimensions of Systems Containing ^{235}U, ^{239}Pu, and ^{233}U. 1986 Revision, LA-10860-MS, July 1987. Los Alamos National Laboratory, Los Alamos (1987).

Peebles, P. J. E. 1993. *"Principles of Physical Cosmology"*. Princeton University Press, Princeton (1993).

Peebles, P. J. E. *et al.* 1991. The case for the relativistic hot Big Bang cosmology. *Nature* **352**, 769-776 (1991).

Peebles, P. J. E. *et al.* 1992. Big Bang contd... *Nature* **357**, 288 (1992).

Peirls, R. 1989. Reflections on the discovery of fission. *Nature* **342,** 852-854 (1989).

Pennisi, E. 1999. Genetic Study Shakes Up Out of Africa Theory. *Science* **283**, 1828 (1999).

Penrose, R. 2005. *"The Road to Reality: A Complete Guide to the Laws of the Universe"*. Alfred A. Knopf, New York (2005).

Perkins, S. 2009a. The iron record of Earth's oxygen. *Science News* **175**, 24-28 (2009).

Perkins, S. 2009b. Pearls Unstrung, When the Great Lakes were lower. *Science News* **176**, 18-21 (2009).

Perlmutter, S. *et al.* 1998. Discovery of a supernova explosion at half the age of the Universe. *Nature* **391**. 51-54 (1998).

Perlmutter, S. *et al.* 1999. Measurements of Omega and Lambda from 42 High-Redshift Supernovae. *Astrophysical Journal* **517**, 565-586 (1999).

Pernetta, J. 1994. *"Philip's Atlas of the Ocean"*. George Philip, London (1994).

Perron, J. T. *et al.* 2007a. Long-Wavelength Shoreline Deformation On Mars Related To True Polar Wander. *Lunar and Planetary Science* **XXXVIII**, 2328.pdf (2007).

Perron J. T. *et al.* 2007b Evidence for an ancient martian ocean in the topography of deformed shorelines. *Nature* **447**, 840-843 (2007).

Perry, K. 1995. Chalk Butte, Inc. Digital Maps, Boulder, WY, personal communication (1995).

Peskin, M. 2002. The matter with antimatter. *Nature* **419**, 24-27 (2002).

Petit, J.–M. *et al.* 2008. The Extreme Kuiper Belt Binary 2001 QW$_{322}$. *Science* **322**, 432-434 (2008).

Petrov, Yu. V. *et al.* 2006. Natural nuclear reactor at Oklo and verification of fundamental constants: Computation of neutronics of a fresh core. *Physical Review C* **74**, 064610-1-17, (2006).

Petrun, N. M. 1965. Respiration through the skin in persons of various ages (from 3-100 years). *Bulletin of Experimental Biology and Medicine* **59**, 26-28 (1965).

Pettini, M., R. Stathakis, P. Molaro, and G. Vladilo 1989. Million Degree Gas in the Galactic Halo and the Large Magellanic Cloud II. The Line of Sight to Supernova 1987A. *Astrophysical Journal* **340**, 256-264 (1989).

Phillips, A. C. 1994. *"The Physics of Stars"*. John Wiley & Sons, Chichester (1994).

Pierson, E. S. 1975. Electromagnetic Self-Excitation in the Liquid-Metal Fast Breeder Reactor. *Nuclear Science and Engineering* **57**, 155-163 (1975).

Pieters, C. M. *et al.* 1986. The Color of the Surface of Venus. *Science* **234**, 1379-1383 (1986).

Pietrzyński, G. *et al.* 2010. The dynamical mass of a classical Cepheid variable star in an eclipsing binary system. *Nature* **468**, 542-544 (2010).

Pilcher, C. B. *et al.* 1970. Saturn's Rings: Identification of Water Frost, *Science* **167**, 1372-1373 (1970).

Pillinger, C. 1987. Nearer yet farther away. *Nature* **326**, 445-447 (1987).

Piperno, D. *et al.* 2009. Starch grain and phytolith evidence for early ninth millennium B. P. maize from the Central Balsas River Valley, Mexico. *Proceedings of the National Academy of Sciences USA* **106**, 5019-5024 (2009).

Plait, P. C. *et al.* 1995. *HST* observations of the ring around SN 1987A. *Astrophysical Journal* **439**, 730-751 (1995).

[Planavsky, N. J. *et al.* 2010. The evolution of the marine phosphate reservoir. *Nature* **467**, 1088-1090 (2010).

Poirier, J.-P. 1991. *"Introduction to the physics of the Earth's interior"*. Cambridge University Press, Cambridge (1991).

Podsiadlowski, Ph. 1991. Running rings around supernova 1987A. *Nature* **350**, 654-655 (1991).

Ponce de León, M. S and C. P. E. Zollikofer 2001. Neanderthal cranial ontogeny and its implications for late hominid diversity. *Nature* **412**, 534-538 (2001) Supplemental Information.

Ponman, T. J. and D. Bertram 1993. Hot gas and dark matter in a compact galaxy group. *Nature* **363**, 51-54 (1993).

Pope, K. 1996. Geology and Climate Change: Mechanisms of Extinction. *Planetary Report* **July/August**, 6 (1996).

Porco, C. C. *et al.* 2006. Cassini Observes the Active South Pole of Enceladus. *Science* **311**, 1393-1401 (2006).

Postberg, F. *et al.* 2011. A salt-water reservoir as the source of a compositionally stratified plume on Enceladus. *Nature* **474**, 620-622 (2011).

Postlethwait, J. H. and J. L. Hopson 1989. *"The Nature of Life"*. McGraw-Hill, New York, (1989).

Potts, R. 1993. Archeological Interpretations of Early Hominid Behavior and Ecology. *"The Origin and Evolution of Humans and Humanness"*. D. T. Rasmussen, Editor. Jones and Bartlett Publishers, Boston (1993).

Powell, C. S. 1991. X-Ray Riddle. *Scientific American* **March**, 26 (1991).

Powell, C. S. 1992. A Second Glance - A dim class of galaxies offers unexpected fascination. *Scientific American* **April 1992**, 36 (1992).

Preston, D. 1995. The Mystery of Sandia Cave. *The New Yorker* **June 12**, 66-83 (1995).

Pringle, H. 1997. Ice Age Communities May Be Earliest Known Net Hunters. *Science* **277**, 1203-1204 (1997).

Pringle, H. 1998. New Women of the Ice Age. *Discover* **April**, 62-69 (1998).

Prueher, L. M. and D. K. Rea 1998. Rapid onset of glacial conditions in the subarctic North Pacific region at 2.67 Ma: Clues to causality. *Geology* **26**, 1027-1030 (1998).

Prussing, J. E. and B. A. Conway 1993. *"Orbital Mechanics"*. Oxford University Press, New York (1993).

Putman, J. J. 1988. The Search for Modern Humans. *National Geographic* **174** (October) 438-477 (1988).

Qiam, S. M., C. H. Wu, and R. Wölfle 1983. ^{3}He Particle Emission in Fast Neutron Induced Reactions. *Nuclear Physics* **A410**, 421-428 (1983).

Quade, J. *et al.* 1998. Black Mats, Spring-Fed Streams, and Late-Glacial-Age Recharge in the Southern Great Basin. *Quaternary Research* **49**, 129-148 (1998).

Quinlan, G. D. 1993. Planet X: A myth exposed. *Nature* **363**, 18-19 (1993).

Rabinowitz, D. 2000. A reduced estimate of the number of kilometer-sized near-Earth asteroids. *Nature* **403**, 165-166 (2000).

Rabinowitz, D. L. *et al.* 2006. Photometric Observations Constraining the Size, Shape, and Albedo of 2003 EL61, a Rapidly Rotating Pluto-sized Object in the Kuiper Belt. *Astrophysical Journal* **639**, 1238-1251 (2006).

Ragettli, R. A. *et al.* 1994. Uranium-xenon chronology: precise determination of λ_{sf} $^{136}Y_{sf}$ for spontaneous fission of ^{238}U. *Earth and Planetary Science Letters* **128**, 653-670 (1994).

Raloff, J. 1995. Caste-Off Orangs. *Science News* **147**, 184-189 (1995).

Rampino, M. 1994. Tillites, Diamictites, and Ballistic Ejecta of Large Impacts. *Journal of Geology* **102**, 439-456 (1994).

Rampino, M. 1996. Impact Process: Reconstructing the Events. *Planetary Report* **Jul/Aug 1996**, 7-8 (1996).

Rampino M. R. and K. Caldeira 1994. The Goldilocks Problem: Climatic Evolution and Long-Term Habitability of Terrestrial Planets. *Annual Reviews of Astronomy and Astrophysics* **32**, 83-114 (1994).

Ramsay, J. G. 2000. A Strained Earth, Past and Present. *Science* **288**, 2139-2140 (2000).

Ramsköld, L. and H. Xianguang 1991. New early Cambrian animal and onychophoran affinities of enigmatic metazoans. *Nature* **351**, 225-228 (1991).

Ratcliffe, H. 2007. *"The Virtue of Heresy"*. AuthorHouse, Central Milton Keynes (2007).

Raup, D. M. 1991. *"Extinction - Bad Genes or Bad Luck"*. W. W. Norton & Company, New York (1991).

R. C. 2001. A new giant in the Kuiper belt. *Science News* **160**, 41 (2001).

Reaves, G. 1997. The Prediction and Discoveries of Pluto and Charon. Pages 3-25 in *"Pluto and Charon"*. S. A. Stern and D. J. Tholen, Editors. University of Arizona Press, Tucson (1997).

Redi, C. A., S. Garagna, and E. Capanna 1990. Nature's experiment with *in situ* hybridization? A hypothesis for the mechanism of Rb fusion. *Journal of Evolutionary Biology* **3**, 133-137 (1990).

Reed, T. C. 2008. The Chinese Nuclear Tests, 1964–1996. *Physics Today* **61**, 47-53 (2008).

Reeves, H. 1994. On the origin of the light elements. *Review of Modern Physics* **66**, 193-216 (1994).

Reich, D. *et al.* 2010. Genetic history of an archaic hominin group from Denisova Cave in Siberia. *Nature* **468**, 1053-1060 (2010).

Reich, E. S. 2010. G-whizzes disagree over gravity. *Nature* **466**, 1030 (2010).

Reich, E. S. 2010b. Antimatter held for questioning. *Nature* **468**, 355 (2010).

Reichen, C.-A. 1968. *"a history of astronomy"* Edito-Service, Geneva (1968).

Reines, F. and C. L. Cowan 1953a. A Proposed Experiment to Detect the Free Neutrino. *Physical Review* **90**, 492 (1953).

Reines, F. and C. L. Cowan 1953b. Detection of the Free Neutrino. *Physical Review* **92**, 830 (1953).

Reines, F. and C. L. Cowan Jr. 1959. Free Antineutrino Absorption Cross Section. I. Measurement of the Free Antineutrino Absorption Cross Section by Protons. *Physical Review* **113**, 273-279 (1959).

Riess, A. G., W. H. Press, and R. P. Kirshner 1995. Using Type Ia supernovae light curves to measure the Hubble Constant. *Astrophysical Journal* **438**, L17-L20 (1995).

Reisz, R. R. and J. J. Head 2008. Turtle origins out to sea. *Nature* **456**, 450-451 (2008).

Reitsma, H. J. 1979. Reliability of Minor Planet Satellite Observations. *Science* **205**, 185-186 (1979).

Renne, P. R. and A. R. Basu 1991. Rapid Eruption of the Siberian Traps Flood Basalts at the Permo-Triassic Boundary. *Science* **253**, 176-179 (1991).

Renne, P. R. *et al.* 1992. The Age of Paraná Flood Volcanism, Rifting of Gondwanaland, and the Jurassic-Cretaceous Boundary. *Science* **258**, 975-979 (1992).

Renne, P. R. *et al.* 1995. Synchrony and Causal Relations Between Permian-Triassic Boundary Crisis and Siberian Flood Volcanism. *Science* **269**, 1413-1416 (1995).

Rest, A. *et al.* 2005. Light echoes from ancient supernovae in the Large Magellanic Cloud. *Nature* **438**, 1132-1134 (2005).

Reynolds, J. H. 1960. Determination of the age of the elements. *Physical Review Letters* **4**, 8-10 (1960).

Ribes, E. *et al.* 1987. Evidence for a larger Sun with a slower rotation during the seventeenth century. *Nature* **326**, 52-55 (1987).

Ribes, E. *et al.* 1991. The Variability of the Solar Diameter. Pages 59-97 in *"The Sun in Time"*. C. P. Sonett, M. S. Giampapa, and M. S. Matthews, Editors. University of Arizona Press, Tucson (1991).

Richards, M. P. *et al.* 2001. Stable isotope evidence for increasing dietary breadth in the European mid-Upper Paleolithic. *Proceedings of the National Academy of Sciences USA* **98**, 6528-6532 (2001).

Richter, F. M., D. B. Rowley, and D. J. DePaolo 1992. Sr isotope evolution of seawater: the role of tectonics. *Earth and Planetary Science Letters* **109**, 11-23 (1992).

Rieppel, O. (2009). How Did the Turtle Get Its Shell? *Science* 325, 154-155 (2009).

Ringwood, A. E. *et al.* 1990. High-pressure geochemistry of Cr, V and Mn and implications for the origin of the Moon. *Nature* **347**, 174-176 (1990).

Rink, W. J. *et al.* 1995. ESR ages for Krapina hominids. *Nature* **378**, 24 (1995).

Roach, J. 2008. First Lungless Frog Found. *National Geographic News* **April 7,** (2008).

Robaudo, S. and C. G. A. Harrisson 1993. Plate Tectonics from SLR and VLBI Global data. Pages 51-71 in *"Contributions of Space Geodesy to Geodynamics: Crustal Dynamics"*. D. E. Smith and D. L. Turcotte, Editors. Geodynamics Series, Volume 23, American Geophysical Union, Washington (1993).

Roberts, M. B. 1994a. A hominid tibia from Middle Pleistocene sediments at Boxgrove, UK. *Nature* **369**, 313 (1994).

Roberts, M. B. 1994b. How old is 'Boxgrove man'? – Roberts replies. *Nature* **371**, 751 (1994).

Roberts, R. G. *et al.* 2001. New Ages for the Last Australian Megafauna: Continent-Wide Extinction About 46,000 Years Ago. *Science* **292**, 1888-1892 (2001).

Robin, E. *et al.* 1993. Evidence for a K/T impact event in the Pacific Ocean. *Nature* **363**, 615-617 (1993).

Robinson, M. S. and P. G. Lucey 1997. Recalibrated Mariner 10 Color Mosaics: Implications for Mercurian Volcanism. *Science* **275**, 197-199 (1997).

Roche, P. F. *et al.* 1989. Old cold dust heated by supernova 1987A. *Nature* **337**, 533-535 (1989).

Rodbell, D. T. 2000. The Younger Dryas: Cold, Cold Everywhere? *Science* **290**, 285-286 (2000).

Rogers, J. J. W. 1993. *"A History of the Earth"*. Cambridge University Press, Cambridge (1993).

Rolfs, C. E. and W. S. Rodney 1988. *"Cauldrons in the Cosmos"*. University of Chicago Press, Chicago (1988).

Roosevelt, A. C. *et al.* 1996. Paleoindian Cave Dwellers in the Amazon: The Peopling of the Americas. *Science* **272**, 373-384 (1996).

Roscoe, S. M. and K. D. Card 1992. Early proterozoic tectonics and metallogeny of the Lake Huron region of the Canadian shield. *Precambrian Research* **58**, 99-119 (1992).

Rose, J. I. 2010. New Light on Human Prehistory in the Arabo-Persian Gulf Oasis. *Current Anthropology* **51**, 849-883 (2010).

Rosenberg, K. R., L. Zuné, and C. B. Ruff 2006. Body size, body proportions, and encephalization in a Middle Pleistocene archaic human from northern China. *Proceedings of the National Academy of Sciences USA* **103**, 3552-3556 (2006).

Ross, D. A. 1995. *"Introduction to Oceanography"*. HarperCollins College Publishers, New York (1995).

Rotaru, M., J. L. Brick, and C. J. Allègre 1992. Clues to early Solar System history from chromium isotopes in carbonaceous chondrites. *Nature* **358**, 465-470 (1992).

Rothery, D. A. 1992. *"Satellites of the Outer Planets: worlds in their own right"*. Clarendon Press, Oxford (1992).

Rothschild, B. M. and P. L. Thillaud 1991. Oldest bone disease. *Nature* **349**, 288 (1991).

Rothschild, B. M., I Hershkovitch, and C. Rothschild 1995. Origin of yaws in the Pleistocene. *Nature* **378**, 343-344 (1995).

Rougier, G. W., M. S. de la Fuente, and A. B. Arcucci 1995. Late Triassic Turtles from South America. *Science* **268**, 855-858(1995).

Rougier, H. *et al.* 2007. Peştera cu oase 2 and the cranial morphology of early modern Europeans. *Proceedings of the National Academy of Sciences USA* **104**, 1165-1170 (2007).

Rowland R. E. *et al.* 1983. Dose-Response Relationships for Radium-Induced Bone Sarcomas. *Health Physics* **44**, 15-31 (1983).

Rowland, R. E. 1994. *"Radium in Humans"*. U. S. Department of Commerce, Springfield (1994).

R. T. 2001. Edge of the Solar System. *Astronomy* **March**, 32 (2001).

R. T. 2001. The Biggest Asteroid Yet. *Astronomy* **December**, 32,34 (2001).

Rudaux, L. and G. De Vaucouleurs 1959. *"Larousse Encyclopedia of Astronomy"*. The Hamlyn Publishing Group, Feltham (1959).

Ruina, J. *et al.* 1980. Ad Hoc Panel on the September 22 Event. http://foia.abovetopsecret.com/VELA_SATELLITE/ THE_VELA_INCIDENT/REPORTS/AD_HOC_REPORT_SEPT_23_1980.pdf (1980).

Ruiz, J. 2001. The stability against freezing of an internal liquid-water ocean in Callisto. *Nature* **412**, 409-411 (2001).

Ruspoli, M. 1986. *"The Cave of Lascaux: the final photographs"*. Harry N. Abrams, New York (1987).

Russell, H. N. 1930. Planet X. *Scientific American* **July**, 20-22 (1930).

Ryder, G. 1988. Quenching and disruption of lunar KREEP lava flows by impacts. *Nature* **336**, 751-754 (1988).

Safronov, V. S. 1969. *"Evolution of the Protoplanetary Cloud and Formation of the Earth and the Planets"*. (translated from the Russian, 1972) NASA TT F-677, NASA Scientific and Technical Information, Linthicum Heights (1972).

Sahu, K. C. *et al.* 1997. Observations of GRB 970228 and GRB 970508 and the Neutron Star Merger Model. *Astrophysical Journal* **489**, L127–L131 (1997).

Sagan, C. 1980. *"Cosmos"*. Random House, New York (1980).

Sagan, C. and C. Chyba 1997. The early faint sun paradox: Organic shielding of ultraviolet-labile greenhouse gases. *Science* **276**, 1217-1221 (1997).

Sagan, C. and G. Mullen 1972. Earth and Mars: Evolution of Atmospheres and Surface Temperatures. *Science* **177**, 52-56 (1972).

Sandage 1961. *"The Hubble Atlas of Galaxies"*. 1984 reprint. Carnegie Institution of Washington, Washington (1984).

Sandage, A. and J. Bedke 1994. *"The Carnegie Atlas of Galaxies"*. Carnegie Institution of Washington with The Flintridge Foundation, Washington (1994).

Sandage, A. *et al.* 1996. Cepheid Calibration of the Peak Brightness of Type IA Supernovae: Calibration of SN 1990N in NGC 4639 Averaged with Six Earlier Type IA Supernovae Calibrations to Give H_0 Directly. *Astrophysical Journal Letters* **460**, L15-L18 (1996).

Sander, P. M. and M. Clauss 2008. Sauropod Gigantism. *Science* **322**, 200-201 (2008).

Sanderson, I. T. 1960. *Saturday Evening Post* January 18, 1960 (1960).

Santoro, R. T. *et al.* 2005. Radiation Transport Calculations for Hiroshima and Nagasaki. Chapter 3 in Dosimetry System 2002 – DS02, Volume 1 (of 2). *"Reassessment of the Atomic Bomb Radiation Dosimetry for Hiroshima and Nagasaki"*. R. W Young and G. D. Kars, Editors. Radiation Effects Research Foundation, Hiroshima/Nagasaki (2005).

Sato, M. *et al.* 2011. Displacement Above the Hypocenter of the 2011 Tohoku-Oki Earthquake. *Science* **332**, 1395 (2011).

Saura Ramos, P. A. 1999. *"The Cave of Altamira"*. Harry N. Abrams, New York (1999).

Scarre, C. 1993. *"Smithsonian Timelines of the Ancient World"*. Dorling Kindersley, London (1993).

Schilling, G. 2001. Radical Theory Takes a Test. *Science* **291**, 579 (2001).

Schilling, J.-G. 1991. Fluxes and excess temperatures of mantle plumes inferred from their interaction with migrating mid-ocean ridges. *Nature* **352**, 397-403 (1991).

Schlosser, W. and W. Bergmann 1985. An early-medieval account on the red colour of Sirius and its astrophysical implications. *Nature* **318**, 45-46 (1985).

Schmitz, B. *et al.* 1997. Accretion Rates of Meteorites and Cosmic Dust in the Early Ordovician. *Science* **278**, 88-90 (1997).

Schmitz, B. *et al.* 2003. Sediment-Dispersed Extraterrestrial Chromite Traces a Major Asteroid Disruption Event. *Science* **300**, 961-964 (2003).

Schombert, J. S. *et al.* 1995. Dwarf Spirals. *Astronomical Journal* **110**, 2067-2074 (1995).

Schramm, D. N. 1985. Nucleosynthetic Interpretations of Isotopic Anomalies. Pages 106-121 in *"Nucleosynthesis: Challenges and New Developments"*. W. D. Arnett and J. W. Truran, Editors. University of Chicago Press, Chicago (1985).

Schramm, D. N. 1988. Supernova 1987A; one year later. LaThuile Proceedings, February 26-March 5, 1988. (1988).

Schubert, G. 1991. The lost continents. *Nature* **354**, 358 (1991).

Schubert, G., T. Spohn, and R. T. Reynolds 1986. Thermal Histories, Compositions and Internal Structures of the Moons of the Solar System. Pages 224-292 in *"Satellites"*. J. A. Burns and M. S. Matthews, Editors. University of Arizona Press, Tucson (1986).

Schubert, G. *et al.* 1996. The magnetic field and internal structure of Ganymede. *Nature* **384**, 544-545 (1996).

Schull, W. J. 1995. *"Effects of Atomic Radiation"*. Wiley-Liss, New York (1995).

Schulte. P. *et al.* 2010a. The Chicxulub Asteroid Impact and Mass Extinctions at the Cretaceous-Paleogene Boundary. *Science* **327**, 1214-1218 (2010).

Schulte, P. *et al.* 2010b. Response. *Science* **328**, 975-976 (2010).

Schwartz, J. H. 1995. *"Skeleton Keys"*. Oxford University Press, New York (1995).

Schwartz, R. D. 1977. A survey of southern dark clouds for Herbig-Haro objects and H-alpha emission stars. *Astrophysical Journal Supplement* **35**, 161-170 (1977).

Science News Staff 2002. Dear Mummy: Rare Fossil Reveals Common Dinosaur Soft Tissue. *Science News* **October 19**, 243 (2002).

Scotese, C. R. 1997. *"PALEOMAP Paleogeographic Atlas"*. University of Texas at Arlington, Arlington (1997).

Scott, W. B. 1997. *Aviation Week & Space Technology* **July 21, 1997**, 33 (1997).

Seaborg, G. T. 1958. *"The Transuranium Elements"*. Yale University Press, New Haven (1958).

Searle, M. 1995. The rise and fall of Tibet. *Nature* **374**,17-18, (1995).

Seeds, M. A. 1992. *"Foundations of Astronomy"*. Wadsworth Publishing Co., Belmont (1992).

Seidelmann, P. K., B. G. Marsden, and H. L. Giclas 1972. Note on Brady's Hypothetical Trans-Plutonian Planet. *Publications of the Astronomical Society of the Pacific* **84**, 858-864 (1972).

Sekaniana, Z. 1987. Nucleus of comet Halley as a torque-free rigid rotator. *Nature* **325**, 326-328 (1987).

Serber, R. 1992. *"The Los Alamos Primer"*. University of California Press, Berkeley (1992).

Serre, D. *et al.* 2004. No Evidence of Neandertal mtDNA Contribution to Early Modern Humans. *PLoS Biology* **2**, 0313-0317 (2004).

Settegast, M. 1986. *"Plato Prehistorian"*. Lindisfarne Press, Hudson (1990).

Seven Days 2010a. Earth-like planet. *Nature* **467**, 638 (2010).

Seven Days 2010b. Missing planet? *Nature* **467**, 889 (2010).

Shang, H. *et al.* 2007. An early modern human from Tianyuan Cave, Zhoukoudian, China. *Proceedings of the National Academy of Sciences USA* **104**, 6573-6578 (2007).

Shanklin, D. R. 1969. A general theory of oxygen toxicity in man. *Perspectives in Biology and Medicine* **13**, 80-100 (1969).

Shapley, H. 1943. *"Galaxies"*. Third edition, revised 1972 by Paul W. Hodge. Harvard University Press, Cambridge (1972).

Shapley , H. (Part I) and H. D. Curtis (Part II) 1921. The Scale of the Universe. *Bulletin of the National Research Council* **2**, 171-217 (1921).

Sharma, S. D. and M. Khanna 1988. Analysis of perihelia of new comets. *Monthly Notices of the Royal Astronomical Society* **235**, 1467-1471 (1988).

Sharp, Z. D. *et al.* 2010. The Chlorine Isotope Composition of the Moon and Implications for an Anhydrous Mantle. *Science* **329**, 1050-1053 (2010).

Sharpton, V. L. *et al.* 1993. Chicxulub Multiring Impact Basin: Size and Other Characteristics from Gravity Analysis. *Science* **261**, 1564-1567 (1993).

Shen, J. J.-S. *et al.* 1994. Lanthanum isotopic composition of meteoritic and terrestrial matter. *Geochimica et Cosmochimica Acta* **58**, 1499-1506 (1994).

Shepard, F. P. 1963. *"Submarine Geology"*, Second edition. Harper & Row, New York (1963).

Shepard, F. P. and R. F. Dill 1966. *"Submarine Canyons and Other Sea Valleys"*. Rand McNally & Company, Chicago (1966).

Sheppard, S. S. and C. A. Trujillo 2006. A Thick Cloud of Neptune Trojans and Their Colors. *Science* **313**, 511-514 (2006).

Sherrill, T. J. 1991. Orbital Science's "Bermuda Triangle". *Sky & Telescope* **81**, 134-139 (1991).

Shigematsu, I. *et al.* 1993. *"A-Bomb Radiation Effects Digest"*. Translated by B. Harrison. Harwood Academic Publishers, Chur (1993).

Shimamura, T. and G. W. Lugmair 1981. U-Isotopic Abundances. *Lunar and Planetary Science XII* **12**, 976-978 (1981).

Shoemaker, E. M. and R. F. Wolfe 1986. Mass Extinctions, Crater Ages and Comet Showers. Pages 338-386 in *"The Galaxy and the Solar System"*. R. Smoluchowski, J. N. Bahcall, and M. S. Matthews, Editors. University of Arizona Press, Tucson (1986).

Showman, A. P. and R. Malhotra 1999. The Galilean Satellites. *Science* **286**, 77-84 (1999).

Shukolyukov, A. and G. W. Lugmair 1993. Live Iron-60 in the Early Solar System. *Science* **259**, 1138-1141 (1993).

Shukolyukov, Y. A. *et al.* 1995. Noble gases and strontium isotopes in the unique meteorite Divnoe. *Meteoritics* **30**, 654-660 (1995).

Sieh, K. *et al.* 2008. Earthquake Supercycles Inferred from Sea-Level Changes Recorded in the Corals of West Sumatra. *Science* **322**, 1674-1678 (2008).

Sime, R. L. 1998. Lise Meitner and the Discovery of Nuclear Fission. *Scientific American* **January,** 80-85 (1998).

Simkin, S. M. *et al.* 1987. Markarian 348: A Tidally Disturbed Seyfert Galaxy. *Science* **235**, 1367-1370 (1987).

Simkin, T. and L. Siebert 1994. *"Volcanoes of the World"*. Geoscience Press, Tucson (1994).

Simone, G. 1998. On velocities beyond the speed of light. arXiv:hep-ph/9722265 v3 22JAN1998, (now missing, July 2010).

Simons, M. and B. H. Hager 1997. Localization of the gravity field and the signature of glacial rebound. *Nature* **390**, 500-504 (1997).

Simons, M. *et al.* 2011. The 2011 magnitude 9.0 Tohoku-Oki Earthquake: Mosaicking the Megathrust from Seconds to Centuries. *Science* **332**, 1421-1425 (2011).

Simonson, T. S. *et al.* 2010. Genetic Evidence for High-Altitude Adaptation in Tibet. *Science* **329**, 72-75 (2010).

Singer, C., J. Shulmeister, and B. McLea 1998. Evidence Against a Significant Younger Dryas Cooling Event in New Zealand. *Science* **281**, 812-814 (1998).

Sinnott, C. 1996. *Astronomy* **October**, 28 (1996).

Sky & Telescope 1991a. No Light from Gamma-Ray Bursters. *Sky & Telescope* **Jan**, 11-12 (1991).

Sky & Telescope 1991b. Dark Matter: WIMP's or MACHO's? *Sky & Telescope* **September**, 237 (1991).

Sky & Telescope 1991c. Planetary Nebulae: Much More than Meets the Eye. *Sky & Telescope* **October**, 347 (1991).

Sky & Telescope 1992. Runaway Pulsar. *Sky & Telescope* **January**, 9-10 (1992).

Sky & Telescope 1993a. Lithium Coming and Going. *Sky & Telescope* **March**, 11-12 (1993).

Sky & Telescope 1993b. Wrong-Way Galaxies. *Sky & Telescope* **May**, 11-12 (1993).

Sky & Telescope 1993c. Early Venus: The Wet Look. *Sky & Telescope* **June**, 9 (1993).

Sky & Telescope 1993d. Gaspra's Magnetic Personality. *Sky & Telescope* **July**, 11-12 (1993).

Sky & Telescope 1994a. Supernova Surprise. *Sky & Telescope* **August**, 13 (1994).

Sky & Telescope 1994b. Are Quasars Blue or Red? *Sky & Telescope* **November**, 10 (1994).

Slade, M. A., B. J. Butler, and D. O. Muhleman 1992. Mercury Radar Imaging: Evidence for Polar Ice. *Science* **258**, 635-640 (1992).

Slater, J. C. and N. H. Frank 1947. *"Mechanics"*. McGraw-Hill, New York (1947).

Sleep, N. H. 1994. Martian plate tectonics. *Journal of Geophysical Research* **99**, 5639-5655 (1994).

Slipher, V. M. 1909. The spectra of the major planets. *Lowell Observatory Bulletin* No. 42, 231-238 (1909).

Slipher, V. M. 1915. Spectrographic Observations of Nebulae. *Popular Astronomy* **23**, 21-24 (1915).

Slipher, V. M. 1919. The Spectrum of the Milky Way. *Popular Astronomy* **27**, 676 (1919).

Sloan *et al.* 1986. Gradual Dinosaur Extinction and Simultaneous Ungulate Radiation in the Hell Creek Formation. *Science* **232**, 629-633 (1986), and Comments and Response *Science* **234**, 1170-1175 (1986).

Smellie, J. 1995. The Fossil Nuclear Reactors of Oklo, Gabon. *Radwaste Magazine* **Mar 1995**, 18-27 (1995).

Smith, B. A. *et al.* 1986. Voyager 2 in the Uranian System. *Science* **234**, 43-64 (1986).

Smith, B. J. 2009. The Spirals, Bridges, and Tails Interacting Galaxy Survey. East Tennessee State University, http://www.etsu.edu/physics/bsmith/research/sg/arp.html (2009).

Smith, D. E *et al.* 1999. The Global Topography of Mars and Implications for Surface Evolution. *Science* **284**, 1495-1503 (1999).

Smith, D. K. 1996. Comparison of the shapes and sizes of seafloor volcanoes on Earth and "pancake" domes on Venus. *Journal of Volcanology and Geothermal Research* **73**, 47-64 (1996).

Smith, F. A *et al.* 2010. The Evolution of Maximum Body Size of Terrestrial Mammals. Science 330, 1216-1219 (2010).

Smith, J. A. *et al.* 2002. The *u'g'r'i'z'* Standard-Star System. *Astronomical Journal* **123**, 2121-2144 (2002).

Smith, M. S., L. H. Kawano, and R. A. Malaney 1993. Experimental, Computational, and Observational Analysis of Primordial Nucleosynthesis. *Astrophysics Journal Supplement* **85**, 219-247 (1993).

Smith, R. 2011. Dark days of the Triassic: Lost worlds. *Nature* **479**, 287-289 (2011).

Smrekar, S. E. *et al.* 2010. Recent Hotspot Volcanism on Venus from VIRTIS Emissivity data. *Science* **328**, 605-608 (2010).

Solomon, S. C. 1993. Keeping that youthful look. *Nature* **361**, 114-115 (1993).

Sonnett, C. P., S. A. Finney, and C. R. Williams 1988. The lunar orbit in the late Precambrian and the Elatina sandstone laminae. *Nature* **335**, 806-808 (1988).

Sparks, W. B., F. Paresce, and D. Macchetto. 1989. Polarization and visual intensity of the inner light echo around SN 1987A. *Astrophysical Journal* **347**, L65-L68 (1989).

Spencer, J. R. *et al.* 2006. Cassini Encounters Enceladus: Background and the Discovery of a South Polar Hot Spot. *Science* **311**, 1401-1405 (2006).

Spitzer, L. 1963. Star Formation. Pages 39-53 in *"Origin of the Solar System"*. R. Jastrow and A. G. W. Cameron, Editors. Academic Press, New York (1963).

Spudis, P. D. and J. E. Guest 1988. Stratigraphy and Geologic History of Mercury. Pages 118-164 in *"Mercury"*. F. Vilas, C. R. Chapman, M. S. Matthews, Editors. University of Arizona Press, Tucson (1988).

Spyromilio, J. *et al.* 1995. The three-dimensional structure of the ISM in front of supernova 1987A. *Monthly Notices of the Royal Astronomical Society* **274**, 256-264 (1995).

Squyres, S. W. and J. F. Kasting 1994. Early Mars: How Warm and How Wet? *Science* **265**, 749 (1994).

Stacey, F. D. 1969. *"Physics of the Earth"*. John Wiley & Sons, New York (1969).

Standage, T. 2000. *"The Neptune File"*. Walker and Company, New York (2000).

Standish Jr., E. M. 1993. Planet X: No dynamical evidence in the optical observations. *Astronomical Journal* **105**, 2000-2006 (1993).

Stassinopoulis, E. G. 1970. *"World Maps of Constant B, L, and Flux Contours"*. NASA SP-3054, National Aeronautics and Space Administration, Washington (1970).

Staveley-Smith, L. *et al.* 1992. Birth of a radio supernova remnant in supernova 1987A. *Nature* **355**, 147-149 (1992).

Steigman, G. and M. Tosi 1992. Galactic Evolution of D and ³He. *Astrophysical Journal* **401**, 150-156 (1992).

Stephens, N. P. and A. R. Carroll 1999. Salinity stratification in the Permian Phosphoria sea; a proposed paleoceanographic model. *Geology* **27**, 899-902 (1999).

Stephens, S. 1994. AAS Meeting. *Mercury* **January/February**, 18-24 (1994).

Sterken, C. and C. Jaschek, Editors 1996. *"Light Curves of Variable Stars"*. Cambridge University Press, Cambridge (1996).

Stern, A. and J. Mitton 1998. *"Pluto and Charon"*. John Wiley and Sons, New York (1998).

Stern, A. 1992. Where has Pluto's family gone? *Astronomy* **September**, 41-47 (1992).

Stern, S. A. 1997. Obituary – Clyde Tombaugh (1906-97). *Nature* **385**, 778 (1997).

Stern, S. A. 2000. Into the Outer Limits. *Astronomy* **September**, 52-55 (2000).

Steves, D. E. and G. A. Agogino undated. *"Sandia Cave: A Study in Controversy"*. Eastern New Mexico University, Portales (   ).

Stevenson, D. J. 1981. Models of the Earth's Core. *Science* **214,** 611-619 (1981).

Stevenson, D. J. 1994. California Institute of Technology, Division of Geological and Planetary Sciences, personal communication to R. J. Tuttle, February 4, 1994 (1994).

Stevenson, D. J. 1995. Personal communication to R. J. Tuttle, January 1995 (1995).

Stevenson, D. J. 1996. When Galileo met Ganymede. *Nature* **384**, 511-512 (1996).

Stevenson, D. 2010. Letter. *Physics Today* **63**, 8 (2010).

Stevenson, D. J. and E. E. Salpeter 1976. Interior Models of Jupiter. Pages 85-112 in *"Jupiter"*. T. Gehrels, Editor. University of Arizona Press, Tucson (1976).

Stewart, J. A. 1990. *"Drifting Continents & Colliding Paradigms: Perspectives on the Geoscience Revolution"*. Indiana University Press, Bloomington (1990).

Stirling, C. H., A. N. Halliday, and D. Porcelli 2005. In search of live ^{247}Cm in the early solar system. *Geochimica et Cosmochimica Acta* **69**, 1059-1071 (2005).

Stofan, E. R. 1993. The New Face of Venus. *Sky & Telescope* **August**, 22-31 (1993).

Storz, J. F. 2010. Genes for High Altitudes. *Science* **329**, 40-41 (2010).

Stowe, K. 1987. *"Essentials of Ocean Science"*. John Wiley & Sons, New York (1987).

Strahler, A. H. and A. N. Strahler 1992. *"Modern Physical Geography"*. Fourth Edition. John Wiley & Sons, New York (1992).

Strahler, A. N. 1981. *"Physical Geology"*. Harper & Row, New York (1981).

Straume, T. *et al.* 2003. Measuring fast neutrons in Hiroshima at distances relevant to atomic-bomb survivors. *Nature* **424**, 539-542 (2003).

Straus, L. G. 1989. Age of the modern Europeans. *Nature* **342**, 476-477 (1986).

Stringer, C. 2000. Coasting out of Africa. *Nature* **405**, 24-27 (2000).

Stringer, C. and W. Davies 2001. Those elusive Neanderthals. *Nature* **413**, 791-792 (2001).

Strom, R. G. 1987. *"Mercury: the elusive planet"*. Smithsonian Institution Press, Washington (1987).

Strom, R. G. and G. Neukem 1988. The Cratering Record on Mercury and the Origin of Impacting Objects. Pages 336-373 in *"Mercury"*. F. Vilas, C. R. Chapman, M. S. Matthews, Editors. University of Arizona Press, Tucson (1988).

Strom, R. G. *et al.* 1994. The global resurfacing of Venus. *Journal of Geophysical Research (Planets)* **99**, 10899-10926 (1994).

Stringer, C. B. *et al.* 1989. ESR dates for the hominid burial site of Es Skhul in Israel. *Nature* **338**, 756-758 (1989).

Stringer, C. and C. Gamble 1993. *"In Search of the Neanderthals"*. Thames and Hudson, New York (1993).

Sturrock, P. A. *et al.* 2011. Concerning the Phases of the Annual Variations of Nuclear Decay Rates. *Astrophysical Journal* **737** 65 doi: 10.1088/0004-637X/737/2/65 (2011).

Sturrock, P. A., E. Fischbach, and J. H. Jenkins 2011. Further Evidence Suggestive of a Solar Influence on Nuclear Decay Rates. *Solar Physics* **272**, 1-10 (2011).

Subcommittee ANS-5 1973. *"Decay heat power in light water reactors"*. October 1973, American Nuclear Society, La Grange Park (1973).

Sulentic, J. W. 1992. Odd Couples. *Astronomy* **November**, 36-41 (1992).

Sullivan, W. 1991. *"Continents in Motion: the new earth debate"*. Second Edition. American Institute of Physics, New York (1991).

Suntzeff, N. B. and P. Bouchet 1990. The Bolometric Light Curve of SN 1987A. I. Results from ESO and CTIO *U* to *QO* Photometry. *Astronomical Journal* **99**, 650-663 (1990).

Suntzeff, N. B. *et al.* 1988. The light echoes from SN 1987A. *Nature* **334**, 135-138 (1988).

Sunyaev, R. *et al.* 1987. Discovery of hard X-ray emission from supernova 1987A. *Nature* **330**, 227-229 (1987).

Suploe, C. 1999. El Niño – Nature's Vicious Cycle. *National Geographic* **March**, 73-95 (1999).

Surovell, T. A. *et al.* 2009. An independent evaluation of the Younger Dryas extraterrestrial impact hypothesis. *Proceedings of the National Academy of Sciences USA* **0907857106** (2009).

Sutherland, B. M. *et al.* 2000. Clustered DNA damages induced in isolated DNA and in human cells by low doses of ionizing radiation. *Proceedings of the National Academy of Sciences USA* **97**, 103-108 (2000).

Svitel, K. A. 2001. It's a Small Solar System After All. *Discover* **April**, 16 (2001).

Swindle, T. D. and M. K. Burkland 1992. I-Xe studies of chondritic metal. *Meteoritics* **27**, 293-294 (1992).

Tademaru, E. and E. R. Harrison 1975. Acceleration of pulsars to high velocities by asymmetric radiation. *Nature* **340**, 676-677 (1975)

Tadhunter, C. N. *et al.* 1987. Detached nuclear-like activity in the radio galaxy PKS2152-69. *Nature* **325**, 504-507 (1987).

Tag's Broadcasting Services 1998. *"The Satellite Encyclopedia"*, http://www.tbs-satellite.com/tse/online/sat_cosmos_1402.html, 11/8/98, (1998).

Talcott, R. 1991. A Burst of Gamma Rays. *Astronomy* **October**, 46-50 (1991).

Talcott, R. 1992. Bursting with Surprises. *Astronomy* **March**, 54-55 (1992).

Talcott, R. 1993. Toutatis Seen with Radar. *Astronomy* **April**, 36-37 (1993).

Tanaka, K. L. 2005. Geology and insolation-driven climate history of Amazonian north polar materials on Mars. *Nature* **437**, 991-994 (2005).

Tanaka, K. L. *et al.* 2003. Resurfacing history of the northern plains of Mars based on geologic mapping of Mars Global Surveyor data. *Journal of Geophysical Research,* **108**, 8043-8075 (2003).

Tang, T. B. 1986. Star colours. *Nature* **319**, 532 (1985).

Tang, T. B. 1991. Did Sirius change colour? *Nature* **352**, 25 (1985).

Tapley, D. F *et al.*, Editors 1989. *"The Columbia University College of Physicians and Surgeons Complete Home Medical Guide"*. Crown Publishers, New York (1989).

Tatsumoto, M. and T. Shimamura 1980. Evidence for live Cm-247 in the early solar system. *Nature* **286**, 118-122 (1980).

Tattersall, I and J.H. Schwartz 1999. Hominids and hybrids: The place of Neanderthals in human evolution. *Proceedings of the National Academy of Sciences USA* **96**, 7117-7119 (1999).

Taylor, E. L. *et al.* 1992. The Present Is Not the Key to the Past: A Polar Forest from the Permian of Antarctica. *Science* **257**, 1675-1677 (1992).

Taylor, H. S. 1921. *"Industrial Hydrogen"*. Chemical Catalog Company, New York (1921).

Taylor, J. H., R. N. Manchester, and A. G. Lyne 1993. Catalog of Pulsars, VII/156A, Astronomical Data Center, University of Maryland (1993).

Taylor, S. R. 1992. *"Solar System Evolution: A New Perspective"*. Cambridge University Press, Cambridge (1992).

Taylor, S. R. 1994. Silent upon a peak in Darien. *Nature* **369**, 196-197 (1994).

Taylor Jr., H. A. and P. A. Cloutier 1986. Venus: Dead or Alive? *Science* **234**, 1087-1093 (1986).

Tedesco, E. F. 1979. Binary Asteroids: Evidence for Their Existence from Lightcurves. *Science* **203**, 905-907 (1979).

Tegler, S. C. and W. Romanishin 2000. Extremely red Kuiper-belt objects in near-circular orbits beyond 40 AU. *Nature* **407**, 979-981 (2000).

television program 1998. Earthquake in New York City. *Science Channel* **May 3, 2001,** 9:00-10:00 pm (1998).

Teske, R. G. 1993. The star that blew a hole in space, Geminga: the gamma-ray pulsar. *Astronomy* **December,** 30-37 (1993).

Théry, I. *et al.* 1995. First use of coal. *Nature* **373**, 480 (1995).

Thiede, D. S. and P. M. Vasconcelos 2010. Paraná flood basalts: Rapid extrusion hypothesis confirmed by new $^{40}Ar/^{39}Ar$ results. *Geology* **38**, 747-750 (2010).

Tholen, D. J. and M. W. Buie 1997. Bulk Properties of Pluto and Charon. Pages 193-219 in *"Pluto and Charon"*. S. A. Stern and D. J. Tholen, Editors. University of Arizona Press, Tucson (1997).

Thomas, D. *et al.* 1993. Primordial Nucleosynthesis and the Abundances of Beryllium and Boron. *Astrophysical Journal* **406**, 569-579 (1993).

Thomas, J. T. 1978. *"Nuclear Safety Guide – TID-7016 Rev 2"*. Union Carbide Corporation, Oak Ridge (1978).

Thomas, J. T., Editor 1978. *"Nuclear Safety Guide"*. NUREG/CR-0095, National Technical Information Service, Springfield (1978).

Thomas, P. G., P. Masson, and L. Fleitout 1988. Tectonic History of Mercury. Pages 692-708 in *"Mercury"*. F. Vilas, C. R. Chapman, and M. S. Matthews, Editors. University of Arizona Press, Tucson (1988).

Thorsett, S. E. 1994. The times they are a-changing (citing M. Bailes, CSIRO). *Nature* **367**, 684-685 (1994).

Tirion, W. 1981. *"Sky Atlas 2000.0"*. Sky Publishing, Cambridge (1981).

Tisoncik, D. D. 1984. Regional Lithostratigraphy of the Phosphoria Formation in the Overthrust Belt of Wyoming, Utah and Idaho. Pages 295-318 in *"Hydrocarbon Source Rocks of the Greater Rocky Mountain Region"*. J. Woodward, F. F. Meissner, and J. L. Clayton, Editors. Rocky Mountain Association of Geologists, Denver (1984).

Todd, N. 1992. Mammalian Evolution: Karyotypic Fission Theory. Pages 275-292 in *"Environmental Evolution"*. L. Margulis and L. Olendzenski, Editors. The MIT Press, Cambridge (1992).

Tolmachoff, I. P. 1929. The Carcasses of the Mammoth and Rhinoceros Found in the Frozen Ground of Siberia. *Transactions of The American Philosophical Society* **XXIII**, Article I (1929).

Tomeoka, K. 1990. Phyllosilicate veins in a CI meteorite: evidence for aqueous alteration on the planet body. *Nature* **345**, 138-140 (1990).

Tonks, W. B. and H. J. Melosh 1992. Core Formation by Giant Impacts. *Icarus* **100**, 326-346 (1992).

Touma, J., J. Wisdom, and W. R. Kuhn 1998. Resonances in the Early Evolution of the Earth-Moon System. *Astronomical Journal* **115**, 1653-1663 (1998).

Tozer, D. C. 1978. Terrestrial Planet Evolution and the Observational Consequences of their Formation. Pages 433-462 in *"The Origin of the Solar System"*. S. F. Dermott, Editor. John Wiley & Sons, New York (1978).

Travis, J. 1994a. Gamma Rays Burst Out of the Galaxy. *Science* **263**, 467 (1994).

Travis, J. 1994b. Dying Stars Give Galactic Theory New Life. *Science* **264**, 906 (1994).

Trefil, J. 1988. *"The Dark Side of the Universe"*. Doubleday, New York (1998).

Trefil, J. 1998. Puzzling Out Parallax. *Astronomy* **September** 46-50 (1998).

Tremaine, S. 1992. The Dynamical Evidence for Dark Matter. *Physics Today* **February 1992**, 28-36 (1992).

Treiman, A. H. *et al.* 1993. Preterrestrial aqueous alteration of the Lafayette (SNC) meteorite. *Meteoritics* **28**, 86-97 (1993).

Trilling, D. E. and R. H. Brown 1998. A circumstellar dust disk around a star with a known planetary companion. *Nature* **395**, 775-777 (1998).

Trimble, V. 1995. The 1920 Shapley-Curtis Discussion: Background, Issues, and Aftermath. *Publications of the Astronomical Society of the Pacific* **107**, 1133-1144 (1995).

Trimble, V. 1996. H_0: The Incredible Shrinking Constant 1925-1975. *Publications of the Astronomical Society of the Pacific* **108**, 1073-1082 (1996).

Trinkaus, E. 2007. European early modern humans and the fate of the Neandertals. *Proceedings of the National Academy of Sciences USA* **104**, 7367-7372 (2007).

Trümper, J. *et al.* 1978. Evidence for Strong Cyclotron Line Emission in the Hard X-Ray Spectrum of Hercules X-1. *Astrophysical Journal* **219**, L105-L110 (1978)

Tuohy, D. R. and A. Dansie 1996. *"An Ancient Human Mummy From Nevada"*. Nevada State Museum, Carson City (1996).

Turcotte, D. L. 1993. An Episodic Hypothesis for Venusian Tectonics. *Journal of Geophysical Research (Planets)* **98**, 17061-17068 (1993)

Tuttle, R. J. 1994. Universal Fission and Adopted Explanations. *Proceedings of the Pacific Division, AAAS* (abstract) **13**, 99 (1994).

Tuttle, R. J. 1997. Mercurian Meteorites? *Mercury* **January-February**, 4 (1997).

Tyson, N. D. *et al.* 1993. On the possibility of a major impact on Uranus in the past century. *Astronomy and Astrophysics* **275**, 630-634 (1993).

Ulrich, R. K. and L. Bertello 1995. Solar-cycle dependence of the Sun's apparent radius in the neutral iron spectral line at 525 nm. *Nature* **377**, 214-215 (1995).

Ulrich, R. K. and A. N. Cox 1991. The Computation of Standard Solar Models. Pages 162-191 in *"Solar Interior and Atmosphere"*. A. N. Cox, W. C. Livingston, and M. S. Matthews, Editors. University of Arizona Press, Tucson (1991).

United Nations Scientific Committee on the Effects of Atomic Radiation 1986. *"Genetic and Somatic Effects of Ionizing Radiation"*. United Nations, New York (1986).

United Nations Scientific Committee on the Effects of Atomic Radiation 1993. *"Sources and Effects of Ionizing Radiation"*. United Nations, New York (1993).

United Nations Scientific Committee on the Effects of Atomic Radiation 1994. *"Sources and Effects of Ionizing Radiation"*. United Nations, New York (1994).

Unknown 1996. *Science News* **April 20**, 244 (1996).

Valdes, P. 1994. Damping seasonal variations. *Nature* **372**, 221, (1994).

Valladas, H. *et al.* 1988. Thermoluminescence dating of Mousterian 'Proto-Cro-Magnon' remains from Israel and the origin of modern man. *Nature* **331**, 614-616 (1988).

Van Allen, J. A. 1990. Magnetospheres, Cosmic Rays, and the Interplanetary Medium. Pages 29-40 in *"The New Solar System"*, 3rd Edition. J. K. Beatty and A. Chaikin, Editors. Sky Publishing Co., Cambridge (1990).

van Altena, W. F., J. T. Lee, and E. D. Hoffleit 1995. *"The General Catalogue of Trigonometric Stellar Parallaxes"*. Yale University Observatory, New Haven (1995).

van der Kruit, P. C., and S. Shostak 1983. Vertical Motion and the Thickness of HI Disks: Implications for Galactic Mass Models. Pages 69-76 in *"Internal Kinematics and Dynamics of Galaxies"*. E. Athanassoula, Editor. International Astronomical Union, Paris (1983).

Van Flandern, T. C. 1978. A Former Asteroidal Planet as the Origin of Comets. *Icarus* **36**, 51-74 (1978).

Van Flandern, T. 1981. Minor Planet Satellites. *Science* **211**, 297-298 (1981).

Van Flandern, T. C. 1993. *"Dark Matter, Missing Planets and New Comets (Paradoxes Resolved, Origins Illuminated)"*. North Atlantic Books, Berkeley (1993).

Van Flandern, T. 1997. Personal communication, date lost, (1997?).

Van Helden, A. 1985. *"Measuring the Universe: Cosmic Dimensions from Aristarchus of Samos to Halley"*. University of Chicago Press, Chicago (1985).

Van Woerkom, A. J. J. 1948. On the origin of comets. *Bulletin of the Astronomical Institute of the Netherlands* **10**, 445-472 (1948).

Velikovsky, I. 1950. *"Worlds in Collision"*. Doubleday & Company, Garden City (1950).

Velikovsky, I. 1952. *"Ages in Chaos"*. Doubleday & Company, Garden City (1952).

Velikovsky, I. 1955. *"Earth in Upheaval"*. Doubleday & Company, Garden City (1955).

Verhoogen, J. 1980. *"Energetics of the Earth"*. National Academy of Sciences, Washington. (Available through NTIS, Springfield, Virginia) (1980).

Verosub, K. L. 1975. Paleomagnetic Excursions as Magnetostratigraphic Horizons: A Cautionary Note. *Science* **190**, 48-50 (1975).

Vickers-Rich, P. and T. H. Rich 1993. Australia's Polar Dinosaurs. *Scientific American* **July**, 50-55 (1993).

Vidale, J. E. and H. M. Benz 1992. A sharp and flat section of the core-mantle boundary. *Nature* **359**, 627-629 (1992).

Vidale, J. E. and H. M. Benz 1993. Seismological mapping of fine structure near the base of the Earth's mantle. *Nature* **361**, 529-532 (1993).

Vigny, C. *et al.* 2011. The 2010 M_w 8.8 Maule Megathrust Earthquake of central Chile, Monitored by GPS. *Science* **332**, 1417-1421 (2011).

Vitaliano, D. B. 1973. *"Legends of the Earth"*. Indiana University Press, Bloomington (1973).

Vogt, S. S *et al.* 2010. The Lick-Carnegie Exoplanet Survey: A 3.1 M_0 Planet in the Habitable Zone of the Nearby M3V Star Gliese 581. *Astrophysical Journal* **723**, 954-968 (2010).

Vonhof, H. B. and J. Smit (1997). High-resolution late Maastrichtian early Danian oceanic $^{87}Sr/^{86}Sr$ record: Implications for Cretaceous-Tertiary boundary events. *Geology* **25**, 347-350 (1997).

Vsekhsvyatski, S. K. 1962. Comets, small bodies, and problems of the solar system. *Publications of the Astronomical Society of the Pacific* **74**, 106-115 (1962).

Wacker, J. F. and E. Anders 1984. Trapping of xenon in ice: implications for the origin of the earth's noble gases. *Geochimica et Cosmochimica Acta* **48**, 2373-2380 (1984).

Waddle, D. M. 1994a. Matrix correlation tests support a single origin for modern humans. *Nature* **368**, 452-454 (1994).

Waddle, D. 1994b. Modern human origins. *Nature* **372**, 229 (1994).

Wagemans, C. 1991. Ternary Fission. Pages 545-584 in *"The Nuclear Fission Process"*. C. Wagemans, Editor. CRC Press, Boca Raton (1991).

Wagman, M. 2003. *"Lost Stars"*. McDonald and Woodward, Granville (2003).

Wagner, J. K. 1991. *"Introduction to the Solar System"*. Saunders College Publishing, Philadelphia (1991).

Wahr, J. 1990. Getting to the core. *Nature* **345**, 476-477 (1990).

Walker, A., M. R. Zimmerman, and R. E. F. Leakey 1982. A possible case of hypervitaminosis A in *Homo erectus*. *Nature* **296**, 248-250 (1982).

Walker A. R., 1988. Calibration of the Cepheid Period-Luminosity Relation. Pages 89-101 in *"The Extragalactic Distance Scale"*. S. van den Bergh and C. J. Pritchett, Editors. Astronomical Society of the Pacific, San Francisco (1988).

Walker, D. A. 1988. Seismicity of East Pacific Rise and Correlation with Southern Oscillation. *Eos* **69**, 857 (1988).

Walker, D. A. 1995. More Evidence Indicates Link Between El Niños and Seismicity. *Eos* **76**, 33 (1995).

Walker, D. A. 1999. Seismic Prediction of El Niño Revisited. *Eos* **80,** 281 (1999).

Walker, G. 2003. *"Snowball Earth"*. Crown Publishers, New York (2003).

Walker, R. J. *et al.* 1995. Osmium-187 Enrichment in Some Plumes: Evidence for Core-Mantle Interaction? *Science* **269**, 819-822 (1995).

Wallace, P. J. and T. M. Gerlach 1994. Magmatic Vapor Source for Sulfur Dioxide Released During Volcanic Eruptions: Evidence from Mount Pinatubo. *Science* **265**, 497-499 (1994).

Walter, R. C. *et al.* 1991. Laser-fusion $^{40}Ar/^{39}Ar$ dating of Bed I, Olduvai Gorge, Tanzania. *Nature* **354**, 145-149 (1991).

Walter, R. C. *et al.* 2000. Early human occupation of the Red Sea coast of Eritrea during the last interglacial. *Nature* **405**, 24-27 (2000).

Wang, Q., T. Hamilton, and D. J. Helfand 1989. The hot interstellar medium toward SN 1987A. *Nature* **341**, 309-311 (1989).

Ward, M. 1988. Reflected glory of active nuclei. *Nature* **334**, 569-570 (1988).

Ward, P. D. 1997. *"The Call of Distant Mammoths: Why the Ice Age Mammals Disappeared"*. Copernicus, New York (1997).

Ward, P. and S. Kynaston 1995. *"Wild Bears of the World"*. Facts on File, New York (1995).

Wark, D. 1987. News from the early Solar System. *Nature* **331**, 387 (1987).

Wasserburg, G. J. 1955. Time interval between nucleogenesis and the formation of meteorites. *Nature* **176**, 130-131 (1955).

Wasserburg, G. J. 1985. Short-lived Nuclei in the Early Solar System. Pages 703-737 in *"Protostars and Planets II"*. D. C. Black and M. S. Matthews, Editors. University of Arizona Press, Tucson (1985).

Wasserburg, G. J. and D. A. Papanastassiou 1982. Some short-lived nuclides in the early solar system - a connection with the placental ISM. Pages 77-140 in *"Essays in Nuclear Astrophysics"*. C. A. Barnes, D. D. Clayton, D. N. Schramm, Editors. Cambridge University Press, Cambridge (1982).

Wasson, J. T. 1985. *"Meteorites: Their Record of Early Solar System History"*. W. H. Freeman & Co., New York (1985).

Watson, L. L. *et al.* 1994. Water on Mars: Clues from Deuterium/Hydrogen and Water Contents of Hydrous Phases in SNC meteorites. *Science* **265**, 86-90 (1994).

Watters, T. R. 1995, *"Planets, A Smithsonian Guide"*. Macmillan, New York (1995).

Way, M. and H. Nussbaumer 2011. Lemaître's Hubble relationship. *Physics Today* **August 2011,** 8 (2011).

Weast, R. C., Editor 1980. *"Handbook of Chemistry and Physics, 60th Edition"*. CRC Press, Boca Raton (1980).

Weaver, A. H. 2005. Reciprocal evolution of the cerebellum and neocortex in fossil humans. *Proceedings of the National Academy of Sciences USA* **102**, 3576-3580 (2005).

Weaver, H. A. *et al.* 2006. Discovery of two new satellites of Pluto. *Nature* **439**, 943-945 (2006).

Webb Society 1979. *"Webb Society Deep-Sky Observer's Handbook"*, Volume 2, Planetary and Gaseous Nebulae. K. G. Jones, Editor. Enslow Publishers, Hillside (1979).

Webster, P. J. and J. N. Palmer 1997. The past and the future of El Niño. *Nature* **390**, 562-564 (1997).

Wegener, A. L. 1928. *"The Origin of Continents and Oceans"*. Translated by John Biram. Dover Publications, New York (1966).

Weinberg, A. M. and E. P. Wigner 1958. *"The Physical Theory of Neutron Chain Reactors."* University of Chicago Press, Chicago (1958).

Weishampel, D. B., P. Dodson, and H. Osmólska 1990. *"The Dinosauria"*. University of California Press, Berkeley (1992).

Weisman, P. R. 1990. The Oort cloud. *Nature* **344**, 825-830 (1990).

Weissman, P. 1995. Bodies on the brink. *Nature* **374**, 762-763 (1995).

Weissman, P. R. 1993. Comets at the Solar System's Edge. *Sky & Telescope* **January**, 26-29 (1993).

Weisman, P. R. 1998. The Oort cloud. *Scientific American* **September**, 84-89 (1998).

Wetherill, G. W. 1989. Origin of the Asteroid Belt. Pages 661-680 in *"Asteroids II"*. R. P. Binzel *et al.*, Editors. University of Arizona Press, Tucson (1989).

Wetherill, G. W. 1992. An Alternative Model for the Formation of the Asteroids. *Icarus* **100**, 307-325 (1992).

Wetherill, G. W. and M. G. Ingrham 1953. Spontaneous Fission in Uranium and Thorium. Page 30-32 in *"Proceedings of the Conference on Nuclear Processes in Geological Settings"*. National Research Council, Washington (1953).

Whitby, J. *et al.* 2000. Extinct ^{129}I in Halite from Primitive Meteorites: Evidence for Evaporite Formation in the Early Solar System. *Science* **288**, 1819-1821 (2000).

White, F. A. 1986. *"Mass Spectrometry in Science and Technology"*. John Wiley & Sons, New York (1986).

White, R. S. and D. P. McKenzie 1989. Volcanism at Rifts. *Scientific American*, **July**, 44-59 (1989).

White, T. D. 1994. Ape and hominid limb length (reply). *Nature* **369**, 194 (1994).

White, T. D. *et al.* 1993. New discoveries of *Australopithecus* at Maka in Ethiopia. *Nature* **366**, 261-265 (1993).

White, T. D. *et al.* 1994. *Australopithecus ramidus*, a new species of early hominid from Aramis, Ethiopia. *Nature* **371**, 306-312 (1994).

Whiteside, J. H. *et al.* 2010. http://www.pnas.org/content/early/2010/03/15/1001706107 - aff-1#aff-1. Compound-specific carbon isotopes from Earth's largest flood basalt eruptions directly linked to the end-Triassic mass extinction. *Proceedings of the National Academy of Sciences USA* **107**, 6721-6725 (2010).

Whitfield, J. 2009. Postmodern evolution? *Nature* **455**, 281-284 (2009).

Whitney, C. A. 1971. *"The Discovery of Our Galaxy"*. Alfred A. Knopf, New York (1971).

Whittet, D. C. B. 1999. A physical interpretation of the "red Sirius" anomaly. *Monthly Notices of the Royal Astronomical Society* 310, 355-359 (1999).

Wicander, R. and J. S. Monroe 1989. *"Historical Geology: Evolution of the Earth and Life Through Time"*, Revised Printing. West Publishing Company, St. Paul (1989).

Wickramasinghe, N. C. 1992. Cosmic microwave background. *Nature* **358**, 547 (1992).

Wickramasinghe, N. C. and H. Okuda 1994. Iron at the Galactic Centre. *Nature* **368**, 695 (1994).

Wikipedia 2009. http://en.wikipedia.org/wiki/Canada-France-Hawaii_Telescope (2009).

Wilbold, M., T. Elliott, and S Moorbath 2011. The tungsten isotopic composition of the Earth's mantle before the terminal bombardment. *Nature* **477**, 195-198 (2011).

Will, C. M. 1986. *"Was Einstein Right?"*. Basic Books, New York (1986).

Williams, G. E. 1986. The Acraman Impact Structure: Source of Ejecta in Late Precambrian Shales, South Australia. *Science* **233**, 200-203 (1986).

Williams, S. N. 1995. Erupting Neighbors-At Last. *Science* **267**, 340-341 (1995).

Williams, S. N. 1996. Double Trouble. *Earth* **August**, 43-48 (1996).

Wills, B. J. 1992. Quasistellar Objects, Spectroscopic and Photometric Properties. Pages 575-578 in *"The Astronomy and Astrophysics Encyclopedia"*. S. P. Maran, Editor. Van Nostrand Reinhold, New York (1992).

Willson, R. C. 1997. Total Solar Irradiance Trend During Solar Cycles 21 and 22. *Science* **277**, 1963-1965 (1997).

Wilson, R. N. 1990. Problems with telescopes. *Nature* **346**, 693(1990).

Witherell, P. W. 1944. Man and His Expanding Universe (I). Pages 2-12 in *"Beyond the Milky Way: Galaxies, Quasars, and the New Cosmology"*. T. Page and L. W. Page, Editors. The Macmillan Company, London (1969).

Witmer, L. W. 2009. Feathered dinosaurs in a tangle. *Nature* **461**, 601-602 (2009).

Witt, A. N. 1992. Diffuse Galactic Light. Pages 171-173 in *"The Astronomy and Astrophysics Encyclopedia"*. S. P. Maran, Editor. Van Nostrand Reinhold, New York (1992).

WoldeGabriel, G. *et al.* 1994. Ecological and temporal placement of early Pliocene hominids at Aramis, Ethiopia. *Nature* **371**, 330-333 (1994).

Wolfe, C. J. *et al.* 2009. Mantle Shear-Wave Velocity Structure Beneath the Hawaiian Hot Spot. *Science* **326**, 1388-1390 (2009).

Wolpoff, M. H. 1981. Allez Neanderthal. *Nature* **289**, 823 (1981).

Wolszczan, A. *et al.* 1989. A 110-ms pulsar, with negative period derivative, in the globular cluster M15. *Nature* **337**, 531-533 (1989).

Wolter, K. and M. S. Timlin 1993. Monitoring ENSO in COADS with a seasonally adjusted principal component index. Proceedings of the 17th Climate Diagnostics Workshop, Norman (1993).

Wolter, K. and M. S. Timlin 1998. Measuring the strength of ENSO events – how does 1997/98 rank? *Weather* **53**, 315-324 (1998).

Wong, K. 2000. Who Were the Neandertals? *Scientific American* **April**, 98-107 (2000).

Wood, C. A. and J. Kienle (Editors) 1990. *"Volcanoes of North America - United States and Canada"*. Cambridge University Press, Cambridge (1990).

Wood, J. A. 1962. Meteorites: Origin and Distribution. Pages 78-109 in *"Physics of the Solar System"*. Virginia Polytechnic Institute, Blacksburg (1962).

Wood, J. 1968. *"Meteorites and the Origin of Planets"*. McGraw-Hill Book Company, New York (1968).

Wood, J. W. *et al.* 1967a. The Growth and Development of Children Exposed in Utero to the Atomic Bombs in Hiroshima and Nagasaki. *American Journal of Public Health* **57**, 1374-1380 (1967).

Wood, J. W. *et al.* 1967b. Mental Retardation in Children Exposed in Utero to the Atomic Bombs in Hiroshima and Nagasaki. *American Journal of Public Health* **57**, 1381-1390 (1967).

Woosley, S. and T. Weaver 1989. The Great Supernova of 1987. *Scientific American* **August,** 32-40, (1989).

Wynn-Williams, G 1992. *"The Fullness of Space"*. Cambridge University Press, Cambridge (1992).

Xu, J., A. P. S. Crotts, and W. E. Kunkel 1994. Two new light echo structures from SN 1987A at large distances. *Astrophysical Journal* **433**, 274-278 (1994).

Xu, J. and A. P. S. Crotts 1995. A three-dimensional study using light echoes of the structure of the interstellar medium in front of SN 1987A. *Astrophysical Journal* **451**, 806-815 (1995).

Xu, X. *et al.* 2010. A basal tyrannosauroid dinosaur from the Late Jurassic of China. *Nature* **439**, 715-718 (2010).

Xue, X.-X. *et al.* 2008. The Milky Way's Circular Velocity Curve to 60 kpc and an Estimate of the Dark Matter Halo Mass from the Kinematics of ~2400 SDSS Blue Horizontal-Branch Stars. *Astrophysical Journal* **684**, 1143-1158 (2008).

Yeomans, D. K. 1991. *"Comets"*. John Wiley & Sons, New York (1991).

Yeomans, D. K. *et al.* 1999. Estimating the Mass of Asteroid 433 Eros During the NEAR Spacecraft Flyby. *Science* **285**, 560-561 (1999).

Yi, X. *et al.* 2010. Sequencing of 50 Human Exomes Reveals Adaptation to High Altitude. *Science* **329**, 75-78 (2010).

Yin, Q., E. Jagoutz, and H. Wänke 1992. Re-Search for extinct ^{99}Tc and ^{98}Tc in the early Solar System. *Meteoritics* (abstract) **27**, 310 (1992).

Yoder, C. F 1995. Astrometric and Geodetic Properties of Earth and Solar System. Pages 1-31 in *"Global Earth Physics"*. T. J. Ahrens, Editor. American Geophysical Union, Washington (1995).

Young, G. M. 1993. Impacts, Tillites, and the Breakup of Gondwanaland: a Discussion. *Journal of Geology* **101**, 675-679 (1993).

Young, E. D. *et al.* 1999. Fluid Flow in Chondritic Parent Bodies: Deciphering the Compositions of Planetesimals. *Science* **286**, 1331-1335 (1999).

Yulsman, T. 199?. Did the dinosaurs suffocate? *Earth* **date unknown,** 12 (199?).

Yun, M. S., P. T. P. Ho, and K. Y. Lo 1994. A high-resolution image of atomic hydrogen in the M81 group of galaxies. *Nature* **372**, 530-532 (1994).

Zahnle, K. J. and N. H. Sleep 1996. Impacts and the Early Evolution of Life. Pages 176-208 in *"Comets and the Origin and Evolution of Life"*. P. J. Thomas *et al.*, Editors. Springer, New York (1996).

Zahnle, K. and D. C. Catling 2009. Our Planet's Leaky Atmosphere. *Scientific American* **May,** (2009).

Zakamska, N. L. 2010. H_2 emission arises outside photodissociation regions in ultraluminous infrared galaxies. *Nature* **465**, 60-63 (2010).

Zashu, S., M. Ozima, and O. Nitoh 1986. K-Ar isochron dating of Zaire cubic diamonds. *Nature* **323**, 710-712 (1986).

Zeilik, M. and E. v. P. Smith 1987. *"Introductory Astronomy and Astrophysics"*, Second Edition. Saunders College Publishing, Philadelphia (1987).

Zensus, J. A. *et al.* 1987. Superluminal motion in the double-lobed quasar 3C263. *Nature* **325**, 36-38 (1987).

Zent, A. P 1996. The Evolution of the Martian Climate. *American Scientist* **84**, 442-451 (1996).

Zhao, J.-x *et al.* 2001. Thermal ionization mass spectrometry U-series dating of a hominid site near Nanjing, China. *Geology* **29**, 27-30 (2001).

Zilhão, J. 2010. Did Neandertals Think Like Us? *Scientific American* **June**, 72-75 (2010).

Zimmer, C. 1999a. The El Niño Factor. *Discover* **January**, 98-106 (1999).

Zimmer, C. 1999b. New Date for the Dawn of Dream Time. *Science* **284**, 1244-1246 (1999).

Zinnand, R. and M. J. West 1984. The Globular Cluster System of the Galaxy III. Measurements of Radial Velocity and Metallicity for 60 Clusters and a Compilation of Metallicities for 121 Clusters. *Astrophysical Journal Supplement Series* **55**, 45-66 (1984).

Zolensky, M. E. *et al.* 1999. Asteroidal Water Within Fluid Inclusion-bearing Halite in an H5 Meteorite, Monahans (1998). *Science* **285**, 1377-1379 (1999).

Index

A

B

T

U

CPSIA information can be obtained at www.ICGtesting.com
Printed in the USA
LVOW121935070612

285160LV00003B/14/P